Pre-ANDA LITIGATION

Strategies and Tactics for Developing a Drug Product and Patent Portfolio

Kenneth L. Dorsney
Editor-in-Chief

ABA Section of Intellectual Property Law
AMERICAN BAR ASSOCIATION

Cover by Best Design Chicago, Inc.

Printed in the United States of America

18 17 16 15 14 5 4 3 2 1

Library of Congress Cataloging-in-Publication Data

Cataloging-in-Publication data is on file with the Library of Congress.
Pre-ANDA Litigation: Strategies and Tactics for Developing a Drug Product and Patent Portfolio
Kenneth L. Dorsney, Editor-in-Chief
ISBN: 978-1-62722-731-5

Discounts are available for books ordered in bulk. Special consideration is given to state bars, CLE programs, and other barrelated organizations. Inquire at Book Publishing, ABA Publishing, American Bar Association, 321 N. Clark, Chicago, Illinois 60610-4714

www.ShopABA.org

CHAPTER AUTHORS

Chapter 1

Mark J. Feldstein, PhD
Courtney B. Casp, PhD
Danielle A. Duszczyszyn, PhD
Finnegan, Henderson, Farabow, Garrett & Dunner, LLP
Washington, DC

Chapter 2

Joseph M. Bennett-Paris, PhD
Richard T. Ruzich
Taft Stettinius & Hollister LLP
Chicago, IL

Chapter 3

Mark J. Feldstein, PhD
Shana K. Cyr, PhD
Finnegan, Henderson, Farabow, Garrett & Dunner, LLP
Washington, DC

Ian Y. Liu, PhD
Idenix Pharmaceuticals, Inc.
Cambridge, MA

Chapter 4

Milin Acharya, PhD
Modus Consultancy
Austin, TX
Ahmedabad, India

Chapter 5

Michael B. Harlin
Thomas J. Campbell, Jr.
Kevin A. O'Connor, PhD
Neal, Gerber & Eisenberg LLP
Chicago, IL

Chapter 6

B. Brett Heavner
Finnegan, Henderson, Farabow, Garrett & Dunner, LLP
Washington, DC

Chapter 7

Patricia A. Carson
Daniel Forchheimer
Kirkland & Ellis LLP
New York, NY

Howard Suh
Holland & Knight, LLP
New York, NY

Chapter 8

Matias Ferrario
Leslie T. Grab, PhD
Kilpatrick Townsend & Stockton LLP
Winston-Salem, NC

Kathryn H. Wade, PhD
Alyson L. Wooten, PharmD
Kilpatrick Townsend & Stockton LLP
Atlanta, GA

Chapter 9

James M. Lennon
Adam W. Poff
Young Conaway Stargatt & Taylor, LLP
Wilmington, DE

Ryan Cox
MacElree Harvey, Ltd.
West Chester, PA

Chapter 10

David S. Bloch
Winston & Strawn LLP
San Francisco, CA

Charles B. Klein
Winston & Strawn LLP
Washington, DC

Scott H. Blackman
Nath, Goldberg & Meyer
Washington, DC

Chapter 11

Earl B. Austin
Sean McDonagh
Jeremy Merling
Baker Botts L.L.P.
New York, NY

Chapter 12

Dennis M. Bissonnette, PhD
Justine A. Gozzi
Sandra S. Lee
Baker Botts L.L.P.
New York, NY

Chapter 13

David P. Frazier, PhD
Adriana L. Burgy
Kimberly D. Braslow, PhD
Finnegan, Henderson, Farabow, Garrett & Dunner, LLP
Washington, DC

Christine M. Hlavka
Bookoff McAndrews PLLC
Washington, DC

Chapter 14

Matthew Avery
Baker Botts L.L.P.
Palo Alto, CA

Paul A. Ragusa
Jennifer C. Tempesta
Baker Botts L.L.P.
New York, NY

Chapter 15

Gregory J. Brodzik
Samantha G. Wilson
James M. Lennon
Young Conaway Stargatt & Taylor, LLP
Wilmington, DE

Chapter 16

H. Keeto Sabharwal
Daniel E. Yonan
Sarah I. Danley
Diane E. Ghrist
Jon Boljesic
Sterne Kessler Goldstein & Fox
Washington, DC

Chapter 17

Lin Hymel, PhD
McLane, Graf, Raulerson & Middleton, PA
Woburn, MA

Chapter 18

Kevin M. Nelson
Patrick C. Gallagher
Duane Morris LLP
Chicago, IL

Matthew C. Mousley
Duane Morris LLP
Philadelphia, PA

Christopher S. Kroon
Duane Morris LLP
Boston, MA

Chapter 19

David S. Bloch
Winston & Strawn LLP
San Francisco, CA

Charles B. Klein
Winston & Strawn LLP
Washington, DC

Scott H. Blackman
Nath, Goldberg & Meyer
Washington, DC

Chapter 20

Karen E. Keller
Stephanie E. O'Byrne
Shaw Keller LLP
Wilmington, DE

Chapter 21

Scott H. Blackman
Nath, Goldberg & Meyer
Washington, DC

Davis S. Bloch
Winston & Strawn LLP
San Francisco, CA

Chapter 22

Stephen R. Auten
Taft Stettinius & Hollister LLP
Chicago, IL

Chapter 23

Brenda J. Huneycutt, PhD, JD, MPH
Avalere Health, LLC
Washington, DC

Chapter 24

Matthew A. Pearson, PhD
Jason Weil, PhD
Barbara L. Mullin
Akin Gump Strauss Hauer & Feld LLP
Philadelphia, PA

Chapter 25

Nadine Martino
Duncan Longstaff
Ashurst LLP
Adelaide, Australia

Grant Fisher
Ashurst LLP
Melbourne, Australia

Gustavo de Freitas Morais
Rodrigo Augusto Oliveira Rocce
Vyctor Taddeucci de Araujo
Dannemann Siemsen Bigler & Ipanema Moreira
Sao Paulo, Brazil

John E. Norman, PhD
Isabel Jaen Raasch
Scott E. Foster
Darren Blimkie
Gowling Lafleur Henderson LLP
Ottawa, ON

Jennifer Jones
Sarah Faircliffe
Bird & Bird LLP
London, UK

Junko Iyoda
Taiyo, Nakajima & Kato
Tokyo, Japan

Contents

Foreword

All pharmaceutical companies, whether they are an innovator or a generic, must navigate the same complex legal and regulatory framework to bring a product to market and fend off competition. The nuances involved in making decisions along the way can be hard to grasp and can involve a multitude of different pathways. Any practitioner would be remiss in not having a guide to a key pathway—ANDA litigations. This book, *Pre-ANDA Litigation: Strategies and Tactics for Developing a Drug Product and Patent Portfolio*, is an in-depth resource for learning about and planning for ANDA litigations and all the different avenues that the pharmaceutical litigants could follow.

From the perspective of an innovator company, which has made the investment in research and development, clinical testing, and regulatory filings, patents are vital to protect these new drug products both to recoup that initial investment and for future investments leading to the next medical breakthrough. For the innovator and patent owner, the patentee must be aware of the risk to those intellectual property rights and be prepared for any patent challenge. The statutory presumption of validity will never be enough. The risks to the market for the innovator product from patent litigation, which could result from the patent being invalidated or rendered unenforceable, needs to be well understood by the innovator company. Generic companies will challenge patents and it is not just a question of which patents will be challenged—most patents will be challenged given the incentives created by the Hatch-Waxman Act—but also when the generic companies could be ready to file an ANDA and begin a patent challenge. It should be a working assumption for the patent professional that any patent covering an innovative drug product will be subject of a challenge by one or more generic companies as soon as permitted under the exclusivity rules. Whether that challenge is to the validity or enforceability of the patent or is based on an alleged non-infringement defense that may require claim construction, it is imperative that the innovator is prepared. Patent litigations and strategies are complex and will require the patent professional to be able to explain complex technical and legal issues to lay persons, both within the organization and to judges and juries. Having a resource that will help formulate a strategy before patent litigation begins is invaluable.

From the perspective of a generic company, one must understand the innovator's thought processes as well as many other factors in order to get a generic drug to market in a way that will maximize revenue. The generic drug planning process begins many years before the drug may be on the market. A generic drug manufacturer, when preparing a Paragraph IV challenge, must consider, *inter alia*, the strength of its patent challenge and the possible responses from the innovator. A generic company must also plan for battling not only innovators but also other generic companies. Then, as time moves on from the initial planning stage to any ensuing litigation or actual market launch, many events may occur (e.g., many manufacturers filed ANDA's, patents are added to the Orange Book, countersuit by another generic, launches at risk) that may cause a generic to shift directions. The generic manufacturer must be able to break the drug planning and litigation processes down into key decision points so that the manufacturer can decide to move forward or adjust its strategy accordingly. This resource will allow generic manufacturers to locate those points and make informed decisions.

The key for any pharmaceutical ANDA litigant wanting to increase market share, whether as an innovator or a generic, is to plan early and to be ready to alter plans as new events occur. This book, *Pre-ANDA Litigation: Strategies and Tactics for Developing a Drug Product and Patent Portfolio*, will aid us all as we look for the best avenue to follow.

Peter Waibel
Vice President, Patent Litigation
Novartis Pharmaceuticals Corporation

John L. Dauer, Jr.
Chief Patent Counsel
Sun Pharmaceutical Industries, Inc.

Preface

After completing our first book on this topic in 2012, *ANDA Litigation: Strategies and Tactics for Pharmaceutical Patent Litigators*, we immediately recognized a need for another book detailing the enormous undertaking employed by innovators and generic pharmaceutical companies, and their associated professionals, to develop drug products and patent portfolios before litigation begins. It is not a straight and easy path from initial drug concept to public use. The road is long, difficult, and expensive. It is an investment of tremendous resources, and a path full of hazards for innovators and generics alike. As consumers, we want it that way, but we also do not.

We want tough patent laws and standards at the U.S. Patent & Trademark Office (PTO) to prevent effortless and inexhaustible patent monopolies. There must be balance, though. The hurdle to patent exclusivity cannot be too difficult so as to stifle the invention it was designed to promote. If it were, there would not be enough incentive to find the next miracle cure to make life better. The cutting edge, frontline of drug development is risky. It is also expensive. Patent exclusivity is the "carrot" that provides the opportunity for innovators to profit and recoup investment. Drug development must be meaningful. Generic pharmaceutical companies, however, drive prices down and create drug market efficiencies. They require accessibility. We, as consumers, want both—patent laws and PTO rules and regulations geared to innovation but designed "[t]o promote the Progress of Science and the useful Arts"[1] for the benefit of all.

We also want tough standards at the Food and Drug Administration (FDA) to ensure the safety and efficacy of the drugs we consume. FDA rules and regulations must be exacting. Although this likely increases the costs borne by the innovator, which are then passed onto consumers, we cannot have it any other way. It is our protection. A lapse in the system can have catastrophic, long-term consequences similar to the birth defects that

1. U.S. Const., Sec. 8, Cl. 8.

resulted from the use of Thalidomide in the late 1950s and early 1960s.[2] The initial introduction of a brand drug is thus expensive. We do not want this period, however, to last forever. It should have an end where market forces and competition can lower the costs of branded drugs and make FDA approved drugs more obtainable. Accordingly, a patent monopoly for brand drugs is of limited duration, and generic pharmaceutical companies can obtain FDA approval for generic drugs at lower costs with an Abbreviated New Drug Application (ANDA) by utilizing the FDA groundwork laid by innovators in having their New Drug Application (NDA) approved.

The process of developing a drug for market and building a patent portfolio involves numerous complex decisions by innovators and generic pharmaceutical companies and many proceedings before the PTO and FDA. This work occurs before ANDA patent litigation and provides much of the foundation on which it rests. It is important to recognize the different alternatives and consequences these companies may make during this period, as it may impact the course of ANDA and other patent litigation, as well as the success of the drug companies involved. Patenting innovative drugs and coordinating an NDA or ANDA drug portfolio strategy with a goal of achieving FDA approval requires a host of expert professionals with varying sets of skills. We attempt here to collect that knowledge and present it from the perspective of both the innovator and generic pharmaceutical company. We examine the different routes they take before the PTO and FDA to give context to the difficult choices they must make along the way. A perspective on how these same processes unfold in foreign countries is also provided at the end of the book for comparison.

As professionals, we want to build the best drug products and patent portfolios for our companies and our clients. We want to understand the process and make it simple. It is not. The rewards, however, exist. A single blockbuster drug can propel an innovator, save millions of lives, and foster an entire new market for generic competition. A portfolio of such drugs literally changes the world. This likely was not lost by Senators Orrin G. Hatch (R-Utah) and Henry A. Waxman (D-CA) when sponsoring The Drug Price Competition and Patent Term Restoration Act, commonly known as the "Hatch-Waxman Act." The Act strikes a compromise in the pharmaceutical industry to balance market competition against patent exclusivity. It does so in one way by allowing an abbreviated FDA review process for generic drugs and creating a right for innovators to initiate litigation before generic

2. The problems encountered with Thalidomide led to the 1962 amendments to the Federal Food, Drug, and Cosmetic Act (FDCA), requiring the FDA to approve all new drugs as safe and effective. Pub. L. No. 87-781, 76 Stat. 780 (1962) (codified at 21 U.S.C. §§ 321, 331-32, 348, 351-53, 355, 357-60, 372, 374, 376, 381).

companies enter the market. These were the subjects of our first work. This book examines the steps innovator and generic pharmaceutical companies take before litigation to patent drugs and to bring them to market.

As with our first book, I hope you too will find this one a welcome addition.

Kenneth L. Dorsney
March 25, 2014
Wilmington, DE[3]

3. I would like to offer special thanks to Tasha L. Ford and Carole Jean Haley at Morris James LLP for their assistance in compiling, managing, and reviewing the manuscript for this book. We would not have completed this work without them.

chapter 1

Coordinating New Drug Application (NDA) and Patent Portfolio Strategy[1]

I. Introduction

The challenges faced by innovators in building their portfolios continue to evolve. Especially with the waning dominance of the "blockbuster" model for drug development, no one-size-fits-all model that marches through scientific, regulatory, and patent stage gates as independent considerations can fully optimize drug portfolio development. Because the innovator's overall success is best advanced by a coordinated strategy covering all aspects of development, the objective of this chapter is to set a common foundation for the specific development phases and considerations addressed in the next few chapters.

The message, in short, is that strategic planning and coordination are necessary across all aspects and stages of pharmaceutical development. For example, there may be little (or no) commercial incentive to spend billions of dollars to develop a drug product for which there would be no period of exclusivity. Consistent with the intended wide audience for this text, strategic planning and coordination are best achieved through a team-based approach that incorporates input and vision from all areas of drug development, commercialization, and intellectual property (IP) protection. Although uncertainty in the pharmaceutical landscape has made strategic planning more difficult, financial constraints have made coordinated planning all the more important.

1. Mark J. Feldstein, PhD, Courtney B. Casp, PhD, and Danielle A. Duszczyszyn, PhD, of Finnegan, Henderson, Farabow, Garrett & Dunner, LLP, Washington, D.C.

II. A Changing Landscape

A. The Impact of Portfolio Planning

Strategic portfolio planning is one of the most impactful steps toward product commercialization and should begin well before the Investigational New Drug (IND) Application and the New Drug Application (NDA) have been developed. Developing a strong intellectual property portfolio directly adds value to a business by preventing others from practicing the claimed subject matter.[2] Studies have also demonstrated a strong positive correlation between the size of the patent portfolio for a product and its sales.[3]

Patent protection for commercial drugs is all but essential. It protects the enormous pharmaceutical development time and costs, which can entail ten or more years of research and $4 billion to $11 billion.[4] With only one compound out of five entering into Phase I testing ever being commercialized, a significant portion of this development expense is associated with unsuccessful drug candidates.[5] Typically, the portfolio strategy for big pharma addresses all the potential candidates, the majority of which do not ultimately succeed. For smaller biotechnology firms built around a single product, protection of that invention is all the more critical to the firm's industry survival.[6]

There are additional benefits attributed to a strategically planned patent portfolio. For example, the true value of patents may lie not in their individual worth but in their aggregate as a portfolio of related patents.[7] Essentially, "*the whole is greater than the sum of its parts.*"[8]

As a threshold matter, the cost to study the large portfolio of a potential competitor may serve as a barrier to entry, especially from smaller firms.[9]

---

2. Ralph Minderop, Arwed Burrichter & Natalie Kirchhofer, *Key Issues in Building a Strong Life Sciences Patent Portfolio*, INTELL. ASSET MGMT., July/Aug. 2012, at 126, 126.

3. Yang Guo, Yuanjia Hu, Mingli Zheng & Yitao Wang, *Patent Indicators: A Window to Pharmaceutical Market Success*, 23 EXPERT OPINION THERAPEUTIC PATS. 765, 765 (2013).

4. Matthew Herper, *The Truly Staggering Cost of Inventing New Drugs*, FORBES (Feb. 10, 2012), http://www.forbes.com/sites/matthewherper/2012/02/10/the-truly-staggering-cost-of-inventing-new-drugs/; Anne Pariser, *Early Communication: A Key to Reduced Drug Development and Approval Times*, FDA VOICE (Feb. 6, 2013), http://blogs.fda.gov/fdavoice/index.php/2013/02/early-communication-a-key-to-reduced-drug-development-and-approval-times/.

5. PHILIP W. GRUBB & PETER R. THOMPSON, PATENTS FOR CHEMICALS, PHARMACEUTICALS AND BIOTECHNOLOGY 425–26 (5th ed. 2010).

6. Minderop, Burrichter & Kirchhofer, *supra* note 2, at 126.

7. Gideon Parchomovsky, *Patent Portfolios*, 154 U. PA. L. REV. 1, 5–6 (2005).

8. *Id.* at 5.

9. Wei Li & Xiang Yu, *Patents: Patent Strategies for Biotechnology Enterprises*, BIOPHARM INT'L (Aug. 1, 2007), http://www.biopharminternational.com/biopharm/Business/Patents-Patent-Strategies-for-Biotechnology-Enterp/ArticleStandard/Article/detail/444984.

Not surprisingly, a large patent portfolio also can serve as a deterrent to litigation.[10] Although generic competition in the Hatch-Waxman ANDA litigation context involves unique considerations, potential competitors understand that the broad scope of protection created by a well-planned portfolio increases the chances of claims covering a competitive product.[11] Indeed, the likelihood of success of a portfolio holder winning an infringement litigation rises as the number of litigated patents increases.[12]

A large and diverse patent portfolio also confers insurance against changes in patent law.[13] Recent Supreme Court cases, such as *Prometheus* and *Myriad*, have arguably rendered certain types of biotechnology method and isolated gene claims unpatentable.[14] A well-planned portfolio protects against unforeseeable changes in patent law by protecting an invention with a wide variety of claim types and prosecution strategies.

Strategic patent portfolios also help leverage business negotiating power. Granting licenses to others can foster a wide variety of business transactions—from creating strategic alliances to preventing litigation through settlement or cross-licensing.[15] Moreover, a large portfolio can stimulate outside interest in the company. For example, investors are more likely to invest when there is a broader, strategic patent portfolio rather than a few individual patents.[16] Comprehensive portfolios are also an attractive asset to larger companies seeking to acquire a new product or technology.[17]

B. The Changing Landscape—View from the Cliff

Recent changes in the political and economic landscapes have forced many companies to adjust their business strategies with respect to portfolio planning. In the past, pharmaceutical companies enjoyed great success with the blockbuster drug model, in which drugs were intended to target diseases affecting large patient populations, like heart disease, to achieve high-volume sales of at least a billion dollars.[18] The success of the blockbuster model was most evident in the early to mid-2000s. Indeed, in 2006,

10. Parchomovsky, *supra* note 7, at 34–35.

11. *Id.* at 34.

12. *Id.* at 35 n.120.

13. *Id.* at 39–41.

14. Mayo Collaborative Servs. v. Prometheus Labs., Inc., 132 S. Ct. 1289 (2012); Ass'n for Molecular Pathology v. Myriad Genetics, Inc., 133 S. Ct. 2107 (2013).

15. Parchomovsky, *supra* note 7, at 36.

16. *Id.* at 37; Danielle T. Abramson, Patent Strategies for Life Sciences Companies to Navigate the Changing Patent Landscape, 17 J. Com. Biotech. 358, 358 (2011).

17. Minderop, Burrichter & Kirchhofer, *supra* note 2, at 126.

18. Roger Collier, *Bye, Bye Blockbuster, Hello Niche Busters*, 183 Can. Med. Ass'n J. E697, E697 (2011).

there were 52 blockbuster drugs that accounted for almost half of all U.S. drug sales.[19]

Over the last few years, however, the number of drugs achieving $1 billion in sales annually has been decreasing.[20] This decrease coincides with the onset of the "patent cliff," the period between 2011 and 2016 when 20 blockbuster drugs face patent expiry.[21] With the onset of the patent cliff, pharmaceutical companies have been forced to reevaluate past business strategies. It is now clear that the blockbuster drug model is no longer viable as the sole business strategy.

A variety of factors contributing to the rising cost of drug development further complicates the situation. For example, the cost of clinical trial testing has dramatically increased—a single clinical trial can cost as much as $100 million.[22] The regulatory process governing Phase III trials is the biggest contributor to this increase.[23] Over the past decade, the average trial length has grown, and the FDA has increased the stringency of its enrollment criteria and trial protocol requirements.[24] Additionally, when these older blockbuster drugs fall off the patent cliff, companies lose revenue streams, and less money is available to be reinvested in new research and development.[25] Further, in the wake of increased scrutiny after commercialized products were withdrawn from the market for safety concerns (think Vioxx), the FDA has been seen as more cautious in authorizing new drugs.[26]

The financial landscape for pharmaceutical and biotech companies has also changed. In 2012, for the first time in history, the U.S. drug market shrank.[27] It is estimated that by 2015 pharmaceutical companies will have lost a combined $170 billion annually from drugs going off patent.[28] This

19. *Id.*

20. *Id.*

21. Elisabeth Fischer, *The Patent Cliff: Rise of the Generics*, DRUG DEV. TECH. (Oct. 5, 2011), http://www.drugdevelopment-technology.com/features/featurethe-patent-cliff-rise-of-the-generics.

22. *How to Survive the Patent Cliff*, MANAGING INTELL. PROP. (Jan. 29, 2012), http://www.managingip.com/Article/2968146/How-to-survive-the-patent-cliff.html; Herper, *supra* note 4.

23. Avik S.A. Roy, *Stifling New Cures: The True Cost of Lengthy Clinical Drug Trials*, MANHATTAN INST. FOR POL'Y RES. (Apr. 2012), http://www.manhattan-institute.org/html/fda_05.htm.

24. *Id.*

25. The loss of revenue from these drugs is significant: in 2010, 130 blockbuster drugs represented 34 percent of the sales of the global pharmaceutical industry accounting for $295 billion of the $860 billion in worldwide sales. Fischer, *supra* note 21.

26. *How to Survive the Patent Cliff*, *supra* note 22; Linda Johnson, *FDA Takes Caution on Approving New Drugs*, WASH. POST (Aug. 17, 2007), http://www.washingtonpost.com/wp-dyn/content/article/2007/08/17/AR2007081701307_pf.html.

27. Doug Horning, *Is the Patent Cliff a Lethal Blow to Big Pharma?*, CASEY RES. (May 23, 2013), http://www.caseyresearch.com/cdd/is-the-patent-cliff-a-lethal-blow-to-big-pharma.

28. Sandra Lee, *Facing the Patent Cliff Head-On*, DRUG DISCOVERY NEWS (May 5, 2013), http://www.bakerbotts.com/file_upload/documents/SLeeDrugDiscoveryarticleMay2013.pdf.

is not surprising as a name-brand drug can lose up to 80 percent of its worldwide market within three months of generic entry.[29] In response to falling revenues, pharmaceutical companies have reduced research budgets, laid off employees, closed research centers, and sold off parts of their businesses.[30] This demonstrates that in order to bring an economically successful drug to market, these companies must evolve their business strategies to reflect the changing conditions. An evolving IP strategy must likewise follow.

C. Strategic Portfolio Planning—Engaging the Parachute

Pharmaceutical business strategies have shifted, and will continue to shift, to respond to the realities that limit reliance on the blockbuster model. For established companies, many of these strategies begin with a reassessment of the existing intellectual property portfolio. For newly formed companies, this provides an opportunity to develop an intellectual property portfolio primed to maximize the potential for growth and development. For both established and nascent companies, however, strategic portfolio planning becomes more important and requires continual vigilance in order to successfully respond to the changing economic and political landscapes.

With decreasing numbers of new drugs in their respective pipelines, many pharmaceutical companies are pursuing products from other firms, either by acquisition or by in-licensing technology.[31] By purchasing intellectual property portfolios from others, larger firms are able to bypass some of the risk involved in research and development by acquiring compounds that have already gone through preliminary clinical testing.[32] Indeed, some growing pharmaceutical companies have relied solely on drug acquisitions for their portfolios[33] and are now themselves potential targets for still larger pharmaceutical companies.[34] Likewise, to expand

29. Martin Austin, Business Development for the Biotechnology and Pharmaceutical Industry 21 (2008).

30. *See, e.g.*, Katie Thomas, *Pfizer Races to Reinvent Itself*, N.Y. Times (May 1, 2012), http://www.nytimes.com/2012/05/02/business/pfizer-profit-declines-19-after-loss-of-lipitor-patent.html?_r=0; Elisabeth Pain, *A Pharma Industry in Crisis*, Sci. Careers (Dec. 9, 2011), http://sciencecareers.sciencemag.org/career_magazine/previous_issues/articles/2011_12_09/caredit.a1100136.

31. In-licensing is the practice of obtaining rights to third-party technology by license, as opposed to developing that technology internally. *See* Dennis P. Schafer, *In-Licensing as a Business Model*, Bioentrepreneur (2013), http://www.nature.com/bioent/2003/030101/full/nbt0602supp-BE36.html.

32. Austin, *supra* note 29, at 25.

33. "Specialty companies, such as Forest, Shire and Reliant, have been built entirely from these sources and are forces to be reckoned with in the marketplace." *Id.* at 24.

34. Allison Connolly, *Shire to Forest on Deal List for AstraZeneca: Real M&A*, Bloomberg (Feb. 12, 2013), http://www.bloomberg.com/news/2013-02-12/shire-to-forest-on-deal-list-for-astrazeneca-real-m-a.html.

their respective portfolios in the changing market, traditionally generic firms have been increasingly acquiring innovator companies[35] to obtain their existing products and pipelines.[36]

These in-licensed compounds might be more expensive in the short term and deliver less profit in the long term, but the reduced development risk can make them a sound business investment.[37] For instance, in contrast to the extremely long odds for an untested drug product, a drug in Phase III testing has an estimated 70 percent chance of developing into a commercially viable drug product.[38] Understandably, licensing costs increase with each stage in development the drug reaches.[39] It is unlikely, however, that such products would be acquired absent a solid patent portfolio.

Likewise, reevaluating existing patent portfolios for repurposed drugs with new indications, formulations, or novel manufacturing techniques has also proven to be a successful business strategy for pharmaceutical companies.[40] One caveat of this strategy is that it requires careful analysis of the existing patent portfolio for the repurposed drug, especially when the new chemical entity (NCE)[41] was developed by another company, in order to evaluate whether the new use is covered by existing patents or whether additional protection is available.[42] Used successfully, this strategy can extract additional value from a company's existing or underutilized intellectual property assets or create new opportunities for other companies.[43]

Another strategy is to target drug development toward smaller patient populations. Known as "niche busters," these drugs include personalized therapeutics and "orphan drugs."[44] In the past, drug companies tended to avoid development of drugs targeted for smaller populations as they were

---

35. Innovator companies are those that historically have focused on the research and development of new drug products, not previously approved by the FDA.

36. Chris V. Nicholson, *Teva to Buy Cephalon for $6.8 Billion*, N.Y. TIMES DEALBOOK (May 2, 2011), http://dealbook.nytimes.com/2011/05/02/teva-to-buy-cephalon-for-6-8-billion/?_r=0; Madhukar Dubey, *How Acquisitions Mean Growth for These Generic Drug Makers*, MOTLEY FOOL (July 22, 2013), http://beta.fool.com/madhudube/2013/07/22/how-acquisitions-means-growth-for-these-generic-dr/40908/.

37. AUSTIN, *supra* note 29, at 25.

38. *Id.* at 106.

39. Sherry M. Knowles et al., *Monetization Models in Biotech and Pharma*, INTELL. ASSET MGMT., Jan./Feb. 2013, at 26, 31.

40. *Repurposed Drugs and Your IP Strategy*, MANAGING INTELL. PROP. (Sept. 1, 2011), http://www.managingip.com/Article/2893351/Repurposed-drugs-and-your-IP-strategy.html.

41. A "new chemical entity" is "a drug that contains no active moiety that has been approved by FDA in any other" new drug application. *See Small Business Assistance: Frequently Asked Questions for New Drug Product Exclusivity*, U.S. FOOD & DRUG ADMIN., http://www.fda.gov/Drugs/DevelopmentApprovalProcess/SmallBusinessAssistance/ucm069962.htm (last visited Apr. 23, 2014).

42. *Id.*

43. *Id.*

44. Discussed *supra* at section II.B.

not likely to yield enough return on investment. New innovations in personalized medicine, however, have increased the commercial viability of some niche-buster products. For example, a growing number of "companion diagnostic" tests has made it possible to predict whether an individual will respond to a therapeutic.[45] Thus, though personalized therapeutics may be useful for a smaller number of patients than would a typical blockbuster-type drug treatment, they may be more effective for the specialized population.[46] Careful portfolio planning is critical with personalized therapeutics—if the company producing the therapeutic also holds patent rights for the diagnostic test, there is an opportunity for an expanded financial benefit and better return on investment.[47]

Orphan drugs, defined as therapies that treat fewer than 200,000 patients, are another type of niche-buster drug growing in popularity post–patent cliff.[48] The U.S. Orphan Drug Act provides benefits for development and commercialization of drugs to treat rare diseases.[49] For example, companies are entitled to tax credits on clinical trial expenses, special grant funding from the FDA, a waiver of usage fees, and, perhaps most enticing, seven years of marketing exclusivity.[50] Additionally, these drugs enjoy a faster track from Phase II testing to market, better odds for FDA approval, premium pricing given the smaller patient populations, and potentially lower development costs.[51] Moreover, notwithstanding their smaller patient populations, orphan drugs have the potential for significant sales.[52] Orphan indications also have the potential for patent protection, including as a new method of use for a previously known drug.

Biologics have also received more attention as an alternative strategy. Biologics are large-molecule drugs that are derived from living cells as

45. Andrew Jack, *Diagnostics Creating "Niche Buster" Drugs*, Fin. Times (July 15, 2012), http://www.ft.com/intl/cms/s/0/536bceae-c359-11e1-ad80-00144feabdc0.html.

46. *Id.*

47. *How to Survive the Patent Cliff*, *supra* note 22.

48. Craig L. Kephart, *Orphan Drugs: Small Markets, Big Opportunity*, Specialty Pharmacy Times (Feb. 20, 2012), http://www.specialtypharmacytimes.com/publications/specialty-pharmacy-times/2012/February-2012/Orphan-Drugs-Small-Markets-Big-Opportunity.

49. Elie Dolgin, *Big Pharma Moves from "Blockbuster" to "Niche Busters,"* 16 Nature Med. 837, 837 (2010).

50. *Id.*

51. Kephart, *supra* note 48; Jonathan D. Rockoff, *Drug Makers See Profit Potential in Rare Diseases*, Wall St. J. (Jan. 30, 2013), http://online.wsj.com/news/articles/SB10001424127887323926104578273900197322758.

52. *Id.* ("Thanks to such high prices, almost a third of orphan drugs notch more than $1 billion in yearly sales, according to a sample reviewed by Thomson Reuters. The category has more than $50 billion in world-wide sales and has been rising more than 20% annually for the last several years.").

opposed to chemical processes like traditional small-molecule drugs.[53] Generally speaking, biologics are more difficult to make than small-molecule drugs and cannot be precisely reverse engineered, making copying more difficult for generic competitors.[54] Biologics offer other unique advantages, such as 12 years of FDA data exclusivity in contrast to the 3 to 5 years normally allotted for small-molecule drugs.[55] As the average breakeven point on research and development investment requires 12.9 to 16.2 years of exclusivity,[56] the automatic 12 years of biologic exclusivity is particularly significant.

Patent protection for biologics, however, is often perceived to be weaker than for small-molecule drugs. This is because, unlike with small-molecule drugs, broad claims to structurally similar drugs are often not granted due to the unpredictable biologic side effects caused by even small structural changes to the biologic.[57] Thus, there is an even greater premium on portfolio planning and protection strategies tailored to the specific case. Trade secret protection of biologics may also be a potentially valuable intellectual property portfolio asset.[58]

III. Strategic Portfolio Planning

A. Enacting a Team-Based Approach

As discussed in more detail earlier in this chapter, strategic portfolio planning is essential to enforce and defend IP rights as well as to enable collaboration and foster further innovation. A team-based approach to portfolio planning and management is most effective for both large and small companies. Fielding a team with multidisciplinary backgrounds provides balanced, comprehensive insight that helps inform decision making on whether to file patent applications. It also provides for greater recognition of unprotected avenues for competitors to skirt one's patent portfolio or curtail future freedom to operate by patenting improvements,

---

53. *How to Survive the Patent Cliff*, *supra* note 22.

54. *Id.*

55. *Id.*; Biologics Price Competition and Innovation Act of 2009 42 U.S.C. 262(k)(7).

56. *How to Survive the Patent Cliff*, *supra* note 22.

57. *Id.*; *In re* Wallach, 378 F.3d 1330, 1335 (Fed. Cir. 2004) (affirming rejection of patent claim to 30 kDa (roughly 250 amino acids) protein where only 10 amino acids were specified because there was "no evidence that there is any known or disclosed correlation between the combination of a partial structure of a protein, the protein's biological activity, and the protein's molecular weight, on the one hand, and the structure of the DNA encoding the protein on the other").

58. *How to Survive the Patent Cliff*, *supra* note 22.

add-ons, or alternatives to those offered in the patent portfolio. The pitfalls of not coordinating technical, business, and IP strategies are many. In the absence of collaboration, the company might pursue business strategies wrongly believing that they have sufficient IP protection, make product changes that adversely impact the company's freedom to operate, or find that the product and patent claims are no longer aligned.[59]

An effective team should include marketing experts who can judge the commercial potential of an invention, technical experts who know the current state of the art and can provide technical insight on novelty and obviousness, and an IP manager who, in consultation with patent counsel and management executives, reviews and mines the patent portfolio and pending applications periodically to ensure they align with the company's objectives. But the teams should not only have individuals familiar with the "big picture." Equally important are team members who are "in the weeds," including contributors from drug discovery, drug formulation, drug regulatory affairs, clinical pharmacokinetics, and licensing that bring to the table specialized knowledge of the project.[60] These individuals can provide helpful insight during prosecution, consider related technologies that may be covered by the claimed invention, and assess the market potential of a technology.[61] The team should aim to meet regularly rather than just in the initial planning phase to be sure that the portfolio adequately reflects any changes in company goals and strategy, the current market, and the law.

IV. Building the Portfolio

A. Cost/Benefit of Filing

Portfolio strategy requires a balance of short-term pressures with long-term planning. Although there are risk assessment models that which can be utilized to determine the probability of success of products in different stages

59. Rachel Teitelbaum & Mark S. Cohen, *Building a Strong Patent Portfolio*, Intell. Asset Mgmt. Mag., IP in the Life Sciences Industries Supplement 2008, at 41, 43, *available at* http://www.iam-magazine.com/issues/article.ashx?g=fd9a3044-e7a4-4787-b108-8f85c2906b44&q=building+a+strong+patent+portfolio#search=%22building+a+strong+patent+portfolio%22.

60. Clifford M. Davidson, *Product Life-Cycle Management Through Patents*, Drug Dev. & Delivery (Mar. 27, 2008), http://www.drugdeliverytech.com/Main/Back-Issues/Product-LifeCycle-Management-Through-Patents-325.aspx.

61. Teitelbaum & Cohen, *supra* note 59, at 43.

of development,[62] patent filing decisions must be made early in a drug's developmental timeline to avoid risking any loss of rights.

On the one hand, patent prosecution and maintenance costs can be expensive and can be especially burdensome for a company that is pre-revenue.[63] Costs can vary dramatically depending on the complexity of the invention, the number of claims, and the country in which the application is filed.[64] Moreover, once the patent has issued, the patentee must pay maintenance fees on both U.S.- and foreign-issued patents.[65]

On the other hand, prosecution and maintenance costs can usually be more than justified for a successful product. Just one extra day of patent protection for a billion dollar drug amounts to nearly $3 million in additional revenues. The problem, however, is that only a small fraction of candidates become commercial products.[66] As a result, the majority of pharmaceutical patents do not cover commercial products.[67] An internal strategy, aligned with business goals and commercial expectations, becomes essential to focus limited IP budgets on maximizing protection and minimizing expenses.

B. Timing of Filing

In addition to deciding where and what to file, timing is an important strategic tool, as discussed in more detail elsewhere in this book. Early in the drug discovery process, it is difficult, if not impossible, to ascertain which drugs may end up being commercialized. Naturally, companies may hesitate to file applications on drug candidates without a better indication of which compound may be successful. Moreover, because the lifespan of a U.S. patent is 20 years from filing, there may also be a temptation to file a patent later in the drug development process, especially because it takes 3 to 6 years to obtain a patent and 10 to 12 years for a drug to reach the market.[68] But delayed filing increases the probability of encountering prior art that will challenge patent validity.

Rather than delaying patent filings, it may be appropriate to consider other means for controlling expenses and maintaining protection for the actual commercial product. For example, Patent Cooperation Treaty (PCT) applications may extend the time from filing until prosecution, thereby

---

62. AUSTIN, *supra* note 29, at 105–06.

63. David Fagundes & Jonathan S. Masur, *Costly Intellectual Property*, 65 VAND. L. REV. 677, 689–90 (2012).

64. *Id.* at 685, 689–90; Robert Silverman, *Patent Filing Strategies for Pharmaceutical Products: A Simple Cost-Benefit Analysis Based on Filing Costs and Pharmaceutical Sale*, 33 AIPLA Q.J. 153, 168, 157–58 (2005).

65. *Id.* at 158; *Maintain Your Patent*, U.S. PAT. & TRADEMARK OFF., http://www.uspto.gov/patents/process/maintain.jsp (last updated Jan. 9, 2013).

66. AUSTIN, *supra* note 29, at 106.

67. Knowles et al., *supra* note 39, at 28.

68. Minderop, Burrichter & Kirchhofer, *supra* note 2, at 128.

deferring some prosecution costs while, at the same time, securing an early priority date. Subject to the many requirements for patentability, applications directed to improvements can be filed later. Or, if the development is unsuccessful while prosecution has been deferred, the patent filing can be abandoned to avoid unnecessary costs.

As for maintaining exclusivity, various regulatory extensions are available. In the United States, for instance, pursuant to 21 U.S.C. § 360cc and 21 C.F.R. § 314.108, respectively, an orphan drug receives seven years of market exclusivity and a new chemical entity (NCE) receives five years of market exclusivity.

While filing early to ensure coverage and to discourage competition is often the best practice to obtain an early priority date and allow time for prosecution to be completed before commercial launch, there may be times when this is not possible. When a patent application cannot be filed early, it may be important to accelerate prosecution of a later filed application. For example, in the Hatch-Waxman context, a 30-month stay of generic approval may not be available if the innovator's patent is listed after the generic's ANDA is filed.[69] Fortunately, the USPTO (U.S. Patent and Trademark Office) has introduced a number of new processes that help to accelerate the patent prosecution process.

First, the accelerated examination program (fast track) promises that final decisions on patentability will be rendered within 12 months of filing if the applicant provides certain additional information and fees.[70] Second, the USPTO has extended its Full First Action Interview Pilot Program.[71] This program allows applicants to interview with the patent examiner assigned to their application before a first office action on the merits issues in a utility patent application.[72] During this interview, written description and patentability issues can be resolved, thereby advancing the patent prosecution process. Third, the USPTO has joined the international Patent Prosecution Highway (PPH).[73] This international program relies on prosecution in a corresponding foreign application filed in a participating country to speed

69. 21 U.S.C. § 355(j)(5)(B)(iii); *see also* U.S. Food & Drug Admin., Guidance for Industry: Listed Drugs, 30-Month Stays, and Approval of ANDAs and 505(b)(2) Applications Under Hatch-Waxman, as Amended by the Medicare Prescription Drug, Improvement, and Modernization Act of 2003, Questions and Answers (Draft Guidance, Oct. 2004), *available at* http://www.fda.gov/downloads/Drugs/GuidanceComplianceRegulatoryInformation/Guidances/ucm072887.pdf.

70. Abramson, *supra* note 16, at 359–60; 37 C.F.R. § 1.102(e)(1); 37 C.F.R. § 1.17(c).

71. *Full First Action Interview Pilot Program*, U.S. Pat. & Trademark Off., http://www.uspto.gov/patents/init_events/faipp_landing.jsp (last updated Nov. 19, 2012) (extended November 16, 2012 for an undecided period of time); Abramson, *supra* note 16, at 360.

72. *Id.*

73. *Patent Prosecution Highway (PPH): Fast Track Examination of Applications*, U.S. Pat. & Trademark Off., http://www.uspto.gov/patents/init_events/pph/ (last updated Mar. 24, 2014); 37 C.F.R. § 1.102(d); 37 C.F.R. § 1.102(a); Abramson, *supra* note 16, at 360.

up examination.[74] PPH offers shortened wait time before examination[75] and high allowance rates.[76] These accelerated processes offer the possibility for increased certainty before investing in further product development.

V. Maintaining the Portfolio

Safeguarding long-term exclusivity or market share requires an active strategy that employs both offensive and defensive approaches. Periodically reevaluating potential weaknesses throughout the portfolio will help to strengthen the patents and the breadth of the protection. This ongoing active process requires constant review of the market, literature, and internal research and development to monitor competitors and identify potential patentable and commercially significant improvements.

For instance, obtaining new patents on improvements, such as new dosage forms or compositions, can provide valuable protection. Other offensive strategies involve amending pending claims to cover products within the scope of the inventive concept that competitors might develop in the future. Publication in scientific literature is another strategy that can be used to prevent competitors from obtaining blocking patents around the same technology. This can be important, especially as small companies and treating physicians have been known to file for patent protection for methods of using another company's product and then later sue that company for infringement.[77]

As discussed earlier, while quantity itself may confer some level of deterrence, it is important to focus on quality as the basic patents that form the cornerstones of the patent portfolio are crucial. Though quick action is initially tempting, long-term interests may be best served by steering clear of short-term monetary savings gained by cutting corners on one's IP portfolio. In other words, investing time, expertise, and effort into planning, drafting, and prosecuting initial cornerstone patent filings

74. *Id.* When an applicant receives a final ruling from a patent office participating in the PPH that at least one claim is allowed, the applicant may request fast track examination of any corresponding claim(s) in a corresponding patent application that is pending in a second patent office (e.g., USPTO) also participating in the PPH. This allows applicants to reach final disposition of a patent application more quickly and efficiently than standard examination processing.

75. PPH applications are typically examined within two to three months of a granted PPH request. Abramson, *supra* note 16, at 360.

76. PPH applications have an allowance rate of over 90 percent compared with 50 percent for non-PPH applications. *Id.*

77. *See* 35 U.S.C. § 102(a)(1) ("A person shall be entitled to a patent unless the claimed invention was patented, described in a printed publication, or in public use, on sale, or otherwise available to the public before the effective filing date of the claimed invention.")

is essential as these will ultimately form the strong foundation for a constantly growing and evolving portfolio.

Another aspect of maintaining one's portfolio is avoiding unnecessary publications that risk diminishing its value. In the scientific community, publications, posters, abstracts, and talks at scientific meetings serve as measures of reputation and success. But these same public disclosures could eventually invalidate or bar a crucial cornerstone patent. This applies to both publications prior to patent filing, which could be used as invalidating prior art, and after-filing publications that might inaccurately recount the inventive process. In addition to involving IP counsel in publication decisions, periodic employee education and training to comply with secrecy requirements and foster IP awareness internally among non-IP personnel should be a priority.

VI. Utilizing the Portfolio

Companies can also stand to benefit from reevaluating their current portfolios with an eye for valuable IP that is currently not being utilized or is being underutilized. While IP monetization, including out-licensing patents to other pharmaceutical or biotechnology firms, has historically not been very common among pharmaceutical companies, the potential is significant. As discussed earlier, for every pharmaceutical patent covering a commercial product there are nine additional patents in a company's patent portfolio covering other patentable inventions.[78]

There are a number of potential strategies aimed at increasing the accessibility of unused intellectual property. One involves using monetization companies that create and maintain a public database of available licensing technologies.[79] A second option would be to encourage the formation of smaller spin-off companies with the sole focus of developing discrete technology areas. In either case, it may be possible to limit the field of use to be noncompetitive, such as for veterinary use or distinct indications.

Patent pools are another patent monetizing strategy used successfully in other industries, especially the IT industry, as a mechanism to facilitate licensing.[80] In the past, patent pools in the biotech and pharmaceutical industry were used for charitable purposes, providing the public access to patents covering technology related to under-researched medical disorders.[81] The creation of commercial patent pools for pharmaceutical and

78. Knowles et al., *supra* note 39, at 31.
79. *Id.* at 33.
80. *Id.* at 35–36.
81. *Id.* at 35.

biotechnology patents, however, might be another mechanism to facilitate the monetization of underutilized IP. Commercial patent pools might prove to be particularly advantageous in the biotechnology industry as a means to facilitate licensing of diagnostics and research tools.[82]

One such pool, MPEG LA's Librassay, was designed to address licensing issues in diagnostics and personalized medicine.[83] This pool contains over 400 patents and facilitates royalty-bearing and royalty-free licensing for a wide variety of technologies and disease states.[84] Another pharmaceutical patent pool is Medicines Patent Pool, which has a pool of patents specifically directed to HIV treatment.[85] These pharmaceutical pools have the potential to provide greater access to potentially beneficial intellectual property, which might otherwise go unused, while allowing patentees to recoup some of their financial investment. If the income generated by participating in these pools is sufficient, these patent pools will attract others to submit their patents, which should simultaneously spur innovation and provide an additional income stream to cash-strapped firms.

VII. Conclusion

All strategic business planning—from research and development to commercialization—relies integrally on the strength of the portfolio protecting a product. This protection is the cornerstone of the business foundation supporting a commercial product. Achieving a comprehensive and diverse patent portfolio is best achieved by a coordinated strategy specific to the product that takes into account all aspects of drug development, regulatory approval, and competition.

A more in-depth explanation of the stages of development of a commercial pharmaceutical product from preclinical research through prelitigation investigations will follow in the next few chapters. Through all of these stages, portfolio planning and maintenance should be an ongoing focus in order to ensure the most comprehensive coverage for any potential commercial product. The next few chapters will further detail the importance of this business strategy as proactive portfolio planning throughout development and commercialization provides the best chances for a return on investment, making it an essential business strategy for pharmaceutical and biotech firms of all sizes.

---

82. *Id.* at 37.

83. *Id.*

84. *Id.*; *see also* Librassay, https://www.librassay.com/.

85. *See* Meds. Pat. Pool, http://www.medicinespatentpool.org/.

chapter 2

Preclinical R&D[1]

The safety and efficacy of a potential drug candidate must be determined long before it is ever tested in humans. This is accomplished in preclinical testing utilizing both in vitro (i.e., in the laboratory or "test tube") and in vivo (i.e., in living cells and animals) studies. These are all part of the drug development effort. Drug development encompasses such endeavors as target validation, drug discovery, development candidate identification, drug safety assessment, regulatory filing, conducting clinical trials, and finally submission to and approval by a regulatory body. Every drug development effort is unique in some way. Each requires its own roadmap, the details of which are highly dependent on the specific issues that arise in the course of discovering and advancing that drug candidate.

This chapter will focus on the preclinical activities that define the safety, efficacy, and characteristics of the potential drug compound that are all incorporated into an Investigational New Drug (IND) application.[2] An IND will summarize all of the preclinical testing studies for the Food and Drug Administration (FDA). The resulting IND mirrors and supports the clinical trial design; that is, a clinical trial involving daily or multi-dose drug administration requires chronic dosing animal toxicity studies.

Preclinical testing makes up just one part of the three major steps of drug development. These steps or phases are drug discovery, preclinical testing, and clinical trials. Drug discovery and preclinical testing often blend together, while there is a clear division between preclinical testing and the beginning of clinical trials. Upon filing and approval of an IND application, clinical trials can begin. A typical preclinical trial phase itself consists of several components, including, but certainly not necessarily limited to: (1) manufacture of the drug, (2) formulation and dosage design, (3) absorption/distribution/metabolism/excretion (ADME)

---

1. Joseph M. Bennett-Paris, PhD, and Richard T. Ruzich of Taft Stettinius & Hollister, LLP, Chicago, Illinois.

2. The IND application will be discussed in more detail in Chapter 7 of this book.

studies, (4) genetic toxicology, and (5) good manufacturing practice (GMP) documentation.[3]

I. Lead Compounds

A. Identification

The very first step in preclinical drug development generally is the discovery, identification, and characterization of a lead compound or drug. The lead compound will become the active pharmaceutical ingredient (API) in the pharmaceutical formulation to be developed. The API is the compound or component that furnishes pharmacological activity within the organism to which the formulation or drug is administered.[4]

The goal of drug discovery is to identify a biologically significant molecular target or biological process that is thought to be causally connected to the medical condition to be addressed. Then, a molecule is identified that somehow affects this biological process or target in the desired manner. That molecule will become the lead compound.

The journey from drug discovery through clinical trials and leading to the ultimate goal of approval is a long and harrowing one, not for the weary. For every 5,000–10,000 molecules initially identified in drug discovery, only one makes it to approval. (See Figure 1.)

3. *See* Loyd v. Allen, Jr., Nicholas G. Popovich, and Howard C. Ansel et al., *New Drug Development and Approval Process, Current Good Manufacturing Processes and Current Good Compounding Practices, in* Ansel's Pharmaceutical Dosage Forms and Drug Delivery Systems 25–91 (Lippincott, Williams & Wilkins eds., 8th ed. 2008).

4. *Id.* at 69.

Figure 1 The Numbers from Discovery to Approval

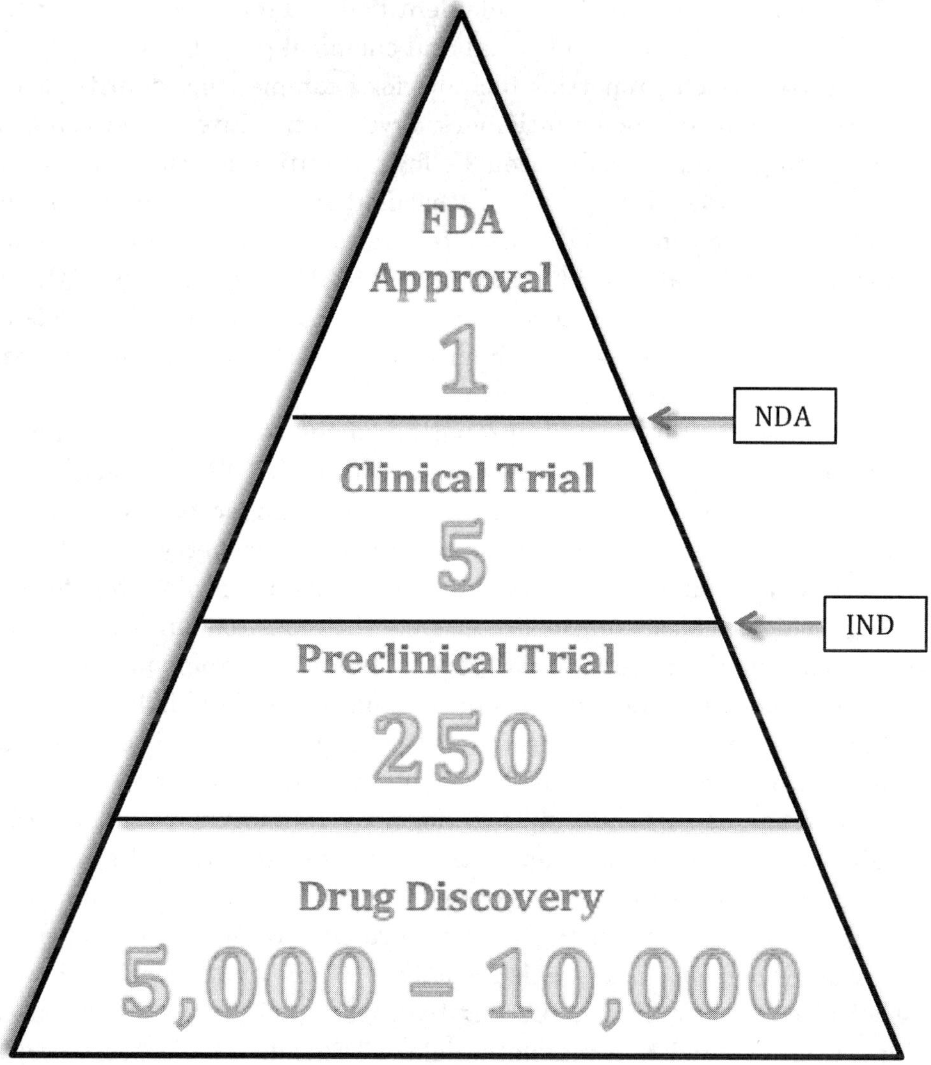

The search for a lead compound can be accomplished by a myriad of tests that can screen molecules for their biological activity. Screens can be high-throughput for evaluating large numbers of potential leads or low-throughput for detailed characterization of leads. High-throughput screens can include in vitro binding studies, X-ray crystallography, and NMR-derived structure assays (to find compounds structurally similar to already known active ones); low-throughput assays include biological (i.e., tissue) or in vivo pharmacology assays.

To qualify for "lead" status, a compound has typically been evaluated in many ways including dose-response experiments, structure-activity relationship (SAR), cellular or tissue assays, and relevant pharmacology models.[5]

After a lead compound has been identified and its activity confirmed in multiple assay systems, its physical and chemical properties must be well characterized. Such properties include, for example, the identification of any present chirality or enantiomers, crystal structure, general appearance, melting point, or salt form. Following initial characterization, the lead compound goes through a process called *optimization* where the compound is prepared in a more cost effective and environmentally friendly way. Optimization allows for tailoring of the ADME properties of the API to reduce both unwanted interactions as well as potential side effects and toxicities. Thus, at this stage, chemistry and biology must work closely together for optimization and testing.

The physical and chemical properties of the lead compound must be understood for any further studies to continue. This characterization takes on several forms. For example, in order to understand the chemical structure and formula of the compound, mass spectrometry is used in conjunction with elemental analysis.[6] Mass spectrometry is a method for identifying molecules based on the detection of the mass-to-charge ratio of ions generated from the molecule by vaporization and electron bombardment of the compound. Deflection of the ions through a magnetic field results in a characteristic pattern that is used to define the structure of the compound of interest. Elemental analysis describes the elemental composition of the compound to determine its empirical formula. In addition to the molecular weight, the molecular formula is then determined. Once the molecular formula is determined, the structure of a compound is then mapped. A key tool in determining chemical structure or how the elements relate to one another is nuclear magnetic resonance (NMR). NMR looks at how the nuclei of particular atoms, such as hydrogen and carbon, in a molecule align when exposed to a magnetic field. Ultraviolet (UV) spectroscopy is also used to detect the presence of conjugated double and triple bonds. Infrared (IR) spectroscopy is different and detects different functional groups that may be present in the compound based on their absorption of infrared radiation. Finally, X-ray crystallography gives any information on the compound's crystal structure, if it exists. All these methods contribute to determining the overall chemical structure.[7]

5. *Id.* at 25–34.

6. *See* Adegoke Adeniji, & Adeboye Adejare, *Chemical and Physical Characterization of Potential New Chemical Entity*, *in* Preclinical Developmental Handbook: ADME and Biopharmaceutical Properties 212–13 (edited by Shayne Cox Gad John Wiley & Sons, Inc., 2008).

7. *Id.* at 213–14.

A compound must be pure before it can be used for testing in either in vitro or in vivo studies. NMR, UV, and IR spectroscopy can detect impurities in a test compound. Impurities can also be detected through differences in melting point (solids) or boiling point (liquids) as compared to the lead compound. Chromatography is a common purification technique that separates these impurities from the lead compound.[8]

Solubility is important not only for formulation and manufacturing considerations, but it also can help to predict the behavior of a compound in the human body.[9] Whether a compound is acidic or basic (pKa or disassociation constant) is also critical to formulation as well as to pharmacokinetics within the body. Insolubility of a drug, or a drug that is not readily soluble in common water or commonly used pharmaceutical solvents, can lead to failure in the very beginnings of preclinical testing.

Lipophilicity, or the tendency of a compound to partition into organic and aqueous environments, is as important as solubility. Commonly measured as the partition coefficient, it is considered one of the most important physiochemical properties of a compound. Lipophilicity can predict how a drug crosses biological membranes, binds to proteins in the blood, and binds to receptors and can even predict its ease of excretion from the body. In other words, determining the partition coefficient and having a measure of lipophilicity are ways to predict the pharmacokinetics of a compound following administration. In the best of situations, an ideal drug candidate will have enough water solubility so that it dissolves readily in bodily fluids, yet have sufficient lipophilicity to easily cross biological membranes.[10]

II. Pharmacokinetics

Pharmacokinetics (PK) deals with the fate of a drug from the moment that it is administered up to the point at which it is completely eliminated from the body. The science of pharmacokinetics describes how the body affects a specific drug after administration through the processes of absorption (the process of a substance entering the bloodstream), distribution (the dispersion or dissemination of substances throughout the fluids and tissues of the body), metabolism (the irreversible transformation of parent compounds into daughter metabolites), and, finally, elimination (the removal of the substances from the body).[11] (See Figure 2.)

8. *Id.* at 214–15.
9. *Id.* at 215–16.
10. *Id.*
11. *See* Allen et al., *Dosage Form Design: Biopharmaceutical and Pharmacokinetic Considerations*, *in* ANSEL'S PHARMACEUTICAL DOSAGE FORMS AND DRUG DELIVERY SYSTEMS,

Figure 2 Pharmacokinectics = ADME

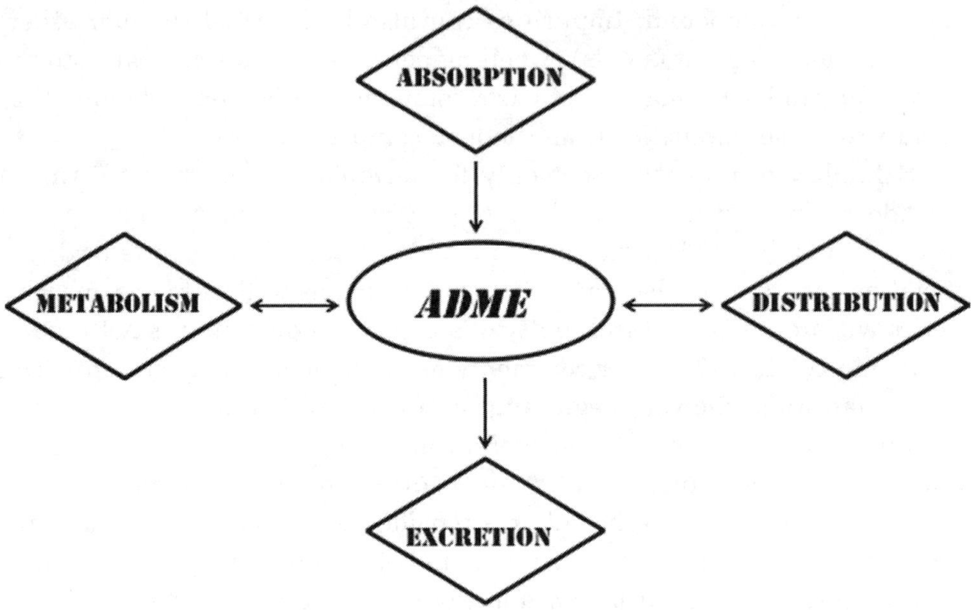

Pharmacokinetic properties determine the amount or the concentration of a drug that is delivered to the appropriate site(s) of action in the body (i.e., the dosage or initial concentration, the extent of absorption and distribution, and the amount and speed of elimination).[12] The process that describes the actions of the active compound (i.e., the response at the appropriate site of action) is called pharmacodynamics.[13]

Absorption is the movement of a drug into the bloodstream and involves several phases. First, the drug needs to be introduced via some route of administration (oral, topical, dermal, etc.) and in a specific dosage form such as a tablet, capsule, solution, and so on. In other situations, such as intravenous therapy, intramuscular injection, enteral nutrition, and others, absorption is even more straightforward; there is less variability in absorption, and bioavailability is often near 100 percent. Typically, intravenous administration does not involve absorption; thus, there is no loss of drug. The fastest route of absorption is inhalation, as there is nearly immediate absorption into the bloodstream.[14]

Knowing how a lead compound or API is distributed following absorption is also critical for further development of a proper dosage form and

supra note 3, at 142–85; Wolfgang .A. Ritschel & Gregory L. Kearns, Handbook of Basic Pharmacokinetics Including Clinical Applications 10–12 (6th ed. 2004).

12. Allen et al., *supra* note 11, at 142.
13. Ritschel and Kearns, at 11.
14. Allen et al., *supra* note 11, at 143–72.

for other preclinical studies. Distribution describes the reversible transfer of drug from one location to another within the body. Once a drug enters into systemic circulation by absorption or direct administration, it must be distributed into interstitial and intracellular fluids. Each organ or tissue can receive different doses of the drug. The drug can also remain in the different organs or tissues for a varying amount of time (i.e., storage). The further distribution of a drug *between* tissues is dependent on such physiological processes as vascular permeability, regional blood flow, cardiac output, and perfusion rate of the tissue. In addition, tissue and plasma binding of the drug plays a key role in distribution as does its lipid solubility and pH partition. For example, a drug is easily distributed in highly perfused organs such as the liver, heart, and kidney, but it is far less easily distributed through less perfused tissues like muscle, fat, and peripheral organs. A drug can also be moved from the plasma to the tissue until an equilibrium is established.[15] The rate and extent of distribution into tissues can affect the therapeutic effects as well as drug toxicity. Should a drug's distribution profile prove undesirable, there are options to manipulate it so as to achieve the desired effects. For example, a delivery system can be utilized to employ an extended release formulation, or the chemical structure of the compound can be modified (e.g., prodrugs[16]).

Metabolism is the mechanism of elimination of foreign and undesirable compounds from the body. Because information on the metabolism of a drug plays a significant role in selection and further characterization of the drug, an in-depth look at the mechanism of drug metabolism is crucial early in preclinical development.

The major site of metabolism in the body is the liver. Metabolism in the liver is a complicated series of processes involving several different enzyme systems and cellular compartments. For example, liver microsomes are in the endoplasmic reticulum of liver cells or hepatocytes and contain a group of enzymes critical to drug metabolism, known as the cytochrome P450 system. These enzymes are responsible for such chemical reactions as aromatic hydroxylation, aliphatic hydroxylation, oxidative N-dealkylation, oxidative O-dealkylation, S-oxidation, reduction, and hydrolysis. These chemical reactions can, for example, make a drug more soluble, facilitating elimination via the kidneys. Alternatively, further conjugation reactions can then occur. These reactions include glucuronidation, sulfation, amino

---

15. *Id.* at 148–50.

16. A prodrug is a drug that has been chemically modified with the addition of a particular substituent or side chain. A prodrug can be modified in such a way as to make it inactive when initially administered, becoming active only upon further metabolism or distribution to a particular site in the body where it is metabolized or enzymatically activated.

acid conjugation, acetylation, methylation, or glutathione conjugation, all of which facilitate elimination via the kidneys.[17]

Clearance, or elimination, is arguably the most important pharmacokinetic process in drug development. Clearance can be the first priority in measuring bioavailability, dosing frequency, and duration of action and in determining a route of administration. If the rate of clearance is not optimal, drug development may not move forward. Clearance allows the body to avoid accumulation of a compound or its by-products. Elimination is typically measured by estimating a compound's half-life of elimination ($t_{½}$), or the time it takes for 50 percent of the compound to be excreted. Half-life of elimination can be either dose-dependent or dose-independent, depending on the particular drug. Alternatively, elimination or clearance can be estimated by measuring the volume from which the entirety of the compound is removed per unit of time, where volume can be blood or urine.[18]

There are three major routes of elimination in the human body. The first of these, and the most important, is renal elimination (i.e., through the kidneys). The kidney is the major organ used for the clearance of drugs in mammals.[19] The kidney mainly eliminates water-soluble products, while lipophilic, or fat-loving, compounds may use alternate routes. However, the metabolites of lipophilic drugs may still undergo renal elimination. In renal disease, a compound or metabolite that is cleared through the kidney will accumulate and cause toxicity.

Many factors can influence the kidney's ability to eliminate drugs, including age, protein binding, increased drug dosage, and drug-drug interactions. Although clearance may also involve other organs than the kidney, it is almost always synonymous with renal clearance or renal plasma clearance. Each substance has a measurable specific clearance that depends on its filtration characteristics. Clearance is a function of glomerular filtration, secretion from the peritubular capillaries to the nephron, and reabsorption from the nephron back to the peritubular capillaries in the kidney.

Another route of elimination is hepatic elimination (i.e., through the liver). Because the liver contains the all-important cytochrome P450 enzymes, many drug-drug interactions and disease-drug interactions can occur in the liver.[20] The liver processes compounds using two processes—biotransformation and biliary excretion. Biotransformation is a metabolic process in which drugs are removed from the enteral route by

17. *See* Sree D. Panuganti & Craig K. Svennsson, *Drug Clearance*, *in* PRECLINICAL DEVELOPMENTAL HANDBOOK: ADME AND BIOPHARMACEUTICAL PROPERTIES, *supra* note 6, at 715–33.

18. *See id.*

19. *Id.* at 720–28.

20. *Id.* at 728–33.

the liver before they become systemically available. This is also referred to as the first-pass effect. In biliary clearance, compounds are directly excreted into bile from the liver.

The last major route of elimination is pulmonary, or via the lungs. Here, some compounds are excreted in exhaled CO_2, which can be directly measured. For example, ethanol is eliminated partly through exhalation from the lungs. In fact, ethanol concentrations in the blood can be measured by determining ethanol concentrations in exhaled CO_2.[21]

A. Safety and Efficacy

Along with characterization of the ADME properties of the lead compound just described, preclinical development includes efficacy, pharmacology, and experimental toxicology studies to help define the dose, route, and frequency required for subsequent studies. There is typically no predetermined order or timeline for all of these studies. Many are conducted simultaneously and the results used to continually modify and optimize how, in what form, and how often the lead compound will be dosed. As optimization progresses, however, safety and efficacy studies play an increasing role in preclinical development.

1. Pharmacology Studies

Pharmacology studies are done using one or more animal models to study the initial efficacy of the lead design. The aim of the studies is to demonstrate the desired effect in vivo. Pharmacology studies identify the ADME characteristics of an investigational drug, while a pharmacokinetics study provides information such as peak plasma concentrations, the half-life of the compound, and clearance. Generally animal studies begin in rodent species (e.g., rat, mouse, hamster, etc.) before progressing to larger or other suitable species (e.g., rabbit, dog, cat, ferret, and primate). Pharmacology studies fall into several categories.

Efficacy: At least one or more pharmacological animal models of the disease are used in initial efficacy studies to demonstrate that treatment with drug candidates has the desired therapeutic effect (efficacy). These studies help to identify the best lead compounds that can be further developed. Although the goal is to test the promising drug candidates in an animal model that best mimics the particular human disease state, there are times when such models do not exist. In those situations, animal models

21. *Id.* at 717.

are chosen that best act as the most efficient drug screening test available that mimics at least one measure of human efficacy used.[22]

Pharmacokinetics (PK): Pharmacology studies are also utilized to address and characterize the properties (described earlier) of absorption, distribution, metabolism, and excretion (ADME) as well as bioavailability of the drugs under development. Bioavailability must be determined in order to examine the percentage of drug absorbed by the body as defined by quantity in plasma. PK studies are critical to characterize how the drug acts within the body and, likewise, how the body affects the drug following administration. Several PK parameters are measured, including maximum attainable plasma concentration (C_{max}), time after dose administration to C_{max} (i.e., T_{max}), mean residence time in the plasma, and renal clearance. PK is critical to the eventual selection of a viable drug or API candidate. Drugs that possess poor PK properties are typically eliminated from further development. Thus, PK has become a predictive screening method for sorting out promising drug candidates. Along with efficacy results, compounds that have favorable PK properties move farther along the preclinical development pathway.[23]

Dose-Range: Early dose-range and toxicity studies will often include single- and multiple-dose administration with several different observation times. Some of the goals of these pharmacology studies are to establish safety parameters such as maximum tolerated dose (MTD), to note any observable signs of toxicity, and to provide a rationale for setting dose levels in future studies. Early range-finding animal studies typically include only a few animals and a single sex per dose level. It is only after a rough suitable dose range is identified that more animals are used per study to enhance statistical comparisons. Typically, only a few variables are measured in these early dose-finding studies as larger numbers of variables will be collected in later studies.

Metabolism: Sometimes following submission of an IND (or during the FDA's pre-submission consultation program described later in this chapter), the FDA may recommend that metabolic studies be performed. These studies usually involve in vitro methods (e.g., hepatic microsomal or culture assays) in order to evaluate possible drug-drug interactions and cytochrome P450 enzyme issues prior to any further animal or human testing. The goal is to examine and identify drug metabolite profiles at significant levels. Certain metabolites are found to be "active" (i.e., possess pharmacological activity) and may require further investigation. Important interspecies

22. *See* Paul Willner, *The Validity of Animal Models of Depression*, PSYCHOPHARMACOLOGY 83:1–16 (1984).

23. *See* Allen et al., *supra* note 3, at 153–82.

differences in how the drug candidate is metabolized are also investigated at this stage.

B. Toxicology

One goal for all pharmacology studies, whether they be for testing efficacy, safety, or toxicology, is to conduct at least initial testing using computer screening methods that avoid the use of costly animal or clinical studies. However, despite the tremendous advances in the fields of pharmacology, pharmaceutics, pharmacology, and toxicology, the application of such computerized techniques has been minimal. Thus, the primary methods used to assess safety remain the tried-and-true tested methods of toxicology studies conducted in various animal (rodent and non-rodent) species.

It is only through very thorough animal studies that safety characteristics of the potential lead compounds, such as no observable adverse effect level (NOAEL) and maximum tolerated doses, are found. Typically, the highest dose levels tested in the toxicological animal studies are based on the maximum tolerated dose determined from the dose-finding studies. These doses almost always differ from the expected dose levels used in subsequent clinical trials. At this point in the drug development scheme, the route and frequency of drug administration, vehicle, and dose formulation that will likely be used in initial human trials is reflected in the definitive animal studies.[24]

Toxicology studies in animals must take into account differences in the species and should be conducted in a species that is particularly sensitive to the toxic effects of the drug in question. Many toxicology studies are conducted in rodents, dogs, or other appropriate species. Typically at least two good laboratory practice (GLP)–compliant safety studies are required by the FDA wherein the route of administration must mimic the proposed clinical route.[25] Thus, for example, even in rodent species, if the proposed route of administration of the drug candidate is oral, the test drug must be administered orally by gastric gavage. Besides route of administration, these studies must also mimic the proposed duration and frequency of dosing and the dose regimen of all the eventual clinical trials.

Later or pivotal safety studies are usually performed with a drug manufactured under good manufacturing practice (GMP), or, at a minimum, a drug product that has been shown to be the same as that used in the animal safety studies. If, for example, the FDA points to significant

24. *See id.* at 153–82.

25. *See* Shayne Cox Gad., *Regulatory Requirements for INDs/FIH (First in Human) Studies, in* PRECLINICAL DEVELOPMENTAL HANDBOOK: ADME AND BIOPHARMACEUTICAL PROPERTIES, *supra* note 6, at 1267–307.

differences that were observed between the materials to be used in the clinical trials and those used in the preclinical safety studies, bioequivalency may be required. Often, in order to preemptively qualify some of the possible manufacturing impurities, such as enantiomers or polymorphs (see later in this chapter) that may be present in large-scale batch production, animal studies may use a smaller batch that contains at least 5 percent impurities.

In addition to the pivotal types of studies described earlier, the International Conference on Harmonisation of Technical Requirements for Registration of Pharmaceuticals for Human Use (ICH) has published several guidelines for regulatory authorities and the pharmaceutical industry concerning safety studies and required ancillary studies to be included in most regulatory packages such as the IND. ICH guidelines include standard procedures for the so-called safety pharmacology core battery, which includes assessments for central nervous, cardiovascular, and respiratory systems. ICH guidelines also provide for genetic toxicology studies required for most small molecules.[26] The FDA requires data from three genetic toxicology assays for most IND applications: gene mutation in bacteria, including four strains of *Salmonella typhimurium* and one strain of *Escherichia coli* (Ames test); an in vitro mammalian cell assay (either evaluation of chromosomal damage in Chinese hamster ovary cells or the mouse lymphoma mutagenesis assay); and an in vivo chromosomal damage in rat or mouse hematopoietic cells (such as the rodent bone marrow micronucleus assay).[27] For drugs that possess immunosuppressive effects or toxicities, additional studies may be required.[28] Similarly, carcinogenicity and reproductive toxicology studies are sometimes necessary.

C. Drug Supply and Scale-Up

At some point during preclinical development, the API or lead drug candidate must be manufactured and supplied on a large-scale basis. No matter how the API is manufactured (e.g., chemical synthesis, fermentation, recombinant DNA, biotechnology methods, isolation/recovery from natural sources, etc.), the API must be well characterized in terms of structure identity (crystalline or polymorphic), salts, impurities, stability, chirality (enantiomers), appearance, solubility, and other chemical and physical properties. These properties are critical to API scale-up and manufacturing.

---

26. *See id.*; *see also* U.S. Food & Drug Admin., Guidance for Industry: S2(R1) Genotoxicity Testing and Data Interpretation for Pharmaceuticals Intended for Human Use (2008).

27. *See generally id.*

28. *See* U.S. Food & Drug Admin., Guidance for Industry: S8 Immunotoxicity Studies for Human Pharmaceuticals (2006).

An adequate and consistent quality drug supply is vital to preclinical (as well as clinical) drug development. Increasing quantities of higher quality APIs are required, from milligram research-grade material early on to kilogram quantities for later studies. During this phase of preclinical development, many preformulation activities begin in order to thoroughly characterize the API and determine its physical and chemical properties, including salt forms, polymorphic forms, solubility, and stability. At this point, a recommended formulation is proposed, and further studies entail large-scale manufacturing efficiency, cost, stability, control of impurities, and batch consistency.

During the scale-up process of preclinical development, the original chemical processes often used for making the API must be changed to accommodate changes in starting materials and reaction materials in order to maximize consistency and yield. The process moves from being able to produce gram quantities to being able to produce kilogram quantities. All of the synthetic steps, along with step-by-step control documentation, are written into the master batch record and included in the IND. API stability and degradation, including identity of major degradation products and stability under a variety of storage conditions, are all documented. Eventually, required batches will be produced along with a Certificate of Analysis or GMP release for each batch. At that point, the batch is ready to be used in GLP safety toxicology studies (described earlier in this chapter) or formulated for clinical use.

1. Salt and Solid Forms

Form selection is a critical step in drug development. Physical properties such as solubility, bioavailability, or toxicology can vary greatly, depending on solid form. The selection of the API, its polymorph, and its salt form is a significant milestone that affects all subsequent drug development.[29] A salt form must be chosen before toxicology and bioavailability studies can be done, and these studies must be repeated if the salt form is changed, resulting in a possible loss of production time and competitive advantage.[30]

The salt form is made by reacting the chosen API with a counter-base or counter-acid.[31] Salt forms can be used to stabilize a lead compound. There are many different considerations when choosing a salt form, chief of which is the delivery system desired of the lead compound. The salt will have different physical properties than the API alone, such as increased water solubility. Choosing a particular salt form over the free compound

29. *See* Howard Y. Ando & Galen W. Radebaugh, *Preformulation*, *in* Remington's the Science and Practice of Pharmacy 701 (Alfonso R. Gennaro ed., 20th ed. 2000).

30. *Id.* at 712.

31. *Id.* at 704.

can be utilized to affect the dissolution rate. For example, a salt will dissolve more rapidly in water due to ionization of the salt species, while the free form does not undergo ionization and, thus, dissolves slower, if at all. The two most common salt forms used are hydrochloride salts and sodium salts, which are both polar groups.[32] Hydrochloride, when used as a salt, is a weak acid and undergoes complete ionization in water. Sodium, when used as a salt, is a weak base.

There are two common approaches to selecting optimal salt forms for an API.[33] One involves sorting salt forms sequentially based upon hygroscopicity, thermal analysis/X-ray diffraction, and accelerated solid-state stability. Hygroscopicity eliminates any salt forms that absorb excessive moisture. Thermal analysis and X-ray diffraction can reveal problems with the crystalline structure when exposed to high moisture and pressure (as in tablet compaction).

Solid-state chemistry of drugs has seen growing importance in the pharmaceutical industry for the development of useful APIs and stable dosage forms. The stability of drugs in solid dosage forms is important because solid dosage forms are the most common pharmaceutical formulations in clinical use. In solid-state stability studies of drugs, an ideal accelerated method (e.g., exposing the solid dosage form to high or low temperatures to measure degradant materials) must not only be selected by different complicated methods but must also detect the formation of degraded product.

The second most common approach to selecting salt forms involves the use of statistics and knowledge of the properties of a chemical structure to predict the behavior of a salt form. Salts are made and then tested for stability in accelerated conditions.

However, it is important to understand that despite all the ways a formulation chemist has of choosing a particular salt form, the world of salt forms from which she or he has to choose is limited and finite. For example, *Remington's The Science and Practice of Pharmacy 20th Edition* lists several salts that are considered to be those which have been deemed "pharmaceutically acceptable" based upon their ease of use, low levels of toxicity, and ability to be adapted to drug formulations.[34] Table 1, adapted from *Remington's*, shows the relative frequency—over a ten-year period examined by those authors—of commonly used salt forms in pharmaceutical formulations.

32. *Id.*
33. *Id.* at 712.
34. *Id.* at 704.

Table 1
Pharmaceutically Acceptable Salts

Salt Form	Frequency
No salt used	390
Hydrochloride	102
Sodium	37
Mesylate	8
Acetate	8
Fumarate	8
Bromide	6
Nitrate	6
Sulfate	5
Tosylate	3
Maleate	3
Phosphate	2
Succinate	2
Besylate	2
Tromethamine	2
Hydrobromide	1
Glucuronate	1
Tartrate	1
Citrate	1
Acistrate	1
Xinafoate	1
Potassium	1
Bismuth	1

2. Enantiomers

Enantiomer identification and isolation also play a pivotal role in preclinical drug development. An enantiomer of an organic chemical compound is one of two stereoisomers[35] that are mirror images of each other that are nonsuperimposable (not identical); for example, one's left and right hands are the same except for opposite orientation.[36] Enantiomers have the same physical properties (except for the direction in which they rotate polarized light and how they interact with different optical isomers of other compounds).

However, different enantiomers of drug compounds may have substantially different biological or toxicological effects. For example, the binding of endogenous substrates or drugs binding to receptors or the active site of enzymes is stereoselective (i.e., preferential for one stereoisomer versus the other).

The biological effects of two enantiomers can be more than innocuous. It is possible for one isomer to have a therapeutically beneficial effect while the other may have severe toxicological effects. Perhaps the most infamous example is the drug thalidomide, a controversial drug that was banned worldwide in 1961. Thalidomide was originally prescribed to ease pain and help pregnant women cope with morning sickness. However, to everyone's horror, the women gave birth to thousands of babies, every one of which had some sort of deformity in their limbs, and some of which were missing limbs completely. It was subsequently found that these horrific effects were due to one of the enantiomers of thalidomide. Thalidomide has a chiral center that allows for two enantiomers, or two slightly different isomers (differing only by the arrangement of their functional groups). The two enantiomers are classified as either "S" or "R." The "R" enantiomer is a relatively safe drug with sedative attributes, while the "S" enantiomer can result in devastating effects.[37]

3. Polymorphs

Polymorphism is the ability of a solid material, such as a drug compound, to exist in more than one form or crystal structure. The presence of polymorphic forms is important in the development of pharmaceuticals. Besides typically receiving eventual regulatory approval for only a single crystal form or polymorph, polymorphism in drugs can also have a direct effect on such things as PK and drug formulation. For example, orally administered drugs are often crystalline solids, and dissolution rates

35. Stereoisomers have the same molecular formula and sequence of bonded atoms but differ only in the three-dimensional orientations of their atoms in space.

36. *See* John E. McMurry & Robert C. Fay, McMurry Organic Chemistry 295 (Brooks/Cole Publ. Co., 4th ed. (2003).

37. Interestingly, one dangerous outcome of thalidomide circulating in the blood is that it racemizes, creating equal amounts of both enantiomers, even if only one was initially present.

in the gastrointestinal tract will depend on the exact crystal form of a polymorph. Furthermore, it is possible for one polymorph to be virtually inactive compared to an alternative crystal form, but the inactive polymorph can subsequently convert the active polymorph into the inactive form on contact (e.g., ritonavir). This is most likely due to its lower energy and greater stability, making spontaneous interconversion energetically favorable. Thus, it is easy to conceive how a few molecules of the lower energy polymorph could convert a manufacturing plant full of ritonavir into a useless, inactive polymorph. Often, these issues of manufacturing can be solved by choosing a different formulation (e.g., tablets versus capsules).

4. Formulations

Formulating a drug product involves combining the API with other inactive chemical ingredients (i.e., excipients) to create a dosage form that can be administered to the organism, absorbed, and distributed to the appropriate site of action at a sufficient concentration over an adequate amount of time to have the necessary therapeutic effect. Creating the proper formulation is often a major hurdle in drug development.

Several key pieces of information must be ascertained during preclinical formulation development. First, it must be known which route of administration is desired for the particular API during the upcoming clinical studies. At this stage, the route of administration intended for the clinic should be clearly identified. Table 2 lists the different and varied routes of administration that drugs may be given and how they would be typically formulated.

Table 2
Routes of Administration

Route	Where Administered	Typical Formulation
Enteric	Oral (swallowed by mouth) Buccal (between the cheek and gum) Sublingual (under the tongue) Rectal (inserted into the anus)	Tablet; capsule; solution; suspension; orally disintegrating film; syrup; solid crystals; natural or herbal plant; seed or food

(continued)

Table 2
Routes of Administration (continued)

Route	Where Administered	Typical Formulation
Injection	Intravenous (into a vein) Intra-arterial (into an artery) Intramuscular (into a muscle) Subcutaneous (under the skin) Epidural (into the epidural space) Intracerebral (into the cerebrum) Intracerebroventricular (into the cerebral ventricles) Intracardiac (into the heart) Intraosseous (into the bone marrow) Intrathecal (into the spinal canal) Intraperitoneal (into the peritoneum) Intravesical (into the urinary bladder) Intravitreal (through the eye) Intracavernous (into the base of the penis) Intravaginal (into the vagina) Intrauterine (into the uterus)	Solution; suspension; suppository; drops; douche; pessary
Topical/ Transdermal	Transdermal (diffusion through the intact skin for systemic rather than topical distribution) Transmucosal (diffusion through a mucous membrane) Topical (for effects directly on the skin)	Paste; powder; lotion, solution; patches; cream; gel; liniment; balm; ointment

Table 2
Routes of Administration

Inhalation	Inhaled or breathed into the lungs Intranasal (into the nose or nasal cavity)	Gas; mists; aerosols

Another consideration to formulating a particular API for a particular route of administration must also be the specific physicochemical properties of the drug. For example, a highly water-soluble drug will be formulated differently (e.g., require different additives or excipients) than a drug that is poorly water-soluble. Thus, excipient selection is critical as well. The particular excipients chosen for a particular formulation must be physically and chemically compatible with the drug in question.

An excipient is generally a pharmacologically inactive substance formulated with the API. Excipients that can be used are bulking agents, fillers, or diluents to allow adequate dissolution of the API from the dosage form. They can also be used to facilitate drug absorption or solubility or for other pharmacokinetic considerations. During preclinical development, excipients are often chosen in order to facilitate the manufacturing process and to aid in the handling of the active substance concerned, such as by facilitating powder flowability or nonstick properties. Furthermore, excipients can aid in long-term stability and shelf life of the formulation (see the following paragraphs).

Formulations (and excipient choice) must also take into account human patient factors, such as tablet/capsule size for ease of swallowing, taste (particularly in drugs to be administered to children), stability, and shelf life.

Drug stability (and maintaining a long shelf life) is one of the most important factors when choosing a formulation.[38] Stability typically refers to the degree to which a drug formulation can withstand both chemical and physical changes. How long a particular formulation remains stable in the container on the pharmacist's or patient's shelf will determine expiration date. Lack of stability within a formulation can lead to degradation of the API or even of the formulation itself (i.e., tablets crumbling, foul odors, discoloration, etc.). The long-term stability of a drug product can also be affected by the introduction of water, change in molecular charge, pH changes, or exposure to air.

38. *See* Alexander V. Lyubimov, *Dosage Formulation*, *in* PRECLINICAL DEVELOPMENT HANDBOOK: ADME AND BIOPHARMACEUTICAL PROPERTIES, *supra* note 6, at 573.

Timing of formulation development during preclinical stages is important as well. For example, optimization often does not occur until after submission of the IND. Therefore, often only a proposed simple formulation is used for the IND submission and/or in initial Phase 1 trials in humans. However, as the clinical trials start to involve more subjects, oral formulations may be prepared in the form of tablets or capsules containing several dosage strengths to allow for ease of administration to study subjects of differing body weights. It is typically not until Phase 2/Phase 3 trials that the final formulation is selected, which will require further clinical trials and eventual product marketing.[39]

III. Summary

The IND application submitted to the FDA (discussed in more detail in Chapter 7) contains all of the components and data gathered during the development of the candidate drug or API. The preclinical development of a drug candidate as described in this chapter is done in order to determine if the product is safe for initial use in humans, if the compound exhibits pharmacological activity, and if the product will lead to further development. When a product is identified as a viable candidate for further development, the IND sponsor then focuses on collecting the data and information necessary to establish that the product will not expose humans to unreasonable risks when used in limited, early-stage clinical studies.

Prior to the submission of any IND to the FDA, communication is done through the agency's Pre-Investigational New Drug Application (Pre-IND) Consultation Program. This program provides early communications between sponsors and divisions in the FDA that are responsible for coordinating new drug development activities. The goal is "to provide guidance on the data necessary to warrant IND submission."[40]

39. Richard Boehner, A Formulation Strategy for Quickly Reaching Clinical Trials 10–14 (BioProcess International 2009).

40. *See Investigational New Drug (IND) Application*, U.S. Food & Drug Admin., http://www.fda.gov/drugs/developmentapprovalprocess/howdrugsaredevelopedandapproved/approvalapplications/investigationalnewdrugindapplication/default.htm (last updated Oct. 18, 2013).

chapter 3

Preclinical Patent Considerations[1]

I. Identifying Intellectual Property (IP) Opportunities and Ensuring Freedom to Operate

Drug discovery and development is an expensive, long, and risky process. Although the most-often quoted estimate for bringing a new compound to market is $1.3 billion,[2] the average cost for developing a drug by a major pharmaceutical company can be as high as $12 billion.[3] It also takes on average 10 to 15 years to develop a drug.[4] At the end of the day, only one out of 5,000 to 10,000 compounds identified during drug discovery and only one out of four to eight compounds entering into clinical trials receives Food and Drug Administration (FDA) approval.[5,6] Of those approved, only about two in ten will recoup research and development (R&D) costs.

1. Mark J. Feldstein, PhD, and Shana K. Cyr, PhD, of Finnegan, Henderson, Farabow, Garrett & Dunner, LLP, Washington, D.C.; and Ian Y. Liu, PhD, of Idenix Pharmaceuticals, Inc., Cambridge, Massachusetts.

2. Joseph A. DiMasi & Henry G. Grabowski, *The Cost of Biopharmaceutical R&D: Is Biotech Different?*, 28 Managerial & Decision Econ. 469 (2007).

3. Matthew Herper, *The Truly Staggering Cost of Inventing New Drugs*, Forbes Feb. 10, 2012), http://www.forbes.com/sites/matthewherper/2012/02/10/the-truly-staggering-cost-of-inventing-new-drugs/2/.

4. Joseph A. DiMasi & Henry G. Grabowski, *The Price of Innovation: New Estimate of Drug Development Cost*, 22 J. Health Econ. 151 (2003); Michael Dickson & Jean Paul Gagnon, *Key Factors in the Rising Cost of New Drug Discovery and Development*, 3 Nature Revs. Drug Discovery 417 (2004); PAREXEL Int'l, PAREXEL Biopharmaceutical R&D Statistical Sourcebook 2010/2011 (2010).

5. Tufts Ctr. for the Study of Drug Dev., *Large Pharma Success Rate for Drugs Entering Clinical Trials in 1993–2004: 16%*, 11 Impact Rep. http://csdd.tufts.edu/files/uploads/summaryjul-aug09.pdf

6. Steve Morgan et al., *The Cost of Drug Development: A Systematic Review*, 100 Health Policy 4 (2011).

Although pharmaceutical companies are investing about 20 percent of their sales[7] into R&D ($48.5 billion in 2012), at the end of the day only two out of ten approved drugs will recoup the average R&D costs.[8]

Innovators who discover, develop, and sell new drugs also face constant competition. Before a new drug product is approved by the FDA, innovators face competition from other innovators. After approval, innovators face further competition from generic manufacturers.

To enter a business that is as expensive, risky, and competitive as the pharmaceutical industry, it is imperative for an innovator to make sure that (1) its technology and/or product is well protected and (2) it is free to conduct R&D and develop products in the field. In practice, the former is achieved by successfully identifying and protecting intellectual property (IP) opportunities in its own R&D, and the latter by constantly monitoring and navigating the patent portfolios of potential commercial competitors and other entities, such as universities.

A. Identifying IP Opportunities

As discussed in Chapter 1, the goal of the preclinical research phase is to identify drug candidates for clinical evaluation. The preclinical phase of drug discovery and development includes a wide spectrum of research activities and can be broadly divided into prediscovery research, drug discovery, and preclinical testing.

When managed properly, preclinical research can provide an innovator company with valuable IP. As discussed in the previous chapter, drug discovery typically starts with a biological target or process causally connected with a medical condition. Biological assays are used to identify a compound or collection of compounds ("lead compound(s)") that can affect the target or process. Medicinal chemistry is employed to modify the lead(s) and to elucidate any structure-activity relationship (SAR). Eventually, a drug candidate is identified for clinical testing. Before the drug candidate enters into clinical trials, preclinical testing, physical characterization, formulation development, and chemical process development are conducted to study and develop physical forms, formulations, and preparations of the drug candidate.

Significant IP value can be found during basic preclinical research in the biological target or process and the method of modulating the target or process. As lead(s) are discovered ("hits to leads," as it is sometimes called), the compound or compound family can also be a significant discovery with

7. Pharm. Research & Mfrs. of Am., PhRMA Annual Membership Survey (2013).

8. John A. Vernon et al., *Drug Development Costs When Financial Risk Is Measured Using the Fama-French Three-Factor Model*, 19 Health Econ. 1002 (2009).

potential IP value. During characterization and development of the drug candidate, its formulation, and chemical preparation, the physical forms (including enantiomers, diastereomers, salts, hydrates, solvates, and polymorphs), formulations, metabolites, methods of preparation, and intermediates that are developed become further discoveries with their own potential IP value. Table 1 lists some of the IP opportunities available in preclinical research.

Table 1
Intellectual Property Opportunities in Preclinical Research

Preclinical Research	Intellectual Property Opportunities
Basic research	Biological targets, processes, assays, therapeutic indications
Drug screening	Compound families (genera)
Hits to leads	Compound selections
Lead development	Active pharmaceutical ingredient (API), enantiomers, diastereomers, salts, hydrates, solvates, polymorphs, particle sizes, crystalline habits
Formulation development	Formulations, dosages, therapeutic regimens
Process development	Methods of preparation, intermediates, impurities
Preclinical testing	Dissolution profiles, pharmacokinetics (PK), pharmacodynamics, metabolites, biomarkers, diagnostics

As discussed in detail in the following sections, identifying and protecting these IP opportunities requires a conscious and concerted effort from all levels in a company.

1. IP Awareness

U.S. patent law, which is continuously evolving, has many nuances that can be traps for the unwary. Raising company-wide awareness of at least the fundamental aspects of IP law can be critical to success.

One fundamental patent law principle is that an invention cannot be patented if it is not new (novel) or if it is obvious. For example, if a claimed invention is published prior to the application date for a patent, it may no longer be considered "new," and the inventor may lose the right to a patent. In many instances, an inventor's own publication has undermined the patentability of her or his invention. The *Klopfenstein*[9] case illustrates this point and highlights why IP awareness throughout the company can be critical.

In October 2000, Carol Klopfenstein and her coworker John Brent applied for a U.S. patent. However, two years before filing their application, they gave a slide presentation on their work at a chemical conference meeting. The presentation was also printed and displayed on a poster board at the meeting. During a subsequent litigation, the court decided that because the inventors, through their disclosures at the meeting, had published the invention more than one year before they applied for a patent, the inventors could not obtain a patent on the invention because it was not new when the patent application was filed.

Thus, raising awareness of at least the fundamental principles of IP law can be very important to an innovator's business. Had these inventors been aware of the potential harm their publications placed, they could have taken different approaches to avoid unintentionally undermining their patents and the companies' business.

2. Implementing a Formalized Corporate Patent System

The primary goals for a formalized corporate patent system are to capture internally generated IP opportunities and incorporate them into the overall business strategy. Two common components of a corporate patent system are invention disclosure and a patent committee.

a. Invention Disclosure

Invention disclosure is used to capture IP opportunities and may include descriptions of

- the invention;
- contributors to the conception of the invention;
- problems that the invention is designed to solve;
- prior approaches to these problems;
- advantages of the invention;
- any possible public uses and sales; and
- any early publications, patents, patent applications, published/patented/marketed technologies, processes, or products.

A sample invention disclosure form is provided as Appendix A. Good practices would entail providing the completed form to IP counsel as a legal communication and for legal review, rather than using it as a substitute for purely technical internal research disclosure.

9. *In re* Klopfenstein, 380 F.3d 1345 (Fed. Cir. 2004).

Improper handling of an invention disclosure may undermine the foundation of any patent. For example, invention disclosure implicates questions of inventorship and, consequently, the duty of disclosure. If not properly addressed, either issue can preclude patent protection or lead to any resulting patent being declared invalid or unenforceable.

First, inventorship in the United States is different from authorship and should preferably be determined by a qualified U.S. patent lawyer. This is important because a patent is invalid unless it lists the correct inventor(s), although errors in inventorship can generally be corrected if made without deceptive intent.[10] Complicating matters, however, the law on inventorship, especially joint inventorship, can be "muddy."[11]

The proper inventors of a U.S. patent are the individuals who conceived the claimed invention.[12] Conception is complete (1) when one has a definite and permanent idea of an operative invention, including every feature of the subject matter claimed, or (2) when "only ordinary skill would be necessary to reduce the invention to practice, without extensive research or experimentation." Conception of a chemical compound generally requires knowledge of both the specific chemical structure of the compound and an operative method of making it.[13]

Joint inventorship exists when two or more individuals contribute to the conception of an invention. Joint inventors do not need to work in the same physical space or work at the same time.[14] Nor must each joint inventor make the same type or amount of contribution to conception of the invention as the other coinventor(s). Rather, a joint inventor needs to at least "(1) contribute in some significant manner to the invention, (2) make a contribution to the claimed invention that is not insignificant in quality, when that contribution is measured against the dimension of the full invention, and (3) do more than merely explain to the real inventors well-known concepts and/or the current state of the art."[15] Thus, a publication coauthor who, for example, does no more than follow another's specific instructions to make a working sample of an invention is not necessarily a coinventor.

Second, each inventor has a duty of candor and good faith to the U.S. Patent and Trademark Office (USPTO), which applies to each claim and exists as long as the patent application is pending. This duty includes a

10. The lack of deceptive intent is no longer required in correcting inventorship under the America Invents Act. *See* 35 U.S.C. §§ 116 and 256 (2011).

11. Mueller Brass Co. v. Reading Indus., 352 F. Supp. 1357, 1372 (E.D. Pa. 1972), *aff'd*, 487 F.2d 1385 (3d Cir. 1973) (joint inventorship is "one of the muddiest concepts in the muddy metaphysics of the patent law").

12. Burroughs Wellcome Co. v. Barr Labs. Inc., 40 F.3d 1223, 1228 (Fed. Cir. 1994).

13. Fina Oil & Chem. Co. v. Ewen, 123 F.3d 1466, 1473 (Fed. Cir. 1997).

14. Pannu v. Iolab Corp., 155 F.3d 1344, 1351 (Fed. Cir. 1998).

15. *Id.*

duty to disclose to the USPTO all information known to the inventor to be material to patentability.[16] Violation of the duty of disclosure with respect to any claim in an application or patent may render the entire application or patent unenforceable[17] and, in limited circumstances, can lead to other related applications or patents in the same technology family being unenforceable.[18]

b. Patent Committee

The primary goal of a patent committee is to implement corporate IP strategies. Typical tasks include the following:

- Reviewing invention disclosures;
- Evaluating each invention in the context of the innovator's own existing products, developing products, and competitors' products;
- Determining whether an invention should be protected by a patent, trade secret, other form of IP, or defensive publishing; and
- Monitoring the implementation of IP protection.

Additional tasks for a patent committee may include implementing an education program, reviewing and approving manuscripts for publications and presentations, making decisions on enforcement or licensing opportunities, and conducting initial reviews on third-party patents and patent applications.

Because an IP strategy should be part of an innovator's overall business strategy and can have important impact on its market share, its optimal implementation requires technical, legal, and marketing expertise. Accordingly, a successful patent committee usually consists of team members from each of these departments.

B. Ensuring Freedom to Operate

Freedom to operate (FTO) is an essential, but sometimes overlooked, requirement for product commercialization. Before discussing the practice of ensuring freedom to operate, it is helpful to discuss what rights a patent affords to the patent owner.

1. Right to Exclude

A U.S. patent gives its owner the right to exclude others from making, using, or selling the patented invention in the United States during the

16. Therasense Inc. v. Becton, Dickinson & Co., 649 F.3d 1276, 1310 (Fed. Cir. 2011) (en banc).

17. *Id.* at 1288 (citing Kingsdown Med. Consultants Ltd. v. Hollister Inc., 863 F.2d 867, 877 (Fed. Cir. 1988)).

18. *Id.* at 1288–89; *see, e.g.*, Consol. Aluminum Corp. v. Foseco Int'l Ltd., 910 F.2d 804, 808–12 (Fed. Cir. 1990).

lifetime of the patent. This exclusive right is rooted in the U.S. Constitution, which gives Congress the power "[t]o promote the progress of science and useful arts, by securing for limited times to . . . inventors the exclusive right to their . . . discoveries."[19] Thus, except for certain circumstances,[20] "whoever without authority makes, uses, offers to sell, or sells any patented invention, within the United States or imports into the United States any patented invention during the term of the patent therefor, infringes the patent."[21]

Other countries afford similar exclusive rights to patent owners.[22] If infringement is found, the infringer can be ordered to stop any infringing activities (called an injunction); pay monetary damages to the patent owner, sometimes up to three times the amount found by a jury or assessed by a court; and/or reimburse reasonable attorney fees to the patent owner if the infringement is found to be willful.[23]

2. Freedom to Operate

FTO needs to be addressed early and often. The analysis involves understanding whether an R&D program may encounter third-party patents and, in doing so, allows a company to develop a course of action to minimize its risk of litigation and minimize any potential damages for infringement. An FTO program usually includes the following three components:

- Identifying third-party patents in FTO searches;
- Analyzing and evaluating risks associated with third-party patents; and
- Managing risks.

a. Identifying Third-Party Patents in FTO Searches

An FTO search starts with defining the scope of the search. As discussed previously, a third-party patent owner can exclude others from making, using, selling, or importing her or his invention. Accordingly, the scope of a third-party patent search should track the innovator company's business plan and, for example, reflect where it conducts R&D, manufactures its products, and makes its sales.

19. U.S. CONST. art. I, § 8.

20. 35 U.S.C. § 271(e), (g) (2012).

21. *See id.* § 271(a); see *id.* § 271(b), (c), (f) for other instances of infringement.

22. *See, e.g.*, Patents Act, 1977, §§ 60, 61 (U.K.); Patentgesetz §§ 139–142b (as amended by the Act on Improvement of Enforcement of Intellectual Property Rights of July 31, 2009, Ger.,); Patent Act of the People's Republic of China arts. 59–74 (Amended Dec. 27, 2008); Patent Act, No. 121, arts. 100–106 (Japan, Apr. 13, 1959); Patent Act, No. 6411, arts. 126–132 (S. Kor. Feb. 3, 2001).

23. *See* 35 U.S.C. §§ 283–285; *see also* Braun Inc. v. Dynamics Corp. of Am., 975 F.2d 815, 822, 829 (Fed. Cir. 1992).

Because drug discovery and development are iterative processes, the results of which cannot be fully predicted in advance, FTO searches might be focused on certain types of patent claims at various stages. However, since the ultimate goal is FTO for a commercial product, it is imperative to keep that end game in mind.

For example, in the preclinical stage, when a new biological target or process is identified, the corresponding search can be focused on the biological targets or processes. When leads are discovered and being developed, the search can be focused on the compound genus. When a clinical candidate is proposed, genus search should be revisited and expanded to address methods of using the clinical candidate in treating identified diseases or other diseases that share similar mechanisms. And when a chemical process of the clinical candidate is being developed, searches on methods of preparing the clinical candidate should be conducted. Again, however, it is not enough to consider short-term FTO; the ultimate commercial product should be part of the overall analysis.

Different databases and different strategies can be used in FTO searches. Commonly used databases include those in the USPTO, the European Patent Office (EPO), the State Intellectual Property Office in China (SIPO), and the World Intellectual Property Organization (WIPO), as well as several commercial databases, including Thomson Innovation and Chemical Abstract Services. Common searching strategies include those based on keywords, inventors, assignees, sequences, formulae, structures, patent classifications, and any combinations of these. Searches can be conducted by in-house librarians, outside law firms, or professional search services.

It is important to consider not only issued patents but also patent applications still being prosecuted at the patent office. If a potentially relevant patent application is identified, diligence may require continuous monitoring of the prosecution to evaluate what claims may issue. Similarly, it may be important to track pending applications related to previously identified issued patents to understand whether the applicant is attempting to, or could later, obtain potentially problematic claims.

b. Analyzing and Evaluating Risks Associated with Third-Party Patents

The second component in ensuring freedom to operate involves comparing an innovator's own product or program against third-party patents and evaluating whether the product or program risks being the subject of a third-party lawsuit. Such a comparison generally simulates an infringement analysis.

The analysis begins by determining the true scope of third-party patents. "It is a 'bedrock principle' of patent law that 'the claim of a patent defines the invention to which the patentee is entitled the right to exclude.'"[24] Properly interpreting claims (called claim construction) is essential for determining whether present or future activity may give rise to liability for patent infringement. Next, the innovator's product is compared to the properly construed claims.[25] A claim covers an accused product if the accused product embodies every element, or limitation, of the claim.[26]

When interpreting claims, the terms "are generally given their ordinary and customary meaning."[27] To determine the ordinary meaning of a claim term, a court will focus on how the patentee used the claim term in the claims, specification, and prosecution history.[28] Other evidence, such as dictionaries, comparable sources, or expert opinions, may be useful to assist in understanding the meaning of a limitation, so long as such sources do not contradict a definition provided by the patent and its prosecution history.

Claim construction can be determinative of infringement and validity. For example, a patent related to Merck's Fosamax® has a single claim:

> A method of treatment of urolithiasis and inhibiting bone reabsorption which consists of administering to a patient in need thereof an effective amount of 4-amino-1-hydroxybutane-1,1-biphosphonic acid.

The common name of the active ingredient, 4-amino-1-hydroxybutane-1,1-biphosphonic acid, is alendronic acid. Teva filed an Abbreviated New Drug Application (ANDA) with the FDA, seeking approval to market a generic version of Fosamax®, and was sued by Merck. Teva argued that, as a matter of claim construction, the claim only covers the free acid form of alendronic acid and, as a result, Teva's monosodium salt does not infringe the patent. After examining the specification and the prosecution history of the patent and listening to expert opinions, both the district court and the Court of Appeals for the Federal Circuit construed the term "4-amino-1-hydroxybutane-1,1-biphosphonic acid" to include the acid salt

24. Phillips v. AWH Corp., 415 F.3d 1303, 1312 (Fed. Cir. 2005) (en banc) (citing Innova/Pure Water Inc. v. Safari Water Filtration Sys. Inc., 381 F.3d 1111, 1115 (Fed. Cir. 2004)).

25. Vulcan Eng'g Co. v. Fata Aluminum Inc., 278 F.3d 1366, 1373 (Fed. Cir. 2002); *see also* Halliburton Energy Servs. Inc. v. M-I LLC, 514 F.3d 1244, 1249 (Fed. Cir. 2008).

26. Read Corp. v. Portec Inc., 970 F.2d 816, 821 (Fed. Cir. 1992).

27. *Phillips*, 415 F.3d at 1312.

28. The prosecution history is the complete record of the proceedings before the USPTO and includes the prior art cited during the examination of the patent.

of alendronic acid and found that administering alendronic acid in its free acid form or acid salt form infringes the patent.[29]

Consequently, an FTO analysis may necessitate considering the impact of multiple different claim constructions. This obviously complicates the analysis and may not be necessary in all cases, but doing so has the potential to avoid surprises at a later date.

c. Managing Risks

When potential FTO issues are identified, there are several potential courses of action. Each has its pros and cons, which will be dictated by the specific facts of the case.

One option would be to simply cancel the R&D project in its entirety. This may be very undesirable for an advanced project, where significant funds have already been spent. However, especially if the FTO issue is identified early, it may be much more desirable to focus limited R&D investments on an alternate project that lacks FTO roadblocks.

Another option is to modify the focus or direction of the project. For example, if an active pharmaceutical ingredient (API) has potential clinical benefits for two different conditions, and there is blocking IP for one, the focus can be shifted to the unencumbered indication. Similar shifts might be envisioned where there are multiple potential lead compounds or formulations, and some carry FTO risks that others do not.

A third option is to maintain the R&D program without change and obtain rights to the potentially blocking IP. If, for example, the IP at issue is owned by a university, it may be possible to work through the university's technology transfer office to obtain a reasonable license. Patents from other entities, such as start-ups, might also be licensed or purchased.

Finally, one might take offensive action to remove potentially problematic patents. The USPTO has recently expanded the ability of third parties to challenge the validity of issued patents in post-grant procedures, such as inter parte review. Third parties can also submit prior art to the USPTO in pending applications, to help insure that obvious or nonnovel patent claims do not issue in the first place. These procedures are significant and less costly alternatives to litigation and provide the potential for clearing an FTO path prior to commercialization.

The options for managing FTO risk are assessed further in the following sections as well as in the next few chapters.

29. Merck & Co. v. Teva Pharm. USA Inc., 347 F.3d 1367, 1372 (Fed. Cir. 2003).

II. Patent Application Filing Strategies

Before discussing patent application filing strategies in detail, it will be helpful to understand what can be patented and what requirements a patent must satisfy.

A. Invention and Patent

1. Patentable Subject Matter

An invention is not patentable simply because it is "good science." Patents are only available for processes, machines, manufactures, compositions of matter, and their improvements.[30] Laws of nature, natural phenomena, and abstract ideas are not patentable.[31]

After three decades of relative stability, a series of landmark cases, including those from the U.S. Supreme Court, has redefined the landscape of patentable subject matter. Cases particularly relevant to the pharmaceutical industry include *Mayo Collaborative v. Prometheus Labs.*[32] and *Association for Molecular Pathology v. Myriad Genetics, Inc.*[33]

The claim in the *Prometheus* case covers a method of optimizing therapeutic efficacy by comparing the level of a drug metabolite with a specified range and adjusting the dosage of the drug according to the comparison. The Supreme Court held that because it covers natural laws and natural phenomena, namely the correlations between the metabolite levels and the efficacy and safety of the drug, and processes that are "well-understood, routine, [and] conventional activity previously engaged in by researchers in the field," the claim is not patentable.[34]

In *Myriad*, the Supreme Court held that "genes and the information they encode are not patent eligible under [35 U.S.C.] § 101 simply because they have been isolated from the surrounding genetic material."[35]

2. Patentability

Patents require utility, novelty, nonobviousness, adequate written description, and enablement. A person is entitled to a patent unless the claimed invention lacks novelty (i.e., "the claimed invention was patented, described

30. 35 U.S.C. § 101 ("Whoever invents or discovers any new and useful process, machine, manufacture, or composition of matter, or any new and useful improvement thereof, may obtain a patent therefor, subject to the conditions and requirements of this title.").

31. Diamond v. Diehr, 101 S. Ct. 1048 (1981); *see also* Bilski v. Kappos, 130 S. Ct. 3218 (2010); Diamond v. Chakrabarty, 100 S. Ct. 2204 (1980).

32. Mayo Collaborative v. Prometheus Labs., 132 S. Ct. 1289 (2012).

33. Ass'n for Molecular Pathology v. Myriad Genetics Inc., 133 S.Ct. 2107 (2013).

34. *Mayo*, 132 S. Ct. at 1294.

35. *Ass'n for Molecular Pathology*, 133 S.Ct. at 2120.

in a printed publication, or in public use, on sale, or otherwise available to the public" before the person applies for a patent, or "the claimed invention was described in a patent" or a published patent application and "the patent or patent application names another inventor" and was filed before the person applies for a patent.)[36]

In addition, if a "claimed invention is not identically disclosed" as just described, but "the differences between the claimed invention and the prior art are such that the claimed invention as a whole would have been obvious to a person with ordinary skills in the art" before the patent application is filed, then a patent for the claimed invention cannot be obtained.[37] Further, if a patent does not sufficiently describe the claimed invention or does not enable a person of ordinary skill in the art to practice the claimed invention, the patent is invalid.[38] The cases in the following sections illustrate how these patent laws have played out in the courts and provide some guidance for preclinical patent considerations.

B. Timing

Timing is critical for obtaining a valid patent and for maintaining market exclusivity, as discussed in more detail in the following two sections. For example, a claim that lacks enablement may be patentable at a later date after additional work and research. On the other hand, wait too long and the subject matter may no longer be new and nonobvious.

As part of the planning process, pharmaceutical companies might also consider the effect of filing patent applications for various aspects of a product in succession rather than all at one time. There can be a balance, though, between the length of patent protection (by filing separate claims to each patentable improvement in succession) and the breadth of patent coverage (by filing broad claims early to avoid the risk of narrow earlier claims rendering later claims obvious). For example, once an API has been patented, that patent becomes a prior art reference. As such, it must be considered for the purposes of assessing novelty and nonobviousness if later seeking additional patent protection on improvements related to that API. Additional patent protection may only be obtained for improvements that were not disclosed or rendered obvious by the original patent.

1. Obtaining a Valid Patent

Under the current law, only the inventor who filed a patent application first can obtain a patent.[39] This law creates a race to file and may counsel

36. 35 U.S.C. § 102.
37. 35 U.S.C. § 103.
38. 35 U.S.C. § 112.
39. 35 U.S.C. § 102.

submitting a patent application as early as possible. In addition, patent law promotes early filing in order to avoid patent ineligibility based on an inventor's own publications or public use. Public disclosure of an invention prior to filing a patent application can jeopardize the validity of a patent claiming the invention, as discussed earlier.[40]

Yet, premature filing can jeopardize the validity of a patent.[41] For example, because of the time gap between the invention of a new molecule and confirmation of its therapeutic value and safety through clinical trials, early filing may result in the inability to obtain a valid patent because the invention cannot be fully described or enabled, or meet the requirements for utility. If a new drug is structurally similar to a prior art molecule, research may be necessary to demonstrate nonobviousness.

The amount of testing necessary to support a patent application depends on the molecule and the utility.[42] In *In re '318 Patent Infringement Litigation*, the Federal Circuit held that claims to methods of treating Alzheimer's disease using galantamine were not enabled by a specification that contained some theory but no animal test results or human clinical test results.[43] In this case, Jansen asserted U.S. Patent No. 4,663,318 ("the '318 patent") against several generic drug companies who had filed ANDAs seeking approval to use galantamine to treat Alzheimer's disease. The '318 patent included only a one-page specification, summarizing six scientific papers in which galantamine had been administered to humans or animals. The summary suggested that galantamine can cross the blood-brain barrier, affect brain wave activity, increase short-term memory in dogs, and reverse amnesia induced by administering acopolamine. It was viewed, however, as lacking any express connection between such research and treating Alzheimer's disease. The Federal Circuit found that the specification provided neither in vitro or animal tests relating to nor sufficient information for a person with ordinary skill in the art to infer the use of galantamine to treat Alzheimer's-like symptoms. Accordingly, the Federal Circuit held the '318 patent invalid for lack of enablement.[44] Of course, galantamine did work as predicted to treat Alzheimer's symptoms (hence Jansen's commercial product and the generic

40. *In re* Klopfenstein, 380 F.3d 1345 (Fed. Cir. 2004).

41. *In re* '318 Infringement Litig., 583 F.3d 1317 (Fed. Cir. 2009); *see also* Centocor Ortho Biotech v. Abbott Labs., 636 F.3d 1341 (Fed. Cir. 2011) (finding the patent-in-suit invalid for lack of written description and reversing a judgment of more than $1.6 billion in damages).

42. *See, e.g.*, Rasmusson v. Smithkline Beecham, 413 F.3d 1318, 1324–25 (Fed. Cir. 2005) (holding that testing was required); *In re* Brana, 51 F.3d 1560 (Fed. Cir. 1995) (reversing rejection for lack of utility of patent application claiming compound that showed anti-tumor activity in vitro).

43. *In re '318 Infringement Litig.*, 583 F.3d at 1325.

44. *Id.* at 1327.

applications), but the animal tests demonstrating efficacy were not obtained until after the patent application was filed.

2. Maintaining Market Exclusivity

As discussed earlier in the chapter, drug discovery and development are long and expensive processes. Studies also show that it typically takes 12 to 16 years of product sales for an innovator to reach the breakeven point recouping R&D expenditures.[45] As a result, market exclusivity becomes critically important for an innovator to recoup investment and to support growth.

Regulatory laws provide drugs with varying lengths of market exclusivity. A new chemical entity (NCE) has market exclusivity of five years; an orphan drug has a market exclusivity of seven years; drugs approved through the supplemental new drug application process, including new salts, new formulations, and new indications, have a market exclusivity of three years; and drugs evaluated for pediatric use have an additional market exclusivity of six months. In comparison to the average breakeven point, these market exclusivities are far from sufficient.

Patent exclusivity provides the additional period that innovator drug companies need to be able to continue to develop new and better cures. A 1986 survey of 100 R&D managers reported that 60 percent of pharmaceutical products would not have been developed had there not been patent protection.[46] A more recent study echoed that the comparative value of each patent is much higher in the life science than in engineering fields.[47]

The unique cost distribution in the pharmaceutical industry determines the importance of a patent in maintaining an innovator's ability to maintain market exclusivity. The majority of an innovator's investment is in drug discovery, development, clinical approval, market development, and lost opportunities. A generic drug company does not have such burdens: the drug has been identified by the innovator, the market has been created, and its value proven. The principal investment a generic drug company has to make is a much simpler and generally less expensive clinical study to show that its drug is bioequivalent to the innovator's drug, a requirement that is even waived by the FDA under certain circumstances.

45. H.G. Grabowski & J. M. Vernon, *Effective Patent Life in Pharmaceuticals*, 19 Int. J. Tech. Mgmt. 98 (2000).

46. Edwin Mansfield, *Patents and Innovation: An Empirical-Study*, 32 Mgmt. Sci. 173 (1986); *see also* Richard C. Levin et al., *Appropriating the Returns from Industrial Research and Development*, 1987 Brookings Papers on Econ. Activity 783–820; Wesley M. Cohen, Richard R. Nelson & John P. Walsh, *Protecting Their Intellectual Assets: Appropriability Conditions and Why U.S. Manufacturing Firms Patent (or Not)* (Nat'l Bureau of Econ. Research, Working Paper No. 7552, 2000), *available at* http://www.nber.org/papers/w7552.pdf.

47. Jay P. Kesan, *Transferring Innovation*, 77 Fordham L. Rev. 2169, 2195 (2009).

Because the cost of goods in this industry is usually a very small portion of the overall expenses, a generic drug company can sell copies of an innovator's drug at a fraction of the innovator's price. This leads to dramatic erosion to the innovator's market.

Although the standard U.S. patent term is 20 years from the filing date, the long drug discovery, development, and clinical approval process gives pharmaceutical patents a much shorter effective life. Recognizing delays resulting from clinical approval, Congress enacted the Drug Price Competition and Patent Term Restoration Act in 1984, authorizing a patent term extension for a pharmaceutical patent for up to five years and setting the remaining patent term after the drug is approved at not more than 14 years. Despite the patent term extension, the average effective patent term for a pharmaceutical patent is about 11 to 12 years.[48] Adjusted by removing antibiotics from the group, the average effective patent term is about nine and a half years.[49] The significant shortening of the effective patent term is largely due to generic companies challenging pharmaceutical patents.[50]

The disparity between the average time to recoup investment in drug discovery and development and the average effective patent term of a pharmaceutical patent shows the importance of considering the effects of exclusivity on the filings of patent applications covering various aspects of a drug product and its use.

This is especially true at the preclinical stages of basic research, drug screening, lead development, process development, formulation development, and preclinical testing, and these processes take place in a somewhat stacked fashion. Preclinical research not only provides an innovator with a wealth of IP opportunities but, when appropriate, also creates opportunities for the innovator to time the patent filing on each new improvement so that the ultimate commercial project can be protected long enough to recoup investment and sustain growth. Whether these opportunities can be captured depends upon various issues discussed in detail later in this chapter.

C. Scope of Offensive and Defensive Filings

An innovator in the pharmaceutical industry faces evolving competition as it develops drug products. As a biological target or process is identified and a screening protocol is validated, new therapeutic indications are being discovered. As indicated earlier in this chapter, an innovator targeting a new indication and a new market will have to face further competition

48. H.G. Grabowski & M. Kyle, *Generic Competition and Market Exclusivity Periods in Pharmaceuticals*, 28 Managerial & Decisions Econ. 491, 497 (2007).

49. *Id.*

50. *Id.*

from other innovators. After a drug is approved and enters the market, the innovator will have to face competition from generic companies. Patents are a key tool for an innovator to fend off these competitors.

The evolving nature of its competition requires an innovator to adopt a patenting strategy that is both offensive and defensive.

1. Offensive Filings

As part of an offensive IP strategy, an innovator might file patent applications to cover all alternatives to the company's R&D and product pipeline. Alternatively, it can also file applications based on its understanding of the market. In either case, by obtaining protection broader than just an anticipated commercial product, the innovator builds a wall to prevent other market entrants from copying its innovations.[51] A strategy including filing offensive patent application is often associated with publication of related discoveries and the use of post-grant proceedings in the PTO to prevent others from improperly obtaining overly broad patents that might limit one's FTO.[52]

2. Defensive Filings

Defensive filings focus on protecting an existing product or R&D program. The pharmaceutical industry is a unique industry in which generic companies are encouraged to challenge innovators' patents—for example, through Hatch-Waxman litigations. Generic companies succeeding in these litigations can be rewarded with comparatively cheap and quick entry to an established market. Accordingly, patents covering existing drug products, particularly successful ones, are likely to be challenged in litigation. The primary concern for defensive filings is the validity of the resulting patents. As a result, claims of these patents are usually narrow and focused.

D. Optimal Patent Protection

Patent protection provides the market exclusivity necessary for companies to recoup their significant R&D costs and provides a foundation for obtaining investments and collaborations. To obtain the optimal protection for a given invention, a company must first understand the invention and foreseeable alternatives, including what product(s) have commercial value to them and to their competitors, how the product(s) are made and used, and

---

51. D. Patrick O'Reilley, *Moving Towards an Integrated IP Strategy*, China IP News (Nov. 2010), http://www.finnegan.com/resources/articles/articlesdetail.aspx?news=e066a9f3-9a69-49af-bf36-764892b84cc4.

52. These proceedings include ex parte reexamination, post-grant review, and inter partes review. The latter two were created under the America Invents Act (AIA).

how easily a third party could design around the patent while still making a commercially viable product.

Especially given the many regulatory considerations and generally long development time before pharmaceutical commercialization, patent strategy is critical from the earliest phases of development and requires more than simply filing applications early or often. A key issue is deciding what to patent, which may seem obvious at first, but is actually a complex question unique to each potential drug product.

One specific consideration, unique to the pharmaceutical industry, is the greater importance given to patents covering drugs and methods of using such drugs.[53] These patents can be listed in the database of Approved Drug Products with Therapeutic Equivalence Evaluations (more often referred to as the Orange Book). As discussed in greater detail in Chapter 5, a company asserting an Orange Book listed patent against a company seeking approval of a generic copy of the protected drug is entitled to a 30-month stay of FDA approval of the generic copy.[54]

Generally, a pharmaceutical company's most valuable patent protection comes from claims directed to the active pharmaceutical ingredient (API) because it is impossible for an AB-rated generic to design around the API itself. To get all the patent protection to which they may be entitled, however, pharmaceutical companies should consider seeking claims beyond the API as further improvements and innovations are developed.

When such continuous improvements are independently patentable, they can provide a company with patent protection better aligned with the actual commercial product and its sales and provide a better chance of recouping investment in the drug discovery and development. An ideal portfolio might include patents covering the genus of compounds in which the API falls; the API itself; any relevant solid, liquid, transdermal, and aerosol forms or formulations of the API; combinations of the API with other pharmaceutically active components; methods of synthesizing the API and intermediates; key scale-up steps; and use of the API for the therapeutic indication alone or in combination with other active agents. Although the ideal portfolio will differ for each product, obtaining the optimal portfolio commonly requires a coordinated understanding of and filing strategy for all different improvements and developments.

53. 21 C.F.R. § 314.53(b)(1) (2012) ("For purposes of this part, such patents consist of drug substance (active ingredient) patents, drug product (formulation and composition) patents, and method-of-use patents.").

54. 21 U.S.C. § 355(j)(5)(B)(iii).

1. Genus/Species

Patenting a genus and a selected number of species is often the first step in protecting an innovator's IP during the R&D stage. Intuition says that broader claims mean broader protection. A balance must be struck, however, because broader claims are at greater risk for being rejected (or invalidated) for obviousness or anticipation. Claiming functional and structural elements of a broader group of compounds may avoid inoperative embodiments and help with overcoming obviousness or anticipation rejections, but this too has special considerations for validity.

As discussed earlier in this chapter, an innovator faces constant yet evolving competition from other innovators and generic companies. The combination of genus patents and species patents provides the innovator with protection from innovator and generic competitors. Before a drug is approved, a genus patent deters other innovators from trying to plow the same field and develop species within the claimed genus.[55] However, the breadth of a genus patent generally renders it more susceptible to a validity attack.

After a drug is approved, generic companies are likely to challenge any patents covering the commercial product because of the potential financial return, especially with the 180-day Hatch-Waxman first-filer exclusivity. In the face of this competition, the narrow scope of a species patent covering the commercial API may be ideal, given its generally better position to withstand a validity challenge.

Thus, claims, or even separate patents, directed to both a genus and a selected number of species should be pursued by the innovator where possible. If these relevant species are not developed until some later time after the genus, a patent directed to the particular species may result in a patent term that is more commensurate with the product's actual commercial lifetime. However, patenting a particular species after the genus is already disclosed may create additional challenges.

For example, in *Eli Lilly & Co. v. Zenith Goldline Pharmaceuticals, Inc.*,[56] Zenith, the generic company, contended that Lilly's patent claiming the olanzapine species was anticipated by a prior Lilly patent and disclosure related to a broader genus. In this case, the generic's anticipation challenge was unsuccessful. The courts held, based on the specific facts of the case, that the prior art genus was not a disclosure of the olanzapine compound.

The principle that allows patenting a later species is that "the disclosure of a genus . . . is not necessarily a disclosure of every species that is a

55. While the genus patent may be a deterrent, it may not actually prevent another entity from developing or patenting a specific species within the genus.

56. Eli Lilly & Co. v. Zenith Goldline Pharm. Inc., 471 F.3d 1369 (Fed. Cir. 2006).

member of that genus."[57] As Judge Giles S. Rich, one of the most influential jurists on U.S. patent law, puts it,

> It is an old custom in the woods to mark trails by making blaze marks on the trees. It is no help in finding a trail or in finding one's way through the woods where the trails have disappeared—or have not yet been made, which is more like the case here—to be confronted simply by a large number of unmarked trees.[58]

When facing a genus encompassing a half million compounds and seeing no guide, he says, "[w]e are looking for blaze marks which single out particular trees" and "[w]e see none."[59] Absent such blaze marks from the prior art genus, these cases show the potential—and importance—of claiming later developed species.

Patenting a species at a later time may also face obviousness challenges. For example, in *Atlanta Pharma AG v. Teva Pharmaceuticals USA, Inc.*,[60] pantoprazole, the drug-at-issue, was found to be obvious over prior art "Compound 12." The compounds were structurally identical, except that the 3-position in pantoprazole is substituted with a methoxy group ($-OCH_3$) while in Compound 12 it is substituted with a methyl group ($-CH_3$).

However, even if structurally similar to prior art compounds, a later developed species may still be nonobvious because obviousness of a compound cannot be based on structural similarity alone. It further requires a motivation to select and modify the prior art compound. Specifically, "[o]bviousness based on structural similarity may be proven by the identification of some motivation that would have led one of ordinary skill in the art to select and modify a known compound in a particular way to achieve the claimed compound."[61]

In cases where the newly developed species is structurally similar to the prior art, it may be important to consider from the start whether there are scientific reasons the later species would not have been obvious. For example, where the prior art indicates that a certain path (e.g., structural modification) would have been undesirable, or otherwise "teaches away"

---

57. Atofina v. Great Lakes Chem. Corp., 441 F.3d 991, 999 (Fed. Cir. 2006).

58. *In re* Ruschig, 379 F.2d 990, 994–95 (C.C.P.A. 1967).

59. *Id.* at 995.

60. Atl. Pharma AG v. Teva Pharm. USA Inc., 566 F.3d 999 (Fed. Cir. 2009); *see also* Aventis Pharma Deutschland v. Lupin Ltd., 499 F.3d 1293 (Fed. Cir. 2007); Procter & Gamble Co. v. Teva Pharm. USA Inc., 566 F.3d 989 (Fed. Cir. 2009).

61. *Atl. Pharma AG*, 566 F.3d at 1007 (citing Eisai Co. v. Dr. Reddy's Labs. Ltd., 533 F.3d 1353 (Fed. Cir. 2008)).

from the later species, the later species may be nonobvious.[62] Unexpected properties are also important to nonobviousness. For example, where the later species has unexpected properties as compared with the prior art, a compound that differs structurally by as little as a methyl versus an ethyl substituent may be nonobvious.[63] Because the knowledge of these considerations may be most fresh during development (although they are sometimes only understood later), it may be important to capture the knowledge early and use it to shape the patenting strategy from the start.

2. Enantiomeric Forms

In several instances where the first generation drug was a racemic mixture, an enantiomer was subsequently found to be more active and/or less toxic than the racemate. Because of the difference in efficacy or toxicity, an enantiomer can be considered a new chemical entity by the FDA.[64] NCE status precludes others from marketing a generic version of the approved enantiomeric drug in the first five years after the approval.[65] Patenting enantiomers, however, has unique challenges. Success or failure often depends on the ability to demonstrate the enantiomer's unexpected properties.

For example, in *Aventis Pharma Deutschland v. Lupin*,[66] a formulation of the 5(S) isomer of ramipril was found to be obvious because, as held by the courts, the prior art pointed to the 5(S) isomer and there was no evidence that separating 5(S) was beyond ordinary skill. On the other hand, AstraZeneca obtained patent protection for its acid-reflux drug Nexium (S-omeprazole) by demonstrating that the (S)-enantiomer had unexpectedly superior efficacy and bioavailability than the racemic mixture in Prilosec (omeprazole).

Consistent with the prior discussion, concerns over obviousness challenges may be addressed by understanding and capturing the advantages of an enantiomer over the racemate. The risk that an enantiomer might be challenged as obvious further underscores the importance of considering claims directed to the other aspects of the product, such as its solid forms or methods of use, as discussed in the following sections. While enantiomer protection can be important to the overall portfolio, there may be no reason to put all of one's eggs in this basket.

---

62. Takeda Chem. Indus. Ltd. v. Alphapharm Pty. Ltd., 492 F.3d 1350, 1359 (Fed. Cir. 2007).
63. Eli Lilly & Co. v. Zenith Goldline Pharm. Inc., 471 F.3d 1369, 1378 (Fed. Cir. 2006).
64. 21 U.S.C. § 355(u).
65. 21 U.S.C. § 355(j)(5)(F)(ii).
66. Aventis Pharma Deutschland v. Lupin, 499 F.3d 1293 (Fed. Cir. 2007).

3. Solid Forms

Potential solid forms of a compound, which might include its polymorphs,[67] amorphous materials, salts, and co-crystals, can be commercially significant because their properties can have significant impacts on solubility, bioavailability, and/or processability of a drug substance. For example, two years after Abbott started to sell its HIV drug Norvir, a previously unknown but more stable polymorph (Form II) of the API was unexpectedly produced, causing the drug to fail in its dissolution assays.[68] The lower solubility of Form II, and inability to reliably generate the prior polymorph, forced Abbott to withdraw this drug from the market until it found ways to reformulate this drug. This disruption cost Abbott at least $250 million.[69]

When seeking patent protection for solid forms, one perspective to keep in mind is that a solid form may not be important solely for what it is (e.g., a new polymorph or salt) but also for what it does, what it enables, and/or why it is an improvement over the prior art. Considering the solid form invention from these perspectives may provide benefits in terms of claiming.

For example, while the besylate salt of the known amlodipine API was found to be obvious (and Pfizer's patent thereon invalid), the record of that case suggests that the besylate salt had greater chemical stability and better processability than a prior maleate salt.[70] Although there is no way to know for sure, the results may have been different if the innovator had claims to an amlodipine product "having a chemical stability of at least x" or "a processability," however defined, "of at least y." Because there is no single perfect way to claim a solid form, it can be important to include a variety of claim types, informed by the importance of the innovation to the product and patient to be treated, that define and protect the discovery from different angles.

As another example, claims related to product purity may also provide protection for discoveries related to solid forms. This was the case in *Glaxo Group Ltd. v. Apotex Inc.*,[71] where the Federal Circuit held that a claim to an amorphous form of the drug product in Ceftin having purity of at least 95 percent was not anticipated by a prior art synthesis of the drug product

67. The term polymorph also includes pseudopolymorph, which may differ not only in the molecules' arrangement but also slightly in the molecular composition. *See* SmithKline Beecham Corp. v. Apotex Corp., 403 F.3d 1331, 1348 (Fed. Cir. 2005). A solvate is a typical pseudopolymorph.

68. John Bauer et al., *Ritonavir: An Extraordinary Example of Conformational Polymorphism*, 18 Pharm. Res. 859–866 (2001).

69. Sherry L. Morissette et al., *Elucidation of Crystal Form Diversity of the HIV Protease Inhibitor Ritonavir by High-Throughput Crystallization*, 100 Proc. Nat'l Acad. Sci. 2180 (2003).

70. Pfizer Inc. v. Apotex Inc., 480 F.3d 1348 (Fed. Cir. 2007).

71. Glaxo Grp. Ltd. v. Apotex Inc., 376 F.3d 1339 (Fed. Cir. 2004).

that failed to necessarily provide the claimed purity.[72] The court also found the claimed amorphous form to be nonobvious based on its unexpectedly better bioavailability and stability compared to the prior art crystalline form.[73] This case illustrates the importance of material properties (bioavailability and stability) to validity and teaches that an innovator should at least consider whether it can claim the solid form based on the novel properties it provides.

There are also unique issues with respect to the enforcement of solid form patents, including issues related to unpredictability in the control of the polymorph formation as was the case for Abbott's original Norvir product discussed earlier. For example, SmithKline argued that the generic company's intended paroxetine hydrochloride (PHC) *an*hydrate product, which was not patent protected at the time, contained at least trace amounts of the patented PHC *hemi*hydrate. The Federal Circuit agreed that trace amounts of the hemihydrate would infringe, because the claim to "[c]rystalline [PHC] hemihydrate" covered even trace amounts of this solid form, regardless of whether it was a commercially significant amount.[74] However, the Federal Circuit held the patent invalid as anticipated, based on its conclusion that the prior art PHC anhydrate would have inevitably produced at least trace amounts of the claimed PHC hemihydrate.[75]

The unpredictability of solid forms[76] is a significant complicating factor in their patenting and enforcement. Effective protection requires careful thought and analysis by all those involved in the project of the unique facts for each, rather than attempting to simply replicate strategies used previously for other products. However, simultaneous consideration of enforcement, validity, and potential design arounds may be particularly important for solid form inventions.

4. Metabolic Products

In appropriate cases, an innovator might seek patent protection for the chemical compounds created in vivo when a drug is metabolized. Patents on metabolites can be commercially significant, as when Ortho's proposed

72. *Id.* at 1348–49.

73. *Id.* at 1349.

74. SmithKline Beecham Corp. v. Apotex Corp., 403 F.3d 1331, 1341 (Fed. Cir. 2005).

75. *Id.* at 1344.

76. JOEL BERNSTEIN, POLYMORPHISM IN MOLECULAR CRYSTALS 241 (Oxford 2002) ("While it may not be surprising that many pharmaceutically important materials have been found to be polymorphic, or that any particular compound may turn out to be polymorphic, every compound is essentially a new situation, and the state of our knowledge and understanding of the phenomenon of polymorphism is still such that we cannot predict with any degree of confidence if a compound will be polymorphic, prescribe how to make possible (unknown) polymorphs, or predict what their properties might be." (citations omitted).)

norgestimate steroid product was found to infringe an American Home Products (now Pfizer) patent because the steroid breaks down in vivo to norgestrel and norgestrel acetate, both of which were claimed.[77] However, there are risks that the claims will be held as not covering the approved product or that the claims are invalid as anticipated or obvious.

In *Mylan Pharmaceuticals, Inc. v. Thompson*,[78] for example, the district court concluded that a metabolite patent for BuSpar (busipirone) could not be listed in the Orange Book because it did not claim the approved product. In *Schering v. Geneva Pharms.*,[79] the Federal Circuit held that a drug patent inherently discloses a metabolite if it teaches administration of the drug to a patient and formation of the metabolite necessarily follows and found the later patent for the metabolite invalid.

Nevertheless, depending on the particular case, it may be possible to protect metabolites. For example, if the in vivo metabolite is discovered to be biologically active, in principle, one could claim a formulation of or method of treating by administering that metabolite. Alternatively, if achieving a certain plasma concentration profile of the metabolite is important and affected by the formulation or method of administering, this profile might be a relevant property to be claimed.

5. Reduced Dosing, Improved Outcomes, or More Favorable Side-Effect Profiles

Patent protection should be sought for inventive formulations involving reduced dosing, improved outcomes, alternative routes of administration, or more favorable side-effect profiles. If all of the patent law requirements are individually met by this type of improvement, a patent may be obtained even after the API is publicly known and may protect the company's further improvements.

A typical formulation in this category is sustained-release/extended-release formulations, which in many instances reduce dosage, side effects, and frequency of dosing and improve patient compliance. For example, Eli Lilly and Company obtained later patent protection for a sustained-release formulation of its antidepressant Prozac product, and Bristol-Myers Squibb obtained later patent protection for an extended-release formulation of its diabetes drug Glucophage that allowed for reduced dosing.

---

77. Ortho Pharm. Corp. v. Smith, 18 U.S.P.Q.2D (BNA) 1977, 1985–86 (E.D. Pa. 1990), *aff'd*, 959 F.2d 936, 939 (Fed. Cir. 1992).

78. Mylan Pharm. Inc. v. Thompson, 139 F. Supp. 2d 1, 19–21 (D.D.C. 2001), *rev'd on other grounds*, 268 F.3d 1323 (Fed. Cir. 2001).

79. Schering Corp. v. Geneva Pharm., 339 F.3d 1373 (Fed. Cir. 2003).

Patenting extended-release formulations has its unique challenges. For example, when it was previously known to make an extended-release formulation of one compound in a given delivery mode (e.g., formulation matrix), patenting an extended-release formulation of a similar compound in a similar delivery mode becomes more difficult. For this reason, success or failure in patenting an extended-release formulation may depend on whether it would have been obvious to select and combine the previously disclosed drug substance with a given extended-release delivery mode.

In *Abbott Laboratories v. Sandoz, Inc.*, for example, the generic side argued that it would have been obvious to make the claimed extended-release clarithromycin antibiotic formulation in view of the FDA regulations for extended-release products and prior extended-release formulations for another antibiotic (azithromycin).[80] The Federal Circuit disagreed because "knowledge of the goal" (i.e., meeting the FDA requirements) "does not render its achievement obvious" and because a person of skill in the field—understanding the differences in the pharmacokinetic, chemical, biological properties, in vitro, and in vivo data of the prior art compound azithromycin and the claimed clarithromycin—"would not have assumed that the two drugs would exhibit similar behavior if placed in the same formulation."[81]

On the other hand, not all later formulation developments are ultimately upheld as patentable inventions. For example, in *Alza v. Mylan*,[82] a patent for a timed-release formulation of oxybutynin was held invalid as obvious because the basis for Alza's formulation was found to be consistent with a previously known correlation between a drug's lipophilicity and its colonic absorptivity. In *Merck v. Teva*,[83] a patent for weekly rather than daily administration of bisphosphonates was held invalid as obvious because similar formulations had been disclosed.

By examining the differences in the chemical, pharmacodynamics, pharmacokinetic, in vitro, and in vivo properties between the claimed compound and its prior art counterpart, and/or distinguishing the claimed formulation from general principles, an innovator can increase its chance of obtaining a valid patent on these formulations. In some cases, the key may relate more to what the formulation achieves (reduced dosing, improved outcomes, or more favorable side-effect profiles) and less to formulation components themselves.

---

80. Abbott Labs. v. Sandoz Inc., 544 F.3d 1341, 1346–47 (Fed. Cir. 2008).
81. *Id.* at 1352.
82. Alza Corp. v. Mylan Labs. Inc., 464 F.3d 1286 (Fed. Cir. 2006).
83. Merck & Co. v. Teva Pharm. USA, 395 F.3d 1364 (Fed. Cir. 2005).

6. Liquid Forms and Formulations

Liquid formulations, such as those used for oral or intravenous administration, present some unique considerations that should be addressed when developing an IP portfolio and strategy. Indeed, to an extent much greater than is generally the case for solid forms, stability can be a significant issue, and the packaging can be an important aspect of the product. Both can be potentially patentable.

For example, sevoflurane, an inhalation anesthetic, degrades in the presence of Lewis acids to produce hydrofluoric acid, which is dangerous to inhale and etches glass containers used to store the product.[84] Abbott discovered that it could reduce sevoflurane degradation and hydrofluoric acid production by deliberately adding water to the composition, a step that was counter to its previous attempts to minimize the water content. As recognized by the courts, the addition of water to reduce degradation was a significant and new discovery. However, Abbott's claims to the composition of sevoflurane and water (or other Lewis acid inhibitor) were held invalid over a prior patent that disclosed a method for purifying sevoflurane, an intermediate step of which involved making a sevoflurane composition saturated with water.

The controlling legal principle was that an old product cannot be patented based on a newly discovered use or property. That is, the newly discovered effect of adding water did not render the old sevoflurane-water composition patentable. The lesson is not that the innovator should forgo seeking patent protection for new discoveries of this type. Instead, the innovator needs to consider the discovery from different angles, such as the practical problems solved, and consider a variety of patent strategies.

For example, a problem solved by Abbott's discovery was that it could store sevoflurane in glass vials without degradation by adding water. Another problem solved was enabling the administration of sevoflurane with reduced risk for exposing the patient to hydrofluoric acid by adding water to the sevoflurane composition. Claims to the composition stored in a glass vial or to the administration of a sevoflurane-water composition may have been novel over the intermediate purification step in the prior art.

7. Inhalation Formulations

Patenting inhalation formulations highlights another interesting consideration at the intersection of patenting strategies and satisfying regulatory requirements. Inhalation formulations include metered dose inhalation

84. Abbott Labs. v. Baxter Pharma Prods. Inc., 471 F.3d 1363 (Fed. Cir. 2006).

(MDI) and dry powder inhalation (DPI).[85] Both formulations include delivery devices. Whether a patent claiming a delivery device is listable in the Orange Book, and therefore available as the basis for a Hatch-Waxman ANDA litigation, determines, in part, the patent's value.

As discussed in greater detail in Chapter 5, a patent is listable in the Orange Book if it claims either the drug in an approved NDA or a method of use of such drug, and can be reasonably asserted against a person not licensed by the owner if the person makes, uses, or sells the drug.[86] For the present purposes, a patent claiming the drug substance, the drug product, or a method of using such drug is listable.[87] In this regard, the FDA has further stated that "[a] device with all of its parts, including any protective packaging (e.g., overwrap), and the formulation together constitute the drug product."[88] However, patents that claim processes of making such a drug, packaging, metabolites, or intermediates are not listable in the Orange Book.[89] Thus, a patent on a delivery device may be most valuable when meeting the FDA's definition of a drug product, with "[t]he key factor [being] whether the patent being submitted claims the finished dosage form of the approved drug product."[90]

However, what about drug delivery device patents? Specifically, is a device patent that discloses but does not claim the active ingredient or formulation of an approved drug product listable in the Orange Book? Or, is a device patent that neither discloses nor claims the active ingredient or formulation of an approved drug product listable? Although the answer may be yes, at least under certain facts, the question has not yet been specifically addressed by the courts or the FDA.

For example, at least one Section 505(b)(2) NDA applicant challenged the listability of a patent that claims an auto-injector but does not claim or disclose the drug substance or drug product used.[91] However, this case was

85. U.S. Food & Drug Admin., Guidance for Industry: Metered Dose Inhaler (MDI) and Dry Powder Inhaler (DPI) Drug Products: Chemistry, Manufacturing, and Controls Documentation (Draft, Oct. 1998); Notice, 63 Fed. Reg. 64,270 (Nov. 19, 1998).

86. 21 U.S.C. § 355(b)(1)(G) (2012); *see also* 21 U.S.C. § 355(c)(2) (requiring an NDA holder to update patent information for patents that are issued after the approval of the NDA).

87. 21 C.F.R. § 314.53(b)(1) (2012) (for purposes of this part, such patents consist of drug substance (active ingredient) patents, drug product (formulation and composition) patents, and method-of-use patents).

88. U.S. Food & Drug Admin., Guidance for Industry: Metered Dose Inhaler (MDI) and Dry Powder Inhaler (DPI) Drug Products, at ll. 113–14 (Nov. 13, 1998).

89. 21 C.F.R. § 314.53(b)(1).

90. 21 C.F.R. § 314.3; *see also* Applications for the FDA Approval to Market a New Drug: Patent Submission and Listing Requirements and Application of 30-Month Stays on Approval of Abbreviated New Drug Applications Certifying That a Patent Claiming a Drug Is Invalid or Will Not Be Infringed, 68 Fed. Reg. 36,676, 36,680 (June 18, 2003).

91. *See* Defendant Intelliject's Answer, Affirmative Defenses, and Counterclaims at 47–49, King Pharm. Inc. v. Intelliject, Inc., No. 1:11-cv-00065-BMS (D. Del. dismissed Feb.

settled, and no guidance was given to the listability issue. In addition, on five separate occasions between 2005 and 2012, NDA sponsors have requested advisory opinions from the FDA on these particular issues.[92] The FDA has provided no public response/opinion to date. However, while seeking guidance from the FDA on whether device patents are listable, Novo Nordisk, Forest, and AstraZeneca listed their device patents in the Orange Book.[93]

Given their potential value, patents covering not only the formulations (including drug substance or drug product) but also devices that deliver the formulation should be considered. Because there are unresolved questions about the listability of certain delivery devices, considering the delivery device from all its aspects, including the problems it solves and how it is used, is again important to obtaining the optimal protection for the invention.

8. Combination Products

Even where the individual drug substances are known, patent protection may be possible for their combinations where they otherwise meet all of the patentability requirements.

For example, Lilly obtained patent protection for a combination product called Symbyax, containing olanzapine (previously marketed as Zyprexa for schizophrenia) and fluoxetine (previously marketed as Prozac for depression), for the treatment of bipolar disorder. Similarly, although the individual components were all previously known, Glaxo-SmithKline obtained patents directed to AZT in combination with lamivudine, marketed as Combivir, and AZT in combination with lamivudine and abacavir sulfate, marketed as Trizivir.

While combination patents have faced battles on the issue of obviousness,[94] many have withstood and survived. For example, in *Knoll Pharm. Co. v. Teva Pharms. USA Inc.*,[95] the Federal Circuit upheld the

16, 2012).

92. *See* Request for Advisory Opinion by GlaxoSmithKline, No. FDA-2005-A-0015 (Jan. 10, 2005); Request for Advisory Opinion by AstraZeneca I, No. FDA-2006-A-0318 (Aug. 10, 2006); Request for Advisory Opinion by AstraZeneca II, No. FDA-2007-A-0261 (June 21, 2007); Request for Advisory Opinion by Forest, No. FDA-2011-A-0363 (May 12, 2011); Request for Advisory Opinion by Novo Nordisk, No. FDA-2012-A-1169 (Nov. 26, 2012).

93. *See, e.g.*, Request for Advisory Opinion by Novo Nordisk, *supra* note 92, at 5; Request for Advisory Opinion by Forest, *supra* note 92, at 5; Request for Advisory Opinion by AstraZeneca II, *supra* note 92, at 3; Request for Advisory Opinion by AstraZeneca I, *supra* note 92, at 3.

94. *See, e.g.*, McNeil-PPC Inc. v. L. Perrigo Co., 337 F.3d 1362 (Fed. Cir. 2003) (finding obvious the combination of antidiarrheal compounds and simethicone to treat accompanying flatulence); Richardson-Vicks v. Upjohn, 122 F.3d 1476 (Fed. Cir. 1997) (finding obvious the combination of ibuprofen and pseudoepedrine).

95. Knoll Pharm. Co. v. Teva Pharm. USA Inc., 367 F.3d 1381 (Fed. Cir. 2004).

validity of a patent on a combination of hydrocodone and ibuprofen based on data from trials after the patent issued showing surprising results. While not required, such evidence of unexpected results is clearly helpful to the patentee. For this reason, consistent with the prior discussions, understanding and capturing the distinguishing features of a combination product may be important to patentability.

9. Methods of Synthesis and Intermediates

Innovators should consider seeking patent protection for methods of synthesizing the API and methods of preparing the various formulations. Key scale-up steps may also be patentable. Even when the API becomes prior art, new methods of synthesis are patentable, especially where they are not simply optimizations or routine improvements of other methods.

However, recent developments on 35 U.S.C. § 271(e)(1),[96] the safe harbor exemption, may have some impact on method patents. In *Momenta Pharms. Inc. v. Amphastar Pharms. Inc.*,[97] Judge Randall Rader, the then Chief Judge of the Court of Appeals for the Federal Circuit, worried that the Federal Circuit's "interpretation of § 271(e)(1) [in *Momenta*] would essentially render manufacturing method patents worthless."[98] A careful reading of the facts of this case and subsequent cases sheds some light on how to obtain useful patent protection on methods.

The drug-at-issue in *Momenta v. Amphastar* was enoxaparin, an oligosaccharide mixture produced from heparin. Unlike a small-molecule drug, analyses of enoxaparin present challenges and, more importantly, opportunities. Momenta Pharmaceuticals[99] obtained a patent that claims methods for analyzing heterogeneous populations of sulfated polysaccharides in enoxaparin and asserted the patent against Amphastar after Amphastar received approval from the FDA to market its generic version of enoxaparin. A representative claim in *Momenta* reads:

---

96. 35 U.S.C. § 271(e)(1) reads: "It shall not be an act of infringement to make, use, offer to sell, or sell within the United States or import into the United States a patented invention (other than a new animal drug or veterinary biological product (as those terms are used in the Federal Food, Drug, and Cosmetic Act and the Act of March 4, 1913) which is primarily manufactured using recombinant DNA, recombinant RNA, hybridoma technology, or other processes involving site specific genetic manipulation techniques) solely for uses reasonably related to the development and submission of information under a Federal law which regulates the manufacture, use, or sale of drugs or veterinary biological products."

97. Momenta Pharm. Inc. v. Amphastar Pharm. Inc., 686 F.3d. 1348 (Fed. Cir. 2012), *cert. denied*, No. 12-1033 (June 24, 2013).

98. *Id.* at 1369 (Rader, J., dissenting).

99. Momenta obtained approval of its generic enoxaparin more than a year before Amphastar.

> A method for analyzing an enoxaparin sample for the presence or amount of a non naturally occurring sugar . . . , comprising:
>
> > providing an enoxaparin sample . . . ;
> >
> > using a separation method to determine, . . . the presence of a structural signature associated with the non naturally occurring sugar associated . . . ; and
> >
> > making a determination about the enoxaparin sample based upon a comparison of the determination of the presence of a structural signature associated with the non naturally occurring sugar . . . ,
> >
> > to thereby analyze the enoxaparin sample.[100]

The Federal Circuit found that Amphastar's use of the claimed method falls within the safe harbor exemption (and therefore was noninfringing) because it is reasonably related to the development and submission of information required by the Federal Food, Drug, and Cosmetic Act even if the use occurs after the ANDA is approved[101] and noninfringing alternatives are available.[102]

After *Momenta v. Amphastar*, the district court issued two opinions on the same patent. In *Momenta v. Teva*,[103] the district court found that the Momenta patent does not claim a product "that is 'made' by the accused tests and sold in the United States."[104] "[W]hile the quality control release testing is a regulatory requirement for sale of enoxaparin in the United States, it is not a method for making enoxaparin."[105] "What is 'made' by the process claimed in the '866 patent is (1) a digested sample of enoxaparin and (2) information about that sample, neither of which is subsequently sold by Teva."[106]

The district court's analyses in the *Momenta v. Teva* case are informative. While mere quality control release testing may be exempted by the safe harbor, methods of making products for commercial purposes or marketing studies may not.[107] Thus, claims to synthesis methods for purposes other than satisfying regulatory requirements may fall outside the safe

---

100. U.S. Patent No. 7,575,886 claim 6.
101. *Momenta*, 686 F.3d at 1357–59.
102. *Id.* at 1359.
103. Momenta Pharm. Inc. v. Teva Pharm. USA Inc., CA-10-12079-NMG (D. Mass. July 19, 2013).
104. *Id.*, slip op. at 7.
105. *Id.*
106. *Id.*
107. *See, e.g.*, Amgen Inc. v. Int'l Trade Comm'n, 565 F.3d 846 (Fed. Cir. 2009).

harbor exemption and should be considered for inclusion in an innovator's IP portfolio.

10. Methods of Use

Pharmaceutical companies often seek patent protection early for multiple methods of using an API, and this clearly has value so long as all of the patent law requirements can be met. Alternatively, filing later applications for newly developed uses may provide companies with late expiring patent protection. For example, Merck originally patented use of finasteride as a treatment for benign prostate enlargement, marketing it as Proscar, and later patented use of the compound to treat male pattern baldness, marketing it as Propecia. Eli Lilly first patented use of atomoxetine as a treatment for depression and later patented its use to treat attention deficit hyperactivity disorder (ADHD), marketing it as Strattera.

Coordination between the indications sought and those patented is also important, as even one labeled but unpatented indication may provide an avenue for generic competition. A recent Supreme Court case, *Caraco Pharm. Lab., Ltd. v. Novo Nordisk A/S*,[108] highlights this connection. Novo's drug, Prandin, was approved for three uses, but its method-of-use patent claims only one of the three.[109] As a result, Caraco, the generic company, was able to apply for marketing a generic version of Prandin for the two uses that are not covered by Novo's patent.

Thus, method-of-use patents are important for an innovator to maximize patent protection of its drug, but care should be taken to pursue claims that are congruent to the intended (or approved) use.

III. Conclusion

The pharmaceutical industry is a unique industry. At preclinical research stage, it creates tremendous amounts of IP, often without creating a single commercial product. Protecting these valuable assets is challenging, particularly because of the unpredictable nature of this business. To do so effectively requires an innovator to take a systematic, proactive, and creative approach.

At the same time, fierce competition is part of the pharmaceutical industry. To avoid falling prey to others' IP traps requires an innovator to be cautious and ensure clearance of any patent hurdle before committing significant investment into drug discovery and development endeavor.

108. Caraco Pharm. Lab. Ltd. v. Novo Nordisk A/S, 132 S. Ct. 1670 (Apr. 17, 2012).
109. *Id.* at 1678–79.

chapter 4

Clinical Trials and Regulatory Considerations[1]

I. Introduction

Design, conduct, and interpretation of clinical trials for any drug or drug product, whether a new molecular entity or an enhancement of existing compounds, are at the core of drug research and development. The current chapter is an attempt to provide an overview of guiding regulatory principles and the legal framework involved with the drug approval process, primarily in the United States but also in other countries using International Conference on Harmonisation (ICH) principles. The chapter also discusses the drug development process by categorizing it in arbitrary steps/phases and briefly summarizes different types of clinical trials commonly required as part of the submission package for Abbreviated New Drug Application (ANDA) or New Drug Application (NDA) submissions.

A. Code of Federal Regulations (CFR)

The Food and Drug Administration (FDA), which regulates the marketing authorizations for drug and food products, falls under purview of Title 21 of the U.S. Code of Federal Regulations (CFR). The mission of the FDA is to enforce laws enacted by the U.S. Congress and regulations established by the agency to protect the consumer's health, safety, and pocketbook. The Federal Food, Drug, and Cosmetic Act is the basic food and drug law of the United States. The law is intended to assure consumers that foods are pure and wholesome, safe to eat, and produced under sanitary conditions; that drugs and devices are safe and effective for their intended uses; that cosmetics are safe and made from appropriate

1. Milin Acharya, PhD, Modus Consultancy, Austin, Texas, and Ahmedabad, India.

ingredients; and that all labeling and packaging is truthful, informative, and not deceptive.[2]

The final regulations published in the Federal Register (the daily published record of proposed rules, final rules, meeting notices, etc.) are collected in the CFR. The CFR is divided into 50 titles that represent broad areas subject to federal regulations. The FDA's portion of the CFR interprets the Federal Food, Drug, and Cosmetic Act and related statutes. Section 21 of the CFR contains most regulations pertaining to food and drugs. The regulations document all actions of all drug sponsors that are required under federal law. For example, some or all of the following regulations apply to the Investigational New Drug (IND) application process:

- 21 CFR Part 312—Investigational New Drug Application
- 21 CFR Part 314—INDA and NDA Applications for FDA Approval to Market a New Drug
- 21 CFR Part 316—Orphan Drugs
- 21 CFR Part 201—Drug Labeling
- 21 CFR Part 50—Protection of Human Subjects
- 21 CFR Part 54—Financial Disclosures by Clinical Investigators
- 21 CFR Part 58—Good Laboratory Practices for Nonclinical Studies
- 21 CFR Part 56—Institutional Review Boards[3]

II. Drug Development Process

Overall, the process of drug development can be divided into four stages: preclinical, clinical, regulatory, and commercial. Each of these stages encompasses several aspects of development efforts for new molecular entities (NMEs), whether chemical or biological products, natural or synthetic.

A. Preclinical Stage of Drug Development

Step 1: Drug Discovery—A sponsor or partner develops drug compounds that have feasible physicochemical properties or that target a specific biological endpoint. Traditionally several thousand compounds were screened using high-throughput technology and using quantitative structure-activity

2. *Federal Food, Drug, and Cosmetic Act (FD&C Act)*, U.S. Food & Admin., http://www.fda.gov/RegulatoryInformation/Legislation/FederalFoodDrugandCosmeticActFDCAct/default.htm (last updated Dec. 5, 2011).

3. *Investigational New Drug (IND) Application*, U.S. Food & Drug Admin., http://www.fda.gov/Drugs/DevelopmentApprovalProcess/HowDrugsareDevelopedandApproved/ApprovalApplications/InvestigationalNewDrugINDApplication/default.htm#Laws, Regulations, Policies and Procedures (last updated Oct. 18, 2013).

relationships (QSARs). Optimal select compounds were then tested against specific targets or proteins in various disease indications to find a good match. In recent years, however, the thinking has evolved toward putting greater effort in understanding the biology of the disease and the underlying mechanisms of pathogenesis. A set of targets or markers are identified that exist in disease states, and then compounds that can regulate the targets are tested and selected—for example, specific upstream or downstream proteins that can be modified to affect eventual translation and expression of certain proteins (like tyrosine kinases, in certain cancer indications).

Step 2: Preclinical/Animal Testing—After in vitro screening against cell lines in the laboratory setting, drugs are tested in vivo in different species of animals (rats, mice, dogs, rabbits, pigs, monkeys, etc.) in different disease models. Testing involves identifying and characterizing short-term and long-term toxicities, pharmacological activity, and target attainment as well as safety and efficacy of compounds. A higher significance is put on characterizations of toxicities and identification of lethal doses. Extrapolations from animals to humans using in vitro in vivo correlations (IVIVC) are used to identify a tolerable starting dose and dosing regimen for further clinical evaluation.

Step 3: IND Application—The sponsor submits an Investigational New Drug (IND) application to the FDA or other regulatory agency based on results from initial testing that includes the drug's composition and manufacturing (chemistry, manufacturing, and controls [CMC]), previous human trial experiences (if any), and a further clinical development plan.

The FDA reviews the IND to ensure that the proposed clinical studies do not place human subjects at unreasonable risk of harm. The FDA also verifies that there are adequate informed consent and human subject protections.

B. Clinical Stage of Drug Development

An IND may be submitted for one or more phases of an investigation. The clinical investigation of a previously untested drug is generally divided into three phases. Although in general the phases are conducted sequentially, they may overlap. These three phases of an investigation are as follows:

Step 4: Phase 1 Studies—Phase 1 includes the initial introduction of an investigational new drug into humans, with primary emphasis on safety. Phase 1 studies are typically closely monitored and may be conducted in patients or normal volunteer subjects. These studies are designed to characterize side effects and to understand the ADME profile (i.e., how the drug is absorbed, distributed, metabolized, and excreted from the body). Other purposes are to identify a safe dose range and to evaluate how long the drug stays in the body and its biomarker engagement. Additionally, there is evaluation of the side effects associated with increasing doses, and, if possible, an attempt to gain early evidence on effectiveness. During Phase 1, sufficient information about the drug's pharmacokinetics and pharmacological effects should be obtained to permit the design of well-controlled, scientifically valid Phase 2 studies. The total number of subjects and patients included in Phase 1 studies varies with the drug but is generally in the range of 20 to 80. Phase 1 studies also include studies of drug metabolism, structure-activity relationships, and mechanism of action in humans, as well as studies in which investigational drugs are used as research tools to explore biological phenomena or disease processes.

Step 5: Phase 2 Studies—Phase 2 includes the clinical studies conducted to evaluate the effectiveness of the drug for a particular indication or indications in patients with the disease or condition under study and to determine the common short-term side effects and risks associated with the drug. In such controlled trials, the test drug is compared to either a placebo or a different drug or a drug that is the current standard of care. Phase 2 studies are typically well controlled, closely monitored, and conducted in a relatively small number of patients, usually no more than several hundred.

At the end of Phase 2, sponsors meet with the FDA to discuss how large-scale Phase 3 studies would be done.

Step 6: Phase 3 Studies—Phase 3 studies are expanded controlled and uncontrolled trials. They are performed after preliminary evidence suggesting effectiveness of the drug has been obtained and are intended to gather the additional information about effectiveness and safety that is needed to evaluate the overall risk-benefit relationship of the drug and to provide an adequate basis for physician labeling. Typically such studies have a sample size in the thousands, with the primary goal of gathering information on safety and effectiveness in different populations and dosage forms and in combinations with other drugs.[4]

C. Regulatory Stage of Drug Development

Step 7: Prefiling Regulatory Review Meeting—The drug sponsor pulls together all the relevant information about the drug using a common technical document (CTD) format under the ICH guidelines and submits the data as a New Drug Application (NDA) or Biologics License Application (BLA). The FDA meets with the sponsor prior to actual NDA submission.

4. 21 C.F.R. § 312.21.

Step 8: NDA Application—The drug sponsor formally asks the FDA to approve the drug. The NDA contains essentially all the drug information from animal and human data, analyses of such data, and information about how the drug behaves in the body as well as manufacturing and formulation details.

The goals of the NDA are to provide enough information to permit the FDA reviewer to reach the following key decisions:

a. Whether the drug is safe and effective in its proposed use(s), and whether the benefits of the drug outweigh the risks.

b. Whether the drug's proposed labeling (package insert) is appropriate, and what it should contain.

c. Whether the methods used in manufacturing the drug and the controls used to maintain the drug's quality are adequate to preserve the drug's identity, strength, quality, and purity.

Step 9: NDA Review, Outcome, and Response—After 60 days from NDA submission, the FDA either accepts or declines to file the NDA, based on the adequacy of what is submitted. Upon acceptance by the FDA, a review team is assigned to evaluate the sponsor's research on the drug's effectiveness and safety. If needed, the FDA sponsors specific meetings to evaluate the research work and to gather opinions from experts in the field by arranging Advisory Committee Meetings (AdComs). The FDA also assigns a date by which it will notify the sponsor of its decision, under the Prescription Drug User Fee Act (PDUFA).

If, after review of the NDA or BLA, the FDA rejects the sponsor's application because of missing information or incomplete data and a lack of confidence that the drug's benefits outweigh the risks, the FDA provides the sponsor a Complete Response Letter (CRL). This document provides the explanation for the rejection along with further regulatory guidance and recommendations. In most cases, sponsors have plans in place for how and when they will approach other countries outside the United States or the EU to seek marketing authorizations. The regulatory requirements, and subsequent hurdles for approval, vary significantly across the globe, and individual countries may require clinical data in their own populations.

Step 10: Drug Labeling—The FDA reviews the drug sponsor's professional labeling and ensures that appropriate information is passed on to health care professionals and users. Key information derived from all clinical studies is summarized in a very structured format. Special precautions and black box warnings are highlighted to allow prescribers to be well informed.

D. Commercial Stage of Drug Development

Step 11: Facility Inspections—The FDA inspects the facilities where the drug will be manufactured to evaluate quality and adherence to good manufacturing practice (GMP).

Step 12: Special Provisions—The *Accelerated Approval* program allows earlier approval of drugs that treat serious, life-threatening diseases and that fill an unmet medical need. Approval is expedited because the FDA bases the drugs' utility on surrogate endpoints or clinical values rather than waiting for final results from a completed clinical study.

The *Fast Track* program helps reduce the review time of products that treat life-threatening diseases. Drug sponsors can submit portions of the application as data becomes available (rolling submission) instead of having to wait until all information is available.

Step 13: Phase 4 Studies (Postmarketing Studies)—Upon scientific review of the NDA, the FDA might recommend one or more follow-up clinical studies postmarketing approval (PMA) of the drug. Such PMA studies need to be completed by the sponsor and the data submitted to the FDA within reasonable timelines. For certain therapeutic areas, the marketing approvals are conditional until further mature clinical data are submitted to the FDA.

The FDA also requires sponsors to submit periodic safety updates to detect any unexpected serious adverse event upon chronic use of the drug in a wider user base. The FDA's *MedWatch* voluntary system makes it easier for physicians and consumers to report adverse events. In certain serious cases, use of the drug may be curtailed or limited to a specific population, or, in very rare circumstances, the drug may even need to be withdrawn from the market.[5]

III. Types of Clinical Trials

Clinical trials are classified mainly according to the time of the study or the key objectives of the study. Focusing on their primary purpose, clinical studies can be broadly classified as human pharmacology, therapeutic exploratory, therapeutic confirmatory, and therapeutic use studies.

5. *Drug Approval Process*, U.S. FOOD & DRUG ADMIN., http://www.fda.gov/downloads/Drugs/ResourcesForYou/Consumers/UCM284393.pdf (last visited Apr. 25, 2014).

A. Human Pharmacology Studies

Key objectives of human pharmacology studies are usually to assess tolerance and get a sense of the pharmacokinetics (PK) and the pharmacodynamic (PD) profile. Such studies could also explore drug metabolism and interactions as well as provide an estimation of pharmacological/biological activity (target attainment).

Typical examples of such studies include the following:

a. *SAD (single-ascending dose)*—Such studies involve a starting dose derived from preclinical studies and administered to subjects as a single dose and can involve different dosing regimen and dose-escalation parameters. The purpose is to find the optimal and safest dose with the highest biological activity. Escalation parameters can be empiric or can have a mathematical basis (e.g., Fibonacci or modified-Fibonacci schemes). Usually key pharmacokinetic parameters are assessed, such as area under the curve (AUC) (exposure to drug, or refers to the extent of bioavailability), C_{max} (peak drug concentration, or refers to the rate of bioavailability), T_{max} (time to reach C_{max}), CL (clearance of drug), Vd (volume of distribution), and so on.

b. *MAD (multiple-ascending dose)*—Such studies involve repeated drug administration over multiple times over a certain period or certain cycles, and the purpose is to identify safe dosing upon increasing exposure to the drug. Accumulation of the drug due to narrow dosing intervals and evaluation of subsequent resulting toxicities can help better optimize a dosing regimen and frequency of drug administration. Both SAD and MAD studies also allow for establishing correlations between dose exposure and dose responses using PK and PD parameters.

c. *Bioavailability (BA)*—Such studies include estimation of drug reach and availability at the target site. For drugs that are given intravenously directly into systemic circulation, by definition, the bioavailability is considered 100 percent. For drugs that are given orally or through other routes of administration (e.g., intramuscular, subcutaneous, rectal, transdermal, sublingual, or pulmonary), there is usually a lag time of absorption, and the drugs also undergo first-pass metabolism through the liver. The absolute bioavailability is measured as a percentage fraction of the drug absorbed through non-intravenous administration compared with the corresponding intravenous administration of the same drug. The comparison must be dose normalized (e.g., account for different doses or varying weights of the subjects); consequently, the amount absorbed is corrected by dividing the corresponding dose administered. Relative bioavailability measures the bioavailability (estimated as the AUC) of a formulation (A) of a certain drug when compared with another formulation (B) of the same drug, usually an established standard, or through administration via a different route. When the standard consists of intravenously administered drug, this is known as absolute bioavailability.

Relative bioavailability is one of the measures used to assess bioequivalence (BE) between two drug products. For FDA approval, a generic manufacturer must demonstrate that the 90 percent confidence interval for the ratio of the mean responses (usually of AUC and the maximum concentration, C_{max}) of its product to that of the name-brand drug is within the limits of 80 to 125 percent.

d. *Bioequivalence (BE)*—The FDA defines bioequivalence as "the absence of a significant difference in the rate and extent to which the active ingredient or active moiety in pharmaceutical equivalents or pharmaceutical alternatives becomes available at the site of drug action when administered at the same molar dose under similar conditions in an appropriately designed study."[6]

Such studies usually compare two products, such as a commercially available name-brand product and a potential, to-be-marketed generic product. Pharmacokinetic studies are conducted in which each of the preparations is administered in a crossover study to volunteer subjects, generally healthy individuals but occasionally patients. Serum/plasma samples are obtained at regular intervals and assayed for parent drug (or occasionally metabolite) concentration. Occasionally, pharmacodynamic endpoints rather than pharmacokinetic endpoints are used for comparison. For a pharmacokinetic comparison, the plasma concentration data are used to assess key pharmacokinetic parameters. Testing is conducted at several different doses, especially when the drug displays nonlinear pharmacokinetics. In addition to data from bioequivalence studies, other supportive data, including analytical method validation and IVIVC studies, may need to be submitted to meet regulatory requirements for establishing bioequivalence.

The FDA considers two products bioequivalent if the 90 percent confidence intervals of the relative mean C_{max}, $AUC_{(0\text{-}t)}$, and $AUC_{(0\text{-}\infty)}$ of the test (e.g., generic formulation) to reference (e.g.,

6. U.S. Food & Drug Admin., Guidance for Industry: Bioavailability and Bioequivalence Studies for Orally Administered Drug Products—General Considerations (Mar. 2003), *available at* http://www.fda.gov/downloads/Drugs/GuidanceComplianceRegulatoryInformation/Guidances/ucm070124.pdf.

name-brand formulation) should be within 80.00 to 125.00 percent in the fasting state.

e. *Food-effect study*—Although there are a few exceptions, generally a bioequivalent comparison of Test to Reference formulations also requires administration after an appropriate meal at a specified time before taking the drug, a so-called fed or food-effect study. A food-effect study requires the same statistical evaluation as the fasting study. Presence of food in the gut delays gastric-emptying time and, hence, affects how long concomitantly administered drugs stay in certain parts of the stomach or intestine and how they get absorbed. Such studies test administration of drug in presence of foods with low-fat or high-fat contents and thus aid in identifying a proper dosing regimen and labeling.

f. *Mass balance*—Mass balance studies investigate the absorption/distribution/metabolism/excretion (ADME) of a drug following a single administration to humans. The drug is often radiolabeled to allow determination of drug disposition and assessment of major metabolic pathways, as well as exposure to the parent drug and its metabolites, and the rate and routes of elimination (urine, sweat, feces, etc.). Characterization of metabolites also helps in identification and optimization of prodrugs.

g. *Drug interactions (DDIs)*— Usually referred to as drug-drug interactions, such studies assess rate and change in exposure and subsequent correlative toxicities due to coadministration of two or more drugs. A substance (usually another drug) when coadministered with the test drug can cause synergistic (when the drug's effect is increased) or antagonistic (when the drug's effect is decreased) effects, or a new effect can be produced that neither produces on its own. Interactions can also occur when certain foods, herbs, medicinal plants, and dietary supplements are coadministered.

Drug interactions may be the result of alterations in the PK properties of a drug, such as alterations in the ADME profile or, alternatively, as a result of modification of the PD properties of the drug (e.g., the coadministration of a receptor antagonist and an agonist for the same receptor). These interactions may occur because of accidental misuse or lack of knowledge about the active ingredients involved in the relevant substances and can either increase the risk of side effects or decrease the therapeutic benefit or neither.

Knowledge of such interactions may even be used to improve the therapeutic effect of certain drugs (e.g., the use of codeine with paracetamol to increase its analgesic effect). Most common pharmaceutical interactions include cytochrome P450 enzymes (CYPs), efflux-transporters, and the like, mediated increase or decrease in exposures of drugs. These are of special interest to the practice of medicine because they primarily have negative effects for the users.

h. *Cardiovascular-safety (QTc)*—The QTc interval represents electrical depolarization and repolarization of the ventricles. A lengthened QTc interval is a marker for the potential of ventricular tachyarrhythmias like torsades de pointes and a risk factor for sudden death. Such cardiovascular safety studies are done in the preclinical setting for most drugs in development, but clinical testing may be necessary where there is incomplete understanding of secondary cardiovascular risks. Prolongation in QTc has been observed in subjects due to adverse drug interactions or underlying pathological conditions. Sometimes genetic predisposition makes certain individuals more susceptible to such cardiovascular changes upon drug administration.

i. *Biomarkers (Phase 0 studies)*—In a shift from classical phases of drug development, in recent times there has been an increased drive to evaluate human PD and/or PK (e.g., bioavailability) of analogs directed at the same target and possessing the same properties in vitro and in animal models to select the most promising candidate for further development using biomarkers or surrogate markers. Such studies are typically done in parallel to Phase 1 or earlier (hence Phase 0). They may include evaluating human biodistribution and binding characteristics using "microdosing" and supersensitive analytical techniques using novel imaging probes. Such studies determine whether a mechanism of action defined in nonclinical models can be observed in humans (e.g., drug binds to or inhibits its alleged target) while simultaneously refining a biomarker assay using human tissue and/or surrogate tissue (e.g., tumors).

The study population varies based on therapeutic areas. In general, these types of studies are carried out in healthy, normal volunteers. However, in certain therapeutic areas where there is high unmet medical need (e.g., oncology), most of these studies for cytotoxic therapies (which involves cell killing) are done in patients who may be refractory to standard available therapies. For cytostatic therapies (which involve cell-signal inhibition or modifications), studies could be done in a healthy, normal population with select predetermined or predefined stratification criteria.

For drugs that are in the postapproval or postmarketing stage, such studies may be done in special patient populations (hepatic- or renal-impairment patients, geriatric patients, pregnant or nursing women, pediatric patients).

B. Therapeutic Exploratory Studies

Key objectives of these studies are to explore the use of targeted indications and to identify or optimize dose and dosing regimen for subsequent large-scale studies. Objectives can include establishing a basis for confirmatory study designs, endpoints, and methodologies for further validation of activity of the drug or drug product.

Typical examples of such studies are the following:

a. *Proof-of-concept*—These are earliest trials with shorter duration in well-defined, narrow patient populations, using surrogate or pharmacological endpoints or clinical measures. Studies can be statistically evaluated for hypothesis testing of the concept (i.e., whether the drug actually has biological activity or not. Testing usually includes a target patient population with a known biomarker in a small, sample-sized patient group. Sometimes the selection of the patient population can be stratified upfront for the presence or absence of certain protein or genetic mutation or pharmacogenetic predisposition.

b. *Dose-response/Exposure-response exploration*—An evaluation of PK-PD correlations using different parameters like dose and exposure and efficacy endpoint is done to better understand the change in effect on an organism caused by differing levels of exposure (or doses) to a stressor (usually a chemical) after a certain exposure time. This may apply to individuals (e.g., a small amount has no significant effect, a large amount is fatal) or to populations (e.g., how many people are affected at different levels of exposure). Dose-response relationships generally depend on the exposure time and exposure route (e.g., inhalation, dietary intake); quantifying the response after a different exposure time or for a different route leads to a different relationship and possibly different conclusions on the effects of the drug. This limitation is caused by the complexity of biological systems and the often unknown biological processes operating between the external exposure and the adverse cellular or tissue response.

c. *Efficacy assessment studies*—Such studies are open label or close label, wherein key assessment of the efficacy profile of the drugs is done. Usually such studies would provide further insight into the correlation of biological activity to effectiveness of the drug in the target population. If statistically powered appropriately to identify clinically meaningful differences with comparator drugs, such studies may be accepted by regulatory agencies for expedited or conditional approvals.

C. Therapeutic Confirmatory Studies

The goals of such studies are to demonstrate/confirm efficacy and to establish safety profiles of drugs or drug products. Other key objectives can include dose-response relationship assessment and validation as well as assessment of risk-benefit to support licensing in various regions.

Types of such studies include the following:

a. *Efficacy studies*—Such studies are done in a target patient population where the clinical benefit is the primary endpoint. Secondary endpoints can be included based on clinical assessment criteria. These are large-scale, large-sample-size, statistically backed studies done in geographically varied patient populations. Usually hypothesis testing is employed in such studies where the test drug is compared head-to-head with either a placebo or existing standard-of-care drugs and drug regimen. For example, in oncology, overall survival is usually a primary endpoint, while progression-free survival (PFS) or reduction in certain biomarkers or the like is used as a secondary endpoint.

b. *Safety studies*—After the drug is approved and marketed, long-term toxicities need to be addressed. Also, in certain pharmacogenetically different patient populations or population subsets with certain underlying disease variance, safety endpoints need to be validated. Primary endpoints of such studies are to characterize not only the primary toxicities but also secondary toxicities due to possible chronic usage.

c. *Randomized parallel dose-response studies*—Such studies are similar to Phase 2 studies but with a larger sample size and with a goal to characterize the dose-response using advanced modeling and simulation techniques to address age effects, gender effects, demographic differences, dosing based on body measures such as body-surface area, and so on.

d. *Outcomes assessment (mortality/morbidity)*—Such studies assess outcomes of quality-of-life measures for drugs indicated for severe, life-threatening illnesses or illnesses with high unmet need.

D. Therapeutic Use Studies

The primary goals of therapeutic use studies are to refine risk-benefit profiles in the general patient population or special patient populations (e.g., pediatric patients, geriatric patients, and patients with organ dysfunction or impairments), to identify less common adverse events due to long-term drug exposure, and to refine dosing recommendations. From a sponsor standpoint, such studies are also done to increase access of the drug or drug product to other countries and to geographically diverse patient populations.

Typically such studies include comparative effectiveness–type studies, expanded-access studies, mortality/morbidity outcome studies, exploratory biomarker or additional endpoints, and pharmacoeconomic studies. Pharmacoeconomic studies can help in understanding market penetration and uptake of drug use by users and prescribers and can help in optimizing pricing in various geographic regions.

E. Classification According to Phases

Clinical trials are often described as four temporal phases. However, a classification using study objectives is technically preferable given that phases are an inadequate basis considering that one trial may occur in several phases. Although very widely used, the phase description is a concept rather than a set requirement.

To simply draw a comparison using the preceding descriptions of types of clinical studies, most human pharmacological studies are Phase 1–type studies, therapeutic exploratory studies are Phase 2–type studies, and therapeutic confirmation and use studies are Phase 3–type studies.

The last classification, Phase 4–type studies, are studies that begin after a drug is approved and the sponsor starts marketing the drug or drug product. These studies are usually in support of approved indications, and the objective is to optimize the drug's use. A lot of drug-drug interaction studies, special population studies, dose-response studies, or safety studies (in addition to those already included in the NDA submission) fall into this category. Hence, a lot of Phase 4 commitments could involve any of the Phase 1, 2, or 3–type studies, depending on the specifics of regulatory requirements or precedence and guidance for that indication, therapeutic area, and drug profile.

An exhaustive list of guidance documents for each of the topics and research components and study designs is available to the research community. These documents are meant to provide directives to sponsors based on current regulatory thinking and expectations.[7]

A harmonized structure of the dataset is required as part of the submission package when a sponsor chooses to file the NDA electronically. Digital Infuzion collaborated with the FDA to design, develop, and implement a scientific reference database with the capabilities to standardize unstructured electronic formats based on Clinical Data Interchange Standards Consortium (CDISC) and Study Data Tabulation Model (SDTM) standards and to query clinical studies by trial design and keywords.[8]

IV. ICH Guidelines

The International Conference on Harmonisation of Technical Requirements for Registration of Pharmaceuticals for Human Use (ICH) brings together the drug regulatory authorities (DRAs) of Europe (EMA), Japan (PMDA), and the United States (FDA), along with the pharmaceutical trade associations from these three regions, to discuss scientific and technical aspects of product registration. ICH's mission is to achieve greater harmonization in the interpretation and application of technical guidelines and requirements for product registration, thereby reducing duplication of testing and reporting carried out during the research and development of new medicines.[9]

A. General Principles

The term "drug" is synonymous to any investigational or medicinal product containing an active pharmaceutical ingredient or biological products or vaccines. The essence of rational drug development is to ask important questions and to answer with appropriate studies. Results of prior studies

---

7. Office of Training & Commc'ns, Center for Drug Evaluation and Research List of Guidance Documents (Apr. 15, 2014), *available at* http://www.fda.gov/downloads/drugs/guidancecomplianceregulatoryinformation/guidances/ucm079645.pdf; http://www.fda.gov/drugs/guidancecomplianceregulatoryinformation/guidances/ucm310704.htm.

8. U.S. Food and Drug Administration (FDA) Inventory of Clinical Trials Protocols and Clinical Study Data (2011), *available at* http://www.fda.gov/downloads/drugs/developmentapprovalprocess/conductingclinicaltrials/ucm309552.pdf.

9. Int'l Conference on Harmonization, The Value and Benefits of ICH to Drug Regulatory Authorities: Advancing Harmonization for Better Health (2010), *available at* http://www.ich.org/fileadmin/Public_Web_Site/News_room/C_Publications/ICH_20_anniversary_Value_Benefits_of_ICH_for_Regulators.pdf.

should influence the plan of later studies, and there should be a scientific approach in design and analysis of all preclinical and clinical data.

Substantial benefits to DRAs resulted when the ICH shifted emphasis on development of a common submission format—the Common Technical Document (CTD)—that greatly influenced regulatory review processes, ultimately leading to harmonized electronic submission and e-review initiatives, which, in turn, have enabled implementation of good review practices. These activities are having a global effect on information review and sharing among drug regulatory authorities.

Harmonizing the differences in these requirements through ICH guidelines helped the industry reduce development times and save resources. To extend the benefits of harmonization, industry participants proposed assembling the building blocks of information intended for inclusion in a submission into a consistent, harmonized format, referred to as the CTD, which would relieve pharmaceutical companies of the time, workforce, and financial burdens of assembling a submission for one DRA and then having to reformat it for another. This new, consistent format has also greatly benefited the U.S. FDA, enabling the agency to establish templates for each of the review disciplines while promoting more consistent review practices and processes.

B. Regulatory Benefits

The CTD has made the exchange of information among drug regulatory authorities easier. For a number of years, the FDA and the EMA have had a confidentiality arrangement in place allowing the sharing of confidential information, greatly increasing interactions between the two agencies. Because submissions are received in the same format and, generally, at the same time, these interactions have become more efficient, facilitating discussions of common concerns as submissions are evaluated. The use of CTD also alleviated the delays in submitting an application caused by format conversions for different ICH regions, which used to delay patient access to new, innovative medicine.

As shown in Figure 1, the CTD is divided into five key modules. Module 1 is not part of the CTD; it represents the administrative information specific to each ICH region. Module 2 is a layering of information and includes an introduction, summaries, and overviews. More complete data are contained in modules 3, 4, and 5. Countries can, in effect, focus on modules of interest. If a regulatory authority is not interested in the complete datasets in modules 3, 4, and 5, the focus can be on modules 1 and 2. This is what some less resourced countries are now doing.

Figure 1 Key Components of NDA

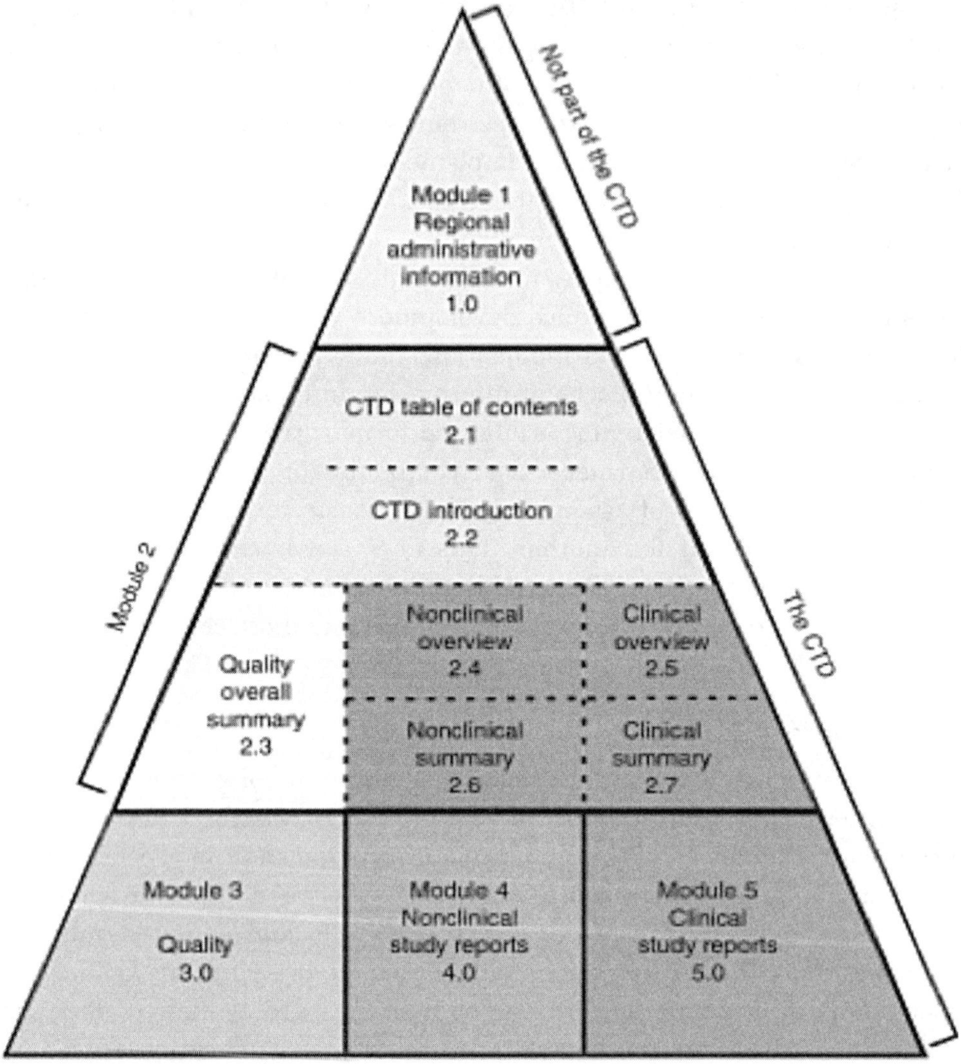

Source: J. A. Molzon et al., "The Value and Benefits of the International Conference on Harmonisation to Drug Regulatory Authorities: Advancing Harmonization for Better Public Health," *Clinical Pharmacology and Therapeutics* 89, no. 4 (2011): 503–12.

Last, and perhaps most important, the CTD has facilitated electronic submissions (the eCTD). The FDA has now implemented the FDA Electronic Submission Gateway, which allows an NDA to be sent electronically, essentially very much like e-mail. After being assessed for completeness, a submission is immediately and fully accessible on the reviewer's desktop. This innovation has alleviated the need for sponsors to create and assemble the many pieces of paper that constituted a traditional paper-based product application, organize the application, box thousands of pages, load the boxes on a truck, and deliver them to the FDA—all before a reviewer could even begin the assessment process.

C. Terms of Reference

- To maintain a forum for a constructive dialogue between regulatory authorities and the pharmaceutical industry on the real and perceived differences in the technical requirements for product registration in the EU, USA, and Japan in order to ensure a more timely introduction of new medicinal products, and their availability to patients;
- To contribute to the protection of public health from an international perspective;
- To monitor and update harmonised technical requirements leading to a greater mutual acceptance of research and development data;
- To avoid divergent future requirements through harmonisation of selected topics needed as a result of therapeutic advances and the development of new technologies for the production of medicinal products;
- To facilitate the adoption of new or improved technical research and development approaches which update or replace current practices, where these permit a more economical use of human, animal and material resources, without compromising safety;
- To facilitate the dissemination and communication of information on harmonised guidelines and their use such as to encourage the implementation and integration of common standards.[10]

D. GCP Guidelines

Adherence to the principles of good clinical practices (GCPs), including adequate human subject protection (HSP), is universally recognized as a critical requirement to the conduct of research involving human subjects. Many countries have adopted GCP principles as laws and/or regulations. The FDA's regulations for the conduct of clinical trials, which have been in effect since the 1970s, address both GCP and HSP.

10. *Vision*, Int'l Conference of Harmonisation, http://www.ich.org/about/vision.html (last visited Apr. 25, 2014).

E. ClinicalTrials.gov

ClinicalTrials.gov is a registry and results database of publicly and privately supported clinical studies of human participants conducted around the world. This website also provides summary information on the clinical trial registration and results submission requirements described in Section 801 of the Food and Drug Administration Amendments Act, known as FDAAA 801. It includes information about Responsible Party, Applicable Clinical Trials, deadlines for submitting required information, and penalties.[11]

V. Summary

In general, the drug development and approval process is extremely time-consuming and tedious and is fraught with significant risks for the sponsors, regulators, and users. The scientific community around the globe continues to strive to make this process shorter, more efficient, and more harmonized to provide easier access to newer and better medicines for general populations and patients.

11. *FDAAA 801 Requirements*, U.S. Nat'l Insts. of Health, http://clinicaltrials.gov/ct2/manage-recs/fdaaa (last updated Nov. 2013).

chapter 5

Clinical Trial–Patent Considerations[1]

I. Introduction

Before pharmaceutical and biotechnology companies can legally market and sell their new products in the United States, the products must go through clinical trials and receive approval from the Food and Drug Administration (FDA) and other regulatory authorities (if marketing outside the United States is desired). As noted in the preceding chapter, regulatory agencies play an important role in this clinical trial paradigm and are the gatekeepers who determine which drugs are safe for human use. While clinical testing must be conducted before product approval, it may or may not constitute a public use or sale of the drug product or its method of use that creates a bar to later filing of a patent application. Clinical trials may also lead to new discoveries that are patentable.

Given the large financial investment in these products before and during clinical trials, it is prudent for pharmaceutical and biotechnology companies to be aware that authorizing the use of their new product in clinical trials may have an impact on the patentability of the product and its use. Although patent applications covering the active compound itself (i.e., compound and/or composition of matter claims) are almost always filed prior to any clinical testing, the clinical trials can be relevant to the patentability of methods of using the product, specific product formulations, or specific salts or polymorphs of the product. When the trial is commenced, the name of the study drug and some details about its use will be disclosed to clinical investigators and patients participating in the study, though specific details such as the formulation may be withheld as part of

1. Michael B. Harlin, Thomas J. Campbell Jr., and Kevin A. O'Connor, Esq, PhD, of Neal, Gerber & Eisenberg LLP, Chicago, Illinois.

the double-blind testing protocol. Thus, composition of matter applications should be filed before the start of clinical trials.

Under the Patent Act of 1952, 35 U.S.C. § 102(b) stated: "A person shall be entitled to a patent unless—(b) the invention was . . . in public use or on sale in this country, more than one year prior to the date of the application for patent in the United States" (2006). The America Invents Act of 2011 (AIA), 35 U.S.C. § 102(a), is the successor to § 102(b), and it now states, "A person shall be entitled to a patent unless (1) the claimed invention was . . . in public use, on sale, or otherwise available to the public before the effective filing date of the claimed invention[.]" The new 35 U.S.C. § 102(a) (effective March 16, 2013) is not limited to public use or sales in this country, and it includes the additional language "or otherwise available to the public." The AIA did not explicitly speak to the nature or extent of disclosure that would rise to "public use" or a "sale," which would imply that case law on such issues retains its precedential value. The impact of a public disclosure will differ depending on which statute governing public use is the effective one, as the America Invents Act significantly altered the statutory framework.

II. Public Use, On Sale, or Otherwise Available to the Public

A claimed invention may lose its novelty by public use or by placing the invention on sale or (under the AIA) otherwise making the invention available to the public. The "public use" and "on sale" bars have often been considered together, though each bar may apply in situations where the other bar does not. Public use has been defined as "any use of that invention by a person other than the inventor who is under no limitation, restriction or obligation of secrecy to the inventor."[2] "[T]he policies underlying the public use bar inform its scope and . . . one such policy is discouraging the removal, from the public domain, of inventions that the public reasonably has come to believe are freely available."[3]

The on-sale bar prevents patenting of an invention if the product was the subject of a commercial offer for sale and the invention was ready for patenting.[4] Under the 1952 act, an invention could be held to have been on sale even if the sale was secret. Under the AIA, it is likely that a secret sale

2. *In re* Smith, 714 F.2d 1127, 1134 (Fed. Cir. 1983) (citing Egbert v. Lippmann, 104 U.S. 333, 336 (1881)).

3. Dey LP v. Sunovion Pharm. Inc., 715 F.3d 1351, 1355 (Fed. Cir. 2013) (internal quotation marks omitted).

4. Pfaff v. Wells Elecs. Inc., 525 U.S. 55 (1998).

or offer for sale does not create a bar to patentability, since § 102(a) has the additional language "or otherwise available to the public," which implies that the on-sale bar also requires that the claimed invention must be available to the public.

For both the public use bar and the on-sale bar, the invention must be "ready for patenting."[5] Where the invention is based on the results of the clinical trial, it is unlikely that the invention is "ready for patenting" before the completion of the trial.

To decide whether a prior use constitutes an invalidating public use, courts must evaluate "whether the purported use: (1) was accessible to the public; or (2) was commercially exploited."[6] This judicial interpretation was articulated before the AIA added the language "otherwise available to the public" to 35 U.S.C. § 102(a). Since similar language was used (available versus accessible), cases considering whether an invention was "accessible to the public" as part of the public use inquiry may be helpful in construing the new statutory language.

A. Public Accessibility

Because clinical testing usually occurs in a public setting (e.g., in one or more hospitals and/or in patients' homes), some have argued that the clinical testing is a public use that bars patenting the product. As a general proposition, there is no requirement that the invention be accessible to the entire general public—access to any member of the public can be enough to bar patenting.[7] "Even limited disclosure to those who are skilled enough to know, understand and 'easily demonstrate the invention to others,' may mean that there was no reasonable expectation of secrecy and that the invention was therefore in public use."[8] That being said, not every use that occurs in public will constitute a public use in the legal sense because the use may not disclose all aspects of the claimed invention or it may be experimental.[9] The issue of public use can be highly fact-specific and depends on what is disclosed and to whom it is disclosed. Factors to consider when evaluating whether certain behavior "was accessible to the public" include: (1) whether the use was for experimentation; (2) the nature of the activity that occurred in public; (3) public access to the use;

5. *Pfaff*, 525 U.S. at 60; Invitrogen Corp. v. Biocrest Mfg. LP, 424 F.3d 1374, 1379 (Fed. Cir. 2005) ("Nonetheless, the Court's analysis of the statutory term 'invention,' or the ready for patenting prong, applies to both of the other parts of section 102(b), 'on sale' and 'public use.'").

6. *Id.* (quoting *Invitrogen*, 424 F.3d at 1380).

7. Egbert v. Lippmann, 104 U.S. 333 (1881).

8. *Dey*, 715 F.3d at 1355–56.

9. *See, e.g., id.* at 1357; (no public use where the descriptions of the patented formulations was incomplete); City of Elizabeth v. Am. Nicholson Pavement Co., 97 U.S. 126, 134–35 (1877) (experimental use).

and (4) confidentiality obligations imposed on members of the public who observed the use.[10]

In *Dey L.P. v. Sunovion Pharms., Inc.*, the Federal Circuit considered whether Dey's patent on a formulation of a drug was invalid based on an alleged public use of the formulation in Sunovion's clinical trial.[11] The court noted that most cases asserting public use based on a clinical trial are based on the patentee's trial rather than a clinical trial conducted by another party. This situation led the court to emphasize that the confidentiality over the use was the focus, rather than a specific obligation or control by the patentee to keep the invention secret. The court considered two groups who were involved with the clinical trial of the formulation: study investigators (physicians) and study participants (patients). The investigators were the most knowledgeable persons involved with the study, but the investigators signed pledges of confidentiality. The participants were not given a complete description of the formulation, but only the name of the active agent and a range of possible dosages in the study. Though the participants could discuss the study with their doctors, and though a small percentage of the study drug was lost rather than taken by participants or returned to the investigators, these facts were insufficient to establish public use. As to the investigators, the public could not "readily discern" the claimed features of the invention due to the participants' limited knowledge about the formulation.

In contrast, a formulation was found invalid based on a public use in *Pronova BioPharma Norge AS v. Teva Pharms. USA, Inc.*[12] Pronova provided a clinician with samples of its formulation, along with a certificate of analysis showing the samples had all the components required by the patent claims. The clinician also analyzed the formulation to confirm its content. The clinician had no restriction on his use or disclosure of the formulation. The court found that this constituted an invalidating public use, as the formulation had been "released into the 'public domain' for purposes of §102(b)."

Thus, a disclosure of some aspects of an invention, but not all, will likely preclude a finding of public use.[13] A court must decide whether the

10. *In re* Omeprazole Patent Litig., 490 F. Supp. 2d 381 (Fed. Cir. 2007); *Invitrogen*, 424 F.3d at 1380; *accord Egbert*, 104 U.S. at 336 (to qualify as "public," a use must occur without any "limitation or restriction, or injunction of secrecy."); Manville Sales Corp. v. Paramount Sys. Inc., 917 F.2d 544, 550 (Fed. Cir. 1990) (use is not likely to be deemed "public" if the inventor has done nothing to make the public reasonably believe that the invention is in the public domain).

11. *Dey*, 715 F.3d at 1353.

12. Nos. 09-CV-0286 and 09-CV-0305, 2013 WL 242727 (Fed. Cir. 2013) (unpublished).

13. *Id.*; *see also*, W.L. Gore & Assocs. Inc. v. Garlock Inc., 721 F.2d 1540, 1549 (Fed. Cir. 1983) (reversing § 102(b) invalidation, in part, because "looking at the machine in operation

"claimed features of the patents [were placed] in the public's possession."[14] Similarly, "an agreement of confidentiality, or circumstances creating a similar expectation of secrecy, may negate a 'public use' where there is not commercial exploitation."[15] The presence or absence of a confidentiality agreement is not dispositive.[16]

In *Janssen Pharmaceutica N.V. v. Eon Labs Mfg., Inc.*,[17] the labels on the boxes of Sporanox stated that it was an investigational drug not for commercial use. Patients did not know any information about the drug other than its name and that it was an investigational antifungal.[18] The patients were not under any confidentiality obligation.[19] Based on these facts, the court found that the confidential and controlled use of the study drug prevented invalidation of the formulation claims based on alleged "public use."[20] Essentially, the lack of information on the investigational drug label supported the confidential nature of the use, despite any formal confidentiality agreement.

An invention may be disclosed to the inventor's employer and coworkers without that disclosure constituting a public use presumably because there is an obligation or expectation of secrecy. Further, the coworkers may use the invention within the confines of the company, so long as the use is secret and not commercial exploitation.[21]

1. Concerns about Publications Arising from the Clinical Trial

In a clinical trial, the study drug is released to investigators and participants, and instructions regarding the use of the drug must be provided. Information about the trial will appear on http://clinicaltrials.gov/ and possibly on the websites of the clinical sites. In most cases, a drug company will be careful to file a patent application before an invention is disclosed in an academic paper or a scholarly journal or in a poster presentation at a scientific conference. However, companies should be cognizant of other publications that may arise in connection with clinical trials. Thus, pharmaceutical and biotechnology companies should be cognizant of what

does not reveal whether it is stretching, and, if so, at what speed. Nor . . . whether the crystallinity and temperature elements of the invention set forth in the claims are involved").

14. Netscape Commc'ns Corp. v. Konrad, 295 F.3d 1315, 1323 (Fed. Cir. 2002).

15. *Invitrogen*, 424 F.3d at 1382.

16. Bernhardt LLC v. Collezione Europa USA Inc., 386 F.3d 1371, 1380–81 (Fed. Cir. 2004).

17. 374 F. Supp. 2d 263 (E.D.N.Y. 2004), *aff'd*, 134 F. App'x 425, 430–31 (Fed. Cir. 2005).

18. *Id.* at 276.

19. *Id.* at 268.

20. *Id.* at 276.

21. *Invitrogen*, 424 F.3d at 1382–83; Moleculon Research Corp. v. CBS Inc., 793 F.2d 1261 (1984).

information is provided on websites, investigators' product labels, brochures, and other product information for the study drug.

No later than 21 days after enrollment of the first patient, information about the clinical trial must be registered by the sponsor or principal investigator so that it can be listed on the ClinicalTrials.gov registry. The information typically appears on ClinicalTrials.gov within a few days after submission (after it has passed review by staff). Pharmaceutical and biotechnology companies may still label the study drug with a drug name, trademark, or some other identifying information. Sometimes an alphanumerical designation is used for the study drug, thereby avoiding disclosure of its chemical name or structure. As one court has stated, such information does not divulge "the alleged innovation" and is "necessary" because "obviously . . . a participant taking a regimen of drugs requires the liberty of discussing same with a spouse, medical personnel in case of emergency, etc."[22]

2. Clinical Trial Patient Agreements

Confidentiality agreements with patients enrolled in clinical trials are infrequent and not necessary to avoid a public use bar. In fact, strict confidentiality agreements with patients preventing any and all disclosure about the study may be inappropriate from a medical stance, as a patient may need to discuss the study drug and his or her reaction to it with a personal physician or other caregivers.[23] Nonetheless, patients often sign consent forms that apprise them of the risks involved with the study drug and that may contain certain confidentiality provisions. The consent forms may instruct the patients that the study drug must be taken as directed and that any unused drug should be returned. Executed patient consent forms can bolster an argument that the use of the patented drug was not in public use because the use was confidential and controlled by the patentee.

Moreover, not all aspects of the trial need to be disclosed to patients. Even when the patient is told what the study drug is, many details may remain confidential so as not to bias the trial. Such additional details may

22. Bayer Schering Pharma AG v. Barr Labs. Inc., No. 05-CV-2308 (PGS), 2008 WL 628592, at *12, *41 (D.N.J. Mar. 3, 2008).

23. *See, e.g.*, *id.* at *38–42 (finding no public use even though "the human patients did not execute confidentiality agreements"); *In re* Omeprazole Patent Litig., 490 F. Supp. 2d 381, 508 (S.D.N.Y. 2007) (dismissing as "meritless" the contention that "inventions were not kept sufficiently confidential" because "patients were not required to sign confidentiality agreements"), *aff'd on other grounds*, 536 F.3d 1361 (Fed. Cir. 2008); Eli Lilly & Co. v. Zenith Goldline Pharm. Inc., 364 F. Supp. 2d 820, 873–75, 912–13 (S.D. Ind. 2005) (rejecting the contention that failing to execute a confidentiality agreement constituted a controlling factor in the public use inquiry), *aff'd*, 471 F.3d 1369, 1380–81 (Fed. Cir. 2006); *Janssen*, 374 F. Supp. 2d at 276.

be patentable. When a dose-finding trial is listed on ClinicalTrials.gov, it may be desirable to identify the amounts as high and low doses, rather than by the numerical amount (such as specific dose amounts).

3. Investigator and Institution Agreements

In contrast to most clinical trial patients, clinical trial investigators have professional expertise with therapeutics and their use, so pharmaceutical and biotechnology companies should control disclosures to these individuals accordingly. "[A disclosure is made] to one highly skilled in the art, with the full ability to know, understand, and fully disclose the invention to others . . . may mean that there is no reasonable expectation of secrecy and that the invention was therefore in public use."[24] Pharmaceutical and biotechnology companies attempting to avoid a public use bar should be sure that the use by the clinical trial investigators is confidential and controlled.

The best practice to demonstrate that the use is confidential and controlled is to secure a confidentiality agreement from each clinical investigator involved in the clinical trials. The confidentiality agreement should be crafted to protect against the unauthorized and ill-timed disclosure of the "invention." This can be accomplished by a blanket confidentiality provision preventing the disclosure of any and all information relating to the study medication or by a more targeted confidentiality provision precluding investigators from disclosing, for example, any pharmacokinetic properties of the study drug. The confidentiality provision should cover any possible new invention that would be supported by the clinical trial data, so that a company has the opportunity to file a patent application on the invention before the data or overall results are publicly disclosed. If publishing information arising from clinical studies is an objective for the investigator, the confidentiality agreement can specify that any publication be provided to the sponsoring company before publication, or even before submission to a journal or medical association. The company can then decide whether a patent filing should be made before the information is published.

B. Commercial Exploitation

Commercial exploitation triggering the public use bar overlaps somewhat with activities that trigger the on-sale bar, which is discussed in the following section. Pharmaceutical and biotechnology companies are in the business of commercializing their inventions. Because the product must

---

24. Pronova Biopharma Norge AS v. Teva Pharm. USA Inc., 549 Fed. Appx. 934, 941, 943 (Fed. Cir. 2013).

undergo clinical testing before it can be commercialized, clinical testing may be viewed as being part of a company's commercialization effort. If the invention is ultimately sold in the marketplace, experiments in furtherance of the invention's development are commercially beneficial in retrospect despite being mandated by the federal government. Thus, the question could be raised, does clinical testing in and of itself constitute a commercialization within the constructs of patent law?

At least one court has said it does not. In *In re Omeprazole Patent Litigation*,[25] Impax, the alleged infringer, asserted that the claimed invention (which was the clinical formulation) had been commercially exploited through the clinical trials before the filing of the patent application. The court rejected that assertion, noting that (1) the patients did not pay for the study medication they received and (2) no evidence was presented to show that the patentee was "testing the market."[26] "Impax merely theorizes that the trials constitute invalidating commercial exploitation, insofar as the trials were a means of obtaining the FDA approval required to enter into the U.S. market. The court rejected this proposition."[27] The court went farther and explained, "Impax's proposed theory, if accepted, would unduly force the hand of inventors of new pharmaceutical formulations to file for patents prior to sufficiently testing the safety and efficacy of the formulation. There is simply nothing in the patent law or its underlying policy which requires or supports this."[28]

C. The On-Sale Bar

As mentioned earlier in this chapter, there is overlap between the on-sale bar and the public use bar, particularly where public use is based on commercial exploitation. "Commercial exploitation is a clear indication of public use, but it likely requires more than, for example, a secret offer for sale."[29] Generally, there will be no sale or offer for sale in a clinical trial since the study drug is provided to investigators and participants at no charge.

---

25. 490 F. Supp. 2d 381 (S.D.N.Y. 2007), *aff'd on other grounds*, 536 F.3d 1361 (Fed. Cir. 2008).

26. *In re Omeprazole Patent Litig.*, 490 F. Supp. 2d at 509.

27. *Id.* The *In re Omeprazole Patent Litigation* decision could be viewed as being at odds with *Pennwalt Corp. v. Akzona Inc.*, 740 F.2d 1573 (Fed. Cir. 1984). In *Pennwalt Corp.*, Pennwalt made a limited distribution of its insecticide under a "temporary permit," issued pursuant to an Environmental Protection Agency (EPA) regulation, which provided that such permits "will be issued only for bona fide experimental programs under the supervision of qualified persons." The Federal Circuit held that Pennwalt's primary motive in obtaining the temporary permit was commercial and thus, the patent was invalid because the claimed invention was in public use under 35 U.S.C. § 102(b).

28. *Id.*

29. Invitrogen Corp. v. Biocrest Mfg. LP, 424 F.3d 1374, 1380 (Fed. Cir. 2005).

D. Experimental Use

If a patent challenger makes a prima facie showing that the public use or on-sale bar negates patentability, the patentee may rebut this showing by demonstrating that the use of the invention was experimental.[30] "Experimental use negates public use; when proved, it may show that particular acts, even if apparently public in a colloquial sense, do not constitute a public use within the meaning of section 102."[31] Experimental use can also negate the on-sale bar, provided the sale was incidental to the primary purpose of experimentation to perfect the invention.[32]

The experimentation must be for developing the invention, which can be moving it toward a reduction to practice; it cannot be for testing the market or customer interest. Once the invention is complete or reduced to practice, further displays or commercial activities are unlikely to be deemed experimental use. "A use may not be experimental if the invention at issue was reduced to practice. For a pharmaceutical composition, a reduction to practice occurs when an inventor 'actually prepared the composition and knew that it would work.'"[33] The Federal Circuit has noted that "when testing is necessary to establish utility, there must be recognition and appreciation that the tests were successful for reduction to practice to occur."[34] However, "further 'experimentation' may constitute a barring public use."[35]

When a pharmaceutical company tests a composition or formulation in clinical trials, it does not know for certain whether the trials will be successful and establish a new way of treating human patients. In *Bayer*, the court noted that European clinical trials were conducted prior to the

30. Lough v. Brunswick Corp., 86 F.3d 1113, 1120 (Fed. Cir. 1996), *cert. denied*, 522 U.S. 806 (1997).

31. Baxter Int'l Inc. v. Cobe Labs. Inc., 88 F.3d 1054, 1059 (Fed. Cir. 1996) (citing TP Labs.Inc. v. Prof'l Positioners Inc., 724 F.2d 965, 971 (Fed. Cir. 1984)), *cert. denied*, 469 U.S. 826 (1984).

32. LaBounty Mfg. v. U.S. Int'l Trade Comm'n, 958 F.2d 1066, 1071 (Fed. Cir. 1992).

33. Bayer Schering Pharma AG v. Barr Labs. Inc., No. 05-CV-2308 (PGS), 2008 WL 628592, at *42 (D.N.J. Mar. 3, 2008) (quoting Estee Lauder Inc. v. L'Oreal SA, 129 F.3d 588, 592 (Fed. Cir. 1997)); Am. Cyanamid Co. v. U.S. Surgical Corp., 833 F. Supp. 92 (D. Conn. 1992).

34. *Estee Lauder Inc.*, 129 F.3d at 594–95. This is a key point in the pharmaceutical/biotechnology context. A claim simply reciting the chemical composition of matter can be adequately supported by in vitro testing. Fujikawa v. Wattanasin, 93 F.3d 1559, 1564–66 (Fed. Cir. 1996). However, for example, the minimum utility for a method claim for treatment of a certain human disease or a formulation claim with certain pharmacokinetic properties may require clinical data to demonstrate the required minimum utility for patentability of that claim.

35. RCA Corp. v. Data Gen. Corp., 887 F.2d 1056, 1061 (Fed. Cir. 1989) (finding that public use is not negated by experiments directed to tangential or unclaimed aspects of the invention).

critical date and that a subsequent U.S. clinical trial was also initiated prior to the critical date.[36] The challenger asserted that the European clinical trial constituted a reduction to practice and that the subsequent U.S. clinical trial could not be viewed as "experimental."[37] The court disagreed and found that the expert testimony regarding the differences in U.S. versus European populations required efficacy confirmation in a U.S. clinical trial.[38] The court found that the *In re* Omeprazole decision supported its conclusion:

> In *In re Omeprazole Patent Litigation*, the court found that the patent challenger had failed to establish by clear and convincing evidence that a reduction to practice occurred where the inventors were still analyzing the results of the clinical trials before the critical date. 490 F. Supp. 2d 381, 507 (S.D.N.Y. 2007). The *In re Omeprazole* court found that a reduction to practice did not occur upon the conception and initial testing of the omeprazole formulation. Rather, the court ruled that the patentee's four clinical trials demonstrated a need to confirm the formulation's effectiveness. Based upon the expert testimony that all results needed to be analyzed before a finding of safety and efficacy, and noting that the patent holder did not have the results from at least one of those clinical trials, the court found insufficient evidence to establish a reduction to practice by clear and convincing evidence.[39]

On the other hand, use in a clinical setting or other regulatory testing does not, per se, establish that the use was experimental. In order for experimental use to negate a public use or on-sale bar, the pharmaceutical or biotechnology company must conduct the activity primarily for the purpose of experiment, and any commercialization must be incidental. In other words, the company's primary purpose must have been to show that the invention was suitable for its intended purpose. This, of course, requires consideration of what the invention is; a chemical compound may be shown to have a useful biological effect without clinical trials, but a formulation designed to achieve certain pharmacokinetic properties in a human patient may require clinical testing.

One strategy to show that uses were experimental is to include them as examples in the patent specification, so that they may be relied upon to

36. *Bayer*, 2008 WL 628592, at *39.
37. *Id.* at *42.
38. *Id.* at *43.
39. *Id.*

show patentability. Such inclusion makes it more likely that they will be viewed as "experimental" rather than a public use.

E. The Grace Period under 35 U.S.C. § 102

Under the 1952 Patent Act, a one-year grace period was provided where the invention was in public use or on sale in this country; the inventor could still file her or his patent application within one year. The AIA also includes a one-year grace period for disclosures of the invention, but only if the disclosures are made by the inventor herself or himself or by another who obtained the disclosed information or material directly or indirectly from the inventor.[40] Accordingly, even if a clinical trial constitutes a public use (or its description on http://clinicaltrials.gov/ is a publication), an inventor may still be able to patent an invention disclosed by the clinical trial within one year. As was true before the enactment of the AIA, this particular grace period exception is peculiar to the United States; most other countries do not provide a grace period for a public use or other public disclosure.

III. Government Funding and Disclosure

The Bayh-Dole Act was passed in 1980 to encourage greater utilization of inventions made by universities and other research organizations that receive federal funding.[41] Various federal agencies such as the National Institutes of Health (NIH) provide funds for research or development by those recipients, and the Bayh-Dole Act allows those recipients to take ownership of any inventions that result. A funding agreement is executed to govern the use of those funds.

The Bayh-Dole Act and implementing regulations require certain provisions in the funding agreement between the government and the recipient. They require the recipient of federal funding to notify the funding agency within two months after the inventor discloses the invention in writing to the recipient.[42] The recipient must also notify the agency if there was any publication, sale, or public use of the invention.[43] The recipient has an obligation to elect in writing whether or not to retain title to the invention by notifying the government agency within two years of that initial disclosure to the agency, though that period may be shortened if there

40. 35 U.S.C. § 102(b)(1)(A).

41. Pub. L. No. 96-517, 94 Stat. 3015 (codified as 35 U.S.C. §§ 200, 211, 301–307 (2000)); H.R. Rep. No. 96-1307, at 3 (1980).

42. 37 C.F.R. § 401.14(c)(1).

43. *Id.*

was a public use, sale, or publication.[44] The regulations require that the recipient file a patent application within one year of the election or before any statutory bar date created by the public use, sale, or publication.[45] Thus, the regulations are cognizant of the possibility of public use, sale, or publication of the invention, and they impose requirements to prevent the loss of patent rights due to them.

One might wonder if the disclosure to the funding agency itself constitutes a public use or otherwise makes the invention available to the public. The Bayh-Dole Act "was not intended to relax the statutory requirements for patentability."[46] However, the funding agency can maintain the disclosure by the inventor in confidence: "Federal agencies are authorized to withhold from disclosure to the public information disclosing any invention in which the Federal Government owns or may own a right, title, or interest (including a nonexclusive license) for a reasonable time in order for a patent application to be filed."[47] The inventor's disclosure is not accessible pursuant to the Freedom of Information Act (FOIA) for a reasonable time in order for a patent application to be filed.[48] No court has held that disclosure to the government pursuant to the Bayh-Dole Act constitutes a public use or other bar to patentability.

Another question is whether the initial request for funding made to a governmental funding agency (such as a grant application) constitutes a public use, offer for sale, or publication. Such funding requests are typically submitted confidentially, and federal regulations authorize an agency to withhold these requests when responding to an FOIA request. However, a grant proposal that is publicly accessible can be an invalidating printed publication under pre- or post-AIA versions of 35 U.S.C. § 102.[49]

The act also requires the funding agreement to provide that the federal agency issuing the funds obtain a nonexclusive, nontransferable, irrevocable, paid-up license to practice or have practiced for or on behalf of the United States any subject invention throughout the world.

44. *Id.* § 401.14(c)(2).

45. *Id.* § 401.14(c)(3).

46. Univ. of Rochester v. G.D. Searle & Co., 358 F.3d 916, 929 (Fed. Cir. 2004).

47. 35 U.S.C. § 205; see also 37 C.F.R. § 401.13(c), which provides that agencies shall not disclose information discussing an invention or patent application pursuant to requests under the Freedom of Information Act for certain periods of time.

48. *Id.*

49. E.I. duPont de Nemours & Co. v. Cetus Corp., No. C 89-2860 MHP, 1990 U.S. Dist. LEXIS 18382 (N.D. Cal. Dec. 11, 1990) (grant applications to the National Science Foundation and the NIM were "sufficiently accessible" to a researcher who exercised "reasonable diligence"; among other factors, the grant applications were identified by their grant numbers in a journal article).

Under the Bayh-Dole Act, an invention is defined as "any invention or discovery which is or may be patentable or otherwise protectable under this title."[50] A "subject invention" is one that is "conceived or first actually reduced to practice in the performance of work under a funding agreement." A company receiving federal funding should be aware that, under these definitions, the statutory scheme applies not just to patentable inventions but also to subject matter that the company intends to protect by keeping it a trade secret and to some inventions where only a small part of the development was supported by federal funding. A company should recognize that using government funding for a clinical trial will result in the government having a royalty-free license to any invention that arises from the trial, and that may include the method of treatment.

Thus, the government has license rights in situations such as (a) where a method of treatment is conceived and developed at private expense, but a sponsor takes government funding to conduct a clinical trial that results in its first actual reduction to practice, (b) where a compound, combination, or diagnostic is conceived while a private party is performing under government funding, but all development and reduction to practice is accomplished at private expense, and (c) in situations falling between these extremes.

A sale or license of intellectual property rights in the invention does not create an on-sale bar.[51] Thus, an assignment from the inventors to the recipient of federal funds does not trigger the on-sale bar.

IV. Expected and Unexpected Results for Patent Considerations

As noted earlier in this chapter, by the time a company initiates a human clinical trial for a new chemical or biological compound, the company has likely filed patent applications disclosing and claiming that compound. Nevertheless, information gleaned from a human clinical trial may provide an impetus to file new patent applications and/or may support previously filed patent applications. Human clinical trials can establish information that eventually appears in the product label, such as detailed methods of use, dosages, contraindications, warnings, and precautions, that may not be available or complete prior to testing in humans. In addition, human clinical trials can yield unexpected results to support patentability of applications directed to formulations, dosing regimens, and combination

50. 35 U.S.C. § 201(d).

51. *See In re* Kollar, 286 F.3d 1326, 1330 n.3, 1330–31 (Fed. Cir. 2002).

treatments. While results from human clinical trials ae generally not necessary to demonstrate patentability,[52] human clinical trial data are often included in new applications. The data also may be submitted via declaration or affidavit to support a previously filed patent application, especially with the U.S. Patent and Trademark Office (USPTO). Thus, in certain instances, findings arising from a human clinical trial provide an avenue for additional patent protection for a known compound.

A. Repurposing Compounds

Human clinical trials may reveal a new use for a known compound and, therefore, give rise to a potentially valuable method-of-use patent.[53] In many instances, discovering a new use for a known compound, or "repurposing" a compound, involves the serendipitous recognition of a side effect having a therapeutic benefit for an unrelated indication. More recently, however, pharmaceutical companies, along with governmental agencies and academia, have consciously engaged in efforts to repurpose compounds that have already been tested in humans.[54] Since the use of the compound is likely to be disclosed on ClinicalTrials.gov and explained to patients, one should consider whether the invention is ready for patenting before the clinical trial (and, accordingly, a patent application should be filed) and/or whether results or data from the clinical trial are needed to complete the invention (so the patent application could not be filed without first conducting the trial).

Pfizer's blockbuster Viagra (sildenafil citrate), which is an inhibitor of a cyclic guanosine monophosphate phosphodiesterase (cGMP PDE) indicated for the treatment of erectile dysfunction, is a classic example of repurposing a compound. Pfizer first synthesized sildenafil in 1989 and

---

52. *See* Manual of Patent Examining Procedure (MPEP) § 2107.03 (rev. Aug. 9, 2012) ("If reasonably correlated to the particular therapeutic or pharmacological utility, data generated using in vitro assays, or from testing in an animal model or a combination thereof almost invariably will be sufficient to establish therapeutic or pharmacological utility for a compound, composition or process.").

53. The value of a method-of-use patent may be tempered by a generic drug manufacturer's ability to "carve out" the patented indication. *See, e.g.*, AstraZeneca Pharm. LP v. Apotex Corp., 669 F.3d 1370, 1379 (Fed. Cir. 2012) ("In other words, the Act allows generic manufacturers to limit the scope of regulatory approval they seek—and thereby forego Paragraph IV certification and a § 271(e)(2) infringement suit—by excluding patented indications from their ANDAs.").

54. Examples of formal units engaged in systemic exploration of repurposing opportunities for existing compounds include Bayer HealthCare's Common Mechanism Research team and Pfizer's New Indications Discovery Unit. Similarly, the National Center for Advancing Translational Sciences (NCATS) launched the Discovering New Therapeutic Uses for Existing Molecules (New Therapeutic Uses) initiative in 2012.

filed patent applications covering the compound.[55] Pfizer's composition of matter patent, U.S. Patent No. 5,250,534, asserts that the disclosed compounds could be used in the treatment of angina, hypertension, heart failure, and atherosclerosis. The '534 patent is listed in the Orange Book for Viagra with an expiration date of September 27, 2012. However, a method-of-use patent, U.S. Patent No. 6,469,012, which covers the use of sildenafil to treat erectile dysfunction, provides Pfizer with exclusivity for Viagra until April 2022 (including pediatric exclusivity).

This inventive use of sildenafil was discovered in the course of a clinical trial. In the early 1990s, Pfizer began clinical development of sildenafil for the treatment of angina, which is a condition caused by insufficient blood flow to the heart.[56] In 1992, Pfizer initiated a multiple-dose study of sildenafil in healthy male volunteers to assess the safety of the drug.[57] At the conclusion of the study, volunteers reported several side effects, including spontaneous erections.[58] Following this revelation, in June 1993, Pfizer filed a patent application in the United Kingdom directed to the use of cGMP PDE inhibitors, including sildenafil, for the treatment of erectile dysfunction.[59] Pfizer subsequently filed a patent application in the United States that claimed priority to the UK application. The U.S. application issued as the '012 patent and provided Pfizer with several additional years of exclusivity for a drug with sales exceeding $1 billion per year in the United States.[60]

Patents containing method-of-use claims covering an FDA-approved indication can be listed in the Orange Book.[61] The FDA requires that a generic drug have the same labeling, including at least one of the approved indications, as the reference drug.[62] Thus, a generic drug manufacturer

55. See, e.g., U.S. Patent No. 5,250,534, which was filed on May 14, 1992 and claims priority to British Patent Application No. 9013750, which was filed on June 20, 1990.

56. *See* Pfizer Inc. v. Teva Pharm. USA Inc., 803 F. Supp. 2d 409, 417 (E.D. Va. 2011).

57. *Id.*

58. *Id.*

59. *See* GB 9311920 ("It has now been found that these disclosed compounds are useful in the treatment of erectile dysfunction.").

60. Pfizer brought litigation against Teva Pharmaceuticals asserting that Teva's ANDA for sildenafil was an infringement of the '012 patent. The parties eventually settled the litigation. As a result of the settlement, Teva will be allowed to launch a generic version of Viagra in the United States on December 11, 2017, or earlier under certain circumstances. Teva will pay Pfizer a royalty for a license to produce its generic version. *See* Press Release, Pfizer, Pfizer Settles Viagra Patent Litigation with Teva (Dec. 17, 2013), *available at* http://www.pfizer.com/news/press-release/press-release-detail/pfizer_settles_viagra_patent_litigation_with_teva.

61. *See* 21 U.S.C. § 355(b)(1)(G) (an NDA holder must identify "any patent . . . which claims a method of using such drug"). The "Orange Book" is the common name for the FDA publication "Approved Drug Products with Therapeutic Equivalence," and it may be accessed at http://www.accessdata.fda.gov/scripts/cder/ob/default.cfm.

62. *See* 21 C.F.R. § 314.127(a)(2), (7).

may be liable for inducing infringement where its proposed label instructs patients to infringe a method-of-use patent.[63]

B. Rebutting a Prima Facie Case of Obviousness

A key aspect of patentability, that an invention is not obvious, was codified in 35 U.S.C. § 103(a) of the 1952 act:

> A patent for a claimed invention may not be obtained . . . if the differences between the claimed invention and the prior art are such that the claimed invention as a whole would have been obvious before the effective filing date of the claimed invention to a person having ordinary skill in the art to which the claimed invention pertains.

The AIA left § 103 essentially intact. Procedurally, the patent examiner (in the context of an application before the USPTO) or the challenging litigant (in the context of a district court proceeding) bears the initial burden of factually supporting any prima facie case of obviousness. If a patent examiner establishes a prima facie case of obviousness, the burden of coming forward with rebuttal evidence or arguments shifts to the patent applicant.[64] Likewise, in a litigation setting, "[t]he patentee . . . is required to come forward with evidence of nonobviousness only after the challenger has successfully made his prima facie case demonstrating that the patent might be obvious."[65] However, this requirement does not shift the burden of persuasion. Indeed, "'because the presumption of validity remains intact . . . throughout the litigation,' the burden of persuasion never shifts to the patentee during the course of a district court obviousness challenge."[66]

To rebut a prima facie case of obviousness, the applicant/patentee may submit evidence of nonobviousness, such as data showing that the claimed invention possesses unexpectedly superior properties as compared to the prior art. Unexpected results may have a foundation in the specification, but the results themselves need not be included in the specification.[67]

---

63. *See* 35 U.S.C. § 271(b) ("Whoever actively induces infringement of a patent shall be liable as an infringer."); *see also* AstraZeneca LP v. Apotex Inc., 633 F.3d 1042, 1060 (Fed. Cir. 2010) ("The pertinent question is whether the proposed label instructs users to perform the patented method.").

64. MPEP, *supra* note 52, § 2142.

65. Novo Nordisk A/S v. Caraco Pharm. Labs. Ltd., 719 F.3d 1346, 1352–54 (Fed. Cir. 2013); *see also* Pfizer Inc. v. Apotex Inc., 480 F.3d 1348, 1359–60 (Fed. Cir. 2007) ("[O]nce a challenger has presented a prima facie case of invalidity, the patentee has the burden of going forward with rebuttal evidence.").

66. *Novo Nordisk*, 719 F.3d at 1352––54 (citing *Pfizer*, 480 F.3d at 1359–60).

67. *See In re* Chu, 66 F.3d 292, 298–99 (Fed. Cir. 1995) ("We have found no cases supporting the position that a patent applicant's evidence and/or arguments traversing a Section 103 rejection must be contained within the specification.").

Indeed, the Federal Circuit has held that evidence of unexpected results may be used to rebut a case of prima facie obviousness even if that evidence was obtained after the patent's filing or issue date.[68]

However, merely demonstrating superior results may not overcome a prima facie case of obviousness. The applicant/patentee is often required to demonstrate that the superior results were surprising or unexpected to a person of ordinary skill in the art. Thus, the applicant/patentee often presents (1) data showing improved results compared with the closest prior art and (2) testimony or other evidence that the results are unexpected.[69] Several recent decisions from the Federal Circuit highlight that the data proffered by the applicant/patentee and the applicant/patentee's characterization of the results as unexpected may be closely scrutinized.

Bayer v. Watson concerned claims 13 and 15 of U.S. Patent Number RE37,564, which was listed in the Orange Book for Bayer's Yaz (drospirenone/ethinyl estradiol) combination oral contraceptive tablets.[70] Claims 13 and 15 of the '564 patent are directed to a combination product for oral contraception comprising 23 or 24 daily dosage units of drospirenone/ethinyl estradiol and 5 or 4 placebo pills, respectively (i.e., a 23/5 or 24/4 dosing regimen).[71]

Bayer's scientists conducted a clinical trial (AA51) to compare a 21/7 dosing regimen to a claimed 23/5 dosing regimen.[72] Study AA51 showed a statistically significant difference in suppressed ovarian activity between the 23/5 dosing regimen and the 21/7 dosing regimen.[73] These data were included in Figure 2 of the '564 patent.[74] The district court credited Bayer's evidence of unexpected results and held that claims 13 and 15 of the '564 were not invalid in view of the prior art.[75]

A Federal Circuit panel disagreed with the district court, finding Bayer's evidence of nonobviousness, including alleged unexpected results, legally insufficient.[76] The court succinctly dismissed Bayer's data

---

68. Genetics Inst. v. Novartis Vaccines, 655 F.3d 1291, 1307 (Fed. Cir. 2011) (citing Knoll Pharm. Co. v. Teva Pharm. USA Inc., 367 F.3d 1381, 1385 (Fed. Cir. 2004)).

69. *See, e.g.*, *In re* Freeman, 474 F.2d 1318, 1324 (C.C.P.A.1973) ("In order for a showing of unexpected results to be probative of nonobviousness, such evidence must at least establish: (1) that there actually is a difference between the results obtained through the claimed invention and those of the closest prior art, and (2) the difference actually obtained would not have been expected by one skilled in the art at the time of the invention." (internal citations omitted)).

70. Bayer Healthcare Pharm. Inc. v. Watson Pharm. Inc., 713 F.3d 1369 (Fed. Cir. 2013).

71. *See* '564 Patent, at cols. 6–8.

72. *See* Bayer Schering Pharma AG v. Lupin Ltd., No. 2:10-cv-01166-KJD-RJJ, slip op. at 17, 29 (D. Nev. Mar. 30, 2012).

73. *See id.*

74. *See* '564 Patent, at fig.2, col. 6.

75. *See* Bayer Schering Pharma AG, No. 2:10-cv-01166-KJD-RJJ, slip op. at 29–30.

76. Bayer Healthcare Pharm. Inc. v. Watson Pharm. Inc., 713 F.3d 1369 (Fed. Cir. 2013).

showing that the 23/5 regimen yielded reduced follicular activity compared to the 21/7 regimen by stating: "[T]hose data merely confirm that administering additional active pills results in additional follicular suppression, which would have been a matter of 'common sense,' as even Bayer's expert agreed."[77]

Novo Nordisk v. Caraco concerned claim 4 of U.S. Patent Number 6,677,358, which was listed in the Orange Book for Novo's Prandin (repaglinide) tablets.[78] Claim 4 of the '358 patent is directed to a method for treating type 2 diabetes by administering two compounds—metformin and repaglinide—in combination.[79]

The '358 patent discussed synergism between repaglinide and metformin.[80] Novo had conducted a clinical trial to determine the effectiveness of the repaglinide/metformin combination for treating patients having type 2 diabetes.[81] In this trial, Novo examined the patients' glucose levels after an eight-hour fast (i.e., fasting plasma glucose, or FPG).[82] The Novo scientists found that the repaglinide/metformin combination was more effective at reducing FPG than either repaglinide or metformin alone.[83] These data were included as example 3 in Novo's patent application.[84]

The obviousness inquiry focused on whether these superior results would have been expected by one of ordinary skill in the art. The district court determined that one of ordinary skill in the art would have expected the repaglinide/metformin combination to yield some synergism.[85] Pointedly, the district court concluded that Caraco had established by "[c]lear and convincing evidence . . . that the results of the claimed combination therapy said by Novo to be unexpected and unexplainable were, to the contrary, expected and explainable in light of the state of the art as of the critical date."[86]

The Federal Circuit panel majority did not find error with the district court's three-step rationale for determining that the metformin/repaglinide combination was obvious: (1) the closest prior art was combination therapy using metformin and an insulin secretagogue in the sulfonylurea class (i.e., glyburide); (2) the prior art combination was well known to produce

---

77. *Id.*
78. Novo Nordisk A/S v. Caraco Pharm. Labs. Ltd., 719 F.3d 1346 (Fed. Cir. 2013).
79. *See* '358 Patent, at col. 10.
80. *See id.*, at col. 3.
81. *Novo Nordisk*, 719 F.3d at 1349.
82. *Id.*
83. *Id.*
84. *Id.* at 1349–50.
85. Novo Nordisk A/S v. Caraco Pharm. Labs., 775 F. Supp. 2d 985, 1012–13 (E.D. Mich. 2011).
86. *Id.* at 1018.

beneficial and even synergistic results in controlling glucose levels in type 2 diabetes patients; and (3) repaglinide was known as an insulin secretagogue having a mechanism of action similar to the sulfonylurea class of secretagogues.[87]

Judge Newman, on the other hand, dissented from the majority's decision that the metformin/repaglinide combination was obvious.[88] She focused on the significant structural and functional differences between glyburide and repaglinide, which, in her opinion, rendered it unreasonable to expect the repaglinide/metformin combination to have synergistic properties superior to the prior art combination of glyburide and metformin.[89]

Allergan v. Sandoz concerned claim 1 of U.S. Patent Number 7,323,463 and claim 4 of U.S. Patent Number 7,030,149.[90] The '463 patent and the '149 patent were listed in the Orange Book for Combigan (brimonidine tartrate/timolol maleate ophthalmic solution). Combigan is indicated for reduction of elevated intraocular pressure in patients with glaucoma or ocular hypertension.[91] The recommended administration is one drop of Combigan in the affected eye(s) twice daily approximately 12 hours apart.[92]

Claim 1 of the '463 patent is directed to a single composition comprising timolol and brimonidine as the sole active agents.[93] Claim 4 of the '149 patent is directed to a method of reducing the number of daily topical ophthalmic doses of brimonidine administered from three to two times a day without loss of efficacy by administering brimonidine and timolol in a single composition.[94]

The '463 and '149 patents included results from a clinical trial comparing the safety and efficacy of twice-daily administration of a fixed-dose combination of brimonidine/timolol with that of concurrent brimonidine three times daily and timolol twice daily.[95] The prior art taught twice-daily dosing of brimonidine as well as concomitant administration of brimonidine and timolol. However, previous attempts to treat patients twice per day with brimonidine resulted in a loss of efficacy eight to nine hours post-administration.[96] This loss of efficacy is referred to as the "afternoon

87. *Novo Nordisk*, 719 F.3d at 1355.
88. *Id.* at 1360 (Newman, J., dissenting).
89. *Id.* at 1364 (Newman, J., dissenting).
90. Allergan Inc. v. Sandoz Inc., 726 F.3d 1286 (Fed. Cir. 2013).
91. Full Prescribing Information for Combigan, ALLERGAN (rev. Nov. 2013), *available at* http://www.allergan.com/assets/pdf/combigan_pi.pdf.
92. *Id.*
93. '463 Patent, at col. 9.
94. '149 Patent, at col. 10.
95. *See, e.g.*, '149 Patent, at col. 4–9.
96. *Allergan*, 726 F.3d at 1293.

trough."[97] Data presented by Allergan in the '463 and '149 patents demonstrated that a twice per day dosage regimen of Combigan did not suffer from the afternoon trough issue.[98]

While the panel majority agreed that this result was unexpected, it treated the formulation claims differently than the method claims. The court found that these unexpected results were insufficient to outweigh the other evidence of obviousness as to the formulation claims. On the other hand, the court emphasized that the method claims included the additional limitation that the daily number of doses of brimonidine be reduced from three to two times a day "*without loss of efficacy*."[99] Sandoz failed to point to evidence in the prior art that would allow for the conclusion that the addition of timolol to brimonidine dosed twice per day would eliminate the afternoon trough.[100] To the contrary, "[t]he record firmly establishes that when brimonidine is dosed twice per day as opposed to three times per day, there is a loss of efficacy in the afternoon—the so called, afternoon trough."[101]

Judge Dyk dissented from the decision upholding claim 4 of the '149 patent. In his opinion, "the different results as between the claims of the '463 patent and claim 4 of the '149 patent cannot be reconciled."[102] He argued that the majority's outcome appears to rest on the notion that claim 4 was not obvious because it claims the result of twice-a-day dosing—avoiding "a loss of efficacy in the afternoon."[103] Judge Dyk reiterated that "a newly-discovered result or property of an existing (or obvious) method of use is not patentable."[104]

C. Compliance with 35 U.S.C. § 112, First Paragraph

In addition to rebutting a prima facie case of obviousness, a patent applicant may use clinical trial data to support an argument that the patent specification complies with 35 U.S.C. § 112, first paragraph.[105] Section 112 requires a patent specification to enable a person of skill in the art to make and use the claimed invention. The enablement requirement "incorporates

97. *Id.*
98. *Id.*
99. *Id.* (emphasis in original).
100. *Id.* at 1294.
101. *Id.*
102. *Id.* at 1295 (Dyk, J., dissenting).
103. *Id.*
104. *Id.* (quoting Bristol-Myers Squibb Co. v. Ben Venue Labs., 246 F.3d 1368, 1376 (Fed. Cir. 2001)).
105. The same section is set forth as 35 U.S.C. § 112(a) pursuant to the AIA.

as a matter of law the requirement of 35 U.S.C. § 101 that the specification disclose as a matter of fact a practical utility for the invention."[106]

"In the context of determining whether sufficient utility as a drug, medicant, and the like in human therapy has been alleged, it is proper for the examiner to ask for substantiating evidence unless one with ordinary skill in the art would accept the allegations as obviously correct."[107] One source of substantiating evidence is information actually submitted to the FDA to obtain approval for clinical trials.[108] In fact, initiation of a human clinical trial has an impact on the utility inquiry under 35 U.S.C. § 101. "[A]s a general rule, if an applicant has initiated human clinical trials for a therapeutic product or process, Office personnel should presume that the applicant has established that the subject matter of that trial is reasonably predictive of having the asserted therapeutic utility."[109] However, the USPTO and the Federal Circuit have made clear that such human clinical data are not required to comply with 35 U.S.C. § 112, first paragraph.[110]

V. Conclusion

Clinical trials conducted before product approval may or may not constitute a public use or sale of the drug product or its method of use. Such trials could create a bar to later filing of a patent application in the United States or abroad. However, clinical trials may also lead to new discoveries that are patentable. It is prudent for pharmaceutical and biotechnology companies to be aware that clinical trials may have an impact on the patentability of the product, methods of using the product, specific product formulations, or specific salts or polymorphs of the product.

---

106. Rasmusson v. SmithKline Beecham Corp., 413 F.3d 1318, 1323 (Fed. Cir. 2005) (internal quotation marks omitted).

107. *Id.*

108. MPEP, *supra* note 52, § 2164 ("In chemical and biotechnical applications, evidence actually submitted to the FDA to obtain approval for clinical trials may be submitted.").

109. *Id.* § 2107.03, at IV.

110. *See id.* ("Office personnel should not impose on applicants the unnecessary burden of providing evidence from human clinical trials."); *see also In re* Brana, 51 F.3d 1560, 1566 (Fed. Cir. 1995) ("FDA approval, however, is not a prerequisite for finding a compound useful within the meaning of the patent laws. . . . The stage at which an invention in this field becomes useful is well before it is ready to be administered to humans.").

chapter 6

Trademark (TM) and Nonproprietary Name Considerations[1]

I. Introduction: The Headache of Pharmaceutical Naming

Pharmaceutical companies face unique branding challenges that are absent from other consumer product and service fields. These challenges arise from the facts that (a) each new pharmaceutical product must have two "names," and (b) those names must be approved by multiple regulatory authorities that use different criteria and do not coordinate with one another. To market a drug in the United States, its manufacturer must create both a proprietary brand, or trademark, for the product that it will use in marketing to the public as well as a nonproprietary, or "generic," name for the product that could be used by competitors for the same formulation. The trademark will need to be approved by the U.S. Food and Drug Administration (FDA), and, to achieve full protection under U.S. law, it will need to be registered with the U.S. Patent and Trademark Office (USPTO). The generic name must be approved by both the United States Adopted Names Council (USAN Council) and the FDA.

Thus, pharmaceutical companies can spend thousands of dollars developing a trademark and generic name for a new drug, only to have one or both of them rejected by any of the three regulating authorities, forcing the pharmaceutical companies to start the name selection and approval

1. B. Brett Heavner of Finnegan, Henderson, Farabow, Garrett & Dunner LLP, Washington, D.C.

process over again.[2] Indeed, that situation is not unusual since the FDA rejects approximately one-third of proprietary brand name applications it receives.[3] Further, if the pharmaceutical company is attempting to develop a worldwide brand for the product, it will also have to satisfy regulatory authorities in other countries, significantly increasing the chances of encountering a roadblock to any proposed trademark or generic name. To account for these challenges, most pharmaceutical companies develop multiple candidates for generic names and trademarks with the hope that at least one generic and one trademark candidate will satisfy all of the relevant authorities. This chapter will outline the general naming standards and procedures that pharmaceutical companies face in navigating the approval process for each of the U.S. regulatory authorities.

II. Pharmaceutical Trademarks

An effective branding strategy is key to the success of any product or service, including pharmaceuticals. Brands or trademarks are symbols used by businesses to identify their products and services and to distinguish them from those of their competitors.[4] Trademarks stand as a guarantee of consistent and reliable quality and thus represent the good reputation a pharmaceutical product enjoys with physicians and patients.

A trademark can be almost anything—letters, words, numbers, sounds, slogans, colors, shapes, logos, packaging, or a combination of all these—used to identify a particular product.[5] For most products and services, the Lanham Act (the statutory basis for federal trademark and unfair competition law) and the USPTO govern the owner's exclusive nationwide right to its trademark.[6] However, because of the unique regulatory position of pharmaceutical products in the United States, both the Lanham Act and FDA regulations govern pharmaceutical trademark protection, and proposed

2. *See* Jeroen Lallemand, *Pharmaceutical Branding: Engineering the Perfect Name*, INTELL. ASSET MGMT. 1 (Sept.–Oct. 2011), http://thomsonreuters.com/products/ip-science/04_085/pharmaceutical-branding-engineering-the-perfect-name.pdf; Gabrielle A. Holley, *Practice Guidelines for Prescription Drug Trademarks*, UPDATE MAG. (July–Aug. 2002), *available at* http://library.findlaw.com/2002/Sep/20/132457.html.

3. *FDA 101: Medication Errors*, FDA CONSUMER HEALTH INFO. 2 (Feb. 20, 2009), http://www.fda.gov/downloads/ForConsumers/ConsumerUpdates/UCM143038.pdf.

4. Two Pesos Inc. v. Taco Cabana Inc., 505 U.S. 763, 768 (1992).

5. 15 U.S.C. § 1127 (2012); *Two Pesos Inc.*, 505 U.S. at 768; *see* Qualitex Co. v. Jacobson Prods. Co., 514 U.S. 159 (1995) (holding that a color alone can function as a trademark).

6. 15 U.S.C. §§ 1051–72.

pharmaceutical trademarks are subject to dual agency review by both the USPTO and the FDA.[7]

A. The USPTO's Role: The Benefits of Federal Trademark Registration

In most cases, trademark rights in the United States are established by being the first to use a mark with a particular product or service. These rights are known as "common law" rights.[8] Trademark owners relying solely on common law rights must prove the validity and ownership of their trademark rights.[9] Further, common law rights are established only in areas where the mark is actually used—a common law mark is not automatically entitled to nationwide protection.[10]

Registration of a mark on the USPTO's Principal Register provides significant advantages and benefits above and beyond common law rights. These include

1. Nationwide priority as of the application filing date over all subsequent users of the trademark and most subsequent applications;[11]
2. Constructive notice of the registrant's claim of ownership in the mark (thus eliminating any good faith defense even against subsequent good faith users);[12]
3. Remedies of injunctive relief as well as profits, damages, and costs in a federal court action, and in certain circumstances, increased damages and attorneys' fees;[13]
4. Prima facie evidence of the mark's validity, the registrant's ownership of the mark, and the registrant's exclusive right to use the mark in connection with the products and services for which it is registered;[14]
5. The right to deposit the registration with the U.S. Customs Service in order to monitor and stop importation of products bearing an infringing mark;
6. Criminal penalties for counterfeiting a registered mark;[15]

---

7. *See id.*; 21 U.S.C. § 352(a) (stating that a drug is misbranded if its labeling is false or misleading in any particular way).

8. *Two Pesos Inc.*, 505 U.S. at 786.

9. Armstrong Paint & Varnish Works v. Nu-Enamel Corp., 305 U.S. 315, 334–35 (1938); House of Westmore v. Denney, 151 F.2d 261, 265 (3d Cir. 1945).

10. United Drug Co. v. Theodore Rectanus Co., 248 U.S. 90, 100 (1918).

11. 15 U.S.C. § 1057(c).

12. *Id.*

13. 15 U.S.C. § 1117.

14. 15 U.S.C. § 1115(a).

15. 18 U.S.C. § 2320.

7. A basis for obtaining trademark registrations in foreign countries;[16] and
8. A deterrent for others interested in adopting similar marks—the application or resulting registration will be discovered in a search, deterring others from adopting the same or a confusingly similar mark.[17]

In short, federal law creates strong incentives to register trademarks with the USPTO. Litigation based on common law trademark rights is inherently more complicated, more costly, and open to more doubt than litigation based on a federally registered trademark. Therefore, obtaining registrations whenever possible is strongly recommended.

B. The FDA's Role: Regulating Direct to Consumer Advertising

The Federal Food, Drug, and Cosmetic Act of 1938 (FDCA) vests the FDA with the authority to approve pharmaceutical products for marketing in the United States.[18] In 1962, Congress granted the FDA statutory authority to regulate prescription drug labeling and advertising.[19] An integral part of this authority is to review drug names and reject any such names that the FDA believes to be misleading.[20]

Under the FDCA and the FDA's implementing regulations, pharmaceutical advertisements and promotional labeling are deemed misleading if they fail to disclose certain information about the product's risk. Generally, promotional pieces: (1) cannot be false or misleading; (2) must reveal material facts about the product being promoted, including facts about the consequences that can result from the use of a product as suggested in the advertisement; and (3) should present information about effectiveness and information about risk in a balanced manner.[21]

Prior to the mid-1980s, the FDA limited approved advertising to medical and pharmaceutical professionals.[22] By 1985, the FDA opened the door

---

16. 15 U.S.C. § 1126(b).

17. Indep. Nail & Packing Co. v. Stronghold Screw Prods. Inc., 205 F.2d 921, 925 (7th Cir. 1953) ("Defendant adopted its logotype emphasizing the word 'Stronghold' with full knowledge of plaintiff's registered trade-mark which featured the same word . . . [and] it did so at its peril.")

18. Pub. L. No. 75-717, 52 Stat. 1040 (1938) (as amended 21 U.S.C. §§ 301–81).

19. The Kefauver-Harris Amendments, Pub. L. No. 87-781, 76 Stat. 780 (1962) (codified in scattered sections of 21 U.S.C. §§ 301–381); *see also Promoting Safe and Effective Drugs for 100 Years*, FDA Consumer Mag. (Jan.-Feb. 2006), *available at* http://www.fda.gov/AboutFDA/WhatWeDo/History/CentennialofFDA/CentennialEditionofFDAConsumer/ucm093787.htm.

20. *See* 21 U.S.C. §§ 352(a), 355(d)(7); 21 C.F.R. §§ 314.105(c), 314.125(b)(6), 314.125(b)(8).

21. 21 U.S.C. § 352(a)–(n), (q)(1); 21 C.F.R. § 202.1(e)(5)(ii)–(iii).

22. Wayne L. Pines, *A History and Perspective on Direct-to-Consumer Promotion*, 54 Food & Drug L.J. 489, 490–91 (1999).

to direct-to-consumer (DTC) advertising, issuing a *Federal Register* notice stating that it had jurisdiction over DTC advertisements and that these advertisements had to meet the same legal standards as those directed at medical professionals.[23] Originally, DTC advertising was primarily limited to print advertising as the volume of information required by the FDA regulations made broadcast media advertising expensive and burdensome.[24] However, in 1997, the FDA issued a draft guidance that allowed companies to advertise pharmaceuticals in broadcast media, so long as the advertisements included a statement of the "major risks" of the product and provided an "adequate provision" that would allow viewers to access the full information about the drug. On January 14, 2014, the FDA published draft DTC guidance directed to online pharmaceutical advertising.[25] The draft guidance deems pharmaceutical brand owners responsible for promotional statements and materials posted on websites that they operate or whose contents they control (such as Facebook and other sponsored social media sites). FDA review of content on such websites would be required before the brand owner could post it. On the other hand, pharmaceutical brand owners are not deemed responsible for online content relating to their products on websites that they do not oversee, edit, or otherwise control (such as third-party blogs and websites). Regarding real-time content (such as online chat and Twitter feeds), the guidance indicates that the pharmaceutical brand owner need not obtain prior FDA approval for the postings, but should periodically (monthly) submit such posting content to the FDA for review. Comments on the draft guidance were to be submitted by April 14, 2014, after which time the FDA commenced work on the final version of the guidance.

The FDA's gradual expansion of the advertising outlets open to pharmaceutical companies has resulted in explosive growth of DTC advertising over the past few decades. In 1989, the U.S. pharmaceutical industry only spent about $12 million on DTC advertising.[26] By 2010, the industry was spending over $4 billion on DTC advertising.[27] Consequently, given the pharmaceutical industry's substantial investment in DTC advertising, the pharmaceutical brand or trademark being promoted is a highly valuable asset for each manufacturer.

---

23. Direct-to-Consumer Advertising of Prescription Drugs; Withdrawal of Moratorium, 50 Fed. Reg. 36,677 (1985).

24. Pines, *supra* note 22, at 495–96.

25. Draft Guidance for Industry on Fullfilling Regulatory Requirements for Postmarketing Submission of Interactive Promotional Media for Prescription Human and Animal Drugs and Biologics, 79 Fed. Reg., 2449 (Jan. 14, 2014).

26. Pines, *supra* note 22, at 493.

27. Beth Snyder Bulik, *Ad Spending: 15 Years of DTC*, AD AGE INSIGHTS (Oct. 17, 2011), http://gaia.adage.com/images/bin/pdf/WPpharmmarketing_revise.pdf.

C. Considerations in Selecting Pharmaceutical Trademarks

When selecting a new pharmaceutical trademark, manufacturers are confronted with the fact that the proposed trademark must satisfy the requirements of both the USPTO and the FDA. Both the USPTO's and the FDA's trademark analyses can be broken down into two separate inquiries, namely (a) does the proposed trademark's inherent characteristics present any problems, and (b) does the proposed trademark present any problems in relation to other, preexisting pharmaceutical trademarks. Unfortunately, the USPTO and the FDA have different concerns and use different tests for conducting each of the two inquiries and, consequently, can reach different conclusions as to whether a proposed pharmaceutical trademark is acceptable. A trademark practitioner can properly advise a client in selecting a mark using USPTO criteria and in fact obtain a federal trademark registration, only to find that the FDA has rejected the proposed trademark under its own, quite different criteria. Similarly, an FDA practitioner may successfully shepherd the proposed pharmaceutical trademark through the FDA approval process, only to find some roadblock to registration of the trademark with the USPTO. Therefore, in developing new pharmaceutical names, it is extremely important that manufacturers consider both USPTO and FDA requirements.

1. Analyzing the Inherent Characteristics of Proposed Trademarks

a. The USPTO's Distinctiveness Inquiry

The first inquiry conducted by the USPTO evaluates whether the proposed pharmaceutical trademark qualifies for trademark protection based on its inherent characteristics. Under U.S. law, a trademark may only be registered with the USPTO if it is "distinctive."[28] A "distinctive" trademark is one that consumers would recognize as identifying a specific manufacturer's product, as opposed to merely being a generic term for a general class of products or merely being a descriptive term that informs the consumer of some quality or characteristic of the product.[29] In short, the distinctiveness inquiry centers on the meaning of the proposed trademark with respect to the underlying goods and whether it can actually function as a trademark.

The USPTO evaluates the distinctiveness of proposed trademarks by placing them along a "spectrum" of meanings comprised of the following categories: fanciful, arbitrary, suggestive, descriptive, or generic.[30] A

28. 15 U.S.C. § 1052 (2012).

29. Two Pesos Inc. v. Taco Cabana Inc., 505 U.S. 763, 768 (1992); Trademark Manual of Examining Procedure § 1212 (Apr. 2013) [hereinafter TMEP].

30. *Two Pesos Inc.*, 505 U.S. at 768; Abercrombie & Fitch Co. v. Hunting World Inc., 537 F.2d 4, 9 (2d Cir. 1976).

mark is fanciful if it serves only the express purpose of functioning as a trademark and has no other meaning or significance, such as Prozac for antidepressant medication or Kodak for cameras. A mark is arbitrary if it is in common use but is arbitrarily applied to goods or services and has no meaning for those goods or services, such as Equate for an antihistamine or Apple for computers. A mark is suggestive if imagination, thought, and perception are required to reach a conclusion as to the nature of the goods based on the mark, such as Apple-a-Day for vitamins, Blink for ophthalmic solutions, or Penguin for refrigerators. A term is descriptive if it immediately conveys an idea of the nature, ingredients, qualities, or characteristics of the product it identifies, such as Clear Eyes for ophthalmic preparations or Apple for apple-flavored gum. Last, a mark is generic if it is the common name for a product, such as Aspirin for acetylsalicylic acid drugs[31] or Apple for the fruit of that name.[32]

The USPTO considers fanciful, arbitrary, and suggestive terms inherently distinctive and entitled to immediate protection and registration.[33] Descriptive terms, on the other hand, are *not immediately* protectable as trademarks because the law presumes that competitors need to use such terms to fairly and accurately describe their goods and services to the public.[34] Consequently, descriptive terms can only be protected if they have acquired sufficient consumer recognition (called "secondary meaning") to distinguish the manufacturer's product from its competitors'.[35] Generic terms can never be protected as trademarks because they will always signify the common genus of a class of products for the consumer.[36] The test used to determine whether a designation is generic boils down to a simple question: "Does the buyer understand the designation primarily to refer to

31. Bayer Co. v. United Drug Co., 272 F. 505 (S.D.N.Y. 1921).

32. *See* J. Thomas McCarthy, McCarthy on Trademarks and Unfair Competition § 15:30 (June 2013).

33. *Two Pesos Inc.*, 505 U.S. at 768.

34. *Id.* at 767; Park 'N Fly Inc. v. Dollar Park & Fly Inc., 469 U.S. 189, 192 (1985); Rodeo Collection Ltd. v. W. Seventh, 812 F.2d 1215, 1218 (9th Cir. 1987).

35. *See* Bos. Duck Tours LP v. Super Duck Tours LLC, 531 F.3d 1, 13 (1st Cir. 2008) (explaining that when the public associates the name "Sports Illustrated" with "a particular sports magazine in addition to its primary meaning as a description of a specific feature or element," the phrase has secondary meaning); SK&F Co. v. Premo Pharm. Labs. Inc., 625 F.2d 1055, 1063–64 (3d Cir. 1980) (maroon and white capsules obtained secondary meaning); Boehringer Ingelheim GmbH v. Pharmadyne Labs., 532 F. Supp. 1040, 1045 (D.N.J. 1980) (orange biconvex tablets obtained secondary meaning); TMEP, *supra* note 29, § 1212; *see* 15 U.S.C. § 1052(f).

36. *Park 'N Fly Inc.*, 469 U.S. at 192; *Two Pesos Inc.*, 505 U.S. at 768; *see* 15 U.S.C. § 1064(3) ("The primary significance of the registered mark to the relevant public . . . shall be the test for determining whether the registered mark has become the generic name of goods or services.").

a general class of goods or services, or to a particular source of that type of good or service?"[37]

Most pharmaceutical trademarks fall in the fanciful category and thus qualify for registration with the USPTO without the need for evidence of consumer recognition.[38] However, any proposed pharmaceutical trademarks that immediately convey to the consumer information about an ingredient, quality, or other characteristic of the product may be deemed descriptive and thus not eligible for USPTO registration without evidence of secondary meaning. Descriptiveness is usually more of an issue for over-the-counter pharmaceuticals and nutritional supplements, which tend to try to convey meaning to consumers with their trademarks.[39] A pharmaceutical company asserting trademark rights in a descriptive term bears the burden of proof in establishing secondary meaning.[40] Section 2(f) of the Lanham Act provides a method for applicants to establish secondary meaning at the time they apply for registration of a trademark.[41] Successfully obtaining a registration under Section 2(f) provides the trademark owner with an enormous benefit in later infringement litigation because a federal registration on the Principal Register constitutes prima facie evidence of the exclusive right to use that mark in connection with the goods and services covered by the registration.[42] Thus, the federal registration of a descriptive mark reverses the burden of proof and forces the party who does not own the registered mark to prove secondary meaning does not exist.

The USPTO and U.S. courts typically look to the following factors when considering whether a mark has acquired secondary meaning:

a. Manner of use;
b. Advertising use;

---

37. *See* Bayer Co. v. United Drug Co., 272 F. 505, 510 (S.D.N.Y. 1921); Magic Wand Inc. v. RDB Inc., 940 F.2d 638, 640 (Fed. Cir. 1991).

38. *See, e.g.*, Pfizer Inc. v. Sachs, 652 F. Supp. 2d 512, 522 (S.D.N.Y. 2009) (Viagra found fanciful and inherently distinctive); Trovan Ltd. v. Pfizer Inc., 107 F. App'x 788, 790 (9th Cir. 2004) (Trovan found fanciful and inherently distinctive); Upjohn Co. v. Am. Home Prods. Corp., No. 1:95CV237, 1996 WL 33322175 (W.D. Mich. Apr. 5, 1996) (Provera found fanciful and worthy of broad protection).

39. *See, e.g.*, *In re* Blue Bonnet Nutrition Corp., No. 77550764, 2010 WL 5522985 (T.T.A.B. Dec. 22, 2010) (Functional Herbs held descriptive for nutritional supplements); *In re* Herbal Dynasty LLC, No. 78058457, 2004 WL 1090656 (T.T.A.B. Apr. 29, 2004) (Herbs and Roots held descriptive for nutritional supplements); *In re* Metabolic Balance GmbH, No. 79009729, 2007 WL 2972231 (T.T.A.B. Sept. 10, 2007) (Metabolic Balance held descriptive of medical instruments and services).

40. *See* Thompson Med. Co. Inc. v. Pfizer Inc., 753 F.2d 208, 217 (2d Cir. 1985); Vision Ctr. v. Opticks Inc., 596 F.2d 111, 118 (5th Cir. 1979); Steem-Elec. Corp. v. Herzfeld-Phillipson Co., 118 F.2d 122, 126 (7th Cir. 1940).

41. 15 U.S.C. § 1052(f).

42. 15 U.S.C. § 1115(a).

c. Unsolicited media coverage;
d. Requests from third parties to license the mark;
e. Intentional copying of the mark by the defendant and other competitors;
f. Survey evidence; and
g. Direct customer testimony.[43]

This list is neither exhaustive nor exclusive. Each factor is not necessarily afforded equal relevance in every case, and there is no single determinative factor.[44]

The USPTO may also reject proposed trademarks if they present other problematic characteristics. For example, the following types of marks are considered to lack "distinctiveness" and can only be registered upon a showing of secondary meaning: (1) marks that are primarily surnames, like Wyeth or Pfizer;[45] (2) marks indicating that goods or services originate from a particular geographic location,[46] such as Boston for ophthalmic solutions[47] or Quaker State for oil made from Pennsylvania-grade crude oil;[48] (3) certain highly laudatory terms and phrases, such as Super Buy,[49] America's Best,[50] or We'll Take Good Care of You;[51] and (4) product shapes, such as the shape and color of a pill.[52]

b. The USPTO's Deceptiveness Test

The USPTO will also refuse registration of a proposed trademark under Section 2(e)(1) of the Lanham Act if the proposed trademark is deceptively misdescriptive of the pharmaceutical product.[53] The U.S. Court of Appeals for the Federal Circuit articulated the following test to determine whether

---

43. *See* Bristol-Meyers Squibb Co. v. McNeil-PPC Inc., 973 F.2d 1033, 1043 (2d Cir. 1992); Thompson Med. Co. v. Pfizer Inc., 753 F.2d 208, 217 (2d Cir. 1985); Spraying Sys. Co. v. Delavan Inc., 975 F.2d 387, 393 (7th Cir. 1992).

44. *Thompson Med.*, 753 F.2d at 213–14 (holding that no one type of evidence is crucial and not every type of evidence need be produced); *see* A&H Sportswear Inc. v. Victoria's Secret Stores Inc., 237 F.3d 198, 212 (3d Cir. 2000) (holding that the district court can consider information outside of the enumerated likelihood of confusion factors).

45. 15 U.S.C. § 1052(e); WYETH, Registration No. 0507245; PFIZER, Registration No. 0501762.

46. 15 U.S.C. § 1052(e).

47. BOSTON, Registration No. 1607115.

48. Quaker State Oil Ref. Co. v. Steinberg, 325 Pa. 273 (1937).

49. *In re* Consol. Cigar Co., 35 U.S.P.Q.2d 1290 (T.T.A.B. 1995).

50. *In re* Wileswood Inc., 201 U.S.P.Q. 400 (T.T.A.B. 1978).

51. Genovese Drug Stores Inc. v. TGC Stores Inc., 939 F. Supp. 340 (D.N.J. 1996).

52. *See, e.g.*, SK&F Co. v. Permo Pharm. Labs. Inc., 625 F.2d 1055 (3d Cir. 1980) (maroon and white capsule is nonfunctional trade dress); A.H. Robins Co. v. Med. Chest Corp., 206 U.S.P.Q. 1015 (E.D. Mo. 1980) (light blue round pill is nonfunctional trade dress); Hoffman La Roche v. Premo Pharm Labs. Inc., 210 U.S.P.Q. 374 (D.N.J.1980) (light green capsule is nonfunctional trade dress).

53. 15 U.S.C. § 2(e)(1).

a mark consists of or comprises deceptive matter: (1) Is the term misdescriptive of a character, quality, function, composition, or use of the goods? (2) Are prospective purchasers likely to believe that the misdescription applies to the goods? and (3) Is the misdescription likely to affect a significant portion of the relevant consumers' decision to purchase the goods?[54]

The first prong of the Federal Circuit's test is met if the ingredient/attribute at issue is not or will not be present in the applicant's goods.[55] Evidence that the ingredient/attribute is actually available in the marketplace satisfies the second prong by showing that consumers would likely believe that the goods contain the key ingredient/attribute, if it is part of the proposed trademark. For pharmaceutical goods, the third prong is generally met through evidence that a significant portion of the relevant consumers believe that the ingredient/attribute provides a health benefit.[56]

The evidence would reflect health claims being touted in connection with the ingredient/attribute. For example, competitors or the applicant may emphasize health claims; mainstream-media references may show a belief by the public in the benefits of the ingredient/attribute in question; or scientific evidence may demonstrate actual benefits from the ingredient/attribute in question.

c. The FDA's Analysis of Misleading Trademarks

Like the USPTO, the FDA considers whether a proposed pharmaceutical trademark has a particular meaning within the pharmaceutical industry. However, rather than focusing on whether the proposed trademark is "distinctive" and capable of functioning as a trademark, or is deceptively misdescriptive, the FDA instead focuses on whether the proposed trademark is likely to cause or contribute to medication errors. If so, the proposed trademark risks FDA rejection and may not be used in U.S. commerce for that drug. Unlike the USPTO, however, the FDA's standards and concerns with potentially misleading trademarks are not specifically articulated in statutory or legal tests. The FDA regulations provide a broad overview of prohibited conduct, such as using a drug trademark that implies "that the drug or ingredient has some unique effectiveness or composition when, in fact, the drug or ingredient is a common substance" and prohibiting the use of labeling that is "false or misleading."[57] However, the regulations do not provide a comprehensive analysis of the factors that determine whether a trademark is misleading. Rather, information regarding the

54. *In re* Budge Mfg. Co., 857 F.2d 773, 775 (Fed. Cir. 1988).

55. U.S. Patent & Trademark Office, Examination Guide 1-11: Section 2(a) Deceptiveness for Class 5 Goods (May 2011).

56. *Id.*

57. 21 C.F.R. §§ 201.10(c)(3), 314.125(b)(6), 314.125(b)(8) (2013).

FDA's "misleading trademark" analysis comes out in anecdotal evidence or informal advisories from the FDA.[58]

The FDA advises that applicants avoid drug names that suggest a drug's composition, dosing interval, dosage form, or route of administration, regardless if that meaning is accurate for the drug or not.[59] Names that exaggerate the drug's effectiveness, superiority, or lack of risk should also be avoided, as the FDA may reject them.[60] Prefixes and suffixes that have pharmaceutical meanings, such as strength, method of administration, or duration, such as "PM" suggesting nighttime administration or "-stat" suggesting fast action, should be avoided because they may cause concern during an FDA review.[61] Also, the FDA is more likely to reject prefixes and suffixes that have multiple meanings depending upon the medical context in which they are used, such as numeric suffixes that may be used to signify both duration and suggested dosage regimens.[62]

The FDA has also been uncomfortable with proposed pharmaceutical trademarks that suggest one or more, but not all, of the drug's ingredients;[63] imply the drug or ingredient has some unique effectiveness or composition;[64] or may be confused with the proprietary name or the established name of a different drug or ingredient.[65]

---

58. *See, e.g.*, Letter from FDA to Nancy Eckert, HPC, Cord Blood BLA Proprietary Name Review Letter (Mar. 22, 2013), *available at* http://www.fda.gov/BiologicsBloodVaccines/CellularGeneTherapyProducts/ApprovedProducts/ucm359530.htm (informal advisory informing applicant that its proposed proprietary name is misleading); Ctr. for Drug Evaluation & Research, Application No. 203568Orig1s000, Proprietary Name Review(s) (Nov. 27, 2012), *available at* http://www.accessdata.fda.gov/drugsatfda_docs/nda/2013/203568Orig1s000NameR.pdf (review of applicant's proprietary name); *see also* Anna Wilde Mathews, *Many Drug Names Fail to Pass FDA Muster*, Sun Sentinel (Mar. 19, 2006), http://articles.sun-sentinel.com/2006-03-19/business/0603171072_1_cholesterol-drug-fda-drug-companies (providing anecdotal evidence that the FDA's name-review procedure for prescription drugs is "black-boxy").

59. U.S. Food & Drug Admin., PDUFA Pilot Project: Proprietary Name Review Concept Paper 12 (Sept. 2008) [hereinafter PDUFA], *available at* http://www.fda.gov/downloads/Drugs/GuidanceComplianceRegulatoryInformation/Guidances/ucm072229.pdf.

60. *Id.*

61. *See* Gerald J. Dal Pan, *Adverse Event Reporting and Medication Safety Considerations: A View from CDER's Office of Surveillance and Epidemiology at the CHPA Regulatory*, Sci. & Quality Conf. (May 3, 2013), http://www.fda.gov/downloads/AboutFDA/CentersOffices/OfficeofMedicalProductsandTobacco/CDER/UCM352766.pdf.

62. *See* Ctr. for Drug Evaluation & Research, Application 21-998, Proprietary Name Review(s) (May 1, 2009), *available at* http://www.accessdata.fda.gov/drugsatfda_docs/nda/2009/021998s000_NameR.pdf (discussing that numerical suffixes in Percocet 5 and Viokase 8 were misinterpreted as pill dosage instead of strength).

63. 21 C.F.R. § 201.6(b) (2013).

64. *Id.* § 201.10(c)(3).

65. *Id.* § 201.10(c)(5).

2. Analyzing Proposed Trademarks with Respect to Existing Trademarks

a. The USPTO's "Likelihood of Confusion" Test

Both the USPTO and FDA will analyze the proposed trademark to determine whether it presents a problem with respect to preexisting pharmaceutical trademarks and names. However, the two agencies employ vastly different analyses. The USPTO employs a "likelihood of confusion" test to reject any proposed mark that consumers will likely confuse with preexisting trademarks.[66] Specifically, Section 2(d) of the Lanham Act forbids registration of a trademark that

> Consists of or comprises a mark which so resembles a mark registered in the Patent and Trademark Office, or a mark or trade name previously used in the United States by another and not abandoned, as to be likely, when used on or in connection with the goods of the applicant, to cause confusion, or to cause mistake, or to deceive.[67]

The USPTO applies the following multifactor balancing test to determine whether confusion between the trademarks in question is likely:

a. Strength or distinctiveness of the preexisting trademark;
b. Similarity of the two marks;
c. Similarity of the goods the two marks identify;
d. Sophistication of purchasers;
e. Similarity of the trade channels through which the parties' goods travel;
f. Similarity of the advertising the two parties use; and
g. Intent of the junior applicant.

The U.S. Court of Customs and Patent Appeals originally articulated these factors in *In re E.I. du Pont de Nemours & Co.*[68] Each factor does not necessarily receive equal relevance in every case, and this list of factors is neither exhaustive nor exclusive.[69] Moreover, the ultimate determination of the highly fact-specific likelihood of confusion test turns on the circumstances of each case.[70] Although the USPTO will consider each of these

---

66. *See* 15 U.S.C. § 1125(a) (2012); FOOD & DRUG ADMIN., GUIDANCE FOR INDUSTRY: CONTENTS OF A COMPLETE SUBMISSION FOR THE EVALUATION OF PROPRIETARY NAMES 3 (Feb. 2010) [hereinafter GUIDANCE], *available at* http://www.fda.gov/downloads/Drugs/GuidanceComplianceRegulatoryInformation/Guidances/ucm075068.pdf.

67. 15 U.S.C. § 1052(d).

68. 476 F.2d 1357 (C.C.P.A. 1973).

69. *Id.* at 1361.

70. *Id.* at 1361–62.

factors where appropriate, the first four factors are the most common ones that arise in the examination process.[71]

To determine a trademark's strength, the USPTO typically looks to the distinctiveness or conceptual strength of a trademark and its commercial strength or marketplace recognition.[72] Courts assess the conceptual "strength" of a trademark using the same spectrum of distinctiveness discussed earlier in this chapter in section II(C)(1)(a), "The USPTO's Distinctiveness Inquiry." Descriptive marks (even those with secondary meaning) tend to be treated as weak and are accorded narrow protection.[73] To ascertain a mark's commercial strength, the USPTO will typically look to the same factors used to determine the existence of secondary meaning (length of use, volume of sales and advertising, media coverage, etc.). Use of the mark by others is also important in assessing the commercial strength of a trademark. Marks that are widely used by numerous other parties will receive narrow protection because such proliferation of the mark in the marketplace likely weakens and dilutes the mark as a single source identifier.[74]

The USPTO determines the similarity factor according to the sight, sound, and meaning trilogy. The degree of similarity between the marks required to prove a likelihood of confusion varies depending on the differences between the goods and services of the parties. As the Trademark Trial and Appeal Board explained, "as the degree of similarity of the goods of the parties increases, 'the degree of similarity [of the marks] necessary to support a conclusion of likely confusion declines.'"[75] Applying this same reasoning, where the parties' respective marks are identical or nearly identical, a likelihood of confusion may be found with a lesser degree of relatedness between the parties' goods and services.[76] For purposes of determining a likelihood of confusion, composite trademarks (made up

71. TMEP, *supra* note 29, § 1207.01.

72. Abbott Labs. v. Pet-Ag Inc., No. 91170148, 2011 WL 6099692 (T.T.A.B. Nov. 22, 2011); Am. Lebanese Syrian Associated Charities Inc., 101 U.S.P.Q.2d 1022 (T.T.A.B. Aug. 30, 2011); Tea Bd. of India v. Republic of Tea Inc., 80 U.S.P.Q.2d 1881, 1899 (T.T.A.B. 2006).

73. *In re* Central Soya Co., 220 U.S.P.Q. 914 (T.T.A.B. 1984); Palm Bay Imps. Inc. v. Veuve Clicquit Ponsardin Mason Fondee en 1772, 396 F.3d 1369, 1373 (Fed. Cir. 2005); *see* TMEP, *supra* note 29, § 1207.01(b)(ix), 1207.01(d)(iii).

74. *Palm Bay Imps. Inc.*, 396 F.3d at 1373; *see, e.g.*, Alaven Consumer Healthcare Inc. v. DrFloras LLC, No. C.A. 109-CV-705-TWT, 2010 WL 481205 (N.D. Ga. Feb. 4, 2010), *aff'd in part*, 399 F. App'x 545 (11th Cir. 2010) (finding mark weak because of extensive use of "Dr." in third-party nutritional supplement registrations); Nature's Best Inc. v. Ultimate Nutrition Inc., 323 F. Supp. 2d 429, 433 (E.D.N.Y. 2004) (finding Isopure mark only moderately strong because of extensive use of similar marks by third parties on dietary supplements).

75. Fossil Inc. v. Fossil Grp., 49 U.S.P.Q.2d 1451, 1456 (T.T.A.B. 1998) (quoting Century 21 Real Estate Corp. v. Century Life of Am., 970 F.2d 874, 877 (Fed. Cir. 1992)).

76. *In re* Opus One Inc., 60 U.S.P.Q.2d 1812 (T.T.A.B. 2001).

of multiple words or graphics) should be viewed in their entireties and, accordingly, should not be "dissected" into their component parts.[77] As the Supreme Court observed: "[t]he commercial impression of a trade-mark is derived from it as a whole, not from its elements separated and considered in detail. For this reason it should be considered in its entirety."[78]

The relatedness of the goods and services offered under the parties' respective marks is critical to determining whether there is a likelihood of confusion. As a rule, goods are sufficiently related to create a likelihood of confusion if the reasonably prudent purchaser believes that they may originate from the same source.[79] Competing goods are always "related" for purposes of a likelihood of confusion analysis.[80] Courts analyze proximity of noncompeting goods using the following factors: whether the goods are complementary, who purchases the goods, where people purchase the goods, how much the goods cost, and how purchasers use the goods.[81] For instance, products likely to be used together, such as syrup and pancake batter or medical devices and pharmaceuticals, have been found sufficiently related to create a likelihood of confusion.[82]

Typically, the USPTO has treated pharmaceutical products as being "related goods" even when the pharmaceutical products treat different conditions.[83] In fact, the USPTO believes that greater care must be taken to avoid confusion between pharmaceutical trademarks because mistakes in the selection and use of these goods may result in serious and harmful

---

77. *In re* Hearst Corp., 982 F.2d 493, 494 (Fed. Cir. 1992); *In re* Hutchinson Tech. Inc., 852 F.2d 552, 554 (Fed. Cir. 1988).

78. Estate of P.D. Beckwith Inc. v. Comm'r of Patents, 252 U.S. 538, 545–46 (1920); *see, e.g.*, Chi. Pharmacal Co. v. Am. Home Prods. Corp., 280 F.2d 148, 150 (C.C.P.A. 1960) (Phenistan and Phenergan confusing when viewed in their entireties); Am. Home Prods. Corp. v. B.F. Ascher & Co., 473 F.2d 903, 904 (C.C.P.A. 1973) (Ayr, Ascher, and Ayerst not confusing when considered in their entireties).

79. Thompson Med. Co. v. Pfizer Inc., 753 F.2d 208, 213 (2d Cir. 1985).

80. *See* TMEP, *supra* note 29, § 1207.01(a).

81. Lambda Elecs. Corp. v. Lambda Tech. Inc., 515 F. Supp. 915, 926 (S.D.N.Y. 1981).

82. Aunt Jemima Mills Co. v. Rigney & Co., 247 F. 407, 410 (2d Cir. 1917); Endo Labs. Inc. v. Fredericks, 199 U.S.P.Q. 824, 828 (T.T.A.B. 1978) (medical spatulas and pharmaceutical preparations are "related goods").

83. *See, e.g.*, Am. Home Prods. Corp. v. USV Pharm. Corp., 190 U.S.P.Q. 357, 359–60 (T.T.A.B. 1976) (finding antidepressives similar to menopause medication); Endo Labs. Inc. v. Mead Johnson & Co., 151 U.S.P.Q. 515 (T.T.A.B. 1966) (finding cold medicine similar to bile medicine because "[t]he products of the parties, although differing in chemical composition and medical uses, are pharmaceutical preparations intended for internal use which would ordinarily be sold through the same sources of drug distribution to the same general class of purchasers"); E.R. Squibb & Sons v. Battista, 53 U.S.P.Q. 606 (Dec. Comm'r Pat. 1942) (holding that cold medicine is similar to "stimulant tonics" although the medicines were applied differently and were intended to treat different conditions); Lehn & Fink Prods. Corp. v. Barrett Prods. Co., 47 U.S.P.Q. 82 (Dec. Comm'r Pat. 1940) (finding mouthwash similar to a disinfectant preparation); Standard Labs. Inc. v. Fried Land, 78 U.S.P.Q. 88 (Dec. Comm'r Pat. 1948) (holding that a mild pain reliever is similar to an iron supplement).

consequences.[84] Thus, the USPTO considers the potential harm to consumers as an additional relevant factor in its analysis and applies a stricter standard that may require a lesser degree of proof to establish a likelihood of confusion.[85]

Another factor the USPTO considers in gauging the likelihood of confusion between products is the sophistication of the target consumers. As a rule, sophisticated purchasers exercise greater care in their purchasing decisions and thus are less likely to be confused.[86] To determine the degree of consumer sophistication, courts look at factors such as whether the consumer is an ordinary purchaser or a professional and whether the goods are expensive, major purchases or cheap, impulse purchases.[87] Where a purchaser is a professional, rather than a member of the general public, courts often presume that the purchaser will exercise additional scrutiny and care. Given that professionals have superior knowledge of the product, courts presume that such purchasers will act with more discrimination than a member of the general public.[88]

The authorities are somewhat split regarding whether prescribing physicians and pharmacists are sufficiently "sophisticated" purchasers to avoid a likelihood of confusion between similar pharmaceutical marks.[89] However, the USPTO has repeatedly held that sophisticated purchasers are not necessarily immune to source confusion if the marks are quite close.[90] Moreover, to the extent that the pharmaceuticals are marketed directly to consumers, the purchasers would likely be unsophisticated.[91]

84. *Am. Home Prods. Corp.*, 190 U.S.P.Q. at 359–60; Clifton v. Plough Inc., 341 F.2d 934, 936 (C.C.P.A. 1965); *In re* Cook Med. Tech. LLC, 105 U.S.P.Q.2d 1377, 1381–82 (T.T.A.B. 2012); Schering Corp. v. Alza Corp., 207 U.S.P.Q. 504, 509 (T.T.A.B. 1980).

85. *See* Glenwood Labs. Inc. v. Am. Home Prods. Corp., 455 F.2d 1384, 1386–87 (C.C.P.A. 1972); Schering Corp., 207 U.S.P.Q. at 509; Ethicon Inc. v. Am. Cyanamid Co., 192 U.S.P.Q. 647, 651–52 (T.T.A.B. 1976).

86. Savin Corp. v. Savin Grp., 391 F.3d 439, 461 (2d Cir. 2004).

87. Kos Pharm. Inc. v. Andrx Corp., 369 F.3d 700, 715–17 (3d Cir. 2004).

88. Ford Motor Co. v. Summit Motor Prods. Inc., 930 F.2d 277, 293 (3d Cir. 1991).

89. Warner-Hudnut Inc. v. Wander Co., 280 F.2d 435 (C.C.P.A. 1960) (physicians constitute "a highly intelligent and discriminating public"). *Compare* Pfizer Inc. v. Astra Pharm. Prods. Inc., 858 F. Supp. 1305 (S.D.N.Y. 1994) (doctors are "as sophisticated a group as one could imagine"), *with* Sanofi-Aventis v. Advancis Pharm. Corp., 453 F. Supp. 2d 834, 851 (D. Del. 2006) ("Trained professionals such as doctors, pharmacists, and nurses are expected to be knowledgeable about and to exercise care in distinguishing between different medicines; however, they are not immune from mistake.").

90. *See In re* Shell Oil Co., 992 F.2d 1204, 1208 (Fed. Cir. 1993) (indicating that "even sophisticated purchasers can be confused by very similar marks"); Top Tobacco LP v. N. Atl. Operating Co., 101 U.S.P.Q.2d 1163, 1170 (T.T.A.B. 2011); *In re* Total Quality Grp. Inc., 51 U.S.P.Q.2d 1474, 1477 (T.T.A.B. 1999); *In re* Decombe, 9 U.S.P.Q.2d 1812, 1814–15 (T.T.A.B. 1988); *In re* Pellerin Milnor Corp., 221 U.S.P.Q. 558, 560 (T.T.A.B. 1983).

91. *Kos Pharm. Inc.*, 369 F.3d at 715 (holding that doctors and professionals are not necessarily the relevant consumer because "drugs are increasingly marketed directly to potential patients through, for example, 'ask-your-doctor-about-Brand-X' style advertising").

b. The FDA's "Sound-Alike/Look-Alike" Test

The FDA's "sound-alike/look-alike" analysis of potentially conflicting pharmaceutical trademarks is quite different from the USPTO's likelihood of confusion test.[92] Put simply, the USPTO's test looks at market conditions to determine whether consumers would be likely to confuse the two proposed trademarks, while the FDA's test looks at the possibility that a physician or pharmacist will accidentally prescribe or dispense the wrong drug. Unlike the USPTO's test, which requires a comparison of marks as a whole, the FDA's analysis typically focuses on similarities at the beginning of the two pharmaceutical trademarks.[93] Specifically, the FDA's analysis focuses on physician handwriting, which it assumes is often difficult to read and may trail off at the ends of words.[94] Similarly, prescriptions are often presented over the telephone to pharmacists, who may only hear the beginning syllables of a trademark.[95] The FDA accordingly presumes that pharmacists may rely more heavily on differences at the beginnings of marks to distinguish drugs than will consumers in other fields. For example, while the marks Serzone and Seroquel would probably pass a trademark "likelihood of confusion" test, the FDA found that they failed the "sound-alike/look-alike" test due to their similar initial syllables.[96]

Under the "sound-alike/look-alike" analysis, the FDA also gives greater scrutiny to marks containing similar-looking letters given the problem of poor physician handwriting.[97] Accordingly, one must be wary of similar marks containing the letters *u*, *v*, *n*, *m*, *w*, *t*, *l*, *i*, and (capital) *I* because the FDA may reject marks that, while pronounced quite differently, might be written in a similar manner.[98] For example, the FDA has been concerned about confusion between the marks Avandia and Coumadin.[99] Although the marks do not look similar when typed or printed and thus would clearly pass the USPTO's test, the names failed the FDA's "look-alike" test because either name might be misread when poorly written in cursive script: the first *A* in Avandia, if not fully formed, can look like a *C*, and the final *a* has appeared to be an *n*.[100] Similarly, while the mark Lodine and the generic

---

92. *See* FOOD & DRUG ADMIN., HOW FDA REVIEWS PROPOSED DRUG NAMES 1 (2010), *available at* http://www.fda.gov/downloads/Drugs/DrugSafety/MedicationErrors/UCM080867.pdf.

93. *See* PDUFA, *supra* note 59, at 15.

94. *Id.* at 17.

95. *Id.* at 14.

96. Carol Rados, *Drug Name Confusion: Preventing Medication Errors*, FDA CONSUMER MAG. (July–Aug. 2005).

97. *Id.*

98. PDUFA, *supra* note 59, at 14.

99. Rados, *supra* note 96.

100. *Id.*

Iodine should pass the USPTO's test, there were problems under the FDA's "look-alike" test because the capital *I* and lowercase *L* look identical.

Many of the factors that are used in the USPTO's likelihood of confusion analysis may be of little importance to the FDA's "sound-alike/look-alike" analysis. For example, the "weakness" factor in the USPTO's analysis is entirely absent from the FDA's analysis. Thus, similar marks that could coexist under the USPTO's analysis (where the field is crowded with equally similar marks or where the two marks are descriptive), may be rejected by the FDA due to the "sound-alike/look-alike" problem. Similarly, factors relating to the sophistication of the consumer and the lack of relationship between the goods will play little or no role in the FDA trademark review. Physician and pharmacist sophistication, and the fact that the pharmaceuticals treat vastly different conditions, will be irrelevant if the individual health care professionals miswrite, mishear, or misread the drug name. These added complexities contribute to the additional costs and time required for trademark owners and practitioners to choose and develop a viable trademark.

3. Clearance Searching for Pharmaceutical Trademarks

Before submitting proposed pharmaceutical trademarks to either the USPTO or the FDA for approval and registration, the manufacturer should complete a clearance search for each of the new trademarks that it is considering. Clearance searching is the best way to predict whether the proposed trademark approval process will be successful. Given the stringent, and potentially conflicting, standards of the USPTO and the FDA just outlined, pharmaceutical manufacturers should have multiple trademark candidates for each pharmaceutical product. Typically, the clearance process involves three steps: (1) an initial inquiry, (2) a preliminary search, and (3) a comprehensive search.

The initial inquiry involves gathering as much information as possible about the proposed trademark from those who have created it, including (a) the products and therapeutic indications with which the trademark will be used; (b) whether the trademark or any of its components have any meaning or significance as applied to the products or therapeutic indications; and (c) the identity of any known identical or similar trademarks. This information will enable counsel to formulate an effective search strategy and perform an accurate analysis of the protectability and availability of the trademark. It also should assist counsel in spotting possibly problematic meanings or suggestions inherent in the mark that might give rise to either a "deceptively misdescriptive" problem with the USPTO or a misleading/misdispensing concern at the FDA.

Next, counsel should conduct a preliminary or "knockout" search of the USPTO's records. This search is typically limited to identical or nearly identical trademarks and is designed to knock out candidates that are clearly unavailable—without having to incur the time and expense of a comprehensive search.

Assuming that a mark clears the preliminary or "knockout" search, the next step is to obtain a "comprehensive" trademark search. At a minimum, this search should encompass the following categories of trademarks and names: (a) federal registrations and applications; (b) state trademark registrations; (c) unregistered trademarks, company names, and business names found in telephone directories, business reporting databases, and various other proprietary databases; and (d) Internet domain names. The search results should be analyzed with an eye toward both the USPTO's likelihood of confusion test and the FDA's "sound-alike/look-alike" test.

In addition, many professional search firms offer pharmaceutical-specific databases that can help spot problems that may arise during the FDA review process. These databases include IMS Health Pharmaceutical In-Use database,[101] USAN Council generic stems,[102] American Drug Index,[103] Physicians' Desk Reference,[104] FDB MedKnowledge,[105] Drug Topics Redbook,[106] The Natural Medicines Database,[107] and the FDA Orange Book.[108] Some search companies even have the ability to search the FDA's Phonetic Orthographic Computer Analysis (POCA) software, which is very useful for spotting potential FDA concerns about similarities in sound and handwriting.[109] The more pharmaceutically focused the clearance process is, the less likely the manufacturer will face surprise rejections of its proposed trademark candidates during the USPTO registration and FDA approval processes.

---

101. *Pharmaceutical In-Use Similarity Search*, THOMSON REUTERS, http://trademarks.thomsonreuters.com/searching/pharmaceutical-searches?id=node/175 (last visited July 31, 2013).

102. *USAN Stem List (2013)*, AM. MED. ASS'N, U.S. ASSIGNED NAME COUNCIL, http://www.ama-assn.org/resources/doc/usan/stem-list-cumulative.pdf (last visited July 31, 2013).

103. NORMAN F. BILLUPS & SHIRLEY M. BILLUPS, AMERICAN DRUG INDEX 2013 (2013).

104. PHYSICIANS' DESK REFERENCE, http://www.pdr.net/ (last visited July 31, 2013).

105. *FDB MedKnowledge*, FIRST DATABANK, http://www.fdbhealth.com/fdb-medknowledge/ (last visited July 31, 2013).

106. *See Red Book*, THOMSON-REUTERS, http://sites.truvenhealth.com/redbook/index.html}} (last visited June 26, 2014).

107. NATURAL MEDS. COMPREHENSIVE DATABASE, http://naturaldatabase.therapeuticresearch.com/ (last visited July 31, 2013).

108. *Orange Book*, U.S. FOOD & DRUG ADMIN., http://www.accessdata.fda.gov/scripts/cder/ob/default.cfm (last visited July 31, 2013).

109. *See, e.g.*, *Pharmaceutical Services*, THOMSON REUTERS, http://trademarks.thomsonreuters.com/sites/default/files/TCMTR_PharmaServices_Final.pdf (last visited July 31, 2013).

D. The Approval Process for Pharmaceutical Trademarks

Once the clearance search process has been completed, and the manufacturer has compiled a list of candidates for the pharmaceutical trademark, it should look toward registering the trademarks with the USPTO and obtaining approval from the FDA. However, the procedures employed by the two agencies are completely different. Furthermore, the two agencies do not coordinate, or even communicate, with respect to pharmaceutical trademark review. Consequently, the pharmaceutical manufacturer will likely need to pursue registration/approval of multiple candidates simultaneously in order to ensure that at least one trademark has been accepted by both agencies at the time the pharmaceutical product has been approved and is ready to be launched in the marketplace.

1. USPTO Registration Procedures

Typically, pharmaceutical manufacturers commence the USPTO registration process first because there is no need to have completed any testing or trials before applying to register the trademark. The following sections outline the basic steps in the USPTO application process.

a. Step 1: Preparation of the Trademark Application

The proper drafting of trademark applications can affect both the ultimate success of the application and the protection accorded the trademark. Therefore, the pharmaceutical manufacturer should exercise a great deal of care in preparing the application. Initially, the applicant must select one of the following statutory bases for the registration of its trademark: (1) actual use of the trademark in U.S. commerce, (2) an intention to use (ITU) the mark in the future, (3) a foreign application for the mark filed within the past six months using the Paris Convention, or (4) an extension of a foreign "home" country application or registration using the Madrid Protocol.[110] Because proposed pharmaceutical trademarks are developed well in advance of the actual launch of the pharmaceutical product, most manufacturers will elect an ITU application or one of the international applications in order to start the USPTO registration as early as possible in order to secure priority in the trademark.

The application must include a drawing of the mark. The drawing is a representation of the mark as actually used, or intended to be used, on the products or services. The drawing may take two forms: (1) typed or block letter form, or (2) a special form depicting logos and stylized script.[111] To obtain the broadest possible protection, applicants will usually apply to register

110. *See* 15 U.S.C. §§ 1051(a)–(b), 1126(a), 1141a(a).
111. 37 C.F.R. § 2.52 (2013).

word marks in a typed or block letter format. When the mark is simply typed in capital letters, the application covers the word mark in block letters *and* in any other stylized format or lettering.[112] A mark registered in typed form thus covers all design variations of that mark. If the applicant develops a logo that accompanies a word mark, it should file a second application for the logo by itself to ensure the broadest possible protection.

One of the most important parts of an application is the description of products and services. Because the description defines the coverage of any resulting registration, care should be exercised in its drafting. Broad descriptions of products and services, to the extent permitted by the USPTO, are generally preferred over narrow, more detailed descriptions. Broad descriptions cover more categories of products and services, thus affording greater protection.[113] However, the USPTO will not accept "pharmaceuticals" as a description of products because it deems that term overly broad and indefinite. Instead, the USPTO usually requires identification of the conditions and diseases that the product is intended to treat.[114] If it is unclear at the time of filing which conditions and diseases the proposed pharmaceutical product will ultimately be approved to treat, the application should include a list of any such conditions that the product may reasonably be viewed as potentially treating.

After all parts of the application have been completed, it is imperative that the application be filed with the USPTO as soon as possible. The application filing date becomes the "constructive date of first use" of the mark throughout the entire United States, provided the mark proceeds to registration on the Principal Register. "Constructive use" means that the mark is considered to have been in use nationwide as of the application filing date, even though actual use may begin after the filing date or may be limited to a particular geographic area. This "constructive first use" benefit thus provides a strong incentive for filing trademark applications as early as possible.

b. Step 2: The Examination Process

Several months after an application is filed, it is assigned to a USPTO Trademark Examining Attorney (Examiner). Examiners review new applications in the order in which the USPTO receives them.[115] Upon receiving an application, an Examiner will review it in three respects. First, he or she will conduct a search of the USPTO records to determine whether

112. *Id.* § 2.52(a); TMEP, *supra* note 29, § 807.03.

113. TMEP, *supra* note 29, § 1402.01.

114. *Trademark Acceptable Identification of Goods & Services*, U.S. PATENT & TRADEMARK OFFICE, http://tess2.uspto.gov/netahtml/manual.html (last visited Apr. 28, 2014).

115. TMEP, *supra* note 29, §§ 702.01, 704.01.

there are any prior registrations or pending applications for confusingly similar marks that would bar the new pharmaceutical trademark application.[116] Second, the Examiner will evaluate the inherent registrability of the mark and will decide whether the mark is generic, merely descriptive, suggestive, and so on.[117] Third, the application will be examined for formalities such as sufficiency of the ownership information, the way the products or services are described, and the form of the application.[118] In most cases, the Examiner will complete the examination of an application within six to eight months after the filing date.

Depending upon the examination results, the Examiner will either approve the application for publication or issue an initial Office Action raising one or more objections of the type discussed earlier in this chapter.[119] An Office Action is the means by which the Examiner communicates the results of his or her examination to the applicant. If the Examiner issues an Office Action, the applicant has six months to file a response.[120] If a response is filed, the Examiner will then review the applicant's response and determine if all the outstanding objections and requirements have been sufficiently addressed to withdraw the refusal. If the applicant does not overcome all of the Examiner's objections, the refusal is typically made final in a second Office Action.[121] From a final refusal, an applicant can either fully comply with the Examiner's objections (if possible) or appeal the refusal to an administrative tribunal.[122]

c. Step 3: Post-examination

If the application is approved by the Examiner without issuing an Office Action, or if an Office Action is issued and any objections raised therein are overcome, the Examiner will approve the mark to be published for opposition in the USPTO's weekly *Official Gazette*.[123] Any person or entity who believes that it will be damaged by registration of the published mark has 30 days to file a Notice of Opposition to registration of the mark or to request an extension of time to oppose.[124] Extensions of time to oppose can be obtained as a matter of right for up to 120 days after the publication

116. *Id.* § 704.01.
117. *Id.*
118. *Id.* §§ 704.01, 802, 1201.
119. *Id.* § 705.
120. *Id.* §§ 705.08, 711.
121. *Id.* §§ 713, 714.03.
122. *Id.* § 715.01.
123. *Id.* § 1603.01.
124. *Id.* § 1503.03.

date.[125] A further 60-day extension can be obtained only with the consent of the applicant.[126]

If a Notice of Opposition is filed, an opposition proceeding is established before the Trademark Trial and Appeal Board, the administrative judicial arm of the USPTO.[127] An opposition, which is very similar to civil litigation in a state or federal court, can be both time-consuming and expensive. The only issue decided in an opposition is whether the applicant is entitled to *register* its mark; the Trademark Trial and Appeal Board is not empowered to enjoin *use* of a mark.[128] The opposing party can interfere with an applicant's *use* of its mark only by obtaining an injunction from a state or federal court.

If no opposition is filed, or if an opposition is filed and the application survives the opposition, the application will proceed toward registration. If the application was filed on a use-basis or under the Paris Convention or Madrid Protocol, the certificate of registration will issue immediately.[129] For intent-to-use applications, the USPTO will instead issue a Notice of Allowance.[130] Following issuance of the Notice of Allowance, an applicant has six months to file either a Statement of Use alleging use of the mark in commerce and a specimen evidencing such use, or a request for extension of time to file the Statement of Use.[131] One of these two filings must be made within six months to avoid abandonment of the application. If a Statement of Use is filed and accepted, the application will be passed to registration, and a Certificate of Registration will issue. If an applicant requests an extension of time, a verified statement alleging that the applicant has a continued bona fide intent to use the mark in commerce must accompany the request.[132] The applicant may obtain extensions of time, in six-month increments, for up to 36 months from the issuance of the Notice of Allowance.[133] Given the length of the FDA approval process for new drugs, it can sometimes take more than the allotted USPTO time to evidence use of the mark. Fortunately for pharmaceutical companies, the USPTO will accept use of the trademark in clinical trials as a "use in

125. *Id.*

126. 37 C.F.R. § 102(c)(3); TMEP, *supra* note 29, § 207.01.

127. 15 U.S.C. § 1063; 37 C.F.R. §§ 2.101–2.107; Trademark Trial & Appeal Board Manual of Procedure § 303 (June 2013) [hereinafter TBMP].

128. Gen. Mills Inc. v. Fage Dairy Processing Indus. SA, 100 U.S.P.Q.2d 1584, 1591 (T.T.A.B. 2011) (holding that the Trademark Trial and Appeal Board has no authority to determine the right to use, or the broader questions of infringement, unfair competition, damages or injunctive relief); TBMP, *supra* note 127, § 102.01.

129. 15 U.S.C. § 1063 (2012).

130. TMEP, *supra* note 29, § 1106.01.

131. 15 U.S.C. § 1051(d); TMEP, *supra* note 29, § 1108.

132. *Id.*

133. *Id.*

commerce" sufficient to allow the applicant to submit a Statement of Use, even if the FDA has not issued a final approval for the pharmaceutical.[134]

2. FDA Approval Procedures

The FDA's Division of Medication Error Prevention and Analysis (DMEPA) has the authority to review proposed pharmaceutical trademarks, both before and after marketing.[135] The FDA encourages manufacturers to submit their proposed pharmaceutical trademarks for DMEPA review early in the drug approval process, preferably around the end of Phase 2 of clinical trials.[136]

DMEPA will review two (and only two) proposed pharmaceutical trademarks during the drug approval process.[137] The manufacturer may indicate to DMEPA which of the two proposed pharmaceutical trademarks it prefers, and DMEPA will review them in order of preference.[138] If the first is rejected, DMEPA will evaluate the second. The manufacturer must also submit the proposed package labeling, including both inserts and outer labeling, and any studies or data evaluating the proposed mark for potential medication error.[139]

DMEPA must complete review of a proposed pharmaceutical trademark within 180 days if the request is submitted during the investigational phase and within 90 days if the request is submitted with an application for a new drug or therapeutic biologic.[140] Proposed trademarks will receive "tentative approval" initially and will then undergo a second evaluation approximately 90 days prior to final approval of the drug or biologic. During that second evaluation, DMEPA will also compare the proposed pharmaceutical trademarks that have been approved since the tentative approval.[141]

In order to determine whether a drug name is misleading, DMEPA collects specific information from the manufacturer about the proposed pharmaceutical trademark, including its intended pronunciation, derivation, modifier meaning such as prefixes or suffixes, and pharmacologic/ therapeutic category; the proposed labeling, including any packaging insert if a prescription drug; the proposed container labels and labeling; and information about the product's dispensing and delivery, including its

134. TMEP, *supra* note 29, § 901.01.
135. 21 U.S.C. § 352(a).
136. Guidance, *supra* note 66, at 15; PDUFA, *supra* note 59, at 7.
137. Guidance, *supra* note 66, at 10; PDUFA, *supra* note 59, at 8.
138. Guidance, *supra* note 66, at 10.
139. *Id.* at 11–12.
140. *Id.* at 8.
141. PDUFA, *supra* note 59, at 8.

likely care environment(s) for dispensing and use, delivery system, and measuring device.[142]

Upon receipt of this information, DMEPA will then use a variety of methods to determine if the name has a problematic meaning or connotation within the pharmaceutical field. It also generates and tests a list of existing drug names that could be confused with the proposed proprietary name, possibly contributing to a medication error. Specifically, DMEPA will conduct a preliminary screening of proposed trademarks that incorporate or suggest, among other things, drug characteristics, dosing intervals, dosage forms/routes of administration, medical/product name abbreviations, and drug compositions.[143] DMEPA will also conduct a United States Adopted Names Council (USAN) stem screening to determine whether the proposed trademark incorporates generic drug stems in a confusing manner. Finally, DMEPA conducts a search of existing drug names in industry databases that are similar to the proposed drug name in spelling, pronunciation, and handwritten appearance using a powerful algorithm called the Phonetic and Orthographic Computer Analysis (POCA). The POCA search generates objective numeric scores that rank the orthographic and phonological similarity between the proposed name and existing drug names. DMEPA flags for further review any proposed trademarks with a POCA score above a specific threshold.[144]

After identifying potentially problematic and similar trademarks, the FDA conducts a failure mode and effects analysis (FMEA), which examines the nomenclature, labeling, and packaging of the drug for possible ways in which an administration error could occur along a drug's route from manufacturer to patient.[145] Part of the FMEA evaluates the name itself for its potential to inadvertently function as a source of error regarding its own ingredients or characteristics. Another part of the FMEA is a "sound-alike/look-alike" test, which compares the proposed trademark to all of the names gathered during the safety review in spelling, appearance, sound, connotation, and commercial impression. Because trademark confusion can occur at any point in the medication-use system, the FDA considers the potential for confusion at each stage of the process, including product procurement, prescribing/ordering, dispensing, administration, and monitoring the effects of a medication.[146] If the FDA determines

142. GUIDANCE, *supra* note 66, at 9–15.

143. *Id.* at 5; PDUFA, supra note 59 at 11-12.

144. PDUFA, *supra* note 59, at 33–34.

145. *Id.* at 21–25.

146. Division of Medical Error Prevention and Analysis (DMEPA) also conducts name simulation studies, which attempt to test the response of practitioners to a proposed name

that under these analyses the proposed trademark could cause medication errors, the trademark will be rejected.

In addition to the safety review, the FDA conducts a promotional review. This review evaluates a proposed trademark to determine if it misleadingly implies unique effectiveness or composition as well as to assess whether the trademark contributes to overstatement of product efficacy, minimization of risk, broadening of product indications, or making of unsubstantiated superiority claims.[147]

3. Conflicting Priority Determinations by the USPTO and FDA

Differences in the FDA and USPTO procedures can create priority battles. Under traditional USPTO trademark principles, rights in a trademark are established through either being the first to use that trademark in interstate commerce or by having the earliest USPTO application filing date for that trademark, whichever is earlier. The FDA, however, does not respect priority established by the USPTO under the Lanham Act.[148] Rather, the FDA insists on approving pharmaceutical trademarks on a first-come, first-served basis as the drug applications wind their way through the drug approval process.

Thus, a pharmaceutical manufacturer who files its USPTO application first may ultimately not be granted the rights to that mark if the FDA approves another company's drug with a similar mark before it approves the pharmaceutical client's drug. In other words, a junior USPTO applicant may be given rights to the mark because the FDA drug application of the senior USPTO applicant moved more slowly through the FDA approval process. The FDA reasons that it cannot hold up the approval of one party's drug merely because another party has expressed a preference for a similar trademark and happened to file its application for that mark with the USPTO first.

This disconnect between the FDA drug approval process and the USPTO's trademark registration process inevitably leads to some uncertainty regarding trademark ownership. Senior USPTO trademark applicants in the pharmaceutical field are constantly under threat of losing their trademark rights to companies whose drug applications move through the FDA at a faster pace. Similarly, companies whose products have been previously approved by the FDA may find that their trademark applications before the USPTO are blocked by prior applicants who do not have

---

by asking them to use the name in simulated real-world conditions. *See* PDUFA, *supra* note 59, at 17–20.

147. Guidance, *supra* note 66, at 6.

148. Nick de la Torre & Jennifer Theis, Pharmaceutical Trademarks 2012, at 62 (World Trademark Review, 3d ed. 2012).

FDA approval for their marks. In a worst-case scenario, conflicting decisions by the FDA and USPTO may preclude either company from using its desired trademark. For example, foreign pharmaceutical company A may be granted a U.S. trademark registration under the Madrid Protocol before it obtains FDA approval of its mark. At the same time, pharmaceutical company B may obtain FDA approval to use a confusingly similar trademark for its drug product, but find itself unable to obtain a U.S. trademark registration for that mark due to the USPTO's prior registration of pharmaceutical company A's mark. Theoretically A could use its registration to prevent B from launching its product, despite the FDA's approval of that product's mark. Likewise, A will find its FDA application blocked by the FDA's previous approval of B's mark. In short, the two pharmaceutical companies would face a stalemate, with no established administrative or legal procedure to help them determine which company has the right to use the mark.

III. Nonproprietary Names

An integral part of a successful drug naming strategy is the selection of a nonproprietary generic name for the new pharmaceutical. After patent protection for a pharmaceutical has expired, competitors are allowed to manufacture generic versions of that pharmaceutical product. If those competitors have no other name to identify the generic product other than the original proprietary brand name, that brand name can become the "de facto" generic name for the product and lose all trademark protection.[149] Such was the case with the now generic term Aspirin, which was once a trademark of Bayer AG.[150] To avoid the risk of their product's valuable brand name becoming generic, and thus available for use by all competitors, pharmaceutical manufacturers must develop a generic name that is available for use by competitors from the moment the new pharmaceutical product is introduced to the public.[151] Therefore, creation of a generic name is vital to the protection of the companion proprietary "brand" name or trademark for the new pharmaceutical.

The nonproprietary generic name for each new pharmaceutical product must be approved during the drug approval process by a regulatory body called the USAN Council.[152] The USAN Council is a private organi-

149. *See* Bayer Co. v. United Drug Co., 272 F. 505 (S.D.N.Y. 1921) (finding Aspirin generic).

150. *Id.*

151. McCarthy, *supra* note 32, § 12:61.

152. *See* U.S. Food & Drug Admin., *supra* note 92, at 2–4; *How to Apply for a Name*, Am. Med. Ass'n, http://www.ama-assn.org/ama/pub/physician-resources/medical-science/united-states-adopted-names-council/how-to-apply-for-usan.page? (last visited July 31, 2013).

zation sponsored by the American Medical Association, the United States Pharmacopeia, and the American Pharmacists Association (formerly the American Pharmaceutical Association) and has been engaged in the assignment of names to drugs since January 1964.[153] The Council consists of a five-person panel that meets three times a year.

A. Selecting Nonproprietary Names

The USAN Council approves nonproprietary names according to principles that ensure "safety, consistency and logic" in the selection of drug names.[154] The USAN Council has set forth several guiding principles to aid in the selection of a nonproprietary name, including the following:

1. A nonproprietary name should be useful to physicians and medical professionals based primarily on the suitability for safe usage, and secondarily on the suitability for educational programs, use in scientific publications, and use internationally.
2. Additional attributes that contribute to usefulness are simplicity, euphony, and ready recognition and recall. The name should generally be a single term with no more than four syllables.
3. The name should reflect terms and characters that will be of practical value to the users, such as a common stem and distinctive terminology.
4. The name should not overlap with existing proprietary and/or nonproprietary names.
5. Whenever possible, the names should follow existing stem and nomenclature schemes established by the USAN Council. The Council assigns a new stem only when it determines that no existing stem is appropriate.[155]

In applying the preceding principles, the USAN Council's analysis is generally a two-part review: first, examining the inherent characteristics of the proposed generic name, and second, examining the proposed generic name in relation to preexisting USAN-approved generic names.

With respect to inherent characteristics, the USAN Council strongly prefers that manufacturers coin generic names for drugs by building off root syllables or "stems" that are shared with other drug formulations composed of the same or related chemical agents.[156] By insisting on these stems as a part of most generic names, the USAN Council has created

153. 21 C.F.R. § 299.4(c) (2013).

154. *Rules for Coining Names*, Am. Med. Ass'n, http://www.ama-assn.org/ama/pub/physician-resources/medical-science/united-states-adopted-names-council/naming-guidelines.page? (last visited July 31, 2013).

155. *See USAN Stem List (2013)*, *supra* note 102.

156. *See Rules for Coining Names*, *supra* note 154.

common "families" of names for similar or related drug groups so that medical professionals and consumers can immediately recognize the drugs' principal ingredients and the conditions that they treat. For example, the USAN Council recommends the stem "-vastatin" for cholesterol drugs called antihyperlipidemics.[157] Several well-known drugs currently on the market use this stem (see Table 1).[158]

Table 1

Brand and Generic Names for Statin Family

Brand Name	Generic Name
Altoprev	Lovastatin
Crestor	Rosuvastatin
Lescol	Fluvastatin
Lipitor	Atorvastatin
Livalo	Pitavastatin
Mevacor	Lovastatin
Pravachol	Pravastatin
Zocor	Simvastatin

The USAN Council publishes a list of common stems that represent various chemical and/or pharmacologic properties to which pharmaceutical manufacturers can refer when developing new generic names.[159]

In addition, the USAN Council generally rejects names that include prefixes that refer to Latin or Greek numbers, prefixes that indicate a medical condition or chemical element, or a variety of homophonic letters, such as the letters *ph* instead of the letter *f*.[160] The Council also recommends distinctive naming schemes for specific drugs or groups, such as biologics, contact lenses, gene therapy, and others.[161]

The USAN Council will also examine a proposed generic name in relation to other, preexisting generic names to ensure that the new name is not too similar.[162] Generic names that are too similar can cause consumer con-

157. *Id.*

158. *High Cholesterol: Medicines to Help You*, U.S. FOOD & DRUG ADMIN. (2009), http://www.fda.gov/forconsumers/ byaudience/forwomen/ucm118595.htm.

159. *See Rules for Coining Names*, *supra* note 154; *New Stems Approved by the USAN Council*, AM. MED. ASS'N, http://www.ama-assn.org/resources/doc/usan/new-stem-list.pdf (last visited July 31, 2013).

160. *Rules for Coining Names*, *supra* note 154.

161. *Id.*

162. *Id.*

fusion and dispensing errors. To better anticipate possible USAN rejections based on preexisting generic names, drug manufacturers are well advised to conduct a clearance search for the proposed generic name. There are searchable databases containing currently approved generic names that can be reviewed by pharmaceutical companies in the process of developing a proposed name.[163]

B. USAN Application and Approval Process

The earliest time at which a pharmaceutical company may request a USAN generic approval is during Phase 1 of the drug development timeline, while the company performs preliminary tests on healthy human volunteers.[164] Most companies, however, apply for USAN approval during Phase 2 of the timeline, while conducting small-scale efficiency studies on patients.[165]

When submitting an application, drug companies may submit for consideration one to three nonproprietary names that adhere to the USAN Council's guiding principles.[166] Once an applicant submits a completed application, the process known as "negotiation" begins. First, the USAN staff processes the application and assigns a lead negotiator to serve as the company's main contact for questions and correspondence. Once the initial review is complete, the USAN negotiator sends a summary of his or her opinion to the USAN Council, which performs additional analyses on the proposed nonproprietary drug names. Once the USAN Council completes its review, it sends its recommendations regarding the generic name to the applicant for acceptance or rejection. If the applicant accepts the recommended generic name, the USAN Council publishes the name on its website as a name under consideration. If the applicant rejects the recommendations, the negotiation process continues with the applicant selecting new generic names or providing additional information for reconsideration.[167]

After the generic name is accepted, the USAN Council requests an international nonproprietary name (INN) from the World Health

163. *See, e.g.*, *Adopted Names*, AM. MED. ASS'N, http://www.ama-assn.org/ama/pub/physician-resources/medical-science/united-states-adopted-names-council/adopted-names.page? (last visited July 31, 2013); RXLIST, http://www.rxlist.com (last visited July 31, 2013).

164. *Drug Name Development Timeline and USAN Review Procedure*, AM. MED. ASS'N, http://www.ama-assn.org/ama/pub/physician-resources/medical-science/united-states-adopted-names-council/drug-name-development-timeline-usan-review.page? (last visited July 31, 2013).

165. *Id.*

166. *How to Apply for a Name: Completing the Application*, AM. MED. ASS'N, http://www.ama-assn.org/ama/pub/physician-resources/medical-science/united-states-adopted-names-council/how-to-apply-for-usan.page? (last visited July 31, 2013).

167. *How to Apply for a Name: Initial Application Processing*, AM. MED. ASS'N, http://www.ama-assn.org/ama/pub/physician-resources/medical-science/united-states-adopted-names-council/how-to-apply-for-usan.page? (last visited July 31, 2013).

Organization, which coordinates drug nomenclature internationally. Following the INN review, the USAN Council sends the applicant a letter of adoption, and then the applicant can begin using the nonproprietary name.[168]

Section 508 of the Federal Food, Drug, and Cosmetic Act authorizes the FDA to designate official names for generic drugs.[169] The FDA, however, formally recognizes the "skill and experience" of the USAN Council in deriving nonproprietary names for drugs.[170] The FDA cooperates with—and is represented on—the USAN Council.[171] The FDA will, however, publish its own name if it deems that the generic name recommended by the USAN Council is unduly complex or not otherwise useful or is identical or confusingly similar to an existing drug name, or if no USAN has been applied to a "medically useful" drug.[172]

IV. Conclusion

While the multiple approval processes may, at times, seem arbitrary and burdensome, it is important to remember that consumer safety is always the highest priority. Medical errors are the eighth leading cause of death in the United States, causing 45,000 to 100,000 deaths annually, with medication errors accounting for as many as 7,000 of those deaths.[173] Each agency has purposefully designed its exhaustive approval process to attempt to prevent the harm to consumers that results from confusing drug names. A drug manufacturer can, however, avoid additional costs and delays with meticulous planning and thorough evaluation of its potential trademarks during the developmental process. By carefully considering the procedural and substantive requirements of both the FDA and USPTO, a manufacturer can greatly improve its chances of getting its mark approved for use in the United States in a timely and relatively painless manner.

---

168. *How to Apply for a Name: Adoption of USAN*, AM. MED. ASS'N, http://www.ama-assn.org/ama/pub/physician-resources/medical-science/united-states-adopted-names-council/how-to-apply-for-usan.page? (last visited July 31, 2013).

169. The Kefauver-Harris Amendments, Pub. L. No. 87-781, 76 Stat. 780 (1962) (codified in scattered sections of 21 U.S.C. §§ 301–81).

170. 21 C.F.R. § 299.4(c) (2013).

171. *Id.* § 299.4(d).

172. *Id.* § 299.4(e)(1)–(3).

173. GUIDANCE, *supra* note 66, at 2–3; U.S. FOOD & DRUG ADMIN., GUIDANCE FOR INDUSTRY: SAFETY CONSIDERATIONS FOR PRODUCT DESIGN TO MINIMIZE MEDICATION ERRORS 2 (Draft Guidance 2012), *available at* http://www.fda.gov/downloads/Drugs/GuidanceComplianceRegulatoryInformation/Guidances/UCM331810.pdf.

chapter 7

Investigational New Drug (IND) and New Drug Applications (NDA)[1]

I. Introduction

Although the path to drug approval begins long before FDA involvement, with scientists working in the laboratory and with animals, FDA approval must be received before the drug is marketed. Three options have been established for seeking drug approval. Newly discovered drugs must go through the New Drug Application (NDA) process. Prior to filing an NDA, the drug sponsor must file an Investigational New Drug application (IND) to begin testing in humans. With IND status in hand, the drug sponsor must conduct extensive clinical trials. The NDA process can be lengthy, arduous, expensive, and fraught with uncertainty.[2] At the opposite end of the spectrum is the abbreviated new drug application (ANDA). This route to approval was established by the Drug Price Competition and Patent Term Restoration Act of 1984, also known as the Hatch-Waxman Act. The Hatch-Waxman Act allows the FDA to approve generic copies of established drugs through a showing of "bioequivalence," without the need for costly and time-consuming clinical trials. The Hatch-Waxman Act also established an "in between" route for approval of drugs that represent a limited change from a previously approved drug. While technically still an NDA, Section 505(b)(2) of the Hatch-Waxman Act allows the 505(b)(2) sponsor to build upon data previously submitted by others, thereby significantly

1. Patricia A. Carson and Daniel Forchheimer of Kirkland & Ellis LLP, New York City, and Howard Suh of Holland & Knight, LLP, New York City.

2. Recent estimates have pegged the cost of developing a new drug at $1.5 billion. Daniel S. Levine, *New Estimate of Drug Development Costs Pegs Total at $1.5 Billion*, Burrill Rep. (Dec. 7, 2012), http://www.burrillreport.com/article-new_estimate_of_drug_development_costs_pegs_total_at_1_5_billion.html.

shortening the development timeline and greatly reducing the expense. The IND process and the NDA routes to drug approval are discussed in this chapter. Also discussed are matters associated with negotiating a product label for NDA drug products, as well as user fees and FDA performance goals associated with the enactment and reauthorization of the Prescription Drug User Fee Act (PDUFA).

II. Investigational New Drug Applications: Types and Requirements

Before a drug company can conduct clinical trials, it must first submit an IND application to the FDA. The literal purpose of the IND application is very specific: the applicant is seeking permission to be exempt from federal laws that prohibit an unapproved drug from entering interstate commerce. See 21 U.S.C. § 355(a).[3] From a practical perspective, clinical trials that form the basis of FDA approval cannot be undertaken unless the drug moves across state lines. Therefore, the IND application allows for the drug company to receive permission to transport its products through the stream of commerce while informing the FDA that based on its successful preclinical program of animal testing, the company intends to advance its product to the next stage of human testing.[4]

There are generally two broad categories of IND, and within each category, there are additional types of IND applications. If the applicant is an individual clinician, then the IND will be categorized as a "research IND." If the applicant is a company or part of the National Institutes of Health, then the IND will be designated a "commercial IND." Within each category, the IND can be further broken down into the following types:

- the Investigator IND, under which a physician is allowed to conduct research and distribute the drug; this is often used when

3. 21 U.S.C. § 355(a) states under the heading "Necessity of effective approval of application" that "No person shall introduce or deliver for introduction into interstate commerce any new drug, unless an approval of an application filed pursuant to subsection (b) or (j) of this section is effective with respect to such drug."

4. *See* Riegel v. Medtronic Inc., 552 U.S. 312, 343 n.15 (2008) ("The process for approving a new drug begins with preclinical laboratory and animal testing. The sponsor of the new drug then submits an investigational new drug application seeking FDA approval to test the drug on humans. See 21 U.S.C. § 355(i); 21 CFR § 312.1 et seq. (2007). Clinical trials generally proceed in three phases involving successively larger groups of patients: 20 to 80 subjects in phase I; no more than several hundred subjects in phase II; and several hundred to several thousand subjects in phase III. 21 CFR § 312.21. After completing the clinical trials, the sponsor files a new drug application containing, inter alia, 'full reports of investigations' showing whether the 'drug is safe for use and . . . effective'; the drug's composition; a description of the drug's manufacturing, processing, and packaging; and the proposed labeling for the drug. 21 U.S.C. § 355(b)(1) (2000 ed. Supp. V).").

the physician would like to apply the drug to a new usage for an approved drug, or in a new subset of patients, or in researching an unapproved drug; this is typically what most research IND are;

- the Emergency Use IND, which can allow for the expedited delivery of an experimental drug or use in dire life-threatening situations where no acceptable treatment options are available; and
- the Treatment IND is used for serious or immediately life threatening conditions while the final clinical work is being conducted and the FDA review takes place.[5]

Finally, there is the Exploratory IND, which, unlike the other types just discussed, seeks approval to conduct very limited clinical trials in small groups of patients where drug safety and efficacy are not the therapeutic endpoint. Instead, an Exploratory IND may want to administer the drug in humans in order to better understand the compound's mechanism of action in the body or its particular pharmacokinetics. This type of IND is contemplating very early clinical testing and is used by companies to select promising drugs for further development.[6]

Under law, an IND is required to contain at least the following information.

- *Animal Pharmacology and Toxicology Information*—this is the standard "preclinical" data that allows the FDA to assess whether the drug is reasonably safe to test in a small group of humans. This section will also include any prior human use outside of the United States.
- *Manufacturing Information*—this section lays out manufacturing details of the drug to ensure that adequate controls are undertaken to create consistent and stable batches of product.
- *Clinical Protocols and Investigator Information*—this section requires detailed protocols for the proposed clinical trials in order to assess any unreasonable risks to the patient population. The qualifications of the investigators are also required as part of this overall safety determination, as well as any commitments to informed consent or review by an institutional review board.[7]

---

5. *See generally Investigational New Drug (IND) Application*, U.S. Food & Drug Admin., http://www.fda.gov/drugs/developmentapprovalprocess/howdrugsaredevelopedandapproved/approvalapplications/investigationalnewdrugindapplication/default.htm (last updated Oct. 18, 2013).

6. *See generally* U.S. Food & Drug Admin., Guidance for Industry, Investigators, and Reviewers: Exploratory ND Studies, (Jan. 2006), *available at* http://www.fda.gov/downloads/Drugs/ GuidanceComplianceRegulatoryInformation/Guidances/UCM078933.pdf.

7. *See generally* 21 C.F.R. § 312.20–.38.

Upon filing an IND with the FDA, the sponsor may begin conducting its clinical trials 30 days after the official FDA receipt date, unless the FDA notifies the sponsor otherwise. In other words, unless the applicant is specifically notified by the FDA within the 30 days, the clinical trials can commence without waiting for actual permission by the FDA. However, a "partial hold" notification within the 30 days allows for only certain portions of the clinical trials to begin while other parts are still being evaluated. A "clinical hold" means that none of the applied-for clinical trials can commence. Under either notification, the FDA usually requires additional safety and/or efficacy data.

In some instances, a third party may partner with the sponsor, at which point some information may need to be shared with the third party. A dilemma arises when the sponsor does not want to disclose all research but must still comply with the content requirements of an IND. In this case, an applicant may submit a Drug Master File (DMF) in order to provide information on "facilities, processes, or articles used in the making, processing, packaging, and storing of the drug."[8] The DMF cannot fully replace an IND application, but given that the information within the DMF remains confidential, it can be a useful tool in the process.[9]

III. New Drug Applications: Types and Requirements

As discussed earlier in this chapter, an NDA must be submitted and approved before a drug can be marketed in the United States. Three key questions are asked throughout this application process: (1) Is the proposed use safe and effective, and do the benefits outweigh the risks? (2) Are the label and package insert appropriate and accurate? and (3) Are there quality controls and methods used to make a sufficient amount of the drug to ensure that the drug's strength, quality, purity, and identity remain constant?[10] NDAs can be broken down into two categories: traditional NDAs (the "505(b)(1)" application) and the paper NDA (the "505(b)(2)" application). While both types of applications end in market approval, the process and requirements for each differ markedly.

8. *Drug Master Files: Guidelines*, U.S. Food & Drug Admin., http://www.fda.gov/Drugs/GuidanceComplianceRegulatoryInformation/Guidances/ucm122886.htm (last visited June 21, 2013).

9. *Id.*

10. 21 C.F.R. § 314; New Drug Applications (NDA), U.S. Food & Drug Admin., http://www.fda.gov/Drugs/DevelopmentApprovalProcess/HowDrugsareDevelopedandApproved/ApprovalApplications/NewDrugApplicationNDA/default.htm (last updated Feb. 21, 2013).

A. The 505(b)(1) Application

The 505(b)(1) application is a massive undertaking, requiring enormous amounts of information and resources. It is used for a drug, the active ingredient of which has never before been approved. See 21 U.S.C. § 355. The drug is classified immediately as "priority" or "standard," the former indicating that the drug has the potential for great improvement over what is currently available in the market (and therefore moving the application into an expedited process).[11] The information contained in these applications is derived directly from the sponsor's efforts rather than from relying on past results or approvals from the FDA. The application must include the following:

> (A) *full reports* of investigations which have been made to show whether or not such drug is safe for use and whether such drug is effective in use; (B) a full list of the articles used as components of such drug; (C) a full statement of the composition of such drug; (D) a full description of the methods used in, and the facilities and controls used for, the manufacture, processing, and packing of such drug; (E) such samples of such drug and of the articles used as components thereof as the Secretary may require; (F) specimens of the labeling proposed to be used for such drug; and (G) any assessments required under section 505B.[12]

Most of the expense in submitting a 505(b)(1) application comes from these "full reports" to assess the safety and efficacy of the drug. Safety reports include things like (1) pharmacological studies that examine possible adverse effects through how the drug acts in the body; (2) toxicology studies, which evaluate the drug's acute, subacute, or long-term chronic toxicity or carcinogenicity; (3) studies that assess any potential adverse effects on human reproduction and the developing fetus; (4) studies on the absorption, metabolism, and excretion of the drug; and (5) any available information regarding interactions with other drugs.[13]

As to the drug's efficacy, the 505(b)(1) application must contain "substantial evidence," which is defined as "consisting of adequate and well-controlled investigations, including clinical investigations, by experts qualified by scientific training and experience to evaluate the effectiveness of the drug involved, on the basis of which it could fairly and responsibly be concluded by such experts that the drug will have the effect it purports or is represented

11. 21 CFR 314.510; 21 CFR 601.41.
12. 21 U.S.C. § 355 (emphasis added).
13. *See generally* 21 C.F.R. § 314.50(d)(2), (3), (5) (2013).

to have under the conditions of use prescribed, recommended, or suggested in the labeling or proposed labeling thereof."[14] Essentially, this means that an NDA requires the large-scale, well-controlled clinical trials that were the subject of the applicant's prior IND applications from which the applicant was allowed to conduct clinical trials in the first place.

The commissioner of the FDA is given significant deference in deciding whether substantial evidence of effectiveness exists and may reject an application or withdraw approval on this basis.[15]

Although the definition of substantial evidence uses the word "investigations" in the plural form, hence suggesting that the FDA requires more than one clinical trial, the law technically allows for drugs to be approved based on one clinical study.[16] In reality, a company will usually submit multiple clinical trials in order to alleviate any safety and efficacy concerns by the agency. These clinical trials constitute a substantial portion of the time and resources for bringing a new drug to market, which as discussed earlier can easily cost a company upward of a billion dollars. The protocols of each study must be submitted, along with the statistical analyses used to determine that particular therapeutic and safety endpoints are being met. These clinical trials are designated by particular phases, ranging from Phases 0 to 4, and are briefly summarized here:

Phase 0 study—this is typically the clinical trial under the Exploratory IND discussed earlier in this chapter. A Phase 0 study may seek to examine the drug's pharmacokinetic profile without determining whether the drug actually has any therapeutic value and is usually conducted in a very small patient population (i.e., fewer than ten patients). Many drugs fail far larger clinical trials because animal studies sometimes cannot predict how the drug works in humans. These Phase 0 studies are designed to

---

14. 21 U.S.C. § 355(d)(7).

15. *See, e.g.*, Warner-Lambert Co. v. Heckler, 787 F.2d 147, 156–61 (3d Cir. 1986) (finding that the uncontrolled use of concomitant medication during a study, including missing case report forms as to the dosage, frequency, or length of administration of the concomitant medication, rendered any evidence from that study unusable and "therefore could not be used as evidence"); Brandenfels v. Heckler, 716 F.2d 553, 555 (9th Cir. 1983) (finding a lack of substantial evidence of effectiveness when a study failed to explain the method of selection of subjects observed who used the treatment, or the methods of observing and recording results, did not provide a system for quantifying symptom severity and did not adequately establish a "control group" for the study); E.R. Squibb & Sons Inc. v. Bowen, 870 F.2d 678, 684–85 (D.C. Cir. 1989) (finding that the FDA was reasonable in withdrawing approval for drug Mysteclin based on a finding by the commissioner that manufacturer failed to show that the claimed suppression effect was of any therapeutic significance); *see also* U.S. Food and Drug Administration, GUIDANCE FOR INDUSTRY, PROVIDING CLINICAL EVIDENCE OF EFFECTIVENESS FOR HUMAN DRUG AND BIOLOGICAL PRODUCTS (May 1998). http://www.fda.gov/downloads/Drugs/.../Guidances/ucm078749.pdf.

16. *See* Pub. L. No. 105-115, § 115 (1997).

ameliorate this concern and streamline the approval process by eliminating unsuccessful drug candidates.

Phase 1 study—this clinical trial is usually conducted in healthy volunteers, typically in the range of 20 to 80 patients. Phase 1 studies are used to assess mostly safety issues, such as common side effects of the drugs and toxicity concerns. Because the Phase 1 trials are assessing "healthy" volunteers, they are not determining the efficacy of the drug for the intended treatment.

Phase 2 study—if the drug is deemed relatively safe based on the Phase 1 studies, the drug sponsor will likely move on to Phase 2 clinical trials, which test the drug's therapeutic efficacy in patients actually experiencing the disease or condition that is intended to be treated. These are larger trials, typically ranging from 60 to 300 patients, and involve controls. For example, in these controlled trials, patients who receive the drug are compared with those who receive a placebo or a different drug that has already been approved to treat the disease or condition. While the overall goal of Phase 2 trials is to assess drug efficacy, these studies also continue to monitor safety issues, like side effects and toxicity.

Phase 3 study—after Phase 2 studies, the drug sponsor will typically meet with FDA officials to design the much larger Phase 3 tests, which continue to assess the effectiveness and safety of the drug, but in larger and more diverse patient populations and often in combination with other drugs. Phase 3 clinical trials continue to be controlled and can range from several hundred to several thousand patients from all age groups and populations throughout the world.

Phase 4 study—this is also referred to as a postmarket study because it is conducted after the drug has been formally approved by the FDA. These studies are either mandated by the agency or done voluntarily by the sponsor in order to evaluate further data on drug efficacy and safety.[17]

As part of the 505(b)(1) submission, the applicant must include the patent number and the expiration date of all patents or patent applications claiming the drug or claiming methods of using the drug.[18] This resulted from the passage of the Hatch-Waxman Act and requires listing of all patents that could form the basis of a patent infringement suit if a generic version of the drug is marketed.[19] The patent listing requirement has

17. *See generally The FDA's Drug Review Process: Ensuring Drugs Are Safe and Effective*, U.S. Food & Drug Admin., http://www.fda.gov/drugs/resourcesforyou/consumers/ucm143534.htm (last updated Apr. 25, 2014).

18. 21 U.S.C. § 355(b)(1)(G).

19. 21 U.S.C. § 355(b)(1); 21 C.F.R. § 314.53(b); *see also* Apotex Inc. v. Thomson, 347 F.3d 1335, 1344 (Fed. Cir. 2003) ("Under that provision, a patent must be listed if it contains a product claim that reads on the drug that is the subject of the NDA or, with respect to a

generated a certain degree of uncertainty, but in general, a drug sponsor must list its patents that cover either the active pharmaceutical ingredient or its formulation of that drug including nonactive excipients.[20] Moreover, while patents that cover the method of using the drug, such as those that contain methods of treating a particular disease state, must be listed, a patent that claims a method of manufacturing the drug cannot be listed.[21] Other types of patents that cannot be listed include those that claim packaging of the drug, claims to metabolites or intermediates of the drug, and those that claim a method of using the drug which is not being sought for approval in the applicant's NDA.[22]

All listed patents in all NDAs are publicly disclosed in the FDA's publication called the *Approved Drug Products with Therapeutic Equivalence Evaluations*, which is universally known in the industry as the "Orange Book." As discussed at length in a subsequent chapter, under the Hatch-Waxman Act, ANDA applicants are required to make one of four patent certifications for Orange Book–listed drugs and uses. Submission of a generic company's ANDA and its patent notice often results in patent litigation brought by the NDA holder against the generic company. In view of this statutory framework, one may think that an NDA applicant who is uncertain as to whether a patent should be listed should nevertheless list it in order to preserve its future rights to litigate. However, NDA holders must be wary since improperly listing a patent with the FDA can potentially lead to antitrust violations.[23]

FDA approval of an NDA can take several months to several years to complete. For an original application with priority designation, the FDA seeks to provide an answer within six months for 90 percent of candidates, and within ten months for standard applications. Of all priority NDAs

method of use claim, if it is reasonable to conclude that a person who makes, uses, or sells the drug would infringe the claim. The listing decision thus requires what amounts to a finding of patent infringement, except that the 'accused product' is the drug that is the subject of the NDA and the 'accused method' is a method that is reasonably likely to be used by a hypothetical infringer. The fact that the issue of patent law that is presented under section 355(c)(2) does not arise in the context of a typical patent infringement action is not surprising, as the Hatch-Waxman Act alters the normal application of patent law principles in various ways. See, e.g., 35 U.S.C. § 271(e)(2) (providing that it is an act of infringement to submit an ANDA 'for a drug claimed in a patent or the use of which is claimed in a patent').").

20. 21 C.F.R. § 314.53(b)(1).

21. *Id.*

22. *Id.*

23. *See In re* Buspirone Patent Litig., 185 F. Supp. 2d 363, 375 (S.D.N.Y. 2002) (holding Bristol Myers Squibb's listing of its '365 Patent' in the Orange Book for patents and subsequent patent infringement litigations were objectively baseless because the patent was no longer enforceable and, therefore, antitrust allegations could not be dismissed).

submitted in 2009 and 2010, 78 percent were reviewed within six months, while 94 percent of all standard NDAs were reviewed within ten months.[24]

B. The 505(b)(2) Application

The Hatch-Waxman Act added a second type of NDA, referred to as a 505(b)(2) application, based on the section of the statute that created it.[25] Section 505(b)(2) provides:

> An application submitted under [section 505(b)(1)] for a drug for which the investigations [into safety and effectiveness] and relied upon by the applicant for approval of the application were not conducted by or for the applicant and for which the applicant has not obtained a right of reference or use[26] from the person by or for whom the investigations were conducted shall also include . . . a certification . . . to each patent which claims the drug for which such investigations were conducted or which claims a use for such drug for which the applicant is seeking approval.

Thus, unlike the traditional NDA application discussed earlier, the 505(b)(2) application can rely upon the safety and efficacy data of someone else's FDA-approved drug. The purpose behind 505(b)(2) applications is to allow companies to develop improved uses and variations of existing drugs without having to invest in costly preclinical and clinical trials. However, 505(b)(2) applications are not ANDAs, which are required by law to be duplicates of the prior approved drug. If, for example, the 505(b)(2) applicant is seeking approval for a new indication of a previously approved drug, the applicant must still provide any additional clinical information to ensure that the new indication is effective, while relying upon preexisting clinical data of the prior approved drug to satisfy safety requirements. In many ways, because the 505(b)(2) application has characteristics of both a regular NDA and an ANDA, this "in between" application shares many attributes of both. For example, because a 505(b)(2) application can

---

24. U.S. Food & Drug Admin., Fiscal Year 2010 Performance Report to the President and Congress for the Prescription Drug User Fee Act (2010), *available at* http://www.fda.gov/downloads/AboutFDA/ReportsManualsForms/Reports/UserFeeReports/PerformanceReports/PDUFA/UCM243358.pdf.

25. Prior to the Hatch-Waxman Act, the FDA allowed "paper" NDAs as a route to approval. This route allowed applicants to use published literature to satisfy the requirement for full reports supporting the safety and efficacy of the drug product.

26. "Right of reference or use" is defined in 21 C.F.R. § 314.3(b) as "the authority to rely upon, and otherwise use, an investigation for the purpose of obtaining approval of an application, including the ability to make available the underlying raw data from the investigation for FDA audit, if necessary."

contain its own new clinical data, its approval can result in market exclusivity, just like a regular NDA.[27] Also, like a regular NDA, the 505(b)(2) applicant must list its patents in the Orange Book, thereby allowing it to sue generic companies that file ANDAs based on the approved 505(b)(2) application.[28] On the flip side, just like an ANDA, the 505(b)(2) application must contain relevant patent certifications to Orange Book–listed patents, which can form the basis of patent litigation brought by the NDA holder.[29] These issues are discussed in more detail in the following section.

C. The FDA Draft Guidance

Although 505(b)(2) applications were established in 1984, they initially were not widely embraced by the industry due in part to uncertainty as to the appropriate types of applications under 505(b)(2) and the process involved. In the face of continued confusion regarding the products covered by Section 505(b)(2) and the procedure for referencing safety and efficacy information to be relied upon, the FDA issued draft guidance in October 1999 (the 1999 Guidance).[30] The 1999 Guidance included information regarding (1) the type of information on which applicants may rely; (2) the types of applications appropriate for filing under Section 505(b)(2); (3) the differences between 505(b)(2) applications and abbreviated new drug applications (ANDAs); (4) patent and exclusivity considerations; and (5) the information to be included in a 505(b)(2) application.

1. The Type of Information a 505(b)(2) Applicant Can Rely On

General references to published information or knowledge will not cause an application to be a 505(b)(2) application. Instead, the relied-upon reference should be to specific information, such as preclinical or clinical trials, "necessary to the approval of the application."[31] In addition, a 505(b)(2) applicant is permitted "to rely on the Agency's finding of safety and effectiveness for an approved drug to the extent such reliance would be permitted under the generic drug approval provisions at section 505(j)."[32]

---

27. *See* 21 U.S.C. § 355(j)(5)(F)(iii), (iv).

28. *See id.* § 355(b)(1)–(2).

29. *See* 21 C.F.R. § 314.50(i).

30. U.S. Food & Drug Admin., Draft Guidance for Industry: Applications Covered by Section 505(b)(2) (1999) [hereinafter 1999 Guidance]. At present, the 1999 Guidance has not been withdrawn. Retrospective Review of Draft Guidance Documents Issued Before 2010: Withdrawal of Guidances, 78 Fed. Reg. 48175-01 (Aug. 7, 2013).

31. *Id.*

32. *Id.* at 3. According to the 1999 Guidance, "[t]his approach is intended to encourage innovation in drug development without requiring duplicative studies to demonstrate what is already known about a drug while protecting the patent and exclusivity rights for the approved drug."

2. The Kind of Application That Can Be Submitted as a 505(b)(2) Application

Applications for new chemical entities/new molecular entities may be submitted under 505(b)(2) "when some part of the data necessary for approval is derived from studies not conducted by or for the applicant and to which the applicant has not obtained a right of reference."[33] The 1999 Guidance notes that for a new chemical entity (NCE), "this data is likely to be derived from published studies, rather than FDA's previous finding of safety and effectiveness of a drug."[34] However, if the applicant has the right of reference to all information necessary for approval, even if not conducted by the applicant, the application is considered a 505(b)(1) application (i.e., a traditional NDA).

Applications reflecting "changes to previously approved drugs" are also appropriate under 505(b)(2). With regard to such applications, the 1999 Guidance states that "an application may rely on the Agency's finding of safety and effectiveness of the previously approved product, coupled with the information needed to support the change from the approved product" and notes that "[t]he additional information could be new studies conducted by the applicant or published data."[35] Referencing 21 C.F.R. 314.101(d)(9), the 1999 Guidance points out that "505(b)(2) applications should not be submitted for duplicates of approved products that are eligible for approval under 505(j)."[36] Applications for changes in the product that would also be eligible for consideration pursuant to a suitability petition under Section 505(j)(2)(C) of the Federal Food, Drug, and Cosmetic Act (FDCA), however, are noted to be eligible as 505(b)(2) applications.[37]

3. Examples of 505(b)(2) Applications

The scenarios in which 505(b)(2) applications should be submitted include: (1) changes in dosage form (e.g., solid oral to a transdermal patch); (2) changes in dosage strength; (3) changes in route of administration (e.g., intravenous to intrathecal); and (4) substitution of an active ingredient

33. *Id.* at 3.

34. *Id.*

35. *Id.* The 1999 Guidance references the letter to the industry from Dr. Paul D. Parkman and states that "[t]his guidance helps to clarify and amplify the approaches stated in the April 10, 1987, letter and in the regulations."

36. *Id.* at 3–4.

37. An ANDA suitability petition is a petition to the FDA to permit the filing of an ANDA for a drug that differs from the reference listed drug. Such changes, however, are often inappropriate in the context of an ANDA because an ANDA can only contain differences that do not need clinical evidence for safety and efficacy. An approved suitability petition is required in order to file an ANDA with the proposed differences.

in a combination product.[38] The 1999 Guidance also provides examples of 505(b)(2) applications that "may be" accepted:[39] (1) formulation where an application "contains a different quality or quantity of an excipient(s) than the listed drug where the studies required for approval are beyond those considered limited confirmatory studies appropriate to a 505(j) application"; (2) change in dosing regimen (e.g., twice daily to once daily); (3) changes to an active ingredient such as a different salt, ester, racemate, or enantiomer of an active ingredient; (4) new molecular entity such as a prodrug or active metabolite of an approved drug; (5) combination product in which the active ingredients have been previously approved individually; (6) previously unapproved indications of listed drugs; (7) switches from prescription indications to over-the-counter (OTC" indications; (8) OTC monograph (i.e., a drug product that differs from a product described in an OTC monograph);[40] (9) naturally derived or recombinant active ingredient (i.e., "drug product containing an active ingredient(s) derived from animal or botanical sources or recombinant technology where clinical investigations are necessary to show that the active ingredient is the same as an active ingredient in a listed drug"); and (10) bioequivalence. As to bioequivalence, the 1999 Guidance explains that applications for pharmaceutically equivalent drug products generally must be submitted under Section 505(j), and bioequivalence to the reference listed drug must be demonstrated.[41] However, the 1999 Guidance singles out applications for proposed drug products where the rate[42] and/or extent[43] of absorption exceed, or are otherwise different from, the 505(j) standards for bioequivalence as eligible under Section 505(b)(2). The potential need for additional clinical studies to document safety and efficacy at a different rate and extent of delivery is noted, as is the need to include such differences in product labeling.[44] Although proposed products do not need to be shown to be clinically better than previously approved products, the 1999 Guidance instructs that "a 505(b)(2) application should not be used as a route of approval for poorly bioavailable generic drug products unable to meet the 505(j) standards for bioequivalence."[45]

The 1999 Guidance also calls out applications that cannot be submitted as 505(b)(2) applications[46]—specifically applications that (1) are duplicates

38. 1999 GUIDANCE, *supra* note 30, at 4.
39. *Id.* at 5–6.
40. 21 C.F.R. § 330.11.
41. 1999 GUIDANCE, *supra* note 30, at 5.
42. 21 C.F.R. § 314.54(b)(2).
43. 21 C.F.R. § 314.54(b)(1).
44. 1999 GUIDANCE, *supra* note 30, at 6.
45. *Id.*
46. *Id.*

of listed drugs and eligible for approval under section 505(j);[47] (2) differ from the reference listed drug only in that the extent to which the active ingredient(s) is absorbed or otherwise made available to the site of action is less than that of the listed drug;[48] or (3) differ from the reference listed drug only in that the rate at which its active ingredient(s) is absorbed or otherwise made available to the site of action is unintentionally less than that of the listed drug.[49]

4. Examples of 505(b)(2) Approvals

In December 2011, the FDA approved Pfizer Consumer Healthcare's 505(b)(2) application for Advil Allergy and Congestion Relief tablets.[50] The application sought approval for a new drug combination—ibuprofen, 200 mg; phenylephrine HCl, 10 mg; and chlorpheniramine maleate, 4 mg—for the relief of symptoms associated with hay fever or other respiratory allergies, or the common cold.[51] The three drugs individually have been on the market for decades. Thus, Pfizer was able to rely on phenylephrine and chlorpheniramine's GRASE[52] classification and the history of extensive use of ibuprofen. To obtain approval under 505(b)(2), Pfizer only had to submit a 90-day toxicology study in rats, and the only clinical study the FDA required was a pharmacokinetic trial.

In November 2011, Antares Pharma Inc. received 505(b)(2) approval for a new formulation of levetiracetam, which is used to treat partial, myoclonic, and primary generalized seizures.[53] Levetiracetam was approved and marketed as Keppra, which came in two formulations: an injection and an oral formulation that required premixing and was only approved for patients 16 years of age and older. Antares's 505(b)(2) application sought approval for three premixed solutions of varying dosages of levetiracetam prediluted in a salt solution and used the already marketed Keppra and Keppra Injection as a reference.[54] Antares was granted a biowaiver so that it did not have to perform new pharmacokinetic studies. No other clinical studies were required because the already approved Keppra and Keppra Injection supported the efficacy and safety of the premixed product.

47. 21 C.F.R. § 314.101(d)(9).

48. *Id.* § 314.54(b)(1).

49. *Id.* § 314.54(b)(2).

50. FDA Approval Letter from Schiffenbauer to Sinclair, dated December 21, 2011, http://www.accessdata.fda.gov/drugsatfda_docs/nda/2011/022113Orig1s000Approv.pdf.

51. *Id.* at 1.

52. "Generally recognized as safe and effective."

53. FDA Approval Letter from Katz to Pizza, dated November 9, 2011, http://www.accessdata.fda.gov/drugsatfda_docs/nda/2011/202543Orig1s000Approv.pdf.

54. *Id.* at 1.

5. Exclusivity and Patent Considerations

The 1999 Guidance also explains the patent and exclusivity provisions that could affect a 505(b)(2) application. An approved 505(b)(2) application may be granted three years of Hatch-Waxman exclusivity if one or more of the clinical investigations, other than bioavailability/bioequivalence studies, was essential to the approval of the application and was conducted or sponsored by the applicant.[55] The application may be eligible for five years of exclusivity if it is a new chemical entity[56] and may also be eligible for orphan drug[57] or pediatric exclusivity.[58]

The 1999 Guidance also notes that, like a 505(j) application, approval or filing of a 505(b)(2) application may be delayed because of patent and exclusivity rights that apply to the listed drug, even if the application also contains clinical investigations supporting approval of the application.[59] Under 21 C.F.R. § 314.50(i), the 505(b)(2) applicant must submit a certification with regard to any patent for a reference drug relied upon by the applicant, stating (1) the patent information was not submitted to the FDA, (2) the patent has expired, (3) the date of expiration of the patent, and (4) the patent is invalid, unenforceable, or will not be infringed by the applicant's drug.[60] Certification under (4) triggers the same deadlines and obligations as a Paragraph IV notice for an ANDA filer.[61] A 505(b)(2) filer certifying under (4) must provide notice to the patent owner and holder of the NDA within 20 days of being informed that the FDA has accepted the application, which in turn triggers a 45-day period in which the NDA holder may file suit for patent infringement under 35 U.S.C. § 271(e)(2). If suit is timely filed, approval is stayed for 30 months, subject to the same events specified for ANDA applicants. Beyond this, the applicant may face a delay in the effective approval date due to market exclusivity on the part of the NDA applicant. If the 505(b)(2) applicant certifies under the third circumstance, the FDA may only grant a tentative approval depending upon the expiration of the patent upon which the applicant relied. Alternatively, the applicant may have to wait until the original drug's patent expires if a court determines that the patent is not invalid or is infringed by the 505(b)(2) application.

55. 21 C.F.R. §§ 314.50(j), 108(b)(4)–(5).

56. *Id.* §§ 314.50(j), 108(b)(2).

57. *Id.* §§ 316.20–36.

58. A 505(b)(2) applicant is not entitled to the 180-day exclusivity afforded to the first ANDA filer.

59. 1999 Guidance at 6.

60. 21 C.F.R. § 314.50(i).

61. *Id.* § 314.107.

6. What Should Be Included in 505(b)(2) Applications

Applications under Section 505(b)(2) should include: (1) "[i]dentification of those portions of the application that rely on information the applicant does not own or to which the applicant does not have a right of reference";[62] (2) specific identification of any and all listed drugs for which the applicant seeks to rely on the FDA's previous finding of safety or efficacy;[63] (3) identification of any patents claiming the drug or proposed use for publication in the Orange Book; (4) information required under 314.50(j), if the applicant believes that it is entitled to marking exclusivity;[64] (5) certification as required under Section 505(b)(2) with respect to relevant patents claiming the listed drug or any other drug or use relied on by applicant;[65] (6) certification stating that the application is for a new, previously unapproved use, if applicable; (7) a statement as to whether any listed drug has received a period of marketing exclusivity; (8) a bioavailability/bioequivalence study comparing the proposed product to the listed drug; and (9) studies to support the change or modification from the listed drug.

7. Legal Challenges to 505(b)(2)

After the 1999 Guidance was issued, a number of pharmaceutical companies, including Pfizer Inc., Pharmacia Corporation, and Amgen Inc., filed citizen petitions challenging the legality of the regulations, asserting, inter alia, that the submissions under Section 505(b)(2) violated their Fourth Amendment rights as an illegal taking of property.[66] In a consolidated response to these petitions, Dr. Janet Woodcock, the director of the Center for Drug Evaluation and Research (CDER), declined to alter the FDA's interpretation of Section 505(b)(2). Dr. Woodcock stated:

> FDA's longstanding interpretation of section 505(b)(2) is intended to permit the pharmaceutical industry to rely to the greatest extent possible under the law on what is already known about a drug. The Agency's approach is to use the 505(b)(2) drug approval pathway to avoid requiring drug sponsors to conduct and submit studies that are not

---

62. *Id.* § 355(b)(2).

63. The 1999 Guidance notes that even if the 505(b)(2) application is based solely upon literature and does not rely expressly on previous safety and efficacy findings, the listed drug(s) on which the studies were conducted, if there are any, must be identified. 1999 Guidance at 1–2.

64. 21 C.F.R. § 314.54(a)(1)(vii).

65. *Id.* § 314.54(a)(1)(vi).

66. Citizen Petition No. 2001P-0323/CPI (July 27, 2001); Comments in support of Citizen Petition No. 2001P-0323/CPI from Ralph Smalling, VP, Regulatory Affairs, Amgen Inc. (Dec. 17, 2001) (on file with author).

> scientifically necessary. The conduct and review of duplicative studies would (1) divert industry resources that could be used to undertake innovative research, (2) increase drug costs, (3) strain FDA review resources, and (4) slow the process for drug approval with no corresponding benefit to the public health.[67]

In denying the petitions, Dr. Woodcock in essence asserted that the pharmaceutical innovators had given up privacy rights in exchange for the patent term extensions provided in the Hatch-Waxman Act.

Several years later, members of the biotechnology industry, including BIO, Genentech, and Pfizer, challenged the applicability of Section 505(b)(2) to therapeutic protein products, asking the FDA to refrain from "approving, under any statute it administers, any application for a therapeutic protein product that does not contain a full complement of original non-clinical and clinical data and that relies on any data or information contained in another applicant's application."[68] Although for the most part, the 505(b)(2) approval pathway is not applicable to biological products[69] approved pursuant to section 351 of the Public Health Service Act, products such as human insulin, human growth hormone, and calcitonin have been approved under section 505 of the FDCA.[70]

The citizen petitions were precipitated by Sandoz's filing of a 505(b)(2) application seeking approval of a follow-on recombinant human growth hormone, relying on the FDA's prior approval of Pfizer's Genotropin product. Among other bases, the petitions asserted that Section 505(b)(2) did not legally permit the FDA to rely on, reference, or otherwise use clinical and manufacturing information contained in Pfizer's Genotropin NDA to

---

67. Letter from Janet Woodcock, MD, Dir., Ctr. for Drug Evaluation & Research, to Petitioners Pfizer Inc., Pharmacia Corp., Biotechnology Industry Org., and Torpharm (Oct. 14, 2003) (on file with author).

68. Citizen Petition No. 2003P-0176/CP1 (Apr. 23, 2003); BIO Citizen Petition Follow-on Therapeutic Proteins at 3.

69. *Transfer of Therapeutic Products to the Center for Drug Evaluation and Research*, U.S. Food & Drug Admin., http://www.fda.gov/AboutFDA/CentersOffices /OfficeofMedicicalProductsandTobacco/CBER/ucm 133463.htm (last updated Feb. 22, 2010).

70. The Biologics Price Competition and Innovation Act of 2009 (BPCIA) was signed into law on March 23, 2010 and created an approval pathway for generic versions of biologic medicines (biosimilars). On April 12, 2012, Abbott Laboratories submitted a Citizen Petition asking the FDA to confirm that it will not accept for filing any application for a biosimilar product citing to Abbott's Humira product as its reference product "or any other product for which the biologics license application (BLA) was submitted to FDA prior to March 23, 2010, the date on which the BPCIA was signed into law." Abbott explained that the FDA's approval of such applications would constitute a taking under the Fifth Amendment, which requires just compensation. Citizen Petition No. FDA-2012-P-0317 (Apr. 2, 2012). As of August 2013, the FDA had not reached a decision on Abbott's petition.

approve Omnitrope and that such reliance is also scientifically unjustified. Specifically, the petitions argued that the use of other companies' data has no assurance of safety and cited the significant differences between therapeutic protein products and chemical drugs in terms of complexity and heterogeneity.

In approving the Omnitrope 505(b)(2) application, the FDA denied the relevant portions of the three petitions, communicating its reasoning in a May 30, 2006, letter (the Response) to BIO, Genentech, and Pfizer.[71] The Response reiterated the FDA's long-standing interpretation of Section 505(b)(2) as being intended to permit the pharmaceutical industry to rely to the greatest extent possible under the law on what is already known about a drug. The Response contends that its review and approval of the Sandoz 505(b)(2) application did not require the use or disclosure of any trade secret or confidential information and, on this basis, does not address the trade secret and constitutional (Fifth Amendment takings clause) arguments made by the petitioners. In explaining its reasoning for approval, the FDA refers to the relative simplicity of the protein, pointing out that it is non-glycosylated and well characterized, with a known mechanism of action and long history of clinical use.[72]

With the uncertainty raised by legal challenges to 505(b)(2) applications resolved and the expiration of patents covering blockbuster drugs, 505(b)(2) applications have gained popularity.[73] While 505(b)(2) applications require information on methods used to ensure identity, strength, quality, and purity of the drug, as well as proper labeling information, the information is tested by the same standards of safety and efficacy that are applied to the 505(b)(1) application. The faster and less costly route to approval offered by 505(b)(2) applications renders them an attractive option for pharmaceutical companies seeking to create new differentiated products.

8. Delisting Counterclaims under 505(c) and (j)

The Medicare Prescription Drug, Improvement, and Modernization Act of 2003 (MMA) added provisions to the Hatch-Waxman Act enabling

71. Letter from Steven Galson, MD, MPH, Dir., Ctr. for Drug Evaluation & Research, to Petitioners Pfizer Inc., Biotechnology Industry Organization, and Genentech (May 30, 2006), *available at* http://www.fda.gov/ohrms/dockets/dockets/04P0231/04P-0231-pdn0001.pdf. The response points out that "[m]any of the legal and regulatory arguments raised by the BIO Petition concerning section 505(b)(2) of the Act were addressed in FDA's October 14, 2003, consolidated response to various citizen petitions." *Id.* at 2 n.5.

72. *Id.* at 5–8.

73. According to data published in October 2012, 505(b)(2) approvals have outnumbered 505(b)(1) approvals since 2006. *The Path of Least Resistance: Repositioned Drugs Surpass New Brands*, PHARM. EXEC. (Oct. 24, 2013), http://blog.pharmexec.com/2012/10/24/the-path-of-least-resistance-repositioned-drugs-surpass-new-brands/.

challenges to patent information at the FDA.[74] These challenges are permitted as a patent delisting counterclaim, added as Section 505(c)(3)(D)(ii)(I) by the MMA and reading as follows:

> If an owner of the patent or the holder of the approved application under [FDC Act Section 505(b)] for the drug that is claimed by the patent or a use of which is claimed by the patent brings a patent infringement action against the applicant, the applicant may assert a counterclaim seeking an order requiring the holder to correct or delete the patent information submitted by the holder under [FDC Act Section 505(b) or (c)] on the ground that the patent does not claim either . . . the drug for which the application was approved, or . . . an improved method of using the drug.[75]

The counterclaim provision references Sections 505(b) and (c), under which an NDA holder must submit with its application and after NDA approval "the patent number and the expiration date of any patent which claims the drug for which the applicant submitted the application or which claims a method of using such drug and with respect to which a claim of patent infringement could reasonably be asserted if a person not licensed by the owner engaged in the manufacture, use, or sale of the drug." Few cases exist in which a defendant has actually asserted a patent delisting counterclaim according to these provisions. In the two major cases in which a patent delisting counterclaim has been asserted in order to seek actual delisting of Orange Book–listed patents, the litigation was ultimately resolved with respect to the counterclaim without a court decision.[76] Relatedly, the Supreme Court recently upheld as proper the use of a patent delisting counterclaim "to force correction of a use code that inaccurately describes the brand's patent as covering a particular method of using a drug" and remanded the case back to the Federal Circuit after the Federal Circuit's initial decision that a patent delisting counterclaim could not be used for this purpose.[77]

---

74. Pub. L. No. 108-173, 117 Stat. 2066 (2003).

75. 21 U.S.C. § 355(c)(3)(D)(ii)(I). Notably, a similar provision in the statute also enables a generic manufacturer in a paragraph IV suit to assert an identical counterclaim. *See* 21 U.S.C. § 355(j)(5)(C)(ii)(I).

76. *See generally* King Pharm. Inc. v. Intelliject Inc., No. 11-065 (BMS) (D. Del. 2011); Purdue Pharma LP v. Ranbaxy Inc., No. 10-cv-03734 (S.D.N.Y. 2010).

77. *See* Caraco Pharm. Labs. Ltd. v. Novo Nordisk A/S, 132 S.Ct. 1670, 1675 (2012); *see also* Hoffmann-La Roche Inc. v. Orchid Chems. & Pharm. Ltd., Nos. 07-4582 (SRC)(MAS), 08-4051(SRC)(MAS), 10-4050(SRC)(MAS), 07-4582, 2011 WL 5508485 (Nov. 7, 2011 D.N.J.) (declining to wait for the Supreme Court to rule on the same issue in Caraco and granting motion to dismiss a patent delisting counterclaim where counterclaim was asserted in order to force a correction of a patent use code).

D. The Mechanics of Preparing and Submitting an NDA

As discussed earlier in this chapter, the NDA is the formal document requesting FDA approval for the sale and marketing of a new pharmaceutical in the United States. The NDA is largely comprised of the data gathered during the animal studies and human clinical trials during the testing phase of the IND, and all documentation produced for the IND becomes part of the NDA and is evaluated by the FDA. Essentially, the NDA includes all information necessary to completely describe the drug and the processes by which it works and is manufactured. [78] The information in the NDA should provide sufficient information to permit the FDA to address the following issues: (1) whether the drug is safe and effective, or whether its benefits outweigh its risks; (2) whether the proposed label is appropriate, and what it should contain; and (3) whether the manufacturing methods are sufficient to maintain quality control of the drug, particularly with respect to consistency in its strength, quality, purity, and identity.[79]

The NDA review process consists of six primary steps: pre-submission activities, submission, planning review of the application, scientific and regulatory review, official action on the disposition of the application, and post-action feedback.[80]

1. Pre-submission Activities

Prior to submitting an application, there are activities an applicant can undertake to improve the NDA and its likelihood of approval. Applicants are encouraged to meet with the appropriate FDA review division in a pre-NDA meeting before submitting an NDA to plan the application process and the content of the application itself. These meetings help ensure that submitted applications are complete and all necessary content is available for review.[81]

At the pre-NDA meeting, applicants and members of the review team discuss the format and content of the application, both the format and quality of the data to be presented in the application, and the expected submission date of the application.[82] Pre-NDA meetings are attended by members of the Office of Surveillance and Epidemiology and may include analysts for Risk Evaluation and Mitigation Strategy and other staff as needed.

---

78. *See generally New Drug Application (NDA)*, U.S. Food & Drug Admin., http://www.fda.gov/Drugs/DevelopmentApprovalProcess/HowDrugsareDevelopedandApproved/ApprovalApplications/NewDrugApplicationNDA/default.htm (last updated Feb. 21, 2013).

79. *See generally id.*

80. CDER 21st Century Review Process: Desk Reference Guide 3 (Sept. 2012), *available at* http://www.fda.gov/downloads/AboutFDA/CentersOffices/CDER/ManualofPoliciesProcedures/UCM218757.pdf.

81. *Id.* at 9.

82. *Id.*

The applicant must provide the following information for the pre-NDA meeting: information about previous contact with the FDA about this drug, including the minutes from previous meetings, and copies of agreements and advice; a summary of the technical data to be included in the NDA, including available data from key trials; information about any potential problems; proposed labeling; the proposed format of the documents to be submitted, including datasets; and an evaluation of the need for Risk Evaluation and Mitigation Strategy analysis or postmarketing safety studies.[83]

2. Submission

When an NDA is submitted, it is assigned to the appropriate review division. The Regulatory Process Manager (RPM) conducts an initial assessment of the application to determine whether regulatory requirements have been met and whether fees have been paid before assigning the application to the appropriate review team.[84]

3. Planning the Review of the NDA

After the review team is assigned, an initial assessment of the application is taken. The review team examines the NDA and the proposed labeling and makes a determination as to the fileability of the application. This determination is made within 45 days of the application's submission date (or 30 days if it is a priority application). If the application is fileable, then the review team plans the actual review of the material in the application.[85]

4. Conduct Review

Reviewers assess their portion of the application and prepare written evaluations to the team leaders of their groups.

5. Official Action

After review of an NDA, there are three possible outcomes. The drug can be designated Not Approvable, and a letter is sent to the sponsor listing the deficiencies in the application. The drug can also be designated Approvable, which indicates that the application has the potential of being approved at a later day if certain deficiencies are corrected. The drug can also be approved following the review process. If the drug is not approved or is designated approvable, applicants may meet with the agency officials to discuss the reasons for the action taken.

83. *Id.*
84. *Id.* at 12–14.
85. *Id.* at 7.

E. Negotiating the Label

Under the FDCA, a "label" is defined as written, printed, or graphic material on the immediate container of the drug. Although "labels" are most commonly thought of as referring to the package insert or prescribing information accompanying a drug, "labeling" includes all labels, as well as all additional written, printed, or graphic material accompanying the drug, and any written, audio, or visual material published by the drug manufacturer. As part of an NDA, drug manufacturers must propose appropriate labeling to be distributed with the product. Each statement proposed for drug labeling must be justified by data and results submitted in the NDA, and the FDA must approve the labeling before the drug enters the market. The specific requirements for the product label are set forth in 21 C.F.R. 201 et seq. and include the following sections in the following order:

Highlights of Prescribing Information
Product Names, Other Required Information
Boxed Warning
Recent Major Changes
Indications and Usage
Dosage and Administration
Dosage Forms and Strengths
Contraindications
Warnings and Precautions
Adverse Reactions
Drug Interactions
Use in Specific Populations
Full Prescribing Information: Contents
Full Prescribing Information
Boxed Warning

1. Indications and Usage
2. Dosage and Administration
3. Dosage Forms and Strengths
4. Contraindications
5. Warnings and Precautions
6. Adverse Reactions
7. Drug Interactions
8. Use in Specific Populations
    8.1 Pregnancy
    8.2 Labor and delivery
    8.3 Nursing mothers
    8.4 Pediatric use
    8.5 Geriatric use

9. Drug Abuse and Dependence
    9.1 Controlled substance
    9.2 Abuse
    9.3 Dependence
10. Overdosage
11. Description
12. Clinical Pharmacology
    12.1 Mechanism of action
    12.2 Pharmacodynamics
    12.3 Pharmacokinetics
13. Nonclinical Toxicology
    13.1 Carcinogenesis, mutagenesis, impairment of fertility
    13.2 Animal toxicology and/or pharmacology
14. Clinical Studies
15. References
16. How Supplied/Storage and Handling
17. Patient Counseling Information[86]

Labeling must also meet the following requirements:

(1) The labeling must contain a summary of the essential scientific information needed for the safe and effective use of the drug.

(2) The labeling must be informative and accurate and neither promotional in tone nor false or misleading in any particular. In accordance with §§ 314.70 and 601.12 of this chapter, the labeling must be updated when new information becomes available that causes the labeling to become inaccurate, false, or misleading.

(3) The labeling must be based whenever possible on data derived from human experience. No implied claims or suggestions of drug use may be made if there is inadequate evidence of safety or a lack of substantial evidence of effectiveness. Conclusions based on animal data but necessary for safe and effective use of the drug in humans must be identified as such and

86. *See* 21 C.F.R. § 201.56; *see also Drug Development and Review Definitions*, U.S. Food & Drug Admin., http://www.fda.gov/drugs/developmentapprovalprocess/howdrugsaredevelopedandapproved/approvalapplications/investigationalnewdrugindapplication/ucm176522.htm (last updated Feb. 22, 2010) (describing general organization of the product label).

> included with human data in the appropriate section of the labeling.[87]

During the NDA review process, agency reviewers evaluate proposed labeling for accuracy based on the data included in the NDA. The FDA also reviews the labeling for consistency with the applicable regulatory requirements. Each individual element of the labeling, including indications, use instructions, and warnings, is reviewed for consistency with the results of the animal and human testing.

The "label negotiation process," through which a drug's final approved labeling is agreed upon, is an extended dialogue between the FDA and the drug manufacturer. According to FDA guidelines, the negotiation process should be scheduled within 45 days of receiving the filed application, as part of the Filing Determination and Review Planning Phase,[88] although the negotiations themselves will likely not occur until later.[89] Even at that initial stage, however, the primary reviewer will make an initial assessment of the proposed labeling to identify potential issues.[90]

Additional activities concerning label negotiation occur throughout the remainder of the FDA review process. For example, in the review phase, the FDA strives to have a mid-cycle meeting by the end of the fifth month, during which high-level discussions of the proposed labeling occur and the FDA assesses the need for additional interactions with the applicant pertaining to the draft labeling. If, at the mid-cycle meeting, sufficient planning for labeling discussions does not occur, the FDA recommends a short labeling meeting within short order (e.g., one week) to allocate responsibility for reviewing the label and to schedule further meetings as appropriate.[91] The FDA notes that label review is an "iterative process."[92] Additional activities in the Review phase may include discussions of major

87. 21 C.F.R. § 201.56.

88. The FDA divides its NDA review process into six phases: (1) pre-submission activities (which, as its name suggests, occurs prior to the NDA filing); (2) process submission; (3) review planning; (4) review phase; (5) action phase; and (6) post-action phase. *See* U.S. FOOD & DRUG ADMIN., GUIDANCE FOR REVIEW STAFF AND INDUSTRY: GOOD REVIEW MANAGEMENT PRINCIPLES AND PRACTICES FOR PDUFA PRODUCTS (Apr. 2005) [hereinafter GOOD REVIEW MANAGEMENT FOR PDUFA], *available at* http://www.fda.gov/downloads/Drugs/ . . . /Guidances/ucm079748.pdf; CDER 21ST CENTURY REVIEW PROCESS DESK REFERENCE GUIDE, *supra* note 80.

89. *See* CDER 21ST CENTURY REVIEW PROCESS DESK REFERENCE GUIDE, *supra* note 80.

90. *See id.* at 7, 18, 22. Even before the application is submitted, the applicant is encouraged to sit for a pre-submission meeting with the FDA and submit the proposed labeling in advance of the meeting so that it can be discussed at the meeting.

91. *Id.* at 33–34.

92. *Id.* at 34.

labeling concerns during a "late-cycle meeting" and the development of the final labeling.[93]

For the action phase, the FDA has implemented the following recommendations for label review:

> Review wrap-up determinations help form the basis for labeling discussions with the applicant to complete the labeling for products whose approval can be anticipated (i.e., approval and most approvable actions). Since essential labeling discussions by necessity occur toward the end of the review cycle when available time is limited, it is important that communication between the FDA and applicants be clear and efficient. Adherence to the review timeline, including completion of primary and secondary reviews well in advance of the PDUFA goal date, allows time to resolve labeling content issues and avoids crisis management of these issues near the PDUFA goal date.[94]

The majority of label discussions and negotiations occur during the review and action phases.[95]

After review, the manufacturer either accepts the FDA suggestions or provides counterproposals until both parties reach an agreement. This process can take a few weeks to many months, largely depending on the number of FDA comments and the manufacturer's willingness to accept the proposed revision. Often, the labeling will undergo several rounds of revision before an agreement is reached and, as discussed in the following section, recent amendments to the law allow for a regulated process for amending product labels.[96]

---

93. *Id.* at 37–38.

94. Good Review Management for PDUFA Products, *supra* note 88, at 21.

95. *Id.* at 23.

96. In recent years, the Supreme Court has issued some notable decisions concerning the adequacy of product labeling. In *Wyeth v. Levine*, 555 U.S. 555 (2009), the Court addressed the issue of whether a state tort claim for inadequate labeling (failure to warn) was preempted by federal regulations when the drug product label complied with FDA requirements. The Court found no preemption, observing that the Vermont law provided for additional consumer protections above and beyond that required by the FDA. In *Mutual Pharmaceutical Co. v. Bartlett*, 133 S. Ct. 2466 (2013), the Court found preemption of state tort law for design-defect claims based on a direct conflict between federal law requiring that generic compounds be chemically identical to the branded product, as well as FDA requirements that generic drug labels mimic that of the corresponding branded drug.

F. Changing the Label after Approval: Section 505(o)(4)

After approval, the drug's label generally cannot be changed except by an amendment submitted by the NDA holder (e.g., for a new approved indication). However, as of 2007, the FDA has the power to require changes for safety purposes. If subsequent testing or new data indicate safety issues that were not understood during the approval process or are not covered by the label, the Food and Drug Administration Amendments Act of 2007 (FDAAA) grants the FDA the ability to mandate a change in labeling. Prior to this act, the FDA historically could not unilaterally require a change in labeling but, rather, relied on a negotiation process with the manufacturer. Unfortunately, this process left the FDA with little negotiating power; the FDA could only request changes and threaten to initiate withdrawal of approval proceedings—an action that could have negative consequences for public health if patients were benefiting from the drug despite the newly discovered safety issues.[97]

Under the FDAAA, Section 505(o)(4) was added to the FDCA, authorizing the FDA to order labeling changes based on new safety information that becomes available after approval. The term "new safety information" means: "information derived from a clinical trial, an adverse event report, a post approval study (including a study under section 505(o)(3)), peer-reviewed biomedical literature, data derived from the post market risk identification and analysis system under section 505(k); or other scientific data deemed appropriate by [FDA]" about:

> A serious risk or an unexpected serious risk associated with use of the drug that [FDA] has become aware of (*that may be based on a new analysis of existing information*) since the drug was approved, since the risk evaluation and mitigation strategy (REMS) was required, or since the last assessment of the approved [REMS] for the drug (emphasis added), *or*
>
> The effectiveness of the approved [REMS] for the drug obtained since the last assessment of [the REMS].[98]

Section 505(o)(4), which only applies to prescription drugs, allows for changes in safety information that would affect the label sections involving

97. *See* http://www.huschblackwell.com/~/media/Files/BusinessInsights/BusinessInsights/2012/01/Issues%20and%20Trends%20in%20Drug%20Product%20Labeling%20Chapt__/Files/Issues%20and%20Trends%20in%20Drug%20Product%20Labeling/FileAttachment/DickensonChapter.pdf

98. *See* U.S. Food & Drug Admin., Guidance for Industry Safety Labeling Changes: Implementation of Section 505(o)(4) of the FD&C Act 3–4 (July 2013) (quoting 21 U.S.C. 355-1(b)), *available at* http://www.fda.gov/downloads/Drugs/GuidanceComplianceRegulatoryInformation/Guidances/UCM250783.pdf.

boxed warnings, contraindications, warnings and precautions, drug interactions, and adverse reactions. Although these sections are the most commonly altered by postapproval safety concerns, the FDA believes there is a broad definition of "new safety information" so as to allow discretion in ordering changes to the label under Section 505(o)(4): "It is FDA's view that the statutory definition of new safety information is broad to enable FDA to require application holders to add information about serious risks to the labeling of a drug when the Agency determines that such information should be included."[99] To date, we have uncovered no litigation surrounding the application or interpretation of Section 505(o)(4).

G. Changing the Label after Approval: Section 506A

When making postapproval changes to an NDA (or ANDA), the FDCA and federal regulations further require that such changes be reported to the FDA, providing for four reporting categories: (1) Major change (prior approval supplement); (2) Moderate change (supplement—changes being effected in 30 days); (3) Moderate change (supplement—changes being effected); and (4) Minor change (annual report).[100] Changes having the substantial or moderate potential to have an adverse effect on the identity, strength, quality, purity, or potency of a drug product require submission of a supplement to the application, as described in the following section.[101] Notably, if an assessment indicates that a change has adversely affected these aspects of the drug product, the FDA recommends that the change be submitted in a prior approval supplement, regardless of the recommended reporting category for the change.[102] Even where changes have a minimal potential to have an adverse effect on these aspects of the drug product, such changes need to be reported in the applicant's next Annual Report to the FDA. The reporting categories for postapproval changes typically concern the following subcategories: (1) Components and composition; (2) Manufacturing sites; (3) Manufacturing process; (4) Specifications; (5) Container closure system; (6) Labeling; (7) Miscellaneous changes; and (8) Multiple related changes.[103]

A major change is a change "that has substantial potential to have an adverse effect on the identity, strength, quality, purity, or potency of a drug product as these factors may relate to the safety or effectiveness of the drug

99. *Id.* at 4.

100. *See* 21 U.S.C. § 356a; 21 C.F.R. § 314.70.

101. *See* Guidance for Industry: Changes to an Approved NDA or ANDA (Apr. 2004).

102. *See id.* at 7; *see also* Guidance for Industry: Changes to an Approved NDA or ANDA, Questions and Answers 2 (Jan. 2001).

103. *See generally* Guidance for Industry: Changes to an Approved NDA or ANDA, *supra* note 101.

product."[104] Major changes require the submission of a supplement called and clearly labeled *Prior Approval Supplement* and approval by the FDA prior to the distribution of the drug product made using that change.[105] The applicant may request expedited review by submitting a supplement called and clearly labeled *Prior Approval Supplement—Expedited Review Requested* for public health reasons or if a delay in making the change "would impose an extraordinary hardship on the applicant."[106] The FDA has released guidance noting that an expedited review request is most likely to be granted for "extraordinary hardship for manufacturing changes made necessary by catastrophic events (e.g., fire) or by events that could not be reasonably foreseen and for which the applicant could not plan."[107] A nonexclusive list of examples of major changes, by subcategory, is provided in Table 1.

Table 1

Changing the Label—Major Changes

Category	Exemplary Major Change(s)
Components/ Composition	Changes in the qualitative or quantitative formulation, including inactive ingredients, as provided in the approved application.[108]
Manufacturing Sites	The FDA recommends that a move to a different manufacturing site, when it is a type of site routinely subject to FDA inspection, be submitted as a prior approval supplement (i.e., a major change) if the site does not have a satisfactory current good manufacturing practice (CGMP) inspection for the type of operation being moved.[109]

(continued)

104. *See id.* at 3–4. Although this Guidance is generally nonbinding, the guidance has binding effect "[i]nsofar as this guidance adjusts reporting categories pursuant to section 506A [21 U.S.C. § 356a] of the [FDCA] and 21 C.F.R. [§] 314.70"—that is, the reporting categories discussed in this section. *Id.* at 3 n.*

105. *See* 21 C.F.R. § 314.70(b).

106. *See* Guidance for Industry: Changes to an Approved NDA or ANDA, *supra* note 101, at 3.

107. *Id.*

108. 21 C.F.R. § 314.70(b)(2)(i).

109. See Guidance for Industry: Changes to an Approved NDA or ANDA, supra note 101, at 9–10.

Table 1

Changing the Label—Major Changes (continued)

Category	Exemplary Major Change(s)
Manufacturing Process	Changes that may affect the controlled (or modified) release, metering, or other characteristics (e.g., particle size) of the dose delivered to the patient, including the addition or deletion of a code imprint by embossing, debossing, or engraving on a modified-release solid oral dosage form.[110] Changes that may affect drug product sterility assurance including, where appropriate, process changes for sterile drug substances and sterile packaging components.[111]
Specifications	All changes in specifications from those in the approved application must be submitted in a prior approval supplement (i.e., as a major change), unless otherwise exempted by regulation or guidance.[112] Establishing a new regulatory analytical procedure including designation of an alternative analytical procedure as a regulatory procedure[113] A change in a regulatory analytical procedure that does not provide for the same or increased assurance of the identity, strength, quality, purity, or potency of the material being tested as the regulatory analytical procedure described in the approved application[114]

110. *See id.* at 12.

111. *Id.*

112. 21 C.F.R. § 314.70(b)(2)(i).

113. See Guidance for Industry: Changes to an Approved NDA or ANDA, *supra* note 101, at 17.

114. *Id.*

Table 1
Changing the Label—Major Changes

Category	Exemplary Major Change(s)
Container Closure System	For liquid (e.g., solution, suspension, elixir) and semisolid (e.g., creams, ointments) dosage forms, a change to or in polymeric materials (e.g., plastic, rubber) of primary packaging components, when the composition of the component as changed has never been used in a Center for Drug Evaluation and Research (CDER) approved drug product of the same dosage form and same route of administration[115] A change in the primary packaging components for any drug product when the primary packaging components control the dose delivered to the patient (e.g., the valve or actuator of a metered-dose inhaler)[116]
Labeling	Any proposed changes in the labeling, except changes designated as moderate or minor by regulation or guidance, must be submitted as a prior approval supplement.[117]
Miscellaneous Changes	Changes requiring completion of studies in accordance with 21 C.F.R. part 320 to demonstrate equivalence of the drug product to the drug product as manufactured without the change or to the reference listed drug[118] Changes to a drug product under an application that is subject to a validity assessment because of significant questions regarding the integrity of the data supporting that application[119]

A moderate change is a change that has "a moderate potential to have an adverse effect on the identity, strength, quality, purity, or potency of the drug product as these factors may relate to the safety or effectiveness of the drug product."[120] The first type of moderate change requires the submission of a supplement to the FDA at least 30 days before distribution

115. See *id.* at 20.
116. *Id.*
117. 21 C.F.R. § 314.70(b)(2)(v)(A).
118. *Id.* § 314.70(b)(2)(i*i*).
119. *Id.* § 314.70(b)(2)(2)(vii*i*).
120. *Id.*

of the drug product made using that change. This type of supplement is called and clearly labeled *Supplement—Changes Being Effected in 30 Days*. The drug product made using this type of moderate change cannot be distributed if the FDA informs the applicant within 30 days of receipt of the supplement that a prior approval supplement is required.[121] Furthermore, the supplement must contain certain information required by the FDA for each change.[122] To the extent the FDA requires additional information within 30 days of receipt of the supplement, distribution of the drug product subject to the change(s) must be delayed until the supplement has been amended to include the additional information.[123]

The second type of moderate change requires submission of a supplement called and clearly labeled *Supplement—Changes Being Effected*.[124] Distribution of the drug product made using that change can occur as soon as the FDA receives this type of supplement.[125] Nevertheless, the FDA may still order the manufacturer to cease distribution if, after receiving the supplement, the agency disapproves the change(s).[126] A nonexclusive list of examples of moderate changes, by subcategory, is provided in Table 2.

Table 2

Changing the Label—Moderate Changes

Category	Exemplary Moderate Change(s)
Manufacturing Sites	A move to a different manufacturing site for the primary packaging of (1) any drug product that is not otherwise listed as a major change and (2) modified-release solid oral dosage form drug products (*supplement—changes being effected in 30 days*)[127] A move to a different manufacturing site for the manufacture or processing of the final intermediate (*supplement—changes being effected*)[128]

121. *Id.* § 314.70(c)(5)(i).
122. *Id.* § 314.70(a)(2), (c)(4).
123. *Id.* § 314.70(c)(5)(ii).
124. *See* Guidance for Industry: Changes to an Approved NDA or ANDA, *supra* note 101, at 3–4.
125. 21 C.F.R. § 314.70(c)(6).
126. *Id.* § 314.70(c)(7).
127. See Guidance for Industry: Changes to an Approved NDA or ANDA, *supra* note 101, at 10.
128. *See id.* at 11.

Table 2
Changing the Label—Moderate Changes

Category	Exemplary Moderate Change(s)
Manufacturing Process	For drug products, any change in the process, process parameters, and/or equipment except as otherwise provided by FDA guidance; for drug substances, any change in process and/or process parameters except as otherwise provided by FDA guidance (*supplement—changes being effected in 30 days*)[129] A change in methods or controls that provides increased assurance that the drug substance or drug product will have the characteristics of identity, strength, quality, purity, or potency that it purports or is represented to possess (*supplement—changes being effected*)[130]
Specifications	Relaxing an acceptance criterion or deleting a test to comply with an official compendium that is consistent with FDA statutory and regulatory requirements (*supplement—changes being effected in 30 days*)[131] An addition to a specification that provides increased assurance that the drug substance or drug product will have the characteristics of identity, strength, quality, purity, or potency that it purports or is represented to possess (e.g., adding a new test and associated analytical procedure and acceptance criterion) (*supplement—changes being effected*)[132]

(continued)

129. *See id.* at 14.
130. *See id.* at 15.
131. 21 C.F.R. § 314.70(c)(2)(iii).
132. See Guidance for Industry: Changes to an Approved NDA or ANDA, *supra* note 101, at 18.

Table 2

Changing the Label—Moderate Changes (continued)

Category	Exemplary Moderate Change(s)
Container Closure System	Changes in the size or shape of a container for a sterile drug substance (*supplement—changes being effected in 30 days*)[133] A change in the size and/or shape of a container for a nonsterile drug product, except for solid dosage forms, without a change from one container closure system to another (*supplement—changes being effected*)[134]
Labeling	A changes-being-effected supplement must be submitted for any labeling change that (1) adds or strengthens a contraindication, warning, precaution, or adverse reaction; (2) adds or strengthens a statement about drug abuse, dependence, psychological effect, or overdosage; (3) adds or strengthens an instruction about dosage and administration that is intended to increase the safe use of the drug product; (4) deletes false, misleading, or unsupported indications for use or claims for effectiveness; or (5) normally requires a supplement submission and approval prior to distribution of the drug product that FDA specifically requests be submitted.[135]
Miscellaneous Changes	Reduction of an expiration dating period to provide increased assurance of the identity, strength, quality, purity, or potency of the drug product. Extension of an expiration date that has previously been reduced should be submitted in a changes-being-effected-in-30-days supplement even if the extension is based on data obtained under a protocol approved in the application.[136]

Finally, a minor change is a change that "has minimal potential to have an adverse effect on the identity, strength, quality, purity, or potency of the drug product as these factors may relate to the safety or effectiveness

133. *See id. at 21.*
134. 21 C.F.R. § 314.70(c)(6)(ii).
135. *Id.* § 314.70(c)(6)(iii).
136. *See* Guidance for Industry: Changes to an Approved NDA or ANDA, *supra* note 101, at 27.

of the drug product."[137] An applicant must describe minor changes in its next Annual Report.[138] A nonexclusive list of examples of minor changes, by subcategory, is provided in Table 3.

Table 3

Changing the Label—Minor Changes

Category	Exemplary Minor Change(s)
Components/ Composition	Deletion or reduction of an ingredient intended to affect only the color of the drug product[139]
Manufacturing Sites	A move to a different manufacturing site for secondary packaging, labeling, manufacture or processing of drug substance interediates other than the final intermediate, or for the ink imprinting of solid oral dosage from drug products[140]
Manufacturing Process	A minor change in an existing code imprint for the dosage form. For example, changing from a numeric to alphanumeric code.[141]
Specifications	For drug substance and drug product, the addition or revision of an alternative analytical procedure that provides the same or increased assurance of the identity, strength, quality, purity, or potency of the material being tested as the analytical procedure described in the approved application or deletion of an alternative analytical procedure.[142]
Container Closure System	A change in the container closure system for a nonsterile drug product, based on a showing of equivalency to the approved system under a protocol approved in the application or published in an official compendium.[143] A change in the size and/or shape of a container for a nonsterile solid dosage form.[144]

(continued)

137. *See id.* at 4.
138. 21 C.F.R. § 314.70(d).
139. *Id.* § 314.70(d)(2)(*ii*).
140. *See* Guidance for Industry: Changes to an Approved NDA or ANDA, *supra* note 101, at 11.
141. *See id.* at 15.
142. *See id.* at 19.
143. 21 C.F.R. § 314.70(d)(2)(v).
144. *Id.* § 314.70(d)(2)(*iv*).

Table 3

Changing the Label—Minor Changes (continued)

Category	Exemplary Minor Change(s)
Labeling	Labeling with editorial or similar minor changes or with a change in the information concerning the description of the drug product or information about how the drug is supplied that does not involve a change in the dosage strength or dosage form.[145]
Miscellaneous Changes	An extension of an expiration dating period based on full shelf life data on production batches obtained under a protocol approved in the application.[146]

IV. Prescription Drug User Fee Act

The Prescription Drug User Fee Act (PDUFA) was originally enacted in 1992 and allowed the FDA to collect user fees in connection with the review and maintenance of drug applications. The original PDUFA authorized the collection of fees for five fiscal years (1993–1997), and the act has been effectively extended by amendment four times since. Essentially, there have been five PDUFAs to date:

PDUFA I: Original legislation enacted in 1992, authorizing user fees for 1993–1997

PDUFA II: Reauthorization of user fees for 1998–2002 via the Food and Drug Administration Modernization Act of 1997

PDUFA III: Reauthorization of user fees for 2003–2007 via the Prescription Drug User Fee Amendments of 2002

PDUFA IV: Reauthorization of user fees for 2008–2012 via the Prescription Drug User Fee Amendments of 2007

PDUFA V: Reauthorization of user fees for 2013–2017 via the Prescription Drug User Fee Amendments of 2012

The notion of collecting user fees, as first put into practice by PDUFA I, arose from a perceived backlog at the FDA that resulted in delayed application review and drug approvals, thus resulting in patients having delayed access to new therapies and, concomitantly, losses in potential revenues

145. *Id.* § 314.70(d)(2)(ix), (*x*).
146. *Id.* § 314.70(d)(2)(v*i*).

for drug manufacturers and sponsors, and a growing loss in patent life.[147] Indeed, in the late 1980s, it was reported that the median approval time for a new application had reached as high as 29 months, and that number hovered around 20+ months until the passage of PDUFA in 1992.[148]

Prompted by the concerns associated with the backlog in the FDA and the inability of the FDA to keep up with the increasing number of drug and biologic applications, Congress in 1992 passed PDUFA I, imposing user fees to increase FDA funding for reviewing and approving applications.[149] PDUFA I established three levels of user fees:

- **Application Review Fees**—Due at the time of submission of an NDA or BLA
- **Establishment Fees**—Annual fees submitted by manufacturers for each manufacturing facility
- **Product Fees**—Annual fees submitted for any product on the market

For fiscal year 1993, the first year the PDUFA was in effect, application review fees were approximately $100,000 per application.[150] Under the PDUFA, the FDA determines fees based on a target revenue figure for the year. Once the target revenue is set, it is divided equally among the three categories of fees, and then the anticipated number of applications, establishments, and products, respectively, are divided into the respective revenue segment to derive the applicable fee. Thus, for example, if the target revenue for a given year was $600 million, the FDA would expect to generate $200 million from each level of fees: application review fees, establishment fees, and product fees. If, then, the FDA anticipated having 200 applications in the given year, it would set the individual application fee at $1 million.[151]

147. *See, e.g.*, Philip J. Hilts, *Plan to Speed Approval of Drugs: Makers Would Pay Fees to U.S.*, N.Y. Times, Aug. 11, 1992, at A1.

148. *FY 1995 PDUFA Performance Report*, U.S. Food & Drug Admin., http://www.fda.gov/AboutFDA/ReportsManualsForms/Reports/UserFeeReports/PerformanceReports/PDUFA/ucm117257.htm (last updated Feb. 4, 2011).

149. The passage of PDUFA and its subsequent reauthorizations have not come without controversy. Many have viewed the act as providing an avenue for pharmaceutical companies to exert substantial influence over the FDA, with the FDA treating drug applicants as its "clients" and leading to rushed and inadequate reviews to meet performance metrics. *See, e.g.*, Jerry Avorn, *Paying for Drug Approvals—Who's Using Whom?*, 356 New Eng. J. Med. 1697 (2007).

150. *See* Susan Thaul, Prescription Drug User Fee Act (PDUFA): 2012 Reauthorization as PDUFA V, at 3 n.10 (Feb. 22, 2013), *available at* http://www.fas.org/sgp/crs/misc/R42366.pdf.

151. *See* Guidance for Industry: User Fee Waivers, Reductions, and Refunds for Drug and Biologic Products 2 n.2 (Sept. 2011).

Fees have generally increased year to year since the inception of the PDUFA. For example, in the most recent fee announcement for 2014, the FDA declared that the application fee for applications with clinical data would be $2,169,100.[152] Relatedly, the establishment fee and product fee for 2014 are, respectively, $554,600 and $104,060.[153]

The purpose of the fees collected pursuant to the PDUFA is to fund the "process for the review of human drug applications," a term that is defined by statute and has been expanded from the enactment of PDUFA I through the enactment of PDUFA V. The current definition of "process for the review of human drug applications" is this:

> (A) The activities necessary for the review of human drug applications and supplements.
>
> (B) The issuance of action letters which approve human drug applications or which set forth in detail the specific deficiencies in such applications and, where appropriate, the actions necessary to place such applications in condition for approval.
>
> (C) The inspection of prescription drug establishments and other facilities undertaken as part of the Secretary's review of pending human drug applications and supplements.
>
> (D) Activities necessary for the review of applications for licensure of establishments subject to section 262 of title 42 and for the release of lots of biologics under such section.
>
> (E) Monitoring of research conducted in connection with the review of human drug applications.
>
> (F) Postmarket safety activities with respect to drugs approved under human drug applications or supplements, including the following activities:
>
> (i) Collecting, developing, and reviewing safety information on approved drugs, including adverse event reports.

152. *See* 78 Fed. Reg. 46,981 (Aug. 2, 2013). The fee for applications without clinical data is half this amount.

153. *Id.*

(ii) Developing and using improved adverse-event data-collection systems, including information technology systems.

(iii) Developing and using improved analytical tools to assess potential safety problems, including access to external data bases.

(iv) Implementing and enforcing section 355 (o) of this title (relating to postapproval studies and clinical trials and labeling changes) and section 355 (p) of this title (relating to risk evaluation and mitigation strategies).

(v) Carrying out section 355 (k)(5) of this title (relating to adverse event reports and postmarket safety activities).[154]

This definition, appearing in PDUFA V, is identical to the definition set forth in PDUFA IV. PDUFA IV (2007) amended the prior version to allow the usage of fees to, inter alia, increase monitoring and data collection relating to adverse events.[155] PDUFA III (2002) amended its predecessor to allow the FDA to utilize user fees to support postmarketing surveillance, develop drug-use databases, and conduct postapproval risk management activities.[156] Finally, PDUFA II (1997) expanded this definition to encompass review of INDs as opposed to just NDAs and/or BLAs.[157]

The PDUFA allows for a number of waivers and reductions in applicable fees. Pursuant to 21 U.S.C. § 379h(d)(1):

> The Secretary shall grant to a person who is named as the applicant in a human drug application a waiver from or a reduction of one or more fees assessed to that person under subsection (a) of this section where the Secretary finds that—
>
> (A) such waiver or reduction is necessary to protect the public health,
>
> (B) the assessment of the fee would present a significant barrier to innovation because of limited resources available to such person or other circumstances,

---

154. 21 U.S.C. § 379g.

155. Pub. L. No. 110-85, 110th Cong. (2007).

156. *See, e.g.*, *Background: PDUFA III*, U.S. Food & Drug Admin., http://www.fda.gov/ForIndustry/UserFees/PrescriptionDrugUserFee/ucm173542.htm (last updated Sept. 10, 2009).

157. Pub. L. No. 105-115, 105th Cong. (1997).

(C) the fees to be paid by such person will exceed the anticipated present and future costs incurred by the Secretary in conducting the process for the review of human drug applications for such person, or

(D) the applicant involved is a small business submitting its first human drug application to the Secretary for review.

The FDA provides nonbinding recommendations regarding the assessment of these waivers and reductions, briefly summarized in the following paragraphs.

Public Health Waiver. A waiver or reduction of fees may be appropriate if it can be shown that the product protects public health *and* the waiver or reduction is *necessary* to the continued protection of public health. Because all approved drug products arguably protect public health to some degree, the inquiry for purposes of waiver is necessarily strict and looks at factors such as whether the drug product is a significant improvement relative to other products, whether treatment alternatives exist, whether the drug product is a priority drug or has been granted fast track or new molecular entity status, whether the drug is intended to treat a serious condition, whether the drug product meets an unmet medical need, and whether the product is designated for a rare disease or condition. As to the second piece of the inquiry, the FDA does *not* consider the resources of the applicant in determining whether a waiver is appropriate.[158]

Barrier to Invention Waiver. This waiver or reduction considers the limited resources of the applicant or other circumstances that justify a waiver or reduction so as to not serve as a barrier to further innovation. The FDA recommends ascertaining whether other products under development by the applicant are innovative and whether the fees would be a *significant barrier* to future innovation by the applicant. As to the first inquiry, the FDA will look at, inter alia, whether the product demonstrates breakthrough research and/or forward thinking, whether the product introduces a unique method of diagnosis and/or treatment, whether the drug has been designated a priority drug or has received fast track or new molecular entity status, and whether the applicant had received any federal grants for innovation.[159]

Fees-Exceed-the-Costs Waiver. The applicability of this waiver requires the FDA to determine its anticipated costs for conducting the

158. *See* Guidance for Industry: User Fee Waivers, Reductions, and Refunds for Drug and Biologic Products 5–6 (Sept. 2011).

159. *Id.* at 6–7.

process for reviewing applications for the waiver applicant. The FDA calculates standard costs on a yearly basis. Once this cost is determined, the FDA will look at the applicant's (including affiliates') submissions pending or received on or after September 1, 1992, and assign a standard cost to each qualifying submission. The sum of these standard costs will be the total anticipated cost. If the total fees exceed the calculated cost, a reduction will be applied.[160]

Small Business Waiver. If the applicant is a small business (employing fewer than 500 employees, including employees of affiliates) and the applicant has no approved drug products and the application is the first human drug application submitted by the applicant, it may qualify for a small business waiver.[161]

The PDUFA also provides additional exemptions from fees for orphan drugs. Specifically, a product designated as a drug for a rare disease or condition (i.e., orphan drug) is not subject to application review fees unless the application also seeks approval for another indication that is not a rare disease or condition.[162] Likewise, an orphan drug is not subject to product and establishment fees if the drug meets the public health requirements contained in the act and so long as the applicant's gross worldwide revenue for the preceding year did not exceed $50 million.[163]

The PDUFA was not simply intended to serve as a revenue-generating scheme for the FDA, but also includes specific performance goals and metrics. PDUFA I, for example, set forth specific performance goals for each of the five subsequent fiscal years covered by the act. For fiscal year 1993, the goals were limited to establishing an industry/FDA working group and initiating a pilot computer-assisted PLA (product license application) review program.[164] The goals for fiscal year 1994 were expanded and included, inter alia, reviewing and acting on 55 percent of complete NDA and PLA/ELA (product license application/establishment license application) submissions received in fiscal year 1994, reviewing and acting on 55 percent of efficacy supplements within 12 months of submission, and reviewing and acting on 55 percent of manufacturing supplements and resubmitted applications within six months of submission.[165] The FDA has set forth new goals for each fiscal year with the passage of each iteration of the PDUFA. In the

160. *See* Guidance for Industry: Fees-Exceed-The-Costs Waivers Under the Prescription Drug User Fee Act (June 1999).

161. *See* Guidance for Industry: User Fee Waivers, Reductions, and Refunds for Drug and Biologic Products, *supra* note 158, at 9–11.

162. 21 U.S.C. § 379h(a)(1)(F).

163. *Id.* § 379h(k).

164. *See FY 1995 PDUFA Performance Report*, *supra* note 148, at app. A.

165. *Id.*

most recent performance report (2012), for example, the FDA set forth goals of reviewing and acting on 90 percent of priority original NDAs and BLAs within six months of receipt (ten months for standard original NDAs and BLAs), as well as separate goals and metrics for meeting management and scheduling, dispute resolution, proprietary name review, and development of topic-specific guidance documents, among numerous other goals and metrics.[166] The FDA publishes yearly performance reports to assess whether goals and metrics are being met. For 2012, the FDA had a 96 percent overall on-time performance record for completed reviews.[167] Performance reviews for other goals and metrics, as well as for other years, are made available on the FDA website.[168]

V. Conclusion

As discussed, drug approval in the United States is difficult and expensive, overseen by an ever-growing body of complicated federal regulations. These regulations have traditionally provided some justification from drug companies of contributing to the high cost of health care.

However, since FDA was established more than a century ago, more regulations have been passed because of the public's need for more streamlined routes to drug approval.

Ultimately, these regulations are FDA's attempts to balance competing interests of drug companies and the public, while ensuring that the country's drug supply is safe, secure, reliable, and innovative.

166. *See* U.S. Food & Drug Admin., FY 2012 Performance Report to the President and Congress app. A, *available at* http://www.fda.gov/downloads/AboutFDA/ReportsManuals Forms/Reports/UserFeeReports/PerformanceReports/UCM359106.pdf.

167. *Id.* at 10.

168. *See* http://www.fda.gov/AboutFDA/ReportsManualsForms/Reports/UserFeeReports /PerformanceReports/ucm2007449.htm

chapter 8

Regulatory Exclusivity[1]

I. Regulatory Exclusivity and Requirements

In 1984, Congress added two more provisions for regulatory exclusivity in the Hatch-Waxman Act,[2] providing five years of exclusivity for new chemical entities not previously approved by the FDA[3] and three years of data exclusivity for changes in a previously approved product requiring new clinical trials to win supplemental NDA approval.[4] These two provisions, discussed more fully in the following sections, prevent competitors from filing Abbreviated New Drug Applications (ANDAs) referencing and/or relying on the safety and efficacy data from a New Drug Application (NDA) during the exclusivity period. The new drug product exclusivity period runs from the date of approval of the first approved drug product. In addition to the exclusivity periods related to new chemical entities and new clinical use, there are a variety of other exclusivity periods that protect investments in drug development, including orphan diseases, pediatric uses, enantiomers, and some antibiotic therapies. The following chapter provides an introduction into what drugs qualify for these exclusivity periods, how the exclusivity periods work, and the ways in which issues relating to these exclusivity periods have been challenged by drug developers or competitors.

A. New Chemical Entities

New chemical entity (NCE) exclusivity is governed by 21 U.S.C. § 355(c)(3)(E) and provides that a pharmaceutical manufacturer may be entitled to exclusivity if the FDA has not previously approved a drug manufacturer's "drug

---

1. Matias Ferrario, Esq. and Leslie T. Grab, PhD., Esq. of Kilpatrick Townsend & Stockton LLP, Winston-Salem, North Carolina; Kathryn H. Wade, PhD., Esq. and Alyson L. Wooten, PharmD., Esq. of Kilpatrick Townsend & Stockton LLP, Atlanta, Georgia.

2. Pub. L. No. 98-417, 98 Stat. 1585 (1984).

3. 21 U.S.C. § 355(c)(3)(E)(ii).

4. *Id.* § 355(c)(3)(E)(iii).

substance." [5] The FDA has issued regulations that provide that a "drug substance" is determined by the product's "active moiety," which is defined as a

> molecule or ion, excluding those appended portions of the molecule that cause the drug to be an ester, salt (including a salt with hydrogen or coordination bonds), or other noncovalent derivative (such as a complex, chelate, or clathrate) of the molecule, responsible for the physiological or pharmacological action of the drug substance.[6]

Thus, in determining whether a drug product qualifies for NCE exclusivity, the FDA looks to the drug product's "active moiety."[7] If the drug product is granted NCE exclusivity, the FDA cannot approve an ANDA for a drug product containing the same active moiety.[8] The exclusivity period for an NCE lasts for five years.[9] The exclusivity period runs from the date of approval and is independent of any patent protection that might be available. A drug's intended use is not a factor in determining NCE exclusivity. In other words, an ANDA will not be considered by the FDA if it contains the same active moiety as an NDA even though the active moiety's intended use is different from the intended use of the active moiety subject to NCE exclusivity.[10] Although the NCE exclusivity period is five years, 21 U.S.C. § 505(c)(3)(E)(ii) provides that a 505(b) application may be submitted after the expiration of four years from the date of approval of the application if that application contains a certification of invalidity or noninfringement under clause (iv) of § 505(b)(2)(A).

Practically speaking, the NCE exclusivity grant can be an effective tool in protecting investment in drug development. Because the average approval process for an ANDA application can be over 30 months, NCE exclusivity can often protect investments beyond the four- to five-year exclusivity period since the FDA cannot accept an ANDA until expiration of the four- to five-year NCE exclusivity period.[11] The Orange Book listing,

5. *Id.* § 355(c)(3)(E); *see also* U.S. FOOD & DRUG ADMIN., DRAFT GUIDANCE FOR INDUSTRY, NEW CHEMICAL ENTITY EXCLUSIVITY DETERMINATIONS FOR CERTAIN FIXED-COMBINATION DRUG PRODUCTS (Feb. 2014) [hereinafter NEW CHEMICAL ENTITY EXCLUSIVITY DETERMINATIONS] (discussing recent change in the interpretation of the term "drug" from "drug product" to "drug substance " or "active ingredient"); *see also infra* Section II.B.1. of this chapter.

6. 21 C.F.R. § 314.108(a).

7. *Id.*

8. 21 U.S.C. § 355(c)(3)(E).

9. *Id.*

10. 21 C.F.R. § 314.108(b).

11. 21 U.S.C. § 355(c)(3)(E).

which can be accessed online at the FDA's website, provides information on exclusivity periods.[12]

B. New Clinical Investigation

Another form of exclusivity provided by the Hatch-Waxman Act is New Clinical Investigation (NCI) exclusivity. NCI exclusivity may be awarded if a drug manufacturer sponsored new clinical trials to support an application for approval of a drug product.[13] The NCI exclusivity period is three years.[14] NCI exclusivity may be granted for existing drug products containing active pharmaceutical ingredients that have been previously approved. The requirements for a drug product to obtain NCI exclusivity are (a) the clinical trials must be new; (b) the clinical trials were "essential to the approval" of the application; (c) the applicant sponsored the clinical trials; and (d) the clinical trial must not be a bioavailability study.[15] Thus, even though a drug product has previously been approved by the FDA, an applicant may apply for NCI exclusivity in a supplemental NDA if the requirements for NCI exclusivity are met.

Although NCI exclusivity provides for a three-year exclusivity period, the FDA may receive and begin to review applications during the exclusivity period.[16] This differs from the exclusivity period related to NCEs (during which the FDA cannot accept ANDAs). The NCI exclusivity period is also limited to the specific indication set forth in the application.[17] Thus, others may submit ANDAs on the same active moiety for other indications.

Unlike NCE exclusivity, the FDA can accept an ANDA and start reviewing the application during the clinical investigation exclusivity period. However, the FDA may not approve the application until the exclusivity period is over.[18] The three-year exclusivity period also only applies to the indication(s) approved in the supplemental application. Approval can still be granted for other indications or uses not covered by the NCI exclusivity period.[19]

---

12. *See Orange Book: Approved Drug Products with Therapeutic Equivalence Evaluations*, U.S. Food & Drug Admin., http://www.accessdata.fda.gov/scripts/cder/ob/default.cfm (last visited June 30, 2014).
13. 21 U.S.C. §§ 355(c)(3)(E)(iii), (iv); 21 C.F.R. § 314.108.
14. *Id.*
15. *Id.*
16. *Id.* § 355(c)(3)(E).
17. *Id.*
18. *Id.* §§ 355(c)(3)(E)(iii), (iv).
19. *Id.* § 355(j)(5)(F).

C. Orphan Drug

Congress passed the Orphan Drug Act in 1983 to encourage drug manufacturers to develop drugs for rare but serious diseases.[20] Because of the high cost of drug development, there was a concern that manufacturers would not invest in drugs to treat rare diseases. The Orphan Drug Act defines a rare disease as (1) a disease that affects 200,000 or fewer people in the United States, or (2) a disease affecting more than 200,000 people in the United States but "for which there is no reasonable expectation that the cost of developing and making available in the United States a drug for such disease or condition will be recovered from sales in the United States of such drug."[21]

The Orphan Drug Act provides manufacturers with seven years of exclusivity.[22] The Orphan Drug Act also provides other monetary incentives, such as research grants and tax credits.[23] The Orphan Drug Act limits exclusivity only for the particular indication approved.[24]

Unlike other exclusivity periods discussed earlier, the FDA will not consider a drug the "same drug" if the follow-on applicant can prove that its drug is "clinically superior" to the previously approved drug.[25] A "clinically superior" drug is defined by the FDA as a drug that provides "a significant therapeutic advantage" over previously approved drugs.[26] Such therapeutic advantages can include (1) greater effectiveness, (2) greater safety, or (3) "otherwise [making] a major contribution to patient care."[27]

Note, however, that parties participating in the Orphan Drug exclusivity program must (a) notify the FDA of the discontinuance of any drug approved under the Orphan Drug exclusivity regulation more than one year before such discontinuance and (b) notify the FDA of the discontinuance of preclinical or clinical investigations that are being conducted under 42 U.S.C. § 262 or 21 U.S.C. § 355(i) more than one year before such discontinuance.[28]

D. Pediatric

Pediatric patients represent another patient population for which Congress determined that special incentives were necessary to encourage research into what may be considered a small target market. Two main regulations deal with pediatric trials and related incentives: the Pediatric Research Equity Act (PREA) and the Best Pharmaceuticals for Children

20. *Id.* §§ 360aa–cc.
21. *Id.* § 360bb(a)(2).
22. *Id.* § 360cc.
23. *See, e.g., id.* § 360ee.
24. *Id.* §§ 360aa–dd.
25. 21 C.F.R. § 316.3(b)(14)(i)(ii).
26. *Id.* § 316.3(b)(3).
27. *Id.*
28. 21 U.S.C. § 360(bb)(b).

Act (BPCA). The PREA requires, except when waived or deferred, that pediatric assessments be conducted and submitted for any new drug application where such drug "may provide health benefits in [the pediatric population]."[29] The BPCA codifies this process for obtaining pediatric exclusivity for pediatric studies.[30] Further, the statute provides that

> The Secretary may, after consultation with the sponsor of [certain applications under 21 U.S.C. § 355], issue to the sponsor or holder a written request for the conduct of pediatric studies for such drug. In issuing such request, the Secretary shall take into account adequate representation of children of ethnic and racial minorities. Such request to conduct pediatric studies shall be in writing and shall include a timeframe for such studies and a request to the sponsor or holder to propose pediatric labeling resulting from such studies.[31]

If the sponsor is granted pediatric exclusivity, such exclusivity lasts six months.[32] The six months of exclusivity is added to the expiration of any other exclusivity period.[33]

To qualify for pediatric exclusivity, "a sponsor must file complete reports of the study(ies) prior to the expiration of patent or exclusivity protection they wish to extend."[34] Additionally, the studies for pediatric exclusivity must be completed at least nine months before any other patent protection or exclusivity runs out.[35]

E. Miscellaneous Exclusivity Provisions

The following subsections describe the exclusivity provided for two particular types of products to encourage development of these products.

1. Enantiomers

Prior to congressional enactment of the Food and Drug Administration Amendments Act (FDAAA), the FDA refused to provide NCE status to enantiomers of previously approved racemic mixtures.[36] Now Section 505(u) of the 2007 FDAAA provides that if certain conditions are met, a manufacturer

29. *Id.* § 355a.

30. Pub. L. No. 108-155 (codified as amended in scattered sections of 21 U.S.C. and 42 U.S.C.).

31. 21 U.S.C. § 355a(d)(1).

32. *Id.* § 355a(b)(1).

33. *Id.*

34. U.S. Food & Drug Admin., Guidance for Industry: Qualifying for Pediatric Exclusivity Under Section 505A of the Federal Food, Drug, and Cosmetic Act (rev. Sept. 1999).

35. 21 U.S.C. § 355a(b)(2).

36. Enantiomers are sometimes referred to as optical isomers or stereoisomers. Regardless of the terminology used, these are molecules that are identical in atomic constitution and bonding but only differ with respect to the three-dimensional arrangements of atoms.

of a single enantiomer of a previously approved racemic drug "may . . . elect to have the single enantiomer not be considered the same active ingredient as that contained in the approved racemic drug."[37] In essence, the amendment provides the same market exclusivity to an enantiomer as is enjoyed by drugs conferred new chemical entity status—namely, they are entitled to five years of market exclusivity.[38] For a single enantiomer to qualify under Section 505(u), the single enantiomer must not have been previously approved except in the approved racemic drug.[39] Additionally, the application for the non-racemic drug must include new clinical data (other than bioavailability studies) that were necessary for the approval of the application and may not rely on clinical studies done in connection with the approval of the racemic drug.[40] The single enantiomer will not receive exclusivity if the application seeks approval for "a condition of use in a therapeutic category in which the approved racemic drug has been approved; or for which any other enantiomer of the racemic drug has been approved."[41] Finally, § 355(u)(2) provides that the non-racemic drug will not be approved "for any condition of use in the therapeutic category in which the racemic drug has been approved."[42] In addition, the statute requires that the labeling contain a statement that the non-racemic drug "is not approved . . . for any condition of use of the racemic drug."[43] Therapeutic category is defined as "a therapeutic category identified in the list developed by the United States Pharmacopeia pursuant to section 1395w-104(b)(3)(C)(ii) of Title 42 and as in effect on September 27, 2007."[44]

2. Qualified Infectious Disease Products

Included in Title VIII of the Food and Drug Administration Safety and Innovation Act (FDASIA) is the Generating Antibiotic Incentives Now Act (GAIN Act), which is intended to encourage the development of antibacterial and antifungal drug products that treat pathogens that cause serious and life-threatening infections.[45] The GAIN Act provides an additional five years of market exclusivity on top of other applicable regulatory exclusivity periods, priority review, and fast-track status for antibiotics to treat serious or life-threatening diseases. For a marketing application to be eligible for the extended exclusivity period, it must be for an antibacterial or antifungal drug that is designated a "qualified infectious

37. 21 U.S.C. § 355(u)(1).
38. *Id.*
39. *Id.* § 355(u)(1)(A)(i).
40. *Id.* § 355(u)(1)(A)(ii).
41. *Id.* § 355(u)(1)(B).
42. *Id.* § 355(u)(1)(2).
43. *Id.*
44. *Id.* § 355(u)(2)–(3).
45. FDASIA, tit. VIII, § 801 (creating new section 505E of the FDC Act) (21 U.S.C. § 355f).

disease product."[46] A qualified infectious disease product is defined as "an antibacterial or antifungal drug for human use intended to treat serious or life-threatening infections, including those caused by (1) an antibacterial or antifungal resistant pathogen, including novel or emerging infectious pathogens; or (2) qualifying pathogens listed by the Secretary under subsection (f)."[47] This type of product is one that is intended to treat a "qualifying pathogen," which is one that "has the potential to pose a serious threat to public health," and includes such pathogens as resistant Gram-positive pathogens, multidrug-resistant Gram-negative bacteria, multidrug-resistant tuberculosis, and *Clostridium difficile*. The U.S. Department of Health and Human Services establishes and maintains a list of qualifying pathogens.[48]

II. Challenges to Exclusivity Determinations

As discussed earlier in this chapter, the Hatch-Waxman Act provides incentives outside the patent system for new-drug manufacturers to engage in research and development by providing market exclusivity for new drug products under certain conditions.[49] This regulatory market exclusivity is independent of and runs concurrently with any related patent term exclusivity.

Instances can arise in which the amount of exclusivity to which a particular drug product is entitled may be unclear. For example, a particular drug product arguably may be entitled to either NCE exclusivity or NCI exclusivity, depending on the interpretation of the language of the relevant statutes and regulations. Other instances arise where exclusivity determinations related to orphan drug status impact a manufacturer's rights. The following sections briefly discuss procedures for challenging exclusivity determinations and identify various instances in which parties have challenged the FDA's determination of exclusivity.

A. Procedural Considerations

A comprehensive discussion of the procedural routes for challenging exclusivity determinations by the FDA is beyond the scope of this chapter. A brief introduction to challenging FDA decisions is provided here; see Chapter 22 for a more detailed discussion of citizen petitions.

46. FDASIA § 505E.
47. 21 U.S.C. § 355f(g).
48. FDASIA § 505E.
49. *See* 21 U.S.C. §§ 355(c)(3)(E)(ii)–(iv).

As provided by 21 C.F.R. § 10.25(a), any "interested person" may petition the FDA "to issue, amend, or revoke a regulation or order, or to take or refrain from taking any other form of administrative action."[50] Such petitions are known as citizen petitions. An "interested person" may be, for example, an NDA applicant seeking marketing exclusivity for which a previous application was denied. An interested person also may be, for another example, a generic company challenging the FDA's decision granting exclusivity to an NDA applicant (or challenging the FDA's rejection of an ANDA as being submitted during a granted exclusivity period). The FDA generally has 180 days to issue a final decision on a citizen petition.[51]

If the applicant or other challenger is not satisfied with the FDA's final decision, that party may bring an action in district court under the Administrative Procedure Act (APA). Under the APA, litigants may challenge an agency action by proving that the agency exceeded its statutory jurisdiction, authority, or limitations.[52] District courts review determinations as to whether an agency has exceeded its statutory authority under the standard set forth in *Chevron v. U.S.A. Inc. v. Natural Resources Defense Council, Inc.*[53] The *Chevron* standard provides that a review of an agency decision must first determine whether Congress's intent was clear; if such is the case, "that is the end of the matter; for the court, as well as the agency, must give effect to the unambiguously expressed intent of Congress."[54] If the statute is ambiguous, "the question for the court is whether the agency's answer is based on a permissible construction of the statute."[55] Where the statute includes ambiguities, the ambiguities "[constitute] an implicit delegation from Congress to the agency to fill in the statutory gaps."[56]

B. New Chemical Entity Challenges

As discussed, an NDA holder that disagrees with the FDA's exclusivity determination may challenge that determination by filing a citizen petition providing the action the NDA holder is requesting that the FDA take,

50. *See* U.S. Food & Drug Admin., Guidance for Industry: Citizen Petitions and Petitions for Stay of Action Subject to Section 505(q) of the Federal Food Drug, and Cosmetic Act (June 2011).

51. FDC Act § 505(q)(2)(A). Section 505(q)(2)(A) also provides that an FDA action is final if 180 days pass without the FDA issuing a final decision.

52. 5 U.S.C. § 706(2)(C).

53. Chevron v. USA Inc. v. Natural Res. Def. Council Inc., 467 U.S. 837 (1984).

54. *Id.* at 842–43.

55. *Id.* at 843.

56. U.S. Food & Drug Admin. v. Brown & Williamson Tobacco Corp., 529 U.S. 120, 124 (2000); Vill. of Barrington, Ill. v. Surface Transp. Bd., 636 F.3d 650, 660 (D.C. Cir. 2011) (noting that court will "defer to the agency's permissible interpretation, but only if the agency has offered a reasoned explanation for why it chose that interpretation").

as well as the factual and legal grounds on which the petition has relied in requesting the action.[57]

1. Combination Drug Products

The differences in statutory and regulatory language interpretation are particularly relevant when dealing with, for example, products that are combination drug products (i.e., drug products that include more than one active pharmaceutical ingredient). Recently, three separate citizen petitions were filed with respect to new drug products that are combination drug products. The FDA can only grant NCE exclusivity to products for which the FDA has not previously approved the active moiety.[58] The FDA regulations define the active moiety as the

> molecule or ion, excluding those appended portions of the molecule that cause the drug to be an ester, salt (including a salt with hydrogen or coordination bonds), or other noncovalent derivative (such as a complex, chelate, or clathrate) of the molecule, responsible for the physiological or pharmacological action of the drug substance.[59]

The FDA has historically taken the position that for a combination drug to be entitled to the five-year NCE exclusivity, each of the active drug moieties must not have been previously approved.[60] Therefore, consistent with this position, the FDA has granted only NCI exclusivity to combination drugs such as Stribild® (Gilead Sciences, Inc.), Prepopik® (Ferring Pharmaceuticals, Inc.), and Natazia® (Bayer Healthcare Pharmaceuticals, Inc.). Each of these drug products contains two or more active pharmaceutical ingredients, at least one of which is a novel active pharmaceutical ingredient, while the remaining active pharmaceutical ingredient or ingredients were previously approved by the FDA.[61] Each of the NDA holders for these drug products filed a separate citizen petition challenging the exclusivity determination and

---

57. *See* 21 U.S.C. § 355; 21 C.F.R. §§ 10.20, 10.30. A patent owner that disagrees with a patent term extension determination also may file a citizens petition to challenge that determination. *Id.*

58. *See, e.g.*, 21 U.S.C. §§ 355(c)(3)(E)(ii), (j)(5)(F)(ii); 21 C.F.R. § 314.08(b)(2).

59. 21 C.F.R. § 314.108(a).

60. *See* New Chemical Entity Exclusivity Determinations, *supra* note 5, at 2, ll. 57–62.

61. Stribild® is a fixed-dose combination of four drugs for the treatment of HIV-1 infection. Two of the drugs, elvitegravir and cobicistat, are new active moieties for which the FDA has not previously approved an NDA. *See* Citizen Petition No. FDA-2013-P-0058. Prepopik is a combination of sodium picosulfate, in combination with magnesium oxide and anhydrous citric acid for cleansing the colon in preparation for a colonoscopy. *See* Citizen Petition No. FDA-2013-P-0119. Sodium picosulfate is an active moiety for which the FDA has not previously approved an NDA. Natazia® is a fixed-dose combination of dienogest and estradiol valerate for use as an oral contraceptive, and dienogest is an active moiety for which the FDA has not previously approved an NDA. *See* Citizen Petition No. FDA-2013-P-0471.

requesting that the FDA instead grant NCE exclusivity to their product and that the FDA not accept any generic drug applications for the drug product until five years from the date of the drug product's first approval.[62]

The position taken by the FDA with respect to Stribild®, Prepopik®, and Natazia® regarding the definition of an active moiety is one that it has taken since the early implementation of the Hatch-Waxman Act, prior to the FDA's extensive rule-making proceeding and prior to advances in medicine that have found that fixed-dose combination therapies are essential to many diseases and conditions.[63] The FDA's interpretation with respect to fixed-dose combination products also does not appear to have been the subject of a public formal analysis or litigation.

The petitioners argued that the FDA's current interpretation is contrary to congressional intent and the agency's regulations.[64] The five-year NCE exclusivity for drug products with new active pharmaceutical ingredients not previously approved by the FDA is complemented by the three-year exclusivity granted for new clinical investigations of new drug products that contain previously approved active pharmaceutical ingredients. This difference in the length of exclusivity reflects the fact that the development of drug products containing novel active pharmaceutical ingredients requires significantly more time and expense than does the development of a drug product containing a previously approved active pharmaceutical ingredient. Therefore, the intent of the five-year NCE exclusivity is to allow the innovator to recoup some of the expense of developing the new drug products with a previously unapproved active ingredient. Each of the petitions cited numerous examples of this interpretation.[65] The petitioners also noted that this interpretation forces new-drug manufacturers to first obtain approval of the new active pharmaceutical ingredient (if possible), followed by obtaining approval of the combination drug, in order to gain the benefit of the five-year NCE exclusivity.[66] For at least these reasons, each of the NDA holders urged the FDA to reconsider and change its interpretation to conform to the congressional intent for the NCE exclusivity provisions in the Hatch-Waxman Act, to conform to the current interpretation of FDA regulations, and based on the current state

62. *See* Citizen Petition Nos. FDA-2013-P-0058, FDA-2013-P-0119, and FDA-2013-P-0471.

63. *See, e.g.*, Citizen Petition No. FDA-2013-P-0058, at 3-11; *see also* New Chemical Entity Exclusivity Determinations, *supra* note 5, at 2, ll. 57–62 & 5–6, ll.157–87.

64. *See* Citizen Petition No. FDA-2013-P-0058, at 16–22; Citizen Petition No. FDA-2013-P-0119, at 8–14; Citizen Petition No. FDA-2013-P-0471, at 5–8.

65. *See* Citizen Petition No. FDA-2013-P-0058, at 20–22; Citizen Petition No. FDA-2013-P-0119, at 8–14; Citizen Petition No. FDA-2013-P-0471, at 7.

66. *See* Citizen Petition No. FDA-2013-P-0058, at 23–25; Citizen Petition No. FDA-2013-P-0471, at 8–10.

of the art of medicine. Furthermore, another unrelated company filed a letter in support of the three citizen petitions' request that the FDA change its position with respect to combination drug products and grant a five-year exclusivity for combination drug products that contain a previously unapproved active pharmaceutical ingredient in combination with one or more previously approved active pharmaceutical ingredients.[67]

The FDA issued a draft regulatory guidance in February 2014 to address the issue of exclusivity for combination drug products.[68] The draft guidance explains that the FDA now intends to interpret the word "drug" in the relevant provisions of the Federal Food, Drug, and Cosmetic Act (FDCA) (i.e., the eligibility clause of the five-year NCE exclusivity statutory provision and the definition of an NCE) to mean "drug substance or active ingredient" rather than "drug product" as it had done previously.[69] In changing its interpretation, the FDA will allow a fixed-combination drug product to be entitled to new chemical entity status and the related five-year exclusivity period, provided that the product contains a drug substance that has not previously been approved either alone or as a fixed combination.[70] The FDA notes that fixed-combination products are of increasing importance in the treatment of serious diseases and conditions, and, therefore, the FDA concludes that the position asserted by the petitioners, as described earlier, would be beneficial to the public health.[71] Notably, however, the draft regulatory guidance is proposed to have a prospective rather than a retrospective effect.[72]. Therefore, the previously approved products, such as Stribild, Prepopik, and Natazia, would not be entitled to the new interpretation or the five-year exclusivity period.

In addition to the citizen petitions and the draft guidance issued by the FDA, Congress is seeking to address the same issues relating to the length of the exclusivity period granted to combination drug products. On August 2, 2013, Representative Jason Chaffetz (R-UT) introduced H.R. 2985, a bill that would amend the NCE exclusivity provisions of the FDCA that affect the submission and approval of ANDA and 505(b)(2) applications.[73] The bill would amend the statute to provide the five-year NCE exclusivity to combination drug products approved after January 1, 2014, that contain a previously unapproved active pharmaceutical ingredient

67. *See* Letter from Novartis Pharmaceutical Corporation to the FDA (July 22, 2013).
68. New Chemical Entity Exclusivity Determinations, *supra* note 5.
69. *See id.* at 2, ll. 57–63.
70. *See id.* at 2, ll. 63–69 & 8, ll. 249–56.
71. *See id.* at 7, ll. 242–44.
72. *See id.* at 1, ll. 31–33.
73. Combination Drug Development Incentive Act of 2013, H.R. 2985, 115th Cong. (2013).

in combination with one or more previously approved active pharmaceutical ingredients.[74] If enacted as currently drafted, the bill goes farther than the pending citizen petitions and draft FDA guidance in that it also provides the five-year NCE exclusivity to combination drug products that contain only previously approved active pharmaceutical ingredients, as long as the specific combination of the active pharmaceutical ingredients has not been previously approved.

2. NCE Presumption of Compounds Not Well Characterized

Another instance in which an FDA action has been challenged is the unique situation involving a previously approved drug that has not been well characterized. In this instance, the FDA granted exclusivity under 505(c)(3)(E)(ii) because it could not determine whether the drug for which approval was sought contained the same active moiety as a previously approved drug.[75] This situation arose with respect to Vitrase (a hyaluronidase compound). The FDA first granted the applicant of Vitrase a three-year period of new clinical investigation exclusivity in September 2004.[76] The FDA reversed its decision in October 2004 and, instead, granted Vitrase a five-year exclusivity period as a new chemical entity.[77] In 2005, a citizen petition was filed by ISTA Pharmaceuticals requesting that the FDA reverse its decision to grant Vitrase a five-year exclusivity marketing period.[78]

In denying the petition, the FDA noted that when the FDA grants a drug application, if the new drug product does not contain a previously approved active ingredient (or any ester or salt of the active ingredient), then the new drug product is granted a five-year marketing exclusivity period.[79] Relative to the compound at issue—hyaluronidase—the FDA noted that such compounds have been approved for certain uses for many years.[80] The FDA further noted, however, that hyaluronidase compounds have never been fully characterized with regard to their chemical structure, pharmacologically active enzymes, or impurities.[81] The FDA found that "the amino acid sequence of hyaluronidase molecules varies based both on the species and the tissue from which they are sourced."[82]

74. *Id.* The bill, if enacted, would not affect the exclusivity of the three drug products that are the subject of the current citizen petitions because the language of the bill relates to combination drug products approved after January 1, 2014.

75. *See* Letter to Mr. Marvin J. Garrett Responding to Citizen Petition (Docket No. 2005P-0134/CP1) 8 (Oct. 25, 2005).

76. *Id.* at 4.

77. *Id.*

78. *See* Citizen Petition of ISTA Pharmaceuticals No. 2005P-0134/CP1 (April 6, 2005).

79. *See* Letter to Mr. Marvin J. Garrett, *supra* note 75, at 1.

80. *Id.* at 3–5.

81. *Id.* at 4.

82. *Id.* at 5.

Accordingly, the FDA concluded that "although the Agency can determine whether a naturally sourced hyaluronidase product contains a member of a class of pharmacologically active enzymes . . . the Agency cannot determine the specific enzyme or enzymes contained in [said product]."[83]

The decision also reviewed the requirements of five-year exclusivity, noting that an NCE is a "drug that contains no active moiety that has been approved by the FDA in any other application submitted under section 505(b) of the [A]ct."[84] Because of the unique nature of the naturally sourced hyaluronidase enzyme in the Vitrase drug product, the FDA concluded that it was appropriate to "presume that a product is an NCE if the FDA cannot determine whether the product contains a previously approved active moiety."[85] Here, the FDA summarized its presumptive rule as follows: "[I]f it is not known whether a product contains a previously approved active moiety, the product also would be treated as an NCE for marketing exclusivity purposes, and, accordingly, granted a five-year exclusivity."[86]

3. Derivative Molecules (aka Prodrugs)

The FDA also has encountered numerous exclusivity challenges involving so-called prodrugs. Prodrugs are drugs that are metabolized or converted in the human body into the active drug.

The FDA has addressed whether prodrugs are entitled to NCE marketing exclusivity on a number of occasions. For example, in 2009, the FDA granted a five-year NCE exclusivity for Emend (fosaprepitant dimeglumine injection).[87] Fosaprepitant is a covalent derivative of the prodrug aprepitant. The FDA previously had approved aprepitant in NDA No. 21-549.

In explaining its decision to grant NCE exclusivity to fosaprepitant, the FDA said:

> Fosaprepitant dimeglumine is a pro-drug phosphoramide derivative of aprepitant. A phosphoramide is a compound in which one or more of the OH groups of phosphoric acid have been replaced with covalent bonds to an amino or substituted amino group. Fosaprepitant dimeglumine is neither an ester, salt (including a salt with hydrogen or coordination bonds), nor other noncovalent derivative (such as a complex, chelate, or clathrate) of the molecule aprepitant. It is a covalent

83. *Id.*

84. *Id.* at 7 (citing 21 C.F.R. § 314.108(a)). Note that the FDA has interpreted "active ingredient (including any ester or sale of the active ingredient) to mean "active moiety."

85. *Id.* at 8.

86. *Id.* at 9.

87. Memorandum from Dr. Nasr to Mr. Gary Buehler on Exclusivity Determination for Enmed (fosaprepitant dimeglumine) for Injection (Nov. 4, 2009).

> phosphoramide derivative. As a non-ester covalent derivative of aprepitant, under 21 CFR § 314.108, fosaprepitant dimeglumine is a new chemical entity entitled to 5 years of exclusivity.[88]

Others have challenged the ability of the FDA to grant five-year exclusivity to prodrugs.[89] In *Actavis*, a generic-drug manufacturer sued the FDA challenging the FDA's award of five-year NCE exclusivity to a prodrug. In *Actavis*, the active ingredient at issue was lisdexamfetamine. The court noted that lisdexamfetamine is comprised of the amino acid lysine covalently linked to dextroamphetamine by an amide (nitrogen) bond. Lisdexamfetamine is a prodrug because when it is ingested, the body converts lisdexamfetamine to dextroamphetamine. Although the FDA previously approved dextroamphetamine, the court held that lisdexamfetamine was still considered an NCE because

> [t]he regulatory definition of active moiety[90] excludes only those appended portions of the molecule that cause the drug to be an ester, salt . . . , or other noncovalent derivative. When the drug molecule is not in the form of an ester, salt, or other noncovalent derivative, the FDA treats the entire molecule as that responsible for the physiological or pharmacological action of the drug substance, and therefore a separate active moiety. Any drug that does not contain a previously approved active moiety is entitled to five-year exclusivity. This leaves certain types of prodrugs eligible for five-year exclusivity—namely, those that are not esters, salts, or other types of noncovalent derivatives.[91]

The FDA's interpretation of "active moiety" under 21 C.F.R. § 314.108(a) as a de-esterified molecule is strictly applied. For example, in 2012, the

88. *Id.* at 4.

89. Actavis Elizabeth LLC v. U.S. Food & Drug Admin., 625 F.3d 760 (D.C. Cir. 2010).

90. Note that FDA has interpreted "active ingredient" to mean "active moiety." The *Actavis* court noted that in the context of Patent Term Adjustments under 35 U.S.C. § 156(a), the Federal Circuit has held that the term "active ingredient" was, in fact, broader than FDA's current interpretation. *See* Photocure ASA v. Kappos, 603 F.3d 1372, 1375–1376 (Fed. Cir. 2010) ("[A] compound can only qualify as the 'active ingredient' of a drug if that compound itself is present in the drug." (citing Glaxo Operations UK Ltd. v. Quigg, 894 F.2d 392, 393 (Fed. Cir. 1990))). *But cf.* Abbott Labs. v. Young, 920 F.2d 984 (D.C. Cir. 1990) (refusing to rule on whether a salt derivative (divalproex sodium) of a previously approved drug (valproic acid) was entitled to five-year exclusivity because the phrase "active ingredient (including any ester or salt of the active ingredient)" is ambiguous, as the phrase can refer to either the active ingredient of the originally approved drug or the active ingredient of the drug at issue).

91. *Actavis*, 625 F.3d at 763 (internal quotation and citations omitted).

FDA denied a request by GlaxoSmithKline (GSK) that Veramyst (fluticasone furoate) be granted five-year NCE exclusivity.[92] In this instance, GSK argued that fluticasone furoate (FF) was comprised of fluticasone with a modified ester bonded appendage that impacted FF's pharmacology. The FDA noted, however, that it had previously approved fluticasone propionate (FP), which was comprised of fluticasone with a different ester appendage.[93] The FDA concluded that for purposes of determining the "active moiety," the FDA has "for eighteen years categorically excluded ester appendages, regardless of the in vivo activity of the particular ester at issue."[94] In making its ruling, the FDA found that the active moiety of FP was fluticasone and that the active moiety of FF was fluticasone. Accordingly, the FDA ruled that FF contained the same active moiety as a previously approved drug and, hence, FF was not entitled to five-year NCE exclusivity.

C. New Clinical Data Challenges

Just as in the NCE context, sponsors have challenged the FDA's denial of the three-year exclusivity determinations under 21 U.S.C. §§ 355(j)(5)(F)(iii) and (iv).

For example, in *Upjohn Co. v. Kessler*, Upjohn sought to overturn the FDA's decision not to grant the company a three-year exclusivity under the new clinical data provisions of the Hatch-Waxman Act for its supplemental NDA related to the sale of over-the-counter Rogaine.[95] In 1988, the FDA approved Upjohn's Rogaine (2 percent minoxidil topical solution). In 1991, Upjohn contacted the FDA to initiate dialogue over Upjohn's desire to sell Rogaine over the counter. The FDA indicated that Upjohn would need to submit a supplemental NDA for approval. The FDA further indicated that the FDA was concerned about the "low tolerance for any risk of serious adverse effects . . . , the need to address cardiovascular risks and the need to address risks associated with gross over use."[96]

Upjohn submitted a supplemental NDA on February 9, 1996. In support of its application, the supplemental NDA included results of a separate study that Upjohn had conducted in support of an application to use minoxidil at 5 percent strength for the treatment of hypertension. Although the FDA granted Upjohn's supplemental NDA, it denied Upjohn's request

---

92. Letter from FDA to GSK on Veramyst New Chemical Entity Exclusivity Determination (June 21, 2012); Letter from Dr. Weber to Mr. Zoffer (May 29, 2012).

93. *Id.*

94. *Id.*

95. 938 F. Supp. 439 (W.D. Mich. 1996).

96. *Id.* at 440.

for three-year exclusivity. Concurrently, the FDA approved several other ANDAs for over-the-counter minoxidil topical solutions.

Upjohn sued the FDA claiming that the FDA's actions were, inter alia, arbitrary and capricious, not in accordance with the law, and in excess of the FDA's statutory authority. At issue in this case was whether the new clinical investigations Upjohn submitted with its supplemental NDA were "essential to the approval of the supplement."[97]

Although this was a highly factual inquiry, the court noted that the FDA never required Upjohn to submit the additional clinical study. In reaching its determination, the court noted that despite the FDA's repeated identification of risks in allowing for the sale of over-the-counter Rogaine, there was "no evidence that the FDA indicated that any additional studies [needed] to be done."[98] On the question as to whether there already was sufficient data (exclusive of the new data) that would have supported Upjohn's supplemental NDA—a position the FDA took in the case—the court deferred, noting,

> When examining scientific determinations within the agency's special expertise, the reviewing court must generally be at its most deferential . . . [The agency's] judgments as to what is required to ascertain the safety and efficacy of drugs fall squarely within the ambit of the FDA's expertise and merit deference from [the judiciary].[99]

Recently, AstraZeneca challenged a decision by the FDA not to grant it three-year NCE exclusivity for a supplemental NDA related to Seroquel, an antipsychotic medication used to treat, among other things, schizophrenia.[100] Seroquel had been approved for use since 1997. In 2008, AstraZeneca submitted two supplemental NDAs for new pediatric indications of Seroquel. In support of its supplemental NDAs, AstraZeneca submitted new clinical study data. AstraZeneca also submitted labeling changes, unrelated to the pediatric use of Seroquel, in its supplemental NDAs. These labeling changes were supported by data from other clinical trials that AstraZeneca conducted for reasons unrelated to the pediatric indication of use. Although the FDA granted AstraZeneca's supplemental NDA for pediatric use, it refused to give AstraZeneca three-year new clinical investigation exclusivity.

AstraZeneca sued the FDA, arguing that the FDA's refusal to grant it three-year exclusivity was not valid. The district court granted the FDA's

---

97. *Id.* at 441.
98. *Id.* at 443.
99. *Id.* at 444 (citations and quotations omitted).
100. AstraZeneca Pharm. LP v. U.S. Food & Drug Admin., 713 F.3d 1134 (D.C. Cir. 2013).

summary judgment motion to dismiss the case. On appeal, the U.S. Court of Appeals for the District of Columbia upheld the district court's decision.[101]

In reviewing the underlying grant of summary judgment, the appeals court noted that the central issue in the case related to the statutory language limiting exclusivity to "a change approved in the supplement" and the requirement that the "supplement contains reports of new clinical investigations . . . essential to the approval of the supplement."[102] Again in a highly factual analysis, the appeals court found that AstraZeneca's submission of the non-pediatric clinical data in support of a labeling change was not "a change approved" in a supplement:

> [T]he administrative record shows that the pediatric supplements were approved on their own merits based upon clinical investigations unrelated to the [labeling change data], which standing alone does not entitle AstraZeneca to exclusivity.[103]

Accordingly, the court of appeals upheld the district court's judgment noting that labeling changes supported by new clinical data did not provide a basis for awardings exclusivity in a supplemental NDA for pediatric indications where the labeling changes (and corresponding clinical data) were unrelated to the pediatric indications.

D. Orphan Drug Challenges

The following subsections provide examples in which challenges have been made to the FDA's approval of ANDAs based on an orphan drug product, or to the FDA's failure to take enforcement action with respect to an orphan drug product.

1. Compounding Pharmacies

Even though a drug product obtains orphan drug status, the FDA's actions may deprive sponsors of any such benefits. In *K-V Pharmaceutical Co. v. United States Food & Drug Administration*, the plaintiff sued the FDA for allowing the importation of an active ingredient and for the FDA's alleged decision not to take action against pharmacies for compounding drugs using the imported active ingredient.[104]

K-V obtained orphan drug status for the drug Makena, which is a hydroxyprogesterone caproate injection used to reduce the risk of preterm birth. As noted by the court, doctors had treated patients at risk of preterm

101. *Id.*

102. *Id.* at 1139 (citing 21 U.S.C. § 355(j)(5)(F)(iv)).

103. *Id.* at 1140 (citing AstraZeneca Pharm. LP v. U.S. Food & Drug Admin., 872 F. Supp. 2d 60, 83 (D.D.C. 2012)).

104. 889 F. Supp. 2d 119 (D.D.C. 2012).

birth for a number of years with versions of hydroxyprogesterone caproate that were "compounded" by certain pharmacies.[105] The plaintiff alleged that the FDA issued a series of press releases indicating that it would not take enforcement action or not prioritize enforcement against compounding pharmacies for making the compound drug (referred to as 17P). The plaintiff alleged that the FDA also failed to prevent the active pharmaceutical ingredient used in 17P from entering the United States. In short, the plaintiff alleged that the FDA's lack of action harmed the plaintiff, which was entitled to a seven-year marketing exclusivity based on its orphan drug status.

The court ruled in favor of the FDA, finding that the agency conduct at issue was not subject to review under the Administrative Procedure Act.[106] Here, the court ruled that the FDA was within its discretion to prioritize enforcement of pharmacies that compounded formulations. Moreover, the court ruled that the FDA could allow compounding pharmacies to formulate drugs so long as such compounding did not exceed the scope of traditional pharmacy compounding (i.e., when such compounding "is necessary for the particular patient and would provide a significant difference for the patient as compared to the FDA-approved commercially available drug product").[107] Thus, under this unique factual situation, the FDA could continue its enforcement policy (or lack thereof) of compounding pharmacies. However, this result may change because on appeal to the U.S. Circuit Court for the District of Columbia, the district court order was vacated and the case remanded in an unpublished decision on January 7, 2014.[108]

---

105. As quoted by the court, "Drug compounding is a process by which a pharmacist or doctor combines, mixes, or alters ingredients to create a medication customized to the needs of an individual patient. Compounded drugs generally are not reviewed or approved by FDA. Compounded versions of 17P were not and are not reviewed or approved by FDA; and, in general, their individual formulations, manufacturing processes, labeling, and adverse-event and treatment-failure histories were and are unknown to FDA. The facilities in which the compounding occurred and continues to occur generally were not and are not registered with or routinely inspected by FDA." *Id.* at 124 (quoting Complaint ¶ 9).

106. *Id.* at 132 ("[A]n agency's decision not to take enforcement action is presumed to be immune from judicial review unless Congress has expressed the intent to circumscribe agency enforcement discretion in the statute . . . ." (citing Heckler v. Chaney, 470 U.S. 821, 825 (1985))).

107. *Id.* at 126.

108. K-V Pharmaceutical Co. v. U.S. Food & Drug Administration, No. 1:12-cv-01105 (D.C. Cir. Jan. 7, 2014) (vacating and remanding for reconsideration in view of recent decision in Cook v. FDA, 733 F.3d 1 (D.C. Cir. 2013) and the recently enacted Drug Quality and Security Act ("DQSA"), Pub. L. No. 113-54, 127 Stat. 587 (2013)). Given that it was only recently enacted, a summary of the Drug Quality and Security Act and its effect on the various exclusivity regimes is beyond the scope of this chapter.

2. Intended Use

In other situations, courts have determined that the orphan drug statutes are limited to particular drugs for particular uses and that "off-label" use should not factor into the FDA's decision in approving generics. For example, in *Sigma-Tau Pharmaceuticals Inc. v. Schwetz*, Sigma-Tau received orphan drug status for the use of levocarnitine to treat a rare inborn metabolic disorder.[109] Sigma-Tau later received orphan drug status for levocarnitine to treat a second rare condition—carnitine deficiency in patients with end-stage renal disease (ESRD). After the expiration of Sigma-Tau's exclusivity period covering the use of levocarnitine for inborn metabolic disorders, the FDA approved various generic forms of levocarnitine for such use. Sigma-Tau sued the FDA, arguing that its approval of generic forms of levocarnitine was improper because there was evidence that the generic forms of levocarnitine were being used to treat ESRD (a use for which Sigma-Tau had marketing exclusivity). Sigma-Tau argued that based on the evidence of intended use, the FDA should have denied the generic applications.

The district court granted summary judgment to the FDA. On appeal, the Court of Appeals for the Fourth Circuit upheld the district court's ruling, noting that § 360cc(a) was disease-specific, not drug-specific. As the court made clear, "the statute as written protects uses, not drugs for any and all uses."[110] The court noted that the FDA approved the generic levocarnitine for people with inborn metabolic disorders (a use for which Sigma-Tau's exclusivity had expired) and that the FDA's approval did not contravene § 360cc(a) because the FDA did not approve the generic's application for treatment of ESRD (a use for which Sigma-Tau maintained exclusivity).[111] Moreover, the court held that the FDA did not violate § 360cc(a) "by relying upon the generic manufacturers' proposed labeling as opposed to the alleged evidence of intended use."[112]

3. Clinically Superior

Although the Orphan Drug Act prohibits the FDA from granting approval to another manufacturer of the same drug to treat the same disease, FDA regulations allow the FDA to approve a drug that treats the same condition if that drug is different from the drug for which exclusivity has been granted. FDA regulations provide that when a drug is "clinically superior," that drug will not be considered the same as a previously approved drug.[113] "Clinically

109. 288 F.3d 141 (4th Cir. 2002).
110. *Id.* at 145.
111. *Id.*
112. *Id.*
113. 21 C.F.R. § 316.3(b)(13)(ii).

superior," as defined by FDA regulations, means that a drug provides "[g]reater safety in a substantial portion of the target populations."[114]

In *Berlex Laboratories, Inc. v. Food & Drug Administration*, a court held that the FDA's determination of clinical superiority based on a single side effect was not arbitrary and capricious.[115] In *Berlex*, the court reviewed the FDA's determination that a new drug was different from a previously approved drug because the new drug showed a decrease in the incidence of injection site necrosis. The court noted that FDA regulations provided "greater safety" and elimination of "an ingredient or contaminant that is associated with relatively frequent adverse effects" as examples of clinical superiority.[116] Hence, the court held that the FDA "had adequate basis upon which to consider [the new drug] clinically superior to [the old drug]."[117]

E. Pediatric Exclusivity Challenges

Challenges to the FDA's determination of pediatric exclusivity are less frequent. Nonetheless, there have been instances in which pediatric exclusivity was an issue in determining effective dates or in which pediatric exclusivity period calculations affected other exclusivity periods.

One such instance occurred in *Altana Pharma AG et al. v. Teva Pharmaceuticals USA, Inc.*[118] In *Altana*, the issue was whether a court had the authority to set the effective date for an ANDA after the expiration of the relevant pediatric exclusivity. In *Altana*, the ANDA holder argued that the court did not have statutory authority to set the period during which the relevant ANDA could not be approved to include the pediatric exclusivity period. The court held, after reviewing other relevant decisions not discussed herein, that "[a]lthough the statute governing pediatric exclusivity, 12 U.S.C. § 355a, does not explicitly grant this Court the authority to reset the effective date of [the defendant's] ANDA, several district courts have granted such relief under § 271(e)(4)(A)."[119] Thus, the court set the effective date at which the defendant's ANDA could be approved to be after expiration of the pediatric exclusivity period.

Similarly, in *Hi-Tech Pharmacal Co. v. United States Food & Drug Administration*, the courts addressed whether the 180-day marketing exclusivity period overlapped with the pediatric exclusivity period.[120] In

114. *Id.* § 316.3(b)(3)(ii).
115. 942 F. Supp. 19 (D.D.C. 1996).
116. *Id.* at 23.
117. *Id.* at 19.
118. No. Civ. A. 04-2355 JLL, 2010 WL 3211126 (D.N.J. Aug. 13, 2010).
119. 2010 WL 3211126, at *3 (citing Alcon Inc. v. Teva Pharm. USA Inc., No. 06-234, 2010 WL 3081327, at *1 (D. Del. Aug. 5, 2010).
120. 587 F. Supp. 2d 13 (D.D.C. 2008).

upholding the FDA's decision that such periods did not overlap, the court held that the 180-day marketing period could not commence until the expiration of the pediatric exclusivity period.[121] Accordingly, the tolling statute at issue in the case (21 U.S.C. § 355a(m)) did not toll the 180-day marketing exclusivity period in view of the pediatric exclusivity period because there was no overlap between the two exclusivity periods.

III. Patent Term Extension

In addition to the marketing exclusivity described in the preceding sections, often a drug may be entitled to an extension of the term of a patent that covers the drug. The following sections briefly discuss the background and the process for applying for a patent term extension.

A. Background

The Drug Price Competition and Patent Term Restoration Act of 1984 (the Hatch-Waxman Act)[122] has the dual purposes of providing incentives to patented-drug manufacturers to engage in new-drug development and of making more low-cost, generic drugs available through the process of generic drug approval under the ANDA.[123] With respect to the first stated goal of incentivizing patented-drug manufacturers to engage in new-drug development, the Hatch-Waxman Act provides relief from the administrative delays incurred in the FDA review process of NDAs. One form of relief is the extension of the patent term of a patent covering a patented drug.[124]

B. Application

The patent term extension provision of the Hatch-Waxman Act provides up to five additional years of patent term to compensate for premarket delays that may result in both the loss of exclusivity and potential profits incurred during the lengthy FDA approval process.[125] The rationale behind the extension in patent term is to provide economic incentive for new drug development—the motivation for which is reduced by the large research and development costs associated with the development of new drugs,

121. *Id.* at 22.

122. Pub. L. No. 98-417, 98 Stat. 1585 (1984) (codified at 21 U.S.C. §§ 355, 360cc; 35 U.S.C. §§ 156, 271), *amended by* the Medicare Prescription Drug Improvement and Modernization Act of 2003, Pub. L. No. 108-173, 117 Stat. 2066 (2003).

123. H.R. Rep. No. 98-857, pt. 1, at 14–15 (1984).

124. 35 U.S.C. § 156 (2006).

125. *Id.*

which can reach into the billions.[126] Under the regulatory scheme, new drugs cannot be marketed until the completion of FDA review, a process that may take many years.

The Hatch-Waxman Act provides for patent term extension for a patent claiming products, methods of manufacturing drugs, methods for use of human and veterinary drugs, medical devices, and food additives. The extension only applies to products and methods that are subject to regulatory approval prior to their first marketing or use.[127] Further, the approval must be the first such approval for the use of the patented product or method. For the patent holder to benefit from the patent term extension, the patent covering the product or method must have issued prior to the completion of the FDA approval process and must not have any additional term extensions. Patentees must file an application for patent extension within 60 days after the product or method has been approved. Further, the total patent term, including the sum of the patent term extension and the amount of patent term remaining after regulatory review, cannot exceed 14 years.[128]

Patent term extension is calculated by the FDA. The calculation includes two separate periods of regulatory review: the testing phase and the approval phase.[129] The testing phase corresponds to the Investigational New Drug (IND) application period whereas the approval phase corresponds to the NDA review period.

Several safeguards are designed to prevent patentees from unfairly extending the life of patents covering new drugs. First, only one patent per drug may be extended through the provisions of the Hatch-Waxman Act.[130] This is designed to prevent patented-drug companies from "layering" a series of patents that may cover various aspects of a single product over time to keep the drug "on patent." Second, only one product may be covered by any patent extension—in other words, the extension does not apply to all products protected by the patent, but only to the product on which the extension was based.[131] Finally, the patentee may not draw out

126. A recent Forbes study suggested that the cost of developing a single new drug for market can be as high as $5 billion, which includes research and development costs as well as the cost of other failed drugs. Matthew Herper, *The Cost of Creating a New Drug Now $5 Billion, Pushing Big Pharma to Change*, FORBES (Aug. 11, 2013, 11:10 AM), http://www.forbes.com/sites/matthewherper/2013/08/11/how-the-staggering-cost-of-inventing-new-drugs-is-shaping-the-future-of-medicine/.

127. Proveris Scientific Corp. v. Innovasystems Inc., 536 F.3d 1256 (Fed. Cir. 2008).

128. 35 U.S.C. §§ 156(c)–(d)1 (2006).

129. *Id.* §§ 156(g)(1)(B)(i)–(ii).

130. *Id.* § 156(g)(4).

131. Merck & Co. v. Kessler, 80 F.3d 1543, 1547 (Fed. Cir. 1996).

the extension process through delay—any lack of due diligence by the patented-drug company will further reduce the term of patent extension.[132]

Patented-drug companies have successfully extended the patent term over certain drugs by temporally layering patents over different forms or applications of one drug. The definition of "product" under the Hatch-Waxman Act is a key factor in the ability of patented-drug companies to layer patents as the act provides that patent term extension only applies to the "first permitted commercial marketing or use of the product."[133] The term "product," as broadly interpreted, would include any derivative form of the active pharmaceutical ingredient whereas a narrow interpretation would limit the product to the exact chemical structure found in the marketed drug compound. Therefore, under a broad interpretation of "product," patent term extensions would be limited since the marketing of a particular parent active ingredient would cover all related or derivative compounds. However, under the narrow reading of "product," extensions of the patent term could be obtained for derivative and related compounds.

Historically, there has been inconsistent treatment of the term "product" within the jurisprudence of the Court of Appeals for the Federal Circuit.[134] In *Pfizer Inc. v. Dr. Reddy's Laboratories, Ltd.*, the Federal Circuit construed the term "product" broadly and found it to include both the active ingredient and its derivatives.[135] Similarly, the court also applied a broad definition of "product" in *Fisons PLC v. Quigg*, finding that "product" should not be restricted to a particular structure rather than its underlying active ingredient.[136] However, just one year after *Fisons*, the court changed course and read the term "product" narrowly to exclude any salts, esters, or other non-covalent derivatives of the active ingredient.[137] These inconsistent determinations have caused unpredictable applications of the Hatch-Waxman Act in the context of litigation.

Recently, the Federal Circuit has again considered the definition of "product" within the context of the act and has twice applied a narrow definition.[138] In *Ortho-McNeil Pharmaceutical, Inc. v. Lupin Pharmaceuticals, Inc.*, the court found that an enantiomer of the previously approved and

---

132. Thomas Chen, Note, *Authorized Generics: A Prescription for Hatch-Waxman Reform*, 93 VA. L. REV. 459, 464 (2007).

133. 35 U.S.C. § 156(a)(5)(A).

134. *See* Pfizer Inc. v. Dr. Reddy's Labs. Ltd., 359 F.3d 1361 (Fed. Cir. 2004); Fisons PLC v. Quigg, 876 F.2d 99 (Fed. Cir. 1989); Glaxo Operations UK Ltd. v. Quigg, 894 F.2d 392 (Fed. Cir. 1990).

135. *Pfizer*, 359 F.3d at 1367.

136. *Fisons PLC*, 876 F.2d at 101.

137. *Glaxo Operations*, 894 F.2d at 399–400.

138. Ortho-McNeil Pharm. Inc. v. Lupin Pharm. Inc., 603 F.3d 1377 (Fed. Cir. 2010); PhotoCure ASA v. Kappos, 603 F.3d 1372 (Fed. Cir. 2010).

marketed drug product was a different "product."[139] Likewise, in *Photocure ASA v. Kappos*, the court held that a methyl ester of the original compound was a separate "product" and not included within the original marketed drug.[140] This recent shift to a narrower definition of "product" opens the door to expansion of patent term extensions for patented-drug companies through the layering of patents.[141]

139. *Ortho-McNeil*, 603 F.3d at 1381.

140. *PhotoCure*, 603 F.3d at 1376.

141. Ann Kotze, *Reining in Patent Term Extensions for Related Pharmaceutical Products Post-*PhotoCure *and* Ortho-McNeil, 106 N.W. UNIV. L. REV. 1419 (2012).

chapter 9

Overview of the NDA Review Process[1]

I. Introduction

The U.S. Food and Drug Administration (FDA) fulfills its mission to ensure the safety and effectiveness of pharmaceutical drugs marketed in the United States through its Center for Drug Evaluation and Research (CDER).[2] Before a new drug is approved for marketing, the company seeking to sell the new drug in the United States—the "sponsor"—must conduct clinical studies (i.e., tests) to demonstrate the drug's effectiveness and define its risks.[3] The CDER monitors this clinical research to ensure the safety of those people who volunteer for these studies and to ensure the quality and integrity of the scientific data collected. As detailed in Chapter 1, sponsors seeking to market a new drug will compile their clinical data into a New Drug Application (NDA) to be submitted to the FDA to prove the new drug's safety and effectiveness for its intended use.[4] The CDER will review submitted NDAs through a multidisciplinary team of reviewers that includes physicians, statisticians, chemists, pharmacologists, and other scientists as well as project managers, the designated liaisons

1. James M. Lennon and Adam W. Poff of Young Conaway Stargatt & Taylor LLP, Wilmington, Delaware; and Ryan Cox, Patent Agent/Law Clerk, MacElree Harvey, Ltd., West Chester, Pennsylvania.

2. *See Development & Approval Process (Drugs)*, U.S. Food & Drug Admin., http://www.fda.gov/Drugs/DevelopmentApprovalProcess/default.htm (last updated Oct. 23, 2013).

3. DHHS (Dep't of Health & Human Servs.), CDER (Ctr. for Drug Evaluation & Research) & FDA (U.S. Food & Drug Admin.), 2007 Center for Drug Evaluation and Research Update: Improving Public Health Through Human Drugs (2007), *available at* http://www.fda.gov/downloads/AboutFDA/CentersOffices/CDER/WhatWeDo/UCM121704.pdf.

4. Although the CDER does conduct limited research on drug quality, safety, and effectiveness, in general, the CDER does not test a drug itself. *See CDER: The Consumer Watchdog for Safe and Effective Drugs*, U.S. Food & Drug Admin., http://www.fda.gov/Drugs/ResourcesForYou/Consumers/ucm143462.htm (last updated Aug. 12, 2011).

between the CDER and the sponsor for each application.[5] An NDA is approved once the CDER determines that the new drug is effective for its intended use, its health benefits outweigh its known risks, and usage and risk details are appropriately described on a proposed drug label. As noted in Chapter 1, there are six major steps in the NDA review process: (1) ensuring readiness of the NDA through pre-submission activities; (2) receiving and processing the NDA submission; (3) planning for review of the NDA; (4) conducting the scientific and regulatory review of the NDA; (5) taking official action on the disposition of the NDA; and (6) providing post-action feedback to the NDA applicant/sponsor.[6] This chapter identifies the written materials governing and guiding the FDA's NDA review process and outlines the elements within the primary steps of review.

II. Governance and Guidance Regarding NDA Procedures

An NDA submitted for review and approval in the United States must adhere to requirements established through a web of federal legislation and administrative policies promulgated over several decades. The framework of laws and regulations governing NDA review is addressed in this section.

A. Laws and Regulations Governing NDA Review

Two sets of iterative federal legislation govern the FDA's NDA review process. These two acts, the first mandating that the FDA approve any new drug and the second establishing a fee charge in order to fund the cost for a proper review by the FDA, are addressed briefly here.

1. FDCA—Mandating FDA Approval of New Drugs

In 1938 the U.S. Congress passed the United States Federal Food, Drug, and Cosmetic Act (FDCA),[7] an overarching set of laws giving authority to the FDA to oversee the safety of food, drugs, and cosmetics, including the mandate for premarket approval of all new drugs by requiring sponsors to

---

5. Office of Inspector General (OIG), DHHS, OEI-01-01-00590, FDA's Review Process for New Drug Applications: A Management Review (2003), *available at* http://oig.hhs.gov/oei/reports/oei-01-01-00590.pdf.

6. CDER 21st Century Review Process: Desk Reference Guide 3 (Sept. 2012), *available at* http://www.fda.gov/downloads/AboutFDA/CentersOffices/OfficeofMedicalProductsandTobacco/CDER/ManualofPoliciesProcedures/UCM218757.pdf.

7. The FDCA is codified in chapter 9 of the 21 U.S.C., available online at http://www.gpo.gov/fdsys/pkg/USCODE-2012-title21/html/USCODE-2012-title21-chap9.htm.

prove to the FDA that a drug is safe and effective before it can be sold.[8] Among many other elements, the FDCA defines "drug"[9] and sets the FDA's mission to enforce federal drug laws and regulations established to protect consumer health and safety.[10] The FDCA has been amended several times over the last century and is considered the most extensive law of its kind in the world.[11] In addition to mandating FDA safety and efficacy approval before a new drug may be brought to market, the FDCA attempts to ensure that an approved drug's labeling and packaging are truthful, informative, and not deceptive.[12]

2. PDUFA—User Fees in Exchange for Review Performance Goals

As noted in Chapter 1, the U.S. Congress first passed the Prescription Drug User Fee Act (PDUFA) in 1992 to authorize a five-year period in which the FDA would collect significant user fees from drug sponsors in order to raise the funds necessary to improve the agency's review process for applications, including NDAs.[13] The PDUFA has been reauthorized four times, each in five-year increments, over the last 20 years.[14] The law was spurred by

---

8. *See FDA History—Part II: The 1938 Food, Drug, and Cosmetic Act*, U.S. Food & Drug Admin., http://www.fda.gov/AboutFDA/WhatWeDo/History/Origin/ucm054826.htm (last updated Sept. 24, 2012).

9. *See* 21 U.S.C. § 321(g)(1) ("The term 'drug' means (A) articles recognized in the official United States Pharmacopoeia, official Homoeopathic Pharmacopoeia of the United States, or official National Formulary, or any supplement to any of them; and (B) articles intended for use in the diagnosis, cure, mitigation, treatment, or prevention of disease in man or other animals; and (C) articles (other than food) intended to affect the structure or any function of the body of man or other animals; and (D) articles intended for use as a component of any article specified in clause (A), (B), or (C). A food or dietary supplement for which a claim, subject to sections 343(r)(1)(B) and 343(r)(3) of this title or sections 343(r)(1)(B) and 343(r)(5)(D) of this title, is made in accordance with the requirements of section 343(r) of this title is not a drug solely because the label or the labeling contains such a claim. A food, dietary ingredient, or dietary supplement for which a truthful and not misleading statement is made in accordance with section 343(r)(6) of this title is not a drug under clause (C) solely because the label or the labeling contains such a statement.").

10. *See New Drug Application (NDA)*, U.S. Food & Drug Admin., http://www.fda.gov/Drugs/DevelopmentApprovalProcess/HowDrugsareDevelopedandApproved/ApprovalApplications/NewDrugApplicationNDA/ (last updated Feb. 21, 2013).

11. *Id.*

12. *Id.*

13. *See PDUFA Legislation and Background*, U.S. Food & Drug Admin., http://www.fda.gov/ForIndustry/UserFees/PrescriptionDrugUserFee/ucm144411.htm (last updated Feb. 22, 2013).

14. PDUFA I established the initial five-year period in which the FDA was authorized to assess user fees from drug sponsors for fiscal years (FY) 1993 through 1997. The user-fee provisions were first reauthorized through the Food and Drug Administration Modernization Act of 1997, authorizing the assessment of user fees for FY 1998 through FY 2002 (PDUFA II). The Public Health Security and Bioterrorism Preparedness and Response Act of 2002 reauthorized the collection of user fees for FY 2003 through FY 2007. The Food and Drug Administration Amendments Act of 2007 reauthorized the FDA's collections of

industry complaints over perceived unpredictability, backlogs, and lengthy delays in the FDA's issuance of decisions on approval of a drug sponsor's NDA, which the agency attributed to insufficient funding for staffing and technology to conduct proper reviews. FDA approval delays eroded the market benefits of limited-term patent exclusivity/protection which is otherwise afforded a patented innovation that does not require governmental premarketing approval.[15] Concerns were raised that unnecessary approval delays for NDAs did not impact only U.S. drug sponsors but U.S. patients as well, whose access to new medicines was found to lag behind other countries.[16] As the name of the act implies, in order to improve the FDA's application review process, Congress authorized the agency to impose significant[17] and increasing[18] fees on sponsors submitting certain applications, including NDAs, to the FDA. These "user fees" are assessed in three ways: application fees—assessed upon application submission—and establishment and product fees—assessed annually for FDA-listed drugs and active establishments.[19] In return, the PDUFA required the FDA to develop a faster and more standardized review process and to commit to progressively increasing goals, culminating with review and action upon 90 percent of all priority NDAs (and certain other applications) within six months of submission by 1997. Through the increased fees imposed on applicants, the PDUFA was intended to give the FDA a necessary revenue source to supplement, not replace, Congress's direct appropriations.[20]

---

user fees for FY 2008 through FY 2012 (PDUFA III). The Food and Drug Administration Safety and Innovation Act of 2012 reauthorized user fees for FY 2013 through FY 2017 (PDUFA V). *See id.*

15. David J. Cantor, Cong. Research Serv., 97-838 E, Prescription Drug User Fee Act of 1992: Effects on Bringing New Drugs to Market 2 (1997), *available at* http://www.law.umaryland.edu/marshall/crsreports/crsdocuments/97-838_E.pdf.

16. U.S. Food & Drug Admin., Prescription Drug User Fee Act (PDUFA): Adding Resources and Improving Performance in FDA Review of New Drug Applications (white paper 2005), *available at* http://www.fda.gov/downloads/ForIndustry/UserFees/PrescriptionDrugUserFee/UCM149130.pdf.

17. For example, under PDUFA I, the standard application fee in FY 1993 was $100,000. Susan Thaul, Cong. Research Serv., RL33914, The Prescription Drug User Fee Act (PDUFA): History, Reauthorization in 2007, and Effect on FDA 3 (June 27, 2008), *available at* http://assets.opencrs.com/rpts/RL33914_20070712.pdf.

18. For FY 2014, the standard NDA application fee rose to $2,169,100. *Prescription Drug User Fee Rates for Fiscal Year 2014*, U.S. Food & Drug Admin., http://www.fda.gov/forindustry/userfees/prescriptiondruguserfee/default.htm (last updated Dec. 16, 2013).

19. *Id.*

20. Susan Thaul, Cong. Research Serv., The Prescription Drug User Fee Act (PDUFA): Background and Issues for PDUFA IV Reauthorization (2007).

B. Code of Federal Regulations (C.F.R.)

The FDA is authorized by Congress to pass regulations interpreting the legislation that governs and empowers it.[21] Final regulations are collected in the *Code of Federal Regulations*, Section 21, which contains all FDA regulations pertaining to food and drugs, including all FDA required actions for drug sponsors. Part 314 of Section 21 of the C.F.R. covers "Applications for FDA Approval to Market a New Drug or an Antibiotic Drug."[22]

III. NDA Review Policy, Procedure, and Guidance Documents

There are several levels of guidance provided by the FDA to aid both reviewers within the agency and NDA applicants. These publications, which range from detailed manuals to broad reference guides, are intended to bring consistency to the process. The following are the primary sources of guidance published by the FDA, directed toward the review process for NDAs.

A. CDER's Manual of Policies and Procedures (MaPPs)

The CDER's Manual of Policies and Procedures (MaPPs) is a collection of documents approved and published by the CDER to provide instruction for its internal practices and procedures followed as well as defining external activities.[23] Among its goals, MaPPs helps standardize the new-drug review process and related CDER activities and allows the public to review and better understand CDER policies, definitions, staff responsibilities, and procedures. MaPPs of particular interest to NDA applicants include: (1) 5015-6—Review of the Same Supplemental Change to More Than One NDA or ANDA in More Than One Review Division; (2) 6010.5—NDAs: Filing Review Issues; (3) 6020.8—Action Packages for NDAs and Efficacy Supplements; (4) 6050.1—Refusal to Accept Application for Filing from Applicants in Arrears; (5) 7211.1—Drug Application Approval 501(b) Policy; and (6) 7600.6—Requesting and Accepting Non-Archivable Electronic Records for New Drug Applications.

---

21. *See, e.g.*, 21 U.S.C. § 301 ("[T]he Secretary is authorized hereby to . . . conduct hearings and to promulgate regulations which shall become effective on or after the effective date of this Act as the Secretary shall direct.").

22. 21 C.F.R. pt. 314, *available at* http://www.accessdata.fda.gov/scripts/cdrh/cfdocs/cfcfr/CFRsearch.cfm?CFRPart=314.

23. *See New Drug Application (NDA)*, *supra*, note 10. [New Drug Application (NDA), U.S. Food & Drug Admin., http://www.fda.gov/Drugs/DevelopmentApprovalProcess/HowDrugsareDevelopedandApproved/ApprovalApplications/NewDrugApplicationNDA/.]

B. CDER 21st Century Review Process Desk Reference Guide (DRG)

CDER publishes guidance for the review activities required for an NDA in a document entitled, *The CDER 21st Century Review Process Desk Reference Guide* (DRG).[24] The DRG describes the review activities required for NDAs and the FDA procedures designed to meet the principles and timelines described in the FDA's *Guidance for Review Staff and Industry: Good Review Management Principles and Practices for PDUFA Products* (GRMP), dated April 2005, as well as the process requirements described in the PDUFA V agreement entitled, *PDUFA Reauthorization Performance Goals and Procedures for Fiscal Years 2013 through 2017*.[25] The DRG is primarily intended for use in the review of original applications for CDER approval, including NDAs, but its principles and procedures also apply to the resubmission of applications. The DRG explains the steps in the review process as well as the center's expectations for the conduct of its reviewers. The DRG also sets out the FDA's timeline goals for completion of various review milestones and the roles of the various participants in the NDA submission and review process.

The DRG describes the latest iteration of the FDA process requirements memorialized in the legislative agreements that have governed the FDA's review process for over 20 years, in PDUFA I through V and associated references.[26]

C. Other FDA Published Guidance

The FDA publishes several guidance documents that "represent the Agency's current thinking on a particular subject."[27] FDA guidance documents are intended to guide both FDA review staff and sponsors. Guidelines are provided for processing, content, and evaluation/approval of applications as well as the design, production, manufacturing, and testing of regulated products.[28] FDA policies articulated through these guidelines are intended

---

24. *See* CDER 21st Century Review Process: Desk Reference Guide, *supra* note 6.

25. *Id.* at 1.

26. Many of the FDA's performance goals for application review were not expressly articulated in the PDUFA legislation but, rather, in public letters expressly referenced in PDUFA.

27. *See New Drug Application (NDA)*, *supra* note 10.

28. Published guidance documents for preparing NDAs include guidelines on (1) "Bioavailability and Bioequivalence Studies for Orally Administered Drug Products"—intended for applicants planning to conduct bioavailability (BA) and bioequivalence (BE) studies during the IND period for an NDA, BE studies intended for submission in an ANDA, and BE studies conducted in the postapproval period for certain changes in both NDAs and ANDAs; (2) "Changes to an Approved NDA or ANDA" and Q&A regarding the same; (3) "Container Closure Systems for Packaging Human Drugs and Biologics"; (4) "Format and Content" for specific sections of an application; (5) "Formatting,

to establish consistency in the agency's procedures for application review, facilities inspection, and enforcement. The FDA's published guidelines are not enforceable, either through administrative actions or through the courts, because they are not regulations or laws. The agency acknowledges that "[a]n alternative approach may be used if such approach satisfies the requirements of the applicable statute, regulations, or both."[29] The FDA encourages the public to contact the originating office for further information on any specific guidance document. The agency's website, http://www.fda.gov, hosts an online *Guidance Index*, which provides a complete and searchable listing of CDER guidance documents.[30]

IV. Primary Steps of the NDA Review Process

A. Pre-submission Meeting to Ensure Application Readiness

As noted in Chapter 1, the FDA encourages prospective applicants to take advantage of certain pre-submission options offered by the agency to improve the quality and content of an NDA (or Biologics License Application [BLA]) prior to the application's submission. Applicants are encouraged to request a pre-submission meeting with the appropriate CDER review division to discuss the planned content of the application. In addition to the interactions sponsors will have with the agency during the IND/clinical trial phases, applicants typically engage in a pre-NDA submission meeting to help ensure that an intended application is complete and *fileable*.[31] During

Assembling and Submitting New Drug and Antibiotic Applications"; (6) "Submitting Supporting Documentation" in applications; (7) "Submitting Samples and Analytical Data for Methods Validation"; (8) "Submitting Supporting Documentation in Drug Applications for the Manufacture of Drug Products"; (9) "NDAs: Impurities in Drug Substances," "Format and Content of the Human Pharmacokinetics and Bioavailability Section of an Application," "Format and Content of the Nonclinical Pharmacology/Toxicology Section of an Application"; (10) "Providing Clinical Evidence of Effectiveness for Human Drug and Biological Products," which describes the quantity of evidence and the documentation of the quality of evidence necessary to support a claim of drug effectiveness; (11) "Drug Master Files"—a Drug Master File (DMF) is a submission to the FDA that may be used to provide confidential detailed information about facilities, processes, or articles used in the manufacturing, processing, packaging, and storing of one or more human drugs; (12) "Required Specifications for FDA's IND, NDA, and ANDA Drug Master File Binders"; (13) "Qualifying for Pediatric Exclusivity"—certain applications may be able to obtain an additional six months of patent exclusivity; (14) "PET Drug Applications—Content and Format for NDAs and ANDAs"; and (15) "Refusal to File"—clarifies CDER's decisions to refuse to file an incomplete application. *See New Drug Application (NDA)*, *supra* note 10.

29. *Id.*

30. *Guidance (Drugs)*, U.S. Food & Drug Admin., http://www.fda.gov/Drugs/GuidanceComplianceRegulatoryInformation/Guidances/default.htm (last updated Apr. 16, 2014).

31. *See* CDER 21st Century Review Process: Desk Reference Guide, *supra* note 6, at 9.

this meeting, sponsors can receive feedback on: (1) the format and content of the anticipated application, including labeling, (2) presentation of data, (3) dataset structure, and (4) acceptability of data for submission. Sponsors are also encouraged to provide the agency with a projected submission date of an application. The CDER recommends that the pre-submission meeting occur sufficiently in advance of a planned submission to allow the sponsor to meaningfully address the agency's feedback, typically not less than two months prior to the planned submission.[32]

B. Recommended Contents of Pre-submission Meeting Request

All meeting requests to the FDA are recommended to include certain standard information.[33] To take advantage of the "Pre-NDA" meetings program, FDA regulations direct an applicant to submit certain information for consideration at the pre-submission meeting at least one month in advance of the meeting.[34] CDER guidance specifically recommends that an applicant provide in advance of the NDA pre-submission meeting: (1) a compilation of prior communications related to the subject matter of the anticipated application;[35] (2) a technical summary of the subject application;[36] (3) highlights of any potential problems;[37] (4) proposed draft labeling; (5) a proposed format for organizing the submission;[38] (6) a discussion of the need for a Risk Evaluation and Mitigation Strategies (REMS) submission or postmarketing study/trial for any safety issue(s) that may have emerged during

32. *Id.* at 9.

33. These include (1) product name and application number (if applicable); (2) chemical name and structure; (3) proposed indication(s); (4) the type of meeting being requested (e.g., Type A, Type B, or Type C); (5) a brief statement of the purpose of the meeting; (6) a list of specific objectives/outcomes expected from the meeting; (7) a preliminary proposed agenda; (8) a draft list of questions, grouped by discipline; (9) a list of all individuals (including titles) who will attend the meeting; (10) a list of agency staff requested by the sponsor or applicant to participate in the proposed meeting; (11) the approximate date on which supporting documentation will be sent to the Division; and (12) suggested dates and times (e.g., morning or afternoon) for the meeting. *See* Judit Milstein, Div. of Special Pathogen & Transplant Products, *Meetings with FDA*, U.S. Food & Drug Admin., http://www.fda.gov/downloads/Drugs/DevelopmentApprovalProcess/SmallBusinessAssistance/UCM185881.pdf (last visited May 2, 2014).

34. *See* 21 C.F.R.§312.47(b)(2), *available at* http://www.accessdata.fda.gov/scripts/cdrh/cfdocs/cfcfr/CFRSearch.cfm?fr=312.47 (last updated Apr. 1, 2013).

35. *See* CDER 21st Century Review Process: Desk Reference Guide, *supra* note 6, at 9 ("The information to be provided by the applicant for the meeting includes . . . [a] timeline and list of prior FDA meetings and agreements, prior FDA advice, and any development changes. This would include meeting minutes and letters sent earlier, regulatory history, issues raised at the end-of-phase 2 meeting, etc."). *Id.* at 9.

36. *Id.* ("including the results of the pivotal trials, if they are available, and the proposed dataset structure").

37. *Id.* ("including product quality issues, data integrity concerns, safety signals, etc.").

38. *Id.* ("including methods for presenting the data and a draft index if available").

development or foreign marketing;[39] and (7) any "other information/issues" that the applicant believes may "require discussion" and the applicant's "[l]ist of questions for FDA."[40]

C. Pre-submission Meeting Participants

Every NDA review, including pre-review activity, is overseen by a CDER Regulatory Project Manager (RPM). The RPM serves as the regulatory leader of the CDER's application review team, manages the review process with a Cross-Discipline Team Leader from the CDER, and coordinates all internal and external communications regarding the application review. The RPM also maintains all CDER documentation associated with the review, conducts an initial Patient Labeling Review of the label format regarding prescribing information, and serves as the primary contact for review-related regulations and policies.[41]

In response to a sponsor's request for a pre-submission meeting, the RPM will evaluate the request, including the submitted package of background information, and schedule the meeting with pertinent reviewers who will also receive and review the background information.[42]

D. Conducting the Pre-submission Meeting

At least 48 hours prior to the pre-submission meeting, the applicant should receive from the RPM the reviewers' comments and draft responses to the applicant's submitted questions.[43]

An applicant should be prepared to negotiate and reach agreement at the pre-submission meeting regarding the required content for a complete application and to engage in preliminary discussion regarding the need for REMS or other risk management actions. During the meeting, the CDER reviewers will provide the applicant with guidance on how the data should

39. Where appropriate, the FDA is authorized to require as part of the NDA review process a sponsor's proposal—and the FDA's approval—of a Risk Evaluation and Mitigation Strategy (REMS) to ensure that the benefits of a drug outweigh its risks. *See* 21 U.S.C. § 355-1. An REMS is a safety strategy to manage a known or potential serious risk associated with a drug and to enable patients to have continued access to the drug by managing its safe use. *See FDA Basics Webinar: A Brief Overview of Risk Evaluation and Mitigation Strategies (REMS)*, U.S. Food & Drug Admin., http://www.fda.gov/aboutfda/transparency/basics/ucm325201.htm (last updated May 2, 2014).

40. *See* CDER 21st Century Review Process: Desk Reference Guide, *supra* note 6, at 9.

41. *Id.* at 3, 5.

42. *Id.* at 9.

43. *Id.* at 10 ("Reviewers should review the meeting package and also be prepared to discuss internally the potential need for an Advisory Committee meeting, REMS, and PMRs. The reviewers draft responses to the pertinent questions submitted by the applicant. The project manager assembles reviewers' comments and sends them to the applicant at least 48 hours prior to the meeting.").

be presented in the application to facilitate review. The FDA will request at the pre-submission meeting that the applicant submit information on all manufacturing facilities and/or "preliminary data on pivotal clinical trials" that would be useful in early determination of sites for inspection.[44] The meeting will also likely include discussion of the data that should be submitted to support the labeling sought by the applicant and of the way that labeling annotations should be linked to the primary supporting information in the Electronic Common Technical Document (eCTD), CDER's standard format for electronic regulatory submissions.[45] All agreements and discussions from the pre-submission meeting will be summarized at the conclusion of the meeting and reflected in meeting minutes prepared and distributed by the RPM.[46]

While major components[47] of the application must be submitted with the original application, under PDUFA V changes, an agreement may be reached at the pre-submission meeting to submit a limited number of application components, for certain applications within the "Program,"[48] not later than 30 days after the submission of the original application.[49] The FDA considers delayable components to "be of a type that would not be expected to materially impact the ability of the review team to begin its review."[50] Applicants may request at the pre-submission meeting that an application receive fast-track review and, in some instances, may submit certain completed sections of the application for FDA review before complete submission of the NDA ("rolling review").[51]

If questions remain regarding technical aspects of the anticipated electronic[52] submission of the NDA, after the pre-submission meeting, an

---

44. *Id.*

45. *Id.*

46. *Id.*

47. For example, the complete study report of a Phase 3 clinical trial or the full study report of required long-term safety data. *Id.*

48. The "Program" refers to an FDA review program under PDUFA V for the review of all new molecular entity new drug applications (NME NDAs), as well as BLAs, received between October 1, 2012, and September 30, 2017. *See Program for Enhanced Review Transparency and Communication for NME NDAs and Original BLAs*, U.S. Food & Drug Admin., http://www.fda.gov/ForIndustry/UserFees/PrescriptionDrugUserFee/ucm327030.htm (last updated Nov. 8, 2012).

49. *See* CDER 21st Century Review Process: Desk Reference Guide, *supra* note 6, at 10.

50. *Id.* "Examples of application components that may be appropriate for delayed submission include updated stability data (e.g., 15-month data to update 12-month data submitted with the original submission) or the final audited report of a preclinical study (e.g., carcinogenicity) where the final draft report is submitted with the original application." *Id.*

51. *See Fast Track, Breakthrough Therapy, Accelerated Approval and Priority Review*, U.S. Food & Drug Admin., http://www.fda.gov/forconsumers/byaudience/forpatientadvocates/speedingaccesstoimportantnewtherapies/ucm128291.htm (last updated June 26, 2013).

52. "For new applications, all electronic submissions must be in eCTD format." *See* Milstein, *supra* note 33.

"electronic pre-submission" meeting may be held at the discretion of the review division, within 30 to 60 days of the intended submission. In some instances, the FDA recommends this meeting.[53] The primary focus of the NDA electronic pre-submission meeting is on electronic site navigation, formatting of electronic files, and the layout of the electronic application.[54]

V. Receiving and Processing the NDA Submission

NDA applications are received and processed by the CDER's document control room staff and then distributed to the review division. The RPM will conduct an initial assessment of the NDA submission to ensure that certain regulatory requirements are met and that the appropriate user fee has been paid, if not waived or exempted. At this time, the CDER will make reviewer assignments for the application.[55] The RPM must send an acknowledgment of receipt letter to the applicant by day 14 after NDA submission.[56]

A. Receipt Marks Beginning of Review

The review timeline governing the NDA review/approval process, pursuant to the requirements of the PDUFA, begins when the FDA receives the application, except for certain extensions within the Program under PDUFA V. Those components of an application that the FDA agreed, pursuant to the Program, could be submitted after the original application must be received not later than 30 calendar days after receipt of the original application. The FDA may refuse to file the application if these components are submitted within the extension period.[57]

B. Determination of Regulatory Compliance

Upon submission of a new NDA, the RPM determines whether the user fees have been paid, waived, or exempted. If the application fee has not been paid within five days of application submission, or if the applicant is on the arrears list, the RPM will send the applicant a letter indicating the submission is "unacceptable for filing," and the review process will be halted until the fees are received.[58] Once the delinquent fees are received, the RPM will send an acknowledgment to the applicant, indicating the new "receipt date," which resets the PDUFA review clock. The RPM will

53. *See* CDER 21st Century Review Process: Desk Reference Guide, *supra* note 6, at 10.
54. *Id.*
55. *See id.* at 13.
56. *Id.* at 14.
57. *Id.* at 13.
58. *Id.*

also notify the review team of the change in the PDUFA date. The RPM will then assess and determine whether the application has been correctly coded in the CDER document room. The RPM will ensure that the application is administratively complete and compliant with content and format regulations and other regulatory requirements.[59]

C. Assessing and Assigning Reviewers

The RPM is responsible for distributing review copies of a submitted application to designated discipline team leaders, so that they may make reviewer assignments for the application. Discipline team leaders will perform a quick review of the contents of an application to determine whether a reviewer assignment is necessary for an original application or efficacy supplement submission and whether the submission contains any obvious issues requiring special attention by the reviewers. Those identified issues are conveyed to the assigned reviewer along with the pertinent application section. The discipline team leader will notify the RPM of the reviewer assignment by day 14 of the submission.[60]

VI. Filing Determination and Review Planning Phase

Once a review team has been assigned to a submitted application, the team will conduct an initial assessment of the NDA and make a recommendation on the *fileability* of the application. The decision on fileability is communicated to the applicant at the filing meeting, to be held within 45 days after submission (or 30 days for priority reviews).[61] Once an application is found to be *fileable*, the FDA will notify the applicant in writing. The application will be deemed "filed" on the date that is 60 days after the FDA received the application. This filing date begins the 180-day period, as specified under the PDUFA legislation,[62] within which the FDA must complete its review of a standard NDA. This 180-day period is called the "filing clock."[63] At the conclusion of the 180-day period, the FDA must send the applicant either an approval letter or a complete response letter outlining the reasons for denying approval.[64]

59. *Id.* at 13–14.
60. *Id.* at 14.
61. *See id.* at 13.
62. 21 U.S.C. § 355(c)(1).
63. *See* 21 C.F.R. § 314.101(a)(1), *available at* http://www.accessdata.fda.gov/scripts/cdrh/cfdocs/cfcfr/CFRSearch.cfm?fr=314.101.
64. *See id.* § 314.101(a)(2).

Upon the determination that an application is "filed," a planning meeting will be held to further address review timelines and activities.

A. Fileability Determination and Start of "Filing Clock"

The FDA must assess the fileability of the submitted application within 60 days of its receipt. A determination that the application has been filed means that the FDA has made a threshold determination that the application is sufficiently complete to permit a substantive review.[65] The FDA also makes a determination on whether the application will be given "priority review" status in conjunction with the fileability determination. A tentative decision on priority should occur by day 14 after submission. The various dates and deadlines for review are dictated by whether the application is given priority status. For example, a filing meeting with the applicant, to communicate the filing determination, must occur by day 45 after a normal NDA submission, but by day 30 for a priority application.[66] A final application priority designation must be communicated to the applicant by day 60.[67]

Fileability is assessed by the RPM and assigned reviewers. The RPM initially determines whether the various sections of the application meet regulatory requirements,[68] including the provision of information concerning the applicant's patents covering the drug or its method of use.[69] Notably, the FDA also imposes upon applicants a duty to amend pending applications to update information concerning later-acquired patents: "If a patent is issued after the application is filed with FDA but before the application is approved, the applicant shall, within 30 days of the date of issuance of the patent, submit the required patent information in an amendment to the application under [21 C.F.R.] 314.60."[70]

As a further element of filing review, each discipline reviewer will assess the application's adherence to technical requirements by comparing the assigned section to a specific discipline filing review template. A reviewer's template-based filing assessment is reviewed by the appropriate discipline team leader and provided to the RPM, preferably before the filing meeting. An application's substantive deficiencies identified during

---

65. *Id.*

66. *See* CDER 21st Century Review Process: Desk Reference Guide, *supra* note 6, at 15.

67. U.S. Food & Drug Admin., Guidance for Review Staff and Industry: Good Review Management Principles and Practices for PDUFA Products 15 (Apr. 2005) [hereinafter Good Review Management for PDUFA], *available at* http://www.fda.gov/downloads/Drugs/.../Guidances/ucm079748.pdf.

68. *See* CDER 21st Century Review Process: Desk Reference Guide, *supra* note 6, at 17; *see also* 21 C.F.R. § 314.50, *available at* http://www.accessdata.fda.gov/scripts/cdrh/cfdocs/cfcfr/CFRSearch.cfm?fr=314.50.

69. 21 C.F.R. § 314.53.

70. *Id.* § 314.53(d)(1).

filing review are noted and communicated in the filing communication, but these issues are considered separately from application filing deficiencies that may serve as the basis for a "refusal-to-file" determination by the FDA.[71] Issues that may result in a refusal-to-file determination by the FDA should be conveyed to an applicant before the filing meeting "so that the review team has adequate time to work with the applicant to try and resolve the deficiencies before the 60-day filing date."[72]

B. Planning a Review of the Filed Application

The RPM and assigned reviewers will also conduct activities to plan for the review of a filed application. This includes holding a review planning meeting with the applicant. Review planning is intended to organize the associated review tasks, minimize review overlap among review disciplines, and establish an internal review timeline.[73] Like the filing meeting, the review planning meeting must be held by day 45 after submission for standard applications. The filing and review planning meetings may be combined into one, but the CDER cautions that sufficient time should be allotted to discuss review planning activities. If meetings are not combined, the CDER still encourages discussion of review planning activities at the filing meeting to take advantage of the review team's collective input and build consensus toward a review plan. Adherence to an established review plan is important to efficiently and effectively review the application to avoid resource-intensive issue resolution that might otherwise occur at the end of the review cycle.[74]

At the conclusion of the review planning meeting, major review milestone timelines are communicated to the applicant. The CDER calls on applicants to be "aware that these timelines are flexible and subject to change based on division workload and other potential review issues (e.g., the need for submission of amendments)."[75] Among other tasks to be addressed at the review planning meeting, participants should: (1) determine the need for consultant reviewers; (2) establish a plan for labeling review; (3) determine the need for an advisory committee for the review; (4) finalize sites for on-site inspection of good laboratory and clinical practices; (5) determine new manufacturing sites for inspections; (6) determine a schedule/timeline for the review; and (7) start the formal Action Package of documentation of the review to be kept by the RPM.[76]

---

71. *See* CDER 21st Century Review Process: Desk Reference Guide , *supra* note 8, at 17.
72. *Id.* at 18.
73. *See id.* at 15.
74. *Id.*
75. *Id.*
76. *See* CDER 21st Century Review Process Desk Reference Guide, *supra* note 6, at 20–24.

VII. Conducting the Scientific and Regulatory Review of the NDA

During the scientific and regulatory review phase of the application process, the submitted application is reviewed in accordance with the established review plan by the review team, including members from the relevant disciplines.[77] The RPM maintains responsibility for the overall review process, including making schedule adjustments, communicating with the applicant, and ensuring regulatory compliance.[78] Once the application reaches the review team, primary reviewers from each discipline assess the contents of the application for regulatory compliance and review the scientific data submitted to identify any deficiencies affecting approvability.[79] During this review process, reviewers may consult with subject matter experts and outside consultants on the review team who provide specific input based on their areas of expertise.[80] As noted, primary reviewers report to the discipline team leader who provides management and guidance for the review team as well as a secondary review of the primary reviewer's findings.[81] A cross-disciplinary team leader may also be involved in the management and review of clinical data from the application.[82] Discipline directors and the Office of New Drugs (OND) division director oversee the process; the OND division director is also responsible for writing the summary review with a recommended action on the application.[83] Finally, the signatory authority, usually an Office of Drug Evaluation (ODE) director, is responsible for conducting a third review of the application and for taking final action.[84]

A. The Review Process

During the primary review process, primary reviewers on the core review team assess the application with respect to their specific discipline areas. Typically, the core review team includes the following disciplines; however, other disciplines may be represented depending on the nature of the application.[85]

77. *See id.* at 25.
78. *Id.*
79. *Id.* at 25.
80. *Id.* at 4.
81. *Id.*
82. *Id.*
83. *Id.*
84. *Id.*
85. *Id.* at 3.

- Medical/Clinical
- Pharmacology/Toxicology
- Product Quality
- Biometrics
- Clinical Pharmacology and Biopharmaceutics
- Clinical Microbiology
- Medication Error
- Risk Management (for New Molecular Entity [NME] or REMS submissions)[86]
- Office of Scientific Investigations
- Office of Compliance's Office of Manufacturing Product Quality

The primary review is focused on validating that the proposed drug or biologic is safe and effective for use.[87] The reviewer is tasked with identifying any approvability issues in the application with respect to the reviewer's specific discipline.[88] During this process primary reviewers may consult with other primary reviewers on the application and report to their discipline team leader, who is responsible for managing the operations within each specific discipline.[89] The primary reviewer and the discipline team leader discuss the application and any issues that arise as the process continues.[90] The core review team also meets regularly during the review to discuss issues and findings and to share information in a cross-disciplinary format.[91] The RPM is responsible for scheduling the meetings, organizing the agenda, and documenting action items arising from the meetings.[92]

In addition to managing review team meetings, the RPM is responsible for communicating with the applicant,[93] setting up other relevant meetings, and communicating with management regarding the application.[94] Communications with the applicant may take the form of information

86. In cases where a product (1) is an NME; (2) is submitted with an REMS; (3) is subject to an existing REMS, but the application is for a new indication; or (4) is part of a class of drugs that already has a REMS, a risk management analyst is assigned to the core review team as a primary reviewer. *Id.*

87. *See Drug Development and Review Definitions*, U.S. Food & Drug Admin., http://www.fda.gov/drugs/developmentapprovalprocess/howdrugsaredevelopedandapproved/approvalapplications/investigationalnewdrugindapplication/ucm176522.htm (last updated Feb. 22, 2010).

88. CDER 21st Century Review Process: Desk Reference Guide, *supra* note 6, at 3, 25–26.

89. *Id.* at 4, 25–26.

90. *Id.* at 26.

91. *Id.* at 26–27.

92. *Id.*

93. 21 C.F.R. § 314.102. The FDA is required to communicate with an applicant during the application review process concerning the review, specifically to identify and provide notice of "easily correctable deficiencies."

94. CDER 21st Century Review Process: Desk Reference Guide, *supra* note 6, at 25.

requests, discipline review letters, meetings, or telephone and e-mail communications.[95] Information requests are sent to the applicant to request additional information or clarification to aid in the review process.[96] Discipline review letters may also be used to communicate to the applicant the results of specific discipline reviews and to identify issues and deficiencies.[97] These processes provide the applicant with notice of potential issues and deficiencies, enabling the applicant to proactively remedy these issues during the review process.

1. The Mid-cycle Meeting

As the review process continues, a mid-cycle meeting is scheduled by the RPM around month 5 for standard reviews and month 3 for priority reviews.[98] The meeting includes all reviewers and team leaders as well as the review division director and deputy director in addition to the deputy director for safety, the safety RPM, and the ODE director if designated as a signatory authority.[99] Other directors and relevant personnel are invited as needed.[100]

The purpose of the mid-cycle meeting is to permit each discipline to present findings and identify issues, give updates from consultants, assess the results of bioresearch audits, review facility inspections, consider labeling, and discuss the need for an Advisory Committee meeting as well as review safety activities and other important issues.[101] The meeting enables the review team and RPM to identify issues affecting approvability and to establish a continuing plan for the review.[102] The meeting also allows for cross-disciplinary interaction and allows for feedback from reviewing authorities including the signatory authority for the application.[103]

Under PDUFA V, the RPM and appropriate team members call the applicant within two weeks of the mid-cycle meeting to inform the applicant of the outcomes of the meeting and address any issues.[104] At this time, the review team can address with the applicant findings and issues as well as new information requests, can update timing for the late-cycle or wrap-up meeting, and can update the schedule, including any Advisory Committee

95. *Id.* at 25–26.
96. *Id.*
97. *Id.* at 36.
98. *Id.* at 27–28.
99. *Id.* at 28.
100. *Id.*
101. *Id.*
102. *Id.*
103. *Id.*
104. *Id.* at 29.

meeting, if necessary.[105] However, for standard applications, it is still recommended that the review team communicate the results of the mid-cycle meeting to the applicant and request necessary information or clarification.[106]

2. Continuing Review after Mid-cycle Meeting

After the mid-cycle meeting, primary reviewers continue the process of reviewing the applicant's submission to determine approvability.[107] In addition, the RPM and review team continue to work with the applicant on labeling.[108] A labeling meeting is held, typically within one week of the mid-cycle meeting, to allow reviewers to discuss and propose labeling, after which the applicant and review team continue to develop the labeling until the final labeling is determined.[109] In addition, postmarket requirements or postmarket commitments as well as REMS are handled in a similar fashion, including meetings of the review team and input from the applicant.[110]

During this time, the review team may also review and incorporate amendments from the applicant as well as seek expert advice to assist in the review of the application.[111] These activities may occur at any time during the review process but may alter the timeline of the review if substantial changes occur; for PDUFA V program applications, this can result in a new goal date and may affect the mid-cycle meeting date if it has not already occurred.[112] The review team may also schedule inspections and compliance reviews to ensure that the applicant adheres to the good clinical practice (GCP), good laboratory practice (GLP), and good manufacturing practice (GMP) requirements.[113]

As the primary reviewers complete their reviews, the reviews are documented in writing and forwarded to the respective discipline team leader for review.[114] The discipline team leader then conducts a secondary review and notes, in writing, any disagreement with the assessments of

105. *Id.* PDUFA V requires that a late-cycle meeting be held between the review team and the applicant no less than 12 days prior to any advisory committee meeting or three months prior to the PDUFA V goal date for standard PDUFA applications. *Id.* at 37.

106. *Id.* at 30.

107. *Id.* at 47.

108. *Id.* at 34–35.

109. *Id.* at 35.

110. *Id.*

111. *Id.* at 30.

112. *Id.*

113. *Id.* at 36. Under PDUFA V, this includes completion of the inspections and compliance reviews within ten months for standard applications and six months for priority applications.

114. *Id.* at 35.

the primary reviewer.[115] Once these reviews are complete, the RPM communicates the findings to the applicant through discipline review letters to provide notice of deficiencies and update the status of the application.[116] The primary and secondary reviews are generally complete by month 8 to month 9 depending on the type of signatory and whether an application is a PDUFA V Program application.[117]

3. Late-Cycle, Wrap-Up, and Advisory Committee Meetings

For PDUFA V program applications, a pre-meeting is held to plan the required late-cycle meeting, allowing the RPM to develop an agenda and identify issues for the late-cycle meeting.[118] The late-cycle meeting must be held no later than three months or two months from the PDUFA V goal date for standard and priority applications, respectively.[119] The meeting must also occur no less than 12 days prior to any Advisory Committee meeting.[120] Prior to the meeting, the RPM sends the applicant an agency briefing package providing an agenda, a list of attendees, an REMS assessment if applicable, and a memorandum from the review team to summarize issues and discipline review letters.[121]

At the late-cycle meeting, major deficiencies are identified and additional application issues are discussed. This may include discussion of safety issues and labeling issues as well as the option for the applicant to provide additional data to remedy deficiencies.[122]

A wrap-up meeting is also conducted among the review team to address remaining issues and to enable planning for the regulatory review of the application and a decision on final action.[123] During the wrap-up meeting, the review team and other personnel from the FDA meet internally to discuss the application and review findings.[124] In addition, an Advisory Committee meeting is held for all new molecular entities (NMEs), for original BLAs, and for other applications for which the review division determines external input would aid in considering the application.[125] Issues that may trigger an Advisory Committee meeting include the use of novel clinical or

---

115. *Id.* at 36. In the case of pharmacology and toxicology, the application is subject to a tertiary review with respect to carcinogenicity by the Executive Carcinogenicity Assessment Committee.

116. *Id.*

117. *Id.* at 47.

118. *Id.* at 36–37.

119. *Id.* at 37.

120. *Id.*

121. *Id.*

122. *Id.*

123. *Id.* at 39.

124. *Id.*

125. *Id.* at 22.

surrogate endpoints in the clinical trial design, significant issues regarding safety or efficacy, and public health questions.[126] During the Advisory Committee meeting, experts provide input on issues in the application to better assist in the review process.[127]

After the review and meetings, the cross-discipline team leader reviews any remaining issues and may meet with review team members to discuss the application and consolidate information for presentation to the signatory authority.[128] Labeling, postmarket commitments and requirements, and REMS are also finalized around the time of the wrap-up meeting, and the RPM completes the regulatory review.[129] Once the review is complete and consolidated by the cross-discipline team leader, it is ready for review by the division director or office director that is the appropriate signatory authority.[130]

VIII. Taking Official Action on the Disposition of the NDA

Once the cross-discipline team leader has completed the review of the application and prepared the application and review materials for the signatory authority, the cross-discipline team leader and the division director, if ODE director sign-off is required, brief the signatory authority on the application and any pertinent issues.[131] The cross-discipline team leader also delivers the action package, which includes all of the review files and letters, and the signatory authority conducts a review of the materials.[132]

After reviewing the action package, the signatory authority writes a review detailing the disposition of issues in the application and determining a course of final action.[133] If the ODE director is the sign-off, the division director also prepares a written summary report.[134] In the case of a BLA, a final facility inspection is requested and should be completed within 30 days of final action on the application.[135]

If the application is approved, the RPM composes an approval letter for review by the review team and signatory authority. The RPM sends the completed letter to the applicant and contacts the applicant to confirm

126. *Id.* at 31.
127. *Id.*
128. *Id.*
129. *Id.* at 38.
130. *Id.* at 41.
131. *Id.* at 41.
132. *Id.*
133. *Id.*
134. *Id.*
135. *Id.*

receipt and then files the approval internally within the FDA within two business days.[136] If, however, the application is not approved, the RPM composes a Complete Response letter detailing the deficiencies and recommended corrective action to place the application for approval.[137]

IX. Providing Post-action Feedback to the NDA Applicant/Sponsor

After action has been taken on the application, feedback meetings and lessons learned meetings may be held to improve the process in the future and to provide feedback to the applicant.[138] In the case of new molecular entities and original BLAs, post-action feedback meetings are available wherein the applicant and the FDA may discuss the process and communication during the application; however, scientific issues are generally precluded.[139] In the case of a Complete Response letter, an End of Review conference may be held to enable the applicant to understand the deficiencies preventing approval and determine a course of action for approvability.[140]

A. Disputes

When an applicant disagrees with the action taken on scientific or technical grounds, the End of Review Conference may provide a venue to discuss and challenge the action.[141] However, if the issue is procedural or administrative in nature, the issue should be addressed to the consumer safety officer for the application and may then be escalated to the ombudsman if the applicant is dissatisfied with the result.[142]

If the applicant is dissatisfied with the outcome of the End of Review conference and has exhausted remedies at the division level, the formal dispute resolution procedure may be followed to appeal the decision.[143] This process permits an applicant to challenge the decision to the next highest level of supervision through the FDA ombudsman.[144] Once the commissioner has rendered a decision, the applicant is considered to have exhausted all administrative remedies and may pursue judicial review of the decision.[145]

136. *Id.* at 41–42.
137. *Id.* at 41; *see also* 21 C.F.R. § 314.110.
138. CDER 21st Century Review Process: Desk Reference Guide, *supra* note 6, at 43.
139. *Id.*
140. *Id.*; *see also* 21 C.F.R. § 314.102(d).
141. 21 C.F.R. § 314.103(c).
142. *Id.* § 314.103(b).
143. *Id.* § 10.75.
144. *Id.*
145. *Id.* § 10.45.

X. Conclusion

As evident from the previous discussion, the CDER's published manuals and guidance documents reflect the mission established by the U.S. Congress, through the FDCA and various iterations of PDUFA, to construct a regulatory process for NDAs, by the FDA, that is robust, fair, and consistent for all applicants. The funding regime established by PDUFA provides the FDA with the resources necessary to conduct a proper review of a new drug's effectiveness and risks. These same publications also define the roles and responsibilities of the CDER's multidisciplinary team of NDA reviewers so that applicants can anticipate and prepare for the various elements of a review and hold the CDER accountable to provide consistency in the review process, while allowing the flexibility demanded by the complexity and importance of this administrative review.

chapter 10

Orange Book Listable Patents[1]

I. Requirements for Orange Book Listing

Proposed new pharmaceuticals for human consumption go through several rounds of regulatory review before an ultimate decision is made about whether to approve the product with labeling appropriate for the indication(s) sought by the sponsor.

In the first round, experts for the FDA review Investigational New Drugs (INDs) to determine if in vitro and animal studies suggest sufficient efficacy and do not display unwarranted toxicity. Once an IND is approved, the sponsor or applicant (a brand pharmaceutical company, in all likelihood, though occasionally a university or other research organization) will undertake human clinical trials. If those human clinical trials are successful, the company will file a New Drug Application (NDA) under the Hatch-Waxman Amendments to the Federal Food, Drug, and Cosmetic Act (FDCA).[2]

NDAs involve years of clinical trials in human subjects and are extremely expensive. Medical and scientific experts for the FDA review the results to determine whether there is a relatively minimal dose that provides efficacy in the human patient while avoiding unreasonable toxicity. Frequently, the experts come back with questions and, sometimes, requests for additional evidence, clinical or otherwise.

Once a drug is approved—including the labeling for the approved indication(s)—the applicant holds the NDA for the Reference Listed Drug (RLD). When an applicant's NDA is accepted by the FDA, the approved

1. David S. Bloch of Winston & Strawn LLP, San Francisco, California; Charles B. Klein of Winston & Strawn LLP, Washington, D.C.; and Scott H. Blackman of Nath, Goldberg & Meyer, Washington, D.C.

2. 21 U.S.C. § 355(b).

product or RLD is listed in the FDA's *Approved Drug Products with Therapeutic Equivalence Evaluations*—the Orange Book. The Orange Book is the FDA's compendium of all RLDs, organized by brand name, by chemical name, and by NDA number. It also contains a listing of all applicable exclusivities, including all patents that the NDA holder identifies to the FDA as covering the RLD or a method of using the RLD.[3] Any live patents on the RLD or an FDA-approved method of using the RLD should be listed in the Orange Book by the applicant, so long as "a claim of patent infringement could reasonably be asserted if a person not licensed by the owner of the patent engaged in the manufacture, use, or sale of the drug product."[4]

Such patents generally are owned or licensed by the applicant. But only patents on approved *active ingredients*, *formulations* and *compositions*, and *methods of use* should be listed in the Orange Book.[5] Patents on processes for manufacture (except for product-by-process patents covering a novel product), packaging, metabolites, or intermediates "must not" be listed.[6] Despite this bright-line restriction, the FDA does not itself confirm or even consider whether any patent submitted by the applicant complies with this limitation.[7] Even a third-party challenge to the FDA regarding such a listing will only result in the FDA requesting the applicant to confirm the listing. Thus, the only way to truly challenge an Orange Book patent listing is through the courts via a counterclaim alleging an improper listing brought under the Hatch-Waxman Amendments.[8]

A patent must be submitted to the FDA for listing in the Orange Book within 30 days of the NDA's approval or 30 days of the patent's issue date, whichever is later.[9] The Hatch-Waxman Amendments require NDA applicants to file with any new drug application "the patent number and the expiration date" of any such patent to be listed in the Orange Book. For method-of-use patents, the NDA holder "shall submit information only on those patents that claim indications or other conditions of use that are described in the pending or approved application" and, further, "shall separately identify each pending or approved method of use and related patent claim."[10]

---

3. *Id.* § 355(j)(7)(A)(i).
4. 21 C.F.R. § 314.53(b)(1).
5. *Id.*; Purepac Pharm. Co. v. Thompson, 238 F. Supp. 2d 191, 193 n.2 (D.D.C. 2002).
6. 21 C.F.R. §§ 314.53(b)(1), 314.53(c)(2)(L).
7. *Purepac Pharm.*, 238 F. Supp. 2d at 196.
8. 21 U.S.C. § 355(j)(5)(C)(ii)(I); Caraco Pharm. Labs. v. Novo Nordisk AS, 566 U.S. ____, 132 S. Ct. 1670 (2012). Along with Winston & Strawn's James Hurst and Steffen Johnson, Messrs. Klein and Bloch represented Caraco throughout this litigation and the related appeals.
9. 21 C.F.R. § 314.53(c)(2)(ii).
10. *Id.* § 314.53(b)(1).

As discussed, the FDA includes in the Orange Book for method-of-use patents a short narrative prepared by the NDA holder that purports to describe each FDA-approved use covered by each such patent. The FDA assigns this patent description a unique numeric code. This code, along with the patent description, is referred to as a "use code."

II. Completion of Forms 3542 and 3542a

The FDA's "submission of patent information" regulation requires "any applicant who submits to the FDA a new drug application or an amendment to it under section 505(b) of the act . . . or a supplement to an approved application under 314.70," which implements § 505(c)(2), to submit patent information to the FDA "for each patent that claims the drug or a method of using the drug . . . and with respect to which a claim of patent infringement could reasonably be asserted if a person not licensed by the owner of the patent engaged in the manufacture, use, or sale of the drug product."[11] This information must be submitted on FDA Forms 3542 or 3542a.[12]

The NDA holder must submit Form 3542a with the filing of an NDA, amendment, or supplement, and Form 3542 after approval of the NDA, amendment, or supplement. The two forms are identical in substance; they differ only in timing—during the pendency, or after approval, of the underlying submission.

According to the FDA, these forms, or declarations, are for use by "NDA applicants or holders or patent owners for submission of patent information at the time of initial NDA or supplement filing, and upon and after NDA or supplement approval." They are standardized and "encompass the required patent declaration information."[13]

At the top of Forms 3542 and 3542a, the NDA holder must list the trade name or proposed trade name of the RLD and the RLD's active ingredients, strengths, and dosage forms. Each form contains six sections, which must be completed by the NDA holder, or completed by the patent owner and submitted by the NDA holder.

A. Section 1—Listed Patents

For each Orange Book–listed patent, the NDA applicant/holder must identify the U.S. patent number, issue and expiration dates, and the name and address of the patent owner. If the patent owner or NDA applicant/holder

11. *Id.* § 314.53(a)(b)(1).
12. *Id.* § 314.53(c).
13. *68 Fed. Reg. 36,686 (June 18, 2003).*

does not reside or have a place of business in the United States, the NDA applicant/holder must identify the name and address of an agent or representative who resides or has a place of business in the United States. The NDA applicant/holder also must state whether or not the patent has been previously submitted for listing in connection with the NDA or any supplement and, if so, whether the patent has a new expiration date (for example, because a term extension has been granted).

B. Section 2—Drug Substance (Active Ingredient)

In Section 2 of Forms 3542a and 3542, the NDA holder must disclose whether a patent listed in Section 1 claims the active ingredient of the NDA. If it does, and the listed patent is a product-by-process patent, the NDA holder must also state whether the product itself is novel.

If the patent only claims a polymorph of the RLD, the NDA holder must certify that it possesses test data showing that the polymorph will perform the same as the RLD and must specify the polymorphic forms claimed by the listed patent. Similarly, the NDA holder must disclose if the patent claims a metabolite or intermediate product of the RLD.

C. Section 3—Drug Product (Composition/Formulation)

The NDA holder must disclose whether the listed patent claims the drug product itself or only an intermediate. If the patent is a product-by-process patent, the NDA holder must disclose whether the product itself is novel.

D. Section 4—Method of Use

In Section 4, the NDA holder must indicate whether any patent listed in Section 1 claims one or more methods of use for which approval is sought. If so, the NDA holder must state the specific claim number and confirm that the NDA holder is seeking (or has obtained) FDA approval for the specific method(s). In that event, the NDA holder must "identify with specificity the use with reference to the proposed labeling for the drug product," and then describe the patented "indication or method of use as identified specifically in the proposed labeling."[14]

The "use code" number itself, assigned by the FDA, corresponds to a particular "use code description," which is a 240-character description of the patented use implicated by the NDA, as supplied by the NDA holder.

This description must be complete and correct, because "accurate and detailed information related to the approved methods of use" is "essential" to the FDA fulfilling its statutory duty to "expedite [its] review of ANDA . . . applications that do not seek approval for all the approved

---

14. *See* 21 C.F.R. § 314.53(c)(2)(ii) (instructions come from FDA Form 3542a at ¶ 4.2a).

uses"[15] (i.e., ANDAs seeking approval under "section viii")[16] In its rule-making findings, the FDA emphasized the critical nature of accurate patent use code information. This is especially true because the FDA's self-appointed role regarding such information is merely ministerial—it does not independently confirm or consider the accuracy of the information submitted by the NDA holder. "To effectively implement the [Paragraph IV] certification and section viii statement provisions . . . it is necessary that an NDA holder submit more specific information on the approved methods of use protected by a submitted patent. Only with this information can [the FDA] determine what submission is required of [applicants for generic versions of the NDA product]."[17]

As the Supreme Court confirmed in *Caraco Pharmaceutical Laboratories v. Novo Nordisk AS*,[18] Congress ratified the FDA's approach and, indeed, went "much further," "provid[ing] a critical complement to the work the FDA has done in clarifying its regulations on patent listing" by empowering courts to resolve disputes concerning erroneous use codes[19] and other patent information submitted under 21 U.S.C. § 355(b) and (c).[20]

E. Section 5—No Relevant Patents

If no patents are to be listed in the Orange Book with regard to the NDA product, the NDA holder must confirm that there are no applicable patents directed to the manufacture, use, or sale of the RLD as covered by the NDA.

F. Section 6—Declaration Certification

Last, an authorized representative of the NDA holder or patent owner must certify that the Form 3542 or 3542a submission is correct and complies with the requirements of the regulation, under penalty for perjury.[21]

III. Confirming or Obtaining Protective Claim Scope

ANDA applicants must always explore whether and to what extent the NDA holder can assert patents to block a new generic version of the RLD.

15. 68 Fed. Reg. at 36,682, 36,685.
16. 21 U.S.C. § 355(j)(2)(A)(viii).
17. 68 Fed. Reg. at 36,682, 36,683.
18. 132 S. Ct. 1670 (2012).
19. *Legislative and Regulatory Responses to the FTC Study on Barriers to Entry in the Pharmaceutical Market-place: Hearing Before S. Comm. on the Judiciary*, 108th Cong. 19 (2003).
20. 21 U.S.C. § 355(j)(5)(C)(ii); *Caraco Pharm.*, 132 S. Ct. 1670, 1686.
21. 21 C.F.R. § 314.53(c)(2)(ii)(R)—for example, FDA Form 3542.

Many relevant patents will be listed by the NDA holder in the Orange Book. Patents with claims covering an RLD or any of its active ingredients will typically be listed without further detail. Patents covering the RLD itself or an active ingredient often provide the greatest degree of protection for the NDA holder, and present the greatest obstacles for the ANDA applicant. For Orange Book–listed method patent claims covering an FDA-approved indication for the RLD, the NDA holder also must include a "use code description," which is supposed to describe the patented method in enough detail to put an ANDA applicant on notice. However, the FDA does not independently confirm any such designations by the NDA holder. If an ANDA applicant believes that a listing is improper, it may challenge the listing by way of a Hatch-Waxman Act counterclaim. In addition, there may be other potentially applicable patents that are not listed in the Orange Book, or that are listed only after the ANDA is filed. We discuss each of these possibilities in this section.

A. Protecting Approved Indications

A method-of-use patent that protects an indication can be listed in Form 3542 (and, thus, in the Orange Book) if, and only if, the FDA has approved that particular indication. Patents that cover unapproved indications cannot be properly listed in the Orange Book.

B. Challenges to Listings under 21 C.F.R. § 314.53(f)

If an ANDA applicant believes a patent use code description is improper, it can object to the listing under 21 C.F.R. § 314.53(f). In response to such a challenge, the FDA will merely ask the NDA holder whether the listing is accurate. The FDA will take no further action and will not require the NDA holder to change it: "Unless the application holder withdraws or amends its patent information in response to FDA's request, the agency will not change the patent information in the [Orange Book] list."[22]

C. Challenges to Listings under 21 U.S.C. § 355(j)

If—and only if—an NDA holder sues an ANDA applicant based on a listed method-of-use patent, the ANDA applicant has a statutory vehicle for challenging the listing pursuant to 21 U.S.C. § 355(j). Section 355(j)'s counterclaim provision allows litigation "to correct or delete the patent information submitted by [the NDA] holder . . . on the ground that the patent does not claim either . . . the drug for which the application was approved; or . . . an approved method of using the drug."[23] The provision

22. 21 C.F.R. § 314.53(f).
23. 21 U.S.C. § 355(j)(5)(C)(ii)(I).

is designed, at least in part, to avoid the problem of overbroad use code descriptions, which the Supreme Court held "throws a wrench into FDA's ability to approve generic drugs as the statute contemplates."[24] The sole remedy for a § 355(j) counterclaim is an order to "correct or delete" the erroneous patent use code information.

D. Coordinating Patent Prosecution and Regulatory Strategies and Submissions

For NDA applicants, patents covering the RLD or its active ingredients are the most far-reaching and valuable forms of patent protection. Such patents typically should be listed on Form 3542a. The patent on a new drug or new chemical entity (NCE) normally will be difficult to invalidate, given the novelty of the subject matter and the statutory presumption of validity.[25] Moreover, it will be extremely difficult—if not impossible—for a subsequent generic filer to engineer around a composition, drug, or NCE patent and still qualify to file an ANDA based on the NDA under the Hatch-Waxman Amendments. As a consequence, many generic applicants opt to submit a Paragraph III Certification with respect to any such compound patent (although there certainly have been challenges to compound patents). If the applicant elects to file a Paragraph IV Certification and is sued, however, the applicant will bear the burden of invalidating the patent by clear and convincing evidence.[26] There is generally little chance of a noninfringement argument under these circumstances.

When such patents are already allowed and pending, and listed in the Orange Book, before any ANDA is filed, the NDA holder/patentee enjoys the greatest amount of regulatory protection. A Paragraph IV Certification to a patent listed in the Orange Book before an ANDA is filed will trigger a mandatory 30-month stay in ANDA approval, if the patentee sues on the basis of the Paragraph IV notice. Until that stay expires or is abrogated by the court, the ANDA applicant has no possibility for an at-risk launch, providing the NDA holder with a guaranteed further period of exclusivity.

Patents issuing and listed in the Orange Book after an ANDA is filed also have significant value, of course, and may raise important issues for the ANDA applicant. However, such a so-called pop-up patent (a patent listed by the NDA holder after the ANDA was filed) does not enjoy the same regulatory advantages that a previously listed patent enjoys. In particular, a pop-up patent does not trigger the mandatory 30-month stay in

24. *Caraco Pharm.*, 132 S. Ct. at 1684.

25. 35 U.S.C. § 282. *But see* Bristol-Myers Squibb v. Teva Pharm., No. 10-805-CJB (D. Del. Feb. 11, 2013) (invalidating patent on new chemical entity).

26. Microsoft Corp. v. i4i Ltd., 131 S.Ct. 2238, 242 (2011); Sciele Pharma Inc. v. Lupin, 684 F.3d 1253, 1260 (Fed. Cir. 2012).

ANDA approval.[27]Nevertheless, an ANDA applicant must promptly file a certification as to any such patents that are timely listed (i.e., filed within 30 days of patent issuance[28]) if the patent was listed after the ANDA was filed but before the date of final approval, and must promptly provide the patentee with the required notice—before the product is launched—if the new certification is a Paragraph IV Certification. To stop a generic launch based on a pop-up patent, therefore, the patentee must obtain preliminary injunctive relief.

Brand companies also frequently pursue method-of-use patents to cover use of the RLD to treat specific indications, particularly for drugs whose chemical structure is unpatented or unpatentable. Method-of-use patents typically issue later in time than compound patents and, hence, can have the effect of prolonging the NDA holder's monopoly on the RLD. If the NDA holder can properly patent *all* of the FDA-approved uses of a particular drug, the generic applicant cannot avoid infringement (under an induced infringement theory) and thus must seek to invalidate the listed patents under Paragraph IV litigation.

When a listed method-of-use patent covers fewer than all of the approved uses, an ANDA applicant can submit a Section viii Statement in lieu of a Paragraph IV Certification so long as the ANDA requests approval only for approved indications that are not implicated by the listed patent(s) as described in the use codes. Thus, as long as there is at least one approved use that is not covered by any method-of-use patent, the ANDA holder potentially may be able to avoid the 30-month stay and the costs of a patent infringement lawsuit. Of course, this assumes, for example, that there are no relevant listed product patents challenged by the ANDA holder.

Lastly, process patents, and patents covering packaging, pharmaceutical metabolites, or intermediates,[29] along with patents covering *unapproved* methods of use,[30] "must not" be submitted to the FDA for listing in the Orange Book. Although a generic applicant might be found to infringe one or more of those types of patents, the NDA holder may not properly use the Hatch-Waxman Amendment's dispute resolution procedures to expedite litigation of those patents or otherwise seek a 30-month stay of ANDA approval based on these patents.

---

27. *See* FDA Draft Guidance for Industry, Listed Drugs, 30-Month stays, and Approval of ANDAs and 505(b)(2) Applications Under Hatch-Waxman, as Amended by the Medicare Prescription Drug, Improvement, and Modernization Act of 2003 at 9.

28. *Id.* § 314.94(a)(12)(vi).

29. *Id.* § 314.53(b)(1).

30. Purepac Pharm. Co. v. Thompson, 238 F. Supp. 2d 191, 196 (D.D.C. 2002).

E. Non–Orange Book Listable Patent Protection

The fact that a patent is not listed in the Orange Book does not necessarily mean that it does not apply to an approved drug. As discussed, certain types of pertinent patents do not need to be listed. And sometimes NDA holders do not properly list all relevant patents.

After a drug is launched commercially, patent owners can and do sue for infringement or induced infringement for patents not listed in the Orange Book (or not previously asserted) based on marketplace activities. At that time, a patent owner is free to sue for infringement of a patent claiming only a metabolite, an intermediate, a product-by-process for a product that is not new, or an infringing package. But for such claims, the patentee generally must wait until the accused product is approved by the FDA and launched commercially—or at least until the controversy becomes sufficiently ripe to support declaratory judgment relief.

F. Delisting

As discussed, a patentee must properly list specific types of patents—and only those patents—in the Orange Book. Only those specified patents are entitled to the benefits of the Hatch-Waxman Amendments. While an erroneous listing for a required patent may be corrected, an improperly listed patent should be withdrawn or "delisted" by the NDA holder by way of a request to the FDA.

Delisting has potentially significant consequences. If a patent is delisted, it can no longer form the basis of a Paragraph IV Certification—including, except as discussed later in this chapter, any pending certifications in an ANDA that was filed but not yet finally approved. However, voluntary delisting by an NDA holder cannot trigger an ANDA first-filer's exclusivity based on a pending Paragraph IV Certification.[31]

However, the FDA cannot remove an improperly listed patent from the Orange Book if removal would impair an ANDA first-filer's eligibility for 180 days of market exclusivity based on a Paragraph IV Certification.[32] If there is a legitimate Paragraph IV first-filer, the listed patent must remain in the Orange Book even after the NDA holder has asked that it be removed. As a consequence, subsequent Paragraph IV filers must continue to certify to the listed patent until such time as the first-filer is able to enjoy its 180 days of market exclusivity. This process helps ensure the integrity of the Hatch-Waxman incentives.

---

31. 21 U.S.C. § 355(j)(5)(D)(i)(I)(CC); Teva Pharm. USA Inc. v. Sebelius, 595 F.3d 1303, 1316-17 (D.C. Cir. 2010).

32. 21 C.F.R. § 314.94(a)(12)(viii)(B). *See generally* Ranbaxy Labs. Ltd. v. Leavitt, 469 F.3d 120, 126 (D.C. Cir. 2006).

G. Nuances of Challenges to Listed Patents

The FDA lists whatever patents are submitted to it by NDA holders using Form 3542 without any substantive confirmation or review. It has long defended its "ministerial" role in the patent-listing process.[33] Anyone can challenge the listing of a particular patent in the Orange Book by writing to the FDA.[34] However, even in the event of a challenge, the FDA will only ask the NDA holder or patentee whether the listing is proper. If the NDA holder affirms the listing, the FDA will take no further action. Section 314.53(f) does not create a private right of action against either the FDA or the listing entity.[35]Thus, an ANDA applicant may need to file a patent certification against a patent that should not be listed in the Orange Book at all. If the patentee sues an ANDA applicant based on a Paragraph IV Certification on an improperly listed patent, however, the defendant can counterclaim to "correct or delete" the erroneous patent listing.[36]

33. aaiPharma Inc. v. Thompson, 296 F.3d 227, 242–43 (4th Cir. 2002); Am. Bioscience, Inc. v. Thompson, 269 F.3d 1077, 1084 (D.C. Cir. 2001); *In re* Buspirone Patent Litig., 185 F. Supp. 2d 363, 371 (S.D.N.Y. 2002); Watson Pharm. Inc. v. Henney, 194 F. Supp. 2d 442 (D. Md. 2001).

34. 21 C.F.R. § 314.53(f).

35. Mylan Pharm. v. Thompson, 268 F.3d 1323 (Fed. Cir. 2001).

36. Caraco Pharm. Labs. Ltd. v. Novo Nordisk AS, 132 S. Ct. 1670, 1680 (2012).

chapter 11

Postmarketing Regulatory Requirements and Life Cycle Management[1]

Before marketing approval of a drug application, the Food and Drug Administration (FDA) evaluates whether the drug is safe and effective. After drug application approval, drug sponsors[2] are subject to comprehensive postapproval FDA surveillance regulations to ensure that the drug's benefits outweigh its risks. Compliance with these postapproval regulatory requirements is necessary to maintain approval of the drug and is an important consideration in the life cycle management of the drug. Additionally, where postapproval regulations apply similarly to generic and branded drugs, branded-drug manufacturers have identified opportunities to extend marketing exclusivity by patenting postmarketing compliance measures that the FDA has determined are required to ensure that the benefits of a drug continue to outweigh its risks.

This chapter provides an overview of (i) the postmarketing surveillance regulatory requirements for human drug products (primarily prescription drugs) and (ii) key considerations in drug life cycle management under the Hatch-Waxman and postmarketing surveillance regulatory frameworks.

I. Postmarketing Regulatory Requirements

In the wake of a number of high-profile drug approval withdrawals due to reports of severe adverse effects in patients that emerged after marketing approval, the U.S. Congress enacted the Food and Drug Administration Amendments Act (FDAAA) of 2007, which significantly expands the FDA's

1. Earl B. Austin, Sean McDonagh, and Jeremy Merling of Baker Botts LLP, New York City.

2. In this chapter, drug sponsors, application holders, and drug manufacturers are referred to interchangeably.

postmarketing surveillance mandate and enforcement capabilities. Title IX of the FDAAA empowers the FDA to require postmarketing clinical studies, changes to labeling, and Risk Evaluation and Mitigation Strategies (REMS) to ensure that the benefits of a drug outweigh the risks to safety associated with use of the drug. Failure to comply with FDA-imposed postmarketing surveillance measures may result in injunction of a drug as misbranded and may subject a drug sponsor to civil, and potentially criminal, penalties.

A. Mandatory Adverse Drug Experience Monitoring and Reporting

The FDA requires sponsors to report adverse effects associated with the use of approved drugs to permit identification of rare, latent, or long-term drug effects not identified during relatively limited (in scale and duration) premarketing approval testing. Pursuant to Section 505(k) of the FDCA and 21 C.F.R. §§ 314.80(a) and 314.98, sponsors of an approved New Drug Application (NDA) or an Abbreviated New Drug Application (ANDA) are obligated to continuously monitor and report to the FDA data concerning "any adverse events associated with the use of [the approved] drug." Such adverse drug events (ADEs) are defined to include adverse events occurring in use in professional practice, accidental or intentional drug overdose, drug abuse, drug withdrawal, and failure of expected pharmacological action.[3]

ADEs are classified as serious or nonserious based upon the outcome of the experience and classified as expected or unexpected based upon whether or not the adverse experience was previously observed (i.e., whether or not the ADE is included in the drug product's current FDA-approved labeling).[4] Serious ADEs include ADEs resulting in death, a life-threatening experience, hospitalization, significant or persistent disability, and congenital anomaly or birth defect.

1. Reporting ADEs

ADEs typically are reported to drug sponsors by health care providers, but include reports received from any source in the context of domestic and foreign marketing experience, postmarketing clinical studies, and the scientific literature.[5] Sponsors of approved drug applications are required to review all ADE information received and submit reports of such ADEs to the FDA.[6] Sponsors are further obligated to establish and maintain records of all ADEs received by the sponsor. ADEs that are both

3. 21 C.F.R. § 314.80(a).
4. *Id.*
5. *Id.* § 314.80(b).
6. *Id.*

serious and unexpected, whether domestic or foreign, must be reported no later than 15 calendar days after initial receipt of the information by the sponsor.[7] The sponsor is further obligated to investigate such ADEs and submit follow-up reports no later than 15 days after receipt of new information based upon the investigation, or as requested by the FDA.[8] These so-called 15-day reports are not required for ADEs received pursuant to postmarketing studies, however, unless the sponsor concludes that there is a reasonable probability that the drug caused the adverse experience.[9]

For nonserious or expected ADEs, sponsors must additionally submit quarterly reports of such experiences for the first three years following application approval and must submit annual reports thereafter. These periodic reports should also contain summaries of all 15-day reports as well as a history of the actions taken by the manufacturer since the prior quarterly report in response to ADEs, such as labeling changes or initiation of a postmarketing study. Nonserious or expected ADEs arising from postmarketing studies, reports in the scientific literature, and foreign marketing experience need not be included in these periodic reports.[10] Sponsors must maintain records of all ADEs known to the sponsor for ten years.[11]

2. FDA Enforcement of Reporting Requirements

A sponsor's failure to comply with ADE reporting requirements constitutes a "prohibited act" under the Federal Food, Drug, and Cosmetic Act (FDCA), exposing the sponsor to civil and criminal penalties and administrative sanctions, including withdrawal of drug approval.[12] The FDCA provides for criminal sanctions of up to one year or up to three years for deliberate or repeated offenses.[13]

ADE reporting enforcement typically commences with advisory enforcement (i.e., "warning letters"), followed by judicial and administrative remedies. Most commonly, FDA enforcement of noncompliance with ADE reporting requirements commences with issuance of a "warning letter" advising the sponsor of the violation and demanding prompt corrective action. Warning letters are addressed to a senior executive of the sponsor and are publicly available on the FDA website. The letters outline with specificity the alleged violations, as well as prior communications with the sponsor about the matter if applicable, and typically conclude by providing

7. *Id.* § 314.80(c)(1)(i).
8. *Id.* § 314.80(c)(1)(ii).
9. *Id.* § 314.80(e).
10. *Id.* § 314.80(c)(2)(iii).
11. *Id.* § 314.80(i).
12. 21 U.S.C. § 331(e); 21 C.F.R. § 314.80(j).
13. 21 U.S.C. § 333(a).

15 days to correct the deficiencies identified under threat of prosecution or injunction. Where identified deficiencies are resolved to the satisfaction of the FDA, the agency will provide a "close-out letter" summarizing the violation and the corrective actions taken by the sponsor.

Where an approved drug sponsor commits serious ADE reporting violations or fails to rectify deficiencies identified in a warning letter, the FDA may pursue formal judicial enforcement. Such enforcement can include legal proceedings to secure injunctions, civil monetary penalties, criminal prosecutions, and seizure of unsafe drugs. For example, in 2011, the Department of Justice announced a $3 billion settlement with a major pharmaceutical company, of which over $240 million was recovered for failure to comply with ADE requirements for a popular diabetes medication.[14]

The FDA may also withdraw approval for a drug for which ADEs were not properly reported. Individuals convicted of felonies relating to failures to report ADEs will also be debarred from providing any services in any capacity to a company with an approved or pending drug application.[15] In addition, individuals and corporations convicted of certain felonies and misdemeanors in connection with ADE reporting noncompliance may also be prohibited from providing services in any capacity to an entity with an approved NDA.[16]

3. Strict Criminal Liability for "Responsible Corporate Officers"

Under the "Responsible Corporate Officer" doctrine, more commonly known as the *Park* Doctrine from the 1973 case *United States v. Park*,[17] a responsible corporate official can be held liable for a first-time misdemeanor, and possible subsequent felony, under the FDCA without proof that the corporate official acted with intent or negligence and even where the corporate official did not participate in or even have knowledge of the specific offense alleged. The *Park* Doctrine essentially imposes strict liability on corporate officials for FDCA violations by their companies, as the government need only prove that the official prosecuted was in a position of authority to prevent or correct an alleged FDCA violation. Conviction under the *Park* Doctrine may also serve as the basis of debarment by the FDA.[18]

Despite the sweeping potential of the *Park* doctrine, the FDA historically has pursued criminal actions against corporate representatives only

14. *See* Press Release, Dep't of Justice, GlaxoSmithKline to Plead Guilty and Pay $3 Billion to Resolve Fraud Allegations and Failure to Report Safety Data (July 2, 2012), *available at* http://www.justice.gov/opa/pr/2012/July/12-civ-842.html.

15. 21 U.S.C. § 335a(a).

16. *Id.* § 335a(b).

17. 421 U.S. 658 (1973). The defendant was fined $50 for each of five violations of the FDCA. *Id.* at 665–66.

18. 21 U.S.C. § 335a(b).

rarely. A notable exception arose in 2007, when the president, general counsel, and medical director of a major pharmaceutical company pled guilty to charges of misbranding of a widely prescribed opiate medication and agreed to pay $34.5 million in criminal penalties.[19] The charges arose from claims by the manufacturer that the opiate was less addictive than competing painkillers.

More recently, the FDA and federal prosecutors have signaled a policy of more aggressive enforcement of the *Park* Doctrine. For example, in a March 2010 public letter to Senator Charles Grassley, the FDA commissioner advised that the FDA recommended "increas[ing] the appropriate use of misdemeanor prosecutions, a valuable enforcement tool, to hold responsible corporate officers accountable."[20] The FDA also recently revised the prosecution guidelines of its Regulatory Procedures Manual, which informs agency inspectors and compliance officers of internal enforcement procedures, to feature factors to consider in recommending *Park* Doctrine prosecutions.[21] These factors include knowledge and actual participation by the corporate official, actual or potential harm to the public, the seriousness of the violation, and whether the violation reflects a pattern of illegal behavior and/or a failure to heed prior warnings.

Significantly, the revised guidelines appear to alter prior practice, which had directed FDA enforcement officers to consider the importance of prior notice and to support prosecution recommendations by facts showing a repeated or ongoing course of violative conduct. Paradoxically, the revised guidelines tout the deterrent value of *Park* Doctrine prosecutions, while emphasizing that knowledge of and participation in violative conduct is not required.

4. Tort Liability for Failure to Report ADEs

In addition to government and agency enforcement, sponsors may face significant tort liability exposure for failure to report ADEs. For example, allegations that a sponsor has failed to satisfy its ADE reporting obligations have been found to give rise to negligence and failure-to-warn product liability claims. For example, a federal district court recently held that withholding adverse drug reports from the FDA can give rise to punitive damages claims under the theory of implied malice.[22] Moreover, the

19. *See* Barry Meier, *Narcotic Maker Guilty of Deceit over Marketing*, N.Y. Times (May 11, 2007), http://www.nytimes.com/2007/05/11/business/11drug.html?pagewanted=all.

20. Margaret A. Hamburg, Comm'r of Food & Drugs, Food & Drug Admin., to the Honorable Charles E. Grassley, Ranking Member, Senate Comm. on Fin. (Mar. 4, 2010).

21. *See* U.S. Food & Drug Admin., Regulatory Procedures Manual § 6-5 (2011), *available at* http://www.fda.gov/ICECI/ComplianceManuals/RegulatoryProceduresManual/default.htm.

22. Wolfe v. McNeil-PPC Inc., 773 F. Supp. 2d 561, 575–76 (E.D. Pa. 2011).

Supreme Court recently held that withholding ADE reports from investors or making misleading statements concerning the significance of ADE reports to investors can give rise to claims for securities fraud.[23] In particular, the Supreme Court held that the public dismissal of ADEs reporting events of anosmia by an intranasal cold remedy manufacturer could give rise to securities fraud claims, even where such ADEs did not rise to the level of statistical significance, where reasonable investors would consider the ADEs material to their trading choices.[24]

On the other hand, courts have recognized that ADE reporting should be encouraged, and because FDA regulations require sponsors to report ADEs without regard to causality, courts generally hold that a plaintiff in a product liability case cannot rely on ADE reports as proof that the drug actually causes the reported adverse effect. For example, in *Nelson v. American Home Products Corp.*,[25] the district court held that proof of causation "must go beyond that which is required by the Food and Drug Administration for reporting adverse drug effects or re-labeling to warn of the potential for adverse drug effects. . . . [I]nformation that defendants are required to report without regard to proof of causation cannot be cited later as an admission of causation."

B. Drug Product Labeling

Drug product labeling is the primary means by which the FDA and the sponsor communicate the risks potentially associated with a drug to health care practitioners in order to assist the prescriber in making therapeutic risk-benefit determinations for individual patients. Drug product labeling is subject to detailed content and formatting regulations, and pursuant to the FDAAA, the FDA may mandate labeling changes based on new safety information.

1. Content and Format

Labeling for a drug product, also referred to as the package insert, is provided by the NDA sponsor and approved by the FDA. Initial marketing approval of an NDA requires approval of the exact text of the proposed label by the FDA.[26] The labeling must be informative, accurate, and "neither promotional in tone nor misleading in any particular."[27]

---

23. *See* Matrixx Initiatives Inc. v. Siracusano, 131 S. Ct. 1309 (2011).

24. *Id.* at 1323.

25. 92 F. Supp. 2d 954 (W.D. Mo. 2000); *accord* Hollander v. Sandoz Pharm. Corp., 289 F.3d 1193 (10th Cir. 2002).

26. *See* 21 C.F.R. § 314.105(b).

27. *Id.* § 201.56(a)(2).

Detailed requirements for the content and format of prescription drug labels are set forth in 21 C.F.R. § 201.57.[28] These requirements were amended in 2006 with the goal of making labeling information more accessible and useful to prescribers and health care practitioners. Pursuant to these 2006 amendments (commonly referred to as the Physician Labeling Rule), all labels for prescription drugs approved after June 30, 2001, must be transitioned to conform to the new formatting requirement.[29] The required date of submission of proposed conforming labeling is based on the approval date of the NDA; however, the latest date for submission of proposed conforming labeling (for NDAs approved between June 30, 2001, and June 29, 2002) was June 30, 2013.

In addition to information concerning the drug name, prescribing, pharmacology, indications, and usage, package inserts must contain both a concise summary and a detailed description of the following:

- each of the drug product's contraindications (i.e., situations in which the drug should not be used because the risk of use clearly outweighs any possible therapeutic benefit);
- warnings and precautions regarding clinically significant adverse reactions associated with use of the drug;
- the overall profile, based on all available safety data, of adverse reactions associated with the drug;[30] and
- information concerning risks of drug abuse and dependence.

For NDAs approved after June 30, 2001, serious risks associated with use of a prescription drug may need to be identified in a boxed warning on the package insert. Boxed warnings, also referred to as black box warnings, are warnings presented in a prominent box near the beginning of the package insert warning of serious potential hazards associated with the drug. Boxed warnings are provided in bold print and give a concise summary of critical information for consideration by a prescriber.[31] A boxed

28. Prescription drugs for which an NDA was approved between June 30, 2001, and June 30, 2006, pending on June 30, 2006, or submitted anytime thereafter are subject to more detailed package insert format and content requirement than prescription drugs approved prior to June 30, 2001. *See* 21 U.S.C. § 201.56(b)(1). Most notably, prescription drugs approved prior to June 30, 2001, are not subject to "Boxed Warning" requirements. *See* 21 C.F.R. § 201.80.

29. 21 C.F.R. § 201.56(b).

30. Adverse reactions are defined for purpose of prescription drug labeling as undesirable effects reasonably associated with use of a drug, which may occur as part of the pharmacological action of the drug or may be unpredictable in its occurrence. The detailed description of the adverse reactions must list the adverse reactions identified from domestic and foreign spontaneous postmarketing reports. 21 C.F.R. § 201.57(c)(7).

31. *Id.* § 201.56(d).

warning is typically used to highlight for prescribers one of the following situations:

- potential adverse reactions so serious (e.g., fatal, life-threatening, or permanently disabling) in proportion to the potential benefit from the drug that it must be considered in assessing the risks and benefits of using the drug;
- potential adverse reactions that can be prevented or reduced in frequency or severity by appropriate use of the drug, such as patient selection, monitoring, avoiding concomitant therapies, or avoiding use in specific clinical situations; or
- the drug was approved with restrictions to ensure safe use because the FDA concluded that the drug can only be used safely with restrictions on distribution or use.[32]

Only those risks with a substantiated association are included in drug labeling; the labeling regulations provide, for example, that only "[k]nown hazards and not theoretical possibilities shall be listed" as contraindications to use in order to prevent dilution of the significance of the labeling.[33] A warning is suitable for inclusion only where "reasonable evidence of an association of a serious hazard with a drug" is identified; importantly, however, "a causal relationship need not have been proved."[34] When substantiated risks are discovered, however, NDA sponsors must alert the FDA and add the new risk information to the drug labeling.[35]

2. Mandatory Labeling Changes

Section 505(o)(4) of the FDCA significantly strengthened the authority of the FDA to impose safety-related labeling changes. Prior to the enactment of that provision, a labeling directive from the FDA to an NDA sponsor typically resulted in negotiations as to the content of the revised labeling, and FDA enforcement options were limited to withdrawing approval or initiating misbranding proceedings, both of which the FDA was often reluctant to

32. *See* U.S. Food & Drug Admin., Guidance for Industry: Warnings and Precautions, Contraindications, and Boxed Warning Sections of Labeling for Human Prescription Drug and Biological Products—Content and Format 11-12 (2011) [hereinafter Warnings and Precautions].

33. 21 C.F.R. § 201.57(d).

34. *Id.* § 201.57(e). Although beyond the scope of this chapter, the import of this provision is a source of frequent dispute in products liability litigation. Because the sponsor's obligation to warn is triggered by "reasonable evidence of an association" even though "a causal relationship need not have been proved," most courts hold that the mere inclusion of a reported adverse effect in a product label is not sufficient by itself to prove that the drug actually causes the effect.

35. *See* 44 Fed. Reg. 37,447 (June 26, 1979).

pursue.[36] Section 505(o)(4),[37] enacted pursuant to Section 901 of the FDAAA, now vests the FDA with the authority to unilaterally require sponsors of approved NDAs for prescription drug products to amend their package inserts to reflect postmarketing safety information. In particular, the FDA can order labeling changes based upon "new safety information," broadly defined as "information derived from a clinical trial, an adverse event report, a post-approval study (including a study under Section 505(o)(3)), peer-reviewed biomedical literature, data derived from the post-market risk identification and analysis system under Section 505(k), or other scientific data deemed appropriate" by the FDA concerning:

- "A serious risk or an unexpected serious risk associated with the use of the drug that [the FDA] has become aware of (that may be based on a new analysis of existing information) since the drug was approved, since the risk evaluation and mitigation strategy ("REMS") was required, or since the last assessment of the approved [REMS] for the drug," or
- "The effectiveness of the approved [REMS] for the drug obtained since the last assessment of [the REMS]."[38]

The FDA may learn about new safety information from many sources, including monitoring or data mining of its Adverse Event Reporting System, ADE inspections, periodic or 15-day alert reports, ADE reports, medical literature submitted by sponsors or third parties or identified by staff, or meta-analyses or new analyses of previously submitted information. New safety information may also be derived by the FDA based on a risk-benefit assessment of a drug relating to a new use or indication of the drug or on use of the drug in a new patient population.[39]

The FDA has indicated in industry guidance that labeling changes will be required when it determines, upon identification and review of new safety information, that changes to the boxed warnings, contraindications, warnings and precautions, drug interactions, or adverse reactions sections of the package insert are warranted. Where changes based on new safety information would be warranted only for the adverse reactions section of the label, the FDA has indicated that it is unlikely to exercise its authority under Section 505(o)(4).[40]

36. *See* U.S. Food & Drug Admin., Guidance for Industry: Safety Labeling Changes—Implementation of Section 505(0)(4) of the FD&C Act 2 (2013).

37. 21 U.S.C. § 355(o)(4).

38. *Id.* § 355-1(b). Postapproval studies and REMS are discussed at Section I-C.

39. *See* Warnings and Precautions, *supra* note 32, at 3.

40. *See id.* at 4.

The FDA initiates a labeling change by issuing a notification letter to the sponsor providing the proposed labeling change as well as a description of the new safety information and the source from which the new safety information was derived. As part of the required labeling changes, the FDA may additionally require the sponsor to create a new Medication Guide to inform patients of serious risks associated with the drug.[41]

After receiving notification of the required safety labeling changes, a sponsor must either submit a supplement with proposed labeling changes to reflect the new safety information or notify the FDA that it does not believe that a labeling change is warranted with a rebuttal statement detailing the reasons why the change is unwarranted.[42] If the sponsor submits proposed labeling changes identical to those included in the notification letter, the application holder can submit the supplement in the form of a Changes Being Effected (CBE) supplement. All other proposed labeling changes must be submitted in the form of a prior approval supplement.

Required safety labeling changes are reviewed on a rapid schedule. A sponsor has 30 calendar days from notification to submit a supplement or rebuttal. The FDA has set a goal of 30 calendar days for completion of review of the response. If the FDA disagrees with the sponsor's supplement or rebuttal, it will initiate a 30-day negotiation period. If no consensus is achieved during negotiations for any part of the required changes, the FDA will issue an order letter ordering the sponsor to file a CBE supplement containing the required labeling changes for which consensus was not achieved. Class-wide order letters will be publicly available, but order letters concerning individual drugs will not be publicly released, although the resulting supplement (if no appeal is taken or appeal is unsuccessful) will be posted.

The sponsor may appeal the order within five calendar days. Appeals are conducted within the FDA according to the dispute resolution procedures of Section 505(o)(4)(F) of the FDCA. No judicial review is provided by statute. Under Section 505(o)(4)(G) of the FDCA, if a sponsor fails to comply with an order letter within 15 days (where no appeal is filed within five days), that sponsor will be in violation of the statute and may be subject to enforcement. Similarly, if the FDA determines at the conclusion of any dispute resolution process that a supplement must be submitted, the sponsor must submit the supplement within 15 days or be subject to enforcement for violation of the statute.

41. *Id.* at 5.

42. 21 U.S.C. § 355(o)(4)(B)(i), (ii).

3. Voluntary Labeling Changes

The prohibitions against drug misbranding require that prescription drug labeling be revised to include warnings about a "clinically significant hazard" as soon as there is "reasonable evidence of a causal association with a drug."[43] Notwithstanding the FDA's expanded authority to mandate labeling changes, sponsors remain obligated to revise drug labeling when they become aware of newly acquired information.

NDA sponsors may submit labeling supplements voluntarily at any time after approval and without prior notification to the FDA. Supplements, including labeling supplements, submitted by a drug sponsor are classified as preapproval supplements under 21 C.F.R. § 314.70(b), which require FDA approval before implementation by a drug sponsor, or as CBE supplements under 21 C.F.R. § 314.70(c)(6), which do not require FDA approval before sponsor implementation. As provided by 21 C.F.R. § 314.70(c)(6)(iii), sponsors may distribute, without prior FDA approval, drugs with labeling supplements based on newly acquired information to add or strengthen a contraindication, warning, precaution, or adverse reaction, a statement about drug abuse, dependence, or overdosage, or an instruction about dosage and administration intended to increase safe use of the drug, or to delete false, misleading, or unsupported indications for use or claims for effectiveness. Changes to labeling the FDA deems "major changes," including labeling changes associated with new indications and usage, changes based on preclinical study data, changes to clinical pharmacology information, revisions of the drug population, and claims of superiority to another drug, must be submitted to and approved by the FDA before inclusion with marketed drug products.[44]

Under current FDA regulations, ANDA holders generally cannot make labeling changes through the formal supplement process in all circumstances in which NDA holders can, because ANDA drug labeling must be the same as its reference NDA drug's labeling.[45] Accordingly, the CBE supplement process is not available to ANDA sponsors except to match a reference drug's labeling or to respond to FDA-mandated labeling changes. ANDA holders are required to provide the FDA with information about labeling concerns, including concerns that new information should be added to a product's labeling,[46] and ANDA sponsors themselves are

43. 21 C.F.R. § 201.57(c)(6).

44. *See* U.S. Food & Drug Admin., Guidance for Industry: Changes to an Approved NDA or ANDA 24–25 (2004).

45. 21 U.S.C. § (j)(2)(A)(v). Certain exceptions are specified at 21 C.F.R. § 314.94(a)(8)(iv).

46. *See* 57 Fed. Reg. 17,950, 17,961 (Apr. 28, 1992).

responsible for ensuring that the labeling contained in the application is identical to the currently approved labeling of the NDA reference drug.[47]

4. Changes Being Effected Supplements and State Law Failure-to-Warn Liability

Under current law, liability for state law failure-to-warn claims differs between branded- and generic-drug manufacturers. Because the obligation to revise product labeling rests with the manufacturer of the branded reference drug, the manufacturer of a generic version of the reference drug generally is not subject to state law failure-to-warn claims. Brand manufacturers, on the other hand, are subject to state law failure-to-warn claims for their own products. Moreover, an evolving body of recent case law suggests that brand manufacturers also may be liable, under certain circumstances, for the products of their generic competitors.

The Supreme Court held in *Wyeth v. Levine* that state law failure-to-warn tort claims associated with allegedly defective branded-drug labeling were not preempted by federal regulations requiring FDA approval of drug labeling.[48] In particular, the Supreme Court held that it was not impossible for the branded manufacturer to comply with both state and federal requirements based on the availability of the CBE labeling supplement process to a branded-drug manufacturer.[49] Subsequently, in *PLIVA Inc. v. Mensing*, a case very similar to *Wyeth* but concerning a generic drug, the Supreme Court held that state failure-to-warn tort claims against generic manufacturers are preempted by federal drug regulations.[50] The ruling was based on the FDA's avowal that a generic drug's label must be identical to that of the reference branded drug, and accordingly, generic manufacturers may not strengthen drug labeling by the voluntary CBE supplement process.[51] In the wake of the *Mensing* ruling, the Supreme Court of Alabama held that under Alabama law, defective labeling of a branded prescription drug product may render that branded manufacturer liable for fraud or misrepresentation, even for injuries caused by its competitors' generic products.[52] Other courts have declined to impose such liability, however, holding that

47. *See* U.S. Food & Drug Admin., Guidance for Industry: Revising ANDA Labeling Following Revision of the RLD Labeling 4–5 (2000).

48. 555 U.S. 555 (2009).

49. *Id.* at 573.

50. 131 S. Ct. 2567 (U.S. 2011).

51. *Id.* at 2575–76.

52. Wyeth Inc. v. Weeks, 2013 WL 135753 (Ala. 2013). Other courts have applied so-called innovator liability to brand drug manufacturers. *See, e.g.*, Conte v. Wyeth Inc., 85 Cal. Rptr. 3d 299 (Ct. App. 2008); Kellogg v. Wyeth Inc., 762 F. Supp. 2d 694 (D. Vt. 2010).

a branded-drug manufacturer owes no duty of care to individuals who have never taken the drug actually manufactured by that company.[53]

In August 2013, the FDA published on the website of the Office of Management and Budget a proposed rule to "create parity between NDA holders and ANDA holders with respect to submission of CBE labeling supplements." This proposed rule could potentially expose generic-drug manufacturers to failure-to-warn liability.

5. Consumer Advertising

Pursuant to 21 U.S.C. § 352(n), a prescription drug shall be deemed misbranded unless the manufacturer or distributor includes in all advertisements a "brief summary relating to side effects, contraindications, and effectiveness" as required by regulation. The implementing regulations provide that advertising causes a prescription drug to be misbranded if it is false or misleading, fails to reveal material facts, or fails to present a fair balance of information.[54] Advertising claims regarding safety, effectiveness, and indications are limited to those appearing in the FDA-approved labeling for the drug.[55] Broadcast prescription drug advertisements must additionally contain a "major statement" conveying all of the product's most important risk information in consumer-friendly language. Pursuant to Section 901(d)(2) of the FDAAA, codified at 21 U.S.C. § 353(b), the FDA is now authorized to require submission of television drug advertisements for FDA preapproval no later than 45 days before dissemination of the advertisement.

C. Risk Evaluation and Mitigation Strategies

Risk Evaluation and Mitigation Strategies (REMS) are submissions to the FDA by drug sponsors that describe measures (beyond professional labeling) taken by the sponsors to identify and minimize the risks associated with the use of a drug. As described in detail in the following paragraphs, these measures typically include one or more of a Medication Guide for patients, a communication plan for health care providers, and Elements To Assure Safe Use (ETASU) such as, for example, training or certification for prescribing health care providers.

Pursuant to the FDA's expanded authority under Section 901 of the FDAAA, which created new Section 505-1 of the FDCA (codified at 21

53. *See, e.g.*, Foster v. Am. Home Prods. Corp., 29 F.3d 165 (4th Cir. 1994); Smith v. Wyeth Inc., 657 F.3d 420 (6th Cir. 2011); Goldych v. Eli Lilly & Co., No. 5:04-CV-1477, 2006 WL 2038436, at *6 (N.D.N.Y. July 19, 2006); Tarver v. Wyeth Inc., No. Civ.A.3-04-2036, 2005 WL 4052382, at *2 (W.D. La. June 7, 2005).

54. 21 C.F.R. § 202.1(e)(5).

55. *See* 64 Fed. Reg. 43,197 (Aug. 9, 1999).

U.S.C. § 355-1), the FDA may require an applicant for new drug approval to submit a proposed REMS as part of an NDA where the FDA determines that a REMS is necessary to ensure that the benefits of the drug outweigh its risks. The FDA may also require approved NDA sponsors to submit a REMS proposal where the FDA identifies new safety information (as defined earlier) concerning a drug. Sponsors seeking approval of new indications for approved NDAs may likewise be required to submit a REMS proposal.[56] The FDA considers the following factors when determining if a REMS is required for a drug:

- the estimated size of the population of likely users of the drug;
- the seriousness of the disease or condition to be treated by the drug;
- the expected benefit of the drug in treating the disease or condition;
- the expected duration of treatment with the drug;
- the seriousness of known or potential adverse events related to the drug and the background incidence of these events in the population likely to use the drug; and
- whether or not the drug is a new molecular entity.[57]

REMS formally supplanted the predecessor Risk Minimization Action Plan (RiskMAP) program initiated in 2005.[58] In contrast to the RiskMAP program, the FDA has broad authority to enforce REMS compliance and to require ongoing assessment of the REMS to verify its effectiveness. A drug is considered misbranded if the sponsor for the drug fails to comply with a requirement of the approved strategy.[59] Additionally, noncompliance with the requirements of an approved REMS can lead to injunction from introduction of the drug into interstate commerce.[60] Sponsors in violation of a REMS requirement may also be subject to civil penalties of up to $250,000 per violation (not to exceed $1 million in a single proceeding). These penalties double every 30 days after the sponsor is notified by the FDA to a maximum of $10 million per proceeding.[61]

---

56. 21 U.S.C. § 355-1.

57. *Id.* § 355-1(a)(1).

58. *See* U.S. Food & Drug Admin., Guidance for Industry: Format and Content of Proposed Risk Evaluation and Mitigation Strategies (REMS), REMS Assessments, and Proposed REMS Modifications 3–4 (2009) [hereinafter Format and Content of REMS]. Of the 30 RiskMAPs in place on the effective date of FDAAA enactment, 14 were not supplanted by REMS and remain in effect. The remaining 16 drugs subject to a RiskMAP were deemed to have an approved REMS in effect, and the sponsors of these drugs were required to submit corresponding REMS proposals.

59. 21 U.S.C. § 352(y).

60. *Id.* § 505(p).

61. *Id.* § 333(f)(4)(a).

1. Elements of REMS

Each REMS may include one or more of the following elements:

- a Medication Guide or patient package insert to inform patients of serious risks associated with a drug;
- a communication plan to educate health care providers about serious risks associated with a drug;
- Elements to Assure Safe Use (ETASU); and
- an Implementation System describing how the Elements to Assure Safe Use will be instituted.

Additionally, all REMS must include a timetable for submission of assessments. The FDA requires all proposed REMS to be submitted with a REMS supporting document that explains the goals, objectives, and assessments for the REMS and describes each of the elements of the REMS in detail.[62]

a. Medication Guides and Patient Package Inserts

Medication Guides are FDA-approved patient labeling provided to educate patients (rather than health care providers) about proper use of the drug, its contraindications, and specific risks associated with the drug. Medication Guides primarily apply to prescription drug products used on an outpatient basis without direct supervision by a health care professional and are distributed by drug dispensers to patients when the drug is purchased. The FDA is authorized to require a Medication Guide if it determines that patient labeling may prevent serious adverse effects of a drug, that the drug has serious risks of which patients should be made aware because these risks could affect patients' decisions to use or continue to use a drug, or that the drug is important to health, and patient compliance with directions for use is crucial to the drug's effectiveness.[63] Medication Guides are provided directly to each patient by authorized drug dispensers when the product is dispensed and may be required independently of a REMS as part of the approved labeling for a drug.[64]

Historically, the FDA approved a large proportion of REMS that required only a Medication Guide and a timetable for assessments. In November 2011, however, the FDA indicated that it generally expects to include a Medication Guide as a part of a REMS only when the REMS includes ETASU.[65] Between the date of enactment of the REMS

---

62. *Id.* § 355-1(c)–(f).

63. 21 C.F.R. § 208.1(c).

64. 21 U.S.C. § 355-1(e)(2)(A); 21 C.F.R. § 208.24.

65. *See* U.S. Food & Drug Admin., Guidance for Industry: Medication Guides—Distribution Requirements and Inclusion in Risk Evaluation and Mitigation Strategies (REMS) 3–4 (2011) [hereinafter Guidance for Industry: Medication Guides].

requirements and March 1, 2011, of approximately 200 approved REMS for drug products, 125 contained only a Medication Guide (and Timeline). As of June 2013, 105 of these REMS were released.

The FDA may additionally require a patient package insert (PPI) for distribution with a drug if it determines that the PPI may help mitigate serious risks of a drug.[66] PPIs are also a component of FDA-approved labeling that may be required independently of a REMS. PPIs contain information similar to that contained in Medication Guides, but unlike Medication Guides, PPIs are included within the physical packaging of a drug product. The FDA has indicated that requiring both a Medication Guide and a PPI is not anticipated to occur frequently.

b. Communication Plans

Communication Plans (i.e., FDA-approved materials used to further a sponsor's implementation of a REMS, to inform health care providers about serious risks associated with a drug, or both) may also be required as an element of a REMS.[67] Communication Plans may consist of "Dear Health Care Professional" letters (commonly referred to as "Dear Doctor" letters),[68] dissemination of information to health care professionals through professional societies, and dissemination of information about the REMS to promote its implementation. Dissemination of information can be achieved by, for example, providing slide decks for access over the Internet from the websites of professional societies or providing links to warnings for health care providers from drug product information websites, or creating informational websites specifically concerning the REMS.

c. Elements to Assure Safe Use

Where a drug has known serious risks such that in the FDA's determination the drug would not be approvable or would be withdrawn without risk mitigation measures beyond a Medication Guide and Communication Plan, the FDA may require "such elements as are necessary to assure safe use of the drug." These Elements to Assure Safe Use (ETASU) must specify one or more goals to mitigate a specific serious risk disclosed in the labeling of the drug. Exemplary ETASUs provided in 21 U.S.C. § 355-1(f) include as potential requirements:

66. 21 U.S.C. § 355-1(e)(2)(B).

67. *Id.* § 355-1(e)(3).

68. For draft guidance on content and format of healthcare provider letters, see U.S. FOOD & DRUG ADMIN., GUIDANCE FOR INDUSTRY AND FDA STAFF—DEAR HEALTH CARE PROVIDER LETTERS: IMPROVING COMMUNICATION OF IMPORTANT SAFETY INFORMATION (2010).

- training or certification for prescribing health care providers to prescribe the drug;
- special certification for dispensaries to dispense the drug;
- restriction of availability of the drug to certain health care settings, such as hospitals;
- documentation of safe-use conditions, such as testing results, with dispensation; and
- monitoring or registration of patients using the drug.

One particularly burdensome set of ETASU for the acne drug isotretinoin, a known teratogen, includes a computerized risk management system that requires prescribers to document each patient's negative monthly pregnancy test results and receive authorization from the computerized system prior to dispensation.

d. Implementation Systems

Where special dispensary certification, setting restrictions on drug availability, or documentation of safe-use conditions is required as an ETASU in a REMS, the REMS may also include a system through which the sponsor can take reasonable steps to monitor and improve the implementation of these elements. Such systems can include, for example, requiring certification of pharmacies and hospitals, or requiring documentation of safe-use conditions, and may also require monitoring and evaluation of implementation of such elements by health care providers, pharmacists, and others responsible for implementation.

e. REMS Supporting Documents

Proposed REMS submissions to the FDA also include a REMS supporting document that details the goals of, and provides additional details regarding, each of the proposed elements of the REMS.[69] A proposed REMS goal is the desired safety-related outcome or the understanding by patients and health care providers of the serious risks targeted by the specific REMS element. Each goal should be associated with specific objectives for achieving the goal. For example, where a REMS goal is to prevent occurrences of a severe adverse event caused by interaction of the sponsored drug with another drug, the objectives could include lowering physician co-prescription rates and pharmacist co-dispensing rates for the drugs. Proposed materials for the REMS, such as proposed educational materials, Medication Guides, enrollment forms, and prescriber and patient agreements, should be appended to the proposed REMS.

---

69. *See* Format and Content of REMS, *supra* note 58, at 9.

REMS supporting documents should also contain a REMS assessment plan, including the proposed evaluation methods for assessing the overall effectiveness of the REMS and the effectiveness of each of the REMS elements. Each of the tools or measures employed in the REMS, such as patient or dispenser registries, surveys, or data systems implemented, should also be evaluated, and the rationales for each of the chosen measures should be explained. The assessment plan should include targeted values for each measure employed in the REMS and the time frame for their achievement and should describe the types of data to be collected, the nature and timing of data collection, and analysis used in assessing the effectiveness of the REMS and its elements.

f. Timelines for Assessment

Every REMS must contain a timetable for the submission of assessments of the REMS. The timetable must include, at a minimum, assessments submitted 18 months, three years, and seven years after initial approval of the REMS, with additional assessments if required.[70] Factors considered by the FDA in determining when more frequent assessments are required include the estimated size of the population of likely users of the drug, the seriousness of known or potential risks associated with the drug, and the extent of the information available about the REMS elements to mitigate the risks of drug use.[71] The FDA may require additional assessments after REMS approval where it determines based on new safety information that a REMS element should be modified. Sponsors seeking approval of new indications of drugs subject to REMS must also submit an assessment of the existing REMS with the application.

Goals are required by statute for each of the Elements to Assure Safe Use required in a REMS, where applicable.[72] Each REMS assessment must contain, for each goal recited in the strategy, an assessment of the extent to which the approved strategy, including each element of the strategy, is meeting the goal or whether one or more goals or elements should be modified.[73] Where applicable, a REMS assessment must also contain the status of all postapproval studies and any difficulties encountered in completing the studies, and/or the status, number of enrolled participants, expected completion date, any difficulties in completion, and registration information for any postapproval clinical trials.[74]

---

70. 21 U.S.C. § 355-1(d).
71. *See* Format and Content of REMS, *supra* note 58, at 15.
72. 21 U.S.C. § 355(f)(3).
73. Format and Content of REMS, *supra* note 58, at 17.
74. 21 U.S.C. § 355-1(g)(3)(B)–(C).

2. Procedures

If the FDA determines that a REMS is required to ensure that the benefits of a drug outweigh its risks, it will notify the drug sponsor by a letter identifying the drug, the specific risks to be mitigated, and the required elements of the REMS. Upon receipt of such notification, the sponsor will have 120 days to submit a proposed REMS, including the text of any required Medication Guides, the content of any Communication Plans, and the details of any mandated or proposed Elements to Assure Safe Use, as well as all goals and implementation and assessment plans for such elements.

A sponsor for NDA approval may dispute the determination of the FDA to require a REMS. Where a REMS is required postapproval, the sponsor may dispute the required REMS content but not the FDA's determination that a REMS is required. All disputes are considered within the FDA, and judicial review is not provided by statute.

3. Modifications to REMS

The FDA may, at any time, require a sponsor to submit a proposed modification to a REMS within 120 days if the FDA determines, on the basis of new safety information, that one or more goals or elements should be added, modified, or removed from the approved strategy to either ensure that the benefits of the drug outweigh its risks or to minimize the burden on the health care system of complying with the strategy.[75] In conjunction with a REMS modification requirement, the FDA may also require safety labeling changes consistent with the modified REMS.

Sponsors of approved NDAs subject to REMS may at any time after approval of the REMS submit a proposal to the FDA to modify the approved strategy, including the addition, modification, or removal of any goal or element of the strategy.[76] Such proposed modifications must be submitted as a prior-approval supplemental application (i.e., an application requiring FDA approval) rather than a CBE supplement and must be submitted with an assessment of the REMS.

Sponsors currently subject to a REMS consisting only of a Medication Guide and timeline may submit a prior approval supplement that proposes REMS modification to eliminate the REMS if they believe that the REMS is not necessary to ensure that the drug's benefits outweigh its risks. Sponsors subject to a REMS consisting only of a Medication Guide and Communication Plan may also submit a proposed modification to remove the Medication Guide from the REMS if they believe it is unnecessary.

75. *Id.* § 355-1(g)(2)(C).

76. *Id.* § 355-1(g)(4).

The proposed modifications must be accompanied by a REMS assessment. If the REMS was previously assessed in the past 18 months, the required assessment may consist of a statement to that effect. If removed, the Medication Guide will continue to be a part of the approved labeling unless the FDA approves a supplement removing the Medication Guide from the approved labeling.[77]

4. REMS for Generic Drugs—Shared REMS

A drug that is the subject of an ANDA is only subject to the Medication Guide and ETASU requirements, if applicable, for the reference branded drug.[78] As provided by 21 U.S.C. § 355-1(i), "[a] drug that is the subject of an abbreviated new drug application and the listed drug shall use a single, shared system under subsection (f)." The statute further provides that the FDA may waive the requirement and permit the ANDA applicant to use a different, comparable aspect of the ETASU if the FDA determines either that

- the burden of creating a single, shared system outweighs the benefit of a single system, taking into consideration the impact on health care providers, patients, the applicant for the abbreviated new drug application, and the holder of the reference drug product; or
- an aspect of the Elements to Assure Safe Use of the applicable listed drug is claimed by a patent or is a method or process that is a protected trade secret for which the ANDA applicant sought and was unable to obtain a license.[79]

The FDCA further provides that "no holder of an approved covered application shall use any [ETASU] required by the Secretary under this subsection to block or delay approval of an application [of an ANDA] or to prevent application of such element [. . .] to a drug that is the subject of an [ANDA]."[80]

There is at present no FDA guidance or rule making addressing the standards and processes for the negotiation of single, shared REMS between NDA and ANDA sponsors. On May 31, 2013, a branded-drug manufacturer filed a citizen petition with the FDA requesting the FDA to initiate notice and comment rule making to establish the standards for shared REMS and waivers to the requirement for shared REMS.[81] The

77. *See* Guidance for Industry: Medication Guides, *supra* note 65, at 8.
78. 21 U.S.C. § 355-1(i).
79. *Id.* § 355-1(i)(1)(B)(i)-(ii).
80. *Id.* § 355-1(f)(8).
81. Citizen Petition No. FDA-2013-P-0572.

citizen petition notes the possible antitrust implications of shared REMS, particularly when the brand and generic manufacturers are engaged in Hatch-Waxman litigation. The petition notes in particular the potential for "pay for delay" allegations, which are especially relevant in light of the subsequent Supreme Court ruling in *FTC v. Actiavis Inc.*,[82] that reverse payment ANDA suit settlements are subject to rule of reason scrutiny. The petition further notes that many REMS programs require substantial investment by innovator companies and argues that statute-mandated negotiations prevent the development of market-based assessment of compensation to innovator companies.

The FDA additionally will seek to consolidate REMS for different drug products of the same drug class pursuant to 21 U.S.C. § 355-1(f)(5) to reduce the burdens of REMS compliance on the health care system, including prescribers, pharmacies, distributors, and patients, where the drug class is subject to particularly burdensome ETASU. There are currently six approved shared system REMS, of which three are opioid drug product class REMS. The shared system REMS for isotretinoin products was converted from the predecessor RiskMAPS program.

D. Postmarketing Studies and Clinical Trials

Approved NDA sponsors often undertake postmarketing (i.e., Phase 4) studies voluntarily, either to identify potential safety issues proactively or in support of marketing approval for additional indications or populations. Although health care providers may prescribe approved drugs for any indication, only approved indications may be included in drug labeling, and accordingly, only approved indications can be promoted to physicians and consumers. Additionally, as discussed in Section IIB, "Data and Marketing Exclusivities," later in this chapter, the Hatch-Waxman scheme provides significant exclusivity incentives for approved NDA holders to seek approval of new indications as well as pediatric population approval.

Prior to enactment of the FDAAA, the FDA was authorized to require postmarketing studies or clinical trials only in specific circumstances where clinical benefit or safety in humans or the pediatric population had not been fully evaluated at the time of approval. In particular, the FDA

82. 133 S. Ct. 2223 (U.S. 2013).

was only authorized to require postmarketing studies for drugs that were granted accelerated approval[83] or subject to deferred pediatric studies.[84]

Section 505(o)(3) of the FDCA now further authorizes the FDA to require sponsors of approved prescription drugs to conduct postmarketing studies and clinical trials for any or all of the following purposes:

- to assess a known serious risk related to the use of the drug;
- to assess signals of serious risk related to the use of the drug; and
- to identify an unexpected serious risk when available data indicate the potential for a serious risk.

The FDA's determination that a study or trial is required must be based on scientific data deemed appropriate by the FDA.[85] Such information can include data regarding chemically related or pharmacologically related drugs. For the purposes of the FDAAA, "clinical trials" are defined as any prospective investigations in which the sponsor/investigator determines the method of assigning the drug product to human subjects, while "studies" are defined as all other investigations, including observational epidemiologic studies, animal studies, and laboratory experiments.[86] NDA sponsors may be required to undertake postmarketing studies or trials at the time of approval or at any time after approval if the FDA becomes aware of new safety information.[87]

The legislation attempts to calibrate the regulatory burden on sponsors according to the necessity of the postmarketing requirement to evaluate serious drug risks. The FDA must find that ADE reporting (and the pharmacovigilance system in development by the FDA) will not be sufficient to assess or identify serious risks before requiring a sponsor to conduct a postmarketing study and must find that a postmarketing study

83. The FDA may grant marketing approval for a new drug product for treatment of serious or life-threatening illnesses and that provide meaningful patient benefit over existing treatments on the basis of clinical trials establishing efficacy of the drug on an indicator of clinical benefit other than survival or morbidity (i.e., on the basis of a "surrogate endpoint"). *See* 21 C.F.R §§ 314.500, .510. The FDA may also grant marketing approval for drugs for treatment of severe toxicity due to exposure to biological, chemical, radiological or nuclear substances on the basis of animal studies which establish that the drug is reasonably likely to produce clinical benefit in humans where human studies would be unethical. *See id.* §§ 314.600, .610.

84. The FDA may, on its own initiative or at a sponsor's request, defer pediatric safety and efficacy studies if, for example, the drug is ready for approval for adult populations before pediatric studies are complete, or where pediatric studies should be delayed until further safety and efficacy data are collected. *See id.* § 314.55.

85. 21 U.S.C. § 355(o)(3)(A).

86. *See* U.S. Food & Drug Admin., Guidance for Industry: Postmarketing Studies and Clinical Trials—Implementation of Section 505(o)(3) of the Federal Food, Drug, and Cosmetic Act 4 (2011) [hereinafter Guidance for Industry: Postmarketing Studies and Clinical Trials].

87. 21 U.S.C. § 355(o)(3)(C).

will not be sufficient to assess and identify these risks before requiring a sponsor to conduct a postmarketing clinical trial.[88] However, the FDA has stated in industry guidance that it believes that the authority to require a sponsor to conduct one or more postapproval studies includes the authority to describe the study or trial to be conducted, including how the study or trial is to be done and the populations and indications to be evaluated.[89]

Postmarketing study and trial requirements are enforceable by the FDA, which can pursue misbranding charges under 21 U.S.C. § 332(z), injunctions against introducing the drug into interstate commerce under 21 U.S.C. § 355(o)(1), or civil monetary penalties according to the schedule outlined earlier in this chapter.

1. Postmarketing Requirements and Commitments

The FDAAA distinguishes between postmarketing requirements (PMRs) and postmarketing commitments (PMCs). Postmarketing requirements now refer to all postmarketing studies or clinical trials required by the FDA, including those required under the FDAAA and those required under the accelerated approval provisions of subpart H of 21 C.F.R. part 314, the Pediatric Research Equity Act, and the Animal Efficacy Rule.[90]

Prior to passage of the FDAAA, postmarketing commitments referred to studies and clinical trials that were either required by the FDA under the limited circumstances noted earlier or that a sponsor agreed to undertake after NDA approval to refine the safety, efficacy, or optimal use of a product or to ensure product quality. Postmarketing commitments now refer to studies and clinical trials that the sponsor has agreed to conduct but that are not required to assess or identify serious risks associated with a drug under Section 505(o)(3)(B) of the FDCA. Under the FDAAA, PMCs, if undertaken, are subject to the input and review of the FDA with regard to their design, conduct, scheduling, and goals.[91]

Section 506B of the FDCA, which was enacted pursuant to the Food and Drug Modernization Act of 1997, gave the FDA the authority to monitor the progress of PMCs by requiring the applicant to provide annual status updates regarding any PMCs, including, if applicable, the reasons for failure to complete the commitment. The FDA has indicated that it will not convert these preexisting PMCs into PMRs unless new safety information emerges relating to a PMC that is not progressing according to schedule.

---

88. *Id.* § 355(o)(3)(D).

89. *See* GUIDANCE FOR INDUSTRY: POSTMARKETING STUDIES AND CLINICAL TRIALS, *supra* note 86, at 6.

90. *See id.* at 3.

91. *Id.* at 5; *see also* 21 U.S.C. § 356b.

As part of FDAAA implementation, the FDA has provided industry guidance regarding the types of studies considered PMRs and those that would be considered PMCs if undertaken. Generally, studies and trials to evaluate specific serious risks or safety endpoints will be considered PMRs. Such studies and trials will include, without limitation:

- observational pharmacoepidemiologic studies designed to assess a serious risk or evaluate factors affecting a risk of serious toxicity associated with drug exposure;
- meta-analyses designed to evaluate a safety endpoint by statistical analysis of data from completed studies or clinical trials;
- clinical trials having a safety endpoint with prespecified assessments and sufficient statistical power to analyze a serious risk identified by the FDA;
- safety studies in animals investigating specific end-organ toxicities (if conducted postapproval);
- in vitro studies to analyze the potential for nonspecific drug activity or the development of drug resistance to antibiotics, or to validate immunogenicity assays used to assess immunologic safety concerns;
- pharmacokinetic studies or trials to evaluate pharmacokinetics in a labeled population or a subpopulation at potential risk of high drug exposure (such as patients with renal or hepatic insufficiency); and
- studies or clinical trials to evaluate bioavailability or potential drug interactions where data indicate a potential serious risk.[92]

In contrast, studies that would not be required under Section 505(o)(3) but that would be considered PMCs if undertaken after agreement between the sponsor and the FDA would include, without limitation:

- drug quality studies, including manufacturing, stability, and immunogenicity studies that do not have a primary safety endpoint;
- pharmacoepidemiologic studies to evaluate a disease or to evaluate the background incidence of adverse events in a population not treated with the drug; and
- clinical trials in which the primary endpoint is related to efficacy.[93]

As discussed later in this chapter, sponsors may additionally or alternatively conduct postapproval studies and clinical trials voluntarily and without agreement between the sponsor and the FDA. Such studies are

92. *See* Guidance for Industry: Postmarketing Studies and Clinical Trials, *supra* note 86, at 7–8.
93. *Id.* at 8–10.

not subject to FDA input, review, or scheduling requirements, but are subject to separate reporting requirements.[94]

2. Reporting

All safety-related postapproval studies and clinical trials are subject to FDA reporting requirements. For studies and trials required by the FDA, the sponsor is subject to periodic reporting requirements as prescribed by the FDA during its review of the PMR, while for PMCs, sponsors must provide annual status reports (506B reports) to the FDA.[95] These reporting requirements are maintained until the FDA notifies the sponsor that it is released from the reporting requirement. PMR updates must additionally be described in REMS assessments. Voluntary safety studies undertaken by the sponsor without agreement with the FDA must be summarized in the annual drug reports required pursuant to 21 C.F.R. § 314.81(b)(2), as discussed in the following section.

E. Additional Postapproval Regulatory Requirements—Annual Reports

Under the comprehensive postmarketing surveillance for approved drugs, in addition to the requirements previously outlined, sponsors of approved drugs are subject to numerous regulatory requirements for the duration of drug approval. Most notably, drug sponsors have a continuing obligation to inform the FDA of significant safety-related information and to maintain drug product quality integrity. The primary means by which the FDA ensures compliance with these requirements is through field reporting and annual reporting requirements.

Drug sponsors (NDA and ANDA) must provide field alert reports of all incidents in which a drug product or its labeling is mistaken for or applied to another drug article and of all incidents of bacteriological contamination, drug product deterioration, or failure of a drug product batch to meet the specifications for the product established in the application for drug approval.[96] FDA regulations require that drug sponsors provide the field reports within three working days of receipt to the FDA field office responsible for the facility involved with the incident.[97] Approved drug sponsors must additionally submit an annual report each year within 60 days of the anniversary date of U.S. approval of the drug application.[98] The annual report must contain, among other things:

94. *See* 21 U.S.C. § 355(o)(3)(E)(ii); 21 C.F.R. § 314.81(b)(2)(viii).
95. 21 U.S.C. § 356(b); 21 C.F.R. § 314.81(b)(2)(vii).
96. 21 C.F.R. § 314.81(b)(1).
97. *Id.*
98. *Id.* § 314.81(b)(2).

- a summary of significant new information concerning the safety, effectiveness, or labeling of the drug product and the actions taken by the applicant as a result of the new information;
- distribution data describing the number of dosage units of each strength distributed for both domestic and foreign use;
- labeling information, including the current professional labeling, patient package inserts if applicable, and a representative sample of the package labels, as well as a summary of all changes to the labeling since the previous annual report;
- clinical data, including published trials of the drug and summaries of completed unpublished trials (including voluntary studies), and analysis of available pediatric safety and efficacy data;
- status reports for each postmarketing study commitment;
- specimens of advertisements and promotional labeling; and
- chemistry, manufacturing, and control changes, including a full description of changes not requiring a supplemental application[99] and reports of new data involving drug properties that may affect the FDA's previous conclusions about the safety or efficacy of the drug.[100]

II. Life Cycle Management in the Regulatory Context

The life cycle of a drug product, including any changes to the drug product and the duration of innovator exclusivity, is managed within the context of both the Hatch-Waxman and postmarketing surveillance regulatory frameworks. The following section outlines the major regulatory considerations in life cycle management and describes strategies employed by branded-drug manufacturers to forestall generic entry based upon the postmarketing surveillance regulatory regime.

A. Postapproval Changes to an NDA

Section 506A of the FDCA and FDA regulations codified at 21 C.F.R. § 314.70 set forth certain requirements for NDA sponsors that plan to make postapproval changes to drugs. Specifically, sponsors must notify the FDA about, and describe fully, each change in each condition established in an approved application beyond the variations already provided for in

99. Discussed at Section II.A.
100. 21 C.F.R. § 314.81(b)(2)(c)(iv).

the application.[101] Depending on the type of change, applicants must notify the FDA either in a supplement or by inclusion of the information in their annual report to the application.[102] As provided by 21 C.F.R. § 314.70, such changes are divided into four reporting categories, including: (1) major changes, (2) moderate changes requiring submission of a supplement 30 days before distribution of the drug, (3) moderate changes requiring submission of a supplement upon distribution of the drug, and (4) minor changes. Each of these reporting categories is discussed in more detail in the following sections. Additionally, the applicant must assess the effects of the change before distributing a drug product made with the change.

1. Major Changes

A major change includes any change in drug substance, drug product, production process, quality controls, equipment, or facilities that has a substantial potential to have an adverse effect on the identity, strength, quality, purity, or potency of a drug product as these factors may relate to the safety or effectiveness of the drug.[103] Examples include, but are not limited to:

- changes in the qualitative or quantitative formulation of the drug product, including inactive ingredients, or in the specifications provided in the approved application;
- changes requiring completion of studies to demonstrate the equivalence of the drug product;
- changes that may affect drug substance or drug product sterility assurance, such as changes in drug substance, drug product, or component sterilization methods or addition, deletion, or substitution of steps in an aseptic processing operation;
- changes in the synthesis or manufacture of the drug substance that may affect the impurity profile and/or the physical, chemical, or biological properties of the substance;
- certain labeling changes;
- changes in a drug product container closure system that controls the drug product delivered to a patient, or changes in the type or composition of a packaging component that may affect the impurity profile of the drug product;
- changes solely affecting a natural product, a recombinant DNA-derived protein/polypeptide, or a complex or conjugate of a drug substance with a monoclonal antibody, including changes in the

101. *Id.* § 314.70(a)(2).
102. *Id.* § 314.70(a)(1)(i).
103. *Id.* § 314.70(b).

virus or adventitious agent removal or inactivation methods, changes in the source material or cell line, and establishment of a new master cell bank or seed; and

- changes to a drug product under an application that is subject to a validity assessment because of significant questions regarding the integrity of data.[104]

When making a major change, applicants must submit a supplement to the FDA, which must be approved by the FDA before distribution of the drug made with the change may begin. The submission for a major change is a Prior Approval Supplement and should be labeled as such. Prior Approval Supplements must include (1) a detailed description of the proposed change; (2) the drug products involved; (3) the manufacturing site or area affected; (4) a description of the methods used and studies performed to assess the effects of the change; (5) the data derived from such studies; (6) for a natural product, a recombinant DNA-derived protein/polypeptide, or a complex or conjugate of a drug substance with a monoclonal antibody, relevant validation protocols and standard operating procedures; and (7) for sterilization processes or test methodologies related to sterilization process validation, relevant validation protocols and standard operating procedures.[105]

If desired, an applicant can request expedited review by the FDA of a Prior Approval Supplement if a delay in making the change would cause extraordinary hardship on the applicant (e.g., based on events that could not have easily been foreseen such as a fire) or for health reasons such as a drug shortage. If expedited review is requested, the submission should be entitled a Prior Approval Supplement—Expedited Review Requested and should be labeled as such.[106]

2. Moderate Changes

A moderate change includes any change in drug substance, drug product, production process, quality controls, equipment, or facilities that has a moderate potential to have an adverse effect on the identity, strength, quality, purity, or potency of a drug product as these factors may relate to the safety or effectiveness of the drug.[107] Examples include, but are not limited to:

- changes in the container closure system that do not affect the quality of the drug product;
- changes solely affecting a natural product, a recombinant DNA-derived protein/polypeptide, or a complex or conjugate of a drug

104. *Id.* § 314.70(b)(2)(i)–(viii).
105. *Id.* § 314.70(b)(3).
106. *Id.* § 314.70(b)(4).
107. *Id.* § 314.70(c)(1).

substance with a monoclonal antibody, including an increase or decrease in production scale during finishing steps that involves different equipment and replacement of equipment with that of a different design that does not affect the process methodology or process operating parameters; and

- relaxation of an acceptance criterion or deletion of a test to comply with an official compendium that is consistent with FDA statutory and regulatory requirements.[108]

When making a moderate change, applicants must submit a supplement to the FDA that provides a full explanation of the basis for the change and the date on which the change is to be made.[109] The supplement must include (1) a detailed description of the proposed change; (2) the drug products involved; (3) the manufacturing site or area affected; (4) a description of the methods used and studies performed to assess the effects of the change; (5) the data derived from such studies; (6) for a natural product, a recombinant DNA-derived protein/polypeptide, or a complex or conjugate of a drug substance with a monoclonal antibody, relevant validation protocols and standard operating procedures; and (7) for sterilization processes or test methodologies related to sterilization process validation, relevant validation protocols and standard operating procedures.[110] Moderate changes are divided into two types—those that require submission of a supplement 30 days before distribution of the drug and those that require submission of a supplement at distribution of the drug.

The submission for the first type of moderate change is called a Supplement—Changes Being Effected in 30 Days.[111] Within a 30-day window after filing the supplement, the FDA may inform the applicant that a Prior Approval Supplement (i.e., the supplement for a major change) is required or that certain information is missing, upon receipt of which notification the applicant must delay distribution of the drug made using the change until the missing information is provided.[112]

The submission for the second type of moderate change is called a Supplement—Changes Being Effected.[113] Applicants of this type of submission may distribute the drug made with the change upon filing the supplement. However, the FDA could disapprove the supplement and order the applicant to stop distribution of the drug made with the change.[114]

108. *Id.* § 314.70(c)(2)(i)–(iii).
109. *Id.* § 314.70(c)(3).
110. *Id.* §§ 314.70(c)(4), 314.70(b)(3).
111. *Id.* § 314.70(c)(3).
112. *Id.* § 314.70(c)(5)(i).
113. *Id.* § 314.70(c)(6).
114. *Id.* § 314.70(c)(7).

Examples of changes that require this type of submission include, but are not limited to:

- additions to a specification or changes in the methods or controls to provide increased assurance that the drug substance or drug product will have the characteristics of identity, strength, quality, purity, or potency that it purports or is represented to possess;
- changes in the size and/or shape of a container for a non-sterile drug product, except for solid dosage forms, without changes in the labeled amount of drug product or from one container closure system to another; and
- certain changes in the labeling.[115]

3. Minor Changes

A minor change includes any change in drug substance, drug product, production process, quality controls, equipment, or facilities that has minimal potential to have an adverse effect on the identity, strength, quality, purity, or potency of a drug product as these factors may relate to the safety or effectiveness of the drug. Examples include, but are not limited to:

- changes made to comply with a change to an official compendium, except the changes previously described for a moderate change, that are consistent with FDA statutory and regulatory requirements;
- deletion or reduction of an ingredient intended to affect only the color of the drug product;
- replacement of equipment with that of the same design and operating principles except those equipment changes previously described for a moderate change;
- changes in the size and/or shape of a container containing the same number of dosage units for a non-sterile solid dosage form drug product, without changes from one container closure system to another;
- changes within the container closure system for a non-sterile drug product, based upon a showing of equivalency to the approved system under a protocol approved in the application or published in an official compendium;
- an extension of an expiration dating period based upon full shelf life data on production batches obtained from a protocol approved in the application;
- addition or revision of an alternative analytical procedure that provides the same or increased assurance of the identity, strength,

115. *Id.* § 314.70(c)(6)(i)–(iii).

quality, purity, or potency of the material being tested as the analytical procedure described in the approved application, or deletion of an alternative analytical procedure;

- addition by embossing, debossing, or engraving of a code imprint to a solid oral dosage form drug product other than a modified release dosage form, or a minor change in an existing code imprint;
- changes in the labeling concerning the description of the drug product, or in the information about how the drug product is supplied, that do not involve a change in the dosage strength or dosage form; and
- editorial or similar minor changes in labeling.[116]

When making a minor change, the applicant does not need to submit a supplement to the FDA. Instead, the applicant must describe each minor change in its next Annual Report for the application.[117] Such disclosure must include: (1) a statement by the holder of the approved application that the effects of the change have been assessed; (2) a full description of the manufacturing and control changes, including the manufacturing site or area involved; (3) the date each change was implemented; (4) data from studies and tests performed to assess the effects of the change; and (5) for a natural product, a recombinant DNA-derived protein/polypeptide, a complex or conjugate of a drug substance with a monoclonal antibody, sterilization process or test methodology related to sterilization process validation, a cross-reference to relevant validation protocols and/or standard operating procedures.[118]

4. Efficacy Supplement for New Indications

Pursuant to 21 C.F.R. § 314.54, new indications or other postapproval modifications that require new investigations, other than bioavailability or bioequivalence studies, for the approval of the change may not be approved via an ANDA. Instead, any person seeking approval of a drug product that is a modification of a listed drug (e.g., new indication, dosage, route of administration, etc.) and for which investigations, other than bioavailability or bioequivalence studies, are essential to approval of the modification may submit a 505(b)(2) new drug application.[119] However, the applicant may not submit an application under 505(b)(2) if the only difference from the listed drug is that (1) the extent to which its active ingredient is absorbed or otherwise made available to the site of action is less than that of the listed drug; or (2) the rate at which its active ingredient is absorbed

116. *Id.* § 314.70(d)(2)(i)–(x).
117. *Id.* § 314.70(d).
118. *Id.* § 314.70(d)(3).
119. *Id.* § 314.54(a).

or otherwise made available to the site of action is unintentionally less than that of the listed drug.[120]

The 505(b)(2) application allows the applicant to rely on the FDA's findings of safety and effectiveness for the listed drug. The 505(b)(2) application need only contain information required to support the modification of the listed drug.[121] Determining specifically what information should be included to meet this standard can be challenging and varies from case to case. FDA draft guidance suggests that an applicant submit a plan, identifying (1) the types of bridging studies to be conducted and (2) the components of its application for which it plans to rely on the FDA's findings for the listed drug, to the appropriate new drug evaluation division before submitting the application.

In 21 C.F.R. § 314.54, the detailed requirements are listed for the contents of the 505(b)(2) application. The applicant must submit a complete archival copy of the application that includes:

- an application form; index; summary; certain technical sections, including the chemistry, manufacturing, and controls and human pharmacokinetics and bioavailability; samples and labeling; and other items as set forth in detail in 21 C.F.R. § 314.50(a), (b), (c), (d)(1), (d)(3), (e), and (g), except that 314.50(d)(1)(ii)(c) shall contain the proposed or actual master production record, including a description of the equipment, to be used for the manufacture of a commercial lot of the drug product;
- certain technical sections, including nonclinical pharmacology and toxicology, microbiology, clinical data, and statistical analyses, and case report forms and tabulations, as set forth in detail in 21 C.F.R. § 314.50 (d)(2), (d)(4) (if an anti-infective drug), (d)(5), (d)(6), and (f), as needed to support the safety and effectiveness of the drug product;
- identification of the listed drug for which the FDA has made a finding of safety and effectiveness and on which the applicant relies in seeking approval by established name, if any, proprietary name, dosage form, strength, route of administration, name of the listed drug's application holder, and the listed drug's approved application number;
- if the applicant is seeking approval only for a new indication and not for the indications approved for the listed drug, a certification so stating;
- patent information with respect to any patent that claims the drug for which approval is sought or a method of using such drug and to

120. *Id.* § 314.54(b).
121. *Id.* § 314.54(a).

which a claim of patent infringement could reasonably be asserted if a person not licensed by the owner of the patent engaged in the manufacture, use, or sale of the drug product;

- a patent certification in accordance with 21 C.F.R. § 314.50(h) with respect to any relevant patents that claim the listed drug or that claim any other drugs on which investigations relied on by the applicant for approval of the application were conducted, or that claim a use for the listed or other drug; and
- information required to show that the applicant is entitled to a period of exclusivity.[122]

The 505(b)(2) application must also include a review copy of the application, including the chemistry, manufacturing, and controls and human pharmacokinetics and bioavailability sections described in 21 C.F.R. § 314.50(d)(1) and (3), except that 314.50(d)(1)(ii)(c) shall contain the proposed or actual master production record, including a description of the equipment, to be used for the manufacture of a commercial lot of the drug product, and the nonclinical pharmacology and toxicology, microbiology, clinical data, and statistical analyses, and case report forms and tabulations, as set forth in detail in 21 C.F.R. § 314.50(d)(2), (d)(4), (d)(5), (d)(6), and (f) when needed to support the modification. Each of the technical sections in the review copy is required to be separately bound with a copy of the application form, index, and summary information required under 21 C.F.R. § 314.50(a), (b), and (c) and a copy of the proposed labeling.

The 505(b)(2) application must also include a field copy of the application, including the chemistry, manufacturing, and controls section described in 21 C.F.R. § 314.50(d)(1), a copy of the application form and summary information required under 21 C.F.R. § 314.50(a) and (c), and certification that the field copy is a true copy of the technical section described in 21 C.F.R. § 314.50(d)(1) contained in the archival and review copies of the application.

As discussed in the following section, approval of the 505(b)(2) application can provide for three years of Hatch-Waxman exclusivity if the application contained reports of new clinical investigations (other than bioavailability studies) that (1) were conducted or sponsored by the applicant and (2) were essential to approval of the application.[123] The investigations must be new and not relied on by the FDA to demonstrate safety and/or effectiveness of a previously approved drug product for any indication. For the investigations to be "essential to approval," there must not be other available data that could have supported approval. Approval of the 505(b)(2) application can also

122. *Id.* § 314.54(a)(1)(i)–(vii).
123. *Id.* § 314.108(b)(4)–(5).

provide for orphan drug and/or pediatric exclusivity, which are discussed in detail in the following section.

B. Data and Marketing Exclusivities

The Hatch-Waxman Act, formally known as the Drug Price Competition and Patent Term Restoration Act of 1984, P.L. 98-417, provides exclusivity incentives to drug sponsors to invest in drug improvements, new approved labeling indications, additional approvals for new patient populations, and additional studies. These can extend marketing exclusivity after patent expiration. In particular, the Hatch-Waxman Act contains exclusivity provisions that prescribe periods of time during which the FDA may not give effective approval to ANDAs submitted according to Section 505(j) of the FDCA or "paper" NDAs[124] submitted under Section 505(b)(2) of the FDCA. Depending on the innovation sponsored by the applicant, either marketing exclusivity or data exclusivity is provided. Data exclusivity prevents generic competitors from relying on data generated by branded-drug manufacturers for regulatory approval, while marketing exclusivity prevents the FDA from approving the same or a highly similar drug product for the period of exclusivity. However, during a term of data exclusivity, a generic competitor is permitted to submit its own NDA based on independently sponsored clinical studies of safety and efficacy.

Each type of exclusivity will be addressed in the following paragraphs. However, generally, the exclusivity periods are:

- New Chemical Entity (NCE): five years of data exclusivity;
- New Clinical Study Exclusivity, or applications for new uses or formulations: three years of data exclusivity;
- Orphan drug exclusivity: seven years of marketing exclusivity after the date of approval of the pioneer NDA; and
- Pediatric exclusivity: six months of marketing exclusivity added to all other exclusivities (including patent exclusivity).

1. New Chemical Entity (NCE) Exclusivity

A New Chemical Entity (NCE) is defined as "a drug that contains no active moiety that has been approved by FDA in any other application submitted" for new drug approval under Section 505(b)(1) of the FDCA.[125] Data exclusivity is available for five years after the date of approval of a pioneer NDA to an NCE, although generic-drug makers may file an ANDA or a

124. New drug applications in which the required safety and effectiveness data are obtained from literature reports of testing by third parties, rather than the original testing by the original NDA sponsor. *See* 21 U.S.C. § 355(b)(2).

125. 21 C.F.R. § 312.108(b).

Section 505(b)(2) application with a Paragraph IV Certification (alleging patent invalidity) one year before expiration of marketing exclusivity.[126]

The five-year data exclusivity is measured from the date of approval of a drug containing an NCE. This exclusivity only applies to the *first* approval of an NCE in any drug product; if another drug product for a different use or indication containing the same NCE is filed, there is no second five-year exclusivity allowed.

2. New Clinical Study Exclusivity

New Clinical Study exclusivity essentially covers new formulations or new indications of an existing approved drug product. New formulation submissions refer to NDA applications for changes to an approved drug product that affect the metabolism of its active ingredient(s), such as new strength, dosage form, route of administration, or conditions of use; new indication submissions, in contrast, refer to NDA or labeling change applications for new therapeutic uses for an existing drug product (in a new or existing formulation).

The requirements for New Clinical Study exclusivity are as follows:[127]

- the study cannot apply to active moiety itself;
- the studies may not be bioequivalency or bioavailability studies;
- the studies must be conducted or sponsored by the applicant;
- the studies must be new; and
- the studies must be essential to approval.

Under New Clinical Study exclusivity, an applicant receives three-year marketing exclusivity, which prevents the FDA from approving a 505(b)(2)[128] application or an ANDA;[129] in contrast to NCE exclusivity, however, New Clinical Study exclusivity does not prevent the FDA from accepting a 505(b)(2) application or an ANDA. Moreover, data exclusivity attaches only if the clinical investigations were essential to the approval of the application containing those changes.

3. Orphan Drug Exclusivity

Orphan drug exclusivity (ODE) is granted to the first approved NDA for a drug intended for the treatment of a rare disease or condition. To qualify for ODE, an NDA must cover a drug for which the sponsor has no reasonable expectation of recouping development costs in the United States or

126. 21 U.S.C. § 355(c)(3)(E)(ii), (j)(5)(F)(ii).
127. 21 C.F.R. § 314.108(a).
128. 21 U.S.C. § 355(c)(3)(E)(iii)–(iv).
129. *Id.* § 355(j)(5)(F)(iii), (iv).

a drug for the treatment of a disease or condition that affects fewer than 200,000 people in the United States.

ODE applies only to the first designated drug to be approved for the orphan indication.[130] Orphan drugs receive seven years of marketing exclusivity, which blocks approval of the same active drug for the same medical indication. Orphan drug status is maintained even if the drug becomes highly successful. It should be noted that another product may be approved if it uses a different drug or if it is clinically superior, offers greater safety, or is a major contribution to patient care.

4. Pediatric Exclusivity

Pediatric exclusivity is granted to applicants who successfully complete FDA-requested clinical trials of a drug in a pediatric population.[131] An applicant for pediatric exclusivity would receive six-month marketing exclusivity added to all other exclusivities and (effectively) to patents in the Orange Book. Pediatric exclusivity does not extend patent term, but it extends the period during which the FDA may not approve competitors' ANDAs or paper NDAs. This applies to any patents listed in the Orange Book for any drug product containing the same active moiety as the drug studied, and, generally, products with no patent life or exclusivity remaining do not qualify. However, pediatric exclusivity may be granted to a product with no remaining exclusivity if the supplemental application itself qualifies for a new exclusivity period under the Hatch-Waxman Act.

5. ANDA First-to-File Exclusivity

In addition to exclusivity for the NDA filer, the Hatch-Waxman Act provides a 180-day marketing exclusivity period (as against other generics) for the first-to-file generic applicant that submits a Paragraph IV Certification.[132] The rules by the FDA are intended to clarify Abbreviated New Drug Application (ANDA) exclusivity rights by providing the following guidelines:

- only the applicant submitting the first substantially complete ANDA (and Paragraph IV Certification) will be eligible for the 180-day exclusivity;
- FDA will consider the first applicant to be eligible for exclusivity even if it is not sued by the patent owner or NDA holder (and thus not penalize an ANDA applicant that designs around a patent in order to avoid an infringement suit);

130. *Id.* §§ 360aa– dd; 21 C.F.R. § 316.
131. *See id.* § 355a.
132. *Id.* § 355(j)(5)(B)(iv).

- if a number of applicants file their ANDAs on the same day, all will be eligible for 180-day exclusivity for the strength of the drug that is the subject of their ANDA;
- each strength of a drug will separately be eligible for exclusivity; and
- the first ANDA applicant will not be eligible for 180-day exclusivity if it is sued by the NDA holder and loses.

The Medicare Prescription Drug, Improvement, and Modernization Act of 2003 provides six different routes whereby 180-day exclusivity can be forfeited. These include:[133]

- failure to market within a specific period by the first applicant;
- withdrawal of the application by the first applicant;
- amendment or withdrawal of the certification by the first applicant;
- failure to obtain tentative approval within 30 months;
- entrance into an agreement that violates antitrust laws; and
- expiration of all of the listed patents that have been certified.

C. Strategies to Forestall Generic Competition

In addition to the exclusivity opportunities created by the Hatch-Waxman regime, the postmarketing surveillance scheme may present barriers by which branded-drug sponsors can exclude generic competitors. In particular, "innovator" drug sponsors have at least temporarily forestalled generic entry by patenting Elements to Assure Safe Use (ETASU) of FDA-mandated REMS. Branded-drug manufacturers have also refused to supply generic-drug manufacturers with drug samples to conduct bioequivalence determinations as required for ANDA approval. Additionally, it is also possible to obtain patents on methods relating to safe administration of a drug product. Where such methods are important to safe administration of a drug, inclusion of the patented information in the drug labeling will not be susceptible to "carve-out" by generic manufacturers and may thus act as a barrier to generic entry for the duration of the patent.

1. Protecting Elements to Assure Safe Use as Intellectual Property

As with any novel, nonobvious, patent-eligible subject matter, it is possible to obtain patents claiming various aspects of Elements to Assure Safe Use of a drug product. As previously discussed in Section IC, "Risk Evaluation and Mitigation Strategies," where an innovator's REMS is covered by patent, the regulations regarding shared REMS provide that the FDA may waive the requirement that generics and innovators shall use a single,

133. *Id.* § 355(j)(5)(D)(i)(1).

shared REMS program where the generic sponsor has sought a license but was unable to obtain one.[134] The innovator is required by statute to negotiate with the generic sponsor regarding a license, but the FDA may not compel the innovator to grant a license, and such negotiations typically result in delayed generic entry. Where a license to a patented (or protected trade secret) REMS is not obtained, the FDAAA indicates that the generic REMS must be comparable to that of the branded drug. The design and implementation of a unique yet comparable REMS are long processes and may be cost-prohibitive to generic manufacturers. Accordingly, although the REMS regulations expressly state that innovators may not use a REMS to "block or delay" ANDA approval or to prevent application of an ETASU,[135] the existence of a protected REMS requirement may nonetheless result in a practical barrier to generic entry.

This strategy has been used with success by branded-drug manufacturers. Most notably, the manufacturer of thalidomide obtained a patent to the ETASU for the REMS program for the drug, which contains elements to severely restrict the distribution of the drug.[136] The manufacturer refused to license the patent and subsequently submitted a citizen petition to the FDA arguing that without an equivalent REMS program, any generic version of the drug would not be safe and requesting that no generic version of the drug be approved.[137] The only ANDA filed for the drug was delayed and ultimately withdrawn as a result of the innovator's patented ETASU.

2. REMS Distribution Controls and Generic Competitor Access

ANDA applicants must conduct bioequivalence testing on a proposed generic product and the branded counterpart before the generic drug can be approved by the FDA. This process necessarily requires obtaining samples of the branded drug product to conduct the required testing. Ordinarily, generic manufacturers obtain the branded samples from pharmaceutical wholesalers, but where the branded drug is subject to restricted distribution under REMS, access by wholesalers and, therefore, by generic manufacturers is prevented. Under these circumstances, generic manufacturers must request the branded drug directly from the innovator company.

An early draft of the FDAAA included provisions that would have required branded manufacturers whose drugs were subject to REMS distribution restrictions to sell ANDA applicants "a sufficient quantity of drug to conduct bioequivalence testing" if the generic manufacturer agreed

134. *Id.* § 355-1(i)(1)(B)(i)–(ii).
135. *Id.* § 355-1(f)(8).
136. *See* U.S. Patent No. 7,141,018, November 28, 2006.
137. *See* Citizen Petition No. FDA-2007-P-0113-0002.

to distribution and safe-use restrictions deemed adequate by the FDA.[138] However, the final version of the FDAAA included only the non-specific provision just noted prohibiting branded manufacturers from using REMS "to block or delay approval of an [ANDA] application."[139] In addition to the confusion previously discussed with respect to the duty of innovators to negotiate single, shared REMS, this language has created uncertainty regarding the scope of a branded manufacturer's duty to supply testing samples to potential generic competitors.

Several branded-drug manufacturers have refused to provide generic-drug makers with testing samples in this context.[140] One notable example also involves thalidomide: in addition to obtaining a patent to protect the ETASU for thalidomide, the manufacturer of thalidomide has refused to provide potential generic competitors with testing samples. In response to the thalidomide manufacturer's refusal to provide samples, one potential generic competitor filed a citizen petition with the FDA, alleging anticompetitive conduct and requesting that the FDA adopt procedures to ensure generic applicants the right to buy sufficient samples for bioequivalence testing of drugs subject to REMS distribution requirements.[141] Thalidomide's manufacturer responded that the legislative history strongly suggested that Congress considered and rejected guaranteed access procedures like the ones proposed and that the FDA thus could not use its administrative power to adopt similar procedures. The branded-drug manufacturer also argued that compelling innovators to supply drugs with known safety risks would impose undue liability risks on the innovator company.

In a separate attack on the thalidomide manufacturer's refusal to provide testing samples, another potential generic competitor initiated antitrust litigation in the Eastern District of Pennsylvania alleging violations of the antimonopolization provisions of the Sherman Act.[142] The generic manufacturer argued that access to thalidomide samples was an "essential facility" for the development of generic thalidomide. The branded-drug manufacturer responded, in a motion for summary judgment, that the

---

138. H.R. 2900, 110th Cong. § 901(b) (2007).

139. 21 U.S.C. § 355-1(f)(8).

140. *See* Katie Thomas, *Drug Makers Use Safety Rule to Block Generics*, N.Y. Times (July 26, 2013), http://www.nytimes.com/2013/04/16/business/drug-makers-use-safety-rule-to-block-generics.html?pagewanted=all&_r=0.

141. Citizen Petition No. FDA-2009-P-0266.

142. Case No. 2:08-cv-3920. Prior to initiating the suit, the generic manufacturer received a letter from the FDA that stated that it would not permit restrictions to prevent access to generic manufacturers—but stated less definitively that the manufacturer of thalidomide was permitted to provide samples once the FDA provided written confirmation that the generic manufacturer's proposed bioequivalence study would be conducted in a manner that would ensure subject safety. The thalidomide manufacturer refused to provide samples even after receipt of FDA confirmation of the bioequivalence study protocol.

Supreme Court had limited the essential facilities doctrine and that under well-established patent law principles, a patentee may refuse to sell its patented invention to rivals, even if the refusal has an anticompetitive effect, provided that the anticompetitive effect does not exceed the scope of the patent. The district court denied the branded-drug manufacturer's motion for summary judgment, and the case ultimately settled before trial.

NDA sponsors with drugs subject to distribution restrictions should carefully consider the benefits and potential liabilities associated with refusing to deal with generics before withholding access to bioequivalence testing samples. Notably, the thalidomide manufacturer's SEC filings show that it has been the subject of significant ongoing antitrust scrutiny by regulators on the basis of its refusal to provide potential competitors with testing samples, including two civil investigative demands from the Federal Trade Commission.

3. Patenting Methods Directed to Safe Use

Patenting discoveries concerning drug safety, including discoveries made in pre- or postapproval safety studies, may provide another avenue to extend branded-drug exclusivity. The FDA will not, as a matter of policy, provide data exclusivity for labeling information concerning drug safety. However, it may be possible to obtain patents directed to methods of safe use based on discoveries in pre- or postapproval studies. Methods identified in postapproval studies in particular may result in significant extensions to marketing exclusivity.

Generic-drug labeling must be identical to the reference branded-drug labeling, except where "aspects of the [branded-drug] labeling are protected by patent or exclusivity and such differences do not render [the generic drug] less safe or effective than the [pioneer] drug for all remaining non-protected conditions of use."[143] While ANDA applicants are frequently permitted by the FDA to omit or "carve out" label instructions pertaining to new indications subject to exclusivity added to branded-drug labeling,[144] the FDA will not permit these carve-outs where the patented information is necessary to ensure that the generic drug is not "less safe or effective" than the branded drug. Patentable discoveries pertaining to safe or effective use and drafted as method claims can be incorporated into drug labeling as a warning, precaution, contraindication, adverse reaction, or dosage or administration instruction. Accordingly, incorporating patented methods of safe use into branded-drug labeling, where such patented information is necessary to

143. 21 C.F.R. § 314.127(a)(7).
144. 21 U.S.C. § 355(j)(2)(a)(viii).

ensure the safe use of the drug, may provide extended patent exclusivity to the branded-drug manufacturer.

4. The 271(e)(1) Safe Harbor May Protect Otherwise Infringing Uses of Protected ETASU and Methods of Safe Use

The safe harbor provision of 35 U.S.C. § 271(e)(1) permits use of a patented invention where the use is "reasonably related to the development and submission of information under a Federal law which regulated the manufacture, use, or sale of drugs [...]." In *Momenta Pharmaceuticals, Inc. v. Amphastar Pharmaceuticals Inc.*,[145] the Federal Circuit held that infringing use by a generic manufacturer of a generic competitor's patented bioequivalence testing methods was protected under the 271(e)(1) safe harbor because the infringing use was reasonably related to information required by the FDA to maintain marketing approval—namely, laboratory determinations of the identity and strength of the drug, as required under 21 C.F.R. § 211.165.[146] Notably, the Federal Circuit held that the safe harbor is available after drug approval, even where noninfringing methods of generating the same information are available.[147] This determination by the Federal Circuit that the safe harbor extends to postapproval activities relating to the submission of FDA-required information may shield potential generic competitors from liability for infringing patents related to ETASU and drug safety labeling.

Although this holding has not been considered in the context of infringing uses of patented postmarketing safety measures, generic manufacturers may be able to argue that use of, for example, patient registries that monitor patient pregnancy status are required for submission of safety information required by the FDA for ongoing drug approval. Likewise, where the infringing use is by inclusion on the drug labeling, submission of the labeling change application itself including the patented safety information would arguably satisfy the safe harbor provision requirements as interpreted by the Federal Circuit in *Momenta*. Branded-drug manufacturers may argue in response, however, at least with respect to patented ETASU, that the REMS implementing regulations of the FDAAA (i.e., 21 U.S.C. § 355-1(i)(1)(B)(i)-(ii)) specifically provide that generic manufacturers must seek a license or develop comparable alternatives to patented ETASU. It does not appear that arguments based on the *Momenta* decision have yet been made in FDA citizen petitions by generic-drug manufacturers seeking a license to patented ETASU.

145. 686 F.3d 1348 (Fed. Cir. 2012).
146. *Id.* at 1358.
147. *Id.* at 1359.

chapter 12

Patent Prosecution[1]

I. Patent Portfolio Strategies

During the postmarketing phase of a pharmaceutical product in connection with an approved NDA, the first series of patent filings directed to the product may be nearing expiration of their patent term. Thus, new patent filing strategies should be considered in order to extend protection for the approved product as well as for product line extensions that are developed during the management of the product's life cycle.

Depending on the patent term remaining for the patent families that cover the approved product, subsequent patent filings may be filed as continuation, divisional, or continuation-in-part applications from a pending member of a patent family.[2] Stand-alone filings may also be considered if appropriate subject matter is available. These specific patent filing strategies are discussed in greater detail in Section II, "Patent Prosecution Strategies," of this chapter.

New patent filings can be pursued to diversify patent protection around the commercialized product and, therefore, extend protection beyond the expiration of the first generation of patents directed to the approved product. The goal of such filings will be to extend the exclusivity period for use of the approved product. The drug discovery process provides many opportunities to file patent applications directed to either the approved commercial product or future follow-on products. Various stages of discovery can provide various patentable options, including the initial discovery of a genus of active chemical agents effective against a particular target, the identification of a lead compound, targeted formulations, and applicable dosages that are approved for commercial use. Such opportunities provide the innovator company with the ability to build a diverse patent portfolio

1. Dennis M. Bissonnette, PhD, Justine A. Gozzi, and Sandra S. Lee of Baker Botts LLP, New York City.

2. Manual of Patent Examining Procedure (MPEP) § 201 (rev. Aug. 9, 2012).

as the active chemical agent is further developed. Thus, while the first generation of patents may be largely focused on the active chemical agent, the second generation of patent filings can be tailored to specific aspects of the approved product, such as formulations, dosages and routes of administration, and purified forms of the approved active ingredient, as well as to improvements to a commercial product.

A. Approved Products and Indications

The first generation of patents covering the approved product is often directed to the active chemical entity present in the approved product. These patents are usually filed during the early phases of development, prior to finalization of the formulation and dosage of the product that is approved by the FDA. For example, first-generation patents may often be directed to the active agent that is the basis of an IND application requesting authorization to administer an investigational drug to humans during early preclinical development.[3] Thus, depending on the broadness of the disclosure of the first generation of patents, various features of the approved product that is on the market may provide additional subject matter that can be pursued in subsequent patent filings.

1. Formulations, Dosages, Routes of Administration, and Salts

The specific formulation(s) of the approved product sold on the market provides a first target for second-generation patent filings, wherein claims are pursued that are tailored to the specific marketed formulation(s). For example, once a particular formulation and dosage of the active ingredient are approved, claims can be pursued to the particular formulation of ingredients, concentration of ingredients, and approved dosages. For such a strategy to be effective, care must be taken when drafting the first generation of patents so that the disclosure of the earlier patent family is not so broad as to anticipate or render obvious any subsequent formulation patents. Thus, strategies for initial filings must be considered with respect to how much broad disclosure is to be included in the application.

Although the specific combination of active ingredient, excipients, salts, stabilizers, and the like of the approved product may not be disclosed in the first-generation patents, claims to such formulations will likely need to be narrowly tailored to the specific formulations and may require the support of secondary considerations to overcome rejections of the claims as being obvious over the first-generation patents. Such secondary considerations include, for example, evidence of commercial success, long-felt but

3. *See, e.g.*, 21 C.F.R. § 312.

unsolved needs, failure of others, and unexpected results.[4] Surprising and unexpected advantages of an approved or new formulation over other formulations, discussed further in the following sections, can provide a good basis for arguing that claims are not obvious over prior formulations. Similarly, claims of the second generation of patents can be drafted to pursue the specific approved dosages of the approved formulation(s).

2. Methods of Treatment

As noted previously, the first generation of patents that cover an approved product are often directed to the active chemical entity present in the product. Thus, in addition to pursuing composition claims directed to features of the marketed product in new patent filings, the second-generation applications can be drafted to pursue the approved method of treatment. However, because the first generation of patents may disclose the approved use as a possible use of the active chemical agent, the claims of the second-generation patent filings may need to include additional combinations of elements not disclosed in the first-generation patents, such as dosages or pharmacokinetic or pharmacodynamic effects. For example, the claims included in the second-generation filings can be drafted to include methods of treating a particular condition by administering the active ingredient in the specific approved dosages. The claimed methods of treatment can be further distinguished from disclosure in any previous filings by including features of the approved product composition not previously disclosed.

For example, the claims can recite methods of treating a condition by administering the approved salt of the active ingredient of the commercial product in combination with other inactive ingredients.[5] Thus, while the first generation of patent filings may disclose numerous uses of the approved active chemical agent, the second-generation filings can be tailored with specificity to the approved use, wherein the active chemical agent is administered in a specific formulation at a specific dosage. Such combination of elements may provide a basis for new patent filings that extend protection of the approved commercial product.

3. Methods of Production

Subsequent to the filings of the first generation of patents directed to the approved active chemical agent, refinement of the method of scaling up production of the approved commercial product may provide additional subject matter that can be pursued in the second-generation patent filings. For example, if the methods used to produce and manufacture the

4. MPEP § 2141.
5. *See, e.g.*, *In re* Wiggins, 488 F.2d 538 (C.C.P.A. 1971).

commercial product include unique processes or techniques that are not disclosed in the first generation of patents or that are not standard in prior art methods of manufacture, such features can be pursued in subsequent filings. Although such claims do not directly cover the approved product or methods of treatment, claims directed to methods of manufacturing the commercial product may provide an alternative means by which to exclude competitors from entering the market with a generic product.

It is important to keep in mind the "safe harbor" provisions when contemplating methods of production. As provided by 35 U.S.C. § 271(e)(1), a "safe harbor" is available for activities that would otherwise be considered to infringe a valid patent. The following is recited in 35 U.S.C. § 271(e)(1):

> It shall not be an act of infringement to make, use, offer to sell, or sell within the United States or import into the United States a patented invention (other than a new animal drug or veterinary biological product (as those terms are used in the Federal Food, Drug, and Cosmetic Act and the Act of March 4, 1913) which is primarily manufactured using recombinant DNA, recombinant RNA, hybridoma technology, or other processes involving site specific genetic manipulation techniques) solely for uses reasonably related to the development and submission of information under a Federal law which regulates the manufacture, use, or sale of drugs or veterinary biological products.

Based on recent Federal Circuit decisions,[6] the safe harbor provision of 35 U.S.C. § 271(e)(1) would apply to methods performed for pre- or postmarket approval to generate data required by the FDA, wherein such methods would otherwise infringe a valid patent.

The Federal Circuit has held in *Classen Immunotherapies, Inc. v. Biogen IDEC* that § 271(e)(1) is directed to activities necessary to obtain regulatory approval and exempted preclinical research and other similar activities, whether or not ultimately submitted to the FDA. Such activities would fall under § 271(e)(1) as long as it was reasonable that they would produce information relevant to an IND or NDA. However, the safe harbor does not apply to information that may be routinely reported to the FDA, long after marketing approval has been obtained.

Momenta Pharmaceuticals, Inc. v. Amphastar Pharmaceuticals, Inc. further extended the safe harbor provision to postapproval testing by holding

---

6. Momenta Pharm. Inc. v. Amphastar Pharm. Inc., 686 F.3d 1348 (Fed. Cir. 2012), *cert. denied*, No. 12-1033, 2013 WL 655193, at *1 (June 24, 2013); Classen Immunotherapies Inc. v. Biogen IDEC, 659 F.3d 1057 (Fed. Cir. 2011), *cert. denied*, 133 S. Ct. 973 (2013).

that postapproval testing falls within the safe harbor of § 271(e)(1) if it is used to satisfy FDA requirements. The court further held that "[P]ost-approval studies that are 'reasonably related to the development and submission of information under a Federal law which regulates the manufacture, use, or sale of drugs' fall within the scope of the § 271(e)(1) safe harbor."[7] According to the court, the "reasonably related" language in § 271(e)(1) does not require submissions to the FDA or activities relevant to filing an ANDA for approval of a generic drug. "As long as the accused infringer 'has a reasonable basis for believing' that use of the patented invention might yield information that 'would be appropriate to include in a submission to the FDA, that use is reasonably related to the development and submission of information under Federal law.'"[8] Thus, postapproval testing that is not routine but is instead required by the FDA, would fall within the scope of § 271(e)(1). Further, the court held that the safe harbor does not mandate the use of a noninfringing alternative when one exists.

Thus, if specific methods to be used in the manufacture of a drug product are necessary to satisfy FDA requirements, such methods would fall within the scope of the safe harbor provision of § 271(e)(1), as the methods would be performed to satisfy FDA requirements. The patenting of such methods would need to be reviewed to determine if the methods can be used outside the context of obtaining data required by the FDA (or any other federal law) and whether performance of the methods for any such uses would not fall within the scope of the safe harbor of § 271(e)(1). If not, then the value of patenting the method may be meaningless.

B. Alternative Formulations

In addition to filing second-generation patent applications directed to the approved commercial product, methods of use, and methods of manufacture, the subsequent filings could focus on derivatives of the active chemical agent, homologs, analogs, new formulations, combination therapies, salts or isomers of the active chemical agent, alternative routes of administration, and alternative means of production.[9] Although such filings may not protect the approved commercial product directly, such filings will help

7. Momenta Pharm. Inc. v. Amphastar Pharm. Inc., 686 F.3d 1348, 1353 (Fed. Cir. 2012).

8. *Id.* at 1356–57.

9. *See, e.g., In re* Elpern, 326 F.2d 762, 767 (C.C.P.A. 1964) (the greater difference in the carbon chain, the better the chance to argue against the presumption of obviousness); *Ex parte* Mowry, 91 U.S.P.Q. 219 (Bd. App. 1950) (claimed isomer cyclohexylstyrene not prima facie obvious over prior art isohexylstyrene); *In re* Baird, 16 F.3d 380, 382 (Fed. Cir. 1994) (the fact that a claimed compound may be encompassed by a disclosed generic formula does not by itself render that compound obvious).

protect and define the landscape for the next-generation follow-on products and improvements of the marketed commercial product.

1. Purified Isomers, Polymorphs, and Metabolites

Although development of a new chemical entity for treating a disease would provide a basis for securing new patents for developing a portfolio around a particular disease state, modifications of an approved chemical entity may also provide patentable subject matter for building a patent portfolio around an existing chemical entity. For example, once an approved product is marketed and commercialized, technical advances resulting from continued development of the approved active chemical entity provide subject matter that can be pursued in the second generation of patent filings. These technical advancements (for example, a more efficacious, more stable, or safer compound) may form the basis for follow-on products of the approved pharmaceutical, which are essential for managing the life cycle of the commercial product.

As the expiration date of the first-generation patents approaches, the second generation of patents can define the product line extensions toward which the innovator can push the market. Thus, preparing and filing these second-generation patents while the first-generation patents are still in force will be important so that the market shift can begin prior to generic entry on the market of the approved product.

One area of subject matter to pursue in the second-generation patents is the active chemical agent itself. First-generation commercial products often include compounds that are mixtures of different isomers (i.e., compounds with the same molecular formula but different structural formulas). Such mixtures can include, for example, structural isomers, wherein the atoms and functional groups are joined in different ways; and stereoisomers, wherein the bond structure of the isomers is the same, but the geometrical positioning of atoms and functional groups in space differs.

In some cases, the different isomers exhibit different levels of activity against the intended target. This often is the case with racemic mixtures of enantiomers (stereoisomers that are mirror images of each other).[10] The structure of an isomer plays an important role in how the isomer interacts with a biological system. Accordingly, one enantiomer of a racemic mixture may be active in a biological system, while the other is not. Thus, a purified form of the active isomer may provide a basis for subsequent patent filings.[11] The purified active isomer can then be prepared in a specific

10. *See, e.g.*, HAWLEY'S CONDENSED CHEMICAL DICTIONARY 441, (R.J. Lewis ed., John Wiley & Sons Inc. 14th ed. 2001).

11. Pfizer Inc. v. Ranbaxy Labs. Ltd., 405 F. Supp. 2d 495, 517 (D. Del. 2005); Ortho-McNeil Pharm. Inc. v. Mylan Labs. Inc., 161 F. App'x 944 (Fed. Cir. 2005).

formulation as a follow-on product of the currently marketed commercial product. The specific formulation of the purified isomer may provide further grounds for additional patent claims.

In addition to purified isomers of a racemic composition, newly developed isomers can also be claimed in the subsequent filings and may be useful when deciding which follow-on products to develop for FDA approval. However, because the FDA requires that a submission of a drug for approval includes a description of the isomeric properties of the drug, extending the life of a product through isomer purification may be more difficult in the future, although the properties of a purified isomer, such as increased efficacy, stability, or safety profile, are factors that will be considered in determining whether such a purified product is patentable.[12]

Similarly, the chemical entities of a drug product may exist in different polymorph forms, for example, different crystalline and amorphous forms. If FDA approval is only granted for a single polymorph of an active ingredient, other polymorphs may provide grounds for later patent filings.[13] If any of the subsequently identified polymorphs are more efficacious than the originally approved polymorph, such evidence may strengthen the second-generation patents covering the new polymorph.

Additionally, prodrugs comprising the active chemical agent of the commercial product can be pursued in the second-generation patent filings. Such compositions would provide a possible alternative product wherein the active chemical agent is a metabolite of the newly developed prodrug. If such compositions provide advantages compared to the marketed product—for example, with regard to shelf life, reduced dosing schedules, bioavailability, or a longer half-life of the active chemical agent—the products may be attractive candidates for use as follow-on products and ones for which a fresh patent term can be obtained.

2. New Salts and Excipients of an Approved Active Ingredient

Reformulation of an approved chemical entity is a convenient strategy for pursuing new second-generation patents to extend the life of an approved product. The initial formulation of the approved chemical entity may likely be a simple formulation of the chemical entity that allows for quick entry into the market. However, technical advances made during continued research on the active chemical agent of an approved commercial product

12. *See, e.g.*, *In re* White, 374 F.2d 1010, 1013–14 (C.C.P.A. 1967); *Ortho-McNeil Pharm.*, 161 F. App'x 944.

13. The U.S. Patent and Trademark Office is content to issue polymorph patent claims. *See, e.g.*, U.S. Patent No. 7,393,874 (July 1, 2008) (titled "Novel Polymorphs of Tolterodine Tartrate").

may identify new patentable features of not only the active chemical agent but also the ingredients it is formulated with.

New salts of the active chemical agent may provide patentable advantages with regard to the storability or bioavailability of the active agent. Similarly, formulating the active chemical agent with alternative excipients, stabilizers, or other inactive ingredients may provide new formulations, such as controlled release dosages, that can be administered less frequently and that may form the basis of second-generation patent claims. However, because many alternative ingredients are known for therapeutic uses, evidence of secondary considerations may be required to meet the threshold of patentability.

Once the patents of the approved simple formulation of the chemical entity are nearing expiration, the more complex reformulations can be introduced into the market (following FDA approval and grant of second-generation patents on such formulations). These second-generation products can be introduced in an attempt to shift the market from the first-generation product to the new and improved second generation. Therefore, if the market is shifted to the new formulation, when the first-generation generics enter the market, the generics will be unable to compete with the new formulations, which would still enjoy exclusivity due to the second-generation patents. However, reformulating the chemical entity using known salts, excipients, stabilizers, fillers, binders, and the like to achieve effects that would have been obvious (as referenced earlier in this chapter) may make it difficult to gain allowance of the second-generation patents absent a surprising and unexpected effect. For example, including a new ingredient with the chemical entity that increases the efficacy of the chemical entity, increases bioavailability, or increases the safety profile of the chemical entity may provide a stronger basis for obtaining allowance of the second-generation patents.

3. Formulations for Alternative Routes of Administration

With the development of new, alternative formulations of the active chemical ingredient, the new formulations can be tested for suitability for use in new dosing regimens and routes of administration. For example, new formulations of an intravenously administered commercial product that are suitable for oral dosing may provide a more attractive follow-on product for patients that can also be protected with new second-generation patent filings. Similarly, new formulations that allow for higher concentrations of the approved active chemical agent compared to the commercial product may allow for a shift of an intravenously administered product to a subcutaneously administered product. Such changes in the route of administration of the active chemical agent may constitute patentable advantages for the follow-on products. Thus, while such advantages are useful for gaining allowance of the second-generation patent claims, they also provide a rational basis for initiating a market shift to the new products as the approved commercial product approaches the end of its exclusivity period.

In addition to new formulations of the commercial product, the second-generation patents can also focus on reformulating the approved product at a new concentration. For example, a commercial product that requires dilution by an end user prior to administration to a patient can be reformulated in a ready-to-use form that would not require any dilution step prior to administration. While providing patentable subject matter for the second-generation patent filings, such reformulations may also provide a more convenient product to the end-user customers and thus help to smoothly shift the market to the reformulated product.

Similarly, repackaging the approved commercial product in an alternative packaging system may provide new patentable subject matter for the second-generation patent filings. For example, repackaging a subcutaneous commercial product in a ready-to-administer syringe, rather than in a vial, may be a sufficient improvement to support a new composition claim. Additionally, if such repackaging provides for unforeseen advantages, for example, with regard to the stability or storability of the active chemical agent, such advantages would also support the new patent claims.

Furthermore, once a new formulation, reformulation, or repackaging of the commercial product has been developed, and claims to such products obtained in the second-generation patent filings, methods of treatment directed to uses of the new formulations, reformulations, or repackaged products can be pursued. Such claims may track the method-of-treatment claims that protected the approved first-generation commercial product, but would utilize the follow-on product and thus be distinct from the first-generation method-of-treatment claims.

C. New Indications and Target Populations

Indication expansion is an effective way to extend the exclusivity period enjoyed by a particular chemical entity.[14] If different diseases share common or overlapping mechanisms of action, then a chemical entity that interacts with the common or overlapping mechanisms may be suitable for use in treating the different diseases. Thus, if approval is obtained for a method of treating one of the diseases, then second-generation patents can be filed, and FDA approval pursued, for methods of treating the other

14. See, e.g., Bimatoprost, marketed by Allegran as Lumigan, an ophthalmic solution for the treatment of high eye pressure. Bimatoprost was subsequently discovered to lengthen eyelashes, and approval for this second indication was obtained from the FDA for the product Latisse.

diseases as the first-generation patents near expiration. Such a strategy would extend the life cycle of a single chemical entity that interacts with multiple disease pathways.

1. Postapproval Studies

Postapproval studies related to a chemical entity protected by first-generation patents may identify different diseases that may also be targeted for FDA approval. These studies may evolve from observations made by physicians who are treating their patients with off-label use or from side effects observed when treating patients for the approved indication. However, the majority of the postapproval studies will likely focus on developing a better understanding of the chemical entity's mechanism of action. Understanding how the chemical entity interacts in a disease pathway—for example, inhibiting a particular enzyme—would allow for the identification of other potentially treatable diseases that share the same enzyme in their mechanism of action. In many instances, research and studies directed to the secondary indications can be developed in parallel with the primary indication, as the mechanism of the chemical's action will often be understood prior to approval.

As the secondary indications are identified, patent protection for methods of treating these diseases can be pursued. The exclusionary effect of such patents would be strengthened if the methods of treating can be combined with additional new elements such as new dosages, formulations, or routes of administration of the chemical entity. When a greater number of new elements are required for treating the secondary indication (and protected by the second-generation patents), it will be less likely that the first-generation generics could be used off-label for the new indication, as they would not be of the appropriate formulation or dosage.

When selecting which secondary indication to pursue, in addition to considering the likelihood of whether the chemical entity will be safe and effective for treating the indication, consideration should be given to the cost of the clinical trials necessary to obtain FDA approval as well as to the size of the target population, whether such a population will grow over time, and whether there is an unmet need for treating such population (i.e., whether the market for treating the condition has been saturated by competitor products).

2. Patient Populations

In addition to undertaking postapproval studies designed to identify new indications to treat, such studies can be focused on identifying biomarkers that can be used in companion diagnostics with the approved primary indication. An approved drug may be more effective for certain subpopulations of the treatable population. Therefore, identifying such a subpopulation would allow for a series of second-generation patents directed to methods of identifying the subpopulation (i.e., methods of identifying patients with the disease who also express the biomarker), in combination with methods of treating the disease with the previously approved chemical entity. Combining the treatment with a diagnostic test may prove to be most effective when an approved chemical entity is not the most effective treatment for the generic population of treatable patients, but is the most effective at treating the identified subpopulation. By specifically targeting the subpopulation, the chemical entity can gain a larger market share for the treatment of this group.

3. Pediatric Uses

The pediatric population, like the subpopulations expressing particular biomarkers, is a subpopulation that can be targeted for extending the exclusivity period of an approved chemical entity.[15] In addition to the pediatric exclusivity period available for extending the term of patents covering the approved product (discussed elsewhere in this book), methods of treating the pediatric subpopulation may provide the basis for new patent filings that may further extend the exclusivity of a chemical entity, at least in the pediatric subpopulation. In addition to patents directed generally to methods of treating pediatric patients, such patients may require novel formulations or dosages that differ from those of the approved product for adult use. Such changes may be required due to differences in pharmacokinetic and/or pharmacodynamic effects of the chemical entity in the pediatric subpopulation.

When determining whether to dedicate resources to developing a patent portfolio around treating the pediatric subpopulation, consideration should be given to how large a subpopulation the pediatric group is and the extent to which differences exist between methods of treating adults and methods of treating pediatric patients with the approved chemical entity. If pediatric subpopulations are successfully being treated with the formulation of the chemical entity that is already approved for adult use, it may be more difficult to obtain new patents covering the pediatric use. However, new formulations that may be more convenient for the pediatric subpopulation, such as an oral liquid formulation rather than a pill or capsule, may provide a basis for developing second-generation patent protection for the pediatric subpopulation.

15. *See* Food and Drug Administration Modernization Act of 1997 § 505(A)..

4. Commercially Relevant Indications of Competitors

Although development of the second-generation patents for the portfolio of an approved chemical entity will likely focus on new formulations, dosages, and indications related to the chemical entity, the portfolio can also include patents directed to commercially relevant indications of competitors. For example, such patents may focus on methods of manufacture or uses of chemical entities that are structurally similar to an approved chemical entity, and thus possibly an alternative to the approved product, which would be of interest to a competitor. Additionally, patents directed to compounds and methods for treating side effects of an approved chemical entity may be pursued. Such compounds and methods may have utility when used as an adjunct to the approved used of the chemical entity. However, the compounds used in such methods may be patentably distinct from the approved chemical entity, and thus provide a basis for new patent filings.

D. Combination Compositions and Alternative Dosing Strategies

As the term of the patents covering an approved chemical entity approaches expiration, attempts to extend protection of the chemical entity can be pursued that are directed to combinations of the chemical entity with one or more different chemical entities or to provision of the approved chemical entity in additional dosage forms that comprise a lower or higher concentration of the chemical entity than is currently on the market. However, when attempting to protect such follow-on products with second-generation patent filings, it will often be necessary that such products provide a benefit other than convenience of use, as such combinations may be held to be obvious by the U.S. Patent and Trademark Office (USPTO).

1. Combination with Second Approved Chemical Entity to Treat More Than One Indication

Combinations of an approved chemical entity with other drugs may provide an additional strategy to extend the life of a product. In some instances, it may be advantageous to combine the approved chemical entity with another drug that is approved to treat the same disease. Such a combination may provide benefits to the patient, as a desired therapeutic effectiveness may be achieved while utilizing lower dosages of each drug than if they were used independently. Although such a combination may provide a basis for a new approval, obtaining allowance of patent claims directed to such combinations may be difficult absent secondary considerations, such as unexpected synergistic effects, if the combination is deemed obvious.

Another option that may enable extending the life of an approved chemical entity may be to combine the approved product with other drugs that are often prescribed to patients receiving the approved chemical entity. For example, if blood pressure lowering medicines are prescribed in conjunction with cholesterol lowering medicines, combining the two drugs as a single fixed dose may provide a more advantageous product, as the patient would be required to take fewer dosages of drugs. Similarly, drug cocktails that are prescribed for treating a single disease may be suitable for combination into a single fixed dosage. For example, combining anti-retroviral therapies for treating AIDS into a single dosage would provide a more convenient and, therefore, more attractive option for consumers.[16]

With regard to patenting such combinations, as described earlier, merely combining two chemical entities into a single fixed dose may be insufficient to obtain allowance of a second-generation patent, absent unexpected benefits with regard to, for example, increased efficacy, bio-availability, or safety profile. However, if one of the chemical entities in the combination remains under patent protection, the combination may provide protection for the off-patent chemical entity if the language of the patent that remains in force is sufficiently broad to encompass combination formulations.

2. New Dosing Regimens

Once a chemical entity has been approved and the first-generation patents are in place protecting that chemical entity, approved formulations, dosages, and methods of using the chemical entity, second-generation patents can be pursued directed to alternative dosing strategies. Such strategies can take the form of new dosage strengths or dosage regimens using the approved formulations.

With regard to alternative forms of dosage strengths, providing the chemical entity in dosage strengths that are lower and higher than the approved dosage will provide physicians with a range of dosage options when prescribing the approved product. These options may be more convenient when treating certain patients for whom only a low dose is necessary. Such an alternative dosage could be prescribed without the need to physically divide dosages. Similarly, a higher dosage alternative would allow a physician to prescribe higher dosages to patients, yet the patients would not need to ingest, for example, more pills. Additionally, providing multiple dosages of the same approved chemical entity would allow a prescribing physician the option to easily titrate an approved chemical entity to a patient. For example, the physician could begin with a low dosage if there is a concern regarding side effects in a particular patient, or alternatively, the patient would have the option to easily increase the dosage if a

16. See, e.g., Combivir, a fixed-dose combination of antiretroviral therapies zidovudine (AZT) and lamivudine (3TC), marketed by GlaxoSmithKline.

standard dose is not sufficiently effective. Thus, more flexible prescribing options would be available to the physician and patient without the need for changing the manner in which the chemical entity is administered.

By providing a chemical entity in multiple dosage amounts, the regimen of administration of the dosage could also be varied for the convenience of the patient, so long as such variations are safe and effective. For example, providing a higher concentration dosage may allow for a patient to change from a twice-daily administration of a product to a once-daily administration regimen. The existence of more convenient dosing regimens may provide for a stronger hold on the market, as patients may be more likely to choose the product with a more convenient dosing option. Additionally, if second-generation patents are granted and directed to the alternative dosage amounts and regimens, generics could not include such alternatives on their label, as such an inclusion could be considered to induce infringement. However, obtaining patent protection for new products that merely provide a patient with a more convenient dosing regimen may be difficult in the absence of secondary considerations, for example, an unexpected increase in effect, bioavailability, or safety profile.

II. Patent Prosecution Strategies

Although Chapter 3 discusses patent application filing strategies during the preclinical phase, certain aspects of filing strategies should be revisited during the postmarketing phase.

A. Continuing Patent Applications

Considerations during the postmarketing phase are whether any patent applications from the original filings are still pending and whether it is possible to file one or more continuing applications based on one of the previously filed pending patent applications.

A continuing patent application is a patent application that has been filed subsequent to and claims priority of an earlier filed original patent application that remains pending.[17] Accordingly, the original patent application can be subsequently used to establish an effective filing date of a later-filed application, which may be valuable with respect to avoiding any intervening prior art since the original filing date.

During the pendency of any patent application, a continuing application may be filed claiming priority to the prior filed parent application.[18]

---

17. *See* MPEP § 201.04.

18. *See* 35 U.S.C. §§ 120, 121, 365(c); 37 C.F.R. § 1.78; *see also* 37 C.F.R. § 1.53 (2012).

The pendency of the patent application begins at the effective filing date of the patent application and ends on the patent grant date of any patent to issue from the patent application (or the date the parent application becomes abandoned).[19] Thus, any continuing applications must be filed prior to the patent grant date of the prior filed parent application.[20]

1. Benefits of Continuing Applications

Continuing applications can have several benefits. One benefit of filing a continuing application is keeping the patent family pending to protect against potential future risks. For example, competitors may attempt to invalidate an existing patent for various reasons. Others may try to design around the protection provided by the patent. By keeping the patent family alive and pending, an applicant can address such issues during the pendency of any continuing applications, as long as there is support in the original disclosure.

Another benefit of filing a continuing application is the potential for broader scope. Existing patented claims with picture claims may cover an immediate ready-to-commercialize product and may have narrow scope or protection. If the patent specification discloses other, broader classes or dosages or alternative uses that could be protected, a continuing application can be filed to protect those disclosures.

The reverse may also be true. If the original patent(s) obtained is directed to broad coverage, without specific embodiments to the commercialized product, a continuation application would allow picture claims to be pursued. Alternative forms of patent protection are always important to contemplate, particularly when the patents may be challenged. Alternative excipients or combination therapies may also be considered in filing strategies, assuming the disclosure supports such claims. The narrower picture claims may withstand such challenge, if broader scope patents are at risk.

It should also be noted that continuing applications can protect against future court decisions that restrict how patents are interpreted

19. *See* MPEP § 706.02.

20. When a patent application has been indicated as allowed, the applicant will receive a notice of allowance letter from the U.S. Patent and Trademark Office that will indicate a date to which the associated fee for granting the patent, such as the issue fee, must be paid. The fee is due three months from the date of the notice of allowance and the applicant can pay the required fee at any time within the three-month window. *See* 37 C.F.R. § 1.311. After the fee is paid, the applicant will next receive an issue notification letter that indicates the date and serial number of the patent to issue from the patent application. In general, it is best practice to file any continuing applications on the date that the applicant pays the issue fee. If an applicant waits until after the issue fee is paid, there may be some risk in missing the opportunity to timely file a continuing application—which could result in loss of rights. Unforeseen circumstances, such as not timely receiving the issue notification letter, could prevent the timely filing of a continuing application. Thus, it is best to file continuing applications early.

and applied.[21] For example, implications of narrowing claim scope during prosecution may be sought as alternatives in a continuation.

In addition, the issued patent may have been examined by an unreasonably difficult examiner who may have only been willing to issue narrow claims. In some instances, although limited, continuing applications are examined by a different examiner than the parent patent. For example, when an original parent patent is directed to product claims but the continuing application is directed to method claims, a new examiner may be assigned. The examiner of the continuing application may be more willing to issue broader claims than the parent.

A further benefit of filing a continuing application, as opposed to another original application, is the consistency in the availability of the prior art. For continuing applications, any new information publicly disclosed since the effective filing date of the parent is not considered prior art of the child continuing application. For example, if a child application is filed ten years after the parent application is filed, prior art available against the child application must have a date prior to the effective filing date of the parent. Thus, for purposes of discussion, intervening art, which is disclosure of new information published after the parent application is filed, is not considered as available prior art against the continuing child application. On the contrary, such disclosure of new information may be available as prior art if the application was filed as a stand-alone original application, as opposed to a child continuing application. Furthermore, when filing an original application as opposed to a child continuing application, the first original application would be considered as prior art against the subsequently filed original application.

2. Caveats of Continuing Applications

There are certain caveats to filing a continuing application that should also be taken into consideration. First, it is important to consider the available patent term of a continuation application. A patent term is limited to 20 years from the filing date of the earliest filed parent application. Thus, a child application will have generally the same patent term as the parent application. If a child application is filed ten years after the effective filing date of the parent application, then an applicant will need to weigh the value in filing a subsequent application as a continuing application because the patent term of the child may be considerably shorter. It may

21. *See, e.g.*, Festo Corp. v. Shoketsu Kinzoku Kogyo Kabushiki Co., 344 F.3d 1359 (Fed. Cir. 2003), *decided on remand from* 535 U.S. 722 (2002) (limits availability of the doctrine of equivalents, which prevents others from avoiding patent infringement by making insubstantial changes to their otherwise infringing products).

be possible, however, to accelerate the prosecution of the application, as discussed further later in this chapter.

If a separate original filing can be done without the parent case anticipating or rendering obvious the new case, this option should be considered as the separate filing will have a patent term of 20 years from its effective filing date. Depending on the circumstances, an application may need the effective filing date of the parent application to overcome recent prior art or other statutory bars, as previously discussed, so for these instances, filing a subsequent original application is not a valid option.

Additionally, to the extent that international filings are contemplated, continuation practice may not be available in all international jurisdictions. An international patent application, such as a Patent Cooperation Treaty (PCT) application or a direct foreign filed application, can be filed claiming priority to an original application only if the original application was filed within a year. Thus, a child continuing application filed after two years of the effective filing date of the parent application is unavailable to be subsequently filed as an international application. Within each specific foreign country, the local rules may limit the ability to file continuations or divisionals, and the parent cases may be prior art to any follow-on applications, regardless of being a direct patent family member.

3. Types of Continuing Applications

Three different types of continuing applications may be filed—each with different uses and benefits. A continuing application can be filed as (1) a *divisional* application, (2) a *continuation* application, or (3) a *continuation-in-part* application. Each of the three types of continuing applications is briefly discussed in the following paragraphs.

A divisional application is a later-filed child application directed to an invention independent or distinct from the parent.[22] A divisional application is often filed as a result of a restriction requirement made by an examiner in a U.S. parent application or as a result of a lack of unity rejection in a foreign application. These are not likely to be applicable in the postmarketing setting, as presumably such divisionals would have already been filed by the postmarket period.

On the other hand, a continuation application may be a viable option during the postmarketing period.[23] As with all continuing applications, the continuation application must be filed before the original parent application becomes abandoned or patented. The disclosure presented in the continuation must be the same as that of the original parent

22. *See* MPEP § 201.06.
23. *See id.* § 201.07.

application and should not include anything that would constitute new matter if inserted in the original parent application.

Additionally, a continuation-in-part (CIP) application may be beneficial as it includes the earlier application and further adds subject matter not previously disclosed.[24] In CIP applications, the priority date is determined claim by claim. Certain of the claims can be directed to the subject matter of the parent application and are accorded the priority date of the parent application, whereas other claims can contain new subject matter and are accorded the filing date of the CIP application. This may be a viable option if the parent case is pending and improvements have been discovered. However, if CIPs are an option, it may be likely that a stand-alone case would be more beneficial, and such options should be carefully weighed.

B. Reissue Patent

As previously mentioned, if a patent issues prior to the filing of a continuing application, an option exists in select circumstances to revive an otherwise dead patent family by way of reissue practice. Whenever any patent is deemed inoperative (at least partly) or invalid by reason of the applicant claiming more or less that she or he had a right to claim in the patent, the applicant can surrender the granted patent and submit an application for reissuance of the patent.[25] In order to file a reissue application, the applicant must submit that the patent is at least partly inoperative or invalid through error without any deceptive intention and pay the applicable fees.[26] Similar to continuation practice, the applicant cannot add new matter to the disclosure of the reissue patent application.

Once the application for reissue is filed and accepted at the U.S. Patent and Trademark Office (USPTO), an applicant can expand the scope of the patent claims if the reissue application is filed within two years of the patent grant date. If the reissue application is applied for beyond two years from the grant of the original patent, then the applicant is foreclosed from broadening the patent claims but can further narrow the claims.

The surrender of the original patent shall take effect upon the issuance of the reissued patent. Every reissued patent shall have the same effect and operation in law as if the reissued patent had been originally granted in the amended form. Insofar as the claims of the original and reissued patents are substantially identical, the reissued patent shall constitute a continuation of the original patent and have effect continuously from the date of the original patent.

24. *See id.* § 201.08.
25. *See* 35 U.S.C. § 251.
26. *See id.* § 251.

It is important to note that once the application for reissue is filed and accepted for broadening reissuance within the two-year window at the USPTO, an applicant can additionally file further broadening continuation reissue applications even if the subsequent continuation reissue applications are outside the two-year window of the patent grant date.[27] Furthermore, the broadening continuation reissue applications can claim features not first claimed in the original patent/first reissue patent. This reissuance practice is a potentially powerful tool for applicants who seek to protect unclaimed subject matter disclosed in a patent that no longer has any pending applications, as long as the first patent issued within the two-year window.

One caution for reissuance practice is the doctrine of intervening rights for potential infringers. The doctrine of intervening rights provides a defense for alleged infringers to be held not liable for infringement of patents that are modified through reissue procedures, or reexamination procedures as discussed in the following section. The doctrine of intervening rights is triggered when the claims of the patent are substantively amended during the reissue proceedings.[28] If the claims are substantively amended, the liability of an alleged infringer may be limited to the date the amended claims issue and not the original issue date of the original patent claims.[29]

To avoid intervening rights with respect to a first reissue patent, the applicant should avoid substantive amendments and preserve claims through argument rather than amendment. The applicant can subsequently file a second continuing reissue patent application with claims directed to more substantive changes, as previously discussed.

---

27. *See In re* Staats, 671 F.3d 1350 (Fed. Cir. 2012) (patentee that filed a first reissue application under § 251 within two years of issuance of its patent could file a second reissue application seeking further broadening corrections to the patent, even though the second application was filed outside the two-year period for filing and covered changes not claimed in the original patent or the first reissue application).

28. *See* 35 U.S.C. §§ 252, 307(b).

29. *See* Marine Polymer Techs. Inc. v. HemCon Inc., 672 F.3d 1350, (Fed. Cir. 2012). The U.S. Court of Appeals for the Federal Circuit in *Marine Polymer Technologies Inc.* considered what constitutes an "amended" claim for purposes of determining if intervening rights apply under 35 U.S.C. § 307(b). During reexamination of the patent at issue, the patentee successfully narrowed the claim scope to avoid prior art without formally amending the remaining claims. This was accomplished through a combination of argument and deletion of several dependent claims. An evenly divided court held that because the remaining claims were not textually "amended" as required by the statute, no intervening rights accrued. If the patentee can preserve claims through argument rather than amendment, an accused infringer is deprived of an intervening rights defense.

C. Reexamination Practice

Reexamination is a process by which the USPTO reviews the validity of an issued U.S. patent upon limited grounds. Almost anyone can file a request for reexamination at any time during the enforceability period of the patent. The requestor must pay a filing fee depending on the type of request, and the request must generally include certain requirements such as (a) a statement pointing out each substantial new question of patentability, (b) identification of every claim for which reexamination is sought, (c) an explanation of the art applied to the claims, (d) a copy of the art relied upon, (e) a copy of the specification and claims, and (f) a certificate that the patent owner has been served.[30] If an existing patent within the patent portfolio of a drug product may warrant reexamination, the patent owner may want to initiate the reexamination to address new prior art that has come to light. Likewise, if a third party challenges the validity of the patent that is listed in the Orange Book, the patent holder may need to seek additional patent scope, as summarized previously. In either case, familiarity with patent prosecution practice is an essential aspect of understanding risks to the portfolio.

Prior to the enactment of the America Invents Act (AIA),[31] two types of reexamination procedures existed: (1) ex parte reexamination and (2) inter partes reexamination. For ex parte reexamination, anyone can file the request for reexamination, including the patent owner. However, for third-party ex parte requests (which can be anonymous), the third party is not permitted to actively participate in the proceedings once the request is filed. Only the patent examiner and the patent owner are involved in the proceedings. In contrast, for inter partes reexamination, only a third party can file a request for inter partes reexamination (not a patent owner), and the third party must be identified. During inter partes reexamination, the third-party requestor may participate in all aspects of the reexamination, including submission of third-party correspondence opposing a patent owner's response to an Office Action.

With the enactment of the AIA, however, inter partes reexamination proceeding was replaced by inter partes review and post grant review, as further discussed in the following sections. Ex parte reexamination remains predominantly unchanged by the AIA.

1. Ex Parte Reexamination

For the USPTO to grant a request for ex parte reexamination, a substantial new question of patentability must be raised by the request.[32] The

30. *See* 37 C.F.R. §§ 1.510, .915.
31. Leahy-Smith America Invents Act, Pub. L. No. 112-29, 125 Stat. 284 (Sept. 16, 2011).
32. *See* 37 C.F.R. § 1.510.

USPTO will grant the request based on patents or publications identified in the request, including those that were previously considered by the USPTO when the patent was originally examined. Other forms of prior art will be considered, such as non-patent literature. The request can be made anonymously by any party, including the applicant/patent owner.[33] However, only the applicant/patent owner may participate in the ex parte reexamination process. Thus, a third-party patent challenger who may have filed the request is excluded from the process once initiated.

The ex parte reexamination proceedings are similar to regular examination procedures in pending patent applications. When the prosecution of an ex parte reexamination proceeding is terminated, a reexamination certificate is issued that indicates the status of all claims following the reexamination. Unless prosecution is reopened by the USPTO, the reexamination proceeding is concluded by the issuance and publication of a reexamination certificate. The Central Reexamination Unit of the USPTO handles ex parte reexaminations, and any party dissatisfied with the examiner's decision can appeal to the Patent Trial and Appeal Board (formerly the Board of Patent Appeals and Interferences).

As similarly noted with respect to reissue practice, one caution for ex parte reexamination practice is the doctrine of intervening rights. The doctrine of intervening rights provides a defense for potential competitors to charges of patent infringement for any patent that is modified through reexamination procedures. Intervening rights is different for ex parte reexamination patents in comparison with reissue patents. For ex parte reexamination patents, the claims must be amended or new to trigger the doctrine of intervening rights for ex parte reexamined patents. Thus, to avoid intervening right defenses of infringers for reexamination patents, applicants should attempt to preserve claims through argument rather than amendment. Alternatively, applicants can try to avoid intervening rights by keeping at least some of the original claims as unamended.[34]

2. Inter Partes Reexamination

Inter partes reexamination was available for cases filed on or after November 29, 1999, wherein the request for inter partes reexamination was filed before September 16, 2012.[35] In addition, only a third party could file a request for inter partes reexamination (i.e., not a patent owner), and the third party must be identified. As previously noted, inter partes reexamination was replaced by other procedures, including inter partes review and

33. *See* MPEP § 2212.

34. *See* Marine Polymer Techs. Inc. v. HemCon Inc., 672 F.3d 1350, (Fed. Cir. 2012).

35. *See* Revision of Standard for Granting an Inter Partes Reexamination Request, 37 C.F.R. § 1.915, *available at* http://www.uspto.gov/aia_implementation/Ip-reaxam-final-rule.pdf.

post grant review, as further discussed later in this chapter.[36] Because inter partes reexaminations are still moving forward for those requests previously filed, the following paragraphs provide highlights of inter partes reexamination procedures.

For requests meeting the preceding conditions for inter partes reexamination, the USPTO initially determines if "a substantial new question of patentability" is presented.[37] If such a new question has been presented, inter partes reexamination will be ordered. Unlike ex parte reexamination, any prior art, including prior art other than patents or publications, can be submitted to the USPTO with the inter partes reexamination request.

The inter partes reexamination proceedings are somewhat similar to regular examination proceedings in pending patent applications. During inter partes reexamination, the third-party requestor may participate in all aspects of the reexamination, including submission of third-party correspondence opposing a patent owner's response to an Office Action. When the prosecution of an inter partes reexamination proceeding is terminated, an inter partes reexamination certificate is issued to indicate the status of all claims following the reexamination and concludes the reexamination proceeding.

Similar to ex parte reexaminations, the Central Reexamination Unit of the USPTO handles inter partes reexaminations. Any party dissatisfied with the examiner's decision can appeal to the Patent Trial and Appeal Board.[38]

One caution for inter partes reexamination practice is the doctrine of intervening rights, as similarly noted and discussed earlier. Intervening rights is different for inter partes reexamination patents in comparison with reissue patents. For inter partes reexamination patents, the claims must be amended or new to trigger the doctrine of intervening rights for inter partes reexamined patents. Thus, to avoid intervening right defenses for inter partes reexamination patents, applicants can try to preserve claims through argument rather than amendment. Alternatively, applicants can try to avoid intervening rights by keeping at least some of the original claims as unamended.[39]

D. Post Grant Review, Inter Partes Review

The AIA enacted two new post-grant review proceedings—*post grant review* and *inter partes review*—to replace inter partes reexamination proceedings.[40] Both new proceedings are effective as of September 16, 2012.

36. *See* MPEP § 2609.

37. *See* 35 U.S.C. § 312(a).

38. *See* 37 C.F.R. § 41.61.

39. *See* Marine Polymer Techs. Inc. v. HemCon Inc., 672 F.3d 1350, (Fed. Cir. 2012).

40. Leahy-Smith America Invents Act, Pub. L. No. 112-29, 125 Stat. 284 (Sept. 16, 2011). The AIA also provided for business method patents to be challenged though the Covered

All patents, regardless of their issue date, are eligible for inter partes review. Therefore, patent holders need to be familiar with these proceedings should their existing patents be called into question. With the exception of business method patents and pending interferences, only patents that contain a claim that has an effective filing date that is on or after March 16, 2013, are eligible for post grant review.

Pending interferences may be stayed or dismissed without prejudice by the filing of a post grant review petition. In addition, the USPTO may limit the number of inter partes reviews and post grant reviews that are accepted until September 16, 2016. The AIA formed a new Patent Trial and Appeal Board (PTAB) to decide post grant review and inter partes review determinations. Any party dissatisfied with a Board decision can appeal directly to the Federal Circuit.

1. Post Grant Review

Post grant review may be petitioned by any third party, exclusive of the patent owner, who has not previously filed a civil action challenging the validity of a claim of the patent. Post grant review petitions are time sensitive and must be requested within nine months of the issue date of the patent. The grounds for filing a petition for post grant review can be based on any ground related to patent invalidity, such as failure to comply with the patentable subject matter requirement or even the written description requirement.[41]

The petition for post grant review has many requirements, such as identification of each claim challenged, the grounds on which the challenge to each claim is based, and the evidence that supports the grounds for the challenge to each claim. Once a petition is filed, the patent owner is notified and has the opportunity to file a preliminary response against the petition. The USPTO has three months to decide whether to grant the post grant review and bases the decision on whether "more likely than not that at least 1 of the claims challenged in the petition is unpatentable" or whether the petition raises a novel or unsettled legal question that is important to other patents or patent applications. The USPTO decision as to whether to grant a post grant review is not currently appealable.

A third party is precluded from filing a petition for post grant review if the third party previously filed a civil action challenging the validity of a claim of the patent. On the contrary, if the third party first or concurrently filed a petition for a post grant review and then filed a civil action

Business Method Patent Review proceedings. More information regarding Covered Business Method Patent Review can be found at *Leahy-Smith America Inventes Act Implementation*, U.S. Pat. & Trademark Off., http://www.uspto.gov/aia_implementation/index.jsp (last visited May 6, 2014).

41. Post-grant review can be filed based on any grounds of 35 U.S.C. § 282 except for failure to comply with the best mode requirement.

challenging the validity of a claim of the patent, the civil action is automatically stayed pending the outcome of the post grant review. However, counterclaims of invalidity do not count as a civil action challenging validity. Further, if the patent owner moves to lift the stay or files an action or counterclaim for infringement, or if the third party moves to dismiss the civil action, the stay will be lifted. Post grant review does allow for discovery limited to evidence directly related to factual assertions advanced by either party in the proceeding.

The third party/petitioner has the burden of proving unpatentability by a preponderance of the evidence. As such, the third party/petitioner has a lower burden of proof than in U.S. litigation practice, which requires proving invalidity by the clear and convincing evidence standard.

With respect to the patentability of any patent claim challenged by the petitioner and any new claim added during the post grant review, estoppel for post grant review becomes effective with a final written PTAB decision.[42] The estoppel furthermore extends to any ground that the petitioner raised or reasonably could have raised during post grant review. Thus, in other words, a third party/petitioner may not request or maintain a subsequent proceeding before the USPTO when estoppel applies. In addition, post grant review estoppel applies both to proceedings before the USPTO and to civil actions and proceedings before the International Trade Commission (ITC).[43] The patent owner, in turn, is estopped from taking action inconsistent with any adverse judgment, including obtaining in a patent a claim that is patentably indistinct from a finally refused or cancelled claim or amending its specification or drawing in a way that was denied during the proceeding.

Upon the joint request of the petitioner and the patent owner for termination of the post grant review, a post grant review shall be terminated unless the office has decided the merits of the proceeding before the request for termination is filed.[44] If the post grant review is terminated with respect to a petitioner under this section, no estoppel shall attach to the petitioner /third party on the basis of that petitioner's institution of that post grant review. If no petitioner remains in the post grant review, the USPTO may terminate the post grant review or proceed to a final written decision.

2. Inter Partes Review

Inter partes review is available to any third party, exclusive of the patent owner, who has not previously filed a civil action challenging the validity of

42. *See* 35 U.S.C. § 328(a) (AIA § 6).
43. *See id.* § 325(e) (AIA § 6).
44. *See id.* § 327(a) (AIA § 6).

a claim of the patent. Inter partes review petitions are time sensitive and can only be filed the latter of (1) 9 months after the grant of a patent/issuance of a reissue of a patent, or (2) the date of termination of any post grant review of the patent. In other words, the time periods for filing a petition for inter partes review and post grant review do not overlap. The grounds for filing a petition for inter partes review are more limiting than those for post grant review and can only be raised under anticipatory grounds or obviousness grounds under 35 U.S.C. § 102 or § 103, respectively, and only on the basis of prior art consisting of patents or printed publications.

Similar to post grant review, the petition for inter partes review has many requirements, such as identification of each claim challenged, the grounds on which the challenge to each claim is based, and the evidence that supports the grounds for the challenge to each claim. Once a petition is filed, the patent owner is notified and has the opportunity to file a preliminary response against the petition. The USPTO has three months to decide whether to grant the inter partes review and bases the decision on whether there is "reasonable likelihood that the petitioner would prevail with respect to at least 1 of the claims challenged in the petition." According to legislative history, the threshold standard for instituting post grant review is slightly higher than the inter partes review threshold standard of a "reasonable likelihood of success."

Similar to post grant review, a third party is precluded from filing a petition for inter partes review if the third party previously filed a civil action challenging the validity of a claim of the patent. On the contrary, if the third party first or concurrently filed a petition for inter partes review and then filed a civil action challenging the validity of a claim of the patent, the civil action is automatically stayed pending the outcome of the inter partes review. As with post grant review, counterclaims of invalidity do not count as a civil action challenging validity. Further, the stay will be lifted under the same parameters as previously explained for post grant review. Inter partes review allows for certain limited discovery. The discovery for inter partes review can include depositions of witnesses submitting affidavits or declarations and what is otherwise necessary in the interest of justice.[45] As with post grant review, the third party/petitioner has the burden of proving unpatentability by a preponderance of the evidence standard. Inter partes review also has similar estoppel effects and termination effects, as previously discussed with respect to post grant review.

45. For more information on Inter Partes Review, visit U.S. PAT. & TRADEMARK OFF., http://www.uspto.gov/aia_implementation/faqs_inter_partes_review.jsp (last visited May 6, 2014).

With the various options available to challenge an issued patent, the timing for new filings may be all the more pressing in the postmarket period. Thus, expediting patent prosecution may be a necessity to ensure the products and methods of use remain protected. In the following section, such options are reviewed and explored.

E. Expedited Prosecution

The USPTO has established procedures and programs under which the examination of a patent application may be accelerated or expedited. Under such procedures and programs, the USPTO can advance a patent application out of turn for examination and/or accelerate the prosecution of the application, if the applicant files a grantable petition under the given program. The following section outlines different programs and procedures instituted by the USPTO for applicants to obtain faster prosecution.

1. Accelerated Examination

The USPTO established the accelerated examination program on August 26, 2006, and retained the program after the enactment of the AIA.[46] The accelerated examination program both advances an eligible patent application out of turn for examination and accelerates the actual prosecution of the application for a final disposition within 12 months.

Although internal preparation can be costly, the accelerated examination program is arguably one of the fastest programs offered by the USPTO to obtain patent protection for eligible patent applications. As with many programs at the USPTO, the accelerated examination program has many requirements and associated fees, as outlined in the Manual of Patent Examining Procedure at Section 708.02(a).

One restriction of entering the accelerated examination program is the limit on the number of claims that can be submitted in the patent application. The application must contain three or fewer independent claims and have a total of 20 or fewer total claims. Further, if the claims are appealed, the applicant agrees not to separately argue the patentability of any dependent claim.

One of the most significant hurdles of the program is the time required to search for and comment on known related prior art. The accelerated examination program shifts the burden from the USPTO to the applicant to explain how the claims of the application are allowable over known related art. Applicants must prepare and submit both a pre-examination search document (PESD) and an accelerated examination support document (AESD).

For the PESD, the applicant must demonstrate that the subject matter of the claims was extensively searched for available prior art and show the

46. *See* MPEP § 708.02(a).

search terms/logic behind the searches conducted. For instance, the applicant must in good faith identify the field of search by U.S. class/subclass, provide the search logic or chemical structure or sequence used as a query, and include the databases/files searched. Furthermore, the conducted searches must involve U.S. patents and patent application publications, foreign patent documents, and non-patent literature. In preparing the PESD forms, the applicant needs to consider and anticipate potential arguments from the USPTO rejecting the claims, because any future amendments to the claims (including any new claim) must be supported by the PESD, otherwise the amendments will be treated as not fully responsive and will not be entered.

In addition to the PESD, the applicant must file the accelerated examination support document (AESD). Basically, the AESD is an advanced and more detailed Information Disclosure Statement (IDS), as used in the regular standard patent prosecution track. However, the IDS of the AESD must cite all references deemed most closely related to the subject matter of each of the claims. In addition, for each reference cited, the AESD must include an identification of all the limitations in the claims that are disclosed by the reference specifying where the limitation is disclosed in the cited reference and must provide a detailed explanation of how each of the claims is patentable over the references.

Once the application is accepted into the program, the prosecution of the application is accelerated. For instance, the response period for a non-final Office Action is one month with no extensions, as opposed to up to six months for regular prosecution. The failure to meet the strict time requirement results in abandonment of the application. Although the consequences for not meeting the deadline are severe in this program, the applicant can have a final disposition within 12 months of the filing date of the patent application.

The decision to enter the accelerated examination program can depend on the urgency and needs of the applicant. For example, if an applicant has a patent that covers the composition of an NDA drug and the patent is going to expire within a year, the applicant may seriously consider the benefits of an accelerated examination for a new patent application directed to the method of administering the drug covered by the patent. For this particular applicant, obtaining a new patent on the method of administering the drug (or other patentable variations) could possibly extend the period of exclusivity for the NDA drug for the life of the new patent.[47]

47. In other areas, such as the computer software field, where the industry moves faster than the regular pendency to obtain a patent, applicants may benefit from the 12-month final disposition of the accelerated examination to obtain patent coverage, prior to the obsoleteness of the patented technology through the standard prosecution track.

One benefit of the accelerated examination program is that it grants the applicant greater control over the direction of the patent application, and the applicant drives the discussion of the patentability of the application with the examiner. In contrast, as made apparent by the description and further discussed in the following section, the prioritized examination program grants more control of the patent application examination to the patent examiner rather than the applicant.

Nevertheless, one caveat of the accelerated examination program is the concern of prosecution history estoppel. Since the applicant has to submit detailed statements regarding the patentability of each claim and arguments with respect to the patentability of each claim over related known prior art, such statements could be damaging to the scope and strength of any patent to be issued from the patent application.

Although the accelerated examination program can be a beneficial tool for expediting the prosecution of a patent application, the prioritized examination program outlined next may be more attractive for some applicants.

2. Prioritized Examination

The AIA established a new accelerated program with the prioritized examination of patent applications. The prioritized examination program solely advances an eligible patent application out of turn for examination. These patent applications are called Track One applications.[48] Unlike accelerated examination, the prioritized examination does not speed up the actual prosecution response periods of the application, although any extension of time declassifies the application to a regular application. Furthermore, there is much less work involved in filing a prioritized examination, as opposed to an accelerated examination.

Unlike accelerated examination, the prioritized examination does not require submission of PESD and AESD documents. Rather, the prioritized examination only requires submission of a simple petition and fee, along with the standard documents for submitting any regular patent application at the USPTO. The prioritized examination fee, however, is not insubstantial and for the year 2013 was $4,800 ($2,400 for small entities) in addition to the regular patent application fees. The goal of the program is to provide a final disposition of the application within 12 months, on average, of prioritized status being granted.[49]

The prioritized examination program does limit the number of claims that can be filed, though the restrictions are not as great as in the accelerated

48. *See* MPEP § 708.02(b).

49. For further information on the AIA implementation of Prioritized Examination, please visit *Patents Examination*, U.S. Pat. & Trademark Off., http://www.uspto.gov/aia_implementation/patents.jsp (last visited May 6, 2014).

examination program. The Track One applications are limited to no more than four independent claims with 30 total claims.

The decision to request prioritized examination can depend on the goals of the applicant. Although the government cost of filing a prioritized examination is greater than the government cost of filing an accelerated examination, the overall cost may be approximately the same, if not less. As explained, applications in accelerated examination must have extensive searches performed and must have detailed documents, such as the PESD and AESD, prepared in advance. The cost of such searches and preparation could potentially outweigh the additional cost of the petition for prioritized examination. Further, for small entities, the prioritized examination program may be most beneficial of all, since the additional fee is 50 percent less than the large-entity fee to enter the program. However, as noted, the prioritized examination program gives more of the control of the patent application examination to the patent examiner, whereas the accelerated examination program gives the applicant arguably greater control over the direction of the patent application.

3. Patent Prosecution Highway (PPH)

The Patent Prosecution Highway (PPH) is a program that permits faster examination of U.S. patent applications than the standard prosecution track. Eligible applications of the PPH are those that have a corresponding foreign application which has at least one claim indicated as patentable or that have a favorable search report from a corresponding PCT international application. If necessary, the U.S. claims must be amended to "sufficiently correspond" to the allowable foreign claims.

Unlike the accelerated and prioritized examination programs, the PPH does not require an additional fee or further support documents other than the filing of a PPH request along with the favorable documents. Once the request to enter the PPH program has been granted, the examination of the patent application occurs within two to three months. Further, applicants can enter the PPH program at *any time* before examination begins—which is in contrast to the accelerated and prioritized examination programs that must be petitioned for upon the application filing.

The USPTO advertises that more than 90 percent of PPH applications are allowed and that PPH applications have fewer actions per disposal on average in comparison to non-PPH applications.[50] It is important to note, however, that not all foreign patent offices partner with the United States

50. For further information on the AIA implementation of Prioritized Examination, please visit *Quick Reference Guide (QRG) for eDAN 2.1*, U.S. Pat. & Trademark Off., http://www.uspto.gov/patents/init_events/pph/pphbrochure.pdf (last visited May 6, 2014).

for the PPH program. As of 2013, the participating partner countries/offices include Australia, Austria, Canada, China, Czech Republic, Denmark, European Patent Office, Finland, Germany, Hungary, Iceland, Israel, Japan, Korea, Mexico, Nordic Patent Institute, Norway, Portugal, Russia, Singapore, Spain, Sweden, Taiwan Intellectual Property Office, and the United Kingdom.[51]

If an applicant has the advantage of having either a corresponding foreign application that has at least one claim indicated as patentable or a favorable search report from a corresponding PCT international application, the applicant should take advantage of the PPH program if seeking to accelerate prosecution. In certain instances, applicants will not want to seek advancement of patent prosecution, such as applicants who have a patent application pending at the USPTO for competitor protection and the like, as previously discussed.

4. Petitions to Make Special

In addition to the preceding programs, a patent application can be advanced out of turn for examination or for further action with the submission of a Petition to Make Special based on certain criteria, such as the applicant's age or health.[52] For petitions based on an applicant's age or health, a fee is not required, but submission of evidence showing the applicant is at least 65 years of age or of very poor health is required, along with the standard documents for submitting any regular patent application at the USPTO. These applications will be advanced out of turn for examination similar to the procedures in the prioritized examination program.

5. Pilot Programs

The USPTO institutes pilot programs that may advance the prosecution of a patent application from time to time. One program that the USPTO instituted in the past was the Green Technology Pilot Program, which advanced green technology–related patent applications out of turn for examination and earlier review without meeting all of the current requirements for the accelerated and prioritized examination programs. Although this program is now closed, the USPTO may institute other advancement

---

51. For further information on PPH programs, please visit *Patent Prosecution Highway (PPH)—Fast Track Examination of Applications*, U.S. PAT. & TRADEMARK OFF., http://www.uspto.gov/patents/init_events/pph/index.jsp (last visited May 6, 2014).

52. *See* 37 C.F.R. § 1.102. Petitions to Make Special based for other reasons, such as the invention will materially (i) enhance the quality of the environment, (ii) contribute to the development or conservation of energy resources, or (iii) contribute to countering terrorism, are also set forth in 37 C.F.R. § 1.102 but are outside the scope of this book.

programs in the future, although such programs may be unlikely due to the opportunities for acceleration just discussed.[53]

Another pilot program that is currently running is the Full First Action Interview Pilot Program, which can expedite the efficiency of the prosecution of an application.[54] Prior to a first Office Action on the merits, an applicant can request entry into the pilot program, which grants an applicant an interview prior to the first Office Action. This pilot program gives applicants a forum to discuss prior art and proposed amendments to the claims with the examiner without the concerns of having a non-final Office Action clock ticking. This program can increase the efficiency of a patent application, which can reduce the pendency of the application.

6. Global Considerations

Although earlier chapters discuss international considerations of patent portfolios, certain aspects of global considerations should be revisited with respect to expedited exams during the postmarketing phase. For example, certain foreign patent offices also reciprocally participate in similar patent prosecution highway programs such that the foreign patent offices expedite prosecution of those patent applications that have a corresponding foreign application with at least one claim indicated as patentable or that have a favorable search report from a corresponding PCT international application. The number of patent offices participating in such programs is continuously growing. The World Intellectual Property Organization (WIPO) maintains a database of the various global patent offices that agree to work with each other in such endeavors.[55]

One consideration to keep in mind upon the filing of any patent application is whether and when to file the application internationally. It is important to remember that foreign or international PCT applications corresponding to a U.S. patent application can only be filed within one year of the effective filing date of the priority application. Another consideration is whether statements made during prosecution of a U.S. patent application can globally impact the prosecution abroad, and vice-versa. Depending on the jurisdictions, statements made during prosecution can have limiting effects on claim scope and patent strength for corresponding patents.

53. For further information on the Green Pilot Program, please visit *Green Technology Pilot Program*, U.S. PAT. & TRADEMARK OFF. http://www.uspto.gov/patents/init_events/green_tech.jsp (last visited May 6, 2014).

54. For further information on the Full First Action Interview Pilot Program, please visit U.S. PAT. & TRADEMARK OFF., http://www.uspto.gov/patents/init_events/faipp_full.jsp (last visited May 6, 2014).

55. For further information on global PPH programs and expedited prosecution internationally, please visit *PCT-Patent Prosecution Highway Pilot (PCT-PCH)*, WORLD INTELL. PROP. ORG., http://www.wipo.int/pct/en/filing/pct_pph.html (last visited May 6, 2014).

This is most important for patent applications entering the accelerated examination program, as the applicant has to submit detailed statements regarding the patentability of each claim and arguments with respect to the patentability of each claim over related known prior art.

III. Conclusion

The various types of patent and prosecution strategies for obtaining the appropriate scope have been covered in this chapter. While not all options may necessarily be available for each postmarket evaluation, keeping the different aspects in mind during the full life cycle of a patent portfolio will be essential to maximize a product's patent protection.

chapter 13

Acquiring and In-Licensing Pharmaceutical Products: Due Diligence Issues[1]

I. Chain of Title/Scope of Rights

In the pharmaceutical context, license agreements enable companies to easily and efficiently expand their intellectual property asset portfolios without exposing themselves to the costs and risks inherent in researching and developing technology in-house. In-licensing, in particular, is a tool to mitigate the costs and risks of drug development while leveraging internal resources and expertise in marketing and distributing new drug products. Similarly, co-development agreements provide opportunities for companies to pool resources and share risk in developing innovations. Such agreements may include licensing provisions setting forth the parties' future rights to any technology created pursuant to the co-development agreement.

Before a company enters into any license agreement, however, it is important that the company conduct a thorough due diligence, as possible problems may arise in a number of areas that could potentially expose the company to costly litigation, loss of its investment, and other liability. In particular, an attorney considering a patent licensing agreement should carefully consider (1) the inventorship of the patent(s) of interest, (2) the transfer of intellectual property (IP) rights from the inventor(s) to the patent-owning company, and (3) the terms of the license and co-development agreements. In addition, the attorney should (4) ascertain that all formalities relating to the patent(s) of interest are in order. Each of these subjects will be discussed in further detail in the following sections.

1. David P. Frazier, PhD, Adriana L. Burgy, and Kimberly D. Braslow, PhD, of Finnegan, Henderson, Farabow, Garrett & Dunner LLP, Washington, D.C.; and Christine M. Hlavka of Bookoff McAndrews PLLC, Washington, D.C.

A. Inventorship

As an initial matter, the diligence attorney weighing a patent license agreement should examine the list of inventors named on the face of the patent and take steps to ascertain that the listed inventorship is correct. In the United States, the named inventors on an issued patent are its presumptive owners.[2] As such, each coinventor presumptively owns a pro rata, undivided interest in the patent as a whole, regardless of each coinventor's actual contribution to the invention.[3] Each coinventor may thus independently use and, perhaps more importantly, independently license the patented invention to third parties, regardless of the wishes or commitments of the other coinventors.[4] To avoid unpleasant surprises, the diligence attorney contemplating an acquisition of patent rights should consider investigating the conception narrative surrounding the patented subject matter as well as any inventorship determinations made on the part of the patent owner. In addition, potential licensors and licensees should seek answers to the following questions:

- What and where is the evidence of conception and reduction to practice?
- Where are the named inventors' lab notebooks?
- Who else was involved in conception and reduction to practice, other than the named inventors? What were their contributions? Where are they now?

1. Legality and *Ethicon*-type Problems

The *Ethicon* case[5] illustrates the adverse consequences of failing to properly investigate inventorship. In *Ethicon*, the patentee and sole inventor Yoon and his exclusive licensee Ethicon sued U.S. Surgical Corp. for infringement of Yoon's patent relating to safety trocars.[6] While the suit was pending, U.S. Surgical Corp. discovered the existence of Choi, an electronics technician who had briefly worked for Yoon.[7] After contacting Choi and confirming his involvement in the development of Yoon's safety trocar, U.S. Surgical Corp. obtained a "retroactive license" from Choi to practice Choi's "trocar related inventions" and Choi's agreement to assist U.S. Surgical Corp. in any suit regarding the specific patent at issue.[8] Thereafter,

2. Israel Bio-Eng'g Project v. Amgen Inc., 475 F.3d 1256, 1263 (Fed. Cir. 2007).
3. Melvin F. Jager, Licensing Law Handbook 76 (2012).
4. *Id.* at 77–78.
5. Ethicon Inc. v. U.S. Surgical Corp., 135 F.3d 1456 (Fed. Cir. 1998).
6. *Id.* at 1458–59.
7. *Id.* at 1459.
8. *Id.*

U.S. Surgical Corp. moved to correct inventorship of the patent-in-suit to add Choi as coinventor of several claims.[9] The district court granted U.S. Surgical Corp.'s motion and furthermore granted U.S. Surgical Corp.'s subsequent motion to dismiss the suit on the basis that Choi, a joint owner of the patent, had granted it a valid license and was unwilling to join the lawsuit.[10] On appeal, the Federal Circuit affirmed.[11]

The *Ethicon* case thus demonstrates how the failure to correctly identify and name the true inventors of patented subject matter can strip a patent license agreement of its value. Ethicon entered litigation as an exclusive licensee but departed as a nonexclusive licensee obligated to share rights to use the patented subject matter with its competitor, U.S. Surgical Corp. To avoid this type of loss, the diligence attorney should carefully investigate inventorship before entering into any patent license agreement.

B. Transfer of Intellectual Property Rights to the Company

The procedure through which companies acquire rights to intellectual property represents another fertile ground for potential problems in patent transactions. Whether a company is a prospective buyer, seller, licensor, or licensee, the diligence attorney should closely examine the transfer of ownership rights from the named inventors to the licensor-company.

1. Company-Developed Intellectual Property

Inventions are conceived by individual inventors, not companies. Companies acquire ownership rights to the intellectual property generated by their inventor-employees via employment agreements and/or formal assignments. Thus, regardless of its role in the licensing context, a company should confirm that it has formalities in place to ensure that its employees timely sign employment agreements pursuant to which the employees assign to it all of their rights in any intellectual property relating to the company's business and resulting from work performed for that company, including all ideas, processes, inventions, discoveries, and business ideas, as well as improvements. A company should further confirm that it has formalities in place to ensure that its inventor-employees timely execute formal assignments of patent applications on which they are named inventors and that these formal assignments are submitted to and recorded by the United States Patent and Trademark Office (USPTO).

9. *Id.*
10. *Id.* at 1459–60.
11. *Id.* at 1459.

A diligence attorney specifically considering licensing one or more of a company's patents to another entity should therefore investigate the following questions:

With regard to the company's employment agreements:

- Are all of the named inventors employees?
- If all named inventors are employees, have they all signed employment agreements under which they are obligated to assign their inventions to their employer?
- Were those employment agreements valid, in force, and applicable to the patented subject matter at the time it was invented?
- If not all named inventors are employees, what relationship do the nonemployees have with the company? Are there valid agreements in place under which the nonemployees are obligated to assign their rights?

With regard to individual assignments:

- Have all of the named inventors signed a formal assignment assigning the patent application(s) to the company?
- Was the executed, formal assignment submitted to and recorded at the USPTO?

2. In-Licensed and Assigned Intellectual Property

Similarly, an attorney considering in-licensing one or more patents from another entity should investigate the preceding questions as well as the following questions:

- Is the potential licensor the same entity as the entity to which the inventor-employee owed an obligation to assign under an employment agreement?
- Is the potential licensor the same entity as the entity named on the recorded assignment?

If the answer to either of these questions is no, the potential licensor may not own the patent(s) of interest and therefore may not possess the power to license the patent(s). In addition, the attorney should investigate what rights any third parties may have in the patent(s) of interest. To that end, the company should investigate the following:

- Do any third parties have any interests, licenses, or other rights in the patent(s) of interests?
- If so, what are the names and addresses of those third parties, and what is the origin, nature, and scope of their interests or rights?

- Were any third-party collaborators involved in research leading to the patent(s)? Were reagents or data from third parties used in the research or incorporated into the patent application?
- If so, were agreements in place to assign any patent rights arising from the collaboration or use of reagents or data?

C. License and Co-development Agreements

An attorney considering a license or co-development agreement must carefully review the terms of the license and/or co-development agreement itself. The terms set forth the rights and responsibilities of both parties to the agreement. Several key contract terms, including transferability, scope, control, and obligation to support, will be discussed in further detail in the following sections. In addition, a discussion of specific antitrust and Federal Trade Commission concerns implicated by license and co-development agreements will be provided. If any of the terms described here are unsatisfactory to a party to the agreement, that party should confer with the other party during contract negotiations to adjust the term(s) as needed to best anticipate and/or prevent future problems.

1. Transferability

A license or co-development agreement's terms of transferability define how access to and use of the patented subject matter may be shared—for example, with a company's affiliates and/or subsidiaries. Thus, an attorney weighing a license or co-development agreement should investigate the terms of the agreement to determine whether the agreement permits the company to share the patented subject matter with others to the extent desired. To that end, the attorney should ask:

- Is access to and use of the patented subject matter limited to the named licensor and licensee?
- Or may the licensor and/or licensee share the patented subject matter with others, such as affiliates or subsidiaries?
- If yes to the latter, is there any limit on the recipients of that transferability (i.e., may the licensor and/or licensee freely share with any affiliate or subsidiary, or with only specific ones)?
- Similarly, are there any limits on the transferability itself (i.e., are affiliates and subsidiaries allowed to access and use the patented subject matter to the same extent as the named licensor and/or licensee)?
- Must any conditions be met or procedures followed before a transfer?

2. Scope

The scope of a license or co-development agreement defines the patented subject matter encompassed by the agreement. In particular, the scope sets forth the patent(s) to which rights are being transferred under the agreement. Products associated with the patent(s) may be listed as well. An attorney contemplating a license should verify that the agreement contains a correct and complete listing of the patent(s) and any associated products of interest. In addition, the attorney should consider whether the patent(s) of interest is part of a larger patent family, which may contain similar patents of slightly different scope or pending applications. To that end, the attorney should investigate the following:

- Does the license agreement set forth a complete and correct listing of the patent(s) to which licensing rights are being sought?
- Is the patent(s) of interest a member of a larger patent family containing other granted patents?
- If yes, what is the scope of those other granted patents relative to the licensed one(s) and how does that scope affect the licensee's ability to use the licensed patented subject matter?
- Does the patent family contain any pending applications? What claims are pending in those applications? Does the license grant the licensee any rights to patents issuing from those applications?

The scope further defines what specific rights to the patented subject matter are being transferred to the licensee and for what period. The attorney should thus investigate the following to ascertain whether, under the terms of the agreement, it is granting or receiving the desired rights for the desired period:

- How is the licensee permitted to use the subject matter under the agreement?
- What is the time period for that use?
- What occurs upon completion of that time period? Do the parties have an option to extend, if desired?
- Is the licensor permitted to grant rights to the patented subject matter to other third parties during the life of the agreement? What are those rights? Are there any limitations on those rights?

3. Antitrust and Federal Trade Commission (FTC) Concerns

License and co-development agreements may implicate a number of antitrust concerns. An attorney considering a license or co-development agreement should therefore consider federal antitrust laws, which are set forth in several statutes summarized in Table 1.

Table 1
Federal Antitrust Laws

Statute	Summary
Sherman Act	Aimed at preserving free and unfettered competition, the act bans contracts that per se (or unreasonably) restrain trade or commerce in the United States and proscribes civil and criminal penalties for violations.[12]
Clayton Act	Prohibits specific practices, such as mergers and acquisitions that lessen competition or create a monopoly and provides civil remedies, including damages and injunction.[13]
FTC Act	Authorizes the FTC to bring actions against those who commit anticompetitive acts and deceptive practices.[14]
Hart-Scott-Rodino Antitrust Improvements Act	Requires prenotification of certain transactions to acquire securities or assets and specifically applies to certain pharmaceutical patent license transactions.[15]

As a further resource, the Department of Justice (DOJ) and Federal Trade Commission (FTC) issued guidelines for the types of antitrust concerns that are raised by a license agreement of intellectual property.[16] In

12. 15 U.S.C.A. §§ 1–2; *The Antitrust Laws*, Fed. Trade Comm'n, http://www.ftc.gov/tips-advice/competition-guidance/guide-antitrust-laws/antitrust-laws (last visited Feb. 26, 2014).

13. 15 U.S.C.A. §§ 12–18. Section 2 of the Clayton Act, 15 U.S.C.A. § 13, further includes the Robinson-Patman Price Discrimination Act. *The Antitrust Laws*, *supra* note a.

14. 15 U.S.C.A. § 45; *The Antitrust Laws*, *supra* note a.

15. 15 U.S.C.A. § 18a (also known as § 7A of the Clayton Act); *see also* Premerger Notification: Reporting and Waiting Period Requirements, 78 Fed. Reg. 68,705 (Nov. 15, 2013) (clarifying types of pharmaceutical transactions that must be reported under Hart-Scott Rodino Act).

16. Fed. Trade Comm'n, Antitrust Guidelines for the Licensing of Intellectual Property (Apr. 6, 1995), *available at* http://www.ftc.gov/sites/default/files/attachments/competition-policy-guidance/0558.pdf.

general, concerns arise when an agreement harms competition of likely competitors in a relevant market. As a starting point under the guidelines, a diligence attorney evaluating a license agreement should identify whether the parties to the agreement are likely or potential competitors and identify the relevant markets of competition. More specific types of questions that may be considered include the following:

- Does the acquisition of IP lessen the competition in a relevant market—for example, would the market have otherwise competed using different technologies?
- Does the agreement merge the research and development (R&D) of two of only a few possible entities to engage in that type of activity?
- Does the agreement place restrictions in one market that harm the competition in another market—for example, by affecting the price or availability of resources in the other market?

An agency analysis will focus on the *effects* of the agreement. For example, the DOJ or FTC may first define the relevant markets in which competition would be affected (e.g., goods, technology, R&D) and consider the significance of market participants.[17] These agencies generally apply a rule of reason approach to assess whether the restraint is reasonably necessary to achieve pro-competitive benefits and whether those outweigh the anticompetitive effects.[18] Antitrust scrutiny may be greater in the context of an exclusive license. Alternatively, if a licensing arrangement is not anticompetitive on its face and the licensor or licensee collectively accounts for no more than 20 percent of each relevant market, the agreement will fall into a "safety zone" at the agency, which recognizes the value of such licenses for the promotion of innovation and competition.[19] During a diligence investigation, conducting this type of market analysis and weighing of competitive factors can help to identify potential antitrust concerns that could have significant consequences for a licensing agreement.

4. Control and Obligation to Support

License and co-development agreements may contain terms of control and/or obligation to support. Terms of control generally appear in a co-development context: they define which party controls any intellectual property conceived as a result of the parties' co-development efforts. Terms setting forth an obligation to support may appear in license or co-development agreements: they define whether either or both parties to the agreement have an obligation

17. *See, e.g.*, *id.* at 9.
18. *Id.* at 16.
19. *Id.* at 22.

to support the other in case of litigation involving the patent(s) of interest, such as patent infringement litigation.

Thus, the attorney considering a license or co-development agreement should investigate the following to verify that any terms of control and/or obligation to support are satisfactory:

- If any innovations are conceived as a result of the co-development efforts pursued under the agreement, which party has control over the innovations? Does that control involve any responsibilities, such as, for example, to pursue patent applications claiming the innovations?
- Does the other party have any rights in the innovations?
- Are there any limitations on what innovations are covered by the co-development agreement, such as subject matter limitations?
- In case of subsequent litigation involving the patent(s) covered by the license and/or co-development agreements, what is each party's obligation to support the other? Are there any limitations on those obligations?

D. Formalities in Order

Finally, the attorney considering a patent licensing agreement should confirm that all formalities with respect to the patent(s) of interest are in order. In particular, the attorney should investigate the following general questions:

- Have all inventors signed an oath or declaration?
- Have all maintenance fees been timely paid to date?
- Was small-entity status established during prosecution of the underlying patent application?

Each of these questions will be discussed in greater detail in the following sections.

1. Declarations

An applicant for a U.S. patent must submit an executed inventor's oath or declaration to the USPTO during prosecution.[20] In the oath or declaration, the inventor attests that he or she is the original inventor of the claimed invention and that he or she authorized the filing of the patent application for the claimed invention.[21] If a patent application lists multiple inventors,

20. 35 U.S.C. § 115.

21. *Id.* § 115. Prior to implementation of the AIA, an inventor was also required (1) to state that he or she was the first inventor of the claimed invention, (2) state that the application filing was made without deceptive intent, and (3) provide his or her country of origin.

the oath or declaration must be signed by every named inventor (although not necessarily on the same sheet). If a continuation or divisional application is filed naming a new inventor not part of the prior inventive entity, a newly executed oath or declaration signed by the new inventor must be submitted.[22] A newly executed oath or declaration also must be filed in any continuation-in-part application, whether the application names all, more, or fewer than all of the inventors named in the prior application.[23]

Thus, an attorney considering a patent license agreement should confirm that all named inventors signed an oath or declaration. In addition, out of an abundance of caution, the attorney should verify that a copy of the executed oath or declaration is stored and accessible somewhere in case problems later arise. In all, the attorney should investigate the following questions:

- Was a proper oath or declaration submitted to the USPTO during prosecution?
- Did all of the named inventors sign it?
- If the patent application was a continuation or divisional and the inventive entity included an inventor (or inventors) not part of the prior inventive entity, was an oath or declaration signed by the new inventor(s) submitted?
- If the patent application was a continuation-in-part, was a newly executed oath or declaration submitted, regardless of how the inventive entity may or may not have changed from the prior application?
- Where is the original copy of the executed oath or declaration?

2. Maintenance Fees

Maintenance fees must be paid in any patent issuing from an application filed on or after December 12, 1980, to maintain the patent in force beyond 4, 8, and 12 years after the date of grant.[24] Maintenance fees are thus due three times during the life of a patent:

- three to three and a half years after the date of issue;
- seven to seven and a half years after the date of issue; and
- eleven to eleven and a half years after the date of issue.[25]

In addition, maintenance fees may be paid with a surcharge during the following "grace periods":

22. 37 C.F.R. § 1.63(d)(5).
23. *Id.* § 1.63(e).
24. Manual of Patent Examining Procedure (MPEP) § 2504 (rev. Aug. 9, 2012).
25. 37 C.F.R. § 1.362(d), MPEP § 2501.

- three and a half years through the date of the fourth anniversary of patent grant;
- seven and a half years through the date of the eighth anniversary of patent grant; and
- eleven and a half years through the date of the twelfth anniversary of patent grant.[26]

If the maintenance fees are not timely paid, the patent will lapse as of the end of the grace period.[27] Thus, as part of any due diligence, the attorney considering licensing a patent should confirm that the patent owner has timely paid all maintenance fees. If not, the company risks licensing a lapsed, and therefore worthless, patent.

3. Small Entity Status

A patent applicant may claim "small entity status" to receive a 50 percent reduction in most USPTO fees for prosecuting and maintaining a patent.[28] To qualify for small entity status, a patent applicant must fall within one of the following categories:

> a *person* (i.e., individual or individuals) who has not granted, conveyed, or licensed, and is under no obligation under contract or law to assign, grant, convey, or license, any rights in the invention; or
>
> a *nonprofit organization* (regardless of size) which has not granted, conveyed, or licensed, and is under no obligation under contract or law to assign, grant, convey, or license, any rights in the invention; or
>
> a *small business concern* which has not granted, conveyed, or licensed, and is under no obligation under contract or law to assign, grant, convey, or license, any rights in the invention.[29]

A "small business concern" is one that has fewer than 500 employees, including affiliates.[30]

Small entity status, once established, must be maintained throughout prosecution of the application, payment of the issue fee, and payment of

26. *Id.* § 1.362(e).
27. *Id.* § 1.362(g).
28. MPEP § 509.02
29. 37 C.F.R. § 1.27. In addition, as part of the implementation of the America Invents Act, effective March 19, 2013, a patent applicant may now claim "micro entity status" to receive a 75 percent reduction in most fees. For more information, see Changes to Implement Micro Entity Status for Paying Patent Fees, 77 Fed. Reg. 75,020 (Dec. 19, 2012) (to be codified at 37 C.F.R. pt. 1).
30. 13 C.F.R. § 121.802.

the maintenance fees.[31] If the applicant's status changes during prosecution, when the issue fee is due, or when the maintenance fees are due such that it no longer qualifies as a small entity, the applicant must timely notify the USPTO of its loss in status.[32] Failure to do so may be considered fraud on the USPTO, resulting in unenforceability of the patent.[33] Thus, an attorney considering a patent license should investigate whether the patent applicant claimed small entity status, and, if so, whether it did so accurately. The attorney should also determine whether the patent applicant timely updated the USPTO of any subsequent change in status.

More importantly, small entity status is lost when a patent is licensed or assigned to an entity that would not itself qualify for small entity status.[34] Thus, depending on the licensee's size, the very license agreement under consideration may potentially change the status of the patent of interest. In such a case, the attorney should work with the patent owner to ensure that any change in small entity status is reported to the USPTO, where appropriate.

II. Asset Evaluation

For the acquiring party, a careful evaluation of the assets associated with the transaction is the central event in the diligence process. Before beginning any evaluation, however, it is worth considering important logistical questions. What is the appropriate evaluation process? What should be its scope and depth? How will the parties communicate, and what feedback will be provided to the seller or licensor? Also important, how will the parties handle privileged information and protect it in the event of ongoing or future litigation? The subsequent sections will address each of these questions in turn.

A. Evaluation Process

Evaluating intellectual property assets in a due diligence setting requires a detailed understanding of the purpose of the underlying transaction. Merely following a generic checklist from another transaction will not suffice. The evaluating attorney needs to know what the transaction is, why the client is pursuing it, and what the perceived value of the deal is. Typically, intellectual property risks are only one part of a larger set of risks that the client is evaluating in a transaction. There may also be risks

31. MPEP § 5.05(VI).
32. 37 C.F.R. § 1.27(g).
33. *Id.* § 1.27(h).
34. MPEP § 509.02(V).

concerning whether the technology will work as planned (for example, whether an early stage compound will survive clinical trials), whether business partners will perform as expected, and whether the ultimate product will be accepted and valued in the market.

Evaluating intellectual property assets without an awareness of how they fit into the overall transaction can lead to either an overly aggressive or an insufficiently detailed assessment. For example, evaluating patents covering an approved pharmaceutical product that is being acquired shortly before a likely generic challenge will be a very different exercise from an evaluation of a patented platform technology that may never itself be marketed. In one case, a diligence investigation could involve interviews with scientists, review of original notebooks and documents, and extensive searching of the prior art. In another, a simple review of issued patents, their prosecution histories, and cited art may provide sufficient information. The details of the transaction will help determine the appropriate budget for the diligence investigation and, correspondingly, its scope and depth.

Of course, communication with the client ideally goes both ways. Clients need to be informed and to understand what will be evaluated and what will not. Budget realities may prevent detailed prior art searching or evaluation of inventor notebooks, but the client should understand that there are unquantifiable risks to so limiting the investigation. Nonetheless, it may be completely appropriate to take those risks, even where they are difficult to quantify, because the value in the transaction may not reside in the ultimate validity or enforceability of the relevant patents.

B. Sharing Information with Seller and Licensor

After an evaluation is completed, or after completing a preliminary assessment, a decision must be made about whether and how to share the findings with the seller or licensor. Again, the attorney should be guided by the overall structure of the transaction and how sharing the intellectual property assessment fits into that structure. Is the goal to use the assessment findings to negotiate a lower price? Is the client ready to walk away from the deal if issues cannot be resolved? Alternatively, is it a certainty that the transaction will be completed regardless of the findings? Can warranties or representations provide sufficient protection to address concerns or risks uncovered in the diligence evaluation? The answers to these questions will help the attorney choose how to communicate and how much to communicate with the target.

Foremost in considering how and what to communicate about diligence findings is the issue of whether the findings, if revealed to third parties, would put the underlying assets at risk or make the situation worse. With that issue in mind, the actual format can be anything from a conference between outside counsel, a "pens down" meeting among inside and outside counsel, a discussion among counsel and businesspeople, or some form of written communication. Each format strikes a different balance between concerns about creating a negative record that could be used by a third party and the need for frank communication that could potentially overcome or ameliorate identified concerns. For example, it may be necessary to have a written communication in order for the information to be effectively communicated from the negotiation team to upper management. Other times, an issue may be more sensitive such that maximizing the chances of maintaining attorney-client privilege or work-product immunity is the overriding concern.

C. Privilege of Counterparty Communications

When the time comes to communicate the results of the diligence with the seller or licensor, a frequent question is whether those communications can be protected from later disclosure by attorney-client privilege or work-product immunity. The short answer is that while courts have granted protection to such communications between parties to a transaction even before the transaction is consummated, the scope of such protection is limited and may be challenged in litigation with sometimes unpredictable results.

Typically, the attorney-client privilege protects communications that are made in confidence, between an attorney and client, for the purpose of securing legal advice.[35] Protection is waived if the confidential communication is shared with a third party.[36] An exception to the waiver principle occurs when privileged communications are shared between parties with common legal interests. There, courts have recognized a "common interest" privilege that, while not an independent privilege, maintains an existing privilege that would otherwise be waived by sharing with a third party.[37] In the patent context, the Federal Circuit has recognized a common interest privilege between licensors and their exclusive licensees, allowing sharing of otherwise privileged communications concerning obtaining and maintaining valid patent rights.[38]

Courts have applied the common interest privilege to protect legal opinions shared between parties to a patent transaction even when the

35. Edna Selan Epstein, The Attorney-Client Privilege and the Work-Product Doctrine 64–66 (5th ed. 2007).

36. *Id.* at 261–66.

37. *Id.* at 274–75.

38. *In re* Regents of Univ. of Cal., 101 F.3d 1386, 1390 (Fed Cir. 1996).

transaction is not completed.[39] But other courts have drawn distinct limits on application of the doctrine in this context.[40] A key consideration is whether the shared interest between the parties is a "legal" interest as opposed to merely a commercial one.[41] Legal interests held sufficient to support a common interest in a transactional context include anticipated or actual litigation and whether products expected to be jointly marketed would infringe third-party patents.[42] Another important factor that courts may examine is whether both parties treated the disclosed information as privileged and confidential. Thus, having a written confidentiality agreement in place before sharing privileged material and limiting the group with access to such material to counsel will improve the chances that a court will recognize the common interest privilege.[43]

Because of the risks associated with sharing opinions of counsel and other privileged information, parties may choose to exchange only factual information and allow their respective counsel to provide analysis and opinions. For example, if the acquiring company is concerned about a particular piece of prior art with respect to a transaction patent, it may choose to have its own counsel analyze the reference rather than seeking any opinions obtained by the target company. Similarly, in requesting information, the acquirer may simply ask for copies of prior art that the target company is aware of or has considered rather than asking for copies of the target's opinions. If no privileged information is actually exchanged with the other side, the risk that evaluations will be discoverable in later litigation is much reduced.

III. Patent Due Diligence

Due diligence can be thought of as a risk assessment. It involves understanding and synthesizing the business, technological, and legal risks associated with acquiring patent rights. The more time and effort placed into the investigation, the more knowledgeable a party will be going forward with the acquisition. Since due diligence is an amalgam of business, technological, and legal aspects, a strategic plan is a requirement. That strategic plan,

39. *See, e.g.*, Hewlett-Packard Co. v. Bausch & Lomb Inc., 115 F.R.D. 308, 312 (N.D. Cal. 1987); Morvil Tech. LLC v. Ablation Frontiers Inc., 2012 WL760603, at *2 (S.D. Cal. Mar. 8, 2012).

40. *See, e.g.*, Nidec Corp. v. Victor Co. of Japan, 249 F.R.D. 575, 579–80 (N.D. Cal. 2007).

41. *See, e.g.*, *id.*

42. *See, e.g.*, *Hewlett-Packard*, 115 F.F.D. 308, 310; Britesmile Inc. v. Discus Dental Inc., 2004 WL 2271589, at *2 (N.D. Cal. Aug. 10, 2004).

43. *See, e.g.*, Tenneco Packaging Specialty & Consumer Prods. Inc. v. S.C. Johnson & Son Inc., 1999 WL 754748, at *2 (N.D. Ill. Sept. 14, 1999).

however, needs to be adaptable and flexible to accommodate the ebb and flow of the usually fast-paced transaction. This strategic due diligence plan can be broken down into the following areas: (a) intellectual property rights to be transferred; (b) patents covering product; (c) validity; (d) patentability searches; (e) freedom to operate (FTO); (f) design around issues; (g) patent term extension (PTE) decision for multi-patent families; and (h) life cycle management. Each of those topics will be discussed in the following sections.

A. Intellectual Property Rights to Be Transferred

A preliminary step in a due diligence is to identify the patents (and, potentially broader, the trade secrets, trademarks, and/or copyrights) to be transferred. Usually, a list of the patents of interest is provided. If the patents (or the intellectual property) have not yet been identified, requests should be made. Example requests seeking this information include the following:

- Provide a list of all pending and issued worldwide patents and applications (including licensed patents and applications) relating to the transaction (or products of interest).
- Provide copies of all applications, including priority documents (whether published or not) and patents (including licensed applications and patents) with file histories relating to the transaction (or products of interest).

Based on this information, the target company's intellectual property should be identified and prioritized. This can be done from various private and public databases, such as the U.S. and European patent office websites. If a pharmaceutical product is part of the transaction, the patents covering the approved product for marking are listed in the Orange Book of the Food and Drug Administration (FDA).[44] With the identification of the intellectual property from the target company, copies of the relevant patents and applications along with the file histories should be obtained.

That list, however, only represents half of the story. To fully understand the rights to be transferred, the attorney needs to first understand the goals of each of the negotiating parties. Perhaps the transaction involves a simple licensing agreement of the manufacture of a single drug. Or, maybe the transfer of the patent rights is part of a larger acquisition of tangible assets—for instance, as part of a merger or part of a sale of a subsidiary. Similarly, the buyer/licensee may have relatively little information about the patent rights associated with the asset simply because the buyer/licensee is more focused on the transfer of the asset itself.

44. *Orange Book: Approved Drug Products with Therapeutic Equivalence Evaluations*, U.S. Food & Drug Admin., http://www.fda.gov/cder/ob (last visited Feb. 23, 2014).

Depending on the goals of the parties involved, the due diligence needs may vary. Relevant questions to consider include the following:

- Why is the particular patent(s) important to this transaction or to the particular parties?
- Does either party lack the resources required to develop an asset covered by the patent (e.g., research and development, implementation, and marketing)? Is the other party a competitor and is threatening expensive litigation otherwise? Does either party have a patent on similar technology and is attempting to strengthen its portfolio and, thus, its proprietary position in the marketplace?
- What are the long-term goals of the parties? For instance, is this an exclusive or nonexclusive license? Is there a potential for third-party transactions in the future? Will the licensor license with multiple partners in different fields of use for a single drug? Will the license agreement account for future development of the drug for new applications? Are there territorial restrictions?
- Is either of the parties a serial acquirer, or is this a first-time acquisition? A party that routinely licenses or acquires may be more knowledgeable in the industry and may have vetted previous legal issues involving patent transactions as opposed to a party facing its first transaction in a litigious industry.
- What are the financial limitations of the parties involved? An acquiring party may want to know, for example, what costs are associated with the asset protected by the patent. How is that cost offset by potential sales? How will remuneration be implemented?

For any transaction of patent rights, whether it is the sole reason for the transaction or merely a smaller part of a larger transaction, understanding the goals of the transaction and of each party should drive the subsequent due diligence analysis. Once the goals are established, the actual investigation should focus on an assessment of the entire portfolio (regardless if just a single patent is to be transferred) and a review of the individual patents within the portfolio.

B. Patents Covering Product

Whether an entire portfolio[45] is the subject of the transfer or the transfer only involves a single patent that is part of a larger portfolio, a portfolio audit is the natural next step in the due diligence investigation. This

45. In this context, a patent portfolio is considered to be a collection or bundle of patents that provides a combination of rights to exclude others from the marketplace. For example, a patent portfolio may encompass patents directed toward varying scopes of related compounds and compositions, processes for making such compounds, methods of use, etc.

involves understanding the big picture of the portfolio—how the individual patents cover the protected asset (if they do) and how they relate to each other within the same portfolio, as well as the details of how the benchtop idea matured into the patent.

The parties should first organize the portfolio in a way that makes logical sense to the parties involved in the transaction. For example, the parties may organize by identifying the core idea in each patent (e.g., which patents cover the broadest genus of compounds to the narrowest genus of compounds, which cover the processes that make the particular compound of interest, the fields of use, etc.) and the scope of protection (i.e., claims). Or, the parties may organize the patents according to their respective value to their company (which may take into account some of the more micro issues discussed next).

Regardless of how the portfolio is organized, an understanding of the products and the patents (or intellectual property) is necessary to determine the relationship between the two. In order to make this determination, the following should be gathered and assessed:

- Confirm all relevant intellectual property of the target (use public sources and verify the list of intellectual property provided; obtain copies of patents, applications, and file histories).
- Obtain detailed information on the processes by which the product(s) is made, including preparation, manufacture, and purification of the product. Are there any third-party licenses needed in order to practice these processes?
- Acquire complete information on the commercial formulation and contemplated routes of administration. Is there any third-party intellectual property for which a license might be needed in order to use such formulations or routes of administration?
- Determine which patents/applications contain at least one claim that literally covers the product(s), bearing in mind potential issues of claim interpretation.

The audit should also reveal if there are any areas relating to the patented product left uncovered (e.g., an unclaimed method of use for the claimed drug). If there is remaining subject matter not yet patented or claimed in a pending application, the party should consider if it is feasible to pursue it in any existing pending applications or if it makes sense to file new related applications. This involves understanding the patent landscape and planning a course of action to fill the gaps of unprotected subject matter in the party's field.

The next step involves assessing the status of each patent in the portfolio. Keeping track of all filing dates and priority dates is key, particularly if any members of the patent portfolio are still under examination. If varying scopes of a compound genus are covered in related patents and patent applications, will any patent be prior art to a related application if their scopes overlap? Expiration dates are also important as patents closer to expiring may be less valuable than one with several years of term remaining. Are any of the patents eligible for patent term extension under 35 U.S.C. § 156(d)? What is the geographic coverage of the patent? Are any subject to any post-grant proceedings (reexamination; past, pending, or threatened litigation; interferences; reissues; oppositions; inter partes review; etc.)?

The information just discussed is also important from a freedom-to-operate standpoint. Freedom-to-operate analysis serves to identify any blocking patents that potentially hinder the party's ability to make, use, and/or sell its products (i.e., the commercialization of its patented product even though the party has the right to exclude others from making it). Parties should perform patent searches to determine if the protected product may be considered to infringe another's patent rights. Ideally, the search should be broad enough to encompass patent literature not within the traditional boundaries of the related industry. For instance, a patent holder of a particular drug may want to search the medical device art, plant patents, design patents, or even business methods. The search results may affect the value of the portfolio and affect the overall negotiation. However, if issues are found, one or both of the parties may further evaluate the blocking patent for invalidity, which may mitigate the issue. Other options include cross-licensing, design around, and post-grant proceedings like the new post-grant review or inter partes review under the America Invents Act (AIA).

C. Validity

Due diligence also requires digging deep into the individual patents and reviewing any validity issues with a fine-tooth comb. This involves conducting a comprehensive investigation into the history of the patent and understanding the competitive landscape that may affect the value of the patent rights to be transferred.

A fundamental question that should be addressed when any patent is sought to be acquired is whether or not the patent is valid. For example, a decision invalidating a patent covering a major pharmaceutical drug may allow generic competitors to enter the marketplace much earlier than expected and dramatically reduce sales by the brand-name pharmaceutical company.

A validity (and also an enforceability) assessment includes evaluating the novelty and nonobviousness of the patent's claims and determining compliance with formal requirements, such as written description, enablement and best mode, and judicially created doctrines such as obviousness-type double patenting. For example, a validity review addresses the following:

1. Anticipation and Obviousness

Most validity analyses begin with a patentability search that includes an investigation of the novelty and nonobviousness of the patent's claims and, particularly, those relevant claims covering the product(s). This analysis may begin with the art cited by the examiner during prosecution and the art provided by the applicant. As the results of the patentability searches are provided, the analysis will be updated and supplemented.

2. Inequitable Conduct

Although inequitable conduct goes to the enforceability of a patent, it is usually when one conducts the validity assessment that issues of inequitable conduct also are flagged and evaluated. The requirements for proving inequitable conduct have recently changed to a but-for materiality standard.[46] Despite these changes, inequitable conduct can still present a significant risk to enforceability under certain circumstances.[47] Unlike a finding of invalidity based on prior art where individual claims may be found invalid, a charge of inequitable conduct can render the entire patent unenforceable. Thus, parties should carefully review statements made during prosecution of the patent, including any declarations submitted to overcome rejections or presented during opposition proceedings. Prosecution files, patent-counsel files, and statements made in related applications should all be reviewed to check for any contradictory statements. While it may be possible to correct mistakes during prosecution,[48] post-issuance corrections—for example, via a reissue application or through the new supplemental examination procedure introduced under the AIA—are likely to be expensive, and it may be difficult to predict the result.

3. Enablement/Written Description

Parties should not only ensure that the patented product is adequately described and enabled by the patent disclosure, they should make sure

46. Therasense Inc. v. Becton, Dickinson & Co., 649 F.3d 1276 (Fed. Cir. 2011) (en banc) (explaining that the defendant must show that the patent applicant (1) withheld or misrepresented so-called material information and (2) did so with the intent to deceive the USPTO).

47. *See, e.g.*, Aventis Pharma SA v. Hospira Inc., 675 F.3d 1324 (Fed. Cir. 2012).

48. *See, e.g.*, Intellect Wireless Inc. v. HTC Corp., 910 F.Supp.2d 1056 (N.D. Ill. Sept. 6, 2012), *aff'd*, 732 F.3d 1339 (Fed. Cir. 2013).

that the claims and the state of the art at the time of filing enable one of ordinary skill in the art to practice the claimed invention.[49] If the patented product covers a drug and the patent claims multiple layers of genus and sub-genus claims, parties should review the patent specification for an adequate description of "structural features common to the members of the genus" or "a representative number of species" within the genus.[50] While broad claims can be helpful for staking out an exclusive space among innovators, in ANDA litigation the most valuable claims are those directed solely to the commercial compound. These claims have the added benefit of reducing potential enablement and written description issues. Thus, part of the diligence review should involve finding a "picture" claim covering only the commercial product.

4. Prior Use or Sale

The due diligence investigation should include a review of the history of the patented product to determine if there are any potential 102(b) on-sale bar issues (for example, if the product was subject to an early contract for its manufacture). Prior use can be problematic for the pharmaceutical industry because it may conflict with clinical trials required for FDA approval. However, as long as the clinical trials occur under formal confidentiality agreements, clinical trials should not be considered a public use and, thus, not invalidate the patent.[51] Parties should take note that the AIA alters "public use" in two situations: (1) a "public use" is no longer limited to the United States but includes any public use anywhere in the world, and (2) no grace period applies to public uses conducted by third parties.[52]

5. Inventorship and Ownership

Inventorship and ownership are two distinct issues, but both are equally important. Errors in inventorship can result in an unenforceable patent, particularly if made with deceptive intent.[53] And for patents, ownership starts with inventorship. A diligence attorney may question the patent owner if too many or not enough inventors are suspected to be named on the patent. The attorney may note in particular those patents arising out

49. Streck Inc. v. Research & Diagnostic Sys. Inc., 665 F.3d 1269, 1285 (Fed. Cir. 2012).

50. Ariad Pharm. Inc. v. Eli Lilly & Co., 598 F.3d 1336, 1350 (Fed. Cir. 2010).

51. Dey LP v. Sunovion Pharm. Inc., 715 F.3d 1351 (Fed. Cir. 2013).

52. George Wheeler, *Patent Portfolio Strategy for the America Invents Act*, 12 J. Marshall Rev. Intell. Prop. L. 289 (2013).

53. Stark v. Advanced Magnetics, 119 F.3d 1551, 1553, 1556 (Fed. Cir. 1997); 35 U.S.C. § 102(f) ("A person shall be entitled to a patent unless . . . he did not himself invent the subject matter sought to be patented"); *see also* Univ. of Colo. Found. Inc. v. Am. Cyanamid Co., 196 F.3d 1366, 1374 (Fed. Cir. 1999) (internal quotations omitted); *see also* 35 U.S.C. §§ 111, 115–16, 256.

of collaborative efforts where inventorship can be a bit murkier. The investigation may include reviewing each named inventor's publications related to similar subject matter as disclosed in the patent and holding discussions with the named inventors themselves, perhaps even holding discussions with upper management to get a feel for the culture of the company (e.g., the company routinely lists directors, managers, and the like as inventors simply out of deference). As part of the due diligence, the attorney may seek the following information:

- A list of all inventors, with their present and past affiliations.
- All notebooks and other documentation (for example, invention disclosure statements, grant applications, etc.) relating to the conception and reduction to practice of inventions.
- A copy of all current, standard company policies, practices, and procedures (and related documents) followed in order to protect trade secrets, inventions, patent rights, disclosure of inventions, notebook record keeping, patenting procedures, and the procedure used to determine inventorship.
- A copy of all company policies, practices, and procedures (and related documents) in place at the time rights to the inventions were obtained in order to protect trade secrets, inventions, patent rights, notebook record keeping, the patenting procedures, and the procedures(s) used to determine inventorship.
- A description of any prior obligation that any employee has to a third party relating to the inventions.
- Copies of all relevant documents relating to the assignments of inventions from the inventors of any patent or application, and a list of any inventor who has not assigned his or her intellectual property rights.

Ownership can also result in an unenforceable patent[54] and, thus, affect its transferability. A prudent investigation involves reviewing the ownership documents associated with each patent:

- Assignments: Check recordings at the USPTO and foreign authorities and compare them with the owner's assignment records. Have all of the inventors transferred rights? Are there any gaps in the chain of title?

---

54. *See, e.g.*, Email Link Corp. v. Treasure Island LLC, No. 2:11-cv-01433-ECRGWF (D. Nev. 2012) (holding a patent subject to a terminal disclaimer unenforceable against the alleged infringers because it was not commonly owned with the related prior patent, even though both patents were owned by subsidiaries that were wholly owned by the same parent company).

- Existing licenses: Is one of the parties actually a third party? The parties must understand if there are any existing patent license agreements and who has the right to enforce the patent under the license (and the nature of the license—exclusive, nonexclusive, compulsory, field of use, territory—duration of the license, covenants not to sue), including any oral license agreements.[55]
- Conditions on ownership: Are any of the patents subject to joint research agreements, distribution agreements, joint ventures, or security agreements, or are any funded by federal/state grants?[56]
- Confidentiality agreements: Are there any confidentiality agreements, nondisclosure agreements, or noncompete agreements associated with the patents?
- Funding: Was the invention funded under federal research and development contracts? Are there Bayh-Dole implications?

Simply put, start at the beginning and follow the paper trail attached to any of the patents to be transferred.

6. Obviousness-type Double Patenting

Double patenting can be in two forms: (1) statutory double patenting under 35 U.S.C. § 101, which bars a later patent claiming the same subject matter of an earlier patent; and (2) obviousness-type double patenting, which is a judicially created doctrine and prevents a later patent from covering an obvious variant of an earlier patent.[57] Within the pharmaceutical context, obviousness-type double patenting should be a consideration. "The doctrine of double patenting seeks to prevent the unjustified extension of patent exclusivity beyond the term of a patent."[58] A terminal disclaimer[59] or having patentably distinct subject matter will overcome a rejection or challenge of obviousness-type double patenting.[60] A terminal disclaimer disclaims or dedicates a portion of the entire term of all the claims of the patent. A reduction in patent term for a pharmaceutical product can significantly affect the value of the intellectual property.

Obviousness-type double patenting can arise among commonly owned patents even when they arise from different families or have different

55. "Only assignments need be in writing under 35 U.S.C. § 261. Licenses may be oral." Waymark Corp. v. Porta Sys. Corp., 334 F.3d 1358, 1364 (Fed. Cir. 2003).

56. For example, 37 C.F.R. § 401 provides that the U.S. government is granted a royalty-free, nonexclusive license to inventions owned by nonprofit organizations and some small businesses that are funded by any federal agency under the Bayh-Dole Act.

57. *See, e.g.*, *In re* Longi, 759 F.2d 887, 892 (Fed. Cir. 1985).

58. MPEP § 804; *In re Longi*, 759 F.2d at 892.

59. In order to file a terminal disclaimer, the conflicting patents or patents and applications must be commonly owned. 37 C.F.R. § 1.321.

60. MPEP § 804.02; *In re Longi*, 759 F.2d at 894.

inventors. Consequently, the diligence attorney will need to consider all of the target company's patents in the relevant subject area to see whether there are double patenting issues across claims in the same or different families.

D. Patentability Searches

Patentability searches identify the closest art or related art within the subject matter of the patent. The search results yield useful information to help determine the novelty and nonobviousness of the patent and/or the patent portfolio. It also can determine whether there is a "knockout" piece of prior art that describes the invention. If that is the case, this information can stop the transaction in its tracks.

A party can become overwhelmed by the sheer number of patent databases and scientific websites when searching for prior art. The key sources for patent searching may include the major patent websites (e.g., USPTO, the World Intellectual Property Organization [WIPO], the European Patent Office [EPO], Google patents) and major scientific literature databases (e.g., PubMed, STN databases, CAS Registry, CAS Marpat Database, Google scholar). Other sources of art may include any New Drug Applications (NDAs) filed with the Food and Drug Administration (FDA), European opposition documents, and U.S. litigation documents, as well as any documents cited during examination of the target patent itself (and possibly any competitors' patents and publications).

In order to identify relevant art, the search terms of the patentability search are instrumental. Time and care should be used to develop these terms in view of the patents and technological landscape.

The patentability search can also encompass inventor patents and publications. Searching for inventor presentations, posters, abstracts, and lecture materials can unearth relevant documents unknown to the target company. Also, inventor publications relating to the patent may be carefully scrutinized to determine if any claimed subject matter was accidentally disclosed prior to the filing date.

Because time may be of the essence for the transaction and searching will take some time to complete depending on the size of the portfolio, outlining and initiating the patentability searches should be started as early as possible. Engaging a competent and reliable search firm that can allocate the time and resources is yet another consideration. Ideally, the search firm engaged provides rolling results in order to concurrently search and review.

Because engaging searchers and analyzing the results of a search can be expensive, parties may choose to rely on prior art searching and analysis done at the patent office during prosecution of relevant patents and prior art cited by the applicant. Of course, such an approach risks missing an important prior art reference. Nonetheless, parties may choose to proceed on lesser information for budgetary reasons, particularly where the art considered during prosecution was relatively close to the claimed invention.

E. Freedom to Operate (FTO)

Determining freedom to operate and identifying risks of third-party infringement suits is often an important aspect of intellectual property diligence. Most fundamentally, the freedom to operate analysis will focus on the particular product to be sold. In the case of small-molecule pharmaceuticals, this will typically involve considerations of patents on compounds and metabolites, methods of use, formulations and excipients, methods of delivery, and solid-state form (e.g., hydrates or polymorphs). Related categories that may be more difficult to search would include aspects like claims to compounds that have a particular mechanism of action or more generic claims that may share only a portion of the structure of the final product. Beyond these basic categories, the attorney may wish to consider potentially relevant manufacturing methods, as well as components like special vials or stoppers, delivery devices, and even methods of dosing or monitoring particular medications. As with other aspects of the analysis, the scope of the freedom-to-operate analysis should be scaled in accordance with factors like the value of the transaction, the availability of workarounds, and the presence or absence of close competitors.

Another important aspect of the freedom-to-operate analysis is deciding on the geographic scope. Extending the scope beyond the United States will significantly increase costs because of the potential need for foreign counsel, translation costs, and logistical issues associated with searching foreign patent office filings. Often the acquiring party wants to build off of freedom-to-operate analyses performed by the target company, but demanding access to opinions of counsel raises the privilege waiver issues discussed earlier. A reasonable compromise is to ask for copies of any patents that the target company has considered from a freedom-to-operate standpoint and have those considered by the acquiring company's counsel. To save time and money, the parties may have a counsel-only discussion about relevant portions of such patents to focus the efforts of counsel performing the analysis.

Freedom-to-operate analyses are necessarily snapshots of a single moment in time and need to be revisited throughout the development of a product. One recurring issue is deciding how to deal with published patent applications in the freedom-to-operate search. Often many parties will have filed applications with what may be considered "aspirational" claims—claims that have practically no chance of issuing because of prior art, but which appear to broadly cover products in the field. Such published applications can be numerous, and analyzing the validity of each can dramatically increase costs. Short of a full validity analysis of such claims, it is often helpful to look at the FTO context on the same timeline as the validity analysis. This combination of prior art and later published claims can make it more immediately apparent when later applications are attempting to cover well-known concepts in the field.

It is also important to keep an open mind when analyzing freedom to operate. While a third-party patent that appears to cover some aspect of a potential product may be initially troubling, it may also present an opportunity to obtain additional exclusive rights. A patent owner who is not a direct competitor may be willing to license on reasonable terms. Such an alternative should be considered before investing in a significant analysis of how to attack an identified patent.

F. Design Around Issues

One possible response to a third-party patent identified in a freedom-to-operate analysis is to consider design around alternatives. But the decision to design around a patent is not one that can be made lightly. The potential for a viable design around is completely dependent on the particular technology at issue. While choosing one form of delivery device over another to avoid a potential infringement concern may be relatively straightforward, substituting different excipients or altering concentrations in hopes of avoiding infringement of a formulation patent is quite another. Likewise, deciding at the diligence phase to pursue lead compounds of one structure as opposed to another for freedom-to-operate reasons is a significant decision. The more ambitious a design around project, the more one is simply shifting one kind of risk (infringement of third-party patents) for another (ability to devise a medically appropriate alternative).

Recognizing that there may be unpredictable technical challenges in carrying out a design around, the legal team should conduct a thorough validity analysis and carefully weigh the legal versus the technical risks before recommending such course of action. Moreover, the question of whether the design around actually avoids the patent claims is something that typically cannot be known until years in the future. There may also be a risk that the patent owner obtains broader claims from a pending application while the design around is proceeding. Particularly in the context of generic pharmaceuticals, the potentially noninfringing design around may end up needing to be closer than originally planned to the patent claims in order to preserve bioequivalence. In such cases, the design around, even if ultimately successful, may end up being litigated. During litigation, the patent owner will typically recast an effort to design around as merely a bad faith attempt to co-opt the benefit of the patent while creating the appearance of noninfringement.

Even if a design-around product is found not to infringe, the ultimate test is in the market, where the question will be whether the design around is truly an adequate alternative to the patented product or component. This is not to say that designing around patents is impossible or that the eventual product will necessarily be inferior. Rather, it is for the attorney involved in the project to lay out the risks and benefits of the individual design around so that the company can compare the legal component of the project with the technical and commercial components.

G. PTE Decision for Multi-Patent Families

When evaluating a portfolio of patents covering an approved or near approval drug product, questions may arise about eligibility for patent term extension (PTE). The PTE statute allows patentees to recover a portion of the term of patent protection lost while a product was under regulatory review.[61] The detailed requirements for obtaining PTE include a requirement that an application be submitted within 60 days of regulatory approval of the corresponding pharmaceutical product, as well as requirements that the patent to be extended has not yet expired and has not previously been extended.[62] Because only one patent may be extended for a given product, if that product is covered by more than one patent eligible for extension, patentees must make a strategic choice about which patent to extend.

Factors to consider include the available term for extension, the scope of rights associated with the extension, and general considerations of the strength of a given patent. The available term is an important consideration because the statute limits the combined term (the remaining term of the patent after the date of approval plus the period of extension for regulatory review) to 14 years beyond the date of regulatory approval.[63] Consequently, if product approval comes very early in the life of a patent and FDA review does not take much time, there may be no extension available. Similarly, on the opposite end of the life of a patent, an extension cannot be obtained after a patent expires, although the USPTO may grant interim extensions if the product is under review and approval is expected after the patent expires.[64]

61. 35 U.S.C. § 156.
62. *Id.* § 156.
63. *Id.* § 156(c)(3).
64. *Id.* § 156(b)(1)(A), (d)(5)(A).

The scope of protection granted under a patent term extension is also a factor in determining the best patent on which to obtain an extension. For example, the rights granted by extension to a compound patent covering a product are limited to uses for which the product was approved prior to patent expiration.[65] A similar limit applies to extended method claims.[66] The effect of such limits is to minimize the difference in protection afforded by product as opposed to method claims. Thus, while compound patent protection is generally considered the most desirable for a pharmaceutical product, it does not necessarily follow that a patent containing a compound claim is the best choice for patent term extension.

In the context of a diligence review, the limited scope of a term extension is also important. The business team working on a deal should understand that exclusivity for a product may include multiple expiration dates. If the primary expected competitors will be generic filers, the extended term may be the most relevant. However, if the question is one of competition by next-generation products or products with related but different indications, the business team may need to focus on the earlier, nonextended expiration date for the relevant patent.

H. Life Cycle Management

Life cycle management considerations in the context of the diligence review typically involve analysis of the extent to which the patentee has obtained or is seeking coverage of all aspects of a pharmaceutical product. A detailed discussion of pharmaceutical product life cycle decisions is beyond the scope of this chapter, but the diligence attorney should have a basic understanding of the expected economics of a particular product based on factors like the available patient population, expected or actual market penetration by the product that is the subject of the diligence review, and the expected number and type of competitors, including expectations about when such competitor products might be available. Analyzing the portfolio of protection for a particular product, the attorney will want to understand the interplay between protection on the relevant pharmaceutical compound and its approved (or expected) uses, physical forms (e.g., salts and polymorphs), formulations, methods of administration, and manufacturing methods. This may involve making predictions 10 to 20 years into the future with corresponding difficulty and uncertainty. Important questions include these: What is the scope of each type of exclusivity, and how long is it expected to last? Is there expected to be a need or demand for a product in the more distant future, or are there demographic limits on the target patient population or limits

65. *Id.* § 156(b)(1).
66. *Id.* § 156(b)(2).

associated with expected transformative therapies (e.g., a vaccine or curative medication rather than a chronic treatment)?

Often, the patent attorney will need to explain or make predictions about the strength of various forms of patent exclusivity. As part of this exercise, it is important to investigate statements that may have been made during prosecution of later-filed applications that may undercut assumptions about what was disclosed or understood in an initial filing. For example, while there may be basic or generic support in an initial application for a particular medical use for a compound, a later-filed application may claim that same use, but based on more extensive data, such as a clinical trial. A review of the file histories may reveal that the patentee has already developed a record suggesting that the initial application lacked adequate support for the particular use. In such a case, it may be difficult or impossible to reverse course and seek protection for that use based on claims in the earlier application.

Thus, evaluating the overall life cycle protection in a diligence requires an understanding not only of the existing patent claims but also of available pending applications and pending claims, as well as the detailed record at the USPTO, to evaluate any limiting admissions or representations. Where frank advice about strengths and weaknesses in the portfolio is to be provided, the diligence attorney must be mindful of attorney-client privilege issues. Advice must be given so that the business team can make informed decisions. Nonetheless, the attorney should be sure that the team understands how to protect such information, including by limiting its dissemination beyond key decision makers, not disclosing it during negotiations, and being mindful of how it is incorporated into any reports or memos being prepared for management.

IV. Trademarks

As defined by 15 U.S.C. § 1127, a trademark is "any word, name, symbol, or device or any combination thereof" used by any person "to identify and distinguish his or her goods, including a unique product, from those manufactured or sold by others and to indicate the source of the goods, even if that source is unknown." Trademarks, like other forms of intellectual property, are intangible assets of a company; they are powerful source identifiers and a means to build public goodwill. From a pharmaceutical perspective, trademarks can, for example, promote brand awareness, enable cost-effective marketing, streamline labeling, minimize confusion between medicines, minimize risk of medication errors, and enhance the ability to monitor the safety of medications.

In the United States, trademark protection is based on federal law, state law, and common law. In particular, federal law is based on the Lanham Act[67] and the common law of unfair competition, and statutory laws are protected under state law.

In conducting an asset evaluation of a trademark portfolio, there are three areas of investigation: (1) validity, (2) scope, and (3) ownership. But before the investigation begins, a preliminary step is to identify the trademark assets used, intended to be used, owned, and/or licensed. Once the trademark assets are identified, the registrations of the marks should be confirmed with the records of the USPTO and in trademark offices of the respective countries. As part of the exercise in identifying the trademark assets, a comprehensive review of the company's publicly available materials should be also conducted to not only further confirm the identified trademark assets but also determine any other marks in use but not registered or applied for. This may include a review of promotional materials, electronic materials, press releases, annual reports, the company's website, and social media. When conducting this preliminary identification step, it is important to keep in mind what can qualify as a trademark.

A. Traditional and Nontraditional

A trademark is a word, phrase, symbol, or design, or a combination thereof, that identifies and distinguishes the source of the goods from one party from those of others.[68] For example, trademarks usually take the form of a word, a logo, or a combination of both. Under common law, trademark rights are acquired through use of the mark in commerce in a particular geographic area and with particular goods and services. These rights must also be maintained through use. Trademark rights are also acquired through federal registration.[69] Although federal registration is not a prerequisite to obtaining rights, there are significant advantages to registration, including the following:

- Public notice of a claim of ownership of the mark;
- Presumption of ownership;
- Federal court jurisdiction to bring an action;
- Use of U.S. registration as a basis to obtain registration in foreign countries;

67. 15 U.S.C. § 1051–1127.

68. U.S. Pat. & Trademark Off., Protecting Your Trademark: Enhancing Your Rights Through Federal Registration (2012), *available at* http://www.uspto.gov/trademarks/basics/BasicFacts.pdf.

69. *See id.* §§ 1051, 1127.

- Ability to record the U.S. registration with the U.S. Customs and Border Protection (CBP) service to prevent importation of infringing foreign goods;
- Right to use the federal registration symbol (®); and
- Listing in the U.S. Patent and Trademark Office's online databases.[70]

Nontraditional trademarks include appearance, motion, shape, sound, scent, taste, and touch.[71] Examples of nontraditional trademarks in the pharmaceutical context include pill color, pill shape, and flavors of medications. Nontraditional trademarks require threshold showings of nonfunctionality, inherent distinctiveness, and/or acquired distinctiveness (secondary meaning) for protection.

B. Validity

If trademarks are considered part of the value of a transaction, a validity analysis should be conducted to examine whether the marks are the subject of registrations or pending applications, whether the marks are registered in the appropriate form, whether the marks are connected with the services/products currently used and/or to be used in the future, and whether the marks are registered or pending in the appropriate countries. In the United States, there are two government agencies that oversee the registration and approval process for pharmaceutical marks: the USPTO and the FDA.[72]

For the USPTO to register a mark, it assesses whether the mark is sufficiently distinctive and whether there will be a likelihood of confusion with respect to other registered marks. In reviewing the federal registration process, consider whether the mark was registered in standard characters, in a stylized form, or with a logo or design mark. Standard character registration provides greater protection, but in some instances, a stylized form may be necessary to overcome the grounds for rejection by the USPTO, such as lack of inherent distinctiveness or a likelihood of confusion. Understanding the rejections and amendments enable a better evaluation of the strength and value of the trademark portfolio.

The FDA's assessment of proprietary drug names includes consideration of both the safety and the promotional aspects of a name. The agency makes this determination for products marketed under an application (i.e., IND, NDA, BLA, and ANDA). The promotional review assesses whether

70. *Frequently Asked Questions About Trademarks*, U.S. Pat. & Trademark Off., http://www.uspto.gov/faq/trademarks.jsp (last visited Jan. 2, 2014).

71. *Nontraditional Trademarks*, Int'l Trademark Ass'n, http://www.inta.org/TrademarkBasics/FactSheets/Pages/NontraditionalTrademarksFactSheet.aspx (last visited Jan. 2, 2014).

72. 15 U.S.C. §§ 1051-1127; 21 U.S.C. § 352(a).

the name is overly fanciful, implying unique effectiveness or composition. The safety review utilizes methods that generate lists of names that could be confused with the proposed proprietary name and tests the likelihood of confusion between these names and the proposed proprietary name. It is important to keep in mind the purpose and approach that each agency takes when reviewing the mark or proprietary name, as one may allow the mark whereas the other may not.

A validity analysis further includes whether the marks are connected with the services/products currently used and/or to be used in the future. This entails reviewing the marks as registered to ensure that they are still in use for the identified goods and services. This confirms that the marks are still in use and will not be vulnerable to cancellation based on nonuse. Similarly, the final step of the validity analysis is determining whether the marks are registered or pending in the appropriate countries. This may entail engaging local counsel in the respective countries of registration in order to confirm and review registration, use, and ownership.

C. Scope

Scope, or the "strength" of a trademark, is determining the relative protection in view of third parties. A "strong" trademark is one that will allow you to prevent a third party from using your mark. Thus, the question for the marks of interest is how to assess the scope or strength.

Courts and the Lanham Act categorize marks and, according to that categorization, the protection afforded to them; the classification is based on increasing distinctiveness: (1) generic; (2) descriptive; (3) suggestive; (4) arbitrary; or (5) fanciful.[73] Classification, however, may not be clear-cut. In the seminal case of *Abercrombie & Fitch Co. v. Hunting World, Inc.*, the Second Circuit noted that

> [t]he lines of demarcation, however, are not always bright. Moreover, the difficulties are compounded because a term that is in one category for a particular product may be in quite a different one for another, because a term may shift from one category to another in light of differences in usage through time, because a term may have one meaning to one group of users and a different one to others, and because the same term may be put to different uses with respect to a single product.[74]

73. *See* Abercrombie & Fitch Co. v. Hunting World Inc., 537 F.2d 4, 9 (2d Cir. 1976).
74. *Id.*

Generally, marks that fall into the categories of fanciful, arbitrary, or suggestive are considered inherently distinctive, and there is no need to show secondary meaning. In contrast, descriptive marks require a showing of secondary meaning. And generic marks are incapable of acquiring distinctiveness, which means that they cannot serve as trademarks. Classifying a mark into one of these categories is a question of fact.[75]

Fanciful marks are terms that are invented or concocted. There is no dictionary definition or commonplace meaning to the term; it has been fabricated. Arbitrary marks are those that already have meaning (e.g., a dictionary definition), but do not describe the goods or services. Fanciful and arbitrary marks are considered the strongest types of marks because they are inherently distinctive, and there is no need to show secondary meaning.

Suggestive marks are "spawned by the felt need to accord protection to marks that were neither exactly descriptive on the one hand nor truly fanciful on the other."[76] One commentator described "suggestive" trademarks as "a term that subtly suggests something about the product."[77] Suggestive marks, like fanciful and arbitrary marks, are also considered inherently distinctive with no need to prove secondary meaning.[78]

Descriptive marks "are merely descriptive of a product [and] are not inherently distinctive."[79] A descriptive mark "tells something about the product . . . [for example,] describes some characteristic, quality, purpose, component, or other property of the product."[80] Descriptive marks that are used to describe a product "do not inherently identify a particular source and . . . may acquire the distinctiveness which will allow them to be protected under the [Lanham] Act."[81] This acquired "distinctiveness" is generally referred to as "secondary meaning."

In order to establish secondary meaning, "an applicant must show that 'in the minds of the public, the primary significance of a product feature or term is to identify the source of the product rather than the product itself.'"[82] Courts consider a number of different factors to determine whether a mark has acquired secondary meaning, such as advertising expenditures and sales success, length and exclusivity of use, unsolicited media coverage, copying of the mark by others, and consumer studies.[83]

75. Bristol-Myers Squibb Co. v. McNeill-PPC Inc., 973 F.2d 1033, 1039–40 (2d Cir. 1992).

76. *Abercrombie*, 537 F.2d at 10.

77. Kane on Trademark Law § 2.1 (6th ed. 2013).

78. *Abercrombie*, 537 F.2d at 10.

79. Two Pesos Inc. v. Taco Cabana Inc., 505 U.S. 762, 769 (1992).

80. Kane on Trademark Law, *supra* note 75, § 2:1.2.

81. *Two Pesos Inc.*, 505 U.S. at 769.

82. Coach Servs. Inc. v. Triumph Learning LLC, 668 F.3d 1356, 1379 (Fed. Cir. 2012) (citing *In re* Dial-A-Mattress Operating Co., 240 F.3d 1341, 1347 (Fed. Cir. 2001)).

83. *Id.* (citing *In re* Steelbuilding.com, 415 F.3d 1293, 1300 (Fed. Cir. 2005)).

A generic term is "the common name for the kind of product."[84] There is no legal protection for generic terms.[85] For example, an applicant attempted to register "Cell Therapeutics Inc." for "laboratory research and development services in the field of biomedical and therapeutic products that affect cellular signaling pathways."[86] Agreeing with the examining attorney, the Trademark Trial and Appeal Board found "cell therapeutics" to be generic for use in connection with the identified goods and services and denied the registration.[87] Generic marks should be avoided.

D. Ownership

As with most types of asset evaluations, one needs to determine who owns assets—in this case, who owns the trademarks. This can be done by reviewing the USPTO assignment database[88] to determine the owner of the mark for those registered marks. The diligence attorney should coordinate efforts to identify if any encumbrances (e.g., security interests or liens) are associated with or may transfer with the marks.

V. Trade Secrets

Trade secrets pose special challenges for the diligence attorney. Part of the economic benefit of patents is that they clearly disclose and define an invention. This allows for efficient identification, evaluation, and transfer of intellectual property rights between companies. By contrast, trade secrets may be hard to identify or define, and evaluating them can be difficult because they are less tangible than patented inventions. The target company may never have attempted to inventory its trade secrets, and even if its secrets have been reduced to writing, they may be buried among other information that is neither secret nor valuable.

Unlike a patent, a trade secret does not have to be new or nonobvious. So trade secret protection can encompass what might be considered routine optimization if the result is something that adds value for the company. While it would be impossible to buy a piece of equipment and then obtain a patent on it, it is common to have a protectable trade secret involving the particular or optimal use parameters of even an off-the-shelf device. Often trade secrets involve detailed procedures or methodologies that have been

84. Kane on Trademark Law, *supra* note 75, § 2:1.

85. *Id.* § 2:1.1.

86. *In re* Cell Therapeutics Inc., 67 U.S.P.Q.2d 1795, 1796 (T.T.A.B. 2003).

87. *Id.* at 1798.

88. *Trademark Assignment Query*, U.S. Pat. & Trademark Off., http://assignments.uspto.gov/assignments/q?db=tm (last visited Jan. 2, 2014).

refined through experience. They are commercially valuable because they may make a process more efficient or reliable.

Trade secrets frequently arise in pharmaceutical diligence evaluations in the context of manufacturing methods. Trade secrets may relate to individual steps in manufacturing of drugs or biologics or to the combined steps of an entire process. Such secrets can include many different parameters: optimum temperatures; time periods for steps like incubations, drying, or stirring; preferred ingredients; or preferred equipment. Other secrets may involve preferred vendors for an active pharmaceutical ingredient (API) or excipients. Although ingredients are usually listed in the package insert for a drug, the characteristics of excipients from certain vendors may be valuable and maintained as secret. The clinical data set associated with a drug can be valuable, confidential information to the extent it has not been published. Finally, a large potential category of trade secret information involves computer databases or analysis tools. Depending on the drug, there may be valuable databases of physician or patient contacts, computer models for "in silico" clinical tests, or other important compilations of data.

A. Enforceability

Unlike patent law, trade secret enforcement is grounded firmly in state law. Federal law makes certain types of trade secret theft a crime, but state law governs private enforcement of trade secret rights.[89] Nearly every state has enacted a version of the Uniform Trade Secrets Act (UTSA).[90] The act defines a trade secret as information that derives economic value from not being generally known or readily ascertainable to third parties who would economically benefit from knowing the secret.[91] Importantly, a party asserting a trade secret right must also show that reasonable efforts were made to maintain its secrecy.[92] Such efforts include, for example, nondisclosure agreements with employees and visitors, and controlled access to facilities.[93] "Information" that can be a trade secret under the UTSA is defined broadly, and includes formulas, patterns, compilations, programs, devices, methods, techniques, or processes

89. *See* 18 U.S.C. § 1831-39 et seq.; *see also* Exec. Office of the President, Administration Strategy on Mitigating the Theft of U.S. Trade Secrets (2013), *available at* http://www.whitehouse.gov/sites/default/files/omb/IPEC/admin_strategy_on_mitigating_the_theft_of_u.s._trade_secrets.pdf (noting that civil enforcement of trade secret rights under federal law is on behalf of the U.S. attorney general).

90. *Trade Secrets Act*, Uniform Law Comm'n, http://www.uniformlaws.org/Act.aspx?title=Trade%20Secrets%20Act (last visited Feb. 23, 2014).

91. Uniform Trade Secrets Act § 1(4) (amended 1985).

92. *Id.* § 1(4)(ii).

93. *Id.* § 1 cmt.

The holder of a trade secret has no independent right to prevent others from performing the secret process or otherwise independently developing the same secret. The only enforceable legal right associated with a trade secret is the right to prevent others from unlawfully obtaining and using or disclosing it.[94] The UTSA includes a right to an injunction against actual or threatened disclosure or unauthorized use of a trade secret, damages for economic loss or unjust enrichment due to such disclosure or use, and attorney's fees for willful and malicious misappropriation.[95] While these limited rights are less potent than those granted by a patent, an important advantage of trade secret protection is that it is unlimited in term. So long as the information continues to meet the definition of a trade secret, including reasonable efforts by the owner to prevent disclosure, it will enjoy the protection of the statute.

B. Process

As discussed, it can be difficult to evaluate or account for trade secrets in a diligence investigation unless the secret is of a type that the company knows is important and valuable (the classic example being the formula for Coca-Cola). A good starting point is to ask whether the company has a trade secret inventory or any other compilation of information it would consider to be its trade secrets. Absent such an inventory, relevant questions are these: Does the company have information that it considers commercially valuable that is not generally known or readily ascertainable? Would the information be valuable to others if it were disclosed? Has the company made reasonable efforts to maintain the secrecy of the information? Of course, not every commercially valuable secret will be worth evaluating, but starting with a broad sweep will spark discussion about what aspects of the company's technology are protected by secrecy rather than through patents.

C. Know How

Trade secret licenses may also be evaluated. The term "Know How" is often used in license agreements to refer to information that is not patented but is otherwise considered proprietary by the licensor. Depending on the agreement, Know How may or may not qualify for the UTSA definition of a trade secret. Nonetheless, Know How terms in a license should alert the attorney for the possible presence of valuable trade secret rights. Know How licenses often arise in development agreements or collaborations where the in-licensing party is seeking more than merely rights under

---

94. *Id.* §§ 2–4.
95. *Id.*

patents. Know How licenses may involve access to laboratory or manufacturing facilities, training or disclosure of detailed operating procedures by the licensor's personnel, or other non-patented components of a product or platform technology. As with most trade secrets, it is unusual for Know How to be precisely defined in license agreements. More often, it is defined broadly as all technical information relating to a particular subject, occasionally with certain exceptions. Such broad definitions, while common, can lead to disputes later between the parties to a license if there is disagreement about the scope of what was provided by the licensor or used by the licensee with respect to Know How.

An important issue for the diligence attorney considering Know How licenses is how the payment term is defined. Because trade secrets have unlimited terms, rights in Know How may never expire absent a specific clause defining an endpoint or declaring any royalty obligation paid in full.[96] Hence, where Know How licenses are in place, the diligence attorney should consider whether there is a provision terminating the obligation to pay for Know How. Moreover, even where there is a clear termination date for Know How payments, the license agreement should be reviewed to be certain that the right to use the Know How is continuing. It is also helpful to determine whether Know How obligations are continuing under a license agreement even if the Know How information becomes public or is independently known or developed by the licensor. For example, agreements may have provisions waiving the obligation to pay for Know How that is in the public domain or that is otherwise known to or developed by the licensee.

D. Valuation

A detailed economic analysis of the valuation of trade secret rights is beyond the scope of this chapter. From the perspective of an intellectual property transaction diligence, the value of the trade secret rights will depend on three factors: (1) the ability to define a specific trade secret that is important in giving value to the transaction; (2) the extent to which the target has cultivated and protected the secrecy of the technology; and (3) the ease or difficulty for others to independently develop the secret.

As discussed, pharmaceutical companies often have unpatented information that would qualify as a trade secret associated with product development and manufacturing, but such secrets are of less value if they cannot be specifically defined or characterized apart from generalized

---

96. *See, e.g.*, Warner-Lambert Pharm. Co. v. John J. Reynolds Inc., 178 F. Supp. 655 (S.D.N.Y. 1959), *aff'd*, 280 F.2d 197 (2d Cir. 1960); Nova Chems. Inc. v. Sekisui Plastics Co., 579 F.3d 319, 327–30 (3d Cir. 2009).

processes. For example, knowing that a particular filtering method will allow faster and more efficient processing of a drug product may be a valuable and defined secret. On the other hand, it may be more difficult to determine the value of knowing the broad steps of a manufacturing process, even if the overall process would be a protectable secret. The value of such secrets may be easier to ascertain where the target company has performed tests or studies to compare the secret step or configuration to other ways of achieving the same result.

Just as a well-defined secret is easier to value, such secrets are more easily analyzed to determine whether they can be effectively implemented by the licensee company. If the acquiring company is planning to set up an identical manufacturing line, it may be relatively easy to incorporate a discrete, secret step that has been identified as optimal or necessary by the licensor. Of course, the acquiring company may not have definite plans, at the diligence stage, about how much of the target's manufacturing methods will be adopted. Similarly, where licenses are to platform technologies as well as particular targets, the valuation will depend on the extent to which the acquiring party needs to continue using the platform.

Protecting the confidentiality of trade secrets is also important to maintaining their value. Frequently, however, there may be pressure to file patents or publish. Patenting an invention destroys the right to maintain it as a protectable trade secret.[97] From a diligence perspective, where the value of a trade secret is important, it can be important to analyze the target's patent filings directed to the same or similar technology to see if they disclose important aspects of the purported "secret." Similarly, companies may publish manufacturing methods or processes for a variety of reasons. Such publications should also be considered to see that care has been taken to preserve secrets that are considered valuable. Other potential disclosures might involve presentations to investors or other potential licensees of the technology as well as communications with collaborators. If the secret in question is important, the attorney will want to investigate and get assurances about all potential disclosure avenues. Secrecy agreements should be in place with all parties to whom a valuable trade secret is disclosed. Consideration must be given to required regulatory disclosures and the likelihood that the secret will inevitably be disclosed when a product is marketed.

A final question in valuing a trade secret is the likelihood that others will independently develop the same knowledge or that they will obtain a similar result through different means. Because trade secret rights cannot be enforced against third parties that learn or develop them independently, the

97. *See* Kewanee Oil Co. v. Bicron Corp., 416 U.S. 470 (1974).

value of a secret depends heavily on how likely it is that others will inevitably learn the secret. One good measure for this is the difficulty experienced by the target company in developing the information. What was required? How long did it take? What did it cost? Answers to these questions can help to distinguish a trade secret that is just providing a head start on optimizing a process or product from one that is truly a competitive advantage. Likewise, the diligence attorney should determine whether there are comparable technologies and whether the target has compared its own secret method to those. To the extent the trade secret is viewed as substantially better than available routine technologies, the attorney should also investigate whether it might still be possible to file a patent application instead of continuing to maintain the invention as secret.

VI. Standing to Enforce Intellectual Property Rights

Where patents are being evaluated for exclusive in-licensing with the goal of ultimately enforcing them against competitors or other infringers, it is important to be sure that the acquiring company obtains the necessary rights for enforcement.

A. Necessary Parties

By statute, patentees have standing to enforce their patents.[98] Absent an agreement to the contrary, all owners of a patent are necessary parties to an infringement action.[99] A "bare licensee" may not enforce a patent.[100] In practice, this means that a nonexclusive licensee has no right to enforce a patent at all, but an exclusive licensee can bring suit in its own name, provided that it joins the patent owner.[101] In rare circumstances, an exclusive licensee can bring suit in its own name by having received "all substantial rights" in the patent via a written license agreement.[102] The "all substantial rights" inquiry is rigorous and involves consideration of the parties' intent, whether the agreement conveys the full right to exclude to the licensee, and the nature of rights retained by the patentee.[103] Ultimately,

98. 35 U.S.C. § 281.

99. Schering Corp. v. Roussell-UCLAF SA, 104 F.3d 341, 344 (Fed. Cir. 1997) (noting that absent agreement, a co-owner may not sue for infringement if another co-owner refuses to join).

100. Prima Tek II LLC v. A-Roo Co., 222 F.3d 1372, 1377, 1379–80 (Fed. Cir. 2000).

101. *Id.*

102. *Id.* (citing Vaupel Textilmaschinen KG v. Meccanica Euro Italia SPA, 944 F.2d 870, 875 (Fed. Cir. 1991)).

103. *Id.*

for the licensee to proceed without the licensor, the transfer of rights must be tantamount to an assignment.[104]

B. Acquirer Rights

If the acquiring company's goal is to be able to enforce exclusive patent rights, the diligence attorney must consider the ownership and license agreements to determine whether sufficient rights will reside in the acquirer to bring suit in its own name, with or without the patent owner as co-plaintiff. Of course, if another party is necessary for litigation, there will be questions about the circumstances in which that party is required to join and whether it may refuse. In some circumstances, control is ceded to the exclusive licensee, and the patent owner must join whenever the licensee asks. In other circumstances, the owner will have some say or some ability to not join a suit, and sometimes control remains entirely with the patent owner. Of course, where the exclusive licensee lacks complete enforcement rights, there is a risk that its business interests may diverge from those of the patent owner, who may not support a decision to enforce a patent. Indeed, where the patent owner's interest is primarily in collecting royalties, it may have little interest in putting the patent at risk or enduring the burdens of litigation. Issues concerning which party chooses counsel and how costs of litigation will be divided are also typically addressed in the license agreement and should be considered by the diligence attorney.

Multiple patent owners present special issues. In addition to the issues discussed earlier, even where a patentee has been granted the right to bring suit without its co-owner, that unjoined co-owner may still grant a license to the target of the suit, if it wishes.[105] This means that the co-owner can effectively derail any effort to sue an infringer by granting a license, regardless of the business objectives of or harm suffered by the suing co-owner. From the diligence perspective, this means that the acquiring company will want to account for all actual and potential owners. "Potential" owners that may be considered include inventors who may not have been included on the patent, or creditors or investors who may have been granted an ownership interest in the patent specifically or in the company's assets more generally.

Another question that may arise concerns the right to sue for past damages. Absent a specific provision, a patent assignment will not include the right to sue third parties for damages accrued prior to the time of the

104. *Id.*; Mentor H/S Inc. v. Med. Device Alliance Inc., 240 F.3d 1016, 1017 (Fed. Cir. 2001).

105. *Schering*, 104 F.3d at 345; *see also* Brian G. Brunsvold, Dennis P. O'Reilley & D. Brian Kacedon, Drafting Patent License Agreements 32–33 (7th ed. 2012).

transfer.[106] If there are current competitors against whom the acquiring company may want to assert the relevant patent rights, past damages should be accounted for.

VII. Regulatory Considerations

In addition to issues squarely within the patent law, such as patent infringement and validity, other important considerations concern regulatory provisions that either overlap with the patent law or stand alone as part of the FDA's power to regulate pharmaceuticals. These provisions include patent term extensions to restore term lost during regulatory approval; regulatory exclusivity periods, during which the FDA cannot approve competing products; and the detailed provisions of the Hatch-Waxman act relating to resolution of patent disputes in the context of generic drugs approval. Although each of these topics could be a chapter on its own, the subsequent sections provide an overview of how they can apply in the diligence context.

A. PTE Candidates

As discussed, where an approved product is available or expected, some consideration of patent term extension strategy is likely to be part of the diligence investigation. From a regulatory perspective, availability of a patent term extension depends on whether the new product is "the first permitted commercial marketing or use of the product."[107] Under the statute, this inquiry is not focused on the drug product, per se, but on the active ingredient.[108] Consequently, the Federal Circuit has upheld denial of patent term extension where the product is a combination of two previously approved products.[109] But patent term extension will be granted where at least one active ingredient has not been previously marketed.[110] Although the statute defines a "product" as including salts and esters of the active ingredient, that definition did not prevent an extension from being granted on the ester of an antibiotic compound where salts of the same compound had previously been marketed.[111] Patent term extension has also been granted where the patent covered an enantiomer active

106. Minco Inc. v. Combustion Eng'g. Inc., 95 F.3d 1109, 1117–18 (Fed. Cir. 1996).

107. 35 U.S.C. § 156(a)(4)–(5).

108. Fisons PLC v. Quigg, 876 F.2d 99, 100–01 (Fed. Cir. 1989).

109. Arnold P'ship v. Dudas, 362 F.3d 1338 (Fed. Cir. 2004).

110. *Id.* at 1341.

111. Glaxo Operations UK Ltd. v. Quigg, 894 F.2d 392 (Fed. Cir. 1990); *see also* PhotoCure ASA v. Kappos, 603 F.3d 1372 (Fed. Cir. 2010) (affirming district court action extending patent on compound over USPTO objection that the approved compound was metabolized in the body into a previously approved "active ingredient").

ingredient even where the racemic mixture of the ingredient had already been marketed.[112]

1. Eligibility of Formulation

The patent term extension requires that the patent in question "cover" the active ingredient subject to regulatory approval.[113] Consequently, merely reformulating an old product will typically not result in eligibility for patent term extension.[114] For example, under this provision, the Federal Circuit upheld denial of term extension where the claims covered a metabolite formed in the body after administration of the product, but did not cover the active ingredient or use of the compound that was actually the subject of regulatory review. [115]

B. Mapping of Market Exclusivity

In addition to patent exclusivity, approved drugs may be eligible for various regulatory exclusivities as well. These exclusivities are designed to incentivize and reward investment in various aspects of pharmaceutical research. The details of the various types of exclusivity are discussed elsewhere in this book. The FDA website is a useful resource when it comes to regulatory exclusivity diligence. The agency maintains a large collection of guidance documents on its website explaining various types of exclusivity available to applicants.[116] In addition, the FDA publishes the exclusivities associated with approved drug products in its Electronic Orange Book.[117]

1. Data and Regulatory Exclusivity

Non-patent data and regulatory exclusivity comes in four broad types. The first is exclusivity that prevents the FDA from accepting new applications for approval for a limited period after approval of a new drug. The Hatch-Waxman Act grants this type of exclusivity to new chemical entities (NCEs).[118] Under NCE exclusivity, the FDA may not accept Abbreviated New Drug Applications (ANDAs) or New Drug Applications relying on data submitted in an approved application (505(b)(2) applications) for five

---

112. Ortho-McNeil Pharm. Inc. v. Lupin Pharm. Inc., 603 F.3d 1377, 1381 (Fed. Cir. 2010).

113. Hoechst-Roussel Pharm. v. Lehman, 109 F.3d 756, 759, 761 (Fed. Cir. 1997).

114. *Fisons PLC*, 876 F.2d at 100–01.

115. *Hoechst-Roussel Pharm.*, 109 F.3d at 759, 761.

116. *Guidances (Drugs)*, U.S. Food & Drug Admin., http://www.fda.gov/Drugs/GuidanceComplianceRegulatoryInformation/Guidances/default.htm (last visited Feb. 26, 2014).

117. *Orange Book: Approved Drug Products with Therapeutic Equivalence Evaluations*, *supra* note 40.

118. 21 U.S.C. § 355(c)(3)(E)(ii), (j)(5)(F)(ii); 21 C.F.R. § 314.108(b)(2).

years after the approval of a drug that has not previously been approved.[119] An exception limits the prohibition to four years if the ANDA or 505(b)(2) application contains a certification challenging the patents covering the approved drug or its use.[120]

Until recently, the FDA denied NCE exclusivity to combination therapies if any active ingredient in the combination had been previously approved. The FDA followed an "umbrella" policy that allowed the NCE exclusivity attached to a new drug to cover combinations or other members of the same product line, but as a rule, if one member of a combination had been previously approved, the combination would not have NCE exclusivity.[121] Recently, however, the agency issued draft guidance indicating its intent to award such exclusivity in the future if any member of the combination has not been previously approved.[122] From a diligence perspective, this change may add more uncertainty about whether certain combinations will be entitled to NCE status, although for approved products, the FDA will continue to publish relevant exclusivities on its website. The new draft guidance should not affect the basic case of a new drug with an active ingredient that has not been previously marketed.

The second type of exclusivity prohibits the FDA from approving an ANDA or 505(b)(2) application for a limited time after approval of a New Drug Application. These exclusivities include the following:

- three-year exclusivity for new clinical investigations[123]
- seven-year exclusivity for orphan drug status[124]

Orphan and new clinical investigation exclusivities are limited to the particular indications or uses for which the New Drug Application was approved, so they will not prevent a competitor from obtaining approval of the same drug for a different indication. This can be important because health care providers may use a competing product "off-label" if it has the same active ingredient as a product approved for an indication. Consequently, while both types of exclusivity are useful, the acquiring company should be aware of the potential for competition even where such exclusivity has been granted.

119. *See* Actavis Elizabeth LLC v. U.S. Food & Drug Admin., 625 F.3d 760 (D.C. Cir. 2010).

120. 21 U.S.C. § 355(c)(3)(E)(ii), (j)(5)(F)(ii).

121. U.S. Food & Drug Admin., FDA-2013-D-1675-0002, Draft Guidance for Industry, New Chemical Entity Exclusivity Determinations for Certain Fixed-Combination Drug Products 6 (Feb. 24, 2014).

122. *Id.* at 1.

123. 21 U.S.C. § 355(c)(3)(E)(iii)–(iv), (j)(5)(F)(iii)–(iv); 21 C.F.R. § 314.108(b)(4)–(5).

124. 21 C.F.R. § 316.3(b)(12); *see also* Orphan Drug Regulations, 78 Fed. Reg. 35117; 35,132–33 (June 12, 2013).

A third type of regulatory exclusivity adds additional time to existing exclusivity. Historically, this category was limited to pediatric exclusivity, which is granted by the FDA if the New Drug Application (NDA) holder performs certain required clinical trials in children.[125] Pediatric exclusivity extends existing exclusivity periods by six months. For example, an approval with five years of NCE exclusivity would be extended to five years and six months by pediatric exclusivity.[126] Recently, Congress created a new type of "add on" exclusivity as an incentive to encourage development of new antibiotic agents.[127] Under the Generating Antibiotic Incentives Now (GAIN) Act, exclusivity periods can be extended by five years for "qualified infectious disease products" designated by the FDA.[128] Qualified infectious disease products include drugs intended to treat certain difficult-to-treat or resistant bacteria.[129] Added to the standard NCE exclusivity, exclusivity from the GAIN Act has potential to provide ten total years of exclusivity to a newly developed antibiotic agent.

The final class of regulatory exclusivity only arises in the context of generic drugs in which there has been a challenge to the patents covering the branded product. The Hatch-Waxman Act grants 180 days of regulatory exclusivity to an ANDA first-filer who lawfully maintains a challenge (a Paragraph IV Certification) against the patents covering a drug product.[130] The exclusivity can be forfeited as a penalty for various actions, such as failing to timely obtain tentative approval of the ANDA or failing to timely market a drug after prevailing in patent litigation.[131] The provisions setting forth these circumstances are complicated and should be given careful consideration if exclusivity of the ANDA product is an important factor in the transaction. Another important consideration is that the 180-day exclusivity applies only to prevent approval of additional ANDA applications. It will not prevent another company from selling a generic product under either the original New Drug Application (a so-called authorized generic) or under a 505(b)(2) application. Thus, even where an ANDA

125. U.S. Food & Drug Admin., Guidance for Industry: Qualifying for Pediatric Exclusivity under Section 505A of the Federal Food, Drug, and Cosmetic Act (Sept. 28, 1999), *available at* http://www.fda.gov/downloads/Drugs/DevelopmentApprovalProcess/DevelopmentResources/UCM049924.pdf.

126. *Id.*

127. Food and Drug Administration Safety and Innovation Act, Pub. L. No. 112-144, § 801, 126 Stat. 993, 1077–79 (2012).

128. *Id.* at 1077 (inserting § 505(E)(a)).

129. *Id.* at 1078–79 (inserting § 505(E)(f)–(g)); *see also FDA Safety and Innovation Act*, U.S. Food & Drug Admin. (June 19, 2013), http://www.fda.gov/downloads/drugs/developmentapprovalprocess/smallbusinessassistance/ucm361320.pdf.

130. 21 U.S.C. § 355(j)(5)(B)(iv).

131. *Id.* § 355(j)(5)(D).

first-filer succeeds in litigation, it may not be able to enjoy the benefit of complete market exclusivity during the 180-day period.

2. Overlap with Patent Coverage

Regulatory exclusivity typically applies independently of patent coverage on a drug product. An exception is pediatric exclusivity, which extends the patent protection associated with a product by six months. The six-month extension is not a true patent term extension, but instead extends the FDA's prohibition on approving either an ANDA or 505(b)(2) application.[132]

C. Timing of Potential ANDA Filings

As discussed, regulatory exclusivity can prevent competitors from filing ANDA or 505(b)(2) applications or delay the FDA's approval of such applications. Nonetheless, the powerful market incentives and potential 180-day exclusivity offered by the Hatch-Waxman Act make such filings inevitable for almost any approved drug and particularly for a highly successful one. Thus, part of the diligence analysis in a transaction involving drugs that are approved or soon-to-be-approved involves estimating the likely date of ANDA filings.

The timing of such filings will depend on a variety of circumstances, including regulatory exclusivity, the number and scope of patents listed in the Orange Book and their expiration dates, the ease or difficulty of making the product, and the value of the product in the market. Where a new drug has been granted NCE exclusivity, it is a safe bet that competitors will file ANDAs on the first possible date (four years after approval, the so-called NCE minus-one date). From the perspective of the generic applicant, there is pressure to file early, because only a "first-filer" can qualify for 180-day exclusivity. There may, however, be more than one first-filer, so even filing on the first possible date will not guarantee an undivided share of market exclusivity. It is not uncommon to have multiple ANDA filers on the NCE minus-one date for valuable drug products.

If 180-day exclusivity is not an issue, the ANDA filer may pay closer attention to the overall patent portfolio. For example, a potential filer could decide that it is unwilling to devote resources to challenging a drug compound patent or a controlling use patent, but might choose to challenge or attempt to design around a later-expiring formulation patent. Thus, there may be an additional window of opportunity for ANDA filers shortly before expiration of certain families of patents. Publicized market entry

132. U.S. Food & Drug Admin., Guidance for Industry: Qualifying for Pediatric Exclusivity under Section 505A of the Federal Food, Drug, and Cosmetic Act (Sept. 28, 1999), *available at* http://www.fda.gov/downloads/Drugs/DevelopmentApprovalProcess/DevelopmentResources/UCM049924.pdf.; Guidance for Industry, *supra* note 53, at 13–15.

dates resulting from a settlement between a brand and first-filers may also encourage additional ANDA filers. Later ANDA filers may conclude that the brand will be more likely to settle with minimal litigation once other generic competitors are already preparing to enter the market.

Sometimes issues affecting the timing of ANDA filings are largely opaque to the brand. The product may turn out to be more difficult to manufacture than the generic company anticipated, a design around may unexpectedly fail, or market priorities may divert resources to other, more valuable products. In such cases, ANDA filings may not materialize as anticipated, or the number of challengers may be smaller.

1. Indications

Analysis of likely dates for an ANDA filing must be done on an indication-by-indication basis. Once initial compound patent coverage expires, different indications may be covered by patents with different expiration dates. In addition, ANDA filers may choose to "carve out" certain indications that remain subject to patent protection if other indications have lost patent protection.[133] As discussed, once a single indication is free of patent protection, there is a market opportunity for generic filers because health care providers may choose to substitute the less expensive drug even if it is not labeled for a particular indication. For similar reasons, there may be little incentive to ever file an ANDA on indications subject to longer patent terms unless the incremental increase in sales solely relating to such indications justifies the expense and investment of resources needed to obtain ANDA approval.

133. *See* AstraZeneca Pharm. LP v. Apotex Corp., 669 F.3d 1370, 1381 (Fed. Cir. 2012).

chapter 14

Pre-litigation Investigations and Due Diligence[1]

I. Introduction

New Drug Application (NDA) owners and patent owners who license to such NDA owners should conduct a pre-litigation fact investigation and due diligence to evaluate potential concerns prior to initiating Abbreviated New Drug Application (ANDA) litigation. Although evidence of infringement will come largely from the generic ANDA filer, other litigation issues—including document preservation, evidence to support the validity of the relevant patents, and the invention "story"—will lie in the hands of the patent and NDA owners. Likewise, the patent owner's own records will set the stage for potential counterclaims (e.g., of invalidity and inequitable conduct). Strategies for managing these issues prior to filing suit, and potentially thorny issues to monitor, are addressed in the following sections.

A. Fact Investigation

1. Document Preservation

Patent files, research and development documents, marketing materials, and financial and other documents relating to the NDA product in the possession or control of the patent and NDA owners are potentially discoverable, and thus should be subject to preservation prior to the initiation of a lawsuit.[2] Once the duty to preserve has been triggered, the failure to do so raises the issue of spoliation of evidence.[3] Spoliation "refers to the destruction or material alteration of evidence or to the failure to preserve property for another's use as evidence in pending or reasonably

1. Matthew Avery of Baker Botts LLP, Palo Alto, California, and Paul A. Ragusa and Jennifer C. Tempesta of Baker Botts LLP, New York City.

2. *See* Zubulake v. UBS Warburg LLC, 220 F.R.D. 212, 216 (S.D.N.Y. 2003).

3. *See id.*

foreseeable litigation."[4] It occurs "when a party has intentionally or negligently breached its duty to preserve potentially discoverable evidence."[5] The failure to take appropriate action to preserve relevant documents can result in consequences that can affect both liability and damages. For example, courts have the discretion to impose sanctions for "the purpose of leveling the evidentiary playing field and for the purpose of sanctioning the improper conduct," such as entering a default judgment against the offending party as to the claims or defenses for which the destroyed relevant documents were material.[6]

A litigant's duty to preserve relevant documents begins when litigation is pending or "reasonably foreseeable."[7] Whether litigation is reasonably foreseeable "is an objective standard, asking not whether the party in fact reasonably foresaw litigation, but whether a reasonable party in the same factual circumstances would have reasonably foreseen litigation."[8] Additionally, as long as the contingencies are to be foreseeably resolved, this standard does not require litigation to be imminent or probable without significant contingencies.[9] However, "this standard does not trigger the duty to preserve documents from the mere existence of a potential claim or the distant possibility of litigation."[10]

Determining when litigation is "reasonably foreseeable" is fact specific and conducted on a case-by-case basis; however, some recent cases provide guidance regarding certain circumstances that trigger the duty to preserve.[11] For example, the destruction of documents as a "litigation strategy,"

---

4. Micron Tech. Inc. v. Rambus Inc., 645 F.3d 1311, 1320 (Fed. Cir. 2011) (quoting Silvestri v. Gen. Motors Corp., 271 F.3d 583, 590 (4th Cir. 2001)).

5. Kounelis v. Sherrer, 529 F. Supp. 2d 503, 519 (D.N.J. 2008); *see also* Sanofi-Aventis Deutschland GmbH v. Glenmark Pharm. Inc., No. 07-CV-5855 (DMC-JAD), 2010 WL 2652412, at *3 (D.N.J. July 1, 2010) (stating that "if the duty to preserve evidence has not been triggered at the time the evidence was destroyed, then there can be no spoliation").

6. Broccoli v. Echostar Commc'ns Corp., 229 F.R.D. 506, 508 (D. Md. 2005) (citing Vodusek v. Bayliner Marine Corp., 71 F.3d 148, 156 (4th Cir.1995)).

7. Trevino v. Ortega, 969 S.W.2d 950, 956 (Tex. 1998) (Baker, J., concurring); *see also* William T. Thompson Co. v. Gen. Nutrition Corp., 593 F. Supp. 1443, 1455 (C.D. Cal. 1984) (stating that "a litigant is . . . under a duty to preserve what it knows, or reasonably should know, is relevant in the action, is reasonably calculated to lead to the discovery of admissible evidence, is reasonably likely to be requested during discovery, and/or is the subject of a pending discovery request").

8. *Micron Tech.*, 645 F.3d at 1320.

9. *See* Hynix Semiconductor Inc. v. Rambus Inc., 645 F.3d 1336, 1345 (Fed. Cir. 2011).

10. *Micron Tech.*, 645 F.3d at 1320.

11. *See, e.g.*, Forest Labs. Inc. v. Caraco Pharm. Labs. Ltd., No. 06-CV-13143, 2009 WL 998402, at *2 (E.D. Mich. Apr. 14, 2009) (holding that the innovator drug manufacturer did not have a duty to preserve electronic evidence about its patented drug product until it received a Paragraph IV Certification letter from the ANDA filer and rejecting the accused infringer's argument that the patentee's alleged destruction of internal e-mails before receiving the letter was spoliation); *In re* NTL Inc. Sec. Litig., 244 F.R.D. 179, 193 (S.D.N.Y.

as opposed to destruction in accordance with a corporate document retention policy, is more likely to result in a court finding that the litigation is "reasonably foreseeable."[12] Additionally, "knowledge of likely infringing activity" by the patentee makes litigation more "reasonably foreseeable," thereby resulting in a duty to preserve.[13]

Once on notice, "the obligation to preserve evidence runs first to counsel, who then has a duty to advise and explain to the client its obligations to retain pertinent documents that may be relevant to the litigation."[14] Document retention policies, which outline the retention and destruction of an organization's records, are "created in part to keep certain information from getting into the hands of others . . . and it is . . . not wrongful for a [company] to instruct [its] employees to comply with a valid document retention policy under normal circumstances."[15] However, "once a party reasonably anticipates litigation, it must suspend its routine document retention/destruction policy and put in place a litigation hold to ensure the preservation of relevant documents"[16] in "existence at the time the duty to preserve attaches, and any relevant documents created thereafter."[17]

To preserve relevant documents, the potential litigant must circulate a document retention memorandum or "litigation hold" notice to key employees and pertinent information custodians and advise information technology (IT) staff to suspend electronic auto-delete functions.[18] To have its intended effect, the litigation hold notice should detail the documents that are to be preserved and their potential sources and inform the key

---

2007) (duty to preserve evidence was triggered by a pre-lawsuit internal memo referencing the heightened risk of litigious activity); Mosaid Techs. Inc. v. Samsung Elecs. Co., 348 F. Supp. 2d 332, 338 (D.N.J. 2004) (holding that duty to preserve, at a minimum, was triggered upon initiation of the lawsuit); *Sanofi-Aventis Deutschland GmbH*, 2010 WL 2652412, at *5 (holding that the defendants' argument that their duty to preserve arose upon the filing of the Paragraph IV letter was invalid because it contradicted their claim of work-product immunity for documents created at an earlier date).

12. *See Micron Tech.*, 645 F.3d at 1322.

13. *See id.* at 1323; *see also* Apple Inc. v. Samsung Elecs. Co., 888 F. Supp. 2d 976, 996–97 (N.D. Cal. 2012) (competitor's duty to preserve evidence relevant to action was triggered after the patent owner gave a presentation to the competitor illustrating the competitor's alleged infringement).

14. Telecom Int'l Am. Ltd. v. AT&T Corp., 189 F.R.D. 76, 81 (S.D.N.Y. 1999).

15. Arthur Andersen LLP v. United States, 544 U.S. 696, 704 (2005).

16. Zubulake, 229 F.R.D. 422, 431 (S.D.N.Y. 2004) (internal quotation marks omitted); *see also Sanofi-Aventis Deutschland GmbH*, 2010 WL 2652412, at *3.

17. *Forest Labs. Inc.*, 2009 WL 998402, at *3 (quoting Zubulake v. UBS Warburg LLC, 220 F.R.D. 212, 218 (S.D.N.Y. 2003)).

18. *See Zubulake*, 229 F.R.D. at 432; *see also Apple Inc.*, 888 F. Supp. 2d at 997 (holding that Apple acted in conscious disregard of its duty to preserve by not issuing a litigation hold notice until after the duty to preserve had arisen, resulting in the destruction of relevant documents).

custodians of such documents.[19] It is "*not* sufficient to notify all employees of a litigation hold and expect that the party will then retain and produce all relevant information."[20] Counsel "must take affirmative steps to monitor compliance so that all sources of discoverable information are identified and searched"[21] and must "monitor[] the party's efforts to retain and produce the relevant documents."[22] In addition, the litigation hold should be periodically reissued to ensure that new employees are aware of it and that employees are reminded of the preservation policy.[23] If a litigation hold notice is insufficient, or its dissemination narrowly limited, the organization should take other steps that may be reasonable to preserve relevant evidence and should be prepared to explain why no litigation hold notice was distributed.

Once a "litigation hold" is in place, a potential litigant and its counsel must make certain that all custodians of potentially relevant information are identified.[24] Although counsel may not be able to locate all sources of such information, counsel must take "*reasonable steps*" to ensure that all sources of relevant information are located.[25] In particular, counsel should identify document custodians, which are typically the client's employees, who can assist in identifying the locations where relevant documents may be stored.

In the context of Hatch-Waxman litigation, documents detailing the research and conception and reduction to practice of the invention at issue and documents detailing the testing and analysis of the invention should be identified and located.[26] Patent prosecution history of the patent(s) covering the invention at issue and any related cases may also provide valuable information and should be identified and located.[27] Other relevant

19. *See Apple Inc.*, 888 F. Supp. 2d at 998 (holding that Apple acted with not just simple negligence but, rather, conscious disregard of its duty to preserve by failing to issue a litigation hold notice to any employees for eight months after its preservation duty had arisen and further delaying issuance of litigation hold notices to several key custodians); *see also* Wiginton v. CB Richard Ellis Inc., No. 02 C 6832, 2003 WL 22439865, at *5 (N.D. Ill. Oct. 27, 2003) (faulting defendant for being too narrow when preservation communication to employees only instructed employees to save documents that "pertained to" the named plaintiff in a putative class action although various other employees and offices were identified in the complaint).

20. *Zubulake*, 229 F.R.D. at 432.

21. *Id.*

22. *Id.*

23. *See id.* at 433.

24. *See id.* at 432.

25. *See id.*

26. *See* Medeva Pharma Suisse AG v. Roxane Labs. Inc., No. 07-5165 (FLW), 2011 WL 310697, at *15–16 (D.N.J. Jan. 28, 2011); *see also Apple Inc.*, 888 F. Supp. 2d at 997 (holding that documents from key witnesses, including inventors and company executives, are relevant documents that must be preserved).

27. *See Medeva*, 2011 WL 310697, at *15–16.

documents that should be identified and located are documents relating to the clinical testing of the invention; documents detailing the testing performed to obtain Food and Drug Administration (FDA) approval and the resulting submissions to the FDA; licensing agreements that cover the patented invention at issue; documents related to marketing, sales, and other commercial activities; and documents detailing the industry recognition of the patented invention.[28]

Once key relevant documents have been identified, these documents should be collected and reviewed. Pre-litigation investigation should include a thorough review of the patents at issue, prosecution history, and relevant evidence, such as the prior art that was cited during the patent application process. Analysis of the prosecution history should be completed to identify any narrowing arguments or admissions that were made during the prosecution of the patent that could limit the scope of the claims. In addition, each patent and application should be analyzed to determine its scope of protection and to ensure that the scope encompasses the product at issue. The rights afforded by a patent are determined by its claims, which are interpreted by its claim language, patent specification, and the prosecution history, all of which should be reviewed.[29]

To ensure proper preservation of relevant documents, counsel must become familiar with the client's document retention policies and the client's data retention architecture.[30] This will invariably include interviewing the client's IT personnel, who will be able to explain the client's "system-wide backup procedures and the actual . . . implementation of the firm's recycling policy."[31] In addition, IT personnel should provide information regarding the sources, custodians, and locations of relevant documents. Counsel should also become familiar with the client's laboratory notebook collection and storage policies. This includes interviewing the individual who oversees the organization and storage of such notebooks in order to determine the compliance with such policies and the locations and sources of the notebooks that may contain data relevant to the litigation.

2. Invention Story

Although documents are the backbone of ANDA litigation, witnesses will be needed to tell the invention story and, regardless, can be extremely valuable in pointing out the key documents among the megabytes of information collected. Interviewing inventors and other key scientists, marketing personnel, declarants, and prosecuting attorneys can provide insight into

---

28. *See id.*
29. *See* Phillips v. AWH Corp., 415 F.3d 1303, 1313–14 (Fed. Cir. 2005) (en banc).
30. *Zubulake*, 229 F.R.D. at 432.
31. *Id.*

the history behind the conception of the invention and the research behind the invention's development. Inventors can explain or clarify the invention story, provide details regarding the state of the art at the time of the invention, and identify leaders in the technological field to potentially serve as expert witnesses. Additionally, inventors can identify key witnesses that may have relevant information regarding the conception and/or the reduction to practice of the invention. Interviewing key witnesses can also help in obtaining facts relevant to the invention story and identifying potential weaknesses that would need to be addressed. For example, interviewing the inventor may reveal information regarding the state of the art at the time of the invention and identify potential prior art that was not cited during patent prosecution. The identification of such prior art before initiation of the litigation will allow early evaluation and strategy with respect to the expected defenses of a potential infringer.

When a potential witness has been identified, cooperation on the part of the witness is essential. In return for their cooperation, potential non-party witnesses may request payment for their time and expenses incurred in testifying and meeting with counsel. In certain circumstances, the agreement to pay for a witness's cooperation can persuade a previously unwilling witness to agree to be interviewed or to testify. These pay-for-cooperation agreements, although permissible, must be for expenses only and cannot constitute a fee for testifying or for the substance of the testimony.[32] For example, payments to former employees who serve as fact witnesses are limited to the purpose of reimbursing the employee for expenses incurred and time lost in preparing for and testifying at the deposition or trial.[33] If the court finds that counsel overstepped the bounds on what is considered acceptable (i.e., compensation is contingent on the outcome of the case), counsel and the client may be subject to civil and criminal sanctions. Possible civil sanctions for paying a witness a fee for testifying or paying a witness an unreasonable amount can include the exclusion of the testimony of the witness who received improper payments, an award of attorney's fees, and disciplinary action by the state bar.[34] Therefore, caution should

---

32. *See* Model Rules of Prof'l Conduct R. 3.4 (proscribes counsel from offering an inducement to a witness that is prohibited by law; statutory and common-law prohibitions exist in one form or another in every jurisdiction); *see also* 18 U.S.C. § 201(d) (1994) (provides that reasonable expenses or the reasonable value of time lost can be paid to witnesses).

33. *Id.*; *see* Cal. Model Rules of Prof'l Conduct R. 5-310(B) (stating that a member of the bar shall not provide "payment of compensation to a witness contingent upon the content of the witness' testimony or the outcome of the case").

34. *See* Model Rules of Prof'l Conduct R. 3.4 cmt. 2; *see also* Rocheux Int'l of N.J. Inc. v. U.S. Merchs. Fin. Grp. Inc., 741 F. Supp. 2d 651 (S.D.N.Y. 2010) (holding that the compensation paid by the plaintiff to the defendant's former employee was unreasonable and excluded the former employee's testimony and sanctioned the plaintiff by requiring it to reimburse the defendant's attorney fee).

be exercised when deciding whether and how much to pay a third-party witness to ensure that the witness's testimony will be allowed during trial.

During the interview of key witnesses and inventors, special attention must be paid to the content of the communications, as that content will determine the communications' privileged status. The so-called subject matter test set forth by the Supreme Court in *Upjohn Co. v. United States* should guide whether discussions between current employees (i.e., potential key witnesses) and an organization's counsel are privileged.[35] Under the subject matter test, the privilege hinges on the subject matter of the communications, rather than on the person with whom the counsel communicated.[36] For example, communications that concern matters within the scope of the current employee's duties at the organization are considered privileged if the employee is aware of the role in which the counsel is acting and if the communications are made at the direction of the organization.[37] With regard to former employees of the client, although not explicitly addressed, *Upjohn* provides some guidance, and courts have continued to follow the rationale set forth in *Upjohn*.[38] Justice Burger, in his concurring opinion, stated that when a former employee speaks with counsel at the direction of the organization's management, the former employee's communications remain privileged.[39] Therefore, during interviews of witnesses, consideration must be given to the content of the communications to ensure that the subject matter relates to the scope of the current or former employees' duties at the organization and that the communications are protected by attorney-client privilege.

3. Secondary Considerations

Patentable inventions must be novel and nonobvious in view of the prior art.[40] Obviousness is a legal conclusion based on the factual inquiries set forth in *Graham v. John Deere Co. of Kansas City*.[41] These inquiries include: (1) the scope of the content of the prior art; (2) the differences

35. 449 U.S. 383 (1981).

36. *Id.*

37. *See id.* at 394–95. Communications are privileged if (1) they were made to the organization's counsel, acting as such; (2) they were made at the direction of organization's superiors, for the purpose of securing legal advice from counsel; (3) they concerned matters within the scope of the employees' duties at the organization; and (4) the employees were sufficiently aware that they were being questioned so the organization could obtain legal advice. *Id.*

38. *See* Peralta v. Cendant Corp., 190 F.R.D. 38, 41 (D. Conn. 1999) (holding that if the communication relates to the former employee's conduct and knowledge during his or her employment, such communication is protected from disclosure by the attorney-client privilege under the *Upjohn* test).

39. *Id.* at 402–03 (Burger, J., concurring).

40. *See* 35 U.S.C. §§ 101–103 (2006).

41. 383 U.S. 1 (1966).

between the claims and the prior art; (3) the level of ordinary skill in the pertinent art; and (4) secondary considerations. Although an invention may appear obvious on its face, secondary considerations, also referred to as secondary indicia of nonobviousness, can be used to establish that the invention was in fact not obvious in light of the prior art.[42] These secondary considerations can serve to protect against the impermissible use of hindsight in determining whether combinations of prior art references would have been obvious to a person of ordinary skill in the art.[43] The Court in *Graham* stated that secondary considerations can include commercial success, long-felt but unsolved need, and the failure of others.[44] Other factors recognized by the Federal Circuit after *Graham* include whether the prior art teaches away from the invention, whether others have copied the invention, and whether the invention has received industry acclaim.[45] In 2007, the Supreme Court in *KSR International Co. v. Teleflex Inc.* reaffirmed the application of the approach articulated in *Graham* to determine nonobviousness, including the analysis of secondary considerations.[46]

When deciding the issue of obviousness, the court must consider any secondary considerations that are relevant to the patent(s) at issue.[47] However, evidence of secondary considerations is not to be given weight if it has no "nexus" to the merits of the claimed invention (e.g., the patentee bears the burden of demonstrating "a legally and factually sufficient connection between the proven success and the patented invention, such that the objective evidence should be considered in the determination of non-obviousness").[48]

Therefore, in addition to determining the invention story behind the patent(s) at issue, as discussed, pre-litigation due diligence should focus on determining the secondary indicia of nonobviousness that may be applicable to the patent(s) at issue.[49] For example, the identification and collection of documents detailing the commercial success of the product and any nexus of that success to the claimed invention will be very helpful in responding to an obviousness argument presented by the ANDA filer.[50]

---

42. *Id.* at 17–18.

43. *See* PharmaStem Therapeutics Inc. v. ViaCell Inc., 491 F.3d 1342, 1377 (Fed. Cir. 2007).

44. 383 U.S. at 17–18.

45. *See* Ecolochem Inc. v. S. Cal. Edison Co., 227 F.3d 1361, 1379–80 (Fed. Cir. 2000), *cert. denied*, 532 U.S. 974 (2001).

46. 550 U.S. 398, 415 (2007).

47. *See* Ruiz v. AB Chance Co., 234 F.3d 654, 667 (Fed. Cir. 2000).

48. *In re* Paulsen, 30 F.3d 1475, 1482 (Fed. Cir. 1994).

49. See Chapter 8, Regulatory Exclusivity.

50. *See* Bristol-Myers Squibb Co. v. Teva Pharm. USA Inc., No. 10-805-CJB, 2013 WL 509152, at *69 (D. Del. Feb. 11, 2013) (holding that commercial success is relevant because

These documents should provide evidence of a nexus between the sales and the patented features of the product and/or show that the sales are not due to other factors, such as marketing and advertising, which would be critical to the success of rebutting a prima facie showing of obviousness.[51] Likewise, as part of their interviews, key witnesses should be asked whether they are aware of articles of other evidence showing that there was a long-felt but unsolved need for the patented invention and/or the failure of others to realize that invention.

4. Initial Disclosures

Initial disclosures can promote the disclosure of important information at the outset of litigation and foster cooperation between the parties on discovery issues.[52] The Federal Rules of Civil Procedure have long required parties to submit their initial disclosures within 30 days of being served or joined in the litigation, unless a different time is set by stipulation or court order.[53] The initial disclosures must provide the opponents with basic factual information, including: (1) the name, address, and telephone number of any witness likely to have discoverable information relevant to disputed facts; (2) a copy or a description by category and location of all relevant documents; (3) a computation of damages (which is typically not relevant in ANDA cases), and (4) any insurance agreement that may be used to satisfy a potential judgment.[54] Because these are due by a date certain, it is prudent to prepare them during the pre-litigation due diligence investigation.

In addition to ensuring that the initial disclosures are prepared by the due date, the initial disclosures should be reviewed for their accuracy and thoroughness, for the failure to disclose all relevant information in the initial disclosure can result in serious consequences. For example, "[i]f a party fails to provide information or identify a witness as required by Rule 26(a) or 26(e), the party is not allowed to use that information or witness to supply evidence on a motion, at a hearing, or at a trial, unless the failure was substantially justified or is harmless."[55] Therefore, the accuracy and completeness of these disclosures should be verified prior to their submission and supplemented in a timely manner to ensure that the necessary

the law presumes that an idea would have been brought to the market sooner, in response to market forces, if the idea was obvious to those skilled in the art).

51. *See id.*

52. *See* David A. Churchill & Sada Manickam, *The Continuing Evolution of the Mandatory Disclosure Requirements of Federal Rule 26*, 35 PROCUREMENT LAW. 3, 3 (1999).

53. FED. R. CIV. P. 26(a)(1)(D).

54. *Id.* at 26(a)(1)(A).

55. *Id.* at 37(c)(1); *see* Gen-Probe Inc. v. Becton Dickinson & Co., 899 F. Supp. 2d 971 (S.D. Cal. 2012) (the court prohibited a witness from testifying at trial because the witness was not identified in the initial disclosures).

witnesses will be permitted to testify and that other potential sanctions will not be levied.

B. Patent Evaluations

All Paragraph IV ANDAs will include a certification that at least one Orange Book–listed patent for the drug the applicant seeks to copy is invalid or will not be infringed by the manufacture, use, or sale of the generic drug for which the application is submitted.[56] In response to a Paragraph IV ANDA, the NDA holder should carefully evaluate each challenged patent claim respecting validity, enforceability, and likelihood of infringement.

Note, however, that the NDA holder does not need to limit its infringement action to Orange Book–listed patents. Instead, the NDA holder may assert relevant patents that are not listed in the Orange Book.[57] Although the following discussion focuses on the evaluation of Orange Book–listed patents that are being directly challenged by a Paragraph IV ANDA, much of the discussion is applicable to any patent the NDA holder might assert against the generic challenger.

II. Claim Construction

The first step of an infringement and/or validity analysis is to determine the meaning of the claims.[58] Only after the boundaries of the claims have been determined can the court decide whether the generic product falls within the scope of the claims and whether the claims are valid.[59] Claim construction is a question of law for the court to decide.[60] The court may consider both intrinsic evidence (i.e., the patent and its prosecution history) and extrinsic evidence (e.g., dictionaries, expert testimony, publications, etc.).

ANDA litigation typically involves claims to a new chemical entity, formulations, and/or methods of making or using such entities or formulations. Typically, the patent holder will argue that the claims should be construed so as to encompass the ANDA applicant's product, while the generic challenger will argue that the drug would not infringe under an alternative construction (usually a narrower construction).

---

56. 21 U.S.C. § 355(j)(2)(A)(vii)(I)–(IV).

57. *See, e.g.*, Cephalon Inc. v. Watson Pharm. Inc., 769 F. Supp. 2d 761, 763 (D. Del. 2011) (where the NDA holder asserted two Orange-Book-listed patents and one non-Orange-Book-listed patent in response to a Paragraph IV ANDA challenge).

58. Abbott Labs. v. Andrx Pharm. Inc., 452 F.3d 1331, 1335–36 (Fed. Cir. 2006).

59. Markman v. Westview Instruments Inc., 52 F.3d 967, 979 (Fed. Cir. 1995) (en banc), *aff'd*, 517 U.S. 370 (1996).

60. *Id.*

A. Intrinsic Evidence

Claim construction begins with the intrinsic evidence of record (i.e., the text of the claims, the specification, and the prosecution history).[61] The words of a claim "are generally given their ordinary and customary meaning."[62] The "ordinary and customary meaning of a claim term is the meaning that the term would have to a person of ordinary skill in the art in question at the time of the invention, *i.e.*, as of the effective filing date of the patent application."[63] Consideration of the intrinsic evidence should begin with the language of the claims at issue.[64]

The court must consider the specification when construing the claims,[65] but should not import limitations from the specification into the claims absent an express disclaimer.[66] In practice, the court will look to the specification for an explanation of the invention and definitions of the terms used in the claims.[67] The context of a claim term, or how the term is used in the claim, should provide guidance as to its meaning. Accordingly, the specification may act "as a sort of dictionary, which explains the invention and may define terms used in the claims."[68] In *CCS Fitness v. Brunswick Corp.*, the Federal Circuit summarized three circumstances where the scope of patent claims may be limited by, among other things, statements made in the specification:

> [A] claim term will not carry its ordinary meaning if the intrinsic evidence shows that [1] the patentee distinguished that term from the prior art on the basis of a particular embodiment, [2] expressly disclaimed subject matter, or [3] described a particular embodiment as important to the invention.[69]

---

61. Phillips v. AWH Corp., 415 F.3d 1303, 1313, 1317 (Fed. Cir. 2005) (en banc); Unique Concepts Inc. v. Brown, 939 F.2d 1558, 1561 (Fed. Cir. 1991).

62. Vitronics Corp. v. Conceptronic Inc., 90 F.3d 1576, 1582 (Fed. Cir. 1996); *see also* Toro Co. v. White Consol. Indus. Inc., 199 F.3d 1295, 1299 (Fed. Cir. 1999); Renishaw PLC v. Marposs Societa' per Azioni, 158 F.3d 1243, 1249 (Fed. Cir. 1998).

63. *Phillips*, 415 F.3d at 1313.

64. Interactive Gift Exp. Inc. v. CompuServe Inc., 256 F.3d 1323, 1331 (Fed. Cir. 2001) ("[T]he analytical focus must begin and remain centered on the language of the claims themselves, for it is that language that the patentee chose to use to 'particularly point out and distinctly claim the subject matter which the patentee regards as his invention.'" (quoting 35 U.S.C. § 112, ¶ 2)).

65. *Phillips*, 415 F.3d at 1313 ("Importantly, the person of ordinary skill in the art is deemed to read the claim term not only in the context of the particular claim in which the disputed term appears, but in the context of the entire patent, including the specification.").

66. *Id.* at 1323.

67. *See Vitronics Corp.*, 90 F.3d at 1582.

68. *Markman*, 52 F.3d at 979; *Vitronics Corp.*, 90 F.3d at 1582 (citing *Markman*, 52 F.3d at 979) (stating that the specification may define terms expressly or by implication).

69. 288 F.3d 1359, 1366–67 (Fed. Cir. 2002) (citations omitted).

Essentially then, "claim terms take on their ordinary and accustomed meanings unless the patentee demonstrated an intent to deviate from [that meaning]."[70] "It is therefore entirely appropriate for a court, when conducting claim construction, to rely heavily on the written description for guidance as to the meaning of the claims."[71]The patentee should also review the file history of each patent to determine any prosecution history estoppel issues that may arise due to statements or amendments made during prosecution. In construing an asserted claim, the court should also refer to the prosecution history, which contains a "complete record of all the proceedings before the Patent and Trademark Office, including any express representations made by the applicants regarding the scope of the claims."[72] Consideration of the prosecution history will make the court aware of any statements made during the prosecution of the patent that limit the available scope of construction.[73] "[T]he prosecution history can often inform the meaning of the claim language by demonstrating how the inventor understood the invention and whether the inventor limited the invention in the course of prosecution, making the claim scope narrower than it would otherwise be."[74]

B. Extrinsic Evidence

Courts may also rely on extrinsic evidence to help construe the claims, though such evidence is generally given less weight than intrinsic evidence. For example, courts may rely on extrinsic evidence "to ensure that the court's understanding of the technical aspects of the patent is consistent with that of a person of skill in the art, or to establish that a particular term in the patent or the prior art has a particular meaning in the pertinent field."[75] Such evidence can include general dictionaries, technical dictionaries, expert testimony, scientific publications, learned treatises, and similar authorities.[76] Of particular importance to ANDA cases may be FDA guidelines and statements by the parties to the FDA related to the drug product.[77]

Dictionaries, and particularly technical dictionaries that are specific to the technical field of the invention, may be useful in understanding the

70. Teleflex Inc. v. Ficosa N. Am. Corp., 299 F.3d 1313, 1327 (Fed. Cir. 2002).

71. *Phillips*, 415 F.3d at 1317.

72. *Vitronics Corp.*, 90 F.3d at 1582.

73. *Id.* at 1583 (noting that "[t]he prosecution history limits the interpretation of claim terms so as to exclude any interpretation that was disclaimed during prosecution" (quoting Southwall Tech. Inc. v. Cardinal IG Co., 54 F.3d 1570, 1576 (Fed. Cir. 1995))).

74. *Phillips*, 415 F.3d at 1317.

75. *Id.* at 1317–18.

76. *Id.* at 1318.

77. *See, e.g.*, Key Pharm. v. Heron Labs. Corp., 161 F.3d 709, 718 (Fed. Cir. 1988).

underlying technology and construing claim terms.[78] However, because such references are not specific to the patent at issue, they can only be relied on to the extent that they do "not contradict any definition found in or ascertained by a reading of the patent."[79]

Similarly, expert testimony may be useful in construing claims "for a variety of purposes, such as to provide background on the technology at issue, to explain how an invention works, to ensure that the court's understanding of the technical aspects of the patent is consistent with that of a person of skill in the art, or to establish that a particular term in the patent or the prior art has a particular meaning in the pertinent field."[80] However, like dictionaries, expert testimony that directly contradicts intrinsic evidence should not be relied upon.[81]

Note that the court may also consider prior constructions of the claims terms by other courts, and such constructions may have a preclusive effect via collateral estoppel or judicial estoppel.

III. Validity/Enforceability

Under 35 U.S.C. § 282, each claim of a patent shall be presumed valid; an accused infringer must prove invalidity by clear and convincing evidence.[82] In the Hatch-Waxman context, this means the generic challenger will always have the burden of proving invalidity. ANDA litigation may involve invalidity arguments under Sections 102, 103, and 112. Because these issues can rely on understanding the state of the art, the patentee should identify and retain potential experts that it can work with to identify and vet these issues.

A. Validity under 35 U.S.C. §§ 102, 103

To show anticipation under 35 U.S.C. § 102, the ANDA applicant will need to show that a single item of prior art discloses each and every limitation of a claim, either expressly or inherently.[83] Because ANDA litigation often involves patent claims to a genus of chemical compounds, it is important to remember that the disclosure of a single species in the prior art is

78. *Id.* at 1321–22.
79. *Id.* at 1322–23.
80. *Id.* at 1318.
81. *Id.* at 1318; *see also* Novartis Pharm. Corp. v. Abbott Labs., 375 F.3d 1328, 1336 (Fed. Cir. 2004).
82. Microsoft Corp. v. i4i LP, 1331 S. Ct. 2238, 2242 (2011).
83. *See* 35 U.S.C. § 102.

sufficient to anticipate an entire genus claim.[84] As for method-of-treatment claims, which are also common in ANDA litigation, a prior disclosure of the compound in a pharmaceutical preparation may inherently anticipate the use of that compound for a specific therapeutic treatment claimed by the patentee.[85]

Under 35 U.S.C. § 103, patent claims will be considered obvious if "the difference between the subject matter sought to be patented and the prior art are such that the subject matter as a whole would have been obvious at the time the invention was made to a person having ordinary skill in the art to which said subject matter pertains."[86] The court must make four factual inquiries to determine whether a claim is invalid under § 103:

1. the scope and content of the prior art;
2. the level or ordinary skill in the art;
3. differences between the prior art and the claims at issue;
4. secondary considerations, such as commercial success, long felt but unsolved needs, failure of others, etc., that may be utilized to give light to the circumstances surrounding the origin of the subject matter sought to be patented.[87]

Under the first prong of the *Graham* test, the scope and content of the prior art under § 103 includes all references and information that qualify as prior art under § 102.[88] The level of ordinary skill in the art is determined by considering many factors, including "type of problems encountered in the art; prior art solutions to those problems; rapidity with which innovations are made; sophistication of the technology; and educational level of active workers in the field."[89] Differences between the prior art and the claims at issue are then evaluated to determine whether the claimed invention would have been obvious to a person of ordinary skill in the art at the time of the invention.[90] Secondary considerations include

84. *See, e.g.*, Eli Lilly & Co. v. Zenith Goldline Pharm. Inc., 471 F.3d 1369 (Fed. Cir. 2006).

85. *See, e.g.*, Merck & Co. v. Teva Pharm. USA Inc., 347 F.3d 1367, 1372 (Fed. Cir. 2003).

86. 35 U.S.C. § 103.

87. Graham v. John Deere Co. of Kan. City, 383 U.S. 1, 17–18 (1966). In *KSR International Co. v. Teleflex Inc.*, 550 U.S. 398 (2007), the Supreme Court reaffirmed the *Graham* test as an appropriate test for obviousness.

88. *See* J.A. LaPorte Inc. v. Norfolk Dredging Co., 787 F.2d 1577, 1580 n.4 (Fed. Cir. 1986).

89. Custom Accessories Inc. v. Jeffrey-Allan Indus. Inc., 807 F.2d 955, 962 (Fed. Cir. 1986). "The person of ordinary skill is a hypothetical person who is presumed to be aware of all the pertinent prior art." *Id.*

90. *See, e.g.*, Yamanouchi Pharm. Co. v. Danbury Pharm. Inc., 231 F.3d 1339, 1342–45 (Fed. Cir. 2000).

commercial success, the copying of the invention by others, or the filling of a long-felt need, among others.[91]

A common obviousness argument in Hatch-Waxman cases is that the claimed chemical compound is obvious in light of a structurally similar compound disclosed in the prior art. When analyzing the obviousness of chemical compounds in this context, the court applies a two-part test to determine: (1) "whether a chemist of ordinary skill would have selected the asserted prior art compounds as lead compounds, or starting points, for further development efforts"; and (2) "whether the prior art would have supplied one of ordinary skill in the art with a reason or motivation to modify a lead compound to make the claimed compound with a reasonable expectation of success."[92]

Another common argument in ANDA litigation is that the claimed invention was "obvious to try." In *KSR*, the Supreme Court explained that a claimed invention can be rendered obvious when it would have been obvious to try a finite number of predictable solutions and where there was a reasonable expectation of success.[93]

B. Validity under 35 U.S.C. § 112

In the *invalidity* defenses asserted in the ANDA, the generic challenger may argue that the patent at issue is invalid for one or more requirements of § 112. That statutory provision requires the specification to "contain a written description of the invention, and of the manner and process of making and using it, in such full, clear, concise, and exact terms as to enable any person skilled in the art to which it pertains, or which it is most nearly connected, to make and use the same." Because both the written description and enablement requirements are based on the state-of-the-art knowledge at the time the patent was filed, the NDA holder may want to identify and retain potential experts who can provide guidance on the state of the art in the field of use of the inventions.

To satisfy the written description requirement, the specification must "reasonably convey[] to those skilled in the art that the inventor had possession of the claimed subject matter as of the filing date."[94] The disclosure

91. Cont'l Can Co. USA Inc. v. Monsanto Co., 948 F.2d 1264, 1273 (Fed. Cir. 1991). While secondary considerations "must be considered, they do not control the obviousness conclusion." Newell Cos. v. Kenney Mfg. Co., 864 F.2d 757, 768 (Fed. Cir. 1988). Rather, "Graham set forth a broad inquiry and invited courts, where appropriate, to look at any secondary considerations that would prove instructive." *KSR Int'l*, 550 U.S. at 415.

92. Otsuka Pharm. Ltd. v. Sandoz Inc., 678 F.3d 1280, 1291–92 (Fed. Cir. 2012) (citing Takeda Chem. Indus. v. Alphapharm Pty. Ltd., 492 F.3d 1350, 1357 (Fed. Cir. 2007)).

93. *KSR Int'l*, 550 U.S. at 402–03.

94. Ariad Pharm. Inc. v. Eli Lilly & Co., 598 F.3d 1336, 1350 (Fed. Cir. 2010) (en banc) (citation omitted).

needed "varies with the nature and scope of the invention at issue."[95] The Federal Circuit has held that an adequate written description in unpredictable arts like biotechnology or chemistry "requires a precise definition, such as by structure, formula, or chemical name" to distinguish the claimed subject matter from other materials.[96]

Claims to a genus of chemical compounds, which are common in ANDA litigation, may be vulnerable to written description attacks. To describe a genus, the specification must disclose a representative number of species in that genus or relevant identifying characteristics of the genus, such as structural features commonly possessed by members of the genus that distinguish them from others.[97] The written description requirement can be met by

> showing that an invention is complete by disclosure of sufficiently detailed, relevant identifying characteristics . . . i.e., complete or partial structure, other physical and/or chemical properties, functional characteristics when coupled with a known or disclosed correlation between function and structure, or some combination of such characteristics.[98]

However, a "functional description can be sufficient only if there is also a structure-function relationship known to those of ordinary skill in the art."[99]

To satisfy the enablement requirement under 35 U.S.C. § 112, first paragraph, a patent specification "must teach those skilled in the art how to make and use the full scope of the claimed invention without undue experimentation."[100] A "reasonable" amount of "routine experimentation" is not considered undue.[101] Whether "undue experimentation" is required to practice the claimed invention is not a factual determination but, rather, a legal conclusion based on factual considerations.[102] Any subject matter experts retained by the NDA holder should be able to analyze the specifica-

---

95. Capon v. Eshhar, 418 F.3d 1349, 1357 (Fed. Cir. 2005). "The purpose of the written description requirement is to prevent an applicant from later asserting that he invented that which he did not; the applicant for a patent is therefore required to 'recount his invention in such detail that his future claims can be determined to be encompassed within his original creation.'" Amgen Inc. v. Hoechst Marion Roussel Inc., 314 F.3d 1313, 1330 (Fed. Cir. 2003).

96. Regents of Univ. of Cal. v. Eli Lilly & Co., 119 F.3d 1559, 1568 (Fed. Cir. 1997) (internal quotation marks and alterations omitted).

97. *Ariad Pharm.*, 598 F.3d at 1350; Billups-Rothenberg Inc. v. Associated Reg'l & Univ. Pathologists Inc., 642 F.3d 1031, 1037 (Fed. Cir. 2011); *Regents of Univ. of Cal.*, 119 F.3d at 1568.

98. Enzo Biochem Inc. v. Gen-Probe Inc., 296 F.3d 1316, 1324 (Fed. Cir. 2002).

99. *In re* Wallach, 378 F.3d 1330, 1335 (Fed. Cir. 2004).

100. Genentech Inc. v. Novo Nordisk AS, 108 F.3d 1361, 1365 (Fed. Cir. 1997) (internal quotation marks omitted).

101. ALZA Corp. v. Andrx Pharm. LLC, 603 F.3d 935, 940 (Fed. Cir. 2010).

102. *In re* Wands, 858 F.2d 731, 737 (Fed. Cir. 1988).

tion of the patent at issue to offer an opinion regarding the level of experimentation that would be necessary to arrive at the claimed invention.

Although a specification need not disclose what is well known in the art, it "must supply the novel aspects of an invention in order to constitute adequate enablement."[103] In addition, when the claimed invention involves the application of an unpredictable technology, a specification must provide a "specific and useful teaching" regarding how to make and use the invention.[104] Because the chemical/pharmaceutical arts can be considered "unpredictable," the patents involved in ANDA litigation may be bound to this "specific and useful teaching" standard, such as experimental results showing that the claimed invention was actually made and used.

C. Inequitable Conduct

An accused infringer can raise a defense that a patent is unenforceable because it was procured from the USPTO using inequitable conduct. In order to prove inequitable conduct, the accused infringer must show with clear and convincing evidence that the patentee either failed to disclose information or presented false information to the patent office that was: (1) material to patentability; (2) withheld with the specific intent to deceive the patent office; and (3) resulted in the patentee receiving an unwarranted claim.[105] If a patent is held unenforceable, every claim of the patent is considered unenforceable.[106]

If the generic challenger raises an inequitable conduct defense, it can be based on the theory that either the inventor or the attorney that prosecuted the patent failed to disclose one or more pieces of prior art to the USPTO. However, the Federal Circuit's recent decision in *Therasense, Inc. v. Becton, Dickinson & Co.* significantly raised the bar on what is needed to prove inequitable conduct.[107] First, the accused infringer has to show that the withheld prior art was but-for material—that is, but for the deception, the USPTO would not have allowed the claims.[108] Second, the accused infringer has to show that the inventors or prosecution attorneys made a

103. *Genentech*, 108 F.3d at 1366; *see also* Auto. Techs. Int'l Inc. v. BMW of N. Am., 501 F.3d 1274, 1283 (Fed. Cir. 2007) ("Although the knowledge of one skilled in the art is indeed relevant, the novel aspect of an invention must be enabled in the patent.").

104. *Genentech*, 108 F.3d at 1367–68; *see also* Chiron Corp. v. Genentech Inc., 363 F.3d 1247, 1254 (same); PPG Indus. Inc. v. Guardian Indus. Corp., 75 F.3d 1558, 1564 (Fed. Cir. 1996) (explaining that, in unpredictable arts, the Federal Circuit has refused to find broad generic claims enabled by specifications that do not specifically demonstrate how to make and use embodiments across the full scope of the claim).

105. *See* Therasense Inc. v. Becton, Dickinson & Co., 649 F.3d 1276 (Fed. Cir. 2011) (en banc).

106. J.P. Stevens & Co. v. Lex Tex. Ltd., 747 F.2d 1553 (Fed. Cir. 1984).

107. *Therasense Inc.*, 649 F.3d 1276.

108. *Id.* at 1291.

deliberate decision to deceive the USPTO.[109] Merely showing that the patentee should have known the reference was material, or even showing that the patentee lacked a good-faith explanation for withholding the reference, is not sufficient to show intent to deceive.

Furthermore, intent to deceive cannot be inferred from the materiality of the reference. But-for materiality and intent to deceive must be proved separately,[110] which will likely be difficult for any generic challenger to show without some egregious admission on the part of the patentee. Consequently, inequitable conduct has rarely been successfully raised as a defense under the new *Therasense* standards.

IV. Infringement

As discussed previously, the Hatch-Waxman Act provides that a Paragraph IV Certification is itself an act of patent infringement.[111] The statute requires all Paragraph IV ANDA applicants to provide notice of the application to the NDA/patent holder, including a detailed factual and legal analysis explaining why the patent is either invalid or not infringed.[112] However, because the proposed generic product will have not received FDA marketing approval, it will not yet be manufactured, imported, marketed, or sold (except as necessary for purposes of obtaining FDA approval) at the time the NDA holder files an infringement action.

This means that the infringement analysis will focus on the details of the ANDA, which describes the applicant's proposed product. Hatch-Waxman requires the generic applicant to offer the NDA holder confidential access to its application so it can determine whether there is possible infringement.[113] If the Paragraph IV Certification asserts that the generic drug does not infringe the patent at issue, the NDA holder should carefully examine the details about the drug product provided in the application to determine whether the NDA holder can actually prove infringement. However, the infringement analysis is not necessarily limited to the scope of approval sought in the ANDA, but should be "based on all of the relevant evidence, including the ANDA."[114]

---

109. *Id.* at 1290.

110. *Id.*

111. 35 U.S.C. § 271(e)(2)(A) ("It shall be an act of infringement to submit . . . an [ANDA] for a drug claimed in a patent or the use of which is claimed in a patent.").

112. 21 U.S.C. § 355(j)(2)(B).

113. *Id.* § 355(j)(5)(C)(i)(III).

114. Glaxo Inc. v. Novopharm Ltd., 110 F.3d 1562, 1567–68 (Fed. Cir. 1997).

An infringement analysis involves two steps: (1) a determination of the scope and construction of the patent claims asserted, and (2) a determination of whether the claims, as so construed, cover the accused product.[115] The second step of an infringement analysis, determining whether the claims as construed cover the accused product, is factual in nature.[116] The patent owner has the burden of proving by a preponderance of the evidence that every limitation of the patent claims asserted to be infringed is found in the accused composition or method, either "literally" or under the "doctrine of equivalents."[117] Consequently, the patentee should consider any difficulties that may be associated with proving infringement—for example, whether testing of the generic product would be needed to prove that certain claim elements are met. This may require the patentee to identify and retain potential experts whom it can work with to identify and vet these issues.

To find literal infringement, an accused composition must possess features that literally correspond to each of the elements set forth in a claim.[118] However, because the ANDA applicant must certify that its generic product is bioequivalent to the NDA holder's product, any patent claims directed to the active pharmaceutical ingredient are likely to be literally infringed. Even when there is no literal infringement, the doctrine of equivalents permits infringement to be found where the differences between an accused and a claimed composition are insubstantial.[119] In applying the doctrine of equivalents, all limitations of a claim are material and must be satisfied equivalently.[120] This is so regardless of whether the limitations are necessary to achieve the claimed result.[121]

The NDA holder should also carefully review the file history of each asserted patent, and any continuations or divisional applications of those patents, for prosecution history estoppel issues. Prosecution history estoppel can limit the scope of equivalents to which a claim is entitled by barring a patentee from recapturing subject matter surrendered during prosecution in order to secure allowable subject matter.[122] Likewise,

---

115. Palumbo v. Don-Joy Co., 762 F.2d 969, 974 (Fed. Cir. 1985).

116. SmithKline Diagnostics Inc. v. Helena Labs. Corp., 859 F.2d 878, 889 (Fed. Cir. 1988).

117. *Id.*

118. Engel Indus. Inc. v. Lockformer Co., 96 F.3d 1398, 1405 (Fed. Cir. 1996); *see also* Pennwalt Corp. v. Durand-Wayland Inc., 833 F.3d 931, 934 (Fed. Cir. 1987) (en banc).

119. Warner-Jenkinson Co. v. Hilton Davis Chem. Co., 520 U.S. 17, 24 (1997).

120. Becton, Dickinson & Co. v. C.R. Bard Inc., 922 F.2d 792, 798 (Fed. Cir. 1990); *Pennwalt*, 833 F.2d at 934-35.

121. *Becton, Dickinson*, 922 F.2d at 798.

122. *See, e.g.*, Festo Corp. v. Shoketsu Kinzoku Kogyo Kabushiki Co., 535 U.S. 722, 737 (2002); *Warner-Jenkinson*, 520 U.S. at 30–31; Athletic Alternatives Inc. v. Prince Mfg. Inc., 73 F.3d 1573, 1582 (Fed. Cir. 1996).

arguments made or positions taken during prosecution can create prosecution history estoppel.[123] Because the patents asserted in Hatch-Waxman cases are often from large patent families, this can multiply estoppel risk, since concessions made in related applications can sometimes be used to limit the scope of other members of the family.[124] Although it is generally too late to consider when responding to an ANDA, the patentee should be mindful that all patents in a family are prosecuted consistently in order to minimize possible prosecution history estoppel issues.

V. Ownership and Standing

Because many NDA holders acquire their products through partnerships or acquisitions,[125] patentees should review their Orange Book–listed patents for any ownership issues. In particular, the NDA holder should review all patent assignments and license agreements covering the drug product at issue.

The NDA holder bears the burden of proof on patent ownership and standing and must establish ownership of enforceable title at the inception of the infringement action.[126] This means the NDA holder must show that it is the sole owner or exclusive licensee of the asserted patents or, if it cannot do this, must show that all co-owners have joined the action at the time of filing. Otherwise, the NDA holder risks having its infringement suit dismissed under Rule 12(b)(1) of the Federal Rules of Civil Procedure for lacking subject-matter jurisdiction.

To establish ownership, the NDA holder must show that there is an unbroken chain of title from the inventors to it. Alternatively, if the NDA holder is not the patent owner, then it must be the exclusive licensee in order to have standing to bring suit. Note, however, that even if the license to the NDA holder does not recite the term "exclusive licensee," whether the NDA holder is considered an exclusive licensee for purposes of standing is judged within the context of the agreement. For example, a license to a particular field of use may be considered "exclusive" if the drug covered by the license is only approved by the FDA for that particular use.[127]

---

123. Viskase Corp. v. Am. Nat'l Can Co., 261 F.3d 1316, 1321–22 (Fed. Cir. 2001); Biovail Corp. Int'l v. Andrx Pharm. Inc., 239 F.3d 1297, 1304 (Fed. Cir. 2001).

124. Baxter Int'l Inc. v. McGaw Inc., 149 F.3d 1321, 1328 (Fed. Cir. 1998).

125. *See, e.g.*, Sergio Garcia & Matthew Avery, *Emerging Trends in Biotech/Pharmaceutical Collaborations*, Licensing J., Oct. 2008, at 1, 1.

126. Abraxis Bioscience Inc. v. Navinta LLC, 625 F.3d 1359, 1364 (Fed. Cir. 2010) ("[T]he plaintiff must demonstrate that it held enforceable title to the patent at the inception of the lawsuit."). After-the-fact or retroactive assignments made after the suit is filed are not sufficient to fix the plaintiff's lack of standing. *Id.* at 1365–68.

127. *See, e.g.*, Sanofi SA v. Med-Tech Veterinarian Prods. Inc., 565 F. Supp. 931, 937 (D.N.J. 1983).

Hatch-Waxman cases often have multiple corporate entities involved as plaintiffs, and the NDA holder must ensure that all needed parties are willing to be part of the litigation against the generic challenger. As discussed, the patent owner or exclusive licensee has standing to sue. However, related corporate entities, like a parent corporation, an agent, or a bare licensee, likely lack standing to sue, even as a co-plaintiff.[128] Consequently, if the NDA holder is one of these entities that lacks standing to sue, it will need to have the cooperation of the patent owner in order to actually sue the ANDA filer for patent infringement. Similarly, if multiple entities have an ownership interest in the patent, then the NDA holder should ensure that all co-owners join as plaintiffs in the infringement suit.

VI. Receiving Paragraph IV Notices

Perhaps more so than in any other field of patent litigation, the receipt of a Paragraph IV notice letter should trigger a prompt reaction from NDA and patent owners to conduct essential diligence. In contrast to most patent cases, little if any information helpful to the patent owner will be publicly available. Instead, the patent owner must rely on a well-thought-out plan to assess whether and on what basis to file suit.

A. What to Look For

Both NDA owners and patent owners who license to NDA owners should develop a process for addressing Paragraph IV notice letters as soon as they contemplate listing patents in the Orange Book. The 45-day period from receipt of the notice letter runs quickly, and suit must be filed within that short period to obtain a statutory 30-month stay of FDA approval of the ANDA corresponding to the notice letter.[129] Strategies for conducting a pre-notice letter investigation, including Orange Book due diligence, handling confidentiality issues, and deciding on where and how to bring suit, are addressed in our first book;[130] here we will address the notice letter itself.

---

128. *See* Spine Solutions Inc. v. Medtronic Sofamor Danek USA Inc., 620 F3d 1305, 1317–18 (Fed. Cir. 2010) (parent company did not have standing to sue); *In re* Rosuvastatin Calcium Patent Litig., 719 F. Supp. 2d 388, 398–99 (D. Del. 2010) (agent did not have standing to sue); Rite-Hite Corp. v. Kelley Co., 56 F.3d 1538, 1553 (Fed. Cir. 1995) (en banc) (bare licensee did not have standing to sue).

129. 21 C.F.R. § 355(c)(3)(c).

130. Kenneth L. Dorsney, ANDA Litigation: Strategies and Tactics for Pharmaceutical Patent Litigators ch. 3, Notice and Pre-Filing Considerations (ABA Publishing 2012). Following the publication of that book, the U.S. Supreme Court unanimously held that an ANDA applicant may employ 21 U.S.C. § 355(j)(5)(C)(ii) and assert a counterclaim seeking correction of a use code listed in the Orange Book that inaccurately describes the brand's patent as covering a particular method of using a drug. Caraco Pharm. Labs. Ltd. v. Novo

Often, the Paragraph IV notice letter will focus on noninfringement of the patent(s) listed in the Orange Book. As noted earlier, claims for literal infringement and/or infringement under the doctrine of equivalents may be pursued by the patent owner and NDA holder. In particular, the doctrine of equivalents can be an important tool. Rarely does an ANDA filer pursue a generic product that is unambiguously covered by the claims of a formulation patent that covers the brand-name drug. Instead, ANDA filers often rely on a combination of claim construction arguments to narrow the Orange Book–listed patent claims and product modifications that fall outside the scope of the purported constructions.[131]

In addition to pursuing a specific active ingredient that is no longer covered by a patent, ANDA filers may strategically pursue specific formulations that have inactive ingredients different from those of the patented product. The ANDA filer may engage in design around efforts to avoid infringement—for example, by using different or additional excipients, including a filler, stabilizer, diluent, binder, or lubricant. The generic company may pursue a different form of administration or a different crystalline form of the active ingredient. Accordingly, it is important for the patent owner to obtain an accurate description and understanding of the composition of the ANDA product.

The Paragraph IV notice letter and its related documents, however, do not necessarily contain sufficient information for evaluating whether the ANDA product infringes the patent. In addition, obtaining and analyzing samples of the proposed ANDA product and/or its constituents (and any intermediary products) may be needed for ascertaining certain properties or functions of the ANDA product that are relevant to the asserted patent claims. This is especially true when the claim limitations are set forth by a function or property of the drug formulation as a whole or of the ingredients, while the corresponding data for the ANDA product are not necessarily available from the ANDA applicant.

In addition to the notice letter itself, publicly available documents may shed light on the ANDA filer's strategy. Often, the generic ANDA filer has filed its own U.S. or foreign patent applications to cover what it perceives to be an alternative formulation. Those patent documents can provide valuable insight concerning the noninfringement strategy. Additionally, the prosecution history of these patent applications may contain arguments to differentiate claims corresponding to the ANDA product from the patent holder's patent and offer further insight into the noninfringement strategy.

Nordisk AS, No. 10-844 (U.S. Apr. 17, 2012). Accordingly, the propriety of the patents listed in the Orange book should be revisited in connection with pre-suit diligence.

131. See Section 14(B)(II) for a discussion of claim construction issues.

Likewise, prior litigations involving the ANDA filer may shed light on its strategy, particularly with regard to drug products having a similar formulation or intended use. A generic may submit multiple ANDAs in certain areas over a period of time. The past conduct of the ANDA applicant may shed light on relevant technical features of the instant ANDA product and may help the patent owner gain insight into the ANDA applicant's strategies and legal theories for escaping patent infringement liability. Whether from the notice itself or from the notice and additional materials, a careful analysis should be conducted to ensure that a bona fide case of patent infringement exists.[132]

The Paragraph IV notice will not always focus on or even address infringement. A frivolous assertion of noninfringement can subject the ANDA applicant to sanctions.[133] Therefore, the ANDA applicant may choose to focus most or all of the notice letter on invalidity, as it may under the controlling statute and regulations.[134] In such cases, the patent owner should be careful to determine whether the claims can be validly construed and, at the same time, maintain breadth to establish infringement.[135] This is important since the ANDA filer may not be precluded from raising additional defenses if litigation is commenced.[136] Where validity is raised, evidence concerning secondary considerations should be analyzed.[137] In addition to the patent owner's own evidence, patent applications filed by the ANDA filer may be fertile ground for statements concerning secondary considerations of nonobviousness of the patent (e.g., by admitting a long-felt need).

B. Evaluation of Concerns

In addition to evaluating infringement, validity, and potential inequitable conduct issues, various antitrust and patent misuse claims can arise from ANDA litigation.[138] With a proper evaluation of these issues in hand, the NDA holder can evaluate how to respond to the Paragraph IV notice, either by preparing to initiate suit, seeking confidential access to the ANDA materials, or ignoring the notice and permitting the generic to launch. If expert analysis is needed to clarify the infringement analysis, a suitable expert should be retained.

The NDA and patent owners should also consider whether the current Orange Book listing is correct and complete and should delist or add

132. See Section 14(B)(IV) for a discussion of infringement.
133. Takeda Chem. Indus. Ltd. v. Mylan Labs. Inc., 549 F.3d 1381 (Fed. Cir. 2008).
134. 21 U.S.C. § 355(j)(2)(B)(iv); 21 C.F.R. § 314.95(c)(6).
135. See Section 14(B)(III)(C) for a discussion of validity and inequitable conduct.
136. See Dorsney, *supra* note 130, at ch. 3(II)(B).
137. See Section 14(A)(3) for a discussion of secondary considerations.
138. See Dorsney, *supra* note 130, at ch 13 for antitrust issues.

patents as needed. Patent ownership should be evaluated to ensure proper standing to bring suit. The NDA and patent owners should evaluate potential forums for litigation in view of the outcome of the infringement and validity due diligence process, as every forum has its own track record on those issues.[139] Finally, the anticipated timeline of events following receipt of the Paragraph IV letter should be considered, including by taking into account any anticipated additional ANDA filers.

VII. Conclusion

The receipt of a Paragraph IV notice letter presents both challenges and opportunities for the NDA and patent owners. Although the lack of publicly available infringing products leaves NDA and patent owners more dependent on information to be obtained from the ANDA applicant, there are steps that can be taken to ensure that meaningful diligence is conducted before a decision is made to file suit.

139. Venue is discussed in Chapter 16 of this book.

chapter 15

Market Entry Business Considerations for Generic Companies[1]

I. Brand-Name Drug Markets and Patent Expiry

Generic manufacturers have penetrated the drug market significantly since the adoption of the Hatch-Waxman Act and particularly over the past two decades.[2] For example, during the 1980s, the percentage of retail pharmacies' generic drug sales almost doubled.[3] Further, it has been estimated that between 2004 and 2010, generic drugs decreased branded drugs' market share by an average of 7.6 percent per year.[4] In 2002, the generic prescription dispensing rate reached 51 percent, and by 2012, this rate increased to 67 percent.[5] In 2012, roughly 84 percent of all prescriptions filled were generic, which accounted for 28 percent of consumer drug

1. Gregory J. Brodzik, Samantha G. Wilson, and James M. Lennon of Young Conaway Stargatt & Taylor LLP, Wilmington, Delaware.

2. *See* Henry G. Grabowksi & John M. Vernon, *Brand Loyalty, Entry, and Price Competition in Pharmaceuticals After the 1984 Drug Act*, 35 Duke J.L. & Econ. 331, 333 (1992) ("As a result of the 1984 law, generic products need only demonstrate bioequivalence to the pioneer's brand, and generic entry has increased significantly.").

3. Richard G. Frank & David S. Salkever, *Generic Entry and the Pricing of Pharmaceuticals*, 6 MIT J. Econ. & Mgmt. Strategy 75, 75 (1997).

4. Somnath Pal, *Shifts in the Generic-Drug Market: Trends and Causes*, U.S. Pharmacist (June 13, 2013), http://www.uspharmacist.com/content/s/253/c/41309/.

5. Murray Aitken et al., *Prescription Drug Spending Trends in the United States: Looking Beyond the Turning Point*, Health Affairs (Dec. 2008), http://content.healthaffairs.org/content/28/1/w151.full.

spending.[6] It is predicted that by 2017, the generic dispensing rate could be as high as 88 percent.[7]

Despite generic firms' continued success and erosion of the branded markets, a particular generic firm's decision to enter a particular market remains complex and depends on consideration of a multitude of factors, which are often interrelated. This section and those that follow attempt to distill and present some of the more vital considerations of a generic firm prior to market entry.

First, though it may seem like a matter of common sense, studies have confirmed that a generic firm's decision to enter a particular market is "*primarily influenced by drug markets' pre-entry revenues*."[8] Indeed, one study has shown that only a 1 percent increase in pre-entry revenue makes it 14 to 18 percent more likely a generic firm will enter a particular market.[9] Given that pre-entry revenue is a primary determinant of whether a generic firm may enter a particular market, it is worth discussing the current and possible future landscapes of various brand-name drug markets.

In 2012, consumer spending on brand drugs was roughly $230.2 billion.[10] Growth has been identified in the markets for drugs treating psychological ailments, nervous system malfunctions, pain, and respiratory disease.[11] On the other hand, markets for drugs providing allergy therapies, antibacterial drugs, and osteoporosis agents have recently encountered declines.[12] Tables 1 and 2 demonstrate the top-selling drugs in the United States in 2012 and the best-selling drugs of all time, respectively.

---

6. Michael Bartholow, *Top 200 Drugs of 2012*, PHARMACY TIMES (July 17, 2013), http://www.pharmacytimes.com/publications/issue/2013/July2013/Top-200-Drugs-of-2012.

7. Adam J. Fein, *A New Generic Wave Update, Courtesy of Express Scripts*, DRUG CHANNELS (Sept. 4, 2012), http://www.drugchannels.net/2012/09/a-new-generic-wave-update-courtesy-of.html.

8. Silvia Appelt, Entry and Competition in the Pharmaceutical Market Following Patent Expiry 34 (June 1, 2011) (dissertation, University of Munich) (emphasis added), *available at* http://edoc.ub.uni-muenchen.de/13108/1/Appelt_Silvia.pdf; *see also* Grabowski & Vernon, *supra* note 2, at 335; Fiona M. Scott Morton, *Entry Decisions in the Generic Pharmaceutical Industry*, 30 RAND J. ECON. 421, 430 (1999).

9. Appelt, *supra* note 8, at 11.

10. Bartholow, *supra* note 6.

11. *Id.*

12. *Id.*

Table 1
Top-Selling Drugs in the United States (2012)[13]

Brand Name	Sales Revenue	Drug Type	Primary Application(s)	Manufac-turer
Nexium	$5,989,000,000	Small Molecule	Erosive esophagitis	AstraZeneca
Abilify	$5,870,000,000	Small Molecule	Antidepressant	Otsuka America
Crestor	$5,092,000,000	Small Molecule	Dyslipidemia	AstraZeneca
Advair Diskus	$4,889,000,000	Small Molecule	Asthma and COPD	GlaxoSmith-Kline
Cymbalta	$4,720,000,000	Small Molecule	Depression and anxiety	Lilly
Humira	$4,609,000,000	Biologic	Rheumatoid arthritis	Abbvie
Enbrel	$4,337,000,000	Biologic	Autoimmune diseases	Amgen
Remicade	$3,876,000,000	Biologic	Autoimmune diseases	Centocor
Copaxone	$3,581,000,000	Small Molecule	Multiple Sclerosis (RRMS)	Teva CNS
Neulasta	$3,460,000,000	Biologic	Neutrophil stimulation after chemotherapy	Amgen
Singulair	$3,300,000,000	Biologic	Asthma	Merck & Comapany
Rituxan	$3,197,000,000	Biologic	Autoimmune diseases	Genetech
Plavix	$2,971,000,000	Small Molecule	Atherosclerosis	Bristol-Myers Squibb/ Sanofi-Aventis
Atripla	$2,899,000,000	Small Molecule	HIV	Bristol-Myers Squibb Gilead
Spiriva Handihaler	$2,833,000,000	Small Molecule	COPD	Boehringer Ingelheim

13. *Id.*; *Small Molecule Drugs Versus Biomolecular Drugs (Biologics)*, PORTFOLIO MGMT. SOLUTIONS, http://www.portfoliomanagementsolutions.com/the-organization-of-pharmaceutical-rd/small-molecule-drugs-versus-biomolecular-drugs-biologics/ (last visited Jan. 12, 2014) [hereinafter *Small Molecule Drugs*).

Table 2
Best-Selling Drugs of All Time[14]

Brand Name	Peak Year Sales Revenue	Drug Type	Primary Application(s)	Peak Year
Lipitor	$13,696,000,000	Small Molecule	Dyslipidemia	2006
Plavix	$9,318,000,000	Small Molecule	Artherosclerosis	2011
Humira	$9,265,000,000	Biologic	Autoimmune diseases	2012
Seretide	$8,148,000,000	Small Molecule	Asthma/COPD	2011
Enbrel	$7,830,000,000	Biologic	Autoimmune diseases	2011
Abilify	$7,363,000,000	Small Molecule	Schizophrenia	2011
Crestor	$7,021,000,000	Small Molecule	Dyslipidemia	2011
Rituxan	$6,798,000,000	Biologic	Autoimmune diseases	2011
Remicade	$6,782,000,000	Biologic	Autoimmune diseases	2011
Losec	$6,260,000,000	Small Molecule	Dyslipidemia	2000
Avastin	$6,216,000,000	Biologic	Colon, lung, rectal cancer	2010
Seroquel	$6,177,000,000	Small Molecule	Schizophrenia	2011
Diovan	$6,053,000,000	Small Molecule	Hypertension	2010
Herceptin	$5,947,000,000	Biologic	HER-2 and breast cancer	2011
Singulair	$5,479,000,000	Biologic	Asthma	2011

As reflected in the tables, several of the top-selling drugs in 2012 and of all time are "biologics." Biologics are large-protein molecules manufactured with DNA recombinant technology.[15] They include, for example, oncologics, which take a targeted approach to hindering the spread of cancer.[16] There

14. Simon King, *The Best Selling Drugs of All Time; Humira Joins the Elite*, FORBES (Jan. 1, 2013), http://www.forbes.com/sites/simonking/2013/01/28/the-best-selling-drugs-of-all-time-humira-joins-the-elite/. These drugs were "best selling" as of January 1, 2013.

15. Aitken et al., *supra* note 5.

16. *Id.*

has been an increasing emphasis on the development and marketing of biologics.[17] The Abbreviated New Drug Application (ANDA) process under the Hatch-Waxman Act, however, applies only to traditional small-molecule drugs and not follow-on biologics.[18] As a result, generic firms lack the incentive and resources to enter biologic markets, as they are unable to gain Food and Drug Administration (FDA) approval inexpensively and expeditiously.[19] Nevertheless, as reflected in Tables 1 and 2, there remain numerous top-selling small-molecule drugs, which continue to provide fertile ground for generic entry under the ANDA process. For example, Plavix, a small-molecule drug that generated $73 billion in revenue over an 11-year period, is predicted to accrue only $16.1 billion between 2013 and 2018 as the result of generic competition.[20]

Another noteworthy trend relates to the number of blockbuster brand-name drugs on the market and forecasted to enter the market. Between 1997 and 2006, the number of blockbuster brand-name drugs (selling in excess of $1 billion)[21] increased more than eightfold—from 6 to 52. In 2007, however, the number of blockbuster drugs began to decline, dropping from 52 to 48.[22] The number of blockbuster drugs on the market and new blockbuster drugs appears to be continuing on a downward trend. That is, while many of the brand-name drugs that appear in Table 2 represent the "golden age" of blockbuster drugs, brand-name manufacturers are not able to develop and bring to market blockbusters with the same consistency as in the past.[23] In fact, one commentator has noted that there appears to be only one new drug in the pipeline that is estimated to accumulate revenue

17. *See* King, *supra* note 14 (noting "robustness" of "biologic franchises" and forecasting Humira and Enbrel, two biologics, to have revenues of $58 billion and $45 billion, respectively, between 2013 and 2017); Aitken et al., *supra* note 5 (noting "real spending" on biologics grew at average annual rate of 19.3 percent between 2002 and 2007).

18. *See* Sanya Sukduang & Jonathan R. Davies, *Follow-On Biologics: A Patent Litigation Perspective*, Finnegan, Henderson, Farabow, Garrett & Dunner LLP (Aug. 29, 2009), http://www.finnegan.com/resources/articles/articlesdetail.aspx?news=1da6f430-bba4-405a-a63d-1c31986e330b ("[T]he [Hatch-Waxman] Act did not create a framework for the approval of follow-on biologics, which are also referred to as 'biosimilars' and 'generic biologics.'").

19. *See* Jeremiah Helm, *The Patent End Game: Evaluating Generic Entry into a Blockbuster Pharmaceutical Market in the Absence of FDA Incentives*, 14 Mich. Telecomm. Tech. L. Rev. 175, 178 (2007) (explaining that requiring a generic firm to conduct its own clinical trials is a "major barrier to generic entry because a generic firm spending the money for a clinical trial would, at best, share duopoly profits" with the brand-name manufacturer); *see also* 21 U.S.C. § 355(j) (providing process for generic firm to gain market approval by demonstrating generic bioequivalence to the brand-name drug).

20. *See* King, *supra* note 14.

21. Aitken et al., *supra* note 5.

22. *Id.*

23. *See* King, *supra* note 14 ("[T]he industry is no longer able to deliver products that hold this level of commercial potential to the market with nearly as much regularity."); Aitken et al., *supra* note 5 ("As more blockbusters go off patent and fewer new ones are developed, the share of sales attributable to blockbuster molecules will likely decline still further.").

of $5 billion a year.[24] While the decline in blockbusters may be a testament to the success of generic firms' continued market penetration,[25] it remains questionable whether this decline may ultimately have a negative impact on the generic industry by limiting potentially large, new markets generic firms could enter and benefit from via the ANDA process.

Although pre-entry revenue has been found to be the most influential factor when determining whether a generic firm will enter a particular market, it may also be necessary to consider this factor in conjunction with the date the patents protecting a particular brand-name drug or pool of drugs will expire. Specifically, if the generic firm does not challenge the validity of the patents in court or claim noninfringement, the generic manufacturer must wait for the patents protecting the brand-name drug to expire to enter the market.[26]

Brand-name drugs' loss of patent protection spiked in 2012, with approximately 40 drugs, representing $35 billion in sales, losing patent rights.[27] In 2013, it was approximated that 14 drugs would lose protection, representing $17 billion in sales.[28] In 2014 and 2015, it has been estimated that brand-name drugs with revenues of $14.8 and $14.1 billion, respectively, would be impacted by a loss of patent rights.[29] From an alternative, broader perspective, between 2012 and 2016, brand-name drugs with revenues of $77.2 billion have lost or will lose patent protection. Interestingly, while 117 drugs will be affected by the loss of patent rights during this period, only ten of these drugs will account for 51 percent of the brand revenues lost.[30]

Although the immediate discussion provides details regarding patent expiration and generic entry, there may exist significant benefits to challenging a brand-name drug's patent prior to expiration, as discussed in the following section.

II. First-Filer and Early Entrant Advantage

The Hatch-Waxman Act permits a generic firm to file an ANDA before a brand-name drug's patent expires if the generic firm submits a Paragraph IV Certification, in which the applicant must certify that the patent at

24. *See* King, *supra* note 14. It was specified that this figure of $5 billion per year was projected for the "medium term." *Id.*

25. *See* Aitken et al., *supra* note 5 (indicating the decrease in blockbusters may be attributable, in part, to generic penetration).

26. *See* 21 U.S.C. § 355(j)(2)(A)(vii)(IV); *see also* Helm, *supra* note 19, at 179.

27. Pal, *supra* note 4.

28. *Id.*

29. Fein, *supra* note 7.

30. *Id.*

issue is either invalid or not infringed by the generic drug.[31] In fact, to foster competition and early generic entry, the Hatch-Waxman Act incentivized the submission of Paragraph IV Certifications by granting the first generic firm to file a successful Paragraph IV Certification (the first-filer) a 180-day "exclusivity" period during which time no other ANDA-filing firm may market the first-filer's drug dose.[32] The frequency of Paragraph IV filings has increased significantly: from 1984 to 1989, only 2 percent of ANDA submissions contained Paragraph IV Certifications, whereas 20 percent of ANDA submissions between 1998 and 2000 included certifications.[33] The number of 180-day exclusivity periods granted has also increased from none between 1992 and 1998 to 31 between 1998 and 2002.[34] Because the competition faced by the first-filer during the 180-day exclusivity period is limited, the first-filer holds the potential to accrue significant revenue during that time.

In addition to the revenues during the exclusivity period, studies have indicated that there may be other substantial and lasting competitive advantages to being an early entrant in the market. That is, if a generic manufacturer is able to gain a several-month lead on its rivals, that manufacturer may enjoy a "first mover" or early entrant advantage, which grants the generic manufacturer the ability to capture and maintain a larger market share with perhaps greater prices.[35] For example, one study demonstrated that of the 18 generic drugs sampled, there was almost a 50 percent difference between the most costly and the cheapest drugs. Even with prices 50 percent higher, the more expensive generics were nevertheless able to capture and maintain a significant share of the generic market.[36] The significant market share maintained—even with higher prices—was explained at least in part by the first mover advantage.[37] A subsequent study, which largely confirmed these results, explained that as a result of a "first mover" advantage, first generic entrants achieved on average a 9.9 percent market share in the generic market segment in the five-year time period studies, whereas subsequent entrants during that time period obtained only a 3.8 percent market share.[38]

This first mover or early entrant advantage has been traced to at least the following concepts: (1) once a pharmacy begins stocking a

31. *See* 21 U.S.C. § 355(j)(2)(A)(vii)(IV).

32. *See id.* § 355(j)(5)(B)(iv).

33. Ernest R. Berndt et al., *Authorized Generic Drugs, Price Competition, and Consumer's Welfare*, 26 Health Aff. 790, 791 (2007).

34. *Id.*

35. Grabowski & Vernon, *supra* note 2, at 346; Appelt, *supra* note 8, at 49.

36. Grabowski & Vernon, *supra* note 2, at 346.

37. *Id.*

38. Appelt, *supra* note 8, at 49.

particular generic drug, it will desire to keep stocking that drug given its identifiable size, shape, and color;[39] (2) patients develop a "habit" to a particular generic the longer and more frequently they receive a drug; and (3) patients and pharmacies have a low sensitivity to price differentials between generic drugs.[40]

Generic firms, however, must weigh the potential exclusivity period benefits and other first mover advantages against possible detriments in the event a brand-name firm has or will release an "authorized generic." Authorized generics are generic products released by brand-name manufacturers, often via a subsidiary or licensing agreement, that rely on that brand manufacturer's New Drug Application (NDA).[41] Authorized generics may be marketed during a first-filer's exclusivity period[42] and may decrease the value of the 180-day exclusivity period for the first-filer. For example, the first-filer may not be able to gain market share as rapidly because substitution laws would permit pharmacies to substitute the brand-name drug with the authorized generic.[43] Further, generic firms would realize a decreased revenue stream during this period.[44] However, studies have confirmed that in general, an authorized generic does not significantly deter generic firms from filing Paragraph IV Certifications.[45] This has been attributed to the fact that in most instances, brand-name manufacturers release authorized generics in markets where there has been a high-grossing brand-name drug.[46] Consequently, even if the authorized generic renders a negative impact on an early entrant's revenue, there nevertheless remains revenue sufficient to justify a generic's patent challenge and early entry.[47]

III. Generic Entry's Effect on Branded and Generic Prices

As discussed, a primary determinant of whether a generic firm will enter a particular market is the market's pre-entry revenue. The more pre-entry revenue a market accrues, however, the greater the number of generic firms

39. Grabowski & Vernon, *supra* note 2, at 346 n.38.
40. Appelt, *supra* note 8, at 73, 75.
41. Berndt et al., *supra* note 33.
42. *Id.*
43. Helm, *supra* note 19, at 189.
44. *See id.*
45. Ernest R. Berndt et al., *Do Authorized Generic Drugs Deter Paragraph IV Certifications? Recent Evidence* 16 (Working Paper 2007), *available at* http://www.analysisgroup.com/uploadedFiles/Publishing/Articles/PhRMA_Authorized_Generic_Entry.pdf.
46. *Id.*
47. *Id.*

that market will attract, which may negatively impact those generic firms' revenues and profits.[48] This resultant decrease in revenue is particularly problematic for generic manufacturers who must, by law, market bioequivalent products and are therefore often already competing on price.[49] Given this landscape, it is important to understand the effect that generic entrants have on both brand-name and generic drug prices in a particular market.

To a certain extent, a brand-name drug's price behavior in response to generic entry is counterintuitive. That is, while there is a decrease in the brand-name drug's market share, the price of branded drugs has been shown to increase in response to generic entry.[50] This phenomenon has been explained with the "segmented-market model."[51] Rather than compete with generic firms on price, the segmented market model provides that brand manufacturers tend to split the market into price-sensitive and price-insensitive consumers.[52] Those price-sensitive consumers will purchase the generic drugs on the market, whereas the price-insensitive (i.e., "brand loyal") consumers will continue to purchase the branded drug at an increased price.[53] Brand manufacturers have found that, in general, they are able to more efficiently recoup revenue from their lost market share by maximizing profits within the price insensitive portion of the market.[54] One study that examined 32 brand-name drugs with an average revenue of $80 million per year prior to patent expiration found that, on average, each new generic entrant equated roughly to an increase in the brand-name price of 0.7 percent.[55] The same study found that five years after generic entry, the price of a brand-name drug increased on average by 50 percent.[56]

By contrast, generic drug market entry eventually leads to significant decreases in generic drug prices.[57] If at first only one generic manufacturer enters the market during a period of exclusivity resulting from its successful Paragraph IV Certification, that sole generic entrant's price is generally only 10 to 20 percent less than that of the branded drug.[58] However, as more generic firms enter the market following the period of exclusivity, additional downward pressure is exerted on the generic price.

48. *See* Morton, *supra* note 8, at 430 ("There is a risk to profits from excessive entry.").

49. *See* Appelt, *supra* note 8, at 7 (explaining that some literature assumes generic firms are "Bertrand competitors," providing an "undifferentiated drug product at identical cost . . . and competing exclusively on price").

50. Frank & Salkever, *supra* note 3, at 76.

51. Grabowski & Vernon, *supra* note 2, at 339.

52. Helm, *supra* note 19, at 185; Grabowski & Vernon, *supra* note 2, at 340.

53. Helm, *supra* note 19, at 185; Grabowski & Vernon, *supra* note 2, at 340.

54. Helm, *supra* note 19, at 194.

55. Frank & Salkever, *supra* note 3, at 86–87.

56. *Id.* at 83.

57. *Id.* at 76.

58. Berndt et al., *supra* note 33, at 792.

For example, an earlier study showed, based on its sample, that for each additional generic entrant, there is a fall in the average price of the generic drug between 5.6 percent and 7.2 percent.[59] The same study additionally concluded that three years after generic entry, the average price of the generic drugs that have entered a particular market is less than 50 percent of the price of the market's brand-name drugs.[60]

Further studies have confirmed the view that the generic market segment "represents a highly competitive fringe to the brand-name drug market,"[61] especially with respect to price, and have further enhanced our understanding of generic entrants' effect on generic price behavior. A relatively recent study, which considered data from1999 to 2003, found that after the fourth or fifth generic entrant, the effect of additional generic entrants on price is generally "negligible."[62] More particularly, this study found that at 24 months subsequent to the first generic entrant, the following price behavior was demonstrated:

> [O]nly one drug with fewer than five generic entrants (out of seven) had a generic-to-brand ratio below .37; all ten drugs with more than five generic entrants had generic-to-brand price ratios falling below .25, and no discernible downward trend in ratios appearing as the number of generic entrants increased further.[63]

This trend—where downward pressure on generic price appears to level off around four to five generic entrants—may help to explain the pattern of generic entry over time that has been observed. Specifically, literature shows that, on average, five generic firms penetrate the market during the first year after a brand-name drug's patent expiry, but in the following three years, only three generic firms in total will enter the same market.[64]

IV. Key Generic Manufacturer Costs

The Hatch-Waxman Act significantly decreased the cost and time for a generic manufacturer to bring a drug to market, allowing the generic manufacturer to rely on the brand-name drug's safety and efficacy data.[65] In general, a brand-name manufacturer must pursue extensive research

59. Frank & Salkever, *supra* note 3, at 88.
60. *Id.* at 83.
61. *Id.*
62. Berndt et al., *supra* note 33, at 792.
63. *Id.* at 792–93.
64. Frank & Salkever, *supra* note 3, at 83–85.
65. *See* Morton, *supra* note 8, at 422.

and development and must complete the expensive and time-consuming clinical trials pursuant to an NDA to gain FDA approval.[66] Indeed, during the 1990s, the average cost of gaining FDA approval was $335 million for drug development and $467 million for clinical and other forms of testing.[67] It has been estimated that the clinical trial process averaged eight years to complete during this time.[68]

By contrast, under the Hatch-Waxman Act, a generic manufacturer need only file an ANDA and demonstrate the generic's bioequivalence to the brand-name drug to gain FDA approval. Some estimates place the typical cost of filing an ANDA and completing its required testing at less than $1 million;[69]other estimates are as high as $5 million.[70] In any event, the typical cost to ANDA filers is relatively low compared to NDA filing, which includes the generic manufacturer's costs to research how to manufacture the drug in addition to conducting studies to demonstrate the drug's bioequivalence to the brand-name drug with respect to strength, dosage form, and route of administration.[71] The ANDA process requires the FDA to inspect a generic manufacturer's facilities and requires that the generic manufacturer provide a number of preliminary batches of the drug.[72] In other words, the generic firm must purchase and verify the operability of new equipment and be prepared to make the product months before it reaches the market.[73] Further, the FDA requires the ANDA applicant to pay certain costs involved with the application process, including, for example, facilities fees (up to $220,000)[74] and a drug master file fee (around $21,000).[75]

Although the Hatch-Waxman Act has drastically decreased the amount of time to bring a generic drug to market, the timeline for ANDA approval can still be "considerable and variable."[76] That is, while bioequivalence

66. David Reiffen & Michael R. Ward, *Generic Drug Industry Dynamics*, 87 REV. ECON. & STAT. 37, 38 (2005).

67. *Id.*

68. *Id.*

69. Berndt et al., *supra* note 33, at 792; *see also* Reiffen & Ward, *supra* note 66, at 38 (providing cost estimate for ANDA in earlier 1990s at $603,000).

70. Jason Chew, *Cubist Pharmaceuticals vs. the Hatch-Waxman Act*, SEEKING ALPHA (June 23, 2010), http://seekingalpha.com/article/211404-cubist-pharmaceuticals-vs-the-hatch-waxman-act; *see also* DAVID HARDIN, GAINING MARKET SHARE IN THE GENERIC DRUG INDUSTRY THROUGH ACQUISITIONS AND PARTNERSHIPS (Dec. 2010), *available at* http://thomsonreuters.com/products/ip-science/04_013/newport-deals.pdf.

71. *See* Morton, *supra* note 8, at 423.

72. *Id.*

73. *Id.*

74. *Facility Fees*, U.S. FOOD & DRUG ADMIN. (Aug. 20, 2013), http://www.fda.gov/ForIndustry/UserFees/GenericDrugUserFees/ucm319566.htm.

75. *Drug Master File Fee*, U.S. FOOD & DRUG ADMIN. (Sept. 9, 2013), http://www.fda.gov/ForIndustry/UserFees/GenericDrugUserFees/ucm319567.htm.

76. Reiffen & Ward, *supra* note 66, at 38.

studies generally require 12 to 24 months, an ANDA submission requires an additional one to two months;[77] the overwhelming majority of ANDA applications are found deficient on their first submission and must be resubmitted with additional testing or information.[78] A generic firm, therefore, must account for resubmission costs. In fact, in many instances, an ANDA is not approved until it is resubmitted two or three times.[79] It was found that between 1984 and 1994, the average time between ANDA submission and approval was 19 months, and it has been advised that a generic firm should plan for two to three years between when it begins ANDA preparation to when its drug may finally reach the market.[80]

It is apparent that before a generic firm may begin marketing its product, it must "sink" significant entry costs.[81] Further, because the FDA does not publicly disclose when a generic firm has submitted an ANDA, generic firms must sink these entry costs before they know how many generic competitors are aiming to enter a particular market.[82] In light of these sunk costs and the uncertainty surrounding market entry, including the uncertainty surrounding the number of generic competitors, it is important for generic firms to enter markets where their entry costs are minimized (or where they may increase the price of their products), as discussed in the next section.

Further, if the FDA requires that a drug have a Risk Evaluation and Mitigation Strategies (REMS) program,[83] then a generic manufacturer must also consider the cost of such a program. REMS programs may require package inserts, medication guides, or communication plans with prescribers to make sure that prescribers and patients are aware of the

77. *ANDA vs. 502(b)(2)—When and Why?*, WEINBERG GROUP (Dec. 21, 2011), http://www.weinberggroup.com/pdfs/ANDA_vs_505b2_When_and_Why_Webinar_Slides.pdf.

78. Reiffen & Ward, *supra* note 66, at 38.

79. *Id.*

80. *Id.*

81. *See* Morton, *supra* note 8, at 423; Reiffen & Ward, *supra* note 66, at 38.

82. *See* Morton, *supra* note 8, at 423; Reiffen & Ward, *supra* note 66, at 38.

83. In 2007, Title IX of Food and Drug Administration Amendments Act (FDAAA) amended federal law to authorize the FDA to require and enforce REMS programs if the FDA deemed they were necessary to ensure that the benefits of a drug outweighed its risks. *See* Kuyler Doyle et al., *REMS: The New Reality*, CAMPBELL ALLIANCE 1, http://www.campbellalliance.com/articles/campbell_alliance_REMS_article.pdf (last visited Jan. 12, 2014). Factors that the FDA may consider in determining whether a REMS program is needed are as follows: "the population size of the users," "the seriousness of the disease," "the anticipated benefit of the drug and its treatment duration," "the severity of adverse events," and "whether the drug is a new molecular entity." Rachel Bohard, *Risk Evaluation and Mitigation Strategies (REMS)*, PURDUE U. C. PHARMACY 7, http://www.pharmacy.purdue.edu/cmsa/education/presentations/REMS.pdf (last visited Jan. 12, 2014) (PowerPoint presentation). Biologics and specialty drugs are most likely to need REMS programs. *See* Heather Morel & Regina Murphy, *Dealing with REMS Challenges in Drug Commercialization*, MCKESSON 3, http://www.mckesson.com/uploadedfiles/mckessoncom/content/manufacturers/_body_components/_right_rails/rems%20article%20reprint-8-09.pdf (last visited Jan. 12, 2014).

risks associated with a product; more aggressive programs may include "Elements To Assure Safe Use (ETASU) such as physician training and certification, patient data collection, and restricted product distribution."[84]

When deciding whether to enter a market, a potential generic manufacturer should research what REMS programs, if any, are in place for a target drug. A generic manufacturer who has submitted an ANDA that is the subject of a product requiring a REMS program, for example, must implement any mandated medication guides, package inserts, or ETASUs.[85] While REMS programs restricting distribution of a product cannot be used to block or delay generic competition,[86] the cost of a REMS program may still discourage generic market entry in a number of ways.

First, REMS programs will increase the cost of manufacturing and distributing a drug. Although it is difficult to quantify the cost of REMS programs,[87] one study estimates that REMS costs for each line item range from $5,000 to $500,000 for manufacturers and from $5,000 to $1,000,000 for distributors.[88] Second, aggressive REMS programs may be particularly labor-intensive for generic firms because where ETASUs are required, brand and generic manufacturers must use a single, shared system of ETASUs, unless the burden of that implementation outweighs its benefits or if the original manufacturer's REMS program involves patent-protected methods.[89] On the other hand, where a single, shared system is not required, "a novel [and patented] REMS program could remain a barrier to entry, point of market differentiation, or source of revenue through license agreements."[90] Similarly, "[a] well designed REMS registry and information network can be an effective means of brand differentiation and could, depending on the level of unmet need, potentially allow for premium pricing."[91] Finally, REMS programs' limits on distribution could impede ANDA filings by making it more difficult for generic competitors to obtain the brand-name product in order to perform comparative evaluations.[92]

---

84. Deloitte Dev. LLC, *Challenge or Opportunity? Proactively Planning and Managing a REMS*, DELOITTE 3 (2011), http://www.deloitte.com/assets/Dcom-UnitedStates/Local%20 Assets/Documents/us_lshc_REMS_031011.pdf.

85. 21 U.S.C. § 355-1(i)(1).

86. Kathleen M. Sanzo, *Regulatory Considerations and Evolving Issues Concerning REMS Programs*, MORGAN LEWIS & BOCKIUS 20 (Aug. 2009), http://www.morganlewis.com/pubs /RegulatoryConsiderationREMSPrograms_Aug2009.pdf (PowerPoint presentation).

87. Bohard, *supra* note 83, at 13.

88. George Miller, REMS Costs Calculated in Study, FIERCE PHARMA MANUFACTURING (June 27, 2011), http://www.fiercepharmamanufacturing.com/story/rems-costs-calculated -study/2011-06-27 (describing study by the Center for Healthcare Supply Chain Research and its partner of the Healthcare Distribution Management Association).

89. 21 U.S.C. § 355-1(i)(1)(B).

90. Doyle et al., *supra* note 83, at 5.

91. Morel & Murphy, *supra* note 83, at 5.

92. Doyle et al., *supra* note 83, at 5.

V. Competitive Advantage: Prior Experience and Branding

A generic firm may gain a cost advantage by having similar past experiences with a particular drug or therapy.[93] Specifically, an ANDA may cost less if the generic firm already has in place the required equipment and procedures.[94] Similarly, if a generic firm has experience with using a particular ingredient, the ANDA's cost may decrease because the firm will already have a supplier certified by the FDA and a team of researchers familiar with the properties of the supplier's ingredient.[95] Moreover, the FDA standards for body absorption to demonstrate a drug's bioequivalence vary per therapeutic class.[96] Thus, a research team's familiarity with a particular therapeutic class may decrease costs.[97] Finally, if a generic firm is already distributing drugs from a particular therapeutic class, costs related to finding customers for a new generic drug falling within the scope of that therapeutic class may be reduced.[98]

Further, a generic firm may be permitted to charge an increased markup on its product as the result of establishing "distribution or reputation economies of scope."[99] More specifically, a firm may charge higher prices by grouping together similar products, taking advantage of a common brand (i.e., "umbrella brands").[100] For example, when a generic firm establishes a common brand or drug family, retail establishments may in turn develop a preference to purchase that single generic firm's complete line, as purchasing agents will already have experience with that generic manufacturer.[101] Further, a common or umbrella brand may positively impact the public's perception of a particular drug falling under that brand and may increase a generic firm's investment in quality.[102] Indeed, studies have demonstrated that the size of a brand's umbrella directly correlates with the price a generic manufacturer is able to charge for a drug falling under that umbrella.[103]

---

93. *See* Morton, *supra* note 8, at 421, 428, 431; Appelt, *supra* note 8, at 111.
94. *See* Morton, *supra* note 8, at 428.
95. *Id.*
96. *Id.*
97. *Id.*
98. *Id.*
99. *Id.* at 428–29.
100. *Id.* at 422, 428–29, 436; Appelt, *supra* note 8, at 111.
101. *Id.* at 428–29.
102. Appelt, *supra* note 8, at 111.
103. *Id.*

chapter 16

Intellectual Property Considerations for Generic Product Opportunities[1]

The Hatch-Waxman Act is a complex federal statute that weaves together the Food and Drug Administration (FDA) generic drug approval processes with the U.S. patent system.[2] While the act seeks to incentivize brand companies (brands) to innovate new products and to create a system providing consumer access to affordable medicine, it also presents a number of potential pitfalls for generic drug applicants (generics) unaware of the statute's intricacies.

This chapter focuses on the interplay between generics, FDA procedures, and U.S. intellectual property law.[3] It first discusses the different kinds of non-patent exclusivities, commonly referred to as Reference Listed Drug (RLD) exclusivities, and how patent law considerations affect RLD exclusivities. It then explains strategies generics should consider in light of RLD exclusivities. Next, it details the types of patents covering RLDs, analyzes these patents' strengths and validity, and provides strategies for potential generics to maximize their product opportunities against brands and other generic competitors. Last, the chapter considers design around opportunities alongside patent office proceedings and examines how the America Invents Act (AIA) affects the relevant Hatch-Waxman provisions.

The Hatch-Waxman Act presents generics with two nonexclusive filing options for an RLD: a 505(b)(2) "paper" application[4] and an Abbreviated New Drug Application (ANDA).[5] Both filing options mirror one another

---

1. H. Keeto Sabharwal, Daniel E. Yonan, Sarah I. Danley, Diane E. Ghrist, and Jon Boljesic of Sterne Kessler Goldstein & Fox, Washington, D.C.

2. 21 U.S.C. § 355 (2012).

3. For a detailed analysis of the requirements of the FDA RLD exclusivity provisions, see Chapter 8.

4. Federal Food, Drug, and Cosmetic Act § 505(b)(2), 21 U.S.C. § 355(b)(2).

5. Abbreviated New Drug Application, 21 U.S.C. § 355(j). The essential differences between the two applications are that the 505(b)(2) application relies on the clinical

and allow the generic filer to challenge the validity of patents covering the RLD. The Hatch-Waxman Act requires both 505(b)(2) and ANDA filers to certify that:

> to the best of [their] knowledge, with respect to each patent which claims the listed drug . . .
>
> (I) that such patent information has not been filed,
>
> (II) that such patent has expired,
>
> (III) of the date on which such patent will expire, or
>
> (IV) that such patent is invalid or will not be infringed by the manufacture, use, or sale of the new drug for which the application is submitted[.][6]

The most litigated certification is Paragraph IV because it directly challenges the patent claims granted by the U.S. Patent and Trademark Office (USPTO) and triggers the artificial act of infringement that gives rise to a cause of action.[7] Therefore, the Hatch-Waxman Act expressly ties the filing of a generic drug application with the FDA to the patent system and its corresponding litigation proceedings.

I. Reference Listed Drug Exclusivities

The Hatch-Waxman Act created non-patent exclusivities, commonly referred to as Reference Listed Drug (RLD) exclusivities, to preserve incentives for brands to continue innovating new products. RLD exclusivities prevent generics from having their products reviewed by or filed with the FDA, which allows an RLD holder to monopolize the market through regulations distinct from, yet parallel to, the patent system.[8] The FDA defines an RLD as an "approved drug product to which new generic

investigations conducted by the RLD holder not the generic, that the applicant did not obtain rights from the RLD holder, and that 505(b)(2) does not require a generic to make an identical product to the RLD (e.g., a different salt may be substituted or the product may comprise a different dosage form). An ANDA, on the other hand, requires a filer to show a nearly identical product to the RLD (i.e., bioequivalence and the same active ingredient, dosage form, strength, and labeling). However, an ANDA need not have the same inactive ingredients as an RLD. 21 U.S.C. § 355(j)(2)(A).

6. 21 U.S.C. § 355(b)(2)(A)(i)–(iv), (j)(2)(A)(vii)(I)–(IV).

7. 35 U.S.C. § 271(e)(2)(A).

8. RLD exclusivities provide statutory exclusivity specific to the FDA and are distinct from patent exclusivities that extend the term of a patent. If obtained, the RLD holder can maintain RLD exclusivity even when a patent is found invalid. Patent term extensions are governed under 35 U.S.C. § 156(g)(6) in part II of the Hatch-Waxman Act and are granted

versions are compared to show that they are bioequivalent"[9] and exhaustively catalogs them in the FDA publication *Approved Drug Products with Therapeutic Equivalence Evaluations* (Orange Book).[10]

Aside from encouraging innovation, the enactment of RLD exclusivities was also justified based on drug safety and effectiveness.[11] The FDA was concerned that adverse events and serious side effects from drugs that did not appear in the clinical trials would best be handled by the brand due to its familiarity with the drug's pre-approval testing. Brands are also obligated to submit adverse reaction reports to the FDA within 15 days of being notified by a patient's doctor, an additional regulatory restriction balanced with additional FDA exclusivity periods. Congress also wanted to provide incentives to innovate drugs when the drug is not patented. This combination of regulatory incentives bolstered the passage of RLD exclusivities. It also limited generics by delaying market entry until the brand drug has been on the marketplace for a number of years.

Generics must be cognizant of RLD exclusivities because RLD exclusivities determine *when* an ANDA can be first reviewed by or filed with the FDA. And timing is critical. Missing the end of RLD exclusivity risks losing the coveted legal status of an ANDA applicant as the first generic to file with the FDA, which later entitles the ANDA holder to 180 days of marketing exclusivity.[12] Although the respective effective dates of RLD exclusivity are listed in the Orange Book for each drug, it is essential to learn how the various categories of RLD exclusivity and the Hatch-Waxman Act interact. Additionally, the various RLD exclusivities can affect a generic company's ANDA or a 505(b)(2) application in different ways. Thus, familiarity with this process enables a generic to best strategize obtaining first to file status. While the Hatch-Waxman statute provides for a generic to receive 180 days of marketing exclusivity, it does not prevent a New Drug Application (NDA) holder from creating a "brand-generic" version of its own drug to market concurrently with the generic.[13]

---

for the duration that the patent was subject to FDA regulatory review, not to exceed five years. For a detailed analysis of patent term extensions, see Chapter 8.

9. *Drugs@FDA Glossary of Terms*, U.S. Food & Drug Admin., http://www.fda.gov/Drugs/InformationOnDrugs/ucm079436.htm#R (last updated Feb. 2, 2012).

10. *Orange Book: Approved Drug Products with Therapeutic Equivalence Evaluations*, U.S. Food & Drug Admin., http://www.accessdata.fda.gov/scripts/cder/ob/default.cfm (last updated May 17, 2013).

11. *Drug Legislation: Hearing on H.R. 3605 Before the Subcomm. on Health & the Env't of the H. Comm. on Energy & Commerce*, 98th Cong. 7 (1983) (statement of Mark Novitch, MD, Deputy Commissioner of the FDA).

12. For a detailed analysis on the marketing considerations of RLD exclusivities, see Chapter 15.

13. Teva Pharm. Indus. v. Crawford, 410 F.3d 51, 55 (D.C. Cir. 2005).

A brand can obtain one or more RLD exclusivities on its drug product through the FDA approval process via an NDA, a supplemental NDA (sNDA) achieved through new clinical investigations,[14] or a citizen petition or letter filed with the FDA seeking administrative action. These options provide brands with a series of options to stall generic competition. RLD exclusivities pose a dynamic obstacle for generics because the exclusivity can be granted throughout the entirety of the drug product's life on the market, including after patent expiry.[15] Brands also have the option to waive an exclusivity, although this is a rarely invoked option. RLD exclusivities take two different forms: the stricter "filing exclusivity" prevents a generic drug application from being reviewed by the FDA, and the "data (or approval) exclusivity" allows a generic drug application to be reviewed by the FDA but prevents that application from receiving final approval.[16]

The current statutorily effective RLD exclusivities are New Chemical Entity (NCE) exclusivity; New Clinical Investigation (NCI) exclusivity; orphan drug exclusivity; and pediatric exclusivity. In the past, certain antibiotics and enantiomers were distinctly granted periods of marketing exclusivity that have since expired. These provisions were allowed to expire because they had been subsumed by the five- or three-year exclusivities provided to 505(b)(2) applications. The following subsections focus on ANDA litigation and provide guidance as to how the FDA grants market exclusivity and how courts treat the exclusivity provisions.

A. New Chemical Entity Exclusivity (Five Years)

New Chemical Entity (NCE) exclusivity provides a five-year filing exclusivity for any drug approved under Section 505(b) of the Federal Food, Drug, and Cosmetic Act.[17] This exclusivity includes both NDAs approved under either 505(b)(1) or 505(b)(2) and applies to both generic 505(b)(2) and ANDA applicants. NCE exclusivities are the strongest form of RLD exclusivity because an NCE "contains no active moiety that has been approved by the FDA in any other application submitted under section 505(b)."[18]

---

14. A supplemental NDA refers to the submission a brand company must make to the FDA to receive FDA approval of label changes for a drug in order to market a new dosage, change in manufacturing, a new patient population, or a new formulation. *Drugs@FDA Glossary of Terms*, *supra* note 9 (last visited July 29, 2013).

15. *See also* ViroPharma Inc. v. Hamburg, 898 F. Supp. 2d 1 (D.D.C. 2012) (implying that a brand has the possibility of obtaining new clinical investigation exclusivity even after a product has been available for generic entry for multiple years).

16. Shashank Upadhye, Generic Pharmaceutical Patent and FDA Law § 12:1, at 660 (2012).

17. 21 C.F.R. § 314.108(b)(2) (2012); 21 U.S.C. § 355(b)(5), (j)(5).

18. 21 C.F.R. § 314.108(a); *see* Actavis Elizabeth LLC v. U.S. Food & Drug Admin., 625 F.3d 760 (D.C. Cir. 2010) (defining what constitutes an NCE exclusivity under the Hatch-

Because new active moieties themselves are often new chemical innovations, they typically result in the strongest compound patents. Therefore, it is difficult for a 505(b)(2) or ANDA applicant to file a Paragraph IV Certification challenging the patent validity of an NCE with the FDA.

However, NCEs are not entirely immune from generic challenges. An NCE exclusivity can be approved by the FDA even when a patent is found invalid or is not granted to the brand by the USPTO.[19] Cases exist in which the claims underlying a patent are weak and vulnerable to attack under provisions of the Patent Act even though the NCE contains a new active moiety. The act incentivizes generics to challenge the underlying patents of NCE exclusivities by slashing the exclusivity by one year if the patent is found invalid.[20] This kind of challenge results in what is commonly referred to as an NCE-1 four-year exclusivity and is accomplished by filing a 505(b)(2) application or ANDA with a Paragraph IV Certification.

Although the NCE-1 exclusivity can be advantageous for a generic, it also has concurrent risks. In particular, if a generic files an application containing a Paragraph IV Certification and the generic is subsequently challenged in district court and subjected to the 30-month stay after the NCE-1 exclusivity expiry date, then the last year of exclusivity will be attached to the end of the 30-month stay. Essentially, if a generic chooses to challenge the patent validity of an NCE-covered drug, the filing exclusivity may ultimately be extended from five to seven and a half years if the brand files an infringement action against the generic. This extension prevents the ANDA from being reviewed by the FDA until after the seven and a half years have expired because of the strict filing exclusivity. However, if the generic waited until the five-year NCE filing exclusivity expired, it would then be able to be tentatively approved by the FDA during the 30-month stay and enter the market at the end of the seven-and-a-half-year period. Yet, to obtain the 180-day exclusivity of a first to file applicant, a generic must file at the NCE-1 date to obtain market advantage over its competitors. This presents a quandary for applicants to decide whether obtaining the 180-day exclusivity period is a worthwhile counterincentive to the typical 15- to 18-month ANDA approval period.

Generics should also be wary of a brand delisting some or all of its patents from the Orange Book. Brands may choose to delist their patents in

---

Waxman Act to be a bright-line rule by defining the active moiety to be the entirety of the molecule affecting the physiological or pharmacological actions).

19. Upadhye, *supra* note 16, § 12:4, at 666 (noting that it is possible for an NDA-approved drug to have NCE exclusivity without a patent because the patent may have been rejected by the USPTO, in effect giving brands additional incentives to innovate regardless of a drug's patent status).

20. 21 U.S.C. § 355(c)(3)(E)(ii), (j)(5)(F)(ii).

order to prevent generics from being able to file Paragraph IV Certification ANDAs with the FDA.[21] This procedure ultimately forces generics to abide by the five-year NCE exclusivity expiration date before filing an ANDA with the FDA. Delisting may occur if the respective NDA asserting an NCE exclusivity is only propped up with weak patents. In this case, ANDA filers may need to amend their application certifications under Paragraph I at the end of the five-year exclusivity period.

B. New Clinical Investigation (or New Indication) Exclusivity (Three Years)

During the legislative approval process, the Hatch-Waxman Act went through a series of iterations that demonstrated the need to allow for a data exclusivity for NDAs based on new clinical investigations (NCIs). The report for the first 1984 version of the House bill noted that the exclusivities were also added in part to assure brands and generics that the FDA dealt solely with the safety and efficacy of the drug rather than with substantive issues of patent law.[22]

The Hatch-Waxman Act provides a three-year data exclusivity for NDAs and sNDAs that meet the following three conditions: (1) the NDA or sNDA contains reports of new clinical investigations (other than bioavailability studies); (2) the new clinical investigations are essential to the approval of the NDA or sNDA; and (3) the new clinical investigations were conducted or sponsored by the NDA applicant or person submitting the sNDA.[23] These three conditions allow an ANDA filer to challenge an RLD exclusivity based on new clinical investigations (NCIs) and arise most frequently with respect to sNDAs. The NCI exclusivity provides an incentive for brands to continue working with existing, FDA-approved products for new applications by rewarding additional testing for new indications and usage. Its intent is to maximize the safe application of an existing drug for all indications for which the drug has a therapeutic effect. NCI studies are

21. Upadhye, *supra* note 16, § 12:5, at 666–67 (citing Iver B. Cooper, Biotechnology and the Law §§ 14:12-14:15 (2012)).

22. H.R. Rep. No. 98-857, pt. 1, at 33 (1984). ("[T]he Committee recognizes that some Paper NDAs for listed drugs will be submitted and ready for approval before the patent on the listed drug has expired. To deal with this situation and to assure that the FDA concerns itself solely with the safety and effectiveness of the generic drug, section 505(c)(3) requires the FDA to approve a Paper NDA but make the approval effective at some later date when appropriate.").

23. 21 U.S.C. § 355(j)(5)(F)(iii)–(iv); 21 C.F.R. § 314.108(b)(4)(iv), (b)(5)(ii); *see* Upjohn Co. v. Kessler, 938 F. Supp. 439 (W.D. Mich. 1996) (holding that the FDA's refusal to grant a new clinical investigation exclusivity was warranted because the supplemental study provided by the brand was not essential to the approval of an sNDA, thereby allowing approved ANDA filers to enter the market).

considered data exclusivities in that the FDA can tentatively approve a generic drug product pending the expiration of the exclusivity.

NCI exclusivities have the ability to reach beyond the approval of new indications to prevent generics from entering the market.[24] The FDA has interpreted this exclusivity to apply to any change to a drug that affects the information relied on by the generic.[25] For example, a brand can submit an sNDA based on new clinical investigations that result in changing the labeling of a brand drug. The labeling change can entitle the brand to an additional three years of exclusivity because ANDA applicants are required to label the generic drug the "same as" the brand-name drug label. This process, in turn, effectively prevents ANDA drugs from receiving final FDA approval.[26] In the case that a generic company expects a product to later be approved for an additional indication, it should consider filing a 505(b)(2) application that does not require identical labeling in lieu of an ANDA, or dual filings, in order to avoid potential labeling exclusivities based on new clinical investigations.

Generics should also remember that an ANDA or a 505(b)(2) application must obtain final approval by the FDA before a brand can assert its exclusivity through litigation. So far, courts have been unwilling to allow for a preliminary injunction preventing a generic from market entry when the FDA has only tentatively approved the generic drug.[27] Therefore, to challenge NCI exclusivity, a generic should submit a citizen petition or letter to the FDA to review the validity of the exclusivity based on the three statutory elements or file a claim under the Administrative Procedures Act against the FDA for an improperly granted exclusivity.[28]

C. Orphan Drug Exclusivity (Seven Years)

The orphan drug (OD) exclusivity passed as part of the Orphan Drug Act of 1983.[29] FDA rules provide that an OD has a seven-year data exclusivity

24. AstraZeneca Pharm. LP v. U.S. Food & Drug Admin., 850 F. Supp. 2d 230, 235 (D.D.C. 2012) ("The FDA has interpreted this language as establishing a relationship between the information obtained from the clinical investigation, the change approved through the brand's supplemental application, and the scope of the information relied upon by a generic competitor in a specific ANDA.").

25. *Id.*

26. ViroPharma Inc. v. Hamburg, 898 F. Supp. 2d 1, 7 (D.D.C. 2012) (discussing the FDA's broad interpretation of the new clinical investigation exclusivity).

27. *AstraZeneca*, 850 F. Supp. 2d 230 (giving deference to the FDA decision not to approve an exclusivity and denying brand a preliminary injunction for possible harm because the FDA merely tentatively approved a generic version of the contested drug).

28. Actavis Elizabeth LLC v. U.S. Food & Drug Admin., 625 F.3d 760 (D.C. Cir. 2010) (challenging the FDA grant of an NCE exclusivity under the Administrative Procedures Act in district court).

29. 21 U.S.C. § 360cc; *see* Orphan Drug Act, Pub. L. No. 97-414 (Jan. 4, 1983).

against subsequent generics sponsoring "the same drug product for the same indication."[30] The Orphan Drug Act aims to provide brands with incentives to research and develop drugs for rare diseases affecting 200,000 people or fewer.[31] The OD exclusivity applies to both NCEs and NCI-type drugs affecting a small patient population, with the logical justification that a brand needs additional time to collect royalties to compensate for its investment. Before the Orphan Drug Act passed in 1983, only a handful of companies were willing to invest in drugs for rare and devastating diseases and conditions due to their low commercial value.[32] The act immediately created incentives for companies to invest in such medications for the public good but has also led to brands earning blockbuster profits by exploiting the seven-year exclusivity on the drug.[33]

The FDA provides OD exclusivity for no two drugs with the same active moiety, unless the second drug is deemed to be "clinically superior."[34] This means that multiple NDA applicants can vie for the seven-year exclusivity for drugs with the same active moiety and different excipient (or inactive) ingredients, with the FDA granting a single OD exclusivity to the drug first obtaining final approval.[35] In other words, the OD exclusivity encompasses only the active and likely patented ingredient(s), rather than the entire drug product. Because OD exclusivity applies to a specific drug tied to a specific indication, brands can obtain subsequent seven-year OD exclusivities for demonstrating a new indication.[36] However, generics are still free to enter the market after seven years for the first indication, and physicians have discretion to prescribe the drug for any off-label uses. This option is possible for generics because the Orphan Drug Act does not mandate identical labeling for all indications like a typical ANDA. Therefore, a generic would benefit from filing a 505(b)(2) application that does not include a labeling requirement for ODs that are likely to have multiple indications approved over a period of time. A 355(j) application would hinder a generic by binding

---

30. 21 C.F.R. § 316.3(b)(12).

31. 21 U.S.C. § 360bb(a)(2).

32. *Orphan Drug Amendments of 1991: Hearing on S. 2060 Before the S. Comm. on Labor & Human Res.*, 102d Cong. 18 (1992) (opening statement of Sen. Orrin G. Hatch, Comm. Member) (noting the success of the Orphan Drug Act by showing that only ten products with orphan drug indications were approved between 1975 and 1983, whereas 60 orphan drugs were approved between 1984 and 1991 with another 350 in the development).

33. *Id.* at 1–4 (prepared statement of Sen. Howard M. Metzenbaum, Comm. Member) (recognizing the prohibitively high prices of orphan drugs due in part to the seven-year exclusivity).

34. Baker Norton Pharm. Inc. v. U.S. Food & Drug Admin., 132 F. Supp. 2d 30, 31 (D.D.C. 2001) (citing 21 C.F.R. § 316.3(b)(13)(i)).

35. *Id.*

36. Sigma-Tau Pharm. v. Schwetz, 288 F.3d 141, 148 (4th Cir. 2002).

the ANDA applicant to its statutory labeling requirement, thereby forfeiting access to the off-label market.[37]

D. Pediatric Exclusivity (Six Months)

Pediatric exclusivity was originally crafted in 1997 under the Food and Drug Administration Modernization Act and made permanent in 2012 with congressional passage of the Food and Drug Administration Safety and Innovation Act.[38] Congress authorized the FDA to provide a pediatric exclusivity so that the agency could encourage brands to invest in pediatric studies for diseases that affect children. Previously, doctors could only prescribe drugs to children for off-label use, forcing these doctors to estimate or even guess whether a drug would help a child or what a drug's appropriate dose would be for a child.[39] These public policy concerns drove Congress to create the pediatric exclusivity to encourage brands to examine the drug's pharmacokinetics and pharmacodynamics in children. The pediatric exclusivity functions differently than all other exclusivities in key ways that make it a complicated hurdle for generic drug applications. These are the four key differences:

- The pediatric exclusivity attaches a six-month data exclusivity to the end of the patent term or other RLD exclusivity.
- The exclusivity is granted merely if pediatric studies are conducted, regardless of whether the drug later obtains a pediatric indication.
- The pediatric exclusivity is not tied to a specific indication; rather it is granted for all dosage forms and indications with the same active moiety as the specific drug studied.
- The pediatric exclusivity does not affect the drug's labeling requirements.[40]

Similar to the other RLD exclusivities, the pediatric exclusivity can be approved for currently marketed drugs but is limited to requests made by the FDA to the NDA holder. Yet in practice, most brands have significant sway in convincing the FDA to request pediatric studies.

---

37. *Id.*

38. Food and Drug Administration Safety and Innovation Act, Pub. L. No. 112-144 (July 9, 2012) (striking the sunset provision for the Pediatric Exclusivity, 21 U.S.C. § 355a(q), from the statute).

39. Susan Thaul, Cong. Research Serv., 7-5700, FDA's Authority to Ensure that Drugs Prescribed to Children are Safe and Effective 1–2 (2008).

40. *Qualifying for Pediatric Exclusivity under Section 505A of the Federal Food, Drug, and Cosmetic Act*, U.S. Food & Drug Admin., http://www.fda.gov/Drugs/DevelopmentApprovalProcess/DevelopmentResources/ucm077915.htm (last visited July 16, 2013).

The most complicated intellectual property aspect to the pediatric exclusivity for ANDA and 505(b)(2) applicants is the brand's ability to artificially extend the patent expiry date as recognized by the FDA. This extension affects generics because a pediatric exclusivity granted near the end of patent expiry can change the certification requirements of the ANDA or 505(b)(2).[41] Brands often use this process to their advantage by obtaining pediatric exclusivities on or near the last day of their patent's expiration date. Thus generics should keep continuously aware of when the FDA has requested pediatric studies through the FDA website.[42]

The pediatric exclusivity has the ability to convert the certification of an ANDA from Paragraph IV to Paragraph II, adding an additional period of exclusivity. The Hatch-Waxman statute requires generics to amend their applications to reflect the current status of the patent. So when a patent expires during Paragraph IV litigation, instead of treating the application as invalid, the FDA itself will de jure treat the certification as Paragraph II.[43] The effect of the conversion from Paragraph IV to Paragraph II certification results in an automatic six-month exclusivity period. This effect complicates the ANDA process for generics because it automatically attaches an additional period of exclusivity even if the patent is found to be invalid or if it expires during litigation. If settling with a brand, a generic should consider requesting a waiver of the pediatric exclusivity or make clear the date of market entry to include the six-month exclusivity to avoid later litigation. Courts have interpreted this exclusivity conversion to apply even when a patent has been found valid and infringed although the brand did not file in the 45-day window to trigger the 30-month stay despite language in § 355a(b)(1)(B)(ii).[44]

The pediatric exclusivity can also convert the final approval of a generic drug to a tentative approval. This conversion means that an ANDA that has been granted final approval after filing a Paragraph IV Certification, either because the patent wasn't asserted during the 45-day period or because the 30-month stay passed, can revert back to tentative approval if the pediatric exclusivity adds six months to the patent expiry date.[45] Courts have also allowed for the conversion of final to tentative approval

---

41. Ranbaxy Labs. Ltd. v. U.S. Food & Drug Admin., 307 F. Supp. 2d 15 (D.D.C. 2004).

42. FDA List of Written Requests Issued for Pediatric Studies, U.S. Food & Drug Admin., http://www.fda.gov/Drugs/DevelopmentApprovalProcess/DevelopmentResources/ucm050002.htm (last visited July 16, 2013).

43. *Ranbaxy*, 307 F. Supp. 2d at 18 (recognizing that Paragraph IV Certification does not remain intact after a patent expires, including during the 30-month stay).

44. Mylan Labs. Inc. v. Thompson, 389 F.3d 1272, 1282–84 (D.C. Cir. 2004).

45. *Id.* at 1281–82 (adding the six months of exclusivity to the brand after the patent was found valid although not asserted in the 45-day period); Barr Labs. v. Thompson, 238 F. Supp. 2d 236 (D.D.C. 2002).

as equitable relief for brands with patents that expired during Paragraph IV litigation. The courts have reasoned that when the patent expired during litigation, it converts the Paragraph IV into a Paragraph II Certification, thereby returning the parties to the status quo before infringement. Therefore, converting the final approval to a tentative approval is a proper construction of the pediatric exclusivity and the artificial act of infringement under 35 U.S.C. § 271(e)(2).[46] Therefore, generics that have seemingly been approved by the FDA may be subject to market entry delay as a result of pediatric studies submitted close to the date of patent expiry. Generics should pay special attention to late changes in the exclusivity and patent status of RLDs before entering the market and becoming targets for litigation.

E. Biologics Price Competition and Innovation Act

Currently new technologies in the biologics market have prompted the FDA to create a Hatch-Waxman–style litigation for biologics under the Biologics Price Competition and Innovation Act of 2009 (BPCI Act).[47] The Affordable Care Act passed during the Obama administration provided for a 12-year period of biologic drug exclusivity. This exclusivity is possibly divided into a four-year filing exclusivity and an eight-year data exclusivity, although these exact numbers have not been clarified by the FDA or courts. The act has also created a pediatric exclusivity that includes an additional one year, rather than six months, of exclusive marketing at the point of patent or listed drug exclusivity expiry.

Distinguishable from an ANDA filing, the BPCI Act allows for a generic that manufactures "a biosimilar that is interchangeable" to file due to the scientific impossibility of identically substituting biologics. Generics must also disclose to the reference product sponsor (as opposed to the reference listed drug) the process by which it makes the biosimilar. However, generic entry into the biologics market poses unique costs and risks to a generic compared to the biochemical market.[48] Brand biosimilars are also aiming to delay generic entry by building massive patent portfolios and launching "biobetter" drugs as a form of evergreening. Although there are a number of similarities to the Hatch-Waxman Act, the BCPI Act is substantially

46. AstraZeneca AB v. Impax Labs. Inc., 490 F. Supp. 2d 368 (S.D.N.Y. 2007).

47. 21 U.S.C. § 379h; *see also* Erwin A. Blackstone & Joseph P. Fuhr, Jr., *The Future of Competition in the Biologics Market*, 31 Temp. J. Sci. Tech. & Envtl. L. 1 (2012) (discussing in detail the effect of the BPCI Act on the generic-drug market and the FDA non-patent exclusivities).

48. Developing a biologic is estimated at $80 million to $100 million as compared with the estimated $5 million development cost of chemical compounds. Biosimilars also pose a challenge to generics because they have no Orange Book listing equivalent.

different, and generics interested in pursuing biologics currently have little judicial certainty. For the time being, generics should proceed with some caution in challenging brand drugs under the BCPI Act. Still, as the BCPI Act is filtered through the courts and clarified by the FDA, generics should familiarize themselves with the biologics framework and its corresponding FDA exclusivities.

II. Identification and Evaluation of Patents on RLDs

Under the Hatch-Waxman Act, a generic can file an Abbreviated New Drug Application (ANDA) by showing bioequivalence to a listed drug.[49] A listed drug is any drug for which a New Drug Application has been approved for safety and efficacy.[50] The generic must generally show that the generic drug has the same active ingredient(s) as, and is bioequivalent to, the Reference Listed Drug (RLD) that the applicant is relying on for approval of the generic version.[51] The FDA generally requires that brands list each patent covering a listed drug in the Orange Book.[52] However, not every patent granted on a drug is listed in the Orange Book. An in-depth review and evaluation of all of the patents for an RLD is critical to the strategic development of a generic drug formulation.[53]

A. Review and Analysis of Orange Book and Related Patents

The brand must identify all of the patents for each RLD, including any formulations (e.g., salts and solid forms), compositions, methods of use/treatment, polymorphs and methods of manufacture patents. For method-of-use patents, FDA regulations provide that only patents claiming "indications or other conditions of use" that either have been approved by the FDA or are in a pending NDA are to be submitted for listing in the Orange Book.[54] These patents are given a use code, which is also listed in the Orange Book.[55] The Orange Book lists all NDAs, over-the-counter (OTC) drugs, and ANDAs. It also includes the expiration dates of patents and exclusivities. It is imperative, however, that the generic perform an

49. 21 U.S.C. § 355(j)(2).

50. *Id.* § 355(j)(7).

51. *Id.* § 355(j)(2)(A)(ii), (iv).

52. *Id.* § 355(b)(1). The FDA regularly updates the Orange Book on its website: http://www.fda.gov/cder/orange/.

53. The type of patent listed and its relative strength to withstand validity challenges could be a key component to the decision to develop a generic drug and will also dictate the manner in which it is brought to market.

54. 21 C.F.R. § 314.53(b).

55. *See id.* § 314.53(c)(2)(ii)(P)(3), (e).

exhaustive search of the landscape for all patents associated with the RLD and not simply rely on the Orange Book. For example, the brand does not necessarily have to list each polymorph in the Orange Book, so the generic must be diligent in its patent search to ensure that it will not infringe a different polymorph. The generic should also note all licensees and assignees to the patent and look at the patents within the same family of drugs. The generic is statutorily permitted to submit a request for disclosure to the NDA holder for a list of all method and process patents.[56]

Before submitting an ANDA, the generic must first make certain required statements about the patent status of the Reference Listed Drug to certify to the FDA that its generic version will not infringe the RLD's patent(s). Depending on the type of patent covering the RLD and its status, the generic may choose from one of several statements. For method-of-use/treatment patents, the generic can submit a Section viii statement asserting that it will market the generic drug only for those methods not covered by the brand's patent.[57] Under this process, the generic should propose a label that carves out the patented method(s) of use based on the RLD's listed use code.[58] The FDA will not approve an ANDA with a Section viii statement whose proposed label overlaps any of the brand's use codes or if removing the carved-out information renders the drug less safe or effective.[59]

Second, the generic can make one of four paragraph certifications that provide either that: (I) no patent information for the brand-name drug has been filed with the FDA; (II) the patent has expired; (III) the patent will expire on a specifically identified date and no marketing of the generic drug will occur before; or (IV) the "patent is invalid or will not be infringed by the manufacture, use, or sale of the new drug for which the application is submitted."[60] Generics most commonly use the Paragraph IV Certification when challenging a patent that has not yet expired.

The FDA mandates that the generic must give notice to the brand of patent noninfringement, unenforceability, or invalidity status.[61] The notice must include, but is not limited to, the following: (1) a statement that the FDA has received an Abbreviated New Drug Application submitted by the

56. 35 U.S.C. § 287(b). A person who receives a response to a request for disclosure under this subsection shall pay to the person to whom the request was made a reasonable fee to cover actual costs incurred in complying with the request, which may not exceed the cost of a commercially available automated patent search of the matter involved, but in no case more than $500.

57. *See* 21 U.S.C. § 355(j)(2)(A)(viii).

58. *See* 21 C.F.R. § 314.94(a)(8)(iv).

59. *See id.* § 314.127(a)(7); *see also* Applications for FDA Approval to Market a New Drug, 68 Fed. Reg. 36,676, 36,682 (June 18, 2003).

60. 21 U.S.C. § 355(j)(2)(A)(vii).

61. *Id.* § 355(j)(2)(B)(i).

applicant containing any required bioavailability or bioequivalence data or information; (2) the ANDA number; (3) the established name, if any, as defined in Section 502(e)(3) of the act, of the proposed drug product; (4) the active ingredient, strength, and dosage form of the proposed drug product; (5) the patent number and expiration date, as submitted to the agency or as known to the applicant, of each patent alleged to be invalid, unenforceable, or not infringed; and (6) a detailed statement of the factual and legal basis of the applicant's opinion that the patent is not valid, is unenforceable, or will not be infringed.[62] The generic shall include in the detailed statement: (i) for each claim of a patent alleged not to be infringed, a full and detailed explanation of why the claim is not infringed; and (ii) for each claim of a patent alleged to be invalid or unenforceable, a full and detailed explanation of the grounds supporting the allegation.[63]

Using the listed patent information, the generic should acquire the product label and file history of each patent found. It is often helpful to review the corresponding European file histories of each patent as well. This information is generally publicly available. When reviewing the file histories, the generic should note what prior art was considered when the patent issued. It is critical to then determine the date each patent expires, being mindful to check for any patent term extensions listed.

B. Evaluation of Strength/Type of Patent Protection

The generic can evaluate the strength of the listed patent based on a variety of factors. One way to determine the patent's strength is to look at the type of the patent listed. Generally speaking, patents claiming compounds for new chemical entities are less susceptible to challenge, because the entity being claimed has typically not been used before (i.e., is not subject to an anticipation analysis). In the spectrum of susceptibility to an invalidity challenge, method-of-use/treatment patents are more vulnerable than patents claiming compounds. Patents covering pharmacokinetics and pharmacodynamics are generally more susceptible to validity challenges because the validity of these types of patents rests on the validity of the molecule itself, because the molecule has inherent properties that typically include the pK parameters. A patent claiming a specific formulation typically includes well-known excipients and, on that basis, is subject to a strong validity challenge or design around that changes the claimed excipient(s).[64]

---

62. 21 C.F.R. § 314.53(c) (2012).

63. *Id.*

64. It should be noted, however, that a patent covering an injectable formulation is less susceptible to validity challenges (and harder to work around) than an oral formulation because the generic must contain the same excipients as the RLD for injectable formulations.

A generic must also pay careful attention to the label of the RLD and make sure that its proposed labeling will not constitute induced infringement. In particular, the generic must look at the "indications and usage" section of the label when asserting a Paragraph IV or Section viii statement to ensure that the proposed indication of the generic version will not infringe.[65] FDA regulations require the indications listed in the "indications and usage" section of the label to be supported by "substantial evidence of effectiveness based on adequate and well-controlled studies."[66] The regulation adds that indications or uses "must not be implied or suggested in other sections of the labeling if not included in this section."[67] These regulations appear to favor the generic by limiting the FDA-approved uses to those listed in the "indications and usage" section of the label.

The generic should also evaluate the risk and profit potential of a potential launch. A key factor to keep in mind is that many generics will file on the same day an NCE expires, thereby reducing the profitability of a generic launch of the compound alone. Consequently, many generics choose to avoid filing Paragraph III Certifications for the single compound and instead focus on another formulation or method of use. On the other hand, if multiple generics file a Paragraph IV Certification on or about the same date, they will often be able to split the litigation costs associated with the process. An interesting example of this process was found during the generic launch of Seroquel. The brand had two listed Orange Book patents—molecule and formulation. Most generics filed a Paragraph III Certification on the molecule (the patent expired in 2012), and some did a Paragraph IV on the formulation, which had different excipients, and gave rise to noninfringement positions. The litigation went past the expiration of the molecule patent, making it an ideal candidate. The market for this drug was between $2 billion and $3 billion, making it a lucrative move for the generics.

C. Evaluation of Patent Validity and Enforceability

Once the generic assesses the relative strength of the patent on the RLD, it must further consider that patent's validity as well as any challenges that can be raised as to its enforceability. Typically, this includes a diligent

65. *See* Barr Schering Pharma AG & Bayer HealthCare Pharm. Inc. v. Lupin Ltd., 676 F.3d 1316 (Fed. Cir. 2012). The ANDA applicants sought approval for a generic form of Yasmin for contraceptive use only, for which there was no valid patent. Bayer argued that because the formulation would necessarily provide the simultaneous effects of contraceptive effect, anti-androgenic effect, and anti-aldosterone effect, which were patent-protected, that the ANDA would infringe. The court held that those three simultaneous effects were not listed in the "indications and usage" section of the label and were not FDA-approved uses and therefore could not be infringed.

66. 21 C.F.R. § 201.57(c)(2)(iv).

67. *Id.*

landscape search of all prior art—including that which is cited within the specification of the referenced patents themselves—to develop potential invalidity positions under 35 U.S.C. §§ 102, 103, and 112. In light of a recent Supreme Court case, the generic may also have a case to use an ineligible subject matter defense under 35 U.S.C. § 101.[68] The search should include any patent family members, any other work by the inventor, and any activity by any assignees.[69] The generic should order the file history, check into the chain of title and standing of the patent, and track down publications and conference materials in which the claimed subject matter may have been disclosed. Additionally, the generic should retain experts and perform tests to corroborate specification findings and testing summaries.

A lesser-used strategy involves an evaluation of the RLD to see if obvious-type double patenting exists. This non-statutory double-patenting rejection is based on a judicially created doctrine grounded in public policy and is primarily intended to prevent prolonging the patent term by prohibiting claims in a second patent that are obvious variations on claims in an earlier patent.[70] This type of double-patenting rejection only requires a single common inventor and does not require common ownership. The terminology "not patentably distinct" or "distinguished" from the earlier claims is often used to describe obviousness-type double patenting. Double-patenting rejections are commonly seen in continuing patent applications, so the generic should note any continuations and evaluate if a double-patenting objection can be raised. Such rejections can usually be overcome if the patent applicant files a terminal disclaimer and disclaims the term of the later-issuing patent that extends beyond the term of the earlier patent. The disclaimer may be filed, however, only if the same inventor or combination of inventors is named in both patents. Even then, the rejections may stand because of unfavorable facts (e.g., the earlier issuing patent has expired).[71]

---

68. *See* Mayo Collaborative Servs. v. Prometheus Labs. Inc., 132 S. Ct. 1289 (2010). The patents concerned a method of determining the proper dosage of thiopurine drugs, which were metabolized differently by different patients. The Court unanimously held that the patents were not patent-eligible since the relationships between concentrations of metabolites in the blood and the likelihood that a thiopurine drug dosage would prove ineffective or cause harm were known laws of nature, and the patents did not describe genuine applications of those laws.

69. The generic must also ascertain whether there is any current opposition to the patent in question, including any foreign litigation.

70. *See In re* Goodman, 11 F.3d 1046 (Fed. Cir. 1993).

71. *See* Boehringer Ingelheim Int'l GmbH v. Barr Labs. Inc., 592 F.3d 1340 (Fed. Cir. 2010). The patentee filed a terminal disclaimer to continue marketing Mirapex, seeking to overcome obviousness-type double patenting. The Federal Circuit affirmed the district court's decision that the terminal disclaimer was ineffective to overcome the obviousness-type double-patenting rejection, because the disclaimer was filed after the earlier patent had

The generic should also consider other standard invalidity defenses, including obviousness, lack of enablement, lack of written description anticipation, and lack of utility. Courts have recently found that patents can be invalidated for lack of enablement based on a lack of utility because relevant animal testing was not complete before the patent issued.[72]

In addition to the statutory patent defenses, generics can assert the equitable defense of inequitable conduct by claiming that the brand obtained the patent by breaching its duties of disclosure or candor to the USPTO when the patent was prosecuted. A successful inequitable conduct defense can render all of a patent's claims unenforceable and thereby serve as a complete defense to infringement. The Federal Circuit recently clarified the standard for inequitable conduct and stated that two elements must be proven for a successful defense: one, intent to deceive the USPTO (infringer must show that the applicant or its counsel knew of the prior art reference, knew that it was material, and made a deliberate decision to withhold the information); and two, that the information was material to the decision (if the USPTO had been made aware of the information, it would not have allowed the claim).[73]

III. Design-Around Opportunities

Generics have several options when attempting to design around a brand patent, but they must first accurately determine what subject matter is claimed. In many cases, an expert is retained to determine the claim coverage by analyzing the claim language. After the claims have been analyzed,

expired. *See also* Sun Pharm. Indus. v. Eli Lilly & Co., 611 F.3d 1381 (Fed. Cir. 2010). The court affirmed the lower court holding that three claims of the patent-in-suit were invalid for obviousness-type double patenting, rejecting the patentee's argument that the district court erred in consulting the specification of the issued prior patent, as opposed to the specification of an earlier application. The patentee could not avoid portions of the prior patent specification by resorting to the specification as originally filed.

72. Janssen Pharmaceutica NV v. Teva Pharm. USA Inc. (*In re* '318 Patent Infringement Litig.), 583 F.3d 1317 (Fed. Cir. 2009). The Federal Circuit affirmed the district court's decision that the patent was invalid for lack of enablement on two distinct grounds: the specification did not demonstrate utility because relevant animal-testing experiments were not finished by the time the patent was allowed, and the specification provided only minimal disclosure of utility. The court alternatively found that the specification and claims did not teach one of skill in the art how to use the claimed method because the application did not provide sufficient dosage information. Although a person of ordinary skill in the art would have viewed the invention as scientifically grounded, that did not demonstrate that a person of ordinary skill in the art would have recognized that the specification conveyed the required assertion of a credible utility. The specification did no more than state a hypothesis and propose testing to determine its accuracy.

73. Therasense v. Becton, Dickinson & Co., 649 F.3d 1276 (Fed. Cir. 2011).

the generic should look to the specification and prosecution history of the file to determine what was "given up" during prosecution. Any elements that were forfeited during prosecution can be incorporated in the ANDA product. This section explains these design-around tactics in greater depth.

A. Prosecution History Estoppel

One of the primary design-around options for generics is the use of prosecution history estoppel. This technique requires the generics to obtain and analyze the file history of the potentially blocking patent. Over the course of patent prosecution, there is a dialogue between the patentee and the examiner about which claim limitations are patentable in view of the prior art that has been submitted by the patentee as well as prior art located by the examiner. As a result, the claims are often amended to overcome the prior art in order to render the claims allowable.

The claim language that has been "disclaimed" can be very useful to a generic in its attempts to develop a design-around position. Because these elements have been acknowledged by the patentee to be present in the prior art and removed from the claims as filed, the generic can typically rely on these representations to avoid infringement. Specifically, the brand will be estopped from asserting those claim limitations forfeited during prosecution. This technique is important for assessing the bounds of the prior art and determining the areas in which the brand would have liked more patent coverage.

B. Disclosed "But Not Claimed" Embodiments

Another option for generics is to examine disclosed but unclaimed embodiments within a brand patent. The Federal Circuit has held that the doctrine of equivalents does not cover embodiments that are disclosed in the patent specification but not claimed.[74] Thus, a generic can argue that it is not infringing, literally or equivalently, any claim of a potentially blocking patent by employing the use of a disclosed embodiment that was not claimed. It can be a useful strategy for generics to take advantage of other functional embodiments that are effective but not protected.

C. Section VIII Statements

Properly listed Orange Book patents can protect both the chemical formulation of the drug as well as the methods of using that formulation.[75] Generics often seek ANDA approval for only some of those methods of use so as not to provoke litigation on the protected uses. This process can occur

74. Johnson & Johnston Assocs. v. RE Serv., 285 F.3d 1046 (Fed. Cir. 2002).
75. Upadhye, *supra* note 16, § 12:1 at 543.

in a situation where a brand has multiple patents on different methods of use where some have expired and some have not.[76] The generic would like to market the drug for only the uses that are no longer protected by patent coverage. A Section viii statement allows a generic to carve out uses from its proposed label so that the drug will only be marketed for specific uses that may not include all of the FDA-approved uses.[77] This carve-out prevents litigation against the protected uses and allows for generic entry into the market for the unprotected indication.

Although not a true design around because it does not affect the formulation, the use of Section viii statements does allow generics to sidestep patented uses of a drug. It can be extremely valuable to carve out protected uses so that the generic will not have to litigate on these matters during the ANDA process. A Section viii statement can only be used if the brand has listed more than one indication on its label.[78] If only one indication is listed, a generic will not be able to get around this use because drug labels must have at least one FDA-approved use.[79]

Although generics rely on Section viii statements, a recent decision by the District of New Jersey has seemingly weakened their effect.[80] In this case, the generic filed a Section viii statement carving out the treatment of "adults" with ADHD from its label.[81] As a result, the generic label would not indicate that the drug is intended for use by adults with ADHD. The brand alleged that although the treatment of adults may not technically appear on the drug label, the label is still causing the generic to induce users to infringe the patented use.[82] The brand argued that adult users would still be likely to use the drug in an infringing way, notwithstanding that the label did not include adult use.[83]

The generic argued that the brand had no claim where the generic was not seeking approval for the patented use.[84] The court was not persuaded and instead held that § 271(e)(2) was open-ended enough to allow for a claim of this type, where the brand had adequately alleged the inducement theory in the pleadings.[85] Although the court did not make any merits determinations, it did decide that this claim could go forward and

76. *Id.* at 552.

77. 21 U.S.C.A. § 355(j)(2)(A)(viii).

78. UPADHYE, *supra* note 16, § 12:1 at 553.

79. *Id.* (explaining that if there was no approved indication on the label, then physicians would not know for what indications the the drug was approved).

80. Shire v. Amneal Pharm., No. 2:11-CV-03781-SRC-CLW, Order No. 218 (D.N.J. 2011).

81. *Id.*

82. *Id.*

83. *Id.*

84. *Id.*

85. *Id.*

be added to the amended complaint.[86] The impact of this decision could be very broad. After this decision, any time a generic carves out patented aspects of a brand's label, a question may exist as to whether the generic has induced infringement of the brand patent.

Allowing inducement claims to go forward in cases where a Section viii statement has been executed may minimize the effectiveness of Section viii. The value of a Section viii statement is to disclaim a protected use on the drug label to avoid litigation with the brand. If the brand can allege that although the generic has disclaimed the use, it still induces customers to use the product in an infringing manner, then litigation will necessarily arise. This type of burdensome litigation is favorable to large brands and prejudicial to generics that are trying to launch a noninfringing product.

IV. AIA Proceedings

The 2011 passage of the America Invents Act (AIA) introduced new contested proceedings at the U.S. Patent and Trademark Office (USPTO). These proceedings allow litigants to challenge various aspects of patent validity in a trial-type setting before administrative patent judges. Although largely untested, these proceedings have the potential to play a major role in the Hatch-Waxman framework.

Inter partes reviews (IPRs) and post-grant reviews are new post-grant proceedings designed to encourage the litigation of validity challenges at the USPTO. Only validity issues can be litigated in these proceedings, and they are designed to be a speedy and cost-efficient alternative to district court litigation.

A. Inter Partes Review

Inter partes reexamination has been replaced at the patent office by the new IPR proceeding. To commence an IPR, a third party must establish a reasonable likelihood that it will prevail in invalidating at least one claim of the challenged patent.[87] The Patent Trial and Appeal Board (PTAB) can grant an IPR on any of the claims requested by the third party.[88] Additionally, only patents and other printed publications are permitted to be considered during the IPR, which only determines issues of obviousness and anticipation.[89] The IPR proceeds much like a mini-trial, with both sides presenting their case to experienced patent law expert judges

86. *Id.*
87. 35 U.S.C. § 314 (2012).
88. *Id.*
89. *Id.* § 311(b).

on the PTAB.[90] A litigant will be estopped from raising any claims that were raised or reasonably could have been raised during the IPR at a later district court or patent office proceeding.[91]

The IPR setting presents several potential benefits to a generic challenger of a brand patent. The first is that there is no presumption of patent validity in an IPR as there is in district court litigation. This lack of presumption is significant for a generic because it significantly lessens its burden when trying to invalidate the brand patent. Similarly, a generic need only show that the patent claims are invalid by a preponderance of the evidence as opposed to the more demanding "clear and convincing" standard required in district court.[92] Both of these lesser burdens make it easier to invalidate patent claims during an IPR as opposed to litigation in district court.

Another benefit is that IPRs at the USPTO are required by statute to be completed within one year of commencement.[93] The validity decision is made much faster in an IPR than in district court, resulting in quicker resolution and greater certainty. Additionally, IPR proceedings are significantly less expensive than district court proceedings because discovery is limited and the duration is much shorter. Obtaining a relatively inexpensive and fast invalidity decision in an IPR is a significant benefit for a cost-conscious generic. For example, generic Apotex made a Paragraph IV Certification against Alcon's Orange Book patents on Vigamox, an antibiotic used in the treatment of conjunctivitis. Alcon had previously won a district court action against another generic on the basis of infringement and the generic's failure to show invalidity.[94] Apotex promptly filed an IPR challenging the validity of the Orange Book–listed patents on Vigamox at the USPTO.[95] The PTAB initiated the IPR by finding that there was a reasonable likelihood that Apotex would succeed in invalidating at least one claim of the challenged patents, despite the holding from the district court in the previous litigation.[96] The IPR provided Apotex with another venue and opportunity to challenge the validity of the blocking patents that it would otherwise not have had.[97]

90. *Id.* § 316(c).
91. *Id.* § 315(e).
92. *Id.* § 316(e).
93. *Id.* § 316(e)(11).
94. *Id.*
95. Brian Murphy, *A New Weapon in Generic Drug Companies' Arsenal*, Law 360 (Apr. 19, 2013), http://www.law360.com/articles/430745/a-new-weapon-in-generic-drug-companies-arsenal.
96. *Id.*
97. *Id.*

IPRs are not without their risks to generics. Generic litigants filing IPRs may create estoppels preventing them from raising invalidity arguments before a district court. Specifically, the petitioner in an IPR that results in a final written decision by the PTAB may not assert in district court litigation any invalidity defenses that were raised or reasonably could have been raised in the IPR.[98] But since IPR is limited to printed publications and patents, any other prior art would still be available to the generic litigant.

In addition, there is still uncertainty surrounding the effect of a final judgment in an IPR (and subsequent Federal Circuit affirmation) on forfeiture of the 180-day exclusivity period.[99] A generic that has obtained ANDA approval and has been awarded a 180-day exclusivity period is subject to the "failure-to-market forfeiture provision" under the Medicare Modernization Act (MMA). The MMA provides that forfeiture of the 180-day exclusivity will occur if the generic has not marketed by "75 days after . . . an infringement action . . . or declaratory judgment action . . . [where] a court enters a final decision from which no appeal (other than a petition to the Supreme Court for a writ of certiorari) has been or can be taken that the patent is invalid or not infringed."[100]

The plain statutory language of the MMA indicates that the failure-to-market provision will only be triggered as a result of a final decision from an "infringement action" or a "declaratory judgment action." An IPR is a validity determination made as a result of an administrative proceeding, not a district court infringement or declaratory judgment action.[101] Additionally, a Federal Circuit affirmation of the PTAB decision would still not render it applicable under the failure-to-market provision.[102] On the other hand, once an invalidity determination has been affirmed at the Federal Circuit, the patent has been found invalid with the same effect as if a district court had so found. It remains unclear exactly how the failure-to-market provision will be interpreted in light of the new AIA proceedings, but these are arguments that will doubtlessly be advanced.

Another issue for generics to consider is the possibility of concurrent litigation in district court and in the USPTO. Once district court litigation has commenced, a party may still request an IPR to contest validity at the USPTO. Unless the district judge grants a stay of the ongoing litigation,

98. 35 U.S.C. § 315(e)(1).

99. Kurt R. Karst, H. Keeto Sabharwal & Dennies Varughese, *How Inter Partes Review Impacts Hatch-Waxman Exclusivity*, Law 360 (Feb. 27, 2013), http://www.law360.com/articles/417119/how-inter-partes-review-impacts-hatch-waxman-exclusivity.

100. 21 U.S.C. § 355(j)(5)(D)(i)(*l*)(bb)(AA).

101. *See* Karst et al., *supra* note 99.

102. *Id.*

the parties will be forced to litigate the same issues on different fronts until entry of a final judgment in one venue becomes preclusive to the other. To date, some district judges have been hesitant to grant stays pending an IPR because of the potentially prejudicial delay to the litigation that would occur.[103] Judges are still attempting to find the right balance between deferring to expert patent judges on validity issues and maintaining a prompt and reasonable schedule in the concurrent litigation. Generics should be mindful that litigation stays are not likely and that the effect of a potential stay on the 30-month stay is unknown at this point.

B. Post-grant Review

Another new proceeding under the AIA is the post-grant review (PGR), which can be used to challenge patent validity within the first nine months of patent issuance.[104] The grounds that can be used to challenge the validity of a patent are much broader in a PGR than in an IPR. In a PGR, almost all validity challenges are permissible, not merely those based on patents or printed publications as in an IPR.

PGR proceedings may be less desirable to generics than IPRs because they need to be filed so quickly after patent issuance. Generics tend to challenge brand patents later in the patent term than a PGR would allow for. This makes the IPR a more attractive option for generics. In addition, since PGR proceedings allow for a wider range of validity challenges, they also create larger estoppels. A litigant will be estopped from raising anything in the district court, USPTO, or International Trade Commission (ITC) that was raised or reasonably could have been raised during the PGR.[105] This estoppel would include virtually all validity challenges, except obviousness-type double patenting, which is excluded from a PGR.

---

103. *See, e.g.*, Universal Elecs. Inc. v. Universal Remote Control Inc., C.A. No. SACV 12-00329 AG 2013 U.S. Dist. LEXIS 64154 (C.D. Cal. May 2, 2013); Davol Inc. v. Atrium Med. Corp., C.A. No. 12-958-GMS 2013 U.S. Dist. LEXIS 84533 (D. Del. June 17, 2013); Automatic Mfg. Sys. v. Primera Tech. Inc., C.A. No. 6:12-cv-1727-Orl-37DAB 2013 U.S. Dist. LEXIS 67790 (M.D. Fla. May 13, 2013). To determine if a stay should be granted, courts will look to a three-part test. District judges will consider whether (1) discovery is complete and whether a trial date has been set, (2) a stay will simplify the issues in question and trial of the case, and (3) a stay would unduly prejudice or present a clear tactical disadvantage to the nonmoving party. Regents of the Univ. of Mich. v. St. Jude Med. Inc., C.A. No. 12-12908 2013 U.S. Dist. LEXIS 76845 (E.D. Mich. May 13, 2013).

104. 35 U.S.C. § 321(c).

105. *Id.* § 235(e).

chapter 17

Preparation of Noninfringement/Invalidity Legal Opinions[1]

I. Patent Legal Opinions in Support of an ANDA Product

A noninfringement or invalidity opinion regarding an Abbreviated New Drug Application (ANDA) product is a crucial component of ANDA strategy, particularly if the related pioneer drug is covered by a patent listed in the Orange Book. If that is the case, then the ANDA filer must include a Paragraph IV Certification stating that it has obtained an opinion of noninfringement and/or invalidity.[2] The ANDA filer also must provide notice to the patentee including a detailed factual and legal basis for noninfringement, invalidity, or unenforceability.[3] The patentee then has 45 days to sue the ANDA applicant for infringement.

Clearly, the opinion for an ANDA Paragraph IV–certified product is not only a statutory requirement, but it will also set the stage for litigation strategy. The noninfringement and/or invalidity opinion may offer relief from damages if litigation is unsuccessful, and it may provide leverage for negotiation and allow litigation to be averted or curtailed. As a key piece of the entire ANDA strategy for a generic product, the opinion must be carefully considered and prepared and integrated into the overall strategy for developing and launching the product.

1. Lin J. Hymel of McLane, Graf, Raulerson & Middleton PA, Woburn, Massachusetts.
2. 21 U.S.C. § 355(j)(2)(A)(vii)(IV).
3. *Id.* § 355(j)(2)(B)(i).

II. Timing and Logistics of Opinion Preparation

The opinion is actually a process rather than simply a document, and as such it must be part of the overall process of developing a generic drug and preparing the ANDA. Research and analysis performed in the course of preparing the opinion can inform the selection of a formulation, the timing of filing the ANDA and product launch, or even whether to go forward with the ANDA product at all.

The process of preparing the opinion can be performed in stages or all at once. If the details of the ANDA product are decided upon and the innovator patent landscape has been characterized before the opinion writer is involved, then clearly the opinion can be prepared all at once. However, if the opinion writer is involved earlier in the process, then the intermediate results of preparing the opinion, such as prior art search results, identification of weaknesses in the innovator patents, or infringement analysis, can be factored into the ANDA development process.

Logistics of preparing the opinion generally require as a first step the identification of the relevant patents covering the intended innovator product. These are identified using the Orange Book, and additional patents can be identified using the usual patent databases. Patent coverage of the innovator product, and the need for preparing a strong opinion, may influence decisions concerning the formulation of the generic product or its method of use, target patient population, mode of administration, and the like.

The second step is to carefully scrutinize the innovator patents with a view to identifying a basis for noninfringement as well as weaknesses that would support an invalidity position. The claims of the patents should be carefully reviewed and compared to the contemplated ANDA product. The specification and drawings should be reviewed for definitions of any claim terms and support for claim limitations. Then, the prosecution history of each patent must be carefully reviewed, again looking for any intrinsic evidence that could be used by a court to interpret the claims. At the same time, the rejections and prior art used during prosecution should be carefully studied, with particular attention to claim amendments and arguments used by the applicant to overcome the rejections. This review process will inevitably point out weaknesses in the patent that can be used to attack its validity.

The next step is to decide whether the opinion will be based on noninfringement, invalidity, or a combination of both. The preparation of the opinion then diverges, depending on this decision.

For an opinion based at least in part on noninfringement, a careful and complete factual description of the generic product should be provided to the opinion writer. Then, claim charts should be prepared for at least all of the independent claims of the relevant innovator patents. The elements of each claim should be listed separately in rows of the chart, with the exact claim language in one column and any corresponding features of the generic product or method of use in an adjacent column. The claim chart will assist in quickly recognizing whether or not there is a viable basis for noninfringement and will also point out where issues of claim construction exist that need to be dealt with.

On the other hand, if the opinion will include an assertion of invalidity of any claim, then at this stage the appropriate prior art will need to be identified to support the arguments for invalidity. In certain cases, such as where it is believed that the patent examiner has erred or where legal standards have changed since the patent was prosecuted, the prior art used by the patent examiner could be sufficient. However, in many cases, the prior art of record may not be the best available and should not be relied upon; in such cases, a fresh prior art search will need to be conducted by the opinion writer, an independent searcher, or by the generic manufacturer itself. The identified references will then have to be carefully reviewed to select the best references or combination of references to support a basis for invalidity. In certain cases, particularly where no adequate prior art can be found, invalidity may be based on another legal argument, such as non-patentable subject matter or insufficient written description or enablement.

Once the appropriate prior art and/or legal arguments have been identified, the opinion writer should discuss the general basis of the opinion with the generic manufacturer and then begin drafting the opinion.

III. Noninfringement versus Invalidity Opinions

While at times it will be clear whether the opinion should be based on noninfringement as opposed to invalidity, there is often a choice to be made. The choice is often determined by the availability or relative strength of noninfringement and invalidity positions for a given patent or claim. However, when both positions are available, their relative merits need to be considered.

On the one hand, since the legal burden of proving noninfringement is lower than that of proving invalidity, preparing a noninfringement opinion may be preferred where the relative merits of noninfringement and invalidity are similar. On the other hand, where there is a strong basis for invalidity, such as highly relevant, previously unconsidered prior art, or a change in legal standards since the patent issued which calls into question the patentability of the claimed subject matter, invalidity may be a preferred basis for the opinion. A strong invalidity opinion can eliminate the need for infringement analysis, which is an attractive option where proving noninfringement is difficult or nuanced. The option is also available to present both invalidity and noninfringement positions.

Prior art may be available to knock out broader claims with some confidence, but less so with certain dependent claims of the same patent. In such cases, a noninfringement position is preferred, which only needs to address the independent claims. When several patents are analyzed in the opinion, some of them might be invalid while others are not infringed; the same can hold true for different independent claims of a single patent. If that is the case, then a mixture of noninfringement and invalidity positions might provide the strongest opinion.

IV. Requirements for a Competent Opinion

The goal of a patent noninfringement or invalidity opinion is not only to satisfy the Paragraph IV requirements and to inform the recipient of the likelihood of a holding of noninfringement or invalidity or of relevant legal issues that might arise during litigation but also to avoid a finding of willful infringement in the event of a holding of infringement. Under 35 U.S.C. §§ 284-285, a court can award up to triple damages and, in exceptional cases, attorney's fees for willful infringement. Incompetent opinions offered in support of an ANDA Paragraph IV Certification have been considered grounds for awarding attorney's fees.[4] In order to avoid triple damages and the award of attorney's fees, the ANDA applicant should obtain and rely upon a competent legal opinion of noninfringement, invalidity, or unenforceability. The key is whether the opinion was "thorough enough, as combined with other factors, to instill a belief in the infringer that a court might reasonably hold the patent is invalid, not infringed, or unenforceable."[5] In general, a competent opinion is one that is thorough, rests upon an adequate foundation, and is not merely conclusory.[6] As such, it should involve a review of the patent, the patent's file history, prior art of record and any additionally identified prior art, applicable legal standards, the details of the allegedly infringing product or method, and the scope and construction of the claims. In the case of a noninfringement opinion,

4. Yamanouchi Pharm. Co. v. Danbury Pharmacal Inc., 231 F.3d 1339 (Fed. Cir. 2000).
5. Ortho Pharm. Corp. v. Smith, 959 F.2d 936, 944 (Fed. Cir. 1992).
6. Westvaco Corp. v. Int'l Paper Co., 991 F.2d 735, 743 (Fed. Cir. 1993).

the analysis should include a comparison of the product or method to each element of each of the claims and a consideration of the legal standards of both literal and equivalents infringement as well as both direct and indirect infringement. In the case of an invalidity opinion, the analysis should include a comparison of the claims to the prior art and any other relevant legal factors under the laws governing patentability. The opinion should discuss in writing the relevant factors in reaching its conclusion.

V. Opinion Format and Content

There is no required format for a noninfringement or invalidity opinion. However, the opinion is typically in the form of a letter from the attorney providing the opinion to the in-house attorney or executive who is receiving the advice. The use of letter format underscores that the advice is personal in nature. Indeed, the opinion letter is typically only the formal representation of a series of conversations that have already transpired between counsel and client on the topic.

The opinion letter is a privileged and confidential attorney-client communication, as well as an attorney work product, and should be boldly labeled as such at the very beginning. The substance of the letter begins with a brief introduction that states the nature and scope of the advice, including the type of opinion, the product or process considered, and the identity of the patents analyzed. A brief executive summary can be included next, which states the conclusion and one or two key points that led to the conclusion. Following the summary, a description of the client's product or process should be provided if the opinion includes noninfringement analysis. The description should include facts that are relevant to the analysis, and it is advisable to have the client review this section before the opinion is finalized. The introductory sections should include a description of any search performed as well as a listing of the important results (if the listing is long, it can be attached as an exhibit). By the end of the introductory sections, it should be clear what the opinion does and does not cover and what methods were used to reach the conclusions.

Next is a summary of the applicable legal standards. This section summarizes the statutes and case law that govern the issues to be considered in the remainder of the opinion. It is neither intended to be a comprehensive law review nor should it provide blanket coverage for all types of patent legal analysis. Rather, it should be brief but thorough enough to provide legal authority for all the issues raised in the opinion. It should also be up to date, including any recent refinements to the relevant law.

The main body of the opinion should follow the presentation of legal standards. This section may be entitled, for example, Analysis and will probably have several subsections with appropriate headings. The analysis should start with a summary or listing of the patent claims under consideration, followed by a brief summary of the major issues and prior art that arose during prosecution, and significant claim amendments and arguments leading to allowance. The relevant terms of the claims should then be construed. The remainder of the analysis section is taken up by the major noninfringement and/or invalidity analysis that forms the basis of the opinion (see details in the following sections).

After the analysis section, the conclusions of the opinion are presented, followed by a discussion of how the opinion should and should not be used. The letter closes with a signature block and a listing of attached exhibits. Typical exhibits might include copies of the patent(s) analyzed, claim listings, claim charts, important excerpts of the prosecution history, descriptive material of the client's product or process relied upon, copies of prior art references, and copies of legal authority important to the case.

VI. Elements of a Noninfringement Opinion

Actions that can be held as patent infringement are defined by federal statute under 35 U.S.C. § 271(a), (b), (c), (e)(2), (f), and (g). Most relevant for an ANDA product is § 271(e)(2), under which it is an act of infringement merely to file an application with the FDA for a drug or use of a drug claimed in a patent prior to the expiration date of the patent. Since filing such an application is likely the intention of the ANDA product manufacturer, the focus of the opinion shifts to whether or not the product or its method of use is "claimed in the patent" (i.e., whether the manufacture, sale, offer for sale, importation, or use of the product would infringe any of the patent claims) and/or whether or not the relevant claims of the patent are valid.

Analysis of patent infringement involves first construing the scope and meaning of the claims and, second, comparing the properly construed claims to the potentially infringing product or process.[7] While claim construction is a matter of law, the comparison of the product or process to the claims is a question of fact.[8] Infringement is analyzed on a claim-by-claim basis. Infringement of a claim occurs only if every element of the claim is found in a corresponding feature of the accused product or process.

7. Markman v. Westview Instruments Inc., 52 F.3d 967 (Fed. Cir. 1995).
8. Cybor Corp. v. AS Etches, 138 F.3d 1448 (Fed. Cir. 1998) (en banc).

A wide variety of evidence can be brought to bear in construing patent claims. The evidence can be divided into intrinsic evidence (the claims themselves, the specification and drawings, and the prosecution history) and extrinsic evidence (dictionary definitions, expert testimony, or any other evidence outside the patent and its prosecution history). The claim language itself should be the starting point for claim construction, though the meaning of claim terms can be interpreted by definitions and usage in the patent specification, as well as statements made during prosecution.[9] Claim terms are presumed to have their plain and ordinary meaning, as understood by persons skilled in the relevant art.[10] Only after establishing the ordinary meaning of a claim term should intrinsic or extrinsic evidence be considered. When extrinsic evidence is introduced (e.g., dictionary definitions), consistency with the intrinsic evidence is given priority. For example, if the specification or prosecution history indicates that a claim term is used in a manner incompatible with a given dictionary definition, then the dictionary definition must be rejected as inapplicable.

Clearly, not every word or term of a claim requires legal analysis and construction. Therefore, the opinion will focus efforts at claim construction where there appears to be ambiguity or uncertainty as to the meaning of a claim term, or uncertainty as to how a particular claim term ought to be applied to an aspect of the potentially infringing product or method. Often, greater effort at claim construction is provided in the opinion for one or more terms that appear not to align with a corresponding aspect of the accused product or process, as there is a greater need for clarity in such areas of the opinion.

One possible outcome of the claim construction exercise is the production of a claim chart. Claim charts are simply tables that divide claim language into individual elements, each in its own row. The chart should compare the actual claim language in one column with a construed meaning in another column, with intrinsic or extrinsic evidence in additional columns as needed. Claim charts are not necessary to perform infringement analysis, particularly with claims that have relatively few elements. However, the more complex the claim and the greater the number of elements, the greater the benefit of using claim charts, as they provide a more readily appreciated visual presentation.

The final step of analyzing literal infringement is to compare the factual details of the potentially infringing product or process with the construed claim. If even a single element of the claim is not present in the product or process, it does not infringe. For this reason, a noninfringement

9. Vitronics Corp. v. Conceptronic Inc., 90 F.3d 1576 (Fed. Cir. 1996).
10. Texas Digital Sys. v. Telegenix Inc., 308 F.3d 1193 (Fed. Cir. 2002).

opinion will often focus on one or more elements of the claim that are missing in the product or process and will omit discussion of remaining claim elements that may or may not be present. This can avoid creating undesirable and unnecessary assertions regarding aspects that are better resolved during litigation. Analysis of noninfringement can also be incorporated into claim charts by adding a column listing the features, or absence thereof, of the accused product or process.

Noninfringement analysis should start with the independent claims. If none of the independent claims of a patent is infringed, then there is no need to analyze any of the dependent claims. The dependent claims recite all the same elements of the respective independent claim and additional elements, so if an element of an independent claim is missing from the product or process, then the same element of any claims depending from that independent claim will also be missing. Thus, noninfringement opinions generally analyze only the independent claims of a patent.

Once it has been shown that none of the claims is literally infringed, because one or more elements are missing in the generic product or its method of use, then possible equivalents infringement must be considered. Equivalents infringement generally focuses on one feature of the product or method of use that is literally absent but that may be present in an equivalent form as defined under the doctrine of equivalents.

The doctrine of equivalents applies one of two possible tests: (1) the "function-way-result" test, where infringement is found if the accused product or process performs substantially the same function, in substantially the same way, with the same result;[11] or (2) the "insubstantial differences" test, where infringement is found if the differences between the claim and the accused product or process are insubstantial.[12] With the function-way-result test, noninfringement is found if any of the three factors is lacking. Application of this test may require significant additional information about how the accused product or process works as well as analysis of the patent specification and file history with regard to any statement about equivalents or the prior art. Where an equivalent is found in the prior art or is described in the patent but not claimed, there is generally noninfringement under the doctrine of equivalents.

The insubstantial differences test focuses on the interchangeability of the claimed element and the feature of the accused product or process. Interchangeability may be indicated if persons of skill in the art would have known of such interchangeability or if there is evidence of copying; however, interchangeability may not exist where efforts were made to

11. Graver Tank & Mfg. Co. v. Linde Air Prods. Co., 339 U.S. 605 (1950).
12. Warner-Jenkinson Co. v. Hilton Davis Chem. Co., 520 U.S. 17 (1997).

design around the claim element or to independently develop the feature of the accused product or process. Although both the function-way-result and insubstantial differences tests are legally acceptable, one or the other may be better suited to a particular technology or claim feature or to the evidence at hand.

The long-standing doctrine of prosecution history estoppel prevents coverage for equivalents not covered by a claim because the applicant has narrowed the claim during prosecution to avoid prior art that otherwise would have barred patentability.[13] This doctrine was refined in the *Festo* case, in which the Supreme Court held that a narrowing amendment made without a clear reason creates a rebuttable presumption that the amendment was made for reasons of patentability;[14] thus, coverage of equivalents is barred unless successfully rebutted by the patentee. The patentee generally will not be successful in rebutting the presumption if the equivalent was foreseeable and could have been covered literally by a claim. Furthermore, nearly any kind of statement or argument made by a patent applicant, apart from a claim amendment, can give rise to an estoppel that bars coverage of equivalents or otherwise limits claim coverage.[15] Such estoppel can arise from statements about claim scope, the nature of the invention, or the prior art. Recapture as equivalents of disclosed but unclaimed subject matter is also barred.[16]

Noninfringement analysis of methods claims also needs to consider both direct and indirect infringement. Direct infringement occurs when a party itself performs an act that violates 35 U.S.C. § 271. However, indirect infringement can be an issue, for example, when a pioneer drug is not covered by a composition-of-matter patent, but its use is covered by a method patent. To the extent that the generic manufacturer does not perform the method, it is not a direct infringer of claims covering the method of use. However, others (e.g., patients, clinics, etc.) may be directly infringing the method-of-use claims. If the manufacturer actively instructs or advertises third parties regarding infringing uses, then the manufacturer can be held to actively induce infringement, and thus itself be an indirect infringer. See 35 U.S.C. § 271(b). Another form of indirect infringement occurs when a party provides a component (e.g., a drug) that is required for use with a patented method or system, where there are no substantial noninfringing uses of the component. This type of indirect infringement is referred to as contributory infringement. See 35 U.S.C. § 271(c). Thus, if method-of-use

13. *Id.*
14. Festo Corp. v. Shoketsu Kinzoku Kogyo Kabushiki Co., 535 U.S. 722, 736 (2002).
15. Abbott Labs. v. Dey, 287 F.3d 1097, 1103 (Fed. Cir. 2002).
16. YBM Magnex , Inc. v. U.S. Int'l Trade Comm., 145 F.3d 1317 (Fed. Cir. 1998).

claims are analyzed in the noninfringement opinion, attention must be given to whether the intended uses of the generic drug could give rise to indirect infringement. Contributory infringement generally can be avoided if a substantial noninfringing use of the drug composition exists. On the other hand, avoiding induced infringement may require careful attention to the indications listed in the instruction sheet and in promotional materials, and appropriate cautionary advice should be included in the opinion in such circumstances.

VII. Elements of an Invalidity Opinion

Because invalidity is an absolute defense to patent infringement, a position of invalidity can be a strong position for an ANDA product, as it avoids the sometimes nuanced analysis of noninfringement, especially where it includes coverage of equivalents or possible indirect infringement. Nevertheless, the presumption of validity of an issued patent must be overcome with clear and convincing evidence.[17] Consequently, a questionable invalidity position should not be the basis of an opinion.

Invalidity must be proved on a claim-by-claim basis, and every claim possibly infringed by the ANDA product or its method of use must be invalidated, including all relevant dependent claims. Any of the typical grounds for lack of patentability that are encountered during patent prosecution can be grounds for invalidity of an issued patent, as outlined in 35 U.S.C. §§ 102, 103, and 112. These commonly include lack of novelty, obviousness, and failure to meet the requirements for written description or enablement. Also becoming more common as a basis for invalidity is lack of patentable subject matter under 35 U.S.C. § 101 and related case law. Additional bases for invalidity or unenforceability can be considered, including lack of best mode, inequitable conduct, failure to name the correct inventors, or even formal defects in the application. However, often the facts required to prove these issues are not available to the opinion writer, though they might surface later, such as during litigation. If suitable facts are available, they should be used in the opinion.

When a prior art basis for invalidity is needed, the prior art can be found either by performing a prior art search and identifying new prior art not considered during prosecution of the patent, or it can be selected from among the references considered by the examiner. However, an issued patent is entitled to a presumption of validity over the references considered by the examiner and listed at the front of the patent under "References

17. 35 U.S.C. § 282; Microsoft Corp. v. i4i LP, 131 S. Ct. 2238 (2011).

Cited." Therefore, if any of those references are used as the basis for invalidity, that presumption will have to be overcome.

A preferred basis for proving invalidity is anticipation (lack of novelty), as it is the simplest to prove, if suitable prior art is available, and its proof is generally more objective than proving obviousness or deficiencies under § 112. A single prior art reference is found that discloses every element of the claim.[18] Additional evidentiary references can be used to prove the inherency of any missing elements or to define terms in the primary reference. As for a noninfringement opinion, claim charts can prove very useful to demonstrate lack of novelty of patent claims, as they link each limitation of the claim to a corresponding disclosure in the reference. When an element of the claim is not explicitly disclosed in the prior art reference, it may be disclosed inherently.[19] For an inherent disclosure, the relevant feature of the claim must necessarily be present in the anticipating reference, but there is no longer a requirement that the inherency be recognized in the prior art.[20]

If new prior art has been found that supports anticipation of the claims, then this can be a strong and straightforward basis for an invalidity opinion. If, on the other hand, the prior art used to support anticipation is prior art that was considered by the examiner, then a convincing explanation must be presented for why the examiner failed to make a suitable rejection or mistakenly removed such a rejection and allowed the claim.

Sometimes lack of novelty exists for an independent claim but not for one or more dependent claims, which recite additional limitations not disclosed by the reference that anticipates the independent claim. In such cases, one or more additional prior art references can be combined with the anticipating reference to support the obviousness of the dependent claims. Many dependent claims merely recite obvious additions to the independent claim, and such dependent claims can often be proved to be obvious with easily identifiable further references or even based on general knowledge in the field that renders obvious the added limitation(s). However, caution must be exercised to evaluate all of the added limitations of the dependent claims.

If no novelty destroying prior art can be found, then invalidity analysis often turns to obviousness. Proving obviousness can be a somewhat subjective exercise. The legal standard is summarized in the "*Graham* factors," which are: the scope and content of the prior art, the differences between

18. Rockwell Int'l Corp. v. United States, 147 F.3d 1358 (Fed. Cir. 1998).
19. Cont'l Can Co. USA Inc. v. Monsanto Co., 948 F.2d 1264 (Fed. Cir. 1991).
20. Schering Corp. v. Geneva Pharm. Inc., 339 F.3d 1373 (Fed. Cir. 2003).

the prior art and the claimed invention, the level of ordinary skill in the art, and objective indicia of nonobviousness.[21]

The scope and content of the prior art can be determined from the patent disclosure, the prosecution history, the references cited on the face of the patent, and from any additional prior art search carried out. The patent claims and any statements of the technical problem solved are also illustrative. A brief discussion of the scope and content of the prior art is usually sufficient to set the stage for selecting the closest and most appropriate prior art for use in the remainder of the analysis.

The differences between the prior art and the claimed invention are discussed next and become the major focus of the obviousness analysis. The closest prior art references should be described in some detail and particular attention given to their disclosure of key limitations in the claim. Usually, a primary reference is selected that discloses the most elements of the claim; it is typically the "closest prior art." Nevertheless, the primary reference will be missing one or more limitations of the claim (otherwise the reference would be anticipatory), which limitations are found in one or more secondary references. After a discussion of the secondary reference(s), the basis for combining the references must be discussed.

The level of skill in the art tends to be a less important factor when analyzing pharmaceutical patents and methods of treating or diagnosing disease, since most practitioners in the field have a PhD or MD degree. The typically high level of skill of both academic and industry scientists involved in drug discovery, and the extensive expertise and technical support in their laboratories, reduces the burden to prove obviousness when appropriate prior art is available.

Secondary indicia of nonobviousness include the following factors: unexpected results associated with the claimed invention, commercial success of the invention, long-felt need for the solution provided by the invention, failure of others to solve the problem solved by the invention, copying of the invention, licensing of the patent, and prior skepticism that a solution could be found to the problem solved by the invention.[22] Any known evidence related to these factors should be included in the analysis. However, such evidence may or may not be available to the opinion writer. The prosecution history should be checked for any presentation of applicable objective evidence of nonobviousness, such as in rebuttal of obviousness rejections made by the examiner. The ANDA product manufacturer also should be questioned for any known evidence of this kind.

21. Graham v. John Deere Co., 383 U.S. 1 (1966).

22. *In re* Rouffet, 149 F.3d 1350 (Fed. Cir. 1998).

The analysis of obviousness is complicated by the lack of firm guidelines to use in determining whether it would have been obvious to a person of ordinary skill in the art at the time of the invention to combine certain features from the prior art to arrive at the invention. The Court of Appeals for the Federal Circuit has devised the "teaching, suggestion, motivation," or TSM, test as one method of analyzing this aspect.[23] In the TSM test, evidence for a teaching, suggestion, or motivation to combine the prior art teachings is sought in the prior art itself, in the nature of the problem solved by the invention, or in the general knowledge of a person of ordinary skill in the art. However, the Supreme Court in *KSR International v. Teleflex, Inc.* cautioned that the TSM test should not be rigidly applied and acknowledged that the ordinary skilled person could be motivated by common sense in making substitutions of known elements for one another, especially where the number of alternative choices is limited and the results are predictable.[24] Thus, under *KSR*, claims that merely substitute one known element for another and yield a predictable result are considered obvious.

The *KSR* holding in 2007 produced a significant change in the legal standard for demonstrating obviousness. Claims reciting combinations that were considered merely "obvious to try" and thus patentable prior to *KSR* may be properly considered obvious after *KSR*. Because Supreme Court decisions have retroactive effect unless they state to the contrary, this has created a significant opportunity for generic drug manufacturers to invalidate patents for obviousness, particularly where claims were allowed based on structural substitutions or modified formulations without demonstration of unexpected properties or other secondary indicia of nonobviousness. Often patents that issued prior to *KSR* can be invalidated for obviousness simply because of the known prior art cited during prosecution, when the standards outlined in *KSR* are applied.

Another significant change in legal standards has occurred in the definition of non-patentable subject matter. Although 35 U.S.C. § 101 broadly defines patentable subject matter, the exceptions thereto defined by case law have grown dramatically. Specifically, the law has changed in two areas that affect patents covering innovator biologic and pharmaceutical products and methods of their use: (1) medical diagnostics that rely on correlations of naturally occurring substances with disease, and (2) the use of genomic DNA for diagnosis or production of therapeutic products.

In *Mayo Collaborative Services v. Prometheus Laboratories, Inc.*, the Supreme Court held that claims directed to methods of diagnosis involving a newly discovered correlation between a disease marker and a medical

23. *See, e.g.*, Al-Site Corp. v. VSI Int'l Inc., 174 F.3d 1308 (Fed. Cir. 1999).
24. KSR Int'l Co. v. Teleflex Inc., 550 U.S. 398 (2007).

condition are not patentable because they read on and preclude the use of a law of nature, where the methods used and the marker are known and merely conventional.[25] The *Mayo v. Prometheus* decision has invalidated large numbers of patents directed to methods of diagnosis, particularly those involving a previously known marker that is detected using previously known analytical methods and components, where the novel and nonobvious aspect was the previously unknown correlation with a disease or physiological state.

In *Association for Molecular Pathology v. Myriad Genetics, Inc.*,[26] the Supreme Court held that isolated, naturally occurring DNA sequences, such as genomic DNA, are not patentable, while cDNA and other non-naturally occurring nucleic acid sequences are still patentable. The *AMP v. Myriad* decision also raises as yet unresolved questions about the patentability of isolated natural products. Method-of-use claims involving natural products, including genomic DNA, were not dealt with in *AMP v. Myriad* and remain patentable. Although a large number of biotechnology patents were rendered invalid by the *AMP v. Myriad* decision, the continued availability of method patents has limited its impact.

The written description and enablement requirements are less often used as the basis for patent invalidity than prior art grounds, due to the difficulty of presenting clear and convincing evidence in many cases. However, in certain cases, these can be effective grounds for invalidity. Lack of written description may be shown where a claimed feature, especially one added during prosecution, is not described at all in the disclosure (i.e., represents "new matter"), and this fact was missed by the examiner. Another basis for inadequate written description can be the lack of any representative species, or the presentation of very few or only inapplicable species, to support a genus, particularly in the biotechnology area.

In order to satisfy the enablement requirement, the disclosure must include sufficient guidance in how to make and use the claimed invention, over the entire scope of the claims, without carrying out undue experimentation. It is important to recognize the level of skill of the ordinary person at the time of the invention in order to determine how much guidance is required and to determine what constitutes undue experimentation (i.e., experimentation that is nonroutine or requires inventive skill). The use of the patent specification and/or references that illuminate the state of the prior art is recommended to provide an objective basis. Careful examination of the disclosures of later filed patents (e.g., continuation-in-part applications) can be informative, because subject matter and claims were

25. Mayo Collaborative Serv. v. Prometheus Labs. Inc., 132 S. Ct. 1289 (2012).
26. Ass'n for Molecular Pathology v. Myriad Genetics Inc., 133 S. Ct. 2107 (2013).

typically added that would not have been enabled by the earlier filed application. Attention should also be given to whether the full scope of the claims is enabled. Often there is unpredictability as to the efficacy of the invention over a fully claimed range; if such unpredictability is not removed with adequate data or examples, the claim may be invalid.

Double-patenting rejections identified in the prosecution history should be examined with care. Because these are often held in abeyance until the end of prosecution, the examiner might have forgotten to insist on a suitable terminal disclaimer, without which the patent is invalid for double patenting.

VIII. Caveats and Use of the Opinion

The opinion should conclude with reminders to the client concerning the limits and use of the opinion. The advice in the opinion is personal and directed from the opinion writer to the client, in part because it is based on facts specific to the client (not all of which are necessarily stated in the opinion). Therefore, it is worth reminding the reader that other parties to whom the opinion is not directed are not entitled to rely on the opinion. Further, the opinion may be limited by certain search criteria or certain subject matter, and therefore its conclusions may be based on certain assumptions, which should be stated. The client also should be reminded that a court could well reach a different view than that expressed in the opinion. Finally, the recipient of the opinion should be reminded of the need to maintain the confidentiality of the opinion.

chapter 18

ANDA Preparation (with an Eye toward Approval and Litigation) and the FDA Review[1]

I. Introduction

As a manufacturer of generic drug products, undoubtedly one of your goals is to prepare an Abbreviated New Drug Application (ANDA) that will pass regulatory scrutiny efficiently. Your ANDA submission is a culmination of time, money, and effort by many in your organization: from those responsible for identifying target products to those responsible for developing those products and, finally, to those responsible for filing the application. This effort can take several years, and even the smallest missteps can result in months of delay and possibly millions in lost revenue and sunk costs. Keeping your goal in mind, every communication that you have with the Food and Drug Administration (FDA) should be made with an eye toward making the FDA's job of reviewing your ANDA easier so that you can get your approval faster.

But as you probably already know, your ANDA can be a technical act of patent infringement that triggers a patent infringement lawsuit from your adversary under the Hatch-Waxman Act. Let's face it: the ANDAs with the most potential reward are the ones most likely to be sued on. Your ANDA filing, therefore, should be made with an eye not only toward regulatory compliance and approval but also toward the litigation that will likely follow. If your team has devoted the effort to developing a Paragraph IV

1. Kevin M. Nelson and Patrick C. Gallagher of Duane Morris LLP, Chicago, Illinois; Matthew C. Mousley of Duane Morris LLP, Philadelphia, Pennsylvania; and Christopher S. Kroon of Duane Morris LLP, Boston, Massachusetts.

strategy, make sure that the entire ANDA submission accurately portrays the proposed ANDA product in light of the potential patent obstacles.

With that in mind, this chapter digs into the ANDA itself—the required components in the initial filing and post-filing submissions you may need to make—and discusses the FDA's review of your application. In the context of this discussion, we will provide some insight on the FDA's review of your application. We also will discuss how to craft certain parts of your submissions with the knowledge that your audience is not only the FDA but also a future litigant, and a federal judge or jury.

Let's begin with a discussion of the driving force behind most business decisions—cost. Specifically, we begin our discussion with the Generic Drug User Fee Act (GDUFA) and how much filing an ANDA will cost.

II. A Fistful of Dollars: The Generic Drug User Fee Act (GDUFA)

Unfortunately, the costs do not stop once you have completed developing your product and preparing your ANDA. An ANDA applicant is also required to pay certain fees associated with filing its ANDA. The types of fees are identified in the Generic Drug User Fee Act, which is commonly known as GDUFA, and the amount of each fee is published annually in the *Federal Register*. GDUFA fees are just one cost component of your ANDA application process, but it is important enough to merit this brief, initial discussion.

The stated purpose of GDUFA is to speed the entry of safe and effective generic drugs to the market and to reduce costs to the industry.[2] In that respect, GDUFA shares a common goal with the Hatch-Waxman Act. For the purpose of this chapter, we will focus on only two of the GDUFA fees: the generic drug submission fee and the facility fees. The backlog fee—a one-time fee applied to ANDAs pending on October 1, 2012, to reduce the FDA's backlog of applications—and the Drug Master File (DMF) fee are outside the scope of this chapter.

A. Generic Drug Submission Fee

The generic drug submission fee can be thought of as your application fee. The submission fee is a one-time fee due on the date of submission of the application. There are two components to the submission fee: a base

2. U.S. Food & Drug Admin., Guidance for Industry, Generic Drug User Fee Amendments of 2012: Questions and Answers 2 (Revision 1, Sept. 2013) [hereinafter GDUFA 2013 Guidance].

submission fee and a fee calculated with reference to active pharmaceutical ingredient (API) manufacture. For fiscal year 2013, the base submission fee was $51,520. The ANDA submission fee for 2014 is $63,860.[3] The base submission fee is adjusted annually for inflation and is based on the number of projected ANDAs that will be filed in the given year. The base submission fee for the year is published in the *Federal Register* no later than 60 days before the start of each fiscal year (i.e., no later than 60 days before October 1 of each year).

1. Base Submission Fee

The FDA will not consider your application received unless the submission fee is paid at the time of the application or within 20 calendar days after filing the ANDA. If you pay the submission fee after 20 calendar days, the FDA will consider the day it receives the submission fee as the "received" date of your application.[4] This delayed received date could mean that your ANDA goes from being a "first-filed application," which could be entitled to 180 days of marketing exclusivity, to an application that is not entitled to "first-filer" status. In other words, missing the payment date by just one day can mean a loss of millions of dollars in revenue and significant market share. If the FDA refuses your application for a reason other than failure to pay fees associated with the ANDA, you are entitled to a refund of 75 percent of the submission fee.

The submission fee also applies to Prior Approval Supplements (PASs) to ANDAs. PASs are supplements to an approved ANDA and will be dealt with in greater detail in Section IV.B.2.b., "Applicant-Initiated Changes," of this chapter. Briefly, a PAS is a major change to an approved ANDA that could have a substantial adverse effect on the identity, strength, quality, purity, or potency of the drug product.[5] A PAS may include a change in manufacturing site or manufacturing process.[6] Importantly, because generic drug product labels must be the "same as" the product label of the appropriate Reference Listed Drug (RLD),[7] a PAS may also include a change to your approved product label as a result of a labeling change by

3. *Abbreviated New Drug Application (ANDA) and Prior Approval Supplement (PAS) Fees*, U.S. Food & Drug Admin., http://www.fda.gov/ForIndustry/UserFees/GenericDrugUserFees/ucm319568.htm (last updated Feb. 5, 2014); *see also* Generic Drug User Fee—Abbreviated New Drug Application, Prior Approval Supplement, Drug Master File, Final Dosage Form Facility, and Active Pharmaceutical Ingredient Facility Fee Rates for Fiscal Year 2014, 78 Fed. Reg. 46,977, 46,978 (Aug. 2, 2013).

4. GDUFA 2013 Guidance, *supra* note 2, at 11.

5. U.S. Food & Drug Admin., Guidance for Industry, Changes to an Approved NDA or ANDA 3 (Apr. 2004); *see also* 21 C.F.R. § 314.70(b).

6. GDUFA 2013 Guidance, *supra* note 2, at 10–13.

7. 21 U.S.C. § 505(j)(2)(A)(v); 21 C.F.R. § 314.94(a)(8)(iii).

that brand RLD.[8] The PAS fee for fiscal year 2013 was \$25,760, and the PAS fee for fiscal year 2014 is \$31,930.[9]

2. API-related Fee

The API-related fee involves some calculation, which might be painful if you dislike math. But we'll get through this together. GDUFA requires that an ANDA applicant pay a one-time fee for each API manufactured in its own facility *for which the applicant has not previously paid an API-related fee* (i.e., payment as part of its DMF submission). This API-related submission fee is calculated based on the number of APIs referenced in the ANDA and the number of additional facilities identified in the ANDA in which those APIs will be manufactured. It is important to note that this one-time fee is different from the annual facility fee that must be paid for each API-manufacturing facility, which will be discussed in more detail later in this chapter. Roughly, the equation for figuring out the API-related submission fee is:

$$((n)\text{APIs referenced in ANDA} + (n)\text{Additional facilities for each API}) \times \text{DMF fee}$$

Remember, the fee is paid only once per API and only once per facility manufacturing that API, but you will have to pay an additional facility fee if you have a different drug product using the same API that is manufactured in a different facility. See, not so bad.[10]

So, with that taken care of, let's move on to facility fees.

B. Facility Fees

Anyone who owns a facility that is identified or intended to be identified in at least one ANDA to produce a finished dosage form or an API is required to pay a facility fee.[11] GDUFA defines a finished dosage form (FDF) facility as one that makes drug products—tablets, capsules, solutions, topicals, or drug products requiring reconstitution.[12] GDUFA defines an API facility as one that makes drug substances that are (1) intended to be used as a component of a drug and that furnish pharmacological activity, diagnosis, cure, treatment, or prevention of a disease, or (2) intended for final crystallization, purification, or salt formation to become a substance described in (1).[13] A facility that manufactures both FDFs and APIs will incur *both*

8. U.S. Food & Drug Admin., Guidance for Industry, Changes to an Approved NDA or ANDA 24–25 (rev. 1 Apr. 2004).

9. GDUFA 2013 Guidance, *supra* note 2, at 5.

10. For further discussion from the FDA, see GDUFA 2013 Guidance, *supra* note 2, at 11–13.

11. *Id.* at 14.

12. *Id.* at 16.

13. *Id.* at 16–17.

an annual FDF fee and an annual API fee.[14] The facility fee is an annual fee. For 2013, the domestic FDF facility fee was $175,389, and the domestic API facility fee was $26,458. For 2014, the domestic FDF facility fee is $220,152, and the domestic API facility fee is $34,515.[15]

A company may have to pay separate facility fees if its facilities are in two different geographic locations. Two locations will not be considered geographically distinct for the purposes of the facility fee, however, so long as the separate buildings are within close proximity, have closely related activities, are under the same local management, and are capable of being inspected by the FDA during a single inspection.[16] Facilities located outside the United States and its territories will be charged a fee not less than $15,000 and not more than $30,000 more than the amount of the fee for a domestic facility.[17] Thus, where you manufacture your FDFs and APIs is an additional component of the total cost analysis.

C. Drug Master File (DMF) Fees and Additional Information

A full discussion of DMF fees is outside the scope of this article. However, a word of caution is in order for those companies filing ANDAs that rely on a DMF of a separate company. If a DMF holder has not paid its GDUFA fees, the DMF will be deemed not available for reference, and no ANDA will be accepted for filing or reviewed unless the fee is paid.[18] There is a 20-calendar-day grace period after the FDA has provided notification of delinquency to pay any late DMF fee.[19] It should go without saying that any failure to timely pay a DMF fee could jeopardize your status as a first-filer ANDA and your place in the exclusivity line. More information on DMF fees, and other GDUFA fees, can be found at the FDA's draft *Guidance for Industry—Generic Drug User Fee Amendments of 2012: Questions and Answers*, Revision 1, September 2013.

D. Some Helpful Hypotheticals

Consider the following hypotheticals to help get the feel of the GDUFA costs associated with filing an ANDA. Assume you own Staley Generic Pharmaceuticals and you intend to file an ANDA seeking approval to market a generic version of helpamide tablets. You are a vertically integrated company and will be making the API for the helpamide tablets in both your Illinois and New Jersey facilities and the tablets in your Illinois

14. *Id.* at 17.
15. *Id.* at 5.
16. *Id* at 18–19.
17. *Id.* at 18.
18. *Id.* at 9.
19. *Id.*

facility. Using 2013 numbers, here is your cost of filing the ANDA for helpamide tablets:

ANDA base submission fee:	$ 51,520
API submission fee (1(API) + 1(Additional API facility)) × $21,340 (DMF fee):	$ 42,680
Facility fee $26,458 (NJ API) + $175,389 (IL FDF):	$201,847
Total	$296,047

If Staley Pharmaceuticals did not make its own API but, instead, simply referenced another company's DMF, it would only pay the ANDA submission fee and the FDF facility fee for its helpamide tablet ANDA (a total of $226,909).

Now let's say that instead of manufacturing its API in New Jersey, Staley Pharmaceutical's API facility is located in India, while the FDF is still manufactured in Illinois. In this situation, the ANDA base submission fee and the API submission fee would remain the same ($51,520 and $42,680, respectively), but, as a result of employing a foreign API facility, the facilities fees will increase an additional $15,000 to $30,000:

ANDA base submission fee:	$ 51,520
API submission fee (1(API) + 1(Additional API facility)) × $21,340 (DMF fee):	$ 42,680
Facility fee: (India API) $26,458 + ($15,000–$30,000):	$ 31,458–56,458
Facility fee: (IL FDF):	$175,389
Total	$311,047–326,047

III. The Initial ANDA Filing

With an understanding of the fees associated with an ANDA filing, we can now move to the nuts and bolts of actually compiling and filing the ANDA.

Before actually filing your ANDA, you should obtain a pre-assigned application number. In fact, you are required to obtain a pre-assigned application number before filing an electronic copy of your ANDA, which, as discussed in the following section, is the FDA's preferred method for receiving applications. You can request a pre-assigned electronic application number by contacting the Office of Generic Drugs (OGD) at the Center for Drug Evaluation and Research (CDER). The FDA's website provides instructions on how to request an electronic application number and lists the information that will be required when making the request. The FDA will provide a pre-assigned application number within three business days after receiving a request, but plan ahead so that you don't miss out on a filing deadline because you are waiting for an application number. Because your six-digit application number ties your entire submission together, it should be included throughout your ANDA submission and on any correspondence with the FDA regarding the application.

A. Paper or Plastic?

The FDA allows you to file your initial ANDA application in paper form or in electronic form, but you should choose to file in electronic form. Myriad are the pleas from the FDA in speeches, presentations, guidance documents, and the like to file in electronic form. Do not ignore these pleas. Like most branches of the federal government, the FDA is seeking to reduce the volume of paper it has to deal with. Additionally, the electronic form makes it easier for the FDA to review and to "share" your ANDA or portions of your ANDA among the departments responsible for reviewing your submission. The federal government is so serious about wanting submissions in electronic, as opposed to paper, format that Congress enacted Section 745A(e) of the Federal Food, Drug, and Cosmetic Act, which requires (among other things) NDAs and ANDAs to be filed in electronic format beginning 24 months after the FDA finalizes its guidance on providing regulatory submissions in electronic format. The FDA issued Revision 3 of the draft guidance in January 2013.[20]

The format for your electronic submission should be in what is called an XML-based eCTD (electronic Common Technical Document). The XML format helps to organize your application by allowing the user to set up folders and subfolders as well as use hyperlinks to navigate the submission.

20. U.S. Food & Drug Admin., Guidance for Industry, Providing Regulatory Submission in Electronic Format—Certain Human Pharmaceutical Product Applications and Related Submissions Using the ECTD Specifications (Draft Guidance, rev. 3 Jan. 2013).

It goes without saying that, because the FDA is the body responsible for approving your ANDA, your goal should be to make the FDA's job easier. If you do so, you will be that much closer to getting your approval quickly. To that end, consider the following general tips for your electronic submission.

- Hyperlink your sections and use bookmarking so that the FDA can more easily navigate your submission.
- Use hyperlinks for cross-referencing to other documents in your submission. (Make sure you are hyperlinking to the correct document and the correct page of the document.)
- Provide a table of contents (hyperlinked, of course) with headings arranged hierarchically.
- Use easy-to-recognize directory and file names so that the reviewers can quickly and easily find the exact section or document that they are looking for.
- Use searchable PDF documents or, if you cannot use a PDF, create scanned searchable text or OCR (optical character recognition) documents rather than documents that are not searchable. Again, this makes it easier for the FDA to find relevant information, and the format allows the FDA to cut and paste materials from your submission into deficiencies, requests for information, or other correspondence.

One last note on electronic submissions: have an additional copy of your electronic submission for your outside counsel handling the notice letter and patent litigation, and make sure that the copy is in the same format that you provided to the FDA. Your outside counsel needs to know the contents of the application on which the notice letter and subsequent litigation are based, and your ANDA (or portions of it) will need to be offered to the NDA holder and patent owner(s) (on a confidential basis) during the 45 days after receipt of your notice letter. From a practical standpoint, you do not want your outside counsel to have to explain to a judge any delay (absent failure to agree on portions of the ANDA that will be disclosed or individuals who can review the ANDA) in providing your ANDA to the other side during that 45-day period.

B. The ANDA Checklist—Managing Your Modules

Now that you have figured out how much the filing fees will be for your ANDA and that you will file in electronic format, it is time to focus on the component parts of the ANDA that you will need to include for your submission. Fortunately, the FDA has made this part easy. The FDA provides an ANDA checklist—the very document that the FDA will use to determine if you have provided all the necessary components for your ANDA and to provide internal comments on the specific ANDA sections. The ANDA checklist is available at the FDA's website as a PDF link from its generic drug user's page.[21] Use it.

Keep in mind, as discussed further in the following sections, that the FDA does not start its substantive review of your ANDA until it has determined that your ANDA is technically complete (i.e., it includes all of the required portions). This determination can typically take around two to three months, but it can take as long as five to six months.

When it comes to small-molecule products, the FDA knows exactly what it wants to see in an ANDA submission. For the most part, the decision of what to include in your ANDA is not an opportunity for creativity. If you do have a unique aspect to your filing, be upfront with the FDA and provide a short but detailed explanation for your deviation. Do not make the FDA search for or question any new or missing aspect. And, if needed, file an appropriate suitability petition. We address specific modules and what they should contain in the following sections.

1. Module 1: Administrative Materials (and the Patent Certification(s))

For the most part, Module 1 is the simplest module, and there are few excuses for having any issue with this module. The required portions for electronic versions of Module 1 include:

1.1	Signed Application form 356h
1.2	Cover letter
1.2.1	Form FDA 3674
1.3.2	Field Certification Form (21 C.F.R. § 314.94(d)(5))
1.3.3	Debarment certification
1.3.4	Financial certification
1.3.5	Patent information
1.4.1	References
1.12.4	Request for comments
1.12.11	Basis for submission
1.12.12	Product comparison
1.12.14	Environmental impact analysis

21. http://www.fda.gov/Drugs/DevelopmentApprovalProcess/HowDrugsareDevelopedandApproved/ApprovalApplications/AbbreviatedNewDrugApplicationANDAGenerics/.

1.12.15 Request for waiver of in vivo BA/BE (bioavailability/bioequivalence) studies

1.14.1 Draft labeling

1.14.3 Listed drug labeling

Of these required components, each of which is important, we will focus on the Form 356h (but only briefly), the cover letter (also briefly), the patent information and certifications, and the labeling sections. It should go without saying that it is imperative to be candid with respect to all sections of your application, including the debarment and financial certifications.

First, Form 356h is used for many different types of submissions to the FDA, including for original ANDA submissions, 505(b)(2) submissions, supplements to pending applications, annual reports, and more. The FDA provides a fillable PDF version of Form 356h at the "Forms" section of its website along with instructions regarding use of the form. You must sign and complete Form 356h.

Every submission to the FDA should also contain a cover letter. If there is anything that is atypical about your submission, use the cover letter to briefly explain the situation to the FDA. Do not bury any issues that your application may have, but instead get your story out front to facilitate conversation with the FDA. Make the FDA's job as easy as possible.

The next major component of Module 1 is the patent information and certification section. In this section, you must include the patent number and expiration date of all of the patents listed in the Orange Book (OB patents) for the appropriate RLD. The patent numbers and expiration dates associated with the appropriate RLD can be found in the FDA's electronic Orange Book.

You must also include a certification for each OB patent. The certification is simply a one- or two-page document that lists the patent number(s), the type of certification (Paragraph I–IV or Section viii), and the date of expiration of the patent(s). Briefly, a Paragraph I Certification states that no patent information is listed in the Orange Book. A Paragraph II Certification states that any OB patents have now expired. With a Paragraph III Certification, the ANDA applicant is certifying that it will not market its product before the expiration of that particular patent or patents identified in the certification. Finally, with a Paragraph IV Certification, the ANDA applicant is certifying that it intends to market its product before expiration of the OB patent(s) and that the patent certified against is invalid and/or will not be infringed by the ANDA applicant's product. With respect to the Paragraph IV Certification, the short certification is not the same as the notice letter that you are required to send to the NDA holder and patent owner detailing why your product, if approved, will not infringe an OB patent or why the OB patent is invalid. Those are two different submissions made to two different entities. Moreover, the decision as to the type of certification you will make has downstream implications that occur after submission of your application and should be made jointly with legal counsel.

The final section of Module 1 we will discuss, the labeling section, requires multiple copies of your proposed draft product label (packaging, product insert, and a side-by-side comparison of labeling of the RLD with annotations) as well as the RLD's labeling and packaging. From a submission standpoint, this section is quite simple. With some exceptions, the generic product's labeling must be the "same as" the approved labeling for the RLD or else the FDA will refuse to approve the ANDA.[22] The exceptions to the "same label" requirement are changes as to the identity of the manufacturer and distributor, the use of the generic name of the product instead of the brand name, changes in the identity of certain inactive ingredients,[23] and "Section viii" carve-outs of additional indications for which the ANDA sponsor is not seeking to market its product.[24] The "same label" requirement allows the FDA to assume that bioequivalent generic drugs are safe and effective without having to demand and review minor changes in product labeling, thus fulfilling the statutory goal of getting lower-cost generics on the market quickly.[25]

Thc proposed product labeling should be a consideration early in product development. Increasing are the number of ANDA cases involving patent claims that involve information in the product label, including many cases involving method-of-use claims. For example, patent holders argue that the label instructs or "induces" a doctor or patient to infringe a method-of-use claim and that the ANDA applicant is therefore responsible for that infringement. It is therefore imperative to know the product label for the RLD during the development stage of your product to determine whether there are indications from the label that are not covered by the patent claims or that can be carved out.

It is also important to know what kinds of changes you can make to your product label. Although frequently asserted during litigation, it is untrue that a generic company should simply petition the FDA to change its label if, for example, the generic product has a "different" pharmacokinetic

22. 21 C.F.R. §§ 314.105(d), .127(a)(7).

23. *Id.* § 314.94(a)(8)(iii), (iv).

24. 21 U.S.C. § 355(j)(2)(A)(viii); 21 C.F.R. § 314.94(a)(8)(iv).

25. SmithKline Beecham Consumer Healthcare LP v. Watson Pharm. Inc., 211 F.3d 21 (2d Cir. 2000).

profile than the Reference Listed Drug. Putting aside changes to the label as to the manufacturer of the product, excipients that are used, or where there is a Section viii statement, the FDA will not approve an ANDA if the generic product label is different from the RLD's label except where such differences are approved after a petition under 21 C.F.R. § 314.93.[26] Although a separate section of the regulations seems to indicate that an ANDA applicant can petition for labeling changes with respect to aspects of its product like its pharmacokinetic profile, that section only allows for differences that are approved under a Section 314.93 petition.[27] And the *only* differences in labeling that are permitted by a Section 314.93 petition are differences in which the ANDA product "is not identical to a listed drug in route of administration, dosage form, and strength, or in which one active ingredient is substituted for one of the active ingredients in a listed combination drug."[28] Most ANDA products have the same active ingredients, "route of administration, dosage form, and strength" as the RLD. Therefore, according to the regulations, an ANDA applicant would not receive permission for a labeling change with respect to, for example, a "different" pharmacokinetic profile.

The label is typically a primary document in any subsequent patent litigation. It is important to recognize what the label says, what can and cannot be changed in the label, and implications for patent infringement arguments in litigation. While there is little merit in the litigation argument that an ANDA applicant can make all sorts of changes to its label, an ANDA applicant must be aware of the information required to be in its product labeling and the potential implications for litigation.

2. Module 2: Quality Overall Summary (QOS) and Clinical Summary

As the title indicates, Module 2 provides summary information for the reviewers and allows for a high-level understanding of an ANDA application and the product described therein. While the documents that comprise Module 2 are summary documents, an applicant still must be careful to properly describe and cross-reference its product and supporting data in this section because even these summary documents will find their way into a subsequent ANDA litigation. Thus, the individual tasked with drafting these summary documents should work with the author of the primary sections (as well as with counsel, if possible) to ensure both completeness and accuracy.

---

26. 21 C.F.R. § 314.127(a)(7).
27. *Id.* § 314.94(a)(8)(iv).
28. *Id.* § 314.93(a), (b).

a. Quality Overall Summary (QOS)

The QOS is the summary or Cliff Notes version of the Chemistry Manufacturing and Controls (CMC) section of Module 3. In fact, a great deal of the information provided in the QOS is put verbatim into the CMC section. So, you may ask, what is the rationale for having a QOS at all? The primary reason for a QOS is to summarize pertinent information about the drug substance and drug product of the ANDA, such as how it is characterized and how it is made, so reviewers (including reviewers that need the information but are not CMC experts) can quickly access certain information about your product without digging into the more detailed CMC section.

The QOS is separated into information regarding the drug substance (Section 2.3.S, with the "S" standing for "substance") and information regarding the drug product (Section 2.3.P, with the "P" standing for "product"). The FDA elicits information regarding the drug substance and the drug product in these sections through a series of standing questions that should be answered by the applicant. As you should now come to expect, the FDA provides these questions on the website of the Office of Generic Drugs and has a sample QOS with answers.

I. 2.3.S Drug Substance (Active Pharmaceutical Ingredient)

In the section on the drug substance, the applicant is asked to provide information on the molecular structure, formula, and weight; the manufacturer of the drug substance and manufacturing controls; characterization of the drug substance; test specifications for the drug substance, including a discussion of any analytical methods that are used; discussion of the certification of reference standards; an explanation of the container closure system used to package and store the drug substance; and stability testing information on the drug substance. Many of the questions in the drug substance section are answered by referring the FDA to the Drug Master File upon which the applicant is relying.

II. 2.3.P Drug Product

In the drug product section of the QOS, the applicant must provide a summary of the components of the drug product and their respective functions; any potential concerns that may be raised from any differences from the RLD; a summary of the development of the product, including alternative formulations and a rationale for the current formulation; the manufacture of the product and scale-up plans; control of excipients and the drug product; reference standards; the container closure system; and stability testing.

As with other sections of your ANDA, draft these summaries cognizant of the way the drug product was developed and with an eye toward possible litigation. While the QOS provides a helpful summary for the ANDA reviewers, if it is not artfully drafted, it could also be useful for a future adversary who takes poorly worded statements out of context in litigation. For example, Section 2.3.P.1 requests information regarding the function of each excipient in your drug product. Many patent claims are drafted such that a certain excipient is functioning in a specific manner in the claimed formulation. If the excipient in your formulation is not functioning in the same way that excipient functions in the RLD (e.g., not acting as an extended-release excipient or extending release in a substantially different way), then you should not characterize it this way merely because the RLD does.

b. Clinical Summary (Bioequivalence)

The information provided in Module 2.7—the clinical summary section—should be primarily in tabular form based on the data from Module 5. The tables should provide highlights of the data, with brief summaries to make the most pertinent information readily accessible to the reviewer. The more detailed information and reports should be reserved for Module 5.

3. Module 3: Drug Substance and Drug Product Information—The Pharmaceutical Development Report (PDR)

Module 3 will contain detailed information regarding the drug product and drug substance relevant to your ANDA. Here, the FDA wants to know about testing specifications and procedures, validation methods, and stability protocols, as well as how the drug substance and drug product are made and will be made on a large scale. From a safety perspective, the FDA will be focusing on controls, validation, and stability. From a manufacturing perspective, the FDA will be focusing on the description of the manufacturing process and process validation for the drug product (Sections 3.2.S.2 and 3.2.P) and the executed batch records for the drug product (Section 3.2.R). But the full story of your product and how it came before the FDA is told in the pharmaceutical development report found at Section 3.2.P.2. That is the story that you will eventually be recounting during litigation, and it should be consistent with your litigation strategy developed at the outset of the project.

a. 3.2.S Drug Substance

The first part of Module 3 contains information about the drug substance—the active pharmaceutical ingredient (API)—in your ANDA product. For much of this information you will be referring to the DMF for the drug substance used in your ANDA product. For certain sections, however, the FDA does not want you to merely refer to the ANDA product. This includes Section 3.2.S.1 with general information about nomenclature, structure, and general properties of the API and Section 3.2.S.5 regarding reference standard or materials. Otherwise, much of the information concerning characterization of the drug substance, identification of impurities, analytical procedures, batch analyses, and stability will likely come from the DMF.

b. 3.2.P Drug Product

The second part of Module 3 contains information about the drug product—the finished dosage form that will be sold to customers. In this section, the FDA will want all of the information concerning development, manufacture, specifications, controls, and analysis for the finished drug product.

c. Pharmaceutical Development Report (PDR)

The pharmaceutical development report is your opportunity to present to the FDA the knowledge that you have gained through the development of the product and its manufacturing process. Your goal in drafting your pharmaceutical development report is to demonstrate to the FDA the scientific rigor that went into developing a bioequivalent drug product that can be reliably manufactured. You should convey an intimate understanding of your drug product and manufacturing process, while explaining any aspect that is unusual or atypical. Again, getting out in front of any potential issues is key to expedient approval. In addition, familiarity with the patent claims covering the formulation of the branded product is important. You may want to stress that the design of your product was deliberately chosen to meet patients' needs and to perform in a particular manner. More on that later.

The following are the key sections of the pharmaceutical development report:

> *The components of the drug product*: This section includes a discussion of properties of the drug substance that can affect performance of the drug product (e.g., solubility, particle size, and biological activity) and information regarding the excipients in the drug product, including any compatibility issues with the drug substance.
>
> *The drug product*: This section should contain a description of the formulation development, including identifying attributes that are *critical to the quality* of the drug product, and should highlight the "evolution of the formulation design from initial concept up to the final design."[29] In this section, you will also want to include information from any comparative in vitro or in vivo studies (with hyperlinked cross-referencing to the actual studies) and any attempts to establish an in vitro/in vivo correlation.

---

29. U.S. Food & Drug Admin., Guidance for Industry: Q8(R2) Pharmaceutical Development 5 (Nov. 2009).

Manufacturing process development: This section should include information on the selection, control, and improvement of the manufacturing process, as well as a discussion of any critical process parameters. You should clearly but succinctly explain any differences between the manufacturing process for batches used in pivotal clinical trials or stability studies and the manufacturing process described in Section 3.2.P.3.3.

Container closure system: This section should provide information regarding the choice and rationale for the selected container closure system for the commercial product.

Microbiological attributes: Where applicable, this section should provide justification for aspects relating to the microbiological aspects of the drug product.

Compatibility: This section should provide supportive information for product labeling, including recommended storage conditions.

While many pharmaceutical development reports tell the story of the drug products described in an ANDA, those same reports can serve as cautionary tales. The pharmaceutical development report is part of your ANDA and will, at some point, be produced to your adversary in litigation. Because the FDA requires so much information regarding the rationale for chosen steps or components and so much information about the results of such steps and components with respect to the performance of your product, care should be taken with how you draft this document. Draft your pharmaceutical development report with the patents that may likely be asserted against you in mind.

The pharmaceutical development report must tell the accurate and complete scientific development story for the proposed ANDA product. All members of the development team should be aware of and on board with any design-around strategies for potential patent noninfringement positions. Sometimes you will develop design-around strategies at the outset of development. Other times, potential noninfringement positions will come about during the development process by finding a better way of making the product using different excipients or different methods of manufacture. Regardless, if you have designed around a patent, and have a potential noninfringement position for a patent, the scientific facts underlying that position should be contained in the pharmaceutical development report. Of course, this report tells a scientific story, not a legal argument. But it is important that the scientific facts underlying that design around are included in the report. At the end of the day, explain the differences between your ANDA product or its components and the RLD, because that is the story that is going to be told in litigation, and it should be a consistent story.

If you are not considering the future patent lawsuit awaiting you when drafting the pharmaceutical development report, you could find yourself fighting against your own written words. In contrast, a well-written pharmaceutical development report not only will assist in obtaining regulatory approval but can also be used as both a shield and a sword in litigation—protecting you from accusations of patent infringement and, at the same time, providing your litigation expert with support from your internal documents for the expert's scientific opinions.

4. Module 5:[30] Clinical Study Reports

Module 5 contains information regarding the supporting clinical studies associated with your ANDA product. This section will include tables, reports, calculations, protocols, and other associated information regarding your bioequivalence studies. Most importantly, this section will provide data supporting the fact that your proposed ANDA product meets the bioequivalence criteria established by the FDA (AUC and possibly C_{max} mean results that are within 80 to 125 percent of those of the RLD at a 90 percent confidence interval), so that the FDA can presume that your ANDA product is safe and efficacious.

IV. All Hands on Deck: Filing Your ANDA

Compiling various sections of the ANDA will be a cross-disciplinary effort. That effort should be shepherded by someone with regulatory experience who can be responsible not only for making sure that all of the technical aspects are present but also for coordinating review of the final components by each department that has main responsibility for each section. After all the time, money, and effort spent to compile your ANDA, now is not the time for shortcuts. Even generally mundane tasks, such as checking spelling or page numbering, are important and should be done with care and attention to detail. Again, these details make the FDA's job easier, which will hopefully bring your approval more quickly with fewer bumps along the road.

Once your ANDA is compiled and reviewed, you are almost ready for filing. The last review of your ANDA, if you are going to submit in electronic form, should be to check all electronic aspects of the file, such as verifying hyperlinks and file names. Many of the FDA reviewers have

30. Module 4 is not applicable to ANDAs.

expressed frustration at small errors in such electronic aspects of the ANDA submissions. These errors are easily found and fixed, and you do not want these errors to be taken by the FDA as a sign of sloppiness or an indication of larger issues with the application.

Remember that ANDA checklist that we discussed in Section III.B of this chapter? The first thing that the FDA will do when it receives your ANDA is to go down that checklist and make sure you have included all of the required sections of the ANDA. Accordingly, the last thing you should do before filing the ANDA is to go through the checklist with the electronic disc that you will be sending to the FDA and make sure that when the FDA opens the disc it will find every item on the checklist in the place it is supposed to be.

Finally, you are ready to file your application. For electronic submissions, you can send a disc containing your ANDA to the OGD's electronic document room.

A. Start the Clock: Acceptance for Filing and Refusal to File

When the FDA receives your ANDA, the reviewer will go through the ANDA checklist to make sure you have included all of the required sections of the ANDA. This is the acceptance for filing review, and the FDA will not substantively review your ANDA until it is satisfied that it meets the technical requirements and has all the required components. The reasons that the FDA will refuse to accept your ANDA for filing are specifically enumerated at 21 C.F.R. § 314.101(d)–(3) and in the FDA's draft *Guidance for Industry, ANDA Submission—Refuse-to-Receive Standards*, October 2013, and include failure to include the required components of an ANDA, failure to properly organize and format an ANDA, failure to pay required GDUFA fees, lack of a U.S. agent, inadequate bioequivalence or stability information, and so forth.

If your ANDA is missing any required components, or if any component is facially deficient, the FDA will refuse to accept the ANDA. That refusal may result in your losing your filing date. One of the purposes in having an initial acceptance for filing review by the FDA is to deter companies from filing incomplete ANDAs as placeholders with the goal of fleshing out issues and fixing them during the course of a complete review. That practice wastes the FDA's time and resources. Acceptance for filing should be an easy hurdle for your application if you provide the FDA what it is looking for or clearly explain any variances at the outset.

Typically, it takes two to three months from the date of filing for the FDA to issue an acceptance for filing or refusal to file letter, but we have heard from companies that are reporting a full five to six months to receive their acceptance for filing. Once the FDA sends you the letter accepting your ANDA for filing, you have 20 days from the date of the postmark on the FDA's letter to send a notice letter to the NDA holder and patent owner(s) with respect to any patents for which you have filed a Paragraph IV Certification.[31]

B. Post-filing Submissions

There are a variety of required and optional submissions that are made after the initial submission of the ANDA, many of which are outside the scope of this chapter. Here, we will focus on the notice letter (which is submitted not to the FDA but to other parties), certifications that must be made with respect to later listed patents, and certain amendments to the ANDA.

1. The Notice Letter

Prudent companies start working on their notice letters before they even start compiling their ANDAs. The notice letter provides the detailed basis for your assertion to the FDA that the OB patents are invalid and/or that your product will not infringe those patents. To be sure, it may take around four months from the time you first file your ANDA until you actually have to send your notice letter. That may seem like sufficient time to develop invalidity and noninfringement arguments, and it may well be. But because you must make the actual noninfringement and invalidity assertions in Module 1 of your ANDA, it is advisable to have a good-faith basis for such a certification at that time. The time between filing your ANDA and serving your notice letter is best spent revising and focusing the arguments in your notice letter.

We will not burden this chapter with a detailed discussion of notice letters. We will, however, provide you with two important practice points. First, recall that the statute requires a detailed basis for your certifications of noninfringement and invalidity, but it does not require that you provide every such basis that you might raise in the context of litigation. In fact, several cases have expressly held that ANDA applicants are not limited to the defenses raised in their notice letter, but may raise additional defenses during the course of the lawsuit.[32] Second, it is important to stay apprised of the Orange Book entry for the RLD applicable to your ANDA, as the sponsoring company may list additional patents for that drug even after you file your ANDA. While those additional patents, in most cases, will not create any additional 30-month stays of approval of your ANDA, you

31. 21 U.S.C. § 355(j)(2)(B)(ii)(I).

32. *See, e.g.*, Abbott Labs. v. Apotex Inc., No. 09 C 7968, 2010 U.S. Dist. LEXIS 68452 (N.D. Ill. July 9, 2010).

are still required to certify to any "later listed" patents[33] as an amendment to your ANDA. You must also send your notice letter with respect to the patent(s) on the same day you send your certification. To avoid being surprised by "later listed" patents, monitor the U.S. Patent and Trademark Office for patents relating to the RLD applicable to your ANDA.

2. Changes: Amendments and Supplements to ANDAs

There are two main situations that may cause you to make changes to your ANDA. First, there are pre-approval changes that you are compelled to make with respect to your ANDA based on a deficiency from the FDA. Second, there are postapproval changes that you make yourself with respect to your ANDA. As in any dealings with the FDA, the key is to be accurate—here, about the category and nature of the change—and to communicate clearly what the change is and why you are making it.

a. Amendments in Response to a Deficiency

It is difficult to pitch a perfect game in baseball; it is nearly impossible to submit an ANDA that is approvable without comment from the FDA. And when the FDA finds fault with your ANDA, you will receive a deficiency letter from the FDA. The various reviewing departments of the FDA have the ability to send deficiencies regarding your ANDA, and often an ANDA will receive deficiencies from multiple departments in one letter. These deficiencies can relate to incomplete information or additional questions regarding chemistry, manufacturing, controls, microbiology, labeling, or bioequivalence. Sometimes the deficiency will request additional information or require additional testing. When responding to a deficiency letter from the FDA, directly address the issue or issues identified in the deficiency letter. Otherwise you risk delaying approval of your application. In other words, answer the question you have been asked, or the FDA will likely just ask it again.

Your response to an FDA deficiency letter will often require an attendant amendment to your ANDA. The FDA will identify in the deficiency letter whether the responding amendment should be categorized as a major, minor, or telephone amendment. Major amendments are required, for example, if the applicant needs to manufacture a new batch of drug product (and with it, new supporting information), if the applicant has to perform a new bioequivalence study, or if the applicant is required

33. There is a distinction between patents that are listed after you file your ANDA, which would be "later listed" as to your ANDA or any other ANDA already on file, and late-listed patents, which are patents for which the sponsor requests listing in the Orange Book more than 30 days after the patent has issued. With respect to the latter, there is no obligation to certify to late-listed patents, but also nothing prohibiting anyone from asserting position on those patents.

to provide new analytical methods or validation data.[34] Minor amendments—where most amendments fall—are for those issues more easily addressed than with a major amendment, including issues where the FDA needs more information or where resolution of the deficiency is outside the control of the applicant.[35] Finally, telephone amendments are reserved for issues of limited complexity that can be resolved more expediently.[36]

Review timelines and review priority vary depending on the type of amendment. Because major amendments are the most complex, they usually have the longest review time frame and have a lower priority than minor amendments or telephone amendments.[37] Major amendments are given the same priority as an original, unreviewed ANDA and, like original ANDA filings, are reviewed on a "first in, first reviewed" basis.[38] The OGD attempts to review major amendments within 180 days.[39]

With minor amendments, the OGD tries to complete its review within 30 to 60 days, but the review can take longer.[40] A categorization of an amendment as minor usually means that the application is closer to approval, and thus such amendments are given higher priority.[41]

Telephone amendments are a reviewer's highest priority, as such amendments typically mean that the application is on the verge of approval with only clarification or postapproval commitments needed.[42] Sometimes a postapproval commitment will be in the form of an update in an annual report or submission of a Changes Being Effected or Changes Being Effected in 30 Days commitment, which are discussed in greater detail in the following section. With a telephone amendment, the FDA and the applicant discuss the issue by phone and agree on the proper resolution. When the FDA requests a telephone amendment, the applicant should respond within ten calendar days of the telephone call.[43] The applicant's response to the telephone amendment is reviewed "upon receipt."[44]

b. Applicant-initiated Changes

The FDA must be informed of any postapproval changes that you make to your product, the process for making your product, or the labeling for your

34. U.S. Food & Drug Admin., Guidance for Industry: Major, Minor, and Telephone Amendments to Abbreviated New Drug Applications 2–3 (rev. 2 Dec. 2001).

35. *Id.* at 3.

36. *Id.* at 4.

37. *Id.* at 2.

38. *Id.*

39. *Id.* at 4.

40. *Id.*

41. *Id.* at 2.

42. *Id.*

43. *Id.* at 4.

44. *Id.*

product. You must, therefore, report any changes that you are making to your ANDA to the FDA, and you must properly categorize your changes so that the FDA can properly evaluate and approve (or disapprove) your changes. Thus, it is important to know the types of changes that can be made to ANDAs and the expected review times for such changes.

Broadly, the changes that can be unilaterally made to an ANDA by the applicant are categorized as major changes, moderate changes, and minor changes.[45] It is imperative that you know the full effect of your proposed changes before implementing them. Accordingly, testing is crucial. Although the title of each category is telling, we provide a bit more substance on each.

> *Major changes*: A major change is one that "has a *substantial* potential to have an adverse effect on the identity, strength, quality, purity, or potency of a drug product as these factors may relate to the safety or effectiveness of the drug product."[46] Major changes include changes to components and composition of the drug product or drug substance, manufacturing sites, manufacturing processes, testing specifications, container closure systems, and labeling. A major change is made through a supplement called a Prior Approval Supplement (PAS).[47] Expedited review of a PAS is available, but the requester must show a public health reason or hardship on the applicant absent expedited review.[48] The applicant cannot make the change requested in the PAS unless the FDA approves the change. Only with approval can the applicant sell any product made using the change requested in the PAS.
>
> *Moderate changes*: A moderate change is one that "has a *moderate* potential to have an adverse effect on the identity, strength, quality, purity, or potency of a drug product as these factors may relate to the safety or effectiveness of the drug product."[49] Moderate changes can include the same type of subject matter as major changes. A moderate change is effected through either a Supplement—Changes Being Effected in 30 Days (CBE 30) or a Supplement—Changes Being Effected (CBE).[50] A CBE 30 is used for changes that the applicant intends on making in 30 days unless the FDA informs the applicant

45. U.S. Food & Drug Admin., Guidance for Industry: Changes to an Approved NDA or ANDA: Questions and Answers 3–4 (Jan. 2001).

46. *Id.* at 3.

47. *Id.*

48. *Id.*

49. *Id.*

50. *Id.*

that a PAS is required in order to make the change. A CBE is for moderate changes that are being made immediately. If the FDA ends up disapproving a CBE 30 or CBE after review, the FDA may order that the applicant cease distributing any product using the disapproved change(s). The less potential there is for the change to have an adverse effect on the drug product, the more likely that the moderate change is a CBE as opposed to a CBE 30. Alternatively, if you are looking for some assurance from the FDA that your moderate change is appropriate before actually implementing it, you should use a CBE 30.

Minor changes: It is not surprising that a minor change is one that "has *minimal* potential to have an adverse effect on the identity, strength, quality, purity, or potency of a drug product as these factors may relate to the safety or effectiveness of the drug product."[51] A minor change is made in the context of an annual report and does not require pre-approval from the FDA. Like major and moderate changes, minor changes can be made with respect to any aspect of the manufacturing process or components of the drug product or its labeling.

When implementing one of the preceding changes, whether through a supplement or an annual report, provide the FDA with a list of all proposed changes in a cover letter (for supplements) or in a summary section (in an annual report).[52] The supplement or annual report should clearly and completely describe the change or changes so that the FDA can quickly determine whether you have used the proper reporting category and can efficiently evaluate the change.

V. Conclusion

Your ANDA should be the result of the collaborative efforts of your scientists, your regulatory affairs professional, your attorneys, and other key decision makers. And at the end of the day, compliance is the key to obtaining swift FDA approval—provide the FDA with what it is looking for in the form it has outlined for you. But compliance with an eye toward preparation for litigation will mean that all of the effort expended in obtaining the FDA approval will not be wasted when the patent lawsuit is filed.

51. *Id.* at 4.
52. *Id.*

chapter 19

Obtaining ANDA Approval and Right to Market[1]

The Drug Price Competition and Patent Term Restoration Act of 1984 (the Hatch-Waxman Act) has competing goals. On the one hand, it is intended to encourage pharmaceutical innovation by providing opportunities for branded pharmaceutical companies to delay generic competition. On the other hand, the act seeks to "get generic drugs into the hands of patients at reasonable prices—fast."[2]

Congress managed to address both goals. For example, the act encourages research and innovation by protecting the patent interests of the patent owner and innovator drug company through exclusivities: patents and patent term extensions; exclusive rights granted by the FDA for pediatric, geriatric, and other applications; and data exclusivity. The act also promotes competition by creating a process to expedite the filing and approval of generic drug applications and to resolve challenges to patents in court before marketing (thus avoiding the risk of a large damages award).[3] The act also entices generics to challenge inapplicable or weak patents by offering 180 days of generic marketing exclusivity under certain circumstances.

I. Exclusivity and Patent Protection for Reference Listed Drugs

Researching and developing new pharmaceuticals for human consumption involves many layers of regulatory review and lots of time and expense. In the first round, the FDA reviews Investigational New Drugs (INDs) to determine if in vitro and animal studies suggest sufficient efficacy, without

1. David S. Bloch of Winston & Strawn LLP, San Francisco, California; Charles B. Klein of Winston & Strawn LLP, Washington, D.C.; and Scott H. Blackman of Nath, Goldberg & Meyer, Washington, D.C.

2. *In re* Barr Labs Inc., 930 F.2d 72, 76 (D.C. Cir. 1991).

3. 68 Fed. Reg. 36,676 (June 18, 2003).

unwarranted toxicity, to warrant further consideration. Once an IND is approved, the applicant will undertake human clinical trials, generally taking up to several years. If those human clinical trials are successful, the company will file a New Drug Application (NDA).[4] Review and approval of NDAs typically require years of consideration, and maintaining the proceeding can involve significant expense. Once a drug is approved, the brand company holds the NDA for the Reference Listed Drug (RLD). In exchange for the enormous investment of time and expense, which otherwise could waste significant portions of a relevant patent's term, an NDA holder enjoys significant additional exclusivities that avoid generic competition for at least some extended period.

A. NDA Exclusivities

An NDA holder may be entitled to one or more additional exclusivities beyond any patent protection it may already have with regard to manufacturing, using, or selling the RLD. These additional exclusivities provide a wide range of additional protection for the RLD product. The FDA grants additional exclusivities when the NDA satisfies the FDA's requirements for new chemical entities, new indications (for an existing drug product), pediatric exclusivity, or orphan drug exclusivity. Each of these is discussed in detail in the following sections.

1. New Chemical Entities

NDAs directed to products containing chemical entities never before approved by the FDA are entitled to five years of absolute exclusivity. The FDA will not accept Abbreviated New Drug Applications (ANDAs) on new chemical entities (NCEs) for four years after approval of the NDA.[5] From year 4 to year 5, the FDA will only accept ANDAs containing Paragraph IV Certifications, but if an infringement lawsuit is filed during that fifth year, the 30-month stay will be extended to such date as to effectively have begun at the end of the fifth year.[6] Thus, upon the filing of such a lawsuit, the ANDA generic drug generally will not be approved until at least seven and a half years following the date the FDA approves a drug subject to NCE exclusivity.[7]

2. New Indications

For NDAs or NDA supplements containing new, essential clinical studies concerning chemical entities that previously have been approved by the

4. 21 U.S.C. § 355(b).
5. 21 C.F.R. § 314.108(b)(2).
6. *Id.* § 314.108(b)(3).
7. 21 U.S.C. § 355(j)(5)(F)(ii).

FDA (e.g., new dosage strength, route of administration, etc.), the NDA holder is entitled to three years of absolute exclusivity.[8]

These "initial exclusivities" are not tied to patent protection. Rather, they are a function of data exclusivity: the NDA holder owns the clinical trial data on which the NDA relies, and no ANDA applicant can rely on the same clinical trial data for five years for new chemical entities, three years for new indications, and 12 years for biologics.[9] An ANDA or subsequent NDA applicant can submit an application for the same substance, but must rely on its own clinical trial data. These initial exclusivities are a reward for the development of new chemical entities and new uses of chemical entities.

3. Pediatric Exclusivity

A brand company can obtain six months of pediatric exclusivity if the FDA makes a "Written Request" for pediatric studies—studies to demonstrate safety and efficacy in patients 12 years old or younger. This is not as random as it may at first appear, because the NDA holder can ask the FDA to issue a "Written Request." In response to this "Written Request," the NDA holder must submit study reports by way of a new NDA, an amendment, or a supplement. If the FDA determines that the report complies with both the time frame and terms specified in the "Written Request," it will grant an additional six months of exclusivity to the NDA holder as of the date the FDA determines that the report complies with the "Written Request."[10]

Importantly, pediatric exclusivity is not confined to doses and dosage forms designed to treat children, and it is not dependent on labeling. Rather, pediatric exclusivity grants six months of additional exclusivity for *all* dosage forms and all indications, adult and child. Further, pediatric exclusivity is an add-on to the end of any existing exclusivities. Thus, for example, it extends the life of each applicable patent by six months.

If a product has no listed patent life or other exclusivity remaining, there is no benefit to obtaining pediatric exclusivity.

4. Orphan Drug Exclusivity

Under the Orphan Drug Act of 1983, the NDA holder can ask the FDA's Orphan Products Board to declare as an "orphan drug" any substance for treatment of a "rare disease or condition." This means it affects fewer than 200,000 people in the United States, or more than 200,000 people in the United States but with no reasonable expectation that sales will cover costs.

8. *Id.* § 355(b)(2), (j)(5)(F)(iii)–(iv).
9. *Id.* § 355(b).
10. *Id.* § 355a.

Orphan drug exclusivity provides seven years of exclusivity from the date of approval of the NDA, notwithstanding patents or other exclusivities.[11]

B. Enforcement of NDA Exclusivities

The FDA Center for Drug Evaluation and Research (CDER) Exclusivity Board is solely responsible for determining whether NCE, new indication, and biologic exclusivities are available. The board does not, however, rule on 180-day generic marketing exclusivity (discussed in the following section) nor on orphan drug or pediatric exclusivities. Pediatric exclusivity is handled by CDER's Pediatric Implementation Team, and orphan drug exclusivity is handled by the FDA Office of Orphan Products Development. The Office of Generic Drugs rules on generic exclusivity.

C. Patent Protection for NDA-Listed Drugs

Of course, it is common for brand companies to obtain patents on RLDs or methods of using RLDs. For patent applications filed on or after June 8, 1995, the term of the resulting patent is 20 years from the date of filing the earliest related patent application.[12] In principle, therefore, a patentee that owns the patent on a particular RLD (or, potentially, a method of using that RLD) should enjoy as much as a 20-year period of patent exclusivity, less only the time it takes to obtain the patent. The reality, however, is far different.

1. Patent Term Extensions

The time it takes to patent a drug usually is far less than the time it takes to get a drug through clinical trials and obtain NDA approval. Because the FDA's NDA approval process can take many years, effectively wasting the first part of the patent's term, the Patent Act extends a drug patent's term for the duration of any period during which the patentee has obtained a patent but is awaiting FDA approval (i.e., the length of the NDA process).[13]

The decision to grant a patent term extension under § 156 is made by operation of law as implemented by the U.S. Patent and Trademark Office (USPTO). In theory, the FDA can reject the extended term if it concludes that the patentee did not act with reasonable diligence during the regulatory process.[14] In practice, however, the term determined by the USPTO generally is accepted by the FDA.

11. *Id.* § 360cc.
12. 35 U.S.C. § 156.
13. *Id.*
14. *Id.*; *see* 37 C.F.R. §§ 1.710–75.

2. Listing Patents in the Orange Book

The Orange Book—*Approved Drug Products with Therapeutic Equivalence Evaluations*—is the FDA's compendium of all RLDs, organized by brand name, by chemical name, and by NDA number. The Orange Book also contains a listing of all applicable exclusivities, including all patents that any party asserts may cover the RLD or a method of using the RLD.[15] Any live patents on the RLD or a method of using the RLD can be listed in the Orange Book. The NDA holder is authorized to list any applicable patent on the RLD or a method of using the RLD. However, only patents on products and *approved* methods of use must properly be listed in the Orange Book.[16] Patents on processes of manufacture (other than for novel products) and patents on packaging, metabolites, or intermediates must not be listed.[17] The entity listing the patent must certify under threat of criminal sanction that it understands the requirements for the submission and that the submission is correct.[18]

A preexisting patent must be listed in the Orange Book within 30 days of the NDA's approval, or within 30 days of the patent's issuance for an NDA already approved, to be considered timely.[19] Generic applicants still must submit certifications to patents listed after the ANDA was filed (if the patent was listed within the 30-day period), but there are no statutory stays in approval with respect to such patents.[20]

When a generic company files an ANDA asking for FDA permission to sell a generic version of the RLD approved pursuant to a brand company's NDA, it must include one of four specific certifications concerning each patent listed in the Orange Book in association with that NDA. The primary certification options are: (I) no such patent information has been submitted to the FDA; (II) the patent has expired; (III) the patent is set to expire on a certain date; or (IV) the patent is invalid or will not be infringed by the manufacture, use, or sale of the new generic drug for which the ANDA is submitted.[21] These are commonly referred to as Paragraph I, II, III, and IV Certifications.

In lieu of one of these certifications, the applicant can file what is known as a "Section viii statement," limiting the ANDA to one or more non-claimed approved uses. This statement, which functions in essence as a fifth kind of patent certification, is available only with respect to listed

15. 21 U.S.C. § 355(j)(7)(A)(i).
16. Purepac Pharm. Co. v. Thompson, 238 F. Supp. 2d 191 (D.D.C. 2002).
17. 21 C.F.R. § 314.53(b)(1).
18. *Id.* § 314.53(c)(2)(ii)(R); *e.g.*, FDA Form 3542.
19. 21 U.S.C. § 355(c)(2).
20. 21 C.F.R. § 314.94(a)(12)(vi).
21. 21 U.S.C. § 355(j)(2)(A)(vii).

method-of-use patents. Where there are multiple approved uses and a listed method patent only covers some of them, the ANDA applicant can submit a Section viii statement instead of a certification under Paragraphs I–IV. A Section viii statement specifies that the ANDA applicant does not plan to practice the patented method of use.[22]

3. Challenges to Patent Listings

The FDA can remove improperly listed patents if requested or confirmed to do so by the NDA holder, but only if the removal will not affect an ANDA holder's eligibility for 180 days of market exclusivity.[23] As a practical matter, it can be challenging to convince the FDA to take such action, because the FDA's role in listing patents is purely ministerial.[24] If an interested party believes a patent is improperly listed in the Orange Book, it can challenge the propriety of the listing under 21 C.F.R. § 314.53(f). But the FDA merely will ask the NDA holder or patentee whether the listing is proper; and if the NDA holder or patentee responds that it is, the FDA's inquiry is over.

Nor do parties generally have a private right of action in court to seek delisting of patents.[25] However, ANDA applicants that have been sued for infringement on the basis of a Paragraph IV Certification for an improperly listed patent can counterclaim to have the patent delisted or the listing corrected.[26]

4. Exception for Research Involving Patented Drugs

Notwithstanding any patent protection or other exclusivities covering an RLD, scientific research involving a new drug—including specifically all research designed primarily for purposes of preparing, filing, and prosecuting an ANDA—is exempted from being deemed an act of patent infringement, as a matter of law.[27] This allows the preliminary research and application work needed for approval of a drug product to take place before the relevant patents expire, expediting years of effort that otherwise could not begin during the term of the patent. In exchange, once the drug application is deemed complete by the FDA, the application for drug

22. *Id.* § 355(j)(2)(A)(viii).

23. 21 C.F.R. § 314.94(a)(12)(viii)(B).

24. aai Pharm. v. Thompson, 296 F.3d 227, 242–43 (4th Cir. 2002); Am. Bioscience Inc. v. Thompson, 269 F.3d 1077, 1084 (D.C. Cir. 2001); *In re* Buspirone Patent Litig., 185 F. Supp. 2d 363, 371 (S.D.N.Y. 2002).

25. Mylan Pharm. v. Thompson, 268 F.3d 1323, 1332 (Fed. Cir. 2001).

26. Caraco Pharm. Labs. v. Novo Nordisk AS, 132 S. Ct. 1670 (2012). Along with Winston & Strawn's James Hurst and Steffen Johnson, Klein and Bloch represented Caraco throughout this litigation and the related appeals.

27. 21 U.S.C. § 355(i); 35 U.S.C. § 271(e)(1).

approval itself is then effectively deemed an act of infringement for certain specific patents that may be applicable. The net result is that most patent infringement actions regarding a prospective new drug product will not occur until the underlying product actually is complete and ready for market, and not just a remote possibility in the producer's forecast.

II. The ANDA Application Process

Under the Hatch-Waxman Amendments, a generic company can file an ANDA, avoiding the costs of human clinical trials on safety and efficacy in favor of in vitro studies that establish the ANDA product is bioequivalent (chemically and morphologically) to the RLD.[28] In addition, excluding trademarks, the ANDA product must be labeled identically to the RLD's approved label (unless the ANDA applicant does not seek approval for a particular FDA-approved use).

Thus, to file a proper ANDA, the applicant must follow a series of precise steps under 21 U.S.C. § 355(j) and 21 C.F.R. § 314.94. The application must show that the proposed drug's prescribed conditions of use,[29] active ingredients,[30] route of administration, dosage form, strength,[31] and labeling[32] all have previously been approved by the FDA and that the active ingredient is bioequivalent to that of the RLD.[33] If the ANDA contains a different active ingredient, route of administration, dosage form, or strength than the RLD, the applicant must request and obtain permission from the FDA within 90 days.[34] The FDA must approve the application unless it finds that additional studies are needed[35] or that the information contained in the ANDA (as opposed to an NDA) is insufficient for evaluation of the proposed new substance.[36] The ANDA must include a listing of all articles used as components for the proposed drug, a full statement of the drug's composition, the manufacturing location, process and manufacturing controls, samples of the proposed labeling, and (if the FDA requests it) the drug itself.[37] Last, as discussed in more depth in the following sections, the ANDA applicant must submit a certification under

28. 21 C.F.R. § 314.92.
29. 21 U.S.C. § 355(j)(2)(A)(i).
30. *Id.* § 355(j)(2)(A)(ii).
31. *Id.* § 355(j)(2)(A)(iii).
32. *Id.* § 355(j)(2)(A)(v).
33. *Id.* § 355(j)(2)(A)(iv).
34. *Id.* § 355(j)(2)(C); 21 C.F.R. § 10.30(e)(4).
35. 21 U.S.C. § 355(j)(2)(C)(i).
36. *Id.* § 355(j)(2)(C)(ii).
37. *Id.* § 355(j)(2)(A)(v), (b)(1)(B)–(F).

21 U.S.C. § 355(j)(2)(A)(vii), or a statement under subsection (viii), concerning each patent listed in the Orange Book with regard to the RLD.

A. Litigation Involving ANDAs

The ANDA process—and particularly the process of certification—is designed with litigation in mind. Challenging patents under the Hatch-Waxman Amendments is often part of a generic drug company's business model. The Hatch-Waxman Amendments are explicit in trying to funnel generic drug disputes into litigation rather than requiring the FDA to adjudicate issues relating to patent law.

But litigation imposes costs of its own. So the Hatch-Waxman Amendments attempt to ensure that generic-brand patent litigation proceeds as efficiently as possible, meaning that the case is resolved before any product launch. Thus, most ANDA litigation proceeds before a judge (i.e., the only relief requested is equitable), not a jury. Damages for infringement generally are not at issue, because Hatch-Waxman litigation generally occurs before launch. ANDA litigation for method patents, where the generic applicant copies the branded product label for the patented indication, most commonly turns on patent validity, an area where by law the patentee has a significant advantage based on an issued patent's presumption of validity.[38]

Under the Hatch-Waxman Amendments, Congress created a series of triggers intended to provoke pre-approval litigation. These triggers ensure that most—though not all—generic drug patent litigation is initiated by the NDA holder before final FDA approval of an ANDA.

B. Certifying to Orange Book–Listed Patents

As part of the ANDA process, the generic company must review the Orange Book to determine if there are any "listed patents." Listed patents are patents identified by the NDA holder as covering certain specific aspects of the RLD.

If the Orange Book identifies any patent(s) for the RLD, the generic applicant must analyze whether its proposed generic product will infringe the patent(s). For patents covering the active ingredient itself, this can be straightforward. But when the patents cover formulations or methods of use, it can be difficult to determine whether the product subject to the ANDA would infringe the patent. Patents covering unapproved methods of use should not be listed in the Orange Book.[39]

---

38. 35 U.S.C. § 282.

39. Purepac Pharm. Co. v. Thompson, 238 F. Supp. 2d 191 (D.D.C. 2002); Warner-Lambert Co. v. Apotex Corp., 316 F.3d 1348 (Fed. Cir. 2003).

The applicant has five options based on its review of the Orange Book patent listing—four "certification" options and one "statement" option. Note that, if a patent is listed in the Orange Book after an ANDA was filed, the ANDA applicant must certify to the newly listed patent (if listed timely), but the "pop-up" patent does not trigger the 30-month stay or the other exclusivities discussed in the following sections.[40]

1. Paragraph I Certifications

In a Paragraph I Certification, the applicant certifies that there are no patents listed in the Orange Book for the RLD. A Paragraph I Certification for a particular patent implies that the ANDA can be approved immediately absent other listed patents—subject, of course, to the FDA's review of non-patent-related issues.[41] A Paragraph I Certification is not an artificial act of infringement.

2. Paragraph II Certifications

In a Paragraph II Certification, the applicant certifies that the listed patent has expired. Like a Paragraph I Certification, a Paragraph II Certification for a particular patent implies that the ANDA can be approved immediately absent other listed patents, again subject to the FDA's review of non-patent issues.[42] A Paragraph II Certification is not an artificial act of infringement.

3. Paragraph III Certifications

In a Paragraph III Certification, the applicant certifies that it will not seek final approval until after the listed patent expires. A Paragraph III Certification thus postpones the FDA's final approval of the ANDA until the Orange Book–listed expiration date of the patent.[43] Because an ANDA with a Paragraph III Certification does not seek authority to market the drug until the patent expires, it also does not create a cause of action for infringement in the federal courts under the Hatch-Waxman Amendments.

4. Paragraph IV Certifications

In a Paragraph IV Certification, the applicant certifies that the listed patent is invalid, is unenforceable, or will not be infringed by the contemplated ANDA product. A Paragraph IV Certification must be accompanied by a detailed statement of reasons, which must be supplied to the NDA

40. 21 C.F.R. § 314.94(a)(12)(vi).
41. 21 U.S.C. § 355(j)(2)(A)(vii)(I).
42. *Id.* § 355(j)(2)(A)(vii)(II).
43. *Id.* § 355(j)(2)(A)(vii)(III).

holder and the patentee[44] either immediately upon filing or within 20 days after the FDA has informed the applicant that its ANDA is substantially complete.[45] The certification must be delivered to the NDA holder and the patentee by certified mail, return receipt requested.[46] Of the four certifications available to an ANDA applicant, only one—the Paragraph IV Certification—directly exposes the generic company to Hatch-Waxman litigation.

a. Contents of Paragraph IV Certification Notice

Notice of a Paragraph IV Certification must state that the application contains bioavailability and bioequivalence studies sufficient to satisfy the ANDA standards,[47] plus "a detailed statement of the factual and legal basis of the opinion of the applicant that the patent is invalid or will not be infringed."[48] If the ANDA applicant wishes to preserve the ability to initiate a declaratory judgment action, it also must offer the NDA holder and the patentee confidential access to relevant portions of the contents of the ANDA.[49]

Despite the fact that the Paragraph IV Certification is merely a piece of paper, the notification is deemed a "highly artificial act of infringement."[50] The act of infringement created by 35 U.S.C. § 271(e)(2) "consists of submitting an ANDA or a paper NDA containing the fourth type of certification."[51] Thus, litigation concerning ANDAs is triggered by a Paragraph IV Certification, which—without more—creates a cause of action for infringement.[52]

b. Timing and Stays

Once the brand company receives notice of a Paragraph IV Certification, it can sue the ANDA applicant for patent infringement.[53] If the NDA holder sues within 45 days of receiving that notice, the law imposes an automatic stay on FDA approval of the ANDA. That stay usually lasts for 30 months or until a final district court judgment—whichever is shorter—but can be shortened or extended by the court in which the litigation is pending.[54]

As a practical matter, this means that most ANDAs including Paragraph IV Certifications will not be approved for at least 30 months after

44. 21 C.F.R. § 314.95(a); 21 U.S.C. § 355(b)(3), (j)(2)(B).
45. 21 U.S.C. § 355(j)(2)(B)(ii); 21 C.F.R. §§ 314.95(b), (d) & .52(b), (d).
46. 21 U.S.C. § 355(j)(2)(B)(iii); 21 C.F.R. §§ 314.52(a), .95(a).
47. 21 U.S.C. § 355(j)(2)(B)(iv)(I).
48. *Id.* § 355(j)(2)(B)(iv)(II).
49. *Id.* § 355(j)(2)(A)(vii)(IV).
50. Eli Lilly & Co. v. Medtronic, 496 U.S. 661, 678 (1990).
51. *Id.* at 678.
52. 35 U.S.C. § 271(e)(2).
53. 21 U.S.C. § 355(j)(5)(B)(iii).
54. *Id.* § 355(c)(3)(C), (j)(5)(B)(iii).

suit is filed. Thus, by following the Hatch-Waxman procedures to the letter, the NDA holder generally can avoid generic competition for at least 31.5 months after it receives a Paragraph IV Certification. The purpose of this stay "is to enable the judicial adjudication upon which the ANDA . . . scheme depends"[55]—that is, to obtain a judgment on the merits of the patent dispute before the generic product launches.

If the NDA holder or patentee wins the infringement suit, the ANDA holder normally must convert from a Paragraph IV Certification to a Paragraph III Certification, and approval of the ANDA is delayed until the expiration of any such patent. But if the first ANDA applicant wins the infringement suit, it is eligible for the highly lucrative benefit of 180 days as the only approved ANDA generic in the market.[56]

The first generic applicant qualifies for this 180-day marketing exclusivity period even where the patentee elects not to sue the ANDA applicant for patent infringement based on the Paragraph IV Certification.[57]

c. Amendments and Pop-up Patents

ANDA applicants can amend their applications to "convert" other certifications into Paragraph IV Certifications. A new Paragraph IV Certification, however, will not trigger an additional 30-month stay.[58]

Also, the 30-month stay only applies in cases where the patent was listed in the Orange Book before the ANDA was filed. Although an ANDA applicant also must file a certification to patents timely listed after the ANDA was filed, such "pop-up" patents do not trigger a 30-month stay under the Hatch-Waxman Amendments.

5. Section VIII Statements

In lieu of a patent certification, an ANDA applicant can submit a Section viii Statement, averring that it does not seek approval to market the ANDA drug for a patented method of use. A Section viii Statement is *not* considered an artificial act of infringement and thus does not give rise to any notification obligations. Under the Section viii Statement process, the label for the ANDA product does not include the patented use or indication. Thus, a Section viii Statement is not available unless the FDA has approved at least two uses or indications for the referenced drug and at least one approved use is not covered by any patent listed in the Orange Book.

"In determining whether an ANDA applicant can 'carve out' the [patented] method of use, rather than certify to the listed patent, [the FDA] will

55. *Eli Lilly & Co.*, 496 U.S. at 678.
56. 21 U.S.C. § 355(j)(5)(B)(iv).
57. Inwood Labs. v. Young M.D., 723 F. Supp. 1523, 1526 (D.D.C. 1989).
58. 21 C.F.R. §§ 314.95(a), .52(a).

rely on the description of the approved use provided by the NDA holder or patent owner in the patent declaration and listed in the Orange Book."[59] Only when NDA applicants or holders "identify specifically the approved uses claimed by the methodofuse patent . . . and declare under penalty of perjury that the patent claims an approved use" may ANDA and 505(b)(2) applicants as well as the FDA "determine whether the applicant must submit a patent certification or may submit a section viii statement."[60]

If approved, the Section viii Statement and resulting carved-out label avoid a claim that the product label will induce infringement.[61] But unlike the Paragraph IV Certification, the Section viii Statement does not provide the ANDA applicant with any market exclusivity. And, the ANDA must include at least one non-patented indication. If the NDA is approved only for patented indications, a Section viii Statement is not available to the generic applicant.

Warner-Lambert and its progeny hold that an ANDA that omits infringing uses from its label does not induce infringement of the patent, even if it is likely that the generic product will be used by physicians and patients in ways that infringe the patent. But this analysis applies only in the context of Hatch-Waxman litigation. That is, a Section viii Statement does not preclude a post-launch lawsuit if the patentee can prove that the generic company is taking other acts, such as through marketing, to induce infringement.

C. Declaratory Judgment Actions

If the NDA holder does not sue within 45 days, the 30-month stay no longer applies. But that does not mean that the patentee has waived its rights. Rather, the patentee can wait until the generic launches, then sue for infringement. To avoid this, the ANDA applicant can file a declaratory judgment action against the NDA holder and/or patentee.

1. Civil Actions to Obtain Patent Certainty

Under certain circumstances, the Hatch-Waxman Amendments grant to each ANDA applicant the statutory right to a determination as to whether its ANDA infringes any valid Orange Book–listed patents to which it made a Paragraph IV Certification. "[T]o prevent patentees from 'gaming' the Hatch-Waxman Act," Congress created a "civil action to obtain patent

59. 68 Fed. Reg. at 36,682.

60. *Id.*

61. Warner-Lambert Co. v. Apotex Corp., 316 F.3d 1348 (Fed. Cir. 2003); Allergan Inc. v. Alcon Labs, 324 F.3d 1322 (Fed. Cir. 2003); AstraZeneca Pharm. v. Apotex, 633 F.3d 1042 (Fed. Cir. 2012) (no infringement claim under 35 U.S.C. § 271(e)(2) if applicants "excluded any patented treatment indications from their ANDAs").

certainty."[62] Under 21 U.S.C. § 355(j)(5)(C), any ANDA applicant can file such an action to validate its Paragraph IV Certification and (if it is a first-filer) preserve its chance at 180 days of generic marketing exclusivity, if neither the NDA holder nor the patentee has filed suit within 45 days of receiving notice of an ANDA containing a Paragraph IV Certification.[63]

As indicated, however, a civil action to obtain patent certainty is available to the generic applicant only if its ANDA was accompanied by an offer of confidential access to its ANDA "for the purpose of determining whether an action [for infringement] should be brought."[64] The offer may be conditioned on the NDA holder and patentee entering into a confidentiality agreement protecting the ANDA applicant's trade secrets and ensuring that the disclosed information is used only for purposes of evaluating (and potentially filing) a Hatch-Waxman infringement claim "and for no other purpose."[65] The ANDA may be redacted to remove information not relevant to the infringement analysis. The implication of this provision is that the ANDA applicant need not offer confidential access to the NDA holder and patentee. But if the ANDA applicant does not offer this information, it runs the risk that the patentee/NDA holder will decline to sue within 45 days of receiving the Paragraph IV Certification and instead try to interfere with the generic launch of the ANDA product.

Under Article III of the U.S. Constitution, declaratory judgment jurisdiction exists whenever there is a live "case or controversy."[66] Thus, the 45-day limitation on declaratory judgment actions in 21 U.S.C. § 355 (j)(5)(B)(iii) is best viewed as a temporary stay to allow the patentee and/or NDA holder to decide whether and where to sue. Whether or not the patentee/NDA holder sues, however, the ANDA is sufficient to create a justiciable case or controversy, as discussed in the following section.

2. Other Declaratory Judgment Claims

The Hatch-Waxman Amendments create a particular form of declaratory judgment—a civil action to obtain patent certainty—in the event an NDA holder or patentee declines to sue a generic within 45 days of receiving proper notice and an offer of confidential access. Broader principles of decisional law suggest that courts can hear actual disputes between brand

62. Teva v. Novartis, 482 F.3d 1330, 1342 (Fed. Cir. 2007).

63. *See generally* Caraco Pharm. Labs. Ltd. v. Forest Labs. Inc., 527 F.3d 1278 (Fed. Cir. 2008); *Teva*, 482 F.3d at 1345; Janssen Pharm. N. v. Apotex Inc., 540 F.3d 1353, 1360 (Fed. Cir. 2008).

64. 21 U.S.C. § 355(j)(5)(C)(i)(III).

65. *Id.*

66. 28 U.S.C. § 2201; Cordis Corp. v. Medtronic, 835 F.2d 859 (Fed. Cir. 1987).

and generic companies even if the process set forth for Paragraph IV Certifications is not followed.

The Patent Act makes it "an act of infringement to submit[] an application under section 505(j) of the [Hatch-Waxman Amendments] . . . for a drug claimed in a patent or the use of which is claimed in a patent."[67] Though Hatch-Waxman repeatedly references Paragraph IV Certifications, the cited provision of the Patent Act states that the infringing act is the "application" alone. Thus, Article III standing may be available even in cases where there is no Paragraph IV Certification, because Article III only requires "injury in fact, causation, and redressability."[68]

Following Article III, Congress has granted the district courts "original jurisdiction of any civil action arising under any Act of Congress relating to patents."[69] This jurisdiction extends to cases in which "federal patent law creates the cause of action or . . . the plaintiff's right to relief necessarily depends on resolution of a substantial question of federal patent law."[70] When determining whether a controversy exists, "the question in each case is whether the facts alleged, *under all the circumstances*, show that there is a substantial controversy, between parties having adverse legal interests, of sufficient immediacy and reality to warrant the issuance of a declaratory judgment."[71] The Federal Circuit explains that "the only limitation on our jurisdiction under the Declaratory Judgment Act" is the Article III requirement of a "justiciable controversy."[72]

A Hatch-Waxman dispute thus is justiciable under Article III where, as here, "(1) the plaintiff has standing, (2) the issues presented are ripe for judicial review, and (3) the case is not rendered moot at any stage of the litigation."[73] Under *MedImmune*, this remains true even if the generic applicant has not yet actually infringed.[74] For this reason, courts repeatedly have held that they have jurisdiction to entertain a Hatch-Waxman case even if an ANDA does not contain a Paragraph IV Certification.[75]

---

67. 35 U.S.C. § 271(e)(2)(A).

68. Steel Co. v. Citizens for a Better Env't, 523 U.S. 83, 103 (1998).

69. 28 U.S.C. § 1338(a).

70. Christianson v. Colt Indus. Operating Corp., 486 U.S. 800, 809 (1988); *see also* 28 U.S.C. § 1331 (vesting "jurisdiction" in the district courts for "civil actions arising under the . . . laws of the United States").

71. MedImmune Inc. v. Genentech Inc., 127 S. Ct. 764, 771 (2007) (quoting Maryland Cas. Co. v. Pac. Coal & Oil Co., 312 U.S. 270, 273 (1941)).

72. Teva Pharm. USA Inc. v. Novartis Pharm. Corp., 482 F.3d 1330, 1340 (Fed. Cir. 2007) (describing the Hatch-Waxman Amendments as an "example[] of legislation that expand[s] standing to constitutional limits").

73. Caraco Pharm. Labs. Ltd. v. Forest Labs. Inc., 527 F.3d 1278, at 1291 (Fed. Cir. 2008).

74. *See* MedImmune Inc., 127 S. Ct. at 775.

75. *See* Glaxo Grp. Ltd. v. Apotex Inc., 376 F.3d 1339, 1350–51(Fed. Cir. 2004) (finding ANDA with no Paragraph IV Certification an act of infringement under § 271(e)(2)); *see also* Novo Nordisk Inc. v. Mylan Pharm. Inc., No. 09-2445,2010 WL 1372437, at *11 (D.N.J. Mar.

To be sure, infringement under 35 U.S.C. § 271(e)(2) typically is triggered by the filing of an ANDA with a Paragraph IV Certification. But the plain language of the statute is not limited to that situation. Section 271(e)(2) "does not require that the applicant have included the drug in its certification, nor does it limit the cause of action to patents listed pursuant to" Paragraph IV.[76] Thus, infringement claims may be viable under 35 U.S.C. § 271(e)(2) even absent a Paragraph IV Certification; it is simply not the case that "a paragraph IV certification is the linchpin to case or controversy jurisdiction under § 271(e)(2)."[77] Even in the absence of a Paragraph IV Certification, an ANDA applicant's declaratory judgment action may "present[] an actual controversy for the purposes of the Declaratory Judgment Act."[78]

The triggering event for jurisdiction is the mere *filing* of an ANDA to market a generic drug claimed by a patent before patent expiry. As the Federal Circuit put it, "once it is clear that a party seeking approval of an ANDA wants to market a patented drug prior to the expiration of the patent, the patent owner can seek to prevent approval of the ANDA by bringing a patent infringement suit."[79] In this sense, the statute "provided patentees with a defined act of infringement sufficient to create case or controversy jurisdiction to enable a court to promptly resolve any dispute concerning infringement and validity."[80] The ANDA constitutes infringement "despite the fact that the generic companies have not yet infringed the patents at issue."[81] At a minimum, "the filing of an ANDA that *should* include, but does not include, a Paragraph IV Certification nevertheless constitutes an 'act of infringement' which confers jurisdiction under Section 271(e)(2)(A)."[82]

31, 2010) ("[T]he filing of an ANDA that should include, but does not include, a Paragraph IV Certification nevertheless constitutes an 'act of infringement' which confers jurisdiction under Section 271(e)(2)(A)."); Glaxo Grp. Ltd. v. Apotex Inc., 272 F. Supp. 2d 772, 779 (N.D. Ill. 2003); Ben Venue Labs. Inc. v. Novartis Pharm. Corp., 146 F. Supp. 2d 572, 582 (D.N.J. 2001) (noting that jurisdiction under § 271(e)(2) "is possible . . . even if the ANDA included no Paragraph IV Certification at all, so long as a Paragraph IV Certification should have been included"); Marion Merrell Dow Inc. v. Hoechst-Roussel Pharm. Inc., 32 U.S.P.Q.2d 1156 (D.N.J. May 5, 1994) (§ 271(e)(2) "does not require that the applicant have included the drug in its certification, nor does it limit the cause of action to patents listed pursuant to" Paragraph IV).

76. *Marion Merrell Dow Inc.*, 32 U.S.P.Q.2d at 1158.

77. *Glaxo Grp. Ltd.*, 272 F. Supp. 2d at 779; *see also Caraco Pharm.*, 132 S. Ct. at 1680 n.5.

78. DuPont Merck Pharm. Co. v. Bristol-Myers Squibb Co., 62 F.3d 1397, 1401 (Fed. Cir. 1995).

79. Bristol-Myers-Squibb Co. v. Royce Labs, 69 F.3d 1130, 1132 (Fed. Cir. 1995).

80. Glaxo Inc. v. Novopharm Ltd., 110 F.3d 1562, 1569 (Fed. Cir. 1997).

81. *Glaxo Grp. Ltd.*, 376 F.3d at 1351 (citing *Eli Lilly*, 496 U.S. 661).

82. *Novo Nordisk Inc.*, 2010 WL 1372437, at *11; *accord Marion Merrell Dow*, 32 U.S.P.Q.2d at 1158 ("[T]he proper inquiry is, should the certification have included the patent and if so, is there an infringement of that patent?").

However, a properly structured covenant not to sue can defeat declaratory jurisdiction by eliminating a live case or controversy.[83]

D. Counterclaims in Hatch-Waxman Cases

Cases and statutes also confirm the existence of several counterclaims available to a Hatch-Waxman defendant. Counterclaims to correct patent information are a part of the Hatch-Waxman statutory scheme; patent misuse and antitrust defenses spring instead from general principles of patent law.

1. Counterclaims to Correct Patent Information

As noted, NDA holders are required to submit "patent information" to the FDA on FDA Form 3542. Certain information from Form 3542 is then published by the FDA in the Orange Book. This information includes patent numbers, expiration dates, and patent descriptions for method patents corresponding to a particular "use code."

The FDA's regulations contain special rules for method patents that claim one or more approved methods of using the listed drug. One of the required pieces of information is a "use code description," which summarizes the scope of any method patent that the brand claims covers the drug.[84] Before 2003, the FDA did not require this information. But in response to anticompetitive strategies by brand companies concerning the listing of method patents,[85] the agency held a notice-and-comment rulemaking and made it mandatory to submit such information.[86] The new system enables the FDA—and ANDA applicants—to determine which labeled use or uses purportedly implicate each listed method patent. An ANDA applicant may then be able to avoid a listed method patent, and Paragraph IV implications, by simply omitting any affected labeled uses, so long as at least one unpatented labeled use remains in the ANDA's proposed product label.

However, the FDA lacks both the authority and expertise needed to verify the patent information submitted by manufacturers, so it defers to their descriptions of the scope of their patents. The regulations note that the act's mandates "do not contemplate a substantive agency review of the scope of the patent"; that the FDA "lack[s] both the resources and the expertise to resolve [patent] matters"; and that "[a] fundamental assumption of the Hatch-Waxman Amendments is that the courts are the

83. Teva Pharm. USA v. U.S. Food & Drug Admin., 182 F.3d 1003 (D.C. Cir. 1999).

84. 21 C.F.R. § 314.53(c)(2)(ii)(P); *see also* 68 Fed. Reg. at 36,682–36,683.

85. *See, e.g.*, Mylan Pharm. Inc. v. Thompson, 268 F.3d 1323 (Fed. Cir. 2001).

86. *See* Applications for FDA Approval to Market a New Drug, 68 Fed. Reg. 36,676, 36,682 (June 18, 2003).

appropriate mechanism for the resolution of disputes about the scope and validity of patents."[87]

This process, in turn, gives brands the potential to block the approval of ANDAs relying on a Section viii Statement by submitting overbroad or inaccurate patent descriptions to the agency. If the FDA cannot determine whether the generic has properly carved out patented indications, it will not approve the ANDA containing a Section viii Statement—effectively extending patents to cover noninfringing uses.[88]

Normally, a generic applicant has no legal route to challenge erroneous Orange Book information, such as an overbroad use code narrative.[89] The applicant can complain to the FDA about an improperly listed patent or a false or misleading use code description under 21 C.F.R. § 314.53(f), but that challenge is really more of a request for confirmation. In response to a challenge under Section 314.53(f), the FDA will ask the NDA holder whether the listing is accurate, but will take no further action, and certainly will not require the NDA holder to change it: "Unless the application holder withdraws or amends its patent information in response to FDA's request, the agency will not change the patent information in the list."[90]

Thus, an ANDA applicant seeking to rely on a Section viii Statement and carve out a patented use may be forced, as a result of an overbroad use code description, to challenge the patent through a Paragraph IV Certification. Although this may appear unfair on its face, various cases confirm that the FDA's role in evaluating patent information is ministerial and that the agency has no obligation to independently evaluate the correctness of information submitted to it.[91] Moreover, while Form 3542 requires that the NDA holder submit information that is true and correct under threat of criminal penalties, we are unaware of any prosecution for violating the submission requirements.

If the NDA holder or patentee later sues based on a Paragraph IV Certification to that erroneously listed or described patent, however, the ANDA applicant has a statutory remedy by way of Hatch-Waxman's counterclaim provision to correct patent information. This counterclaim provision allows litigation "to correct or delete the patent information submitted by [the NDA holder] on the ground that the patent does not claim either

87. *Id.* at 36,682.

88. *See generally Caraco Pharm.*, 132 S. Ct. 1670.

89. *Mylan Pharm. Inc.*, 268 F.3d 1323.

90. 21 C.F.R. § 314.53(f).

91. *E.g.*, Apotex Inc. v. Thompson, 347 F.3d 1335, 1347 (Fed. Cir. 2003) ("[T]he regulation codifies the FDA's position that its duties with respect to Orange Book listings are purely ministerial.").

(aa) the drug for which the application was approved; or (bb) an approved method of using the drug."[92]

The Supreme Court has held that the counterclaim provision is designed, in part, to avoid the problem of overbroad use codes, which the Court held "throws a wrench into FDA's ability to approve generic drugs as the statute contemplates."[93] As the Supreme Court stated:

> The statutory scheme . . . contemplates that one patented use will not foreclose marketing the generic drug for other unpatented ones. . . . [T]he counterclaim naturally functions to challenge the brand's assertion of rights over whichever discrete use (or uses) the generic company wishes to pursue. That assertion, after all, is the thing blocking the generic drug's entry on the market. The availability of the counterclaim thus matches the availability of FDA approval under the statute: A company may bring a counterclaim to show that a method of use is unpatented because establishing that fact allows the FDA to authorize a generic drug via section viii.[94]

The sole remedy promised by this counterclaim is an order to "correct or delete" the erroneous information; damages are not available. The court can provide specificity as to how patent information should be changed, but it cannot dictate the precise language of the replacement use code description.[95]

2. Patent Misuse/Antitrust Counterclaims

In addition to the statutory counterclaim to correct or delete erroneous patent information, positive principles of patent law can reach patent-related misconduct by patentees and NDA holders.

The judicially created defense of patent misuse occurs where a patentee impermissibly broadens either the coverage or the duration of its patent grant with anticompetitive effect.[96] As the Federal Circuit explained:

> The concept of patent misuse arose to restrain practices that did not in themselves violate any law, but that drew anticompetitive strength from the patent right, and thus were deemed to be contrary to public policy. The policy purpose was to prevent a patentee from using the patent

92. 21 U.S.C. § 355(j)(5)(C)(ii)(I).
93. *Caraco Pharm.*, 132 S. Ct. 1670.
94. *Id.*
95. Novo Nordisk v. Caraco Pharm., 688 F.3d 766 (Fed. Cir. 2012).
96. Windsurfing Int'l Inc. v. AMF Inc., 782 F.2d 995, 1001 (Fed. Cir. 1986).

> to obtain market benefit beyond that which inheres in the statutory patent right.[97]

A patent is unenforceable where the patentee has used its "patent to obtain market benefit beyond that which inheres in the statutory patent right."[98] The remedy for patent misuse is a declaration that the patent is unenforceable until such time as the patentee has abandoned the impermissible conduct and the effects of the misuse have dissipated.[99]

Several cases hold that the submission of inaccurate patent information to the FDA can constitute patent misuse.[100] Similarly, there is case support for the idea that serially listing invalid patents in the Orange Book violates antitrust law and thus constitutes patent misuse as well.[101] There is also support for the notion that a patentee who improperly lists patents may be subject to infringement defenses that "include unclean hands, unenforceability of the patent for fraud and inequitable conduct, misuse, and delay in filing suit resulting in laches or estoppel."[102]

III. Approval of ANDAs

Approval of an ANDA by the FDA is, of course, critical to the applicant, but it does not necessarily indicate that marketing may proceed without any further concerns. Final approval of an ANDA indicates that the regulatory path is clear for the ANDA applicant to launch its product for commercial sale. However, the NDA holder still may be able to attempt to interfere with the launch. At the same time, a cautious ANDA first-filer that moves ahead slowly for purposes of marketing its product may risk forfeiting its first-filer exclusivity.

Tentative approval of an ANDA indicates only that regulatory review is likely complete, but it does not yet confer any right to market the underlying product until final approval is obtained.

---

97. Mallinckrodt Inc. v. Medipart Inc., 976 F.2d 700, 704 (Fed. Cir. 1992); *see also* Zenith Radio Corp. v. Hazeltine, 395 U.S. 100, 136–37 (1969) (the doctrine reflects the "principle that a patentee may not use the power of his patent to levy a charge for making, using, or selling products not within the reach of the monopoly granted by the Government").

98. Princo Corp. v. Int'l Trade Comm'n, 616 F.3d 1318 (Fed. Cir. 2010) (en banc).

99. *See* BAB Chem. Co. v. Ellis, 314 U.S. 495, 498 (1942).

100. Novo Nordisk v. Caraco Pharm., 649 F. Supp. 2d 661 (E.D. Mich. 2009); Astra Aktiebolag v. Kremers Urban Dev. Co., 61 U.S.P.Q.2d 1767 (S.D.N.Y. 2001) (allegations that the NDA holder falsely certified to the FDA that the patent covered the approved product were "sufficient to state a claim of patent misuse").

101. Xechem v. Bristol-Myers Squibb, 372 F.3d 899 (7th Cir. 2004).

102. *Mylan Pharm. Inc.*, 268 F.3d at 1331.

A. Final Approval

An ANDA applicant can launch its product only after it receives "final approval" from the FDA. By statute, the FDA must approve or deny an ANDA within 180 days.[103] But that rarely happens, because the effective date of approval is determined by the ANDA's patent certification and NDA exclusivities. While a patent is in effect, and even if an infringement lawsuit is ongoing, the FDA's scientific and technical review of an ANDA proceeds.

1. Applicant May Launch Product

In Paragraph IV litigation, after any stays and exclusivities have expired, the ANDA applicant may obtain final approval from the FDA to launch its product—even if the lawsuit is ongoing. Once an applicant obtains final approval, it *may*—but need not—launch "at risk." This means that the generic applicant can launch its generic product (and, if a first-filer, enjoy 180 days of marketing exclusivity) even if it has not yet defeated the patent-in-suit.[104]

2. Risks and Benefits of "At Risk" Launches

When the first generic alternative for a drug product is launched, it can be expected to capture a very large share of the overall market for the drug in question. This is the result of price sensitivity as well as federal and private insurance regulations that often mandate the use of cheaper generic drugs in place of higher-priced brand pharmaceuticals. Thus, an at-risk launch is potentially quite lucrative and puts maximum pressure on the brand company—the brand's market share will drop precipitously, and the brand may need to launch an "authorized generic" (a generic-label drug manufactured and sold at lower prices under the brand company's NDA) in order to mitigate damages.

Launching as early as possible also reduces the risk that the patentee (which very often will have additional patents pending) will obtain additional intellectual property (IP) rights that will prevent generic entry.

Also, even if a generic applicant receives a favorable district court judgment and then launches, there is always the possibility that the Federal Circuit will find the patent valid and infringed. If the generic launches at risk and then ultimately loses on infringement and validity, it will be held liable for not less than a reasonable royalty, and possibly the full amount of the brand's lost profits during the at-risk launch period—an amount that could swamp the generic's profits during that same span.

103. 21 U.S.C. § 355(j)(5)(A).
104. *Id.* § 355(j)(5)(B)(iv).

3. Maintaining or Forfeiting First-Filer Exclusivity

An ANDA first-filer may be eligible for a period of generic exclusivity that begins when it first markets the ANDA product. But the exclusivity is not absolute—the first-filer may forfeit its exclusivity under certain circumstances.

a. Timing of Launch

An ANDA first-filer for a particular product is eligible to receive 180 days of generic exclusivity, at least as to other ANDA filers that will use the same labeling as the RLD. When a generic has preserved its 180-day marketing exclusivity, this means that the FDA will not approve subsequent ANDAs with Paragraph IV Certifications until 180 days after it issued final approval of the first-filer's ANDA and that first-filer launches its generic product.[105] (There is no exclusivity as to authorized generics or generic applicants that filed a New Drug Application under Section 505(b)(2) of the Federal Food, Drug, and Cosmetic Act. Also, ANDAs containing Section viii Statements can be approved at any time, notwithstanding the possibility of another company's Paragraph IV exclusivity.)

But the time to enjoy that exclusivity is not unlimited and cannot be "parked" indefinitely. For example, the benefit of 180-days of market exclusivity can be forfeited based on a "failure to market."[106] The first-filer forfeits exclusivity under this provision if it fails to market by the *later* of:

1. The earlicr of
   - 75 days after final approval[107] or
   - 30 months after the date the ANDA was submitted;[108] or
2. With respect to the first-filer or another ANDA applicant that has received tentative approval, 75 days after the date as of which, as to each patent with respect to which the first-filer submitted and lawfully maintained a Paragraph IV Certification qualifying the first-filer for the 180-day marketing exclusivity, at least one of the following has occurred:
   - a court (appellate court but not the Supreme Court) enters a final judgment that the patent to which the ANDA applicant filed a Paragraph IV Certification is invalid, unenforceable, or not infringed;[109]

105. *Id.* § 355(j)(5)(B)(iv)(I).
106. *Id.* § 355(j)(5)(D)(i)(I).
107. *Id.* § 355(j)(5)(D)(i)(I)(aa)(AA).
108. *Id.* § 355(j)(5)(D)(i)(I)(aa)(BB).
109. *Id.* § 355(j)(5)(D)(i)(I)(bb)(AA).

- the court enters a settlement or consent decree that the patent is invalid or not infringed;[110] or
- the NDA holder withdraws the listed patent from the Orange Book.[111]

In the event of Paragraph IV litigation, the appellate process can push the final approval date out for several months or even years: under this provision, a judgment is "final" only if there is no available avenue of appeal "other than a petition to the Supreme Court for a writ of certiorari."[112]

There are additional ways the first-filer can forfeit its 180-day marketing exclusivity—for example, if the applicant (1) withdraws its application, (2) withdraws its Paragraph IV Certification, (3) fails to obtain tentative approval within 30 months of the ANDA filing, or (4) is subject to a final decision of the Federal Trade Commission or a court holding that a settlement violated the antitrust laws, or if (5) all patents supporting the exclusivity have expired.[113]

b. Second-Filers Triggering First-Filer Exclusivity

A successful civil action to obtain patent certainty by "any other applicant" can trigger forfeiture of the first-filer's exclusivity.[114] For example, a second-filer (i.e., an ANDA applicant that is not a first-filer) can trigger a first-filer's 75-day forfeiture clock by obtaining a final decision of noninfringement, unenforceability, or invalidity after obtaining tentative approval of its own ANDA.[115]

c. Implications of Forfeiture

An ANDA holder that forfeits its exclusivity does not lose its right to launch; it merely loses its right to 180 days of marketing exclusivity before additional ANDAs containing Paragraph IV Certifications are approved. If a sole first-filer forfeits its exclusivity, then no generic applicant is entitled to exclusivity, and the FDA is free to approve any and all subsequent ANDA filings.[116]

B. Tentative Approval

If the FDA concludes that the ANDA is proper from a scientific and technical perspective before all relevant exclusivities or stays expire, it will issue "tentative" approval rather than final approval. Tentative approval

110. *Id.* § 355(j)(5)(D)(i)(I)(bb)(BB).
111. *Id.* § 355(j)(5)(D)(i)(I)(bb)(CC).
112. *Id.* § 355(j)(5)(D)(i)(I)(bb)(AA).
113. *Id.* § 355(j)(5)(D)(i)(II)–(VI).
114. *Id.* § 355(j)(5)(D)(i)(I)(bb)(AA).
115. *Id.* § 355(j)(5)(B)(iv)(II).
116. *Id.* § 355(j)(5)(D)(ii).

denotes that the FDA's technical and scientific review of an ANDA is complete and the generic application is in order, but the NDA holder's patent rights, exclusivities, or litigation stays prevent final approval.[117]

1. Paragraph I, II, and III Certifications

An ANDA with a Paragraph I or Paragraph II Certification may be approved effective immediately.[118] If the ANDA instead contains a Paragraph III Certification, approval will be effective upon the date the listed patent expires.[119] If the FDA's scientific review is completed before that patent expires, the FDA will grant tentative approval, with final approval to follow once the patent expires.

2. Paragraph IV Certifications and the 30-Month Stay

If an ANDA contains a Paragraph IV Certification, the application can be approved immediately unless the patentee files a Hatch-Waxman Act lawsuit within 45 days of receiving the required Paragraph IV Certification notice.[120] If a lawsuit is filed, final approval of that ANDA follows a complex formula.

By default, final approval is delayed for 30 months from the date the notice was received or such longer or shorter period as the court may direct.[121] However, if the patent is held invalid or not infringed by the trial court before the 30-month period expires, approval is effective on the date on which the court enters judgment, a settlement order, or a consent decree—whichever occurs first.[122] And if the trial court holds that the patent is valid and infringed but is overturned on appeal before the 30-month stay expires, then final approval is effective on the date the court of appeals rules that the patent is invalid or not infringed or the court of appeals enters a consent decree or settlement order.[123]

The first ANDA filer generally must obtain tentative approval within 30 months or forfeit its 180 days of generic exclusivity.[124] (Under Section 1133(a) of the FDA Safety and Innovation Act, or FDASIA, for Paragraph IV Certifications submitted during the period January 9, 2010, through July 9, 2012, the time to obtain timely tentative approval is 40 months during July 9, 2012, through September 30, 2015, and 36 months during October 1, 2015, through September 30, 2016.)

117. *Id.* § 355(j)(5)(B)(iv)(II)(dd).
118. *Id.* § 355(j)(5)(B)(i).
119. *Id.* § 355(j)(5)(B)(ii).
120. *Id.* § 355(j)(5)(B)(iii).
121. *Id.* § 355(j)(5)(B)(iii).
122. *Id.* § 355(j)(5)(B)(iii)(I)(aa), (bb).
123. *Id.* § 355(j)(5)(B)(iii)(II)(aa).
124. *Id.* § 355(j)(5)(D)(IV).

3. Pediatric Exclusivity

Pediatric exclusivity confers six months of additional exclusivity for a brand company that undertakes acceptable pediatric studies in response to the FDA's written request.[125] Although this additional exclusivity period is triggered by pediatric studies, it applies to all uses of the RLD following the last applicable patent's expiration. Pediatric exclusivity does not itself provide an independent exclusivity, but instead extends all other exclusivities by six months.

4. Orphan Drug Exclusivity

Orphan drug exclusivity confers seven years of exclusivity from the date of approval of the NDA, regardless of whether there are any other applicable exclusivities.[126]

5. 180-day Marketing Exclusivity

As discussed, a generic ANDA filer may be blocked by the 180-day marketing exclusivity enjoyed by a successful ANDA first-filer. Under 21 U.S.C. § 355(j)(5)(B)(iv), the first ANDA filer whose ANDA includes a Paragraph IV Certification is eligible to receive 180 days of marketing exclusivity. Thus, unless that exclusivity is forfeited (or the first-filer loses its litigation), subsequent ANDA filers who file identical applications are not barred by the NDA holder's Orange Book–listed patents, but can only obtain tentative (not final) approval until 180 days after the first-filer launches its generic product.

Of note, exclusivities are determined by strength. For example, if a generic applicant were a first-filer for two of three approved strengths for a particular product, that applicant's 180-day exclusivity would apply only to those two strengths. In other words, the exclusivity would not delay a launch of the third strength by other generic applicants.

Also, a first-filer's 180-day exclusivity does not prevent the brand company from launching an "authorized generic." Nor does it prevent the FDA from approving a product with a Section viii carve-out.

C. Delaying ANDA Approvals

An NDA holder may try to delay approval of an ANDA. For example, the NDA holder may try to manipulate the Orange Book patent listing or product labeling while an ANDA is pending, or it may file a Citizen Petition with the FDA raising safety issues regarding a particular ANDA.

---

125. *Id.* § 355a.
126. *Id.* § 360cc.

1. Manipulation of the Orange Book or Product Labels

Approval can be delayed if the NDA holder manipulates the Orange Book's listings or product labeling, requiring the patentee to certify to patents that should not be listed. Amendment by the NDA holder to the listings or labeling is allowed—indeed, encouraged—under the applicable regulations.[127] The ANDA applicant's remedies for such misconduct are, first, a challenge letter under 21 C.F.R. § 314.53(f), and—if the NDA holder does not voluntarily correct the listing—a counterclaim under 21 U.S.C. § 355(j)(5)(C)(ii)(I).[128] Alternatively, the ANDA applicant might be able to use a Section viii Statement to avoid labeling language that could otherwise block FDA approval.

2. Citizen Petitions

Under 21 U.S.C. § 355(q) and 21 C.F.R. § 10.30, any person can file a citizen petition with the FDA raising issues concerning approval of a particular NDA or ANDA application. The FDA can only delay the approval process if it concludes that "a delay is necessary to protect the public health."[129] Brand and generic companies (but mostly brand companies) have used this process to interfere with drug approvals, though the use of this vehicle should give companies cause for caution, in that any submission must be provided under penalty of criminal sanction.[130]

A citizen petition can toll the 30-month forfeiture period.[131] And the FDA must respond to a citizen petition within 180 days of submission or else the citizen petition is deemed denied.[132]

127. 21 C.F.R. § 314.70.

128. *Caraco Pharm. Labs.*, 132 S. Ct. 1670 ("We accordingly hold that Caraco may bring a counterclaim seeking to 'correct' Novo's use code 'on the ground that' the '358 patent 'does not claim . . . an approved method of using the drug'—indeed, does not claim two.").

129. 21 U.S.C. § 355(q)(1)(A)(ii).

130. *Id.* § 355(q)(1)(H), (I).

131. *Id.* § 355(q)(1)(G).

132. *Id.* § 355(q)(2); 21 C.F.R. § 10.30(e).

chapter 20

Providing Notice to the NDA Holder[1]

I. Introduction

Upon the submission of an infringing Abbreviated New Drug Application (ANDA) to the Food and Drug Administration (FDA), an ANDA applicant is required by statute to provide notice to the New Drug Application (NDA) holder. This notice letter alerts the NDA holder to the potential infringement of its patents and triggers a deadline within which the NDA holder must file a patent litigation in which the ultimate issues will be resolved. The system provides that the FDA will not approve a drug until the culmination of suit, if one is filed. The notice letter is a crucial part of the ANDA process due to its role in triggering critical dates and in framing the issues to be resolved in litigation and, possibly, its role in the assessment of exceptional case fees.

II. Infringing and Noninfringing ANDAs

The FDA's approval of an ANDA, discussed earlier in this text, carries no legal effect of its own. Rather, it is the submission of the ANDA for approval that has the potential to be an act of patent infringement. As discussed in prior chapters, NDA applicants must include information on any patents claimed for the drug or method of using the drug "with respect to which a claim of patent infringement could reasonably be asserted if a person not licensed by the owner of the patent engaged in the manufacture,

1. Karen E. Keller and Stephanie E. O'Byrne of Shaw Keller LLP, Wilmington, Delaware.

use, or sale of the drug product."[2] The identified patents are published in the FDA's *Approved Drug Products with Therapeutic Equivalence Evaluations*, commonly referred to as the Orange Book.[3]

Later-filed ANDAs must certify that any patents listed in the Orange Book are expired,[4] provide a date upon which the patents are set to expire,[5] or provide what is called a Paragraph IV Certification, whereby the ANDA applicant certifies that such patents are "invalid or will not be infringed by the manufacture, use, or sale of the new drug for which the [ANDA] is submitted."[6] The Paragraph IV Certification may be made in the original ANDA, or it may be made by amendment or supplement to the ANDA.[7]

In addition to the four Paragraph certifications, where the Orange Book lists only method-of-use patents, an ANDA applicant may certify in its ANDA that it seeks approval for a drug for a different use than that claimed by the patent(s).[8] The FDA requires that the ANDA applicant submit either this Section viii Statement or a Paragraph IV statement; it may not submit both.[9]

III. Requirement to Provide Notice

The requirement to provide notice is triggered by patent infringement.

A. ANDA with Method-of-Use Certification

Filing an ANDA with a method-of-use certification is noninfringing; thus, no notice letter is required. Applicants submitting a Section viii Statement have no statutory obligation to provide notice to the NDA holder.[10] That is, if no infringing use may be approved by the FDA, there is no corresponding duty to alert the NDA holder for the purpose of initiating patent litigation. In contrast to the Paragraph IV Certification, a Section viii Statement does not trigger a stay of FDA approval pending the outcome of underlying

2. 21 C.F.R. § 314.53(b) (2011). For method-of-use patents, the "artificial" infringement claim provided by § 271(e)(2)(A) only arises against a patented use that has been approved by the FDA. *See* Warner-Lambert Co. v. Apotex Corp., 316 F.3d 1348, 1356 (Fed. Cir. 2003). That is, the statute refers only to the "use" for which the FDA has granted an NDA. *Id.*

3. 21 U.S.C. § 355(b)(1), (c)(2), (j)(7)(A)(iii) (2011); 21 C.F.R. § 314.53(e).

4. 21 U.S.C. § 355(j)(2)(A)(vii)(II) (commonly, a "Paragraph II Certification").

5. *Id.* § 355(j)(2)(A)(vii)(III) (commonly, a "Paragraph III Certification").

6. *Id.* § 355(j)(2)(A)(vii)(IV).

7. *Id.* § 355(j)(2)(B)(ii)(II).

8. *Id.* § 355(j)(2)(A)(viii) (commonly, a "Section viii" Statement).

9. "[E]ither the applicant is seeking approval for the use claimed in the patent, or it is not." Torpharm Inc. v. Thompson, 260 F. Supp. 2d 69, 77 (D.D.C. 2003) (internal quotation omitted).

10. Purepac Pharm. Co. v. Thompson, 354 F.3d 877, 880 (D.C. Cir. 2004).

patent litigation.[11] The FDA is free to approve a Section viii application immediately; however, there is no exclusivity period awarded to the successful applicant.[12]

B. ANDA with Paragraph IV Certification

Filing an ANDA with a Paragraph IV Certification is infringing and requires a notice letter. As discussed earlier, a Paragraph IV Certification avers that the Orange Book–listed patents are either invalid or not infringed. The filing of an ANDA with a Paragraph IV Certification is an act of patent infringement.[13] This type of infringement is "highly artificial," given that at the time of the ANDA filing, no commercial manufacture, use, or sale of the generic drug (i.e., actual infringement) has yet occurred.[14,15] This provision exists to vest the court with statutory jurisdiction to adjudicate the underlying patent infringement dispute.[16]

The submission of the Paragraph IV Certification triggers a cascade of deadlines pursuant to the statutory regime. The Paragraph IV Certification to the FDA first creates an obligation to give notice of the filing (i.e., the act of infringement).[17] Notice must be provided to each owner of the patent that is the subject of the Paragraph IV Certification or to a representative of the owner that has been designated to receive such notice.[18] The holder of the approved NDA must also be notified, if it is not also a patent owner.[19]

Notice must be provided promptly. When the Paragraph IV Certification is made, the ANDA applicant agrees to provide notice to the patent holder no later than 20 days after the date of the postmark on the notice provided by the FDA that the ANDA has been filed.[20] If the original ANDA did not contain a Paragraph IV Certification as to a particular patent, but one is subsequently made by amendment or supplement to the ANDA, immediate notice is required. In that case, the applicant is required to give

11. *Id.*

12. *Id.* (citing Purepac Pharm. Co. v. Thompson, 238 F. Supp. 2d 191, 195 (D.D.C. 2002)).

13. 35 U.S.C. § 271(e)(2)(A) (2010).

14. *See* Eli Lilly & Co. v. Medtronic Inc., 496 U.S. 661, 678 (1990); *accord* Teva Pharm. USA Inc. v. Novartis Pharm. Corp., 482 F.3d 1330, 1334 (Fed. Cir. 2007) (describing Paragraph IV Certifications as "technical infringement" under 35 U.S.C. § 271(e)(1)).

15. As stated, only "commercial" use may be infringing. There is a FDA experimental use exception, under which a drug manufacturer may conduct otherwise-infringing research so long as it is "solely for purposes reasonably related to the submission of information" to the FDA. *See* 35 U.S.C. § 271(e)(1).

16. *Eli Lilly & Co.*, 496 U.S. at 678.

17. 21 U.S.C. § 355(j)(2)(B)(ii)(I).

18. *Id.* § 355(j)(2)(B)(iii)(I).

19. *Id.* § 355(j)(2)(B)(iii)(II).

20. *Id.* § 355(j)(2)(B)(ii)(I).

notice at the time of the submission. This immediate notice is required even if notice was previously given vis-à-vis a Paragraph IV Certification previously made on another patent.[21]

The ANDA applicant's Paragraph IV notice letter, in turn, serves to trigger a 45-day period in which the NDA holder may bring its patent infringement suit.[22] Commonly, the patentee brings its infringement suit regarding the Orange Book–listed patents that are the subject of the notice letter within the 45-day window.[23] When this occurs, it is the ANDA applicant—not the patentee or NDA holder—that is required to notify the FDA of the suit. The ANDA applicant has until the later of (1) the expiration of the 45-day period or (2) the completion of the FDA's review of the application to provide notice of the suit.[24]

The initiation of suit also serves to trigger a 30-month stay under which, without a court judgment, the FDA may not proceed to process and approve the ANDA.[25] The 30-month stay is effectuated by the filing of suit and is calculated from the date the Paragraph IV notice letter is received by the patentee or NDA holder.[26]

If the FDA receives no notice of suit, the FDA may approve the ANDA.[27] Per the statute, the ANDA will be made effective immediately upon the latter of (1) the close of the 45-day window or (2) the agency's review and approval of the ANDA.[28] Given the extensive review required by the FDA, as well as the FDA's backlog in ANDAs to review, approval will not occur within this 45-day window.[29]

If an ANDA with Paragraph IV Certification is filed, the patentee or NDA holder does not file suit, and the FDA approves the ANDA, the ANDA holder may still be liable for damages for patent infringement

21. *Id.* § 355(j)(2)(B)(ii)(II).

22. *Id.* § 355(j)(5)(B)(iii).

23. The statute requires that the NDA holder "shall file" any patent that claims a drug or method of using a drug "to which a claim of patent infringement could reasonably be asserted" against the subject of the ANDA. *See* 35 U.S.C. § 355(c)(2) (2013). Ideally, the patent holder has not listed patents in the Orange Book that it does not intend to assert against others in litigation—but the broad language of the listing statute may not practically reconcile with this result in all cases.

24. 21 C.F.R. § 314.107(f)(2) (2012).

25. 21 U.S.C. § 355(j)(5)(B)(iii).

26. *Id.*

27. *Id.* § 355(j)(5)(B)(iii).

28. 21 C.F.R. § 314.107(f)(2).

29. In February 2013, FDA Commissioner Margaret A. Hamburg represented to the Annual Meeting of the Generic Pharmaceutical Association (GPhA) that the current backlog of ANDAs exceeds 2,500, stretching the median ANDA approval time to 31 months. *See* U.S. Food & Drug Admin., http://www.fda.gov/NewsEvents/Speeches/ucm340870.htm (last visited July 30, 2013).

upon its launch of the generic drug.[30] To protect its interests, the ANDA applicant may seek a declaratory judgment that the Orange Book–listed patent or patents are invalid and not infringed.[31] A declaratory judgment suit may only be initiated after the close of the 45-day window in which the NDA holder may bring its suit.[32] When a declaratory judgment suit is brought, it is again the ANDA applicant's responsibility to alert the FDA about the litigation.[33]

IV. Notice Letters and Exclusivity

The Hatch-Waxman Act further provides that, after a first ANDA applicant has submitted an ANDA containing a Paragraph IV Certification, additional ANDAs containing Paragraph IV Certifications may not be made effective until "180 days after the date of the first commercial marketing of the drug (including the commercial marketing of the listed drug) by any first applicant."[34] This 180-day exclusivity incentivizes challenges to Orange Book patents "by enabling a generic company a period to recover its investment in these challenges."[35]

As noted, an ANDA applicant must provide its notice letter regarding its Paragraph IV Certification no later than 20 days after the date the original ANDA was filed with the FDA. This requirement was added to the Hatch-Waxman Act in 2003.[36] The requirement that notice be immediately given upon the amendment of an ANDA to add a Paragraph IV Certification was in place both before and after this 2003 amendment.

30. The "technical" act of infringement of submitting the ANDA with Paragraph IV Certification does not trigger damages, but damages may be warranted for sales of the generic prior to resolution of the infringement issue at trial, if they are later found to infringe. *See, e.g.*, Biovail Labs. Inc. v. Torpharm Inc., No. 01-C-9008, 2002 WL 1732372, at *2 (N.D. Ill. July 25, 2002).

31. 21 U.S.C. § 355(j)(5)(C)(II).

32. *Id.*; *see also* Paddock Labs. v. Ethypharm SA, No. 09-3779-GEB, 2011 WL 149860 (D.N.J. Jan. 18, 2011) (declining its discretionary jurisdiction over a suit filed by the ANDA applicant within the 45-day window, given that the Hatch-Waxman Act intended that the NDA holder has the first opportunity to sue).

33. 21 C.F.R. § 314.107(f)(2).

34. 21 U.S.C. § 355(j)(5)(B)(iv)(I). This "'180-day exclusivity period' means the 180-day period ending on the day before the date on which an application submitted by an applicant other than a first applicant could become effective under this clause." *Id.* § 355(j)(5)(B)(iv)(II)(aa). To be deemed a "first applicant," the statute requires that the ANDA is "substantially complete" or that it is sufficiently complete to permit the FDA's substantive review. *Id.* §§ 355(II)(bb), (cc).

35. *See* Janssen Pharmaceutica NV v. Apotex Inc., 540 F.3d 1353, 1361–62 (Fed. Cir. 2008) (citations omitted) (collecting authority).

36. *See* 149 Cong. Rec. S8898-03 (daily ed. June 26, 2003); *see also* 149 Cong. Rec. H6007-02 (daily ed. June 26, 2003).

Failure to promptly comply with the statute's requirement to immediately notify the NDA holder of a new Paragraph IV Certification in an existing ANDA could result in the loss of generic exclusivity. This is best illustrated through an exclusivity challenge between two ANDA applicants, Purepac Pharmaceutical Company (Purepac) and TorPharm Inc. (TorPharm). Both companies submitted amended ANDAs with Paragraph IV Certifications for the Orange Book–listed patent covering Neurontin (gabapentin) with the FDA in 2000. Purepac's Paragraph IV Certification was received by the FDA on May 26, 2000, and Purepac provided its notice letter to the NDA holder (Warner-Lambert) on June 13, 2000.[37] TorPharm mailed its amended ANDA to the FDA on June 13, 2000, and on the same day provided notice to Warner-Lambert—though its ANDA was not received by the FDA until June 16, 2003.[38] The FDA determined that "[b]ecause Purepac did not give notice when it submitted the amendment to the FDA, [the] FDA will not treat the original receipt date as the relevant date for exclusivity purposes. Instead, [the FDA] will look to the date that Purepac actually sent the required notice, since this is the date upon which Purepac effectively met the statutory requirements by having both a submitted Paragraph IV certification *and* sent notice of the submission[—]June 13, 2000."[39] In this case, Purepac sent its notice letter before the FDA received TorPharm's ANDA, so Purepac remained eligible for the 180-day exclusivity as the first ANDA filer.[40]

In sum, the notice letter not only serves to apprise the NDA holder of its rights, but it may also be used to establish priority between ANDA filers in analogous situations.

V. Disclosure of Contentions

The ANDA applicant's contentions must be disclosed at the outset. Notice to the patentee/NDA holder must contain more than a statement that an ANDA with a Paragraph IV Certification has been filed and accepted by the FDA. The ANDA applicant is required to "include a detailed statement of the factual and legal basis of [its] opinion that the patent is invalid or will not be infringed" by the ANDA product (or method).[41] Thus, the

37. *See* TorPharm, Inc. v. Thompson, 260 F. Supp. 2d 69, 79–80 (D.D.C. 2003).

38. *Id.*

39. *Id.* at 78 (citing agency record) (emphasis in original).

40. *Id.* at 82 (upholding FDA decision upon TorPharm's challenge), *aff'd*, Purepac Pharm. Co. v. Thompson, 354 F.3d 877, 890 (D.C. Cir. 2004).

41. 21 U.S.C. § 355(j)(2)(B)(iv)(II).

statutory scheme anticipates the patent infringement litigation that will likely ensue.

As part of providing its notice, the potential defendant (the ANDA filer) is required to inform the potential plaintiff (the patentee or NDA holder) of its noninfringement and invalidity contentions immediately or within 20 days of filing the ANDA. While filing the ANDA with a Paragraph IV Certification is, by statute, an act of infringement, the noninfringement contention concerns the merits of the underlying dispute, not the filing itself. That is, the ANDA applicant is required to provide a "detailed statement" of its contentions with respect to the ANDA product or method that may potentially infringe the Orange Book–listed patents upon the launch of an approved generic drug.[42]

VI. The Notice Letter

The notice letter shapes the litigation.

A. Plaintiff's Pre-filing Investigation

Upon receiving the Paragraph IV notice letter, the NDA holder must initially determine whether it has enough information to accuse the sale of the infringing generic drug product in its complaint or whether it will rely on the filing of the ANDA itself as the operative act of infringement.

Accusing the drug product or method itself, however, invokes the requirement of Federal Rule of Civil Procedure 11(b) that an attorney conduct a "reasonable inquiry" into the law and facts before filing a complaint and certifying that the claims therein are not frivolous, legally unreasonable, without factual foundation, or asserted for some improper purpose. The Federal Circuit has interpreted Rule 11 to require, at a minimum, that counsel interpret the asserted patent claims and compare the accused drug(s) or method(s) with those claims before filing a claim alleging patent infringement.[43] At a minimum, the plaintiff must put forth a broad claim interpretation that follows the standard canons of claim construction and that is reasonably supported by the intrinsic record; a formal claim chart is not required.[44] Through discovery, the ANDA filer may inquire as to the

---

42. *See* 21 U.S.C. § 355(j)(2)(B)(iv)(II) (2011).

43. *See* Q-Pharma Inc. v. Andrew Jergens Co., 360 F.3d 1295, 1300–01 (Fed. Cir. 2004) (collecting authority).

44. *Id.*; *accord* View Eng'g Inc. v. Robotic Vision Sys. Inc., 208 F.3d 981, 986 (Fed. Cir. 2000) ("The presence of an infringement analysis plays the key role in determining the reasonableness of the pre-filing inquiry made in a patent infringement case under Rule 11.").

nature of the NDA holder's pre-suit investigation and, if insufficient, seek sanctions pursuant to Federal Rule of Civil Procedure 11(c).[45]

In contrast, the NDA holder may rely on the filing of the ANDA with Paragraph IV Certification as the act of infringement alleged in the complaint, not the manufacture or sale of the generic drug product. In the typical case, it is unlikely that the NDA holder has been given access to the ANDA filer's ANDA prior to the expiration of the 45-day window. Notwithstanding, a reasonable and competent inquiry into the act of infringement is simply an investigation of whether a relevant ANDA has been filed.[46] The notice letter provides sufficient basis for the filing of the complaint in this regard.[47] Put another way, because "a reasonable plaintiff in a Hatch-Waxman case would be expected to know few details about the accused product at the outset of litigation[,] plaintiff's counsel may reasonably rely on discovery to learn the material details."[48]

As the case commences, the NDA holder is likely to seek discovery into the ANDA filer's pre-filing investigation—that is, the bases of its noninfringement and invalidity positions underlying its Paragraph IV Certification. In addition to factual details about the ANDA filer's method of production, drug formulation, and the like, the NDA holder may obtain opinion letters detailing the ANDA filer's possible noninfringement and invalidity arguments, as well as particular claim constructions.[49]

If it was provided with detailed contentions in the Paragraph IV notice letter, the patentee can pursue more streamlined and tailored discovery at the outset of litigation.

B. Challenge of Placeholder Notice Letters

There is no statutory mechanism to challenge placeholder notice letters. The notice statute contemplates that the patentee, armed with detailed noninfringement and invalidity contentions at the outset, may incorporate this detail into its complaint for patent infringement. The process is aimed at streamlining the litigation such that resolution can occur within the 30-month stay.[50]

45. *Q-Pharma Inc.*, 360 F.3d at 1300–01.

46. Celgene Corp. v. KV Pharm. Co., No. 07-4819(SDW), 2008 WL 2856469, at *3 (D.N.J. July 22, 2008).

47. *Id.*

48. AstraZeneca AB v. Mylan Labs. Inc., No. M-21-81, MDL No. 1291, 2010 WL 2079722, at *4 (May 19, 2010).

49. *See, e.g.*, Astrazeneca AB v. Mut. Pharm. Co., 221 F. Supp. 2d 528, 532 (E.D. Pa. 2002).

50. While there is a "paucity" of legislative history on the Hatch-Waxman Act, the generally accepted purpose of the 30-month stay is the protection of NDA holders in this manner. *See* Gerald J. Mossinghoff, *Overview of the Hatch-Waxman Act and Its Impact on*

Importantly, however, there is no statutory mechanism for challenging the sufficiency of the Paragraph IV notice letter.[51] As a corollary, the ANDA applicant is not limited, in the ensuing litigation, to those defenses outlined in its notice letter to the patentee.[52]

At least one NDA holder that was presented with vastly different theories in suit than it anticipated prior to filing attempted to traverse precedent on policy grounds. If unchecked, an ANDA applicant may be tempted to "instigate premature litigation" by serving a Paragraph IV notice letter on an NDA holder that does not contain the ANDA applicant's defenses.[53] In this manner, an ANDA applicant essentially has filed a "placeholder notice letter" as it seeks to "free-ride[] on the efforts of subsequent filers," depriving the patentee of the full benefit of the 30-month stay.[54] Until Congress provides a remedy for nondisclosure, however, the sufficiency of the Paragraph IV letter may only be directly challenged in an action under the Administrative Procedure Act.[55] As described later in this chapter, however, a baseless Paragraph IV notice letter may form a basis for a motion for exceptional case fees pursuant to Section 285 after the close of proceedings.

C. Resolution of Claims

A detailed Paragraph IV notice letter may enable the NDA holder to plead its claims of infringement with equal specificity, but the NDA holder is not required to do so to satisfy the liberal requirements of Federal Rule of Civil Procedure 8.[56] Given that the ANDA applicant has triggered suit against

---

the Drug Development Process, 54 FOOD & DRUG L.J. 187, 187 (1999); *see also* Teva Pharm. USA Inc. v. Novartis Pharm. Corp., 482 F.3d 1330, 1340 n.5 (Fed. Cir. 2007) (citing 21 U.S.C. § 355(j)(5)(B)(iii)) (stay is offered to patent holders that reasonably cooperate in expediting the actions challenging their patents).

51. *See* Abbott Labs. v. Lupin Ltd., No. 09-152-LPS, 2011 WL 1897322, at * 7 (D. Del. May 19, 2011) ("The Court has no authority to penalize deficiencies in Paragraph IV notice letters; the Hatch-Waxman Act creates no private cause of action for such notice letter shortcomings." (citing Minn. Mining & Mfg. Co. (*3M*) v. Barr Labs. Inc., 289 F.3d 775, 777 (Fed. Cir. 2002))). The sufficiency of the Paragraph IV letter may only be challenged in an action under the Administrative Procedure Act. *3M*, 289 F.3d at 783.

52. *Abbott Labs.*, 2011 WL 1897322, at *7 (denying motion to strike defenses not contained in defendant's Paragraph IV notice letter); *accord* Abbott Labs. v. Apotex Inc., 2010 WL 2899113, at *3 (N.D. Ill. July 9, 2010); Takeda Chem. Indus. Ltd. v. Mylan Labs. Inc., 2005 WL 1457696, at *1 (S.D.N.Y. June 15, 2005); Astra Aktiebolag v. Kremers Urban Dev. Co., 2000 WL 257125, at *1 (S.D.N.Y. Mar. 8, 2000).

53. *Abbott Labs.*, 2011 WL 1897322, at *6.

54. *Id.*

55. *3M*, 289 F.3d at 783.

56. Rule 8(a)(2) requires that pleadings contain "a short and plain statement of the claim showing that the pleader is entitled to relief," FED. R. CIV. P. 8(a)(2), such that the opposing party is given "fair notice of what the claim is and the grounds upon which it rests." Bell Atl. Corp. v. Twombly, 550 U.S. 544, 545 (2007) (internal quotations omitted) (interpreting FED.

itself through the filing of its ANDA with Paragraph IV Certification and serving its Paragraph IV notice letter to the patentee, a motion to dismiss the patentee's claim of direct infringement for failure to satisfy the liberal pleading requirements of Rule 8 is highly improbable.

Moreover, the ANDA applicant cannot easily obtain an early dismissal or judgment on the pleadings on the basis that the ANDA product cannot infringe the asserted patent(s) in the manner described by the complaint. Courts have declined to review the ANDA itself on Federal Rule of Civil Procedure 12 motions[57] and often require the presentation of expert testimony prior to resolving the infringement issue.[58] Thus, an ANDA applicant may find itself limited in merit-based dismissal options regardless of whether it has served a "placeholder" or a highly substantive Paragraph IV letter.

D. Confidentiality of Notice Letters

At least one court has been asked to consider whether the Paragraph IV notice letter is a confidential document and has answered the question in the negative. In that case, Tolmar Inc. (Tolmar) submitted to NDA holder Nycomed U.S. Inc. (Nycomed) a Paragraph IV letter notifying Nycomed of Tolmar's ANDA. Tolmar's letter referenced an attached Offer of Confidential Access (OCA),[59] along with a consecutively paginated Detailed

---

R. Civ. P. 8(a)). "When there are well-ple[d] factual allegations, a court should assume their veracity and then determine whether they plausibly give rise to an entitlement to relief." Ashcroft v. Iqbal, 129 S. Ct. 1937, 1950 (2009). The Federal Circuit has held that Form 18, a sample form for direct patent infringement in the Appendix of Forms to the Federal Rules of Civil Procedure, is the measure of specificity required to determine whether a pleading meets the *Twombly/Iqbal* standards. *See In re* Bill of Lading Transmission & Processing Sys. Patent Litig., 681 F.3d 1323, 1334–35 (Fed. Cir. 2012). In this regard, Form 18 requires only "(1) an allegation of jurisdiction; (2) a statement that the plaintiff owns the patent; (3) a statement that defendant has been infringing the patent 'by making, selling, and using [the device] embodying the patent'; (4) a statement that the plaintiff has given the defendant notice of its infringement; and (5) a demand for an injunction and damages." *Id.* at 1334 (citing McZeal v. Sprint Nextel Corp., 501 F.3d 1354, 1357 (Fed. Cir. 2007)).

57. *See* Cima Labs Inc. v. Actavis Grp. HF, No. 06-1999(DRD), 2007 WL 1672229, at *4–5 (D.N.J. June 7, 2007) (where plaintiff did not "explicitly rely" on the ANDA in their complaint, it would not be considered on a motion to dismiss, but would not be considered determinative in any event); Lupin Atlantis Holdings v. Ranbaxy Labs. Ltd., No. 10-3897, 2011 WL 1540199, at *3 (E.D. Pa. Apr. 21, 2011) (holding that the court cannot consider defendant's ANDA for the truth of its contents, given that defendant may have "drafted the ANDA . . . with the specific intention of avoiding or forestalling any infringement action").

58. *See Lupin Atlantis Holdings*, 2011 WL 1540199, at *3 (citing Astra Aktiebolag v. Andrx Pharm. Inc., 222 F. Supp. 2d 423 (S.D.N.Y. 2002)) (noting that an improvident result may occur if infringement is adjudicated without the benefit of expert testimony).

59. The Hatch-Waxman Act provides for an "offer of confidential access" to the ANDA, to which access may be provided under confidentiality restrictions, so that the NDA holder may determine whether an infringement action should be brought. *See* 21 U.S.C. § 355(j)(5)(C)(i)(III).

Statement of its factual and legal bases for its positions.[60] A dispute arose as to whether the Detailed Statement portion of the letter's attachment was confidential or a public disclosure. The court noted that "the FDA addressed concerns of potential disclosure of confidential information by declining to give the notice letter any specific protection, not requiring disclosure of any particular information, and leaving decisions about the contents and the sufficiency of the details to the ANDA applicant, the patent owner, and the NDA holder."[61] The court concluded that the Detailed Statement was part of the notice letter, as proscribed by the statute, and it was not part of the separately described OCA.[62]

In resolving the issue of whether the Paragraph IV notice letter is a public disclosure, the court deferred to the FDA, which has determined on two separate occasions that notice letters are public documents.[63] The first FDA letter opinion, issued in 2003, stated that "[b]ecause the notice to the NDA holder or patent owner of the [P]aragraph IV certification is considered a public disclosure after notice has been given, the certification is available under FOIA" (Freedom of Information Act).[64] The Nycomed court also relied upon a 2010 informal decision by the FDA, in which the FDA denied the ANDA filer's request to remove a citizen petition and the accompanying Paragraph IV letter submitted therewith from public view. According to the FDA, "the statutory scheme contemplates that the [P]aragraph IV notice letter is the first step in an inherently public process."[65] In making that determination, however, the FDA also noted that the ANDA applicant's decision to market its generic drug in that case before the expiration of the patent at issue (an "at risk launch") also evidenced a "deliberate decision . . . to publicly disclose certain information about its ANDA so

60. *See* Nycomed U.S. Inc. v. Tolmar Inc., No. 10-2635(KSH), 2011 WL 1675027 (D.N.J. Apr. 28, 2011). The "detailed statement of the factual and legal basis of the opinion of the applicant that the [drug's] patent is invalid or will not be infringed" is a statutorily required part of the Paragraph IV notice letter—separate and apart from the "confidential access" to the ANDA provision. *See* 21 U.S.C. § 355(j)(2)(B)(iv)(II); *see also* 21 C.F.R. § 314.95(c)(6) (a "full and detailed" explanation is required).

61. Nycomed U.S. Inc. v. Tolmar, 2011 WL 1675027, at *4 (D.N.J. Apr. 28, 2011) (citing 59 Fed. Reg. at 50,350, 50,342).

62. *Id.*

63. The court applied deference to the determination of the FDA pursuant to *Chevron USA Inc. v. Natural Resources Defense Council Inc.*, 467 U.S. 837 (1984), while noting that letter rulings are not necessarily entitled to this level of deference. *Nycomed*, 2011 WL 1675027, at *6 (citing Madison v. Res. for Human Dev. Inc., 233 F.3d 175, 185 (3d Cir. 2000) (no Chevron deference required for informal rulings); Mylan Labs. Inc. v. Thompson, 389 F.3d 1272, 1279–80 (D.C. Cir. 2004) (applying *Chevron* deference to FDA letter decision)).

64. *Nycomed*, 2011 WL 1675027, at *6 (citing 68 Fed. Reg. at 36,690).

65. *Id.* The 2010 FDA letter decision was of record in the *Nycomed* case and not otherwise quoted by the *Nycomed* court.

that it may be eligible for FDA approval of its generic drug product prior to the expiration of the relevant patent."[66]

The Nycomed case reveals a possibility that a court may decide not to apply deference to the FDA's letter rulings on the issue and determine that a Paragraph IV notice letter is confidential. In *Christensen v. Harris County*, the Supreme Court specifically articulated that "[i]nterpretations such as those in opinion letters—like interpretations contained in policy statements, agency manuals, and enforcement guidelines, all of which lack the force of law—do not warrant Chevron-style deference."[67] While not controlling, however, agency opinions "do constitute a body of experience and informed judgment to which courts and litigants may properly resort for guidance" and which "give it power to persuade, if lacking power to control."[68] Thus, an ANDA applicant seeking to keep its Paragraph IV notice letter to the NDA holder outside the public purview, for reasons of protecting its trade secrets or otherwise, may find it difficult to convince a court to counter the FDA's determinations.

E. Damages Issues

As discussed earlier, there is no mechanism for challenging the sufficiency of the Paragraph IV notice letter. Thus, despite the statute's instruction to disclose all of its positions, an ANDA applicant may serve a placeholder letter in lieu of a substantive disclosure. Whether substantive positions are omitted for lack of prior formulation or later proven to be frivolous, the NDA holder may elect to seek enhanced damages for willful infringement and exceptional case fees in the case where discovery reveals that the ANDA applicant lacked a good faith basis of noninfringement or invalidity.

1. Enhanced Damages for Willful Infringement

In the non-pharmaceutical context, a patentee may establish willful infringement by showing, by clear and convincing evidence, that: (1) the infringer acted despite an objectively high likelihood that its actions constituted infringement of a valid patent (a threshold objective prong); and (2) this objectively defined risk was known or so obvious that it should have been known to the accused infringer (a subjective prong).[69] As discussed, the filing of an ANDA with a Paragraph IV Certification is an act

66. *Id.*

67. 529 U.S. 576, 587 (2000) (collecting authority).

68. Skidmore v. Swift & Co., 323 U.S. 134, 140 (1944).

69. *See In re* Seagate Tech. LLC, 497 F.3d 1360, 1371 (Fed. Cir. 2007); *see also* Bard Peripheral Vascular Inc. v. W.L. Gore & Assocs. Inc., 682 F.3d 1003, 1005 (Fed. Cir. 2012), *cert. denied*, 133 S. Ct. 932 (2013).

of patent infringement.[70] The Hatch-Waxman regime, which incentivizes challenges to Orange Book patents,[71] specifically disallows damages or other relief based upon noncommercial activity—that is, the "artificial" act of infringement of filing an ANDA.[72] Thus, the Federal Circuit has made clear that "the mere fact that a company has filed an ANDA application or certification cannot support a finding of willful infringement for purposes of awarding attorney's fees pursuant to 35 U.S.C. § 271(e)(4)."[73]

2. Exceptional Case Status and Enhanced Damages under Section 285

In contrast to the foregoing, Section 271(e)(4) of the Hatch-Waxman Act does permit an award of attorney's fees in exceptional cases on the basis of an ANDA filing.[74] The Hatch-Waxman Act imposes a "duty of care" on an ANDA applicant, and a case initiated by a Paragraph IV Certification may become "exceptional" if the ANDA applicant makes baseless allegations.[75]

The statute does more than lay out a "simple negligence" standard; rather, applicants only fail to meet their "duty of care" when they file "baseless" litigations.[76] While the totality of the circumstances is reviewed in this regard,[77] in at least one case, the sufficiency of the ANDA applicant's Paragraph IV notice letters were described as contributing to the court's finding of an exceptional case.

In *Takeda Chemical Industries, Ltd. v. Mylan Laboratories, Inc.*, the district court discussed Paragraph IV notice letters filed by Mylan Laboratories Inc., Mylan Pharmaceuticals Inc., and UDL Laboratories Inc. (collectively,

70. 35 U.S.C. § 271(e)(2)(A) (2010).

71. *See Janssen Pharmaceutica NV*, 540 F.3d at 1361–62 (citations omitted) (collecting authority).

72. 35 U.S.C. § 271(e)(4)(C) (providing that courts may only award monetary damages when commercial activity has occurred); *see also* Glaxo Grp. Ltd. v. Apotex Inc., 376 F.3d 1339, 1350–51 (Fed. Cir. 2004).

73. Glaxo Grp. Ltd., 376 F.3d at 1350–51. In so holding, the Federal Circuit adopted a holding first suggested in *Yamanouchi Pharm. Co. v. Danbury Pharmacal Inc.*, 231 F.3d 1339, 1346 (Fed. Cir. 2000), that a baseless and "wholly unjustified" Paragraph IV Certification, when combined with litigation misconduct, warranted an exceptional case finding. *Id.* (citing *Yamanouchi*, 231 F.3d at 1346).

74. 35 U.S.C. § 271(e)(4) states: "For an act of infringement described in paragraph (2) [ANDA filings] . . . a court may award attorney fees under section 285." In turn, § 285 provides that the court may, in exceptional cases, award reasonable attorney fees to the prevailing party. 35 U.S.C. § 285; *see also Yamanouchi*, 231 F.3d at 1346.

75. *Yamanouchi*, 231 F.3d at 1347.

76. Takeda Chem. Indus. Ltd. v. Mylan Labs Inc., 549 F.3d 1381, 1387 (Fed. Cir. 2008) (holding that the district court did not commit clear error in finding the ANDA applicant's conduct "baseless" where it "fail[ed] to present even a prima facie case of invalidity in filing the [P]aragraph IV [C]ertification," and awarding fees based on its finding of litigation misconduct).

77. *Yamanouchi*, 231 F.3d at 1347 (citing Kaufman Co. v. Lantech Inc., 807 F.2d 970, 978–79 (Fed. Cir. 1986)).

Mylan) and by Alphapharm Pty. Ltd. and Genpharm Inc. (collectively, Alphapharm).[78] In this regard, the court found that "[v]iewed in its totality, Alphapharm's [Paragraph IV notice letter] [was] so devoid of merit and so completely fail[ed] to establish a prima facie case of invalidity that it must be described as 'baseless.'"[79] While some bases were simply abandoned as insupportable, the court explained that the letter's deficiency with respect to an explanation for why one skilled in the art would be motivated to modify a lead compound to arrive at the patented compound was not an error due to "mere sloppiness, [but] in fact the error is more insidious [as it] underscores that Alphapharm was grasping at straws and did not act with due care or in good faith."[80]

In concluding that the NDA holder was entitled to attorney's fees from Alphapharm, the *Takeda* court stated that Alphapharm's Paragraph IV notice letter was "deeply flawed, filed in bad faith, and fail[ed] to present even a prima facie case of invalidity."[81] Alphapharm waited two years to file its Paragraph IV notice letter,[82] yet provided no indication for "why its [Paragraph IV notice letter] was so flawed[, or] why its description of obviousness went through such a dramatic evolution" between the notice and trial.[83] In support for its conclusion that it was presented with an exceptional case, the court also described Alphapharm's pattern of misconduct spanning the entire litigation, which it described as a "constant[ ] shift[ ] [in] its theory of obviousness in a futile effort to locate a coherent argument."[84]

The invalidity claims set forth in Mylan's Paragraph IV notice letter were "deeply flawed and never defended during discovery."[85] Mylan's notice letter focused on obviousness based on one compound described in a prior art reference (no. 14), and, during discovery, Mylan wholly abandoned this theory for a theory of obviousness based on another compound of that reference (no. 57)—notwithstanding that its 30(b)(6) designee testified that the latter compound would have been ruled out as a lead compound.[86] Mylan's failure to provide any rationale for why one skilled in the art would have

78. Takeda Chem. Indus. Ltd. v. Mylan Labs Inc., 459 F. Supp. 2d 227, 231 (S.D.N.Y. 2006).

79. *Id.* at 235.

80. *Id.* at 238, 237.

81. *Id.* at 245.

82. This case predates the most recent changes to the statute, requiring prompt notice.

83. *Id.* at 242–43.

84. *Id.* at 245. As described by the court, Alphapharm's misconduct included (1) "woefully short" evidence of "a credible claim of obviousness," (2) invalidity expert's positions that were poorly developed and "eviscerated" at deposition and trial, (3) "constantly shifting its theory," and (4) misrepresenting the record in the litigation. *Id.* at 244–45.

85. *Id.* at 245.

86. *Id.* at 246.

been motivated to modify compound 14, and its change of theory notwithstanding its admission, evidenced both baselessness and bad faith.[87]

Notably, the *Takeda* court stated that "Mylan's ANDA filing and Statement were baseless and filed in bad faith. On this ground alone, Takeda would be entitled to an award of attorney's fees."[88] Notwithstanding, the court proceeded to discuss litigation misconduct on the part of Mylan in both pursuing a baseless inequitable conduct claim and disobeying court orders, noting that each example would also, taken alone, support attorney's fees.[89]

The Federal Circuit affirmed on appeal, stating that "[w]ell-supported findings challenging the validity and infringement of patents owned by an NDA holder should not raise the specter of an unjustified holding of an exceptional case."[90] In affirming the award of attorney's fees, the Federal Circuit cited both Alphapharm's and Mylan's notice letters as well as the parties' litigation misconduct and held that the district court correctly found that the circumstances justified the award of fees against each party.[91]

Despite the district court's pronouncement in *Takeda*, it is unclear whether a baseless Paragraph IV notice letter could, alone, justify an award of attorney's fees given the broad, "totality of the circumstances" analysis required by the court in the Section 285 analysis.[92] Parties should be aware, however, that the Paragraph IV notice letter will be viewed with importance by both the NDA holder and by the court, and proceed accordingly.

VII. Conclusion

The notice letter not only serves to advise the patent holder of the submission of the Paragraph IV certification, but the cascade of deadlines triggered therefrom, including the 45-day window within which the patent holder must bring suit. As discussed above, the notice letter may also be used to establish priority between ANDA filers for purposes of determining generic exclusivity. The ANDA filer's descriptions of its possible noninfringement and invalidity contentions in the Paragraph IV notice letter shapes the

87. *Id.* at 247–48.

88. *Id.* at 248. Earlier in the opinion, the *Takeda* court recognized that, pursuant to Federal Circuit precedent, "[f]iling a baseless Paragraph IV [C]ertification *and* proceeding to challenge a patent's validity despite glaring weaknesses in the theory of invalidity constitute litigation misconduct." *Id.* at 232 (citing *Glaxo Grp.*, 376 F.3d at 1350) (emphasis added).

89. *Id.* at 250–51.

90. *Takeda Chem. Indus. Ltd.*, 549 F.3d at 1387.

91. *Id.* at 1388.

92. *Yamanouchi*, 231 F.3d at 1347 (citation omitted); *accord* Knorr-Bremse Systeme Fuer Nutzfahrzeuge GmbH v. Dana Corp., 383 F.3d 1337, 1342–43 (Fed. Cir. 2004) (collecting authority in the non-ANDA context).

ensuing litigation. The FDA has determined that notice letters are public documents. While there is no court mechanism for the patent holder to challenge the sufficiency of the notice letter, a "placeholder" notice letter may contribute to a finding that a case is "exceptional" under Section 285.

chapter 21

Evaluating Scenarios for Launch, Including Marketing and Manufacturing Developments[1]

I. Determining Launch Date

For a new generic product coming to market, selecting a launch date is critical. Because drugs have limited shelf lives that start to run as soon as a product is packaged, identifying the specific launch date drives substantial investments in time, labor, and cost. A series of pre- and post-launch milestones flow from that one corporate decision. A generic company must work backward from the projected launch date to create packaging, publicize the launch, and produce and test full-scale batches, and then work forward to ensure that anticipated demand is met and finished product is supplied to the company's sales channels in time to be prescribed.

External matters can affect the company's target launch date. Some of these matters are within the company's control; others are not. Decisions may need to be made—is the prospective launch date too soon? Too late? In particular, if there are potential or real patent issues that are not yet finally resolved, the applicant may need to determine whether to launch "at risk"—that is, at risk of infringing a patent. If the company miscalculates, an at-risk launch can be a catastrophe, leading to an injunction (possibly killing a first-filer's 180 days of exclusivity), compensatory damages, and even treble damages and attorney's fees for willful infringement.[2] But balanced against that are the expense and delay created by waiting until a

1. Scott H. Blackman of Nath, Goldberg & Meyer, Washington, D.C., and Davis S. Bloch of Winston & Strawn LLP, San Francisco, California.

2. 35 U.S.C. §§ 283–85.

pending case is finally resolved—and the possibility that another applicant will find a way to launch first.

Either way, mandatory or discretionary delay of the launch may result in forfeiture of a first-filer's exclusivity period.

A. At-risk Launch Scenarios

An Abbreviated New Drug Application (ANDA) applicant must launch its generic drug product soon after it receives final approval from the Food and Drug Administration (FDA). Sometimes, however, an immediate launch can carry significant downside risk. Launch of a commercial generic product upon final approval by the FDA is "at risk" if there is still a known, significant risk that a lawsuit could be filed by the patent owner or the owner of the Reference Listed Drug New Drug Application (NDA). Although the Hatch-Waxman Amendments create significant incentives for NDA holders to sue within 45 days of receiving an ANDA Paragraph IV notice, they do not require it. A patentee can instead lie in wait until the ANDA holder launches, and then sue based on the manufacture, use, or sale of the product itself rather than on the ANDA. In that event, however, the suit will be a regular patent infringement case and not one filed under the Hatch-Waxman Act. No matter when a non-Hatch-Waxman Act suit is filed, there is no 30-month stay of the FDA's final approval, which can issue at any time the FDA completes its review and approval process. Without the stay, the court may issue a preliminary injunction preventing launch of the product, even if a final decision on validity and infringement has not yet been determined.

Such a post-launch lawsuit could disrupt sales of the product, lead to an injunction, and require the generic company to disgorge its profits *or pay the brand company's lost profits*—potentially a crushingly larger damages figure. Of course, if the plaintiff does not sue upon notice of the Paragraph IV Certification, but instead chooses to sue later, there may be a strong equitable argument that the plaintiff should not be entitled to injunctive relief post-launch. Nevertheless, there is the possibility that any remaining goods would have to be pulled from the market and held until the patent expires (assuming the product does not expire before then).

To prevent this, the Hatch-Waxman Amendments allow the ANDA holder to sue for declaratory judgment that the patent is invalid and/or not infringed.[3] But that, too, is a risk—it requires the ANDA applicant to spend very significant sums on patent litigation over a patent that the NDA holder might never assert. Even launching after prevailing in dis-

3. 21 U.S.C. § 355(c)(3)(D)(i).

trict court but before all appeals are exhausted is "at risk" in the sense that there is still the possibility that the Federal Circuit will overturn a favorable decision on appeal.

B. Multiple First-filers

The Hatch-Waxman statute clearly contemplates situations where there is more than one first-filer for a particular drug. FDA will not approve later-filed ANDAs until 180 days after the date of first commercial marketing of the drug by *any* first applicant.[4] Likewise, a first applicant is *any* filer that, "on the first day on which a substantially complete application containing a [patent certification that at least one Orange Book patent listed for the reference drug is invalid or not infringed] is submitted for approval of a drug, submits a substantially complete application that contains and lawfully maintains [such] a certification."[5] Multiple first-filer situations are not uncommon. This is particularly true where there is a statutory bar preventing ANDA filings before a particular date—for example, when the reference drug is afforded New Chemical Entity (NCE) exclusivity. In that situation, multiple generic companies may file on the first permissible date, resulting in multiple first-filers.

In that situation, each first-filer—which is to say, all of the applicants who filed on that very first day—enjoy the same (shared) 180 days of exclusive marketing, beginning on the first date that *any first-filer* starts commercial marketing of the drug. Any one first-filer can trigger the 180-day exclusivity for all first-filers. If the other first-filers elect not to launch (because they do not want to face an at-risk situation) or are not able to launch (because they are not ready), they will lose some or all of the 180-day exclusive marketing period to which they otherwise would be entitled. If all first-filers forfeit their entitlement to the 180-day exclusivity, for any reason, then no other applicant shall be entitled to such an exclusivity period.[6]

II. Forfeiture Scenarios

To what extent would delay forfeit entitlement to the exclusivity period? Are the potential rewards from marketing promptly, including potentially taking advantage of the 180-day exclusivity, worth the potential harm from having to pull the product from the market and having to pay damages of one form or another to the NDA holder/patentee? These are examples of

4. *Id.* § 355(j)(5)(B)(iv)(I).
5. *Id.* § 355(j)(5)(B)(iv)(II)(bb).
6. *Id.* § 355(j)(5)(D)(iii).

issues that must be taken into account in determining whether to begin commercially marketing a product with a potential patent issue lurking in the shadows, like a sword of Damocles hanging overhead.

Being cautious may cost too much. A first applicant who delays launch of the product may end up expressly or inadvertently forfeiting its entitlement to the 180-day exclusivity period. Because the period of exclusivity begins when *any* first applicant launches, a particular first applicant may not be able to control when the 180-day period begins. And even if a particular applicant is able to control when the exclusivity period begins (if, for example, it is the only first-filer), the statute provides certain limitations to prompt timely marketing at risk of forfeiting the applicant's exclusivity.

By statute, the 180-day exclusivity window is forfeited upon any of the following:

- the first-filer fails to market the drug in a timely manner;
- the first-filer withdraws the application for approval;
- the first-filer amends or withdraws the patent certification for all patents that qualified the application for the exclusivity originally;
- the first-filer fails to obtain tentative approval from the FDA in a timely manner;
- the first-filer enters into certain agreements with another applicant, the listed drug application holder, or a patent owner; or
- all patents that originally qualified the application for exclusivity expire.[7]

We discuss each of these forfeiture events next, in turn.

A. Failure to Market[8]

The statute notes that failure to market by a first applicant takes place when the first applicant fails to market the drug by the later of—

> (aa) the earlier of the date that is—
>
> > (AA) 75 days after the date on which the approval of the application of the first applicant is made effective under subparagraph (B)(iii); or
> >
> > (BB) 30 months after the date of submission of the application of the first applicant; or
>
> (bb) with respect to the first applicant or any other applicant (which other applicant has received tentative approval), the date that is 75 days after the date as of which, as to each of the patents with

7. *Id.* § 355(j)(5)(D)(i).
8. *Id.* § 355(j)(5)(D)(i)(I).

respect to which the first applicant submitted and lawfully maintained a certification qualifying the first applicant for the 180-day exclusivity period under subparagraph (B)(iv), at least 1 of the following has occurred:

> (AA) In an infringement action brought against that applicant with respect to the patent or in a declaratory judgment action brought by that applicant with respect to the patent, a court enters a final decision from which no appeal (other than a petition to the Supreme Court for a writ of certiorari) has been or can be taken that the patent is invalid or not infringed.
>
> (BB) In an infringement action or a declaratory judgment action described in subitem (AA), a court signs a settlement order or consent decree that enters a final judgment that includes a finding that the patent is invalid or not infringed.
>
> (CC) The patent information submitted under subsection (b) or (c) is withdrawn by the holder of the application approved under subsection (b) of this section.[9]

We can simplify that convoluted analysis. The forfeiture takes place upon failure by the first applicant to bring its product commercially to market by the later of two dates, (aa) and (bb). Date (aa) is itself the date that is the *earlier* of two other dates, (aa/AA) or (aa/BB). Date (aa/AA) is the date 75 days after final approval as governed by 21 U.S.C. § 355(j)(5)(B)(iii); and date (aa/BB) is the date 30 months after the first applicant's submission of a substantially complete application. Date (aa) is the earlier of dates (aa/AA) and (aa/BB).

Date (aa) then still needs to be compared to date (bb) to determine the deadline for marketing. The later of dates (aa) or (bb) is the deadline for a first-filer to commence commercial marketing of the product without forfeiting the 180-day exclusivity.

Date (bb) is determined as follows. Date (bb) is the date 75 days after at least one of the following has happened for each patent subject to the qualifying certification: (bb/AA), (bb/BB), or (bb/CC). Date (bb/AA) is the date the court of appeals (but not the Supreme Court) decides that the patent is invalid or not infringed; (bb/BB) is the date of a settlement order or consent decree entered by the court of appeals (but not the Supreme Court) stating that the patent is invalid or not infringed; and (bb/CC) is the date

9. *Id.*

the reference drug application holder removes the patent from the Orange Book listing.

Thus, to determine date (bb), we must first determine for each patent whether there is a date (bb/AA), (bb/BB), or (bb/CC). As long as there is such a date for each patent at issue, the last of these dates for the patents, plus 75 days, is the (bb) date. This sequence may be depicted graphically as shown in Figure 1.

Figure 1 Requirements for Promptly Launching a Product to Avoid Forfeiting Market Exclusivity

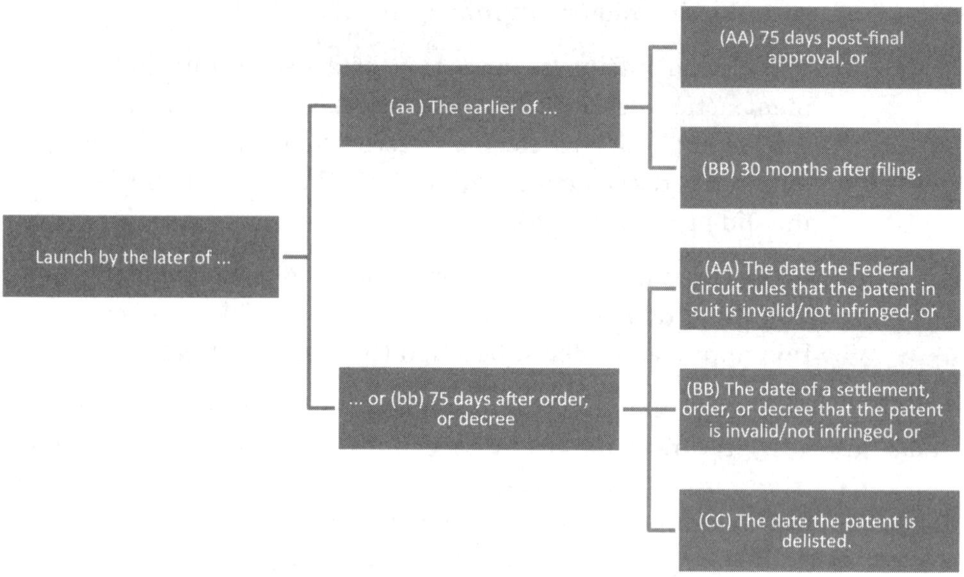

An apparent anomaly here is that if any patent does not have at least one of these (bb) candidate dates—(bb/AA), (bb/BB), or (bb/CC)—by definition, there can be no (bb) date. As a result, there is literally no way to determine a "later" date as between (aa) and (bb), ostensibly because there is no certainty that the (aa) date is later than a potential (bb). *Thus, so long as at least one patent does not have a (bb) date, there can be no deadline for failure to market under this provision.* FDA guidance on this point suggests that the Hatch-Waxman Amendments thus continue to offer at least one vehicle for "parking" generic exclusivity:

> Inherent in the structure of the "failure to market" forfeiture provisions is the possibility that a first applicant would be able to enter into a settlement agreement with the NDA holder or patent owner in which a court does not enter a final judgment of invalidity or non-infringement (i.e., without a forfeiture event under subpart (bb) occurring), and

that subsequent applicants would be unable to initiate a forfeiture with a declaratory judgment action. This inability to force a forfeiture of 180-day exclusivity could result in delays in the approval of otherwise approvable ANDAs owned by applicants that would market their generic drugs if they could but obtain approval. This potential scenario is not one for which the statute currently provides a remedy.[10]

B. Withdrawal of Application[11]

If the "first applicant withdraws the application or the Secretary considers the application to have been withdrawn as a result of a determination . . . that the application does not meet the requirements for approval," the 180-day exclusivity period is forfeited as to that first applicant.[12] Thus, if the application is withdrawn or rejected, there will be no approved drug product and no entitlement to the exclusivity period.

C. Amendment of Certification[13]

If the first applicant withdraws its Paragraph IV Certification for all listed patents, or amends its certification to remove all Paragraph IV Certifications, the first applicant loses its entitlement to the exclusivity period. No subsequent filers will be entitled to recapture the exclusivity period.

The relevant regulations set forth three situations when amendment to the patent certification is mandatory: if a first applicant is found by a court to infringe a valid patent addressed by the certification; if a patent addressed in the certification is removed from the Orange Book; or if the applicant learns that the submitted certification is no longer accurate.[14]

1. Final Judgment of Infringement[15]

When a first applicant is sued for infringement of an Orange Book–listed patent within 45 days of giving notice of the Paragraph IV Certification, the outcome of the lawsuit can affect the certification as to that patent and, thus, may also affect the entitlement to any 180-day exclusivity period based on that patent. If the first applicant loses the infringement suit (thus, the patent is held valid and infringed), it must amend its patent certification from Paragraph IV (asserting that the patent is invalid or not infringed) to Paragraph III (setting forth the expiration date of the patent and asking

10. Citizen Petition Response, No. FDA-2007N-0389, at 5 n.6 (Jan. 17, 2008).
11. 21 U.S.C. § 355(j)(5)(D)(i)(II).
12. *Id.*
13. *Id.* § 355(j)(5)(D)(i)(III).
14. 21 C.F.R. § 314.94(a)(12)(viii).
15. *Id.* § 314.94(a)(12)(viii)(A).

that final approval be made effective on or after that date).[16] Thus, the amendment removes any entitlement and acts as to forfeiture of any right to the exclusivity period based on the original patent certification.

2. Removal of a Patent from the Orange Book[17]

If a patent is removed from the Orange Book, any applicant with a pending application that contains a Paragraph IV Certification to that patent must amend its ANDA to delete the certification. If there are no other listed patents, the ANDA certification should be converted from Paragraph IV to Paragraph I (no listed patents). The reason for the change in certification—that the patent is being or has been removed from the Orange Book—must be included in the amendment.[18]

If the patentee delists a patent that is in litigation against an ANDA applicant, the delisting will not be effective as to any subsequent ANDA filers until the FDA determines that delisting will not alter the first-filer's 180-day exclusivity period.[19] Thus, voluntary delisting by an NDA holder cannot trigger an ANDA first-filer's exclusivity.[20]

3. Other Amendments for Accuracy[21]

Finally, an applicant must amend its Paragraph IV Certification if, at any time before the final effective date of approval, the applicant becomes aware that the submitted certification is no longer accurate.[22] Once final approval is granted and the product is launched, there is no further obligation to ensure the continued accuracy of the certification.[23]

D. Failure to Obtain Tentative Approval[24]

If a Paragraph IV first-filer fails to obtain tentative approval from the FDA within 30 months of the filing date, it forfeits its entitlement to the exclusivity period, unless the failure is caused by a change or review of the requirements for approval imposed after the filing date.[25]

Similarly, if the FDA delays approval of a first-filer's application while considering a citizen petition filed by another party, the 30-month period is extended so that the petition process does not by itself cause forfeiture,

16. *Id.*
17. *Id.* § 314.94(a)(12)(viii)(B).
18. *Id.*
19. *Id.*
20. 21 U.S.C. § 355(j)(5)(D)(i)(I)(CC); Teva Pharm. USA Inc. v. Sebelius, 595 F.3d 1303 (D.C. Cir. 2010).
21. 21 C.F.R. § 314.94(a)(12)(viii)(C).
22. *Id.* § 314.94(a)(12)(viii)(C)(1).
23. *Id.* § 314.94(a)(12)(viii)(C)(2).
24. *Id.* § 355(j)(5)(D)(i)(IV).
25. *Id.*

regardless of the outcome of the petition.[26] Thus, the 30-month period is extended by a period equal to the time interval between the date the FDA receives the citizen petition and the date of final agency action on the petition.[27] Such a petition may delay a generic product's launch, but will not by itself suffice to cause forfeiture of the applicant's exclusivity period, if any. If the Secretary determines that the citizen petition or supplement thereto is submitted to the FDA with the primary purpose of delaying the approval of an application, and if the petition does not raise valid scientific or regulatory issues, the Secretary may deny the petition at any point for that reason.[28]

E. Agreement with Other Applicant, Listed Drug Application Holder, or Patent Owner[29]

This forfeiture provision covers certain situations in which a Paragraph IV first-filer enters an agreement with another applicant, the holder of the application for the Reference Listed Drug (RLD), or the patentee. If the Federal Trade Commission (FTC) or the attorney general successfully challenges such a settlement agreement through to a final decision from which no appeal has been or can be taken (other than a petition to the Supreme Court for a writ of certiorari), the first-filer shall be deemed to forfeit its rights to exclusivity.[30]

F. Expiration of All Patents[31]

In a Paragraph IV situation, if the listed patents expire before final approval of the ANDA, the applicant loses its basis for the Paragraph IV Certification and thus forfeits any entitlement to the 180-day exclusivity period.[32] Thus, to maximize the advantages of a generic filing, an ANDA applicant should time its filings for at least 30 months before at least one listed patent expires.

III. De Facto Exclusivity Longer Than 180 Days

There are circumstances that may lead to a de facto period of market exclusivity longer than the statutory 180 days. In particular, a noninfringement

26. *Id.* § 355(q)(1)(G).
27. *Id.*
28. *Id.* § 355(q)(1)(E).
29. *Id.* § 355(j)(5)(D)(i)(V).
30. *Id.*
31. *Id.* § 355(j)(5)(D)(i)(VI).
32. *Id.*

finding specific to a Paragraph IV applicant's ANDA may not translate to subsequent filers whose products may be different in patent-relevant ways. By contrast, invalidating a patent has the effect of ending all litigation triggered by Paragraph IV filings against that patent and (absent an appeal) starts the 180-day clock ticking for all first-filers. So a Paragraph IV first-filer that can differentiate itself from other first-filers by mounting a noninfringement challenge may be able to obtain its 180 days of statutory exclusivity, plus the extra time it takes for the other applicants to establish their own noninfringement defenses, invalidate the patent, or launch at risk.

IV. Settlement Scenarios

Settling a Paragraph IV lawsuit without a final court decision on validity and infringement issues may provide significant advantages to both parties.

The party defending the patent (the patentee or RLD owner) may be able to maintain the patent—and thus its own period of exclusive marketing—at least with regard to other non-first-filer ANDA applicants. Even other first-filers could possibly be delayed by a settlement with one or more first-filers, so long as the settlement does not declare the invalidity of the patent in suit. To the extent that the applicant is able to enter the market under an "authorized generic" arrangement, it may be able to launch even before the 30-month stay expires and, thus, may enjoy well more than 180 days of exclusivity—especially if the patentee continues to sue subsequent ANDA applicants.

Frequently, "reverse payment" settlements are a mechanism for settling ANDA litigation. That type of settlement typically involves some form of payment by the patentee or RLD owner to the ANDA filer. In exchange for a payment, the generic company agrees to stay off the market for an agreed period. That period usually expires earlier than the regular expiration date of at least one relevant patent, providing a second benefit to the ANDA filer (and arguably a benefit to consumers as well). And if there are other Orange Book–listed patents in play (for example, patents subject to a Paragraph III Certification), the settlement can give the generic entrant a very significant head start.

"Reverse payment" settlements have faced increasing scrutiny from the courts and the Federal Trade Commission (FTC). In one recent case, the U.S. Supreme Court ruled that "reverse payment" settlements were subject to antitrust scrutiny under a "rule of reason" analysis.[33] The Court held that "[i]n sum, a reverse payment, where large and unjustified, can bring with

33. Fed. Trade Comm'n v. Actavis, Inc., 133 S. Ct. 2223 (2013).

it the risk of significant anticompetitive effects; . . . a court, by examining the size of the payment, may well be able to assess its likely anticompetitive effects along with its potential justifications without litigating the validity of the patent."[34] The Court rejected the FTC's assertion that reverse payment settlements should be deemed presumptively unlawful, as well as the Eleventh Circuit's decision providing "near-automatic antitrust immunity to reverse payment settlements."[35]

With this in mind, the parties to a reverse payment (or other reverse value transfer) settlement need to be able to defend the settlement from claims that the agreement is anticompetitive in intent or effect. As the Supreme Court explained, the "likelihood of a reverse payment bringing about anticompetitive effects depends upon its size, its scale in relation to the payor's anticipated future litigation costs, its independence from other services for which it might represent payment, and the lack of any other convincing justification. The existence and degree of any anticompetitive consequence may also vary as among industries."[36]

V. Licensing Scenarios

Especially in light of the increasing antitrust scrutiny brought to bear against settlements, licensing options may be a preferable option. License terms, with significant benefits flowing both ways, also may be used to justify a reverse-payment or other benefit flowing from the patentee to the ANDA filer. For example, the ANDA filer could be licensed by the patentee as a manufacturer or supplier of the active ingredient or the final drug product. In this manner, the ANDA filer profits from sales of the RLD product much sooner than it otherwise could sell its own version of the drug directly to the public.

Alternatively, the ANDA filer could be licensed as an "approved generic"—a product sold under the NDA with different packaging than the reference product, effectively as a "generic" version. An ANDA filer acting as the approved generic source would be preparing and selling the product under the NDA rather than under its own ANDA.

Relationships like these benefit both parties and help to avoid antitrust problems that otherwise could nullify the settlement. Any other resolution that similarly provides reasonable benefits to both parties should also provide a good chance of avoiding antitrust problems.

34. *Id.* at 19–20.
35. *Id.* at 20–21.
36. *Id.* at 20.

chapter 22

Citizen Petitions[1]

Citizen petitions are strategic tools for both branded and generic drug companies alike. Because private entities, including individuals, are generally exempt from antitrust liability for petitioning the U.S. government for a certain action, even if ultimately anticompetitive,[2] such petitions have been used in attempts to extract a market advantage over competitors. Although an exhaustive list of petition topics is not feasible due to space constraints, this chapter shall examine some of those more common to ANDA litigants, such as to:

1. prevent Section viii ANDA/505(b)(2) label carve-outs;
2. require a split patent certification or recertification;
3. impose a second 30-month stay;
4. designate a second Reference Listed Drug (RLD);
5. require a new bioequivalence or other clinical study; and
6. deem that a first-applicant has forfeited any 180-day exclusivity.

We begin with the general definition and authority for most of the citizen petitions relevant to ANDA litigation.

I. Definition

A citizen petition is a mechanism by which a person (not necessarily a citizen of the United States) or an entity may request that the FDA issue, change, amend, uphold, or stay a policy decision that is of interest to the petitioner. Citizen petitions must be submitted to the FDA in writing and signed.[3] A petition may request the FDA to issue, amend, or revoke a regulation or an order, or to take or refrain from taking any other form of administrative action,[4] or to stay the effective date of any administrative

1. Stephen R. Auten of Taft Stettinius & Hollister LLP, Chicago, Illinois.

2. *See, e.g.*, E. R.R. Presidents Conference v. Noerr Motor Freight Inc., 365 U.S. 127 (1961); United Mine Workers v. Pennington, 381 U.S. 657 (1965).

3. 21 C.F.R. § 10.20.

4. *Id.* § 10.30.

action for either a specific or an indefinite period.[5] The right to petition the government with regard to an agency decision is conferred by the First Amendment, as part of the U.S. Constitution's Bill of Rights, and certain petitions directed to drug applications, specifically ANDA and 505(b)(2) applications, are governed by Section 505(q) of the Federal Food, Drug, and Cosmetic Act.[6]

II. The 505(q) Citizen Petition

Section 505(q) of the Federal Food, Drug, and Cosmetic Act was enacted as part of the 2007 Food and Drug Administration Amendments Act (FDAAA).[7] As mentioned, Section 505(q) applies to petitions that request the FDA to take action in connection with an ANDA or a 505(b)(2) application. In June 2011, the FDA issued a final guidance regarding its interpretation of 505(q), particularly specifying what petitions the statute applies to, stipulating that the targeted application(s) must be pending at the time the petition is submitted, and identifying necessary elements for the statutory certification and verification (discussed later in this section).[8]

As initially enacted, the statute provided that:

> The Secretary shall not delay approval of a pending application submitted under subsection (b)(2) or (j) because of any request to take any form of action relating to the application, either before or during consideration of the request, unless—
>
> (i) the request is in writing and is a petition submitted to the Secretary pursuant to section 10.30 or 10.35 of title 21, Code of Federal Regulations (or any successor regulations); and
>
> (ii) the Secretary determines, upon reviewing the petition, that a delay is necessary to protect the public health.[9]

---

5. *Id.* § 10.35.
6. 21 U.S.C. § 355(q).
7. Pub. L. No. 110-85 (as amended by Pub. L. No. 110-316).
8. U.S. Food & Drug Admin., Guidance for Industry: Citizen Petitions and Petitions for Stay of Action Subject to Section 505(q) of the Federal Food, Drug, and Cosmetic Act 6 (June 2011) [hereinafter Guidance for Industry: Citizen Petitions] (stating that FDA interprets "section 505(q) to apply only to petitions for which, at the time the petition is submitted, at least one ANDA or 505(b)(2) application related to the subject matter of the petition is pending").
9. 21 U.S.C. § 505(q)(1)(A)(i).

Section 505(q) was subsequently amended on July 9, 2012, when the Food and Drug Administration Safety and Innovation Act (FDASIA)[10] was signed into law. Under FDASIA, the scope of 505(q) was broadened to include petitions regarding applications for biosimilar and interchangeable biological products filed pursuant to Section 351(k) of the Public Health Service Act.[11]

Prior to FDASIA, Section 505(q) required that the FDA "shall take final agency action on a petition not later than 180 days after the date on which the petition is submitted." FDASIA reduced the FDA's window to take action from 180 days to 150 days.[12] Further, the FDA shall not extend the period to take final agency action for any reason, including any determination that the approval of an application is to be delayed for the public health, the submission of any comments relating to the petition or supplemental information supplied by the petitioner, or by the consent of the petitioner.[13] As such, the FDA cannot extend the 180-day action window for pre-FDASIA petitions or the 150-day action window for post-FDASIA petitions, as applicable.

If the FDA determines that a delay of the approval of the particular ANDA, 505(b)(2), or 351(k) biosimilar/interchangeable application is necessary to protect the public health,[14] the FDA must provide the applicant, within 30 days of making the determination: (1) notification that such a determination has been made, (2) any clarification or additional data that the applicant should provide to the FDA to review the petition promptly, and (3) a brief summary of the specific substantive issues raised in the petition that form the basis for the FDA's determination.[15] The FDA is to provide this information either in a document or a meeting with the applicant,[16] and such information is considered part of the application and thus subject to public disclosure requirements.[17] But importantly, the FDA cannot delay final agency action on a related petition.

As noted, there are separate certification and verification requirements, which are strikingly similar.[18] Generally speaking, the certification requirement applies to any new petition, and the verification requirement applies to any supplemental information or comments on a pending petition. In the certification, the petitioner avers that the petition contains all

10. Pub. L. No. 112-144, 126 Stat. 993 (2012).
11. 42 U.S.C. § 262(k).
12. 21 U.S.C. § 355(q)(1)(F).
13. *Id.* § 355(q)(1)(F)(i)–(iii).
14. *Id.* § 355(q)(1)(A)(ii).
15. *Id.* § 355(q)(1)(B)(i)–(iii).
16. *Id.* § 355(q)(1)(C)(i)–(ii).
17. *Id.* § 355(q)(1)(D).
18. *Id.* § 355(q)(1)(H)–(I).

the information that is relied upon, including unfavorable information, and that reasonable steps were taken to have all such information disclosed.[19] A petitioner must also certify as to the date on which the information became known to him or her, and whether any payment or other consideration was received or expected, as well as from whom.[20]

The requirements in the verification essentially mirror those in the certification, particularly as to timing and consideration, while further averring that there was no intentional delay in the timing of the submission (to avoid outright gamesmanship).[21] If the FDA determines a petition or its supplement was submitted with the primary purpose of delaying the approval of an application and does not on its face raise valid scientific or regulatory issues, the FDA may deny the petition at any time.[22] Finally, the FDA's related guidance cautions that the certification and verification must parrot the statutory language verbatim and contain a specific month, date, and year regarding when the information first became known, or otherwise the submission will be found deficient.[23]

A. Notable Exceptions

The statute explicitly sets forth those petitions to which Section 505(q) does not apply.[24] Excluded are those petitions that relate "solely to the timing of the approval of an application pursuant to subsection (j)(5)(B)(iv)"[25] or the 180-day exclusivity for a so-called first-applicant, and petitions that are submitted by an ANDA or 505(b)(2) sponsor that seek "only to have the [FDA] take or refrain from taking any form of action with respect to that application."[26] Finally, also exempted, specifically as to 351(k) biosimilar and interchangeable biological product applications, is the statutory section regarding the exhaustion of administrative remedies, which must be done before seeking redress in federal court.[27]

19. *Id.* § 355(q)(1)(H)(a)–(c).
20. *Id.* § 355(q)(1)(H)(c).
21. *Id.* § 355(q)(1)(I)(a)–(b).
22. *Id.* § 355(q)(1)(E).
23. Guidance for Industry: Citizen Petitions, *supra* note 8, at 9–12.
24. 21 U.S.C. § 355(q)(4).
25. *Id.* § 355(q)(4)(A)(i).
26. *Id.* § 355(q)(4)(A)(ii).
27. *Id.* § 355(q)(4)(B).

III. Specific Topics in Past Citizen Petitions

A. To Prevent Section viii Label Carve-outs

It is important to keep in mind that the purpose of a Section viii carve-out is a strategic option for an ANDA or a 505(b)(2) applicant so as to avoid having to file a Paragraph I to IV Certification as to an Orange Book–listed patent. If an ANDA or a 505(b)(2) applicant seeks to omit an approved method of use covered by a listed patent, it need not file a Paragraph IV Certification and, hence, remove any barriers that the patent claiming the protected method of use may impose.[28] A Section viii Statement requires the ANDA applicant to omit the protected use from its labeling.[29] In other words, the listed patent will not serve as a barrier to approval, thereby allowing the ANDA applicant to get to market immediately upon FDA approval. Accordingly, the NDA holder usually takes issue with this strategy, using a petition to thwart an applicant's avoidance of a patent certification. Because a carve-out is permissible only when the labeling "does not render the proposed drug product less safe or effective than the listed drug for all remaining, non-protected conditions of use," petitions directed to carve-outs generally revolve around these issues.[30]

For example, UCB filed in October 2010 a petition aimed at preventing ANDA applicants from carving out Xyzal's (levocetirizine dihydrochloride) indications for seasonal allergic rhinitis and perennial allergic rhinitis, leaving in the ANDA label only the indication relating to chronic idiopathic urticaria.[31] According to UCB, the rhinitis indications were "primary" and the urticaria indication "secondary" because "nearly all of the clinical data was obtained" on the former indications, and thus carving out the rhinitis clinical trial information would render the ANDA label not approvable for the urticaria indication.[32]

In February 2011, the FDA denied the petition.[33] While the FDA did agree that certain safety information conveyed in the clinical trial section of Xyzal's labeling was needed, the FDA disagreed that all rhinitis-related information had to be removed for an approvable carve-out. The FDA noted that the regulations permitted differences in a generic product's labeling to prevent disclosure of information in the RLD labeling that is

28. *See* FDCA 21 U.S.C. § 355(j)(2)(A)(viii).

29. *See* 21 C.F.R. §§ 314.94(a)(12)(iii), .95(a)(1).

30. *See id.* § 314.127(a)(7).

31. Docket No. FDA-2010-P-0545.

32. *Id.* at 1.

33. *Id.* (decision letter dated February 24, 2011, from the FDA to R. Trainor, Executive Vice President and General Counsel, UCB Inc.).

protected by patents or exclusivity.[34] Such differences may include deleting certain words or phrases, rather than entire sections of information, as well as other minor changes after which the label still reads properly. Here, that was accomplished merely by removing "allergic rhinitis" from the relevant sections in the label, leaving the relevant safety information in the ANDA label while ensuring that the patent-protected information was not disclosed.

That FDA denied UCB's petition was not unexpected. The FDA has historically denied such petitions while granting only a handful. By way of example, the FDA has allowed a label carve-out relating to the following products (and, where appropriate, denied a related petition that had requested such a carve-out not be allowed): Actos (pioglitazone HCl) and Actoplus Met[35](pioglitazone HCl; metformin HCl), Altace[36] (ramipril), Camptosar[37] (irinotecan HCl), Ethyol[38] (amifostine), Lyrica[39] (pregabalin), Marinol[40] (dronabinol), Naropin[41] (ropivacaine HCl monohydrate), Oxandrin[42] (oxandrolone), Prandin[43] (repaglinide), Pulmicort Respules[44] (budesonide inhalation suspension), Rebetol[45] (ribavirin), Seroquel[46] (quetiapine fumarate), Ultram[47] (tramadol HCl), Vancocin[48] (vancomycin HCl) Capsules, and Xibrom/Bromday[49] (bromfenac).

It is much more rare for the FDA to not allow a label carve-out, which has occurred on at least two occasions relating to the products Rapamune[50] (sirolimus) and Colcrys[51] (colchicine). Regarding the former, Wyeth petitioned the FDA to refrain from approving any ANDA for sirolimus with a label that omitted the then-exclusivity-protected information regarding cyclosporine withdrawal, as the product was originally approved for use in combination with cyclosporine and corticosteroids. After Rapamune was

34. *Id.* at 10 (citing 21 C.F.R. § 314.94(a)(8)(iv)).
35. Docket No. FDA-2009-P-0411 (decision letter Mar. 15, 2010).
36. Docket No. FDA-2008-P-0304 (decision letter June 18, 2008).
37. Docket No. FDA-2008-P-0069 (decision letter July 28, 2008).
38. Docket No. FDA-2006-P-0274 (decision letter Mar. 13, 2008).
39. Docket No. FDA-2010-P-0087 (decision letter July 30, 2010).
40. Docket No. FDA-2007-P-0169 (decision letter Apr. 25, 2008).
41. Docket No. FDA-2009-P-0601 (decision letter June 17, 2010).
42. Docket No. FDA-2005-P-0368 (decision letter Dec. 1, 2006).
43. Docket Nos. FDA-2008-P-0343 and FDA-2008-P-0411 (decision letter June 16, 2009).
44. Docket No. FDA-2006-P-0073 (decision letter Nov. 18, 2008).
45. Docket No. FDA-2003-P-0074 (decision letter Apr. 6, 2004).
46. Docket Nos. FDA-2011-P-0662 and FDA-2011-P-0663 (decision letter Mar. 27, 2012).
47. Docket Nos. 2001P-0495, 2002P-0191, and FDA-2002-P-0003 (decision letter June 11, 2002).
48. Docket No. FDA-2006-P-0007 (Apr. 9, 2012).
49. Docket No. FDA-2011-P-0128 (decision letter May 11, 2011).
50. Docket No. FDA-2003-P-0002 (decision letter Sept. 20, 2004).
51. Docket No. FDA-2010-P-0614 (decision letter May 25, 2011).

first marketed, it was found that the combination of that product with cyclosporine was associated with the risk of increased renal function impairment, and the results of a clinical study showed that the benefits of withdrawing cyclosporine outweighed the risks of immune system reactions. Wyeth was granted three years of data exclusivity regarding the cyclosporine withdrawal trial and thus sought to prevent any related carve-outs. Because the FDA determined that the inclusion of the cyclosporine withdrawal information was necessary for the safe and effective use of the drug, the protected information could not be carved out from the label.[52]

The facts on the Colcrys petition are similar, as Mutual successfully petitioned the FDA to block the carve-out of any information related to a clinical trial showing that a lower-dose regimen is adequate to treat an acute gout flare, which may occur during chronic use of the product and for which Mutual had been granted three years of data exclusivity.[53] Notably, these two instances in which the FDA did not allow a label carve-out relate to clinical data regarding a change in the administration of the product, but not to the deletion of a previously approved indication, which the FDA seemingly views as quite permissible.

B. To Require a Split Patent Certification or Recertification

There have been a few instances in which an entity has petitioned the FDA to require an ANDA or 505(b)(2) applicant to file a new Paragraph IV Certification to an Orange Book–listed patent, the consequence of which may be to restart the 30-month stay. One such scenario involves petitioning the FDA to require a so-called split certification to an Orange Book–listed patent, whereby a patent certification is submitted for certain claims of a listed patent and a Section viii Statement pertains to others.[54] If the ANDA has to be amended to include a new Paragraph IV Certification, even as to one patent claim, then a new 30-month stay would likely result.

For example, in August 2009, Sandoz filed a petition with respect to Actos and Actoplus Met, requesting that the FDA not approve any ANDAs with only Section viii Statements to two listed patents unless a separate patent certification was also received.[55] This is because, as Sandoz argued, the two listed patents contained product claims and were not limited to

52. Docket No. FDA-2003-P-0002 (decision letter Sept. 20, 2004), at 4.

53. Docket No. FDA-2010-P-0614 (decision letter May 25, 2011), at 24.

54. The FDA allows so-called split certifications to the claims in a single patent, allowing a patent certification to some claims and a Section viii Statement as to others. *See* Docket No. FDA-2009-P-0411 (decision letter Mar. 15, 2010), at 7.

55. Docket No. FDA-2009-P-0411, arguing that U.S. Patent Nos. 5,965,584 and 6,329,404 each required a patent certification.

one or to methods of use that could be carved out under Section viii.[56] At the relevant time, due to technical limitations, the Orange Book listed the two patents with only a use code and did not designate the patents as claiming the drug product or drug substance (as is commonplace now), which would have signaled that the patents contained product claims and thus required a patent certification in addition to any possible Section viii Statement. That the patents were not limited to method claims was apparent upon reading the claims of the patents.

The FDA, however, remains consistently ministerial in its role with respect to patents. To confirm whether the listed patents contained product claims, and thus required a patent certification, the FDA reviewed not the patents themselves but the related declarations that Takeda had submitted to the FDA with the Orange Book listing request.[57] Because Takeda's submission included a designation that the two patents claimed a drug product (in addition to a method of use), Takeda had a right to receive a patent certification in addition to any Section viii Statement.[58] The FDA thus granted Sandoz's petition,[59] and an ANDA applicant who had not previously filed a patent certification to those two patents was now required to do so, thereby possibly triggering a new 30-month stay as to those applicants.[60]

Aside from split certifications, another instance for a possible patent recertification occurs when an ANDA or 505(b)(2) sponsor changes the formulation in the application. Does such a change require a new patent certification? Does it matter if the reason behind an original Paragraph IV Certification may have shifted from noninfringement to invalidity, or vice versa? The law in this particular area is unsettled, as there does not

56. Unrelated to the petition issue, Watson later sued the FDA over its decision on the 180-day exclusivity for this product, which was dismissed as moot in June 2013 after the exclusivity period had expired. Watson Labs., Inc. v. Sebelius, Docket No. 12-5332 (D.C. Cir., October 24, 2012).

57. For clarity, the Orange Book listing information does not trigger an applicant's duty to file a patent certification, as the Orange Book may contain errors. The FDA's view is that the certification requirement is triggered "where the patent (or patent claim) claims a listed drug, and where the NDA holder is required to submit and has submitted that patent information to the FDA. This obligation to certify attaches regardless of whether that submission is accurately reflected in the Orange Book." Docket No. FDA-2009-P-0411 (decision letter Mar. 15, 2010), at 10 (citing FDA, Proposed Rule, Abbreviated New Drug Application Regulations, 54 Fed. Reg. 28872 (July 10, 1989).

58. Docket No. FDA-2009-P-0411 (decision letter Mar. 15, 2010), at 10–11.

59. Even though Sandoz's petition did not explicitly apply to the 180-day exclusivity, which is excepted from the then applicable 180-day decision deadline under 505(q), the FDA nevertheless decided the petition shortly after that time frame.

60. The FDA granted a similar petition on Prandin (repaglinide), as the listed patent was directed to a drug substance, drug product, and a method of use. FDA Docket Nos. 2008-P-0343 and 2008-P-011 (decision letter Dec. 4, 2008), at 18.

appear to be any clear obligation in this scenario to file a new Paragraph IV Certification despite the FDA's early guidance document shortly after the Hatch-Waxman Amendments were passed.

The authority begins in a 1986 FDA Letter to Industry from Dr. Paul D. Parkman, Deputy Director and then Acting Director of the FDA's Center for Drugs and Biologics, which states:

> The agency has considered various ways to ensure that patent certifications in supplements are asked for only when necessary and appropriate. The agency has concluded that supplements submitted under sections 505(b)(2) and 505(j) for the following types of change should contain an appropriate certification as specified by sections 505(b)(2)(A) and 505(j)(2)(A)(vii).
>
> - formulation changes
> - new uses or other conditions of use, e.g., route of administration
> - strength changes
>
> October 31, 1986, FDA Letter to Industry at p. 5[61] (1986 Parkman Letter)

The 1986 Parkman Letter states that a certification "should" be submitted when there is a formulation change, not that one "shall" be filed. Moreover, it does not state what types of formulation changes are contemplated. There could be a change in the salt form of the active ingredient (which would likely require a new application altogether), or an adjustment to an impurity specification, or an alternation to a dissolution rate, or anything else imaginable. The types of changes are nearly limitless, and the 1986 Parkman Letter does not firmly outline when a patent recertification would be required. In practice, the FDA has not seemingly sought to officially enforce the 1986 Parkman Letter, which pre-dates the FDA's implementing regulations of Hatch-Waxman, leaving applicants to police (and petition) the issue themselves.

Lupin Atlantis Holdings S.A. filed one such petition in October 2010, asking the FDA to rescind acceptance of any ANDA that does not have the same capsule dosage form as its RLD Antara (fenofibrate) capsules, and to require a new patent certification and notice letter relating to any resulting dosage form change.[62] Lupin apparently wanted the FDA to

---

61. *See Guidance for Industry*, U.S. Food & Drug Admin. (Oct. 31, 1986), http://www.fda.gov/downloads/Drugs/GuidanceComplianceRegulatoryInformation/Guidances/ucm074994.pdf.

62. Docket No. FDA-2010-P-0561.

require these things of Paddock, as Lupin contended that Paddock's drug product was for a capsule with the active ingredient on the outside, which is "not even a dosage form recognized to exist by the FDA" and which Paddock later amended "to be an actual capsule product."[63] The FDA never issued its decision letter, however, as Lupin withdrew the petition.[64] The withdrawal presumably happened because Lupin had recently received its requested relief. That is, in the related ANDA litigation where Lupin was a party, the district court noted a month earlier that Paddock filed a major amendment[65] regarding a dosage form change and that "FDA contacted Paddock . . . and required it to file a new Paragraph IV Certification for the new formulation."[66] Thus, a change in the dosage form via a major amendment may require a patent recertification, though we do not have the FDA's public pronouncement on this petition.

Pfizer, on the other hand, was able to garner the FDA's response, doing so as to a petition it filed in June 2012 regarding its Lyrica (pregabalin) product.[67] There is one patent[68] relevant to the petition, which is directed to a purified enantiomer of the active pharmaceutical ingredient. Not surprisingly, at issue in the ANDA litigation was whether the purity level in the products of several ANDA sponsors infringed the listed patent. When those ANDA sponsors amended their respective purity specifications, perhaps in response to Pfizer's claim construction position and/or the court's then-imminent ruling in the ANDA litigation, Pfizer petitioned the FDA to require new Paragraph IV Certifications (which presumably would trigger new notice letters and new 30-month stays). The FDA disagreed, stating that it will not:

> refuse to accept any amendment that would change the [enantiomer] impurity specification unless the amendment includes a new patent certification. FDA does not require recertification for minor changes in formulation that are presented within the context of an ANDA. In this case, FDA does not believe that an amendment that revises the [enantiomer] impurity specification is a change in formulation that

---

63. Docket No. FDA-2010-P-0561 (Lupin's petition Oct. 22, 2010), at 16.

64. Docket No. FDA-2010-P-0561 (Lupin's letter to the FDA Feb. 17, 2011).

65. For what the FDA considers a major amendment, see U.S. Food & Drug Admin., Guidance for Industry: Major, Minor, and Telephone Amendments to Abbreviated New Drug Applications (Dec. 2001), *available at* http://www.fda.gov/downloads/Drugs/ . . . /Guidances /ucm072888.pdf.

66. Paddock Labs Inc. v. Ethypharm SA, Civil Action No. 09-3779, 2011 WL 149860, at *6 (D.N.J. Jan. 18, 2011).

67. Docket No. FDA-2012-P-0583.

68. U.S. Patent No. 6,197,819.

> would require a new patent certification under 21 U.S.C. 355(j)(2)(A)(vii)[69]

That reasoning raises the question, what kind of change does require a recertification? In this case, is it because an original specification was amended, as opposed to an entirely new parameter or element being introduced, that resulted in no new patent certifications being required?[70] Or, perhaps, do only those changes that constitute a major amendment require a recertification? What about a change to the drug product's formulation and related packaging, or to a container's closure system? Such changes may require new patent certifications, as we will see in the next section.

Like the FDA, the district courts have not seen this issue a lot. At least one district court has found that compliance with one's discovery obligations in an ANDA litigation suffices for notification purposes and that a separate patent recertification and notice letter was not required.[71] In that case, Novartis argued that Ben Venue was required to submit a new patent certification in view of a formulation change. The court disagreed, stating that:

> once the initial Paragraph IV Certification has been filed and litigation commences, a material change to an ANDA triggers a duty on the part of the ANDA filer to inform the patentee of the material change to the ANDA, but does not necessarily trigger a duty to file a new Paragraph IV Certification. While Ben Venue did not inform Novartis directly of the material amendment to its ANDA, it did inform Novartis's outside counsel in a timely matter [*sic*]. Since litigation was on-going at the time of the amendment and discovery had commenced, the Court concludes that . . . such notice was adequate.[72]

69. Docket No. FDA-2012-P-0583 (decision letter Nov. 30, 2012), at 7. The FDA's response continued: "However, we note that an applicant that submits an amendment to a pending ANDA is required to amend its patent certification if, at any time before approval, the applicant learns that the previously submitted patent certification is no longer accurate with respect to the pending application, as amended (see 21 C.F.R. 314.94(a)(12)(viii)(C)). The Agency does not generally assess the claims of listed patents to determine which patent may or may not potentially be infringed by an ANDA or to determine the accuracy of patent certifications." *Id.*

70. Sunovion Pharmaceuticals Inc. petitioned the FDA on nearly an identical issue involving its enantiomer purity patent for Lunesta (eszopiclone), in which the FDA declined to require a new patent certification. Docket No. FDA-2012-P-0943 (decision letter Jan. 23, 2013), at 6.

71. Ben Venue Labs. Inc. v. Novartis Pharm. Corp., 146 F. Supp. 2d 572, 580 (D.N.J. 2001).

72. *Id.* at 580.

Because of the uncertainty in petitioning for a patent recertification, particularly where there is already ongoing, related ANDA or 505(b)(2) patent litigation, one may request relief more directly and seek the imposition of a second 30-month stay.

C. To Impose a Second 30-month Stay

A second 30-month stay is what Genzyme successfully imposed on the first applicant Pentech/Cobrek in its Paragraph IV challenge to Hectorol (doxercalciferol) injection.[73] Genzyme's NDA for Hectorol was first approved in April 2000, and the product was packaged in ampules. Genzyme listed three patents in the Orange Book at that time.[74] In October 2007, Pentech[75] filed the first Paragraph IV ANDA on the ampule product, with a Paragraph IV Certification to all three patents.[76] Meanwhile, Genzyme filed a supplemental NDA, changing the formulation as well as the packaging from ampules to vials, which was approved in December 2008.[77] Genzyme then submitted the same three ampule patents for listing in the Orange Book as to the vial, while also adding a fourth patent (vial patent),[78] followed by discontinuing the ampule formulation.

Cobrek, now owning Pentech's ANDA, filed a Paragraph IV Certification as to the vial patent in the pending ampule ANDA and notified Genzyme, who—understanding that the ANDA was directed to the ampule—did not sue. Thereafter, the FDA advised Cobrek that the ampule ANDA would not be approved because it was not qualitatively and quantitatively the same as to the new vial RLD, and that a reformulation was recommended.[79] Taking that recommendation, Cobrek reformulated to the vials, but the FDA would not accept the amendment, reasoning that Cobrek had to provide new patent certifications, including as to the vial patent.[80] Cobrek did so, notifying Genzyme in November 2009 that the

---

73. Docket No. FDA-2010-P-0223 (decision letter Oct. 19, 2010).

74. Genzyme later requested delisting of two of the three patents, though that is irrelevant for present purposes.

75. Cobrek later acquired an interest in Pentech's ANDA.

76. Docket No. FDA-2010-P-0632 (decision letter Sept. 20, 2011), at 2.

77. Docket No. FDA-2010-P-0223 (decision letter Oct. 19, 2010), at 2.

78. U.S. Patent No. 7,148,211.

79. Alternatively, Cobrek could have continued prosecuting the ampule ANDA with the original patent certifications, as the now discontinued ampule was determined not to have been removed for safety or efficacy but instead decided to match Genzyme's new vial packaging and formulation.

80. It is unclear whether Cobrek ever argued that notice in the context of the ANDA litigation was sufficient, following the reasoning of *Ben Venue v. Novartis*, 146 F. Supp. 2d 572. Cobrek did argue that new patent certifications should not have been required, but the FDA responded that a petition comment was an inappropriate mechanism for that argument, and in any event the relevant time had passed. *See* Decision Letter (Oct. 19, 2010), at 7 n.13.

ANDA was directed to vials and that it had submitted new certifications as requested by the FDA. This time, Genzyme sued within 45 days and, thus, wanted a new 30-month stay.

Enter Genzyme's petition, which asked the FDA to restart Cobrek's 30-month stay from the receipt of the November 2009 notice letter, arguing that regardless of whether pre-[81] or post-MMA (Medicare Prescription Drug, Improvement, and Modernization Act) law is applied, a "second" stay is warranted.[82] As to the latter characterization, the FDA's decision letter says that "[i]t might, however, be more appropriately referred to as a superseding stay. The first stay was related to a formulation covered by the ANDA as initially submitted. As discussed in the text, however, that formulation was subsequently abandoned and the stay at issue here relates to the revised formulation."[83] As to the former argument, the FDA said it need not resolve which law applies because Genzyme wins either way because:

> [o]nce Cobrek [had] made the paragraph IV certification, and Genzyme subsequently sued Cobrek for infringing the [listed] patent, the statutory requirements for a 30-month stay with respect to this paragraph IV certification were met, as the information concerning the [listed] patent [had been] submitted to FDA before either the original submission of the ANDA to FDA in October 2007, or the submission of the ANDA amendment in 2009.[84]

The FDA thus imposed a second (superseding) 30-month stay, beginning with Cobrek's notice letter regarding the ANDA amendment to the vials.[85] Or, in other words, the FDA did not allow Cobrek to relate its 2009 amendment and patent certifications for the vials back to the original 2007 ampule patent certification, at least for purposes of determining whether to impose a 30-month stay.[86]

81. Cobrek also argued that to grant Genzyme's petition would impermissibly result in a multiple 30-month stay situation, which the Medicare Prescription Drug, Improvement, and Modernization Act sought to eliminate. *See* Cobrek's Comment Letter (May 20, 2010). On that, the FDA said that "[t]here is no evidence, from the statutory structure or otherwise, that Congress ever considered the particular set of circumstances presented here, in which the two stays at issue related to separate formulations of the product. Ultimately, Cobrek chose to seek approval of a copy of the reformulated product rather than follow the alternative route, offered by the FDA, of seeking approval of the original formulation. Based on that choice, Cobrek filed the [P]aragraph IV [C]ertification at issue here. Once it did so . . . the statute dictates the result." Decision Letter (Oct. 19, 2010), at 8.

82. Docket No. FDA-2010-P-0223, at 6–8.

83. Docket No. FDA-2010-P-0223 (decision letter Oct. 19, 2010), at 1 n.1.

84. *Id.* at 7–8.

85. *Id.* at 8–9.

86. Based on this reasoning, Sandoz separately petitioned the FDA to deem that Cobrek forfeited its 180-day exclusivity. See Section 2(f).

Consistent with (and citing) this decision, the FDA imposed a second 30-month stay in favor of Baxter Healthcare Corporation and against Minrad Inc. regarding the latter's Paragraph IV challenge to Suprane (desflurane). The patent in that case was directed to a container for the active agent.[87] When Minrad changed the container closure system for its drug product, it submitted a second Paragraph IV Certification and notified Baxter, who timely sued. Because of those facts, "the statutory requirements were met, [and] a [30]-month stay ensue[d]. This would occur whether or not [Baxter's] Petition had been filed."[88] Whether notification in the ANDA litigation context would suffice, and without the submission of a new patent certification, remains unanswered.

D. To Designate a Second Reference Listed Drug

For any ANDA product to be approved as therapeutically equivalent, there needs to be a Reference Listed Drug (RLD) against which to measure bioequivalence.[89] The FDA designates which drug products will be RLDs, as outlined in the 1992 preamble to its final regulations when implementing some of the Hatch-Waxman Amendments.[90] For a single source drug product, which is generally the case for any brand product in ANDA litigation, the RLD will be the respective NDA drug product. In some instances, however, there is no RLD (such as when a product is discontinued due to a dosage form switch),[91] and an ANDA sponsor must know against which drug product to test bioequivalence. As the Orange Book states, a sponsor "wishing to market a generic version of an NDA listed drug that is not designated as the reference listed drug may petition the Agency through the Citizen Petition procedure."[92] This normally is a very routine procedure, with the FDA granting the vast majority of such petitions.[93] An ANDA sponsor then has certainty regarding what drug product against which to test bioequivalence. In July 2013, however, the FDA surprisingly denied such a petition.

In that petition, Lupin Pharmaceuticals Inc. asked the FDA to establish a second RLD for calcium acetate capsules, 667 milligrams, because

87. U.S. Patent No. 5,617,906.

88. Docket No. FDA-2011-P-0127 (decision letter June 7, 2011), at 7.

89. *See* 21 C.F.R. §§ 314.3, .94(a)(3).

90. 57 Fed. Reg. 17,950, 17,958 (Apr. 28, 1992).

91. Even an ANDA can be designated as an RLD. For example, for the Doryx (doxycycline hyclate) delayed-release drug product, the capsules were discontinued in favor of the tablets. The related capsule NDA (No. 05082) is no longer designated the RLD but is now instead the 100 milligram drug product in Medicis's ANDA No. 065281.

92. *Preface* to APPROVED DRUG PRODUCTS WITH THERAPEUTIC EQUIVALENCE EVALUATIONS (32nd ed. 2012).

93. *See, e.g.*, Docket Nos. FDA-2012-P-1127, FDA-2012-P-1043, FDA-2011-P-0606, and FDA-2011-P-0059.

commercial samples available of the first RLD, Fresenius Medical Care North America's Phoslo Gelcaps, had not been available since November 2012.[94] There was thus seemingly no RLD product against which to test bioequivalence. There were, however, three approved ANDAs that referenced Fresenius's NDA, which were held by Roxane, Paddock, and Invagen Pharmaceuticals. Of those, Lupin contended that only Roxane's drug product was commercially available and thus requested that it be designated as a second RLD.[95] On July 3, 2013, the FDA denied Lupin's petition for not having sufficient grounds to establish the need for a second RLD, as the NDA drug product was still being marketed as an authorized generic.[96] So long as the proper documentation is maintained, an ANDA sponsor may use the authorized generic product to conduct its bioequivalence testing.

Even with the 150-day decision deadline established under 505(q), that is critical time to lose for those developing generic drug products. An ANDA sponsor must consider whether the authorized generic of the NDA drug product is available. One place to start is FDA's list of authorized generics, which is publicly available on its website.[97]

E. To Require a New Bioequivalence or Other Clinical Study

There are situations when a branded company will petition the FDA to require a generic sponsor to perform additional bioequivalence studies or even one or more comparative clinical trials comparing the proposed generic drug product to the RLD. The reasoning is typically that there is some unique property about either the RLD (such as a modified release profile or localized site of action), the underlying disease state, or the potential side effects of the active pharmaceutical ingredient or final drug product, that arguably requires further clinical studies beyond traditional bioequivalence testing. The pragmatic effect is that approval of generic drug products could be delayed.

For example, Teva Pharmaceutical Industries Ltd. (or a related company) has filed at least four citizen petitions specifically concerning its flagship branded drug product Copaxone (glatiramer acetate)[98] injection, or approval of any New Drug Application or Biologics License Application

94. Docket No. FDA-2013-P-0040 (Lupin's petition Jan. 2, 2013), at 1.

95. *Id.* at 1–2.

96. Docket No. FDA-2013-P-0040 (decision letter July 3, 2013), at 1–2.

97. FDA's list of authorized generics may be found at *FDA Listing of Authorized Generics as of April 28, 2014*, U.S. Food & Drug Admin., http://www.fda.gov/downloads/AboutFDA/CentersOffices/OfficeofMedicalProductsandTobacco/CDER/UCM183605.pdf.

98. Docket No. FDA-2012-P-0555 (decision letter Nov. 30, 2012); Docket No. FDA-2010-P-0642 (decision letter June 8, 2011); Docket No. FDA-2009-P-0555 (decision letter May 11, 2010); Docket No. FDA-2008-P-0529 (decision letter Mar. 25, 2009).

(BLA) similarly indicated for the treatment of multiple sclerosis.[99] To date, all have been denied. In each of the petitions involving Copaxone, Teva has essentially requested the same thing—additional clinical testing prior to approval of any generic drug product. In its most recent petition, Teva requested that the FDA:

1. Refrain from approving any ANDA that references Copaxone unless the application contains the "[r]esults of comparative clinical investigations in RRMS [relapse-remitting multiple sclerosis] patients using relevant safety and effectiveness endpoints" that demonstrate "that the proposed generic drug is bioequivalent" to, and as safe and effective as, Copaxone; and
2. Refrain from approving any ANDA that references Copaxone unless the application contains the "[r]esults of non-clinical and clinical investigations demonstrating that the immunogenicity risks associated with the proposed . . . product are no greater than the risks associated with Copaxone, including a demonstration that the risks of alternating or switching between use of the proposed product and Copaxone are not greater than the risks of using Copaxone without such alternation or switching; . . . ."[100]

Teva argued several grounds why such action is required, including that: (1) Copaxone is a highly complex mixture of proteins and polypeptides, such that its exact constituents cannot be individually and completely characterized; (2) the drug product is colloidal and thus not a true solution; (3) its precise mechanisms of action are not well understood; and (4) as an immunomodulator, the drug product may potentially alter naturally occurring immune responses.[101] In other words, Teva argued that because of the drug product's complexity, "it currently is not technologically possible for an ANDA applicant to synthesize a generic version of glatiramer acetate that is structurally and compositionally *identical* to the active ingredient in Copaxone."[102] Typical pharmacokinetic studies would be inadequate to determine bioequivalence, Teva argued, because the drug product acts locally and thus does "not produce measurable concentrations of the active ingredient, active moiety, or metabolites in an accessible biological fluid, and even when measurable concentrations are produced, there is a lack of evidence of any correlation between these systemic concentrations and concentrations at the site of drug action."[103]

---

99. Docket No. FDA-2013-P-0025 (decision letter Mar. 27, 2013).
100. Docket No. FDA-2012-P-0555, Teva's Citizen Petition (June 4, 2012), at 3.
101. *Id.* at 4.
102. *Id.* at 9 (emphasis in original).
103. *Id.* at 15–16 (internal quotation marks omitted).

In its repeated denials of Teva's petitions, the FDA has consistently adopted a "wait-and-see" approach. For example, in the most recent denial, the FDA advised that:

> [A]ny decision regarding bioequivalence studies would be highly specific to the active ingredient, taking into account relevant physico-chemical and other quality attributes . . . [Drugs like] Copaxone . . . present unique challenges; thus it is reasonable and appropriate for the Agency to take into account each product's specific attributes in conjunction with sameness and other data provided by an ANDA applicant as it assesses whether a biowaiver or any other requirement for approval is appropriate for any specific ANDA. Thus, as with the issue of sameness, the Agency's position is that any decision regarding a biowaiver is likely [to] be informed by our review of a specific ANDA before us, and must be based on relevant scientific information specific to each active ingredient.[104]

FDA's approach will thus evaluate not just the nature of the drug product but also the complete data set (bioequivalence, sameness, immunogenicity, etc.) submitted by the ANDA sponsor.[105] Given that 505(q) requires that the FDA issue a final decision within 150 days, and that its typical practice is to rule on pending petitions and ANDAs essentially simultaneously, if not the same day, the FDA's view is not terribly surprising.

As noted, drug products with a modified release profile are often targets for a citizen petition requesting additional data beyond that of traditional bioequivalence. Such petitions typically ask the FDA to require studies under various conditions or with partial AUC measurements. Examples include petitions relating to several drug products indicated for attention deficit–related disorders, such as Adderall XR[106] (mixed amphetamine salts) extended-release capsules, Concerta[107] (methylphenidate hydrochloride) extended-release tablets, and Metadate CD[108] (methylphenidate hydrochloride) extended-release capsules as well as for insomnia: Ambien CR[109] (zolpidem tartrate) extended-release tablets. Other examples exist as well.

---

104. Docket No. FDA-2012-P-0555 (decision letter Nov. 30, 2012), at 10.

105. The FDA also adopted the same approach in denying Teva's petition regarding drug products indicated to treat multiple sclerosis, evaluating each product on a "case-by-case basis" as it deemed "patient access to a variety of safe and effective treatment options important to public health." Docket No. FDA-2013-P-0025 decision letter Mar. 27, 2013), at 6–7.

106. Docket No. FDA-2005-P-0120 (decision letter June 22, 2012).

107. Docket No. FDA-2004-P-0151 (decision letter July 19, 2012).

108. Docket No. FDA-2004-P-0290 (decision letter July 19, 2012).

109. Docket No. FDA-2007-P-0182 (decision letter Oct. 13, 2010).

The drug products indicated for attention-deficit disorders are quite interesting from the perspective of the time required from the first Paragraph IV ANDA filing until final approval. In the case of Adderall XR, the first such ANDA filing was in or about November 2002,[110] with the first ANDA final approval (by a different filer!) on June 22, 2012[111]—nearly ten years later. The first generic equivalent to Concerta took slightly less time at almost seven and one-half years after the first Paragraph IV ANDA filing.[112] In both cases, while the FDA was not bound by the 150-day decision deadline as now exists under 505(q), the FDA did depart from conventional bioequivalence metrics in granting the respective petitions' request that partial AUC parameters be required.[113] There is no doubt that these additional requirements contributed to some of the longest, if not the longest, ANDA approval timelines.[114] Any innovative drug maker would be remiss in not examining any legitimate issue to petition for the FDA's consideration.

F. To Deem That a First-applicant Has Forfeited Any 180-day Exclusivity

There are six distinct scenarios in which a first-applicant will forfeit its 180-day exclusivity. Given the value of that exclusivity, and the market opportunity should that exclusivity no longer exist, it is not surprising that non-first-applicants have petitioned the FDA to deem the exclusivity forfeited. As noted, petitions pertaining to the 180-day exclusivity are

---

110. *See Barr Confirms Shire Suit Initiating Adderall XR(TM) Patent Challenge*, PR Newswire, http://www.prnewswire.com/news-releases/barr-confirms-shire-suit-initiating-adderall-xrtm-patent-challenge-74477227.html (last visited May 14, 2014) (noting that Barr Laboratories Inc. claimed to have filed then).

111. See the FDA's approval letter to Actavis Elizabeth, LLC, which is available at http://www.accessdata.fda.gov/drugsatfda_docs/appletter/2012/077302s000ltr.pdf, which has the same date as the FDA's decision letter on Shire's petition, Docket No. FDA-2005-P-0120. Separately, Actavis petitioned the FDA over the 180-day exclusivity, which was withdrawn before any decision letter. *See* Docket No. FDA-2010-P-0188.

112. The FDA's website shows a first-applicant having filed on July 19, 2005, *Paragraph IV Patent Certifications*, U.S. Food & Drug Admin. (May 6, 2014), http://www.fda.gov/downloads/Drugs/DevelopmentApprovalProcess/HowDrugsareDevelopedandApproved/ApprovalApplications/AbbreviatedNewDrugApplicationANDAGenerics/UCM293268.pdf. Mallinckrodt's ANDA No. 202608 received the first final approval on December 28, 2012, with FDA's letter found at http://www.accessdata.fda.gov/drugsatfda_docs/appletter/2012/202608Orig1s000ltr.pdf. As to both Concerta and Adderall XR, the 180-day exclusivity is governed by pre-MMA rules, is quite complicated, and is in any case irrelevant to petition issue on bioequivalence.

113. *See* Docket No. FDA-2005-P-0120 (decision letter June 22, 2012); Docket Nos. FDA-2004-P-0151 and FDA-2004-P-0290 (joint decision letter July 19, 2012).

114. The FDA does not always require partial AUC measurements for extended-release products. *See, e.g.*, Docket No. FDA-2010-P-0082, decision letter on Kadian (morphine sulfate) extended-release capsules (decision letter July 21, 2010).

excepted from 505(q)'s 150-day decision deadline.[115] In that case, whatever time it takes the FDA to respond to a petition (whether grant or deny, in whole or in part) is added to the 30-month period (or 36 or 40 months, as applicable under FDASIA) to garner tentative approval and not forfeit the 180-day exclusivity.[116]

Sandoz filed such a petition regarding the exclusivity for Hectorol (doxercalciferol) injection.[117] As discussed in Section III.C of this chapter, Genzyme reformulated the product from ampules to vials, and first-applicant Cobrek[118] correspondingly amended its ANDA product to be quantitatively and qualitatively identical to the RLD. Upon doing so, Cobrek was required to submit new patent certifications, triggering a superseding 30-month stay. Sandoz thus argued two separate grounds for forfeiture of the 180-day exclusivity: (1) amendment or withdrawal of the "original" certifications under forfeiture condition III,[119] and (2) failure to obtain tentative approval within 30 months under forfeiture condition IV.[120]

On the first ground, the FDA determined "the recertification to any patent based on a formulation change is not an 'amendment' or a 'withdrawal' of the prior certification."[121] The FDA found nothing in the statute or regulations that "stipulate[s] that a resubmission of patent certifications because of a change in formulation mandate[s] the loss of any 180-day exclusivity rights stemming from the original certifications that qualified a first applicant for exclusivity."[122] Thus, no forfeiture under condition III.

On the second ground, the FDA noted that Cobrek had not yet received tentative approval of its ANDA despite passage of more than 30 months. However, there was no forfeiture in this case because the statute explicitly excuses such failure if there is "a change in or a review of the requirements for approval of the application imposed after the date on which the application is filed."[123] Applied here, the FDA stated:

> [t]his determination reflects the requirement that to match a reference listed drug that is an injectable product, the generic drug is expected to be Q1/Q2 identical to the reference listed drug. That would require a reformulation of the

115. 21 U.S.C. § 355(q)(1)(F).

116. *Id.* § 355(q)(1)(G).

117. Docket No. FDA-2010-P-0632.

118. For simplicity, we ignore Pentech's involvement with the ANDA.

119. 21 U.S.C. § 355(j)(5)(D)(i)(III).

120. *Id.* § 355(j)(5)(D)(i)(IV).

121. Docket No. FDA-2010-P-0632 (decision letter Sept. 20, 2011), at 8. Consistent with its practices, the FDA decided the petition nearly simultaneously as acting on a pending ANDA (approving Roxane's No. 091433).

122. *Id.* at 7.

123. 21 U.S.C. § 355(j)(5)(D)(i)(IV).

> product in a pending ANDA after the listed drug referenced reformulated its own product, or alternatively, a request by an ANDA applicant to determine that the original innovator formulation had not been withdrawn for reasons of safety or effectiveness and a request for a waiver of the requirement of Q1/Q2 identity. Either case would constitute a change in the requirements for approval of an ANDA for a generic version of this reference listed drug, and a company's decision to adopt the former option rather than the latter is not determinative. In those situations in which there has been such a change in requirements, the failure to obtain a tentative approval within the 30-month period does not lead to forfeiture under section 505(j)(5)(D)(i)(IV) of the FD&C Act.[124]

Cobrek thus did not forfeit its 180-day exclusivity, as the new patent certifications that triggered a superseding 30-month stay were not themselves superseding for forfeiture purposes.

Sandoz argued the same forfeiture condition IV against Ranbaxy's presumed 180-day exclusivity on AstraZeneca's Nexium (esomeprazole) capsules, but for a different reason.[125] Ranbaxy's ANDA was filed on August 5, 2005, and was presumably the first ANDA with a Paragraph IV Certification.[126] According to the FDA's public records, that ANDA received tentative approval on February 5, 2008.[127] Sandoz thus argued that tentative approval was not received "within" 30 months as required by the statute, but instead on the first day of the thirty-first month.[128] Naturally, Ranbaxy opposed the petition, arguing that the so-called anniversary date is included in the timeline calculation, and there is no forfeiture.[129]

As of this writing, the FDA has not publicly issued a decision letter.[130] Further, the FDA may not likely issue a formal decision letter, as it has already tacitly rejected Sandoz's reasoning in its approval of Caraco Pharmaceutical Laboratories Ltd.'s ANDA No. 077571 for generic Prandin

124. Docket No. FDA-2010-P-0632 (decision letter Sept. 20, 2011), at 9.

125. Docket No. FDA-2012-P-0661 (Sandoz's petition filed June 19, 2012).

126. *Id.* at 2.

127. *See* www.accessdata.fda.gov.

128. Docket No. FDA-2012-P-0661 (Sandoz's petition), at 14.

129. Docket No. FDA-2012-P-0661-0004 (Ranbaxy's comments Jan. 8, 2013).

130. Though timing remains critical, as shown by the FDA's approval letter to Anchen Pharmaceuticals Inc. on fluvoxamine maleate extended-release capsules. The FDA determined that Anchen forfeited the 180-day exclusivity on one dosage strength, but not another, due to a change in the label requirements that fell between the respective 30-month dates for the two strengths. *See* Letter from U.S. Food & Drug Admin. to David Quiggle, Dir., Regulatory Affairs (Mar. 13, 2013), *available at* http://www.accessdata.fda.gov/drugsatfda_docs/applette r/2013/091476Orig1s000ltr.pdf. Ironically, the two, 30-month dates were only a week apart.

(repaglinide) tablets, which upheld Caraco's 180-day exclusivity under similar facts.[131]

Not all petitions regarding a first-applicant's 180-day exclusivity concern forfeiting for failure to timely obtain tentative approval. At least one petition argued that the first-applicant was not entitled to the 180-day exclusivity because its patent certification strategy was a sham. In October 2010, Lupin petitioned the FDA on Combivir (lamivudine and zidovudine) tablets, arguing that Teva, who was the first-applicant, should not be awarded the 180-day exclusivity because it did not file a Paragraph IV Certification to the last-expiring, listed patent.[132] At the time Teva filed its ANDA in June 2006, there were four patents listed in the Orange Book, of which only two were relevant and expiring in May 2012 and May 2016, respectively.[133] Teva apparently filed a Paragraph IV Certification against the earlier-expiring patent, but a Paragraph III to the last-to-expire patent. Later, in October 2008, Teva amended its ANDA with a Paragraph IV Certification to the last-to-expire patent.

Meanwhile, between Teva's original filing and Paragraph IV amendment, Lupin filed its ANDA with a Paragraph IV Certification to both patents. Because Lupin was blocked by Teva's 180-day exclusivity, Lupin argued that it and not Teva should be awarded the 180-day exclusivity. Lupin argued three separate grounds: (1) that under the plain language of Hatch-Waxman, Teva is not a "first applicant" because its original ANDA did not contain a valid Paragraph IV Certification to the last-expiring patent, (2) that Teva's ANDA violates the congressional intent behind the 180-day exclusivity, and (3) that Teva did not "lawfully maintain" its Paragraph IV Certification to the earlier-expiring patent in view of the Paragraph III Certification to the later-expiring patent.[134]

The FDA denied the petition in May 2011, rejecting all three grounds.[135] The FDA noted that Lupin's entire claim "rests on the fact that Teva did not simultaneously challenge" the two last-to-expire patents.[136] As a threshold matter, regarding Teva's amendment to convert its patent certification as to the last-expiring patent, the FDA said that "[a] paragraph IV certification submitted after the first day will not qualify for 180-day exclusivity, even if it is the first paragraph IV certifi-

131. ANDA No. 077571, FDA's approval letter to Caraco dated July 11, 2013, awarding the 180-day exclusivity for its ANDA filed February 10, 2005, and tentatively approved on August 10, 2007.

132. Docket No. FDA-2010-P-0549 (petition filed Oct. 18, 2010).

133. U.S. Patent Nos. 5,859,021 and 5,905,082, with the latter last to expire.

134. Docket No. FDA-2010-P-0549 (petition filed Oct. 18, 2010), at 7–11.

135. Docket No. FDA-2010-P-0549 (decision letter May 25, 2011).

136. *Id.* at 6.

cation to that patent."[137] Practitioners thus must note that only the original Paragraph IV Certification(s) that is filed with the ANDA impinges on the 180-day exclusivity; later certifications are irrelevant. Next, the FDA noted that "[t]he plain language of the MMA requires 'a' certification; it does not require a first applicant to be the first to challenge with a paragraph IV certification more than a single patent or even the latest expiring patent."[138] As such, and because Teva was the first ANDA sponsor to submit a substantially complete application with at least one Paragraph IV Certification, Teva met the statutory requirements to be entitled to the 180-day exclusivity, and the FDA "need not second guess the particular strategy Teva has employed in its patent challenge."[139] Finally, the FDA notes that had Teva not amended its certification to the last-expiring patent, its 180-day exclusivity would have been forfeited under condition VI (expiration of all patents)[140] and in any case would not transfer to Lupin or any other applicant.[141]

As noted in this chapter, there are several grounds on which ANDA litigants, brand and generic sponsors alike, may consider to petition the FDA. The statute dictates those where the FDA is required to issue a final decision within 150 days, though petitions on the 180-day exclusivity are specifically exempt from this rule. Given the dollars at stake in ANDA litigation, and the relatively low antitrust risk due to *Noerr-Pennington* immunity,[142] one should be well poised to imagine what topic may be petitioned next. In the next chapter, we consider petitions on biological products.

137. *Id.* at 5.
138. *Id.* at 6.
139. *Id.*
140. 21 U.S.C. § 355(j)(D)(i)(VI).
141. Docket No. FDA-2010-P-0549 (decision letter May 25, 2011), at 6 n.9.
142. *See supra,* note 2.

chapter 23

Expedited Regulatory Pathways[1]

A number of the Food and Drug Administration's (FDA's) programs were created to reduce the time for drug development and/or regulatory review of a New Drug Application (NDA) for drugs that are intended to treat serious conditions.[2] Currently, the four formal expedited programs—fast track, priority review, accelerated approval, and breakthrough therapies—are referred to as the FDA's "expedited programs." The use of these programs has been extensive; for example, in 2013, over a third of the new molecular entities (NMEs) were approved under priority review, over a third received fast track designation, and almost half of NMEs approved in 2013 received the benefit of one or more of the FDA's expedited programs (some received multiple designations).[3]

A drug must be intended to treat a serious condition to be eligible for any of the four expedited programs. The FDA defines "a serious condition (disease or illness)"[4] in the context of an expedited pathway as:

> a disease or condition associated with morbidity that has substantial impact on day-to-day functioning. Short-lived

1. Brenda J. Huneycutt of Avalere Health LLC, Washington, D.C.

2. These programs apply to both drugs (regulated under the Federal Food, Drug, and Cosmetic Act) and biologics (regulated under the Public Health Service Act), but for the purposes of this chapter, the focus is on drugs.

3. John Jenkins, U.S. Food & Drug Admin., CDER New Drug Review: 2013 Update (Dec. 11, 2013), *available at* http://www.fda.gov/downloads/AboutFDA/CentersOffices/OfficeofMedicalProductsandTobacco/CDER/UCM378227.pdf.

4. U.S. Food & Drug Admin., Draft Guidance for Industry: Expedited Programs for Serious Conditions—Drugs and Biologics (June 2013) [hereinafter Draft Guidance for Industry: Expedited], *available at* http://www.fda.gov/downloads/Drugs/GuidanceComplianceRegulatoryInformation/Guidances/UCM358301.pdf ("All conditions meeting the definition of life-threatening as set forth at 21 CFR 312.81(a) would also be serious conditions."); After drafting and before publication of this chapter, on May 29, 2014, the FDA issued its final expedited programs guidance, available at http://www.fda.gov/downloads/Drugs/GuidanceComplianceRegulatoryInformation/Guidances/UCM358301.pdf (the final guidance does not substantially differ from the draft guidance, but it clarifies a number of details).

> and self-limiting morbidity will usually not be sufficient, but the morbidity need not be irreversible if it is persistent or recurrent. Whether a disease or condition is serious is a matter of clinical judgment, based on its impact on such factors as survival, day-to-day functioning, or the likelihood that the disease, if left untreated, will progress from a less severe condition to a more serious one.[5]

As described in the following sections, the four programs differ in benefits to drug sponsors, eligibility requirements, and procedures for requesting designation or use of the programs by drug sponsors, but they share some terminology and consistency in definitions.

In addition to these formal programs, the Agency has the flexibility to accelerate development and approval in other ways. Subpart E, as it is called, became a part of the FDA's regulations in 1988 "to establish procedures designed to expedite the development, evaluation, and marketing of new therapies intended to treat persons with life-threatening and severely-debilitating illnesses, especially where no satisfactory alternative therapy exists."[6] Under these regulations: (1) sponsors of these medicines can request (and, to the extent FDA resources permit, will be granted) early consultation with the FDA to discuss preclinical and clinical trial design,[7] (2) sponsors' investigational drugs may be granted expanded access for treatment use,[8] and (3) the Agency will use a risk-benefit analysis for approving these medicines, taking into consideration the severity of the disease and the lack of satisfactory treatments.[9]

It is important to recognize that none of these expedited programs or actions changes the statutory standards of demonstrating safety and effectiveness for drug approval. All drugs are held to the same standard

---

5. This definition comes from the FDA's expanded access regulations, 21 C.F.R. § 312.300(b). *See also* Draft Guidance for Industry: Expedited, *supra* note 4; *For Consumers: Fast Track, Breakthrough Therapy, Accelerated Approval and Priority Review*, U.S. Food & Drug Admin., http://www.fda.gov/ForConsumers/ByAudience/ForPatientAdvocates/SpeedingAccesstoImportantNewTherapies/ucm128291.htm (last visited March 1, 2014) ("AIDS, Alzheimer's, heart failure and cancer are obvious examples of serious conditions. However, diseases such as epilepsy, depression and diabetes are also considered to be serious conditions.").

6. 21 C.F.R. § 312.80; Food and Drug Administration, Interim Rule, Investigational New Drug, Antibiotic, and Biological Drug Product Regulations; Procedures for Drugs Intended to Treat Life-Threatening and Severely Debilitating Illnesses, 53 Fed. Reg. 41,516 (Oct. 21, 1988).

7. 21 C.F.R. § 312.82. Early consultation includes pre-investigational new drug (IND) meetings to discuss animal and Phase 1 study design and end of Phase 1 meetings to discuss Phase 2 and pediatric (if required) study design.

8. *Id.* § 312.83.

9. *Id.* § 312.84.

for approval, regardless of regulatory pathway or benefit of acceleration of drug development or regulatory application review. As the Agency's regulations explain:

> while the statutory standards of safety and effectiveness apply to all drugs, the many kinds of drugs that are subject to them, and the wide range of uses for those drugs, demand flexibility in applying the standards. The [FDA] has determined that it is appropriate to exercise the broadest flexibility in applying the statutory standards, while preserving appropriate guarantees for safety and effectiveness. These procedures reflect the recognition that physicians and patients are generally willing to accept greater risks or side effects from products that treat life-threatening and severely-debilitating illnesses, than they would accept from products that treat less serious illnesses. These procedures also reflect the recognition that the benefits of the drug need to be evaluated in light of the severity of the disease being treated.[10]

I. Accelerated Approval

A. Benefits of the Accelerated Approval Program

In 1992, the FDA instituted the accelerated approval program to allow the use of surrogate endpoints (defined in the following paragraphs) to support approval of certain drugs intended to treat serious diseases.[11] In 2012, the Food and Drug Administration Safety and Innovation Act (FDASIA), among other changes, broadened the scope of the accelerated approval program by clarifying that, in addition to surrogate endpoints, "intermediate clinical endpoints" (defined in the following paragraphs) can be used to support accelerated approval of a drug and give the FDA more flexibility to take into account the existence or lack of available therapies when determining a drug's eligibility for accelerated approval.[12]

Accelerated approval is important for serious conditions where the disease course is long and assessment of clinical benefit (e.g., survival or morbidity measures) takes many years, if not decades. If a validated surrogate

10. *Id.* § 312.80.

11. 21 U.S.C. § 506(c); 21 C.F.R. pt. 314(H); Final Rule, New Drug, Antibiotic, and Biological Drug Product Regulations; Accelerated Approval, 57 Fed. Reg. 58942 at 58,948 (Dec. 11, 1992).

12. FDASIA § 901.

endpoint or an intermediate clinical endpoint exists, then collection of data to support an earlier approval can be done to bring these drugs to patients more quickly than if studies were limited to the long-range clinical outcome. For example, accelerated approval has been used for drugs to treat conditions such as HIV/AIDS and cancer.

The Agency defines the following terms in the context of accelerated approval in its 2013 draft guidance:[13]

Clinical endpoint—a characteristic or variable that directly measures a therapeutic effect of a drug: an effect on how a patient feels (e.g., symptom relief), functions (e.g., improved mobility), or survives.

Clinical benefit—a positive therapeutic effect that is clinically meaningful in the context of a given disease. The clinical benefit must be weighed against a treatment's risks to determine whether there is an overall benefit for patients (i.e., a positive benefit-risk profile).

Surrogate endpoint—a marker, such as a laboratory measurement, radiographic image, physical sign, or other measure, that is thought to predict clinical benefit, but is not itself a measure of clinical benefit.

Intermediate clinical endpoint—a measurement of a therapeutic effect that is considered reasonably likely to predict the clinical benefit of a drug, such as an effect on irreversible morbidity or mortality.[14]

Only certain surrogate endpoints and intermediate clinical endpoints can be used to support an accelerated approval:

- a surrogate endpoint that is considered reasonably likely to predict clinical benefit; and
- a clinical endpoint that can be measured earlier than an effect on irreversible morbidity or mortality (IMM) and that is reasonably likely to predict an effect on IMM or other clinical benefit.[15]

Essentially, for surrogate endpoints, a study for which the outcome tested is a clinical benefit (or is a marker that has been shown to predict clinical benefit—a validated surrogate endpoint), accelerated approval does not apply because traditional approval can be used.[16] For a study for which the outcome is not a clinical benefit but, rather, a surrogate "reasonably likely to predict a drug's intended clinical benefit," accelerated approval can

13. Agency guidance documents, whether draft or final versions, are not legally binding but describe the Agency's current thinking on a topic and are generally followed by FDA review staff.

14. Draft Guidance for Industry: Expedited, *supra* note 4.

15. *Id.*

16. *Id.*

be used (e.g., tumor shrinkage as a surrogate for survival).[17] Finally, a surrogate without sufficient evidence demonstrating that it is reasonably likely to predict a drug's intended clinical benefit cannot be used to support either traditional or accelerated approval.[18] The Agency decides whether to accept the drug sponsor's proposed surrogate based on the evidence demonstrating that it predicts the drug's intended clinical benefit.

For intermediate clinical endpoints, if the clinical effect is not a clinical benefit (and, therefore, cannot support a traditional approval), and the endpoint can be shown to be "reasonably likely to predict the ultimate clinical benefit of a drug," then it may be used to support an accelerated approval.[19] Because the use of these endpoints is relatively new (since FDASIA in 2012), the Agency envisions intermediate clinical endpoints being used in the following types of contexts:

- The study results for a clinical endpoint demonstrate a therapeutic effect that would not support traditional approval because:
    - The effect is not a clinical benefit, and
    - The effect is only a modest benefit within the context of the disease that alone would not justify the risks associated with the drug, but there is an evidentiary basis to conclude that the effect is reasonably likely to predict an effect on IMM or other clinical benefit that would be a basis for traditional approval;
- A clinical endpoint demonstrates a relatively short-term clinical benefit in a chronic disease setting in which it is essential to confirm longer-term durability of the clinical benefit for traditional approval, but the short-term benefit is reasonably likely to predict long-term benefit;
- A clinical endpoint demonstrates a clinical benefit that is reasonably likely to predict an effect on IMM in a disease setting in which it is essential to confirm the effect on IMM (e.g., because available therapy has established effects on IMM).[20]

A showing that an endpoint is "reasonably likely to predict clinical benefit" requires clinical evidence to support a biologically plausible contention.[21] The FDA considers "epidemiological, pathophysiological, therapeutic, pharmacologic, or other evidence developed using biomarkers, for example, or other scientific methods or tools" and makes a determination

17. *Id.*
18. *Id.*
19. *Id.*
20. *Id.*
21. *Id.*; Food and Drug Administration, Final Rule, New Drug, Antibiotic, and Biological Drug Product Regulations; Accelerated Approval, 57 Fed. Reg. 58,942 (Dec. 11, 1992).

on a case-by-case basis.[22] Important factors in the Agency's assessment include what is known about the disease process and how well the disease is understood, and what is understood about how the drug's effect on the surrogate endpoint relates to the disease process.[23] "FDA considers all relevant evidence and weighs the uncertainty against the severity of the disease to be treated and the lack of available therapy."[24] Sponsors should submit to the FDA "evidence that a surrogate or intermediate clinical endpoint is reasonably likely to predict the intended clinical benefit of a drug."[25]

B. Eligibility for the Accelerated Approval Program

The Federal Food, Drug, and Cosmetic Act (FDCA) provides that the FDA may grant accelerated approval to:

> a product for a serious or life-threatening condition . . . upon a determination that the product has an effect on a surrogate endpoint that is reasonably likely to predict clinical benefit, or on a clinical endpoint that can be measured earlier than irreversible morbidity or mortality, that is reasonably likely to predict an effect on irreversible morbidity or mortality or other clinical benefit, taking into account the severity, rarity, or prevalence of the condition and the availability or lack of alternative treatments.[26]

The FDA regulations allow accelerated approval only for those drugs "that provide meaningful therapeutic benefit to patients over existing treatments (e.g., ability to treat patients unresponsive to, or intolerant of, available therapy, or improved patient response over available therapy)."[27] The FDASIA amendments to the FDCA added a requirement that when determining whether a drug provides a "meaningful therapeutic benefit," the Agency must take into account whether there are available treatments for the condition or whether there are no alternative treatments.[28]

The Agency interprets the FDASIA changes to "broaden[] use of the accelerated approval pathway to cases in which the advantage of a new drug over available therapy may not be a direct therapeutic advantage, but is a clinically important improvement from a patient and public health

22. 21 U.S.C. § 506(c)(1)(B).
23. Draft Guidance for Industry: Expedited, *supra* note 4.
24. *Id.*
25. *Id.*
26. 21 U.S.C. § 506(c)(1)(A).
27. 21 C.F.R. § 314.500.
28. FDASIA § 901.

perspective."[29] The Agency's 2013 draft guidance offered examples of how a drug could show a "meaningful advantage over available therapy":

- Has an effect on a serious outcome of the condition that is not known to be influenced by available therapy (e.g., progressive disability when the available therapy has shown an effect on symptoms but has not shown an effect on progressive disability).
- Has an improved effect on a serious outcome(s) of the condition compared to available therapy (e.g., superiority of the new drug used alone or in combination with available therapy in an active- or historically-controlled trial assessing an endpoint reflecting mortality or serious morbidity).
- Has a benefit for patients who are unable to tolerate available therapy or whose disease has failed to respond to available therapy, or the treatment can be used effectively with other critical agents that cannot be combined with available therapy.
- Provides efficacy similar to those of available therapy, while (1) avoiding serious toxicity that occurs with available therapy, (2) avoiding less serious toxicity that is common and causes discontinuation of treatment of a serious condition, or (3) reducing the potential for harmful drug interactions.
- Provides similar safety and efficacy as available therapy but with another documented benefit, such as improved compliance, that is expected to lead to an improvement in serious outcomes.
- Addresses an emerging or anticipated public health need, such as a drug shortage.[30]

"Available therapies" (used synonymously with "existing treatments" in the context of expedited programs) are those FDA-approved for the same indication (as the new drug) that are "relevant to current U.S. standard of care (SOC) for the indication" at the time of decision on the specific designation request.[31] Therapies not FDA-approved, FDA-approved but not used, and approved under the accelerated approval program but not

29. Draft Guidance for Industry: Expedited, *supra* note 4.

30. These are the same examples used for how to show a drug addresses an unmet medical need in the context of fast track designation. *See* Draft Guidance for Industry: Expedited, *supra* note 4.

31. *Id.*

yet clinically verified are not generally considered available therapies in this context.[32] It is important to note that, even though the Agency is relying on different endpoints (surrogate or intermediate clinical rather than direct clinical benefit as in traditional approval), drugs granted accelerated approval are still held to the same statutory standards for safety and effectiveness as any other FDA-approved drug.[33]

C. Conditions of Accelerated Approval

Unlike the other three expedited programs, for which a drug receives a designation and the attendant benefits, accelerated approval is an approval pathway. As such, no formal request need be filed, but drug sponsors need the FDA's agreement to the full drug development plan.[34] Drug sponsors should discuss with the FDA any potential use of accelerated approval early in the drug development process. It is critical to do this early as the drug sponsor and the Agency can then discuss proposed surrogate or intermediate clinical endpoints (and their validation status or plans), and clinical trial designs for both pre- and postapproval studies.[35] In addition, applicants must submit copies of all promotional materials (e.g., labeling and advertisements) for the Agency's consideration prior to approval.[36]

Furthermore, use of accelerated approval is generally contingent on the drug sponsor agreeing to conduct postmarketing studies (Phase 4 confirmatory trials) to verify the drug's clinical benefit or effect on IMM (e.g., confirm that tumor shrinkage does predict that a cancer patient will survive longer).[37] The FDA generally requires that these studies measure the clinical benefit, but sponsors may be allowed other designs, such as assessing the surrogate endpoint for a longer time, for example.[38] In addition, the Agency requires that these studies be conducted in a timely manner (i.e., "with due diligence"), and, ideally, these trials would commence prior to drug approval, if not prior to the New Drug Application (NDA) filing.[39] In some instances, such as some cancers, the FDA may allow the confirmatory trial to be conducted in a different study population than that used for approval of the drug (because of issues with recruiting patients on the now-approved drug to join a study of that same drug) or allow accelerated

32. *Id.*
33. 21 U.S.C. §§ 506(f)(2); 505(d).
34. Draft Guidance for Industry: Expedited, *supra* note 4.
35. *Id.*
36. 21 U.S.C. § 506(c)(2)(B); 21 C.F.R. § 314.550.
37. 21 U.S.C. § 506(c)(2)(A); 21 C.F.R. § 314.510.
38. Draft Guidance for Industry: Expedited, *supra* note 4.
39. 21 U.S.C. § 506(c)(3)(A); 21 C.F.R. § 314.510.

approval based on interim analysis of the surrogate endpoint data and thus allow continuation of the same trial to be used as confirmatory.[40]

If the subsequent studies confirm the drug's clinical benefit, then the Agency grants the drug traditional approval.[41] If, on the other hand, the Phase 4 confirmatory studies fail to confirm the drug's clinical benefit (including showing enough benefit to justify the drug's risks), the Agency can pull the drug from the market[42] or the sponsor can voluntarily request that the Agency withdraw the approval.[43] The Agency can also withdraw the drug's accelerated approval if, for example, other data show the product is not safe or effective, the confirmatory studies are not done in a timely manner, or the drug sponsor circulates false or misleading promotional materials.[44]

The FDA does not make public information on discussions with drug sponsors regarding accelerated approval; only at approval will the FDA note that the drug was approved under the accelerated approval pathway (unless, of course, the drug sponsor makes the information public itself).

II. Fast Track

A. Benefits of the Fast Track Program

In 1997, the Food and Drug Administration Modernization Act (FDAMA) created the fast track program, which aims to shorten both drug development time *and* Agency review time for certain drugs that are intended to treat serious diseases.[45] A fast track designation applies to a drug for a specific indication (and, therefore, a new use for an already-approved drug can also receive a fast track designation).[46]

A fast track designation offers two main benefits: (1) the opportunity to accelerate drug development because of more frequent interactions with the Agency to discuss drug development plans, including face-to-face meetings between the drug sponsor and the FDA; and (2) the potential to submit portions of the NDA by rolling review.[47] More frequent interactions

40. Draft Guidance for Industry: Expedited, *supra* note 4.

41. 21 C.F.R. § 314.560.

42. 21 U.S.C. § 506(c)(3); 21 C.F.R. § 314.530.

43. 21 C.F.R. § 314.150(c).

44. 21 U.S.C. § 506(c)(3).

45. *Id.* § 506(b), as added by section 112 of the Food and Drug Administration Modernization Act of 1997, Pub. L. No. 105-115, and amended by FDASIA § 901, Pub. L. No. 112-144.

46. Draft Guidance for Industry: Expedited, *supra* note 4.

47. *Id.*; 21 U.S.C. § 506(d)(1) ("If [FDA] determines, after preliminary evaluation of clinical data submitted by the sponsor, that a fast track product may be effective, the [Agency] shall evaluate for filing, and may commence review of portions of [a marketing application] before the sponsor submits the complete application.").

with the Agency benefit the applicant in that any potential issues and challenges to the development and studies of the drug can be identified and resolved quickly to make drug development and clinical trials of a drug as efficient as possible. Rolling review benefits applicants in that the FDA review of their NDA does not have to wait until the full NDA is compiled; rather, the Agency will begin review of the submission before it is complete (however, the Agency review goal period does not start until receipt of the complete application).[48]

B. Eligibility for the Fast Track Program

A drug is eligible for the fast track program "if it is intended, whether alone or in combination with one or more other drugs, for the treatment of a serious or life-threatening disease or condition, and it demonstrates the potential to address unmet medical needs for such a disease or condition."[49] In addition, a drug designated a qualified infectious disease product (QIDP) is also eligible for fast track designation.[50]

In addition to the requirement of treating a serious condition (defined earlier), qualifying for the fast track program requires the drug sponsor to demonstrate "the potential of a drug to address an unmet medical need" with either nonclinical or clinical data, including theoretical rationale.[51] As the Agency explained in its 2013 draft guidance:

> The type of information needed to demonstrate the potential of a drug to address an unmet medical need will depend on the stage of drug development in which fast track designation is requested. Early in development, evidence of activity in a nonclinical model, a mechanistic rationale, or pharmacologic data could be used to demonstrate such potential. Later in development, available clinical data should demonstrate the potential to address an unmet medical need.[52]

The FDA defines an unmet medical need in this context as "a condition whose treatment or diagnosis is not addressed adequately by available

48. 21 U.S.C. § 506(d)(1)–(2).
49. *Id.* § 506(b)(1).
50. *Id.*; *see also* DRAFT GUIDANCE FOR INDUSTRY: EXPEDITED, *supra* note 4 ("Title VIII of FDASIA entitled 'Generating Antibiotic Incentives Now (GAIN)' provides incentives for the development of antibacterial and antifungal drugs for human use intended to treat serious and life-threatening infections. Under GAIN, a drug may be designated as a qualified infectious disease product (QIDP) if it meets the criteria outlined in the statute. A drug that receives QIDP designation is eligible under the statute for fast track designation and priority review.").
51. DRAFT GUIDANCE FOR INDUSTRY: EXPEDITED, *supra* note 4.
52. *Id.*

therapy.[53] An unmet medical need includes an immediate need for a defined population (i.e., to treat a serious condition with no or limited treatment) or a longer-term need for society (e.g., to address the development of resistance to antibacterial drugs)."[54] The FDA recognizes that a serious condition without any available therapy is clearly an unmet medical need.[55] In addition, however, the availability of previously approved therapies for a serious condition does not prevent a determination of unmet medical need, and a drug "generally would be considered to address an unmet medical need if the treatment":

- Has an effect on a serious outcome of the condition that is not known to be influenced by available therapy (e.g., progressive disability when the available therapy has shown an effect on symptoms but has not shown an effect on progressive disability).
- Has an improved effect on a serious outcome(s) of the condition compared to available therapy (e.g., superiority of the new drug used alone or in combination with available therapy in an active- or historically-controlled trial assessing an endpoint reflecting mortality or serious morbidity).
- Has a benefit for patients who are unable to tolerate available therapy or whose disease has failed to respond to available therapy, or the treatment can be used effectively with other critical agents that cannot be combined with available therapy.
- Provides efficacy similar to those of available therapy, while (1) avoiding serious toxicity that occurs with available therapy, (2) avoiding less serious toxicity that is common and causes discontinuation of treatment of a serious condition, or (3) reducing the potential for harmful drug interactions.

---

53. *Id.* Available therapies are those FDA-approved for the same indication (as the new drug) that are "relevant to current U.S. standard of care (SOC) for the indication" at the time of decision on the specific designation request. Therapies not FDA-approved, FDA-approved but not used, and approved under the accelerated approval program but not yet clinically verified are not generally considered available therapies in this context.

54. *Id.*

55. *Id.* ("FDA recognizes, as a general matter, that it is preferable to have more than one treatment approved under the accelerated approval provisions because of the possibility that clinical benefit may not be verified in post-approval confirmatory trials. FDA may therefore consider products as addressing unmet medical need notwithstanding the availability of therapies with accelerated approval.").

- Provides similar safety and efficacy as available therapy but with another documented benefit, such as improved compliance, that is expected to lead to an improvement in serious outcomes.
- Addresses an emerging or anticipated public health need, such as a drug shortage.[56]

C. Designation Request Procedure

Drug sponsors can submit a request to the FDA for a fast track designation any time between when the Investigational New Drug (IND) is submitted and the NDA is filed.[57] To optimize the benefits from a fast track designation, the earlier in the drug development process the request is filed, the better. Therefore, as soon as the criteria for a fast track designation are met (either with nonclinical or clinical data), a request should be submitted. A request is sent to the appropriate review division or office at the FDA either with the IND or as an amendment to the IND and includes: (1) the sponsor's contact information; (2) the IND number (if an amendment); (3) the proprietary name (if available) and the active ingredient; (4) the division or office where the IND is or is being submitted; (5) the proposed indication(s); and (6) a summary of information supporting the request for fast track designation.[58] A separate request must be filed for each proposed development program for a different use (i.e., developing the drug to treat a different condition).[59]

The FDA will respond to requests within 60 calendar days of receipt with either a designation letter or a nondesignation letter, explaining the reasons for denial.[60] The Agency does not publicize either that a request has been submitted or whether it is granted or denied, although drug sponsors often do publicize grants of fast track designations.[61] Importantly, a fast track designation can be withdrawn by the FDA if the drug no longer meets the eligibility criteria.[62] This could happen if subsequent data on the drug do not show the anticipated advantage over existing therapies, or if

---

56. These are the same examples used to illustrate when a drug may show a meaningful advantage over available therapy in the context of eligibility for accelerated approval.

57. 21 U.S.C. § 506(b)(2).

58. Draft Guidance for Industry: Expedited, *supra* note 4.

59. *Id.*

60. 21 U.S.C. § 506(b)(3); Draft Guidance for Industry: Expedited, *supra* note 4.

61. 21 C.F.R. § 312.130. Because requests for fast-track designation are filed either with or as an amendment to an IND, the FDA cannot disclose a request unless it has been made public.

62. Draft Guidance for Industry: Expedited, *supra* note 4.

a new product subsequently gets approved for the serious condition and, therefore, the drug is no longer addressing an unmet medical need.[63]

III. Breakthrough Therapies

A. Benefits of the Breakthrough Therapies Program

In 2012, the Food and Drug Administration Safety and Innovation Act (FDASIA) created the breakthrough therapies program.[64] Similar to the fast track designation, the breakthrough therapies program is meant to expedite both drug development and regulatory review for certain drugs that are intended to treat serious diseases.[65] The breakthrough designation program was created to speed up the development and approval of drugs that showed large effects early on through preliminary clinical evidence (breakthroughs) to get them to patients suffering from serious diseases faster.

Because a drug that meets the requirements of breakthrough therapy would also qualify for a fast track designation, breakthrough therapy–designated drugs receive all the benefits just described for fast track drugs (e.g., enhanced communication with the Agency and the potential to use rolling review).[66] Further, a breakthrough therapy designation gives the drug's sponsor unprecedented access to senior FDA staff and intensive guidance on developing the most efficient drug development plan.[67] These development plans could be considerably shorter than for other drugs treating the same condition, since breakthrough therapies, as described next, generally show a significantly large treatment effect early in development. As such, the clinical trial data burden may be reduced as fewer and/or smaller clinical trials may meet the standard for approval.[68] In fact, it is conceivable that these products, if the treatment effect is

---

63. *Id.*

64. FDASIA § 902, Pub. L. No. 112-144.

65. 21 U.S.C. § 506(a).

66. DRAFT GUIDANCE FOR INDUSTRY: EXPEDITED, *supra* note 4.

67. *Id.*; 21 U.S.C. § 506(a)(3)(B).

68. *See, e.g.*, *FAQ: Breakthrough Designation*, U.S. FOOD & DRUG ADMIN., http://www.fda.gov/regulatoryinformation/legislation/federalfooddrugandcosmeticactfdcact/significantamendmentstothefdcact/fdasia/ucm341027.htm (last visited Mar. 3, 2014) (Q8 answer: "After a development program is designated as a breakthrough therapy, the FDA will work closely with the sponsor to ensure that the design of the clinical trials is as efficient as practicable, when scientifically appropriate. An efficient trial design typically would help minimize the number of patients exposed to a potentially less efficacious treatment, which could translate to fewer patients enrolled in clinical trials that support marketing approval. Regulatory standards to demonstrate safety and efficacy must still be met.").

impressive enough, may be able to be approved on Phase 2 data (e.g., phases of traditional drug development may even be skipped).

The FDA takes an "all-hands-on-deck" approach to working with a breakthrough therapy sponsor, including being available, responsive, and proactive, and involving senior Agency staff at all stages of the drug development, including the earliest stages. As such, the Agency works closely with drug sponsors to streamline studies and thereby shorten drug development by giving timely advice to help the drug sponsor design and conduct an efficient drug development program.[69] As the Agency described in its 2013 draft guidance, "FDA intends to expedite the development and review of a breakthrough therapy by, where appropriate, intensively involving senior managers and experienced review staff in a proactive collaborative, cross-disciplinary review."[70]

It is important to note that for such an accelerated drug development plan, sponsors must also be ready to accelerate other aspects of drug development, particularly scale-up manufacturing processes, in the event that clinical trials are no longer the rate limiting step in drug development timelines. In other words, the drug sponsor must also follow an "all-hands-on-deck" approach to match the speed of the Agency's development and review advice.

B. Eligibility for the Breakthrough Therapies Program

The FDA can designate a drug as a breakthrough therapy "if the drug is intended, alone or in combination with 1 or more other drugs, to treat a serious or life-threatening disease or condition and preliminary clinical evidence indicates that the drug may demonstrate substantial improvement over existing therapies on 1 or more clinically significant endpoints, such as substantial treatment effects observed early in clinical development."[71] Similar to a fast track designation, a breakthrough therapy designation applies to a drug for a specific indication (and, therefore, a new use for an already-approved drug can also receive a breakthrough therapy designation).[72]

Unlike fast track designation, which does not require clinical evidence to support a request, sponsors asking for a breakthrough therapy designation for their drug must have preliminary *clinical* evidence of a treatment effect (conceivably from Phase 1 trials, but most likely from Phase 2 trials).[73] The Agency explained that:

---

69. Draft Guidance for Industry: Expedited, *supra* note 4.
70. *Id.*
71. 21 U.S.C. § 506(d)(1).
72. Draft Guidance for Industry: Expedited, *supra* note 4.
73. *Id.*

> Assessment of the treatment effect for the purposes of breakthrough therapy designation will be based on preliminary clinical evidence, which could include early clinical evidence of both clinical benefit and an effect on a mechanistic biomarker (generally derived from Phase 1 and Phase 2 trials). Nonclinical information could support the clinical evidence of drug activity. In all cases, preliminary clinical evidence demonstrating that the drug may represent a substantial improvement over available therapy should involve a sufficient number of patients to be considered credible. However, FDA recognizes that the data cannot be expected to be definitive at the time of the designation. Ideally, preliminary clinical evidence would be derived from a study that compares the investigational drug to an available therapy (or placebo, if there is no available therapy) in clinical testing and shows superiority, or from a study that compares the new treatment plus SOC [standard of care] to the SOC alone.[74]

What is a "substantial improvement over existing therapies[75] on 1 or more clinically significant endpoints"? The FDA determines what is "substantial" on a case-by-case basis and takes into account, for example, the magnitude and duration of the effect (as compared to what is currently available to treat the condition), the amount of data available, and the importance of the clinical outcome observed, noting that "[i]n general, the preliminary clinical evidence should show a clear advantage over available therapy."[76] The Agency, in its 2013 draft guidance, listed a number of approaches that it envisions may demonstrate this:

- Direct comparison of a new drug to available therapy (or to no treatment if none exists) showing a much greater or more important response (e.g., complete response where the control treatment results in partial response). Such a trial could be conducted in treatment naïve patients or in those whose disease failed to respond to available therapies either as a comparison with the failed therapy (if ethically acceptable) or as a no-treatment controlled study.
- The new drug added to available therapy results in a much greater or more important response compared to

74. *Id.*

75. Existing therapies is used synonymously with available therapies in the expedited programs context.

76. Draft Guidance for Industry: Expedited, *supra* note 4.

available therapy in a controlled study or to a historical control. This trial also could be conducted in treatment naïve patients or in those whose disease failed to respond to available therapies.
- The new drug treats the underlying cause of the disease, in contrast to available therapies that treat only symptoms of the disease, and preliminary clinical evidence shows significant efficacy. In this case, the treatment effect is entirely new (i.e., has not been observed with available therapies). For example, a drug that targets a defective protein that is the underlying cause of a disease (whereas current therapies only treat the symptoms of the disease).
- The new drug reverses disease progression, in contrast to available therapies that only provide symptomatic improvement.
- The new drug has an important safety advantage that relates to serious adverse events compared to available therapies and has similar efficacy.[77]

For purposes of breakthrough therapy designation, FDA considers *clinically significant endpoint* generally to refer to an endpoint that measures an effect on irreversible morbidity or mortality (IMM) or on symptoms that represent serious consequences of the disease. It can also refer to findings that suggest an effect on IMM or serious symptoms, including:

- An effect on an established surrogate endpoint.
- An effect on a surrogate endpoint or intermediate clinical endpoint . . . considered reasonably likely to predict a clinical benefit (i.e., the accelerated approval standard).
- An effect on a pharmacodynamics biomarker(s) that does not meet criteria for an acceptable surrogate endpoint, but strongly suggests the potential for a clinically meaningful effect on the underlying disease.
- A significantly improved safety profile compared to available therapy (e.g., less dose-limiting toxicity for an oncology agent), with evidence of similar efficacy.

77. *Id.*

In a breakthrough therapy designation request, the sponsor should provide justification for why the endpoint, biomarker, or other findings should be considered clinically significant.[78]

C. Designation Request Procedure

A drug sponsor can submit a request for breakthrough therapy designation with or any time after the IND submission.[79] However, given that the request needs to contain the preliminary clinical evidence required by the standard, requests are anticipated after some preliminary evidence has been collected. As such, the request is generally filed as an amendment to the IND (although it can be filed with the IND). As mentioned earlier in the context of requesting a fast track designation, in order to take full advantage of the benefits of the breakthrough therapy designation, drug sponsors should submit their request as early as possible in drug development. The request is submitted to the appropriate review division or office at the FDA and includes: (1) the sponsor's contact information; (2) the IND number (if an amendment); (3) the proprietary name (if available) and the active ingredient; (4) the division or office where the IND is or is being submitted; (5) the proposed indication(s); and (6) a summary of information supporting the request for breakthrough designation.[80] A separate request must be filed for each proposed development program for a different use (i.e., developing the drug to treat a different condition).[81]

In addition, "[i]f a sponsor has not requested breakthrough therapy designation, FDA may suggest that the sponsor consider submitting a request if: (1) after reviewing submitted data and information (including preliminary clinical evidence), the Agency thinks the drug development program may meet the criteria for breakthrough therapy designation; and (2) the remaining drug development program can benefit from the designation."[82]

All breakthrough therapy designation requests are evaluated by the FDA's Center for Drug Evaluation and Research's (CDER) Medical Policy Council to ensure consistent decisions across requests.[83] The Medical Policy Council consists of high-level FDA officials as well as on-the-ground

78. *Id.* (emphasis in original).

79. 21 U.S.C. § 506(a)(2).

80. Draft Guidance for Industry: Expedited, *supra* note 4.

81. *Id.*; *FAQ: Breakthrough Designation*, *supra* note 68.

82. Draft Guidance for Industry: Expedited, *supra* note 4.

83. *See* Richard Moscicki, The State of Rare Disease Drug Development: An FDA Perspective (Jan. 2014), *available at* http://www.fda.gov/downloads/Drugs/NewsEvents/UCM381085.pdf.

product reviewers.[84] The FDA responds to requests within 60 calendar days of receipt with either a designation letter or a nondesignation letter, explaining why the request was denied.[85] Which drug sponsor submitted a request and whether it is granted or denied is not made public by the Agency, although many (but not all) drug sponsors have made public that they have received breakthrough designations.[86] The FDA has, however, kept track of the number of requests and the number of grants and denials.[87] As of February 28, 2014, the Center for Drug Evaluation and Research (CDER) had received 129 requests for breakthrough therapy designations, granted 39, and denied 66.[88] Common reasons for denial include:

- Evidence does not include clinical data.
- Evidence is too preliminary to be considered reliable; e.g., very small number of patients treated, anecdotal case reports.
- Failure to demonstrate "substantial" improvement over available therapy.
- Reliance on a novel biomarker or surrogate endpoint without sufficient evidence to support benefit to patient.
- Post-hoc analyses of failed studies that identify a subset that may benefit.[89]

As with a fast track designation, if a drug no longer qualifies for the breakthrough designation, it may be withdrawn by the FDA.[90] This may happen when, for example, subsequent study data show a treatment effect that is substantially smaller than that seen in the preliminary clinical evidence used to support the grant of the designation.[91] It can also happen when another drug is subsequently approved for the same serious disease, and the breakthrough therapy designated drug cannot show substantial improvement over the approved drug.[92]

---

84. *See, e.g.*, CDER, (MAPP) 4301.1, MANUAL OF POLICIES AND PROCEDURES (Mar. 19, 2013), *available at* http://www.fda.gov/downloads/AboutFDA/CentersOffices/OfficeofMedicalProductsandTobacco/CDER/ManualofPoliciesProcedures/UCM344444.pdf.

85. 21 U.S.C. § 506(a)(3)(A); DRAFT GUIDANCE FOR INDUSTRY: EXPEDITED, *supra* note 4.

86. 21 C.F.R. 312.130(a); *FAQ: Breakthrough Designation*, *supra* note 68 (Q10 answer: "FDA will not disclose information regarding sponsors who submitted requests for or who have been granted or denied breakthrough therapy designation. Breakthrough therapy designation request are typically submitted to an IND, and the FDA cannot disclose the existence of an IND, or any submissions that have been submitted to the IND, unless it has previously been publically disclosed or acknowledged per 21 CFR 312.130(a).").

87. *FAQ: Breakthrough Designation*, *supra* note 68.

88. *Id.*

89. Moscicki, *supra* note 83.

90. DRAFT GUIDANCE FOR INDUSTRY: EXPEDITED, *supra* note 4.

91. *Id.*

92. *Id.*

IV. Priority Review

A. Benefits of the Priority Review Program

In 1992, the Prescription Drug User Fee Act (PDUFA) created the priority review program to shorten NDA review times for certain drugs.[93] Under the current program, the FDA evaluates every marketing application (NDA or efficacy supplement) when submitted and designates each as either priority review or standard review. The Agency informs applicants within 60 calendar days of submission of the application when it grants priority review.[94] If the application is designated priority review, the FDA has a review goal of six months to take action on the application, compared with a goal date of ten months for an application designated standard review. Given the shortened review time frame, the Agency applies attention and resources to evaluating priority applications.

B. Eligibility for the Priority Review Program

To qualify for a priority review designation, the application must be for a drug "that treats a serious condition and, if approved, would provide a significant improvement in safety or effectiveness."[95]

To be eligible for priority review, the drug must be intended to treat a serious condition (defined earlier), and the FDA has interpreted this to require that the:

> drug must be intended to have an effect on a serious aspect of a condition, such as a direct effect on a serious manifestation or symptom of a condition, or other intended effects, including:
>
> . . .
>
> - A product intended to improve or prevent a serious treatment-related side effect (e.g., serious infections in patients receiving immunosuppressive therapy)
> - A product intended to avoid a serious adverse effect associated with available therapy for a serious condition (e.g., less cardiotoxicity than available cancer therapy)[96]

---

93. Prescription Drug User Fee Act , Pub. L. No. 102-571, Title I, 106 Stat. 4491 (1992).

94. If the Agency determines the application does not receive priority review, it will inform the applicant within 74 days of the application's submission. *See* DRAFT GUIDANCE FOR INDUSTRY: EXPEDITED, *supra* note 4.

95. *Id.*

96. *Id.*

In addition, the FDA determines on a case-by-case basis whether the drug "would provide a significant improvement in safety or effectiveness." In its 2013 draft guidance document, the Agency outlined examples of how to consider this standard:

- Evidence of increased effectiveness in treatment, prevention, or diagnosis of a condition.
- Elimination or substantial reduction of a treatment-limiting drug reaction.
- Documented enhancement of patient compliance that is expected to lead to an improvement in serious outcomes.
- Evidence of safety and effectiveness in a new subpopulation.[97]

This can be demonstrated by clinical evidence comparing the drug to a marketed drug, but clinical evidence is not necessarily required.[98]

In addition, there are other ways for a drug's application to be designated as priority review:

- Any supplement that proposes a labeling change pursuant to a report on a pediatric study under 505A.[99]
- An application for a drug that has been designated as a qualified infectious disease product.[100]
- Any application or supplement for a drug submitted with a priority review voucher.[101]

C. Request Procedure

Even though the FDA automatically evaluates each application for review status, an applicant can request priority review from the Agency. To do

97. *Id.*

98. *Id.*

99. *Id.* ("Any supplement to an application under section 505 of the FD&C Act that proposes a labeling change pursuant to a report on a pediatric study under this section shall be considered to be a priority review supplement per section 505A of the FD&C Act as amended by section 5(b) of the Best Pharmaceuticals for Children Act.").

100. *Id.* ("Title VIII of FDASIA entitled 'Generating Antibiotic Incentives Now (GAIN)' provides incentives for the development of antibacterial and antifungal drugs for human use intended to treat serious and life-threatening infections. Under GAIN, a drug may be designated as a qualified infectious disease product (QIDP) if it meets the criteria outlined in the statute. A drug that receives QIDP designation is eligible under the statute for fast track designation and priority review.").

101. *Id.* ("Any application or supplement that is submitted with a priority review voucher will be assigned a priority review. Priority review vouchers will be granted to applicants of applications for drugs for the treatment or prevention of certain tropical diseases, as defined in section 524(a)(3) and (4) of the FD&C Act and for treatment of rare pediatric diseases as defined in section 529(a)(3) of the FD&C Act.").

so, the applicant submits a request for priority review with its NDA (or efficacy supplement) submission that includes a cover letter containing: (1) the drug sponsor's contact information; (2) the proprietary name (if available) and the active ingredient; (3) the proposed indication; and (4) a summary of the information supporting the request for priority review.[102] Whether a drug is granted priority review or standard review is not made public by the Agency, although drug sponsors often publicize receipt of priority review designations.

102. *Id.*

chapter 24

Biologic Drugs and the Biologics Price Competition and Innovation Act[1]

I. Introduction

In 2010, when Congress passed the Patient Protection and Affordable Care Act,[2] public attention was focused on the changes that were made to health care laws. But the Affordable Care Act also made substantial changes to federal regulations and laws concerning the approval of biologic treatments, or "biologics." That legislation is known as the Biologics Price Competition and Innovation Act of 2009 (BPCIA).[3] The intent of the BPCIA was to do for biological products what the Hatch-Waxman Act was intended to do for small-molecule drugs: create a statutory framework that purports to balance competing interests by (1) providing a way for lower-cost, generic-like competition in the market by allowing follow-on manufacturers to seek approval based on a comparison with an already-approved drug and (2) protecting incentives for the substantial investment required to develop new drugs by providing some period of market exclusivity. In general, the BPCIA mechanisms correspond to the familiar Hatch-Waxman framework. But, as we discuss here, there are some important differences.

---

1. Matthew A. Pearson, Jason Weil, and Barbara L. Mullin of Akin Gump Strauss Hauer & Feld LLP, Philadelphia, Pennsylvania.

2. Patient Protection and Affordable Care Act, Pub. L. No. 111-148, 124 Stat. 119 (2010).

3. *Id.* §§ 7001–03, 124 Stat. at 804–21.

II. The Market for Biological Products

A. What Is a "Biologic"?

First, to put the differences between the BPCIA framework and Hatch-Waxman into context, we need to consider what makes a "biologic" drug different from a small-molecule drug.

Small-molecule drugs are chemical compounds that are often synthesized in a laboratory. In contrast, a biologic, or biological product, is a medicine that is made in a living organism.[4] These include proteins or other molecules synthesized by a living cell that are harvested and used to treat various human conditions.

Congress has defined a "biological product" in a way that potentially covers a wide range of products—with the common thread being that they are all produced using living cells or viruses. Specifically, a biologic product is

> a virus, therapeutic serum, toxin, antitoxin, vaccine, blood, blood component or derivative, allergenic product, protein (except any chemically synthesized polypeptide), or analogous product, or arsphenamine or derivative of arsphenamine (or any other trivalent organic arsenic compound), applicable to the prevention, treatment, or cure of a disease or condition of human beings.[5]

The use of biologics to treat human disease is not new. Doctors have administered vaccines since the 1700s,[6] and the government has regulated biologics since the beginning of the 1900s.[7] But, more recently, in the latter half of the 20th century, scientists developed advanced genetic engineering techniques, and the biologics industry has blossomed into providing substantial pharmaceutical products. Biological products currently regulated by the Food and Drug Administration (FDA) include some allergenic extracts (e.g., allergy shots), vaccines, monoclonal antibodies, cytokines, growth factors, and enzymes.[8]

Sales of these products have been trending upward. In 2002, six biologics generated sales of over $1 billion. Just five years later, 22 biologics

---

4. *See* Wendy H. Schadt & John R. Thomas, Cong. Research Serv., Follow-On Biologics: The Law and Intellectual Property Issues 1 (Dec. 15, 2011) [hereinafter CRS Report].

5. 42 U.S.C. § 262(i)(1).

6. Krista H. Carver et al., *An Unofficial Legislative History of the Biologics Price Competition and Innovation Act of 2009*, 65 Food & Drug L.J. 671, 682 (2010).

7. Following incidents in which contaminated vaccines led to fatalities, Congress passed the Biologics Act of 1902. *Id.*; *see also* Pub. L. No. 57-244, 32 Stat. 728 (1902).

8. *FDA 101: Biological Products*, Food & Drug Admin. 1–2 (July 25, 2008), http://www.fda.gov/downloads/ForConsumers/ConsumerUpdates/ucm048367.pdf [hereinafter *FDA 101*].

topped the $1 billion sales mark.[9] In 2007, worldwide sales of biologics reached $75 billion, up 12.5 percent from the year before.[10] And there is no sign that this trend is slowing. Halfway through 2012, analysts had already predicted that sales of three competing biologics—antibodies against TNFα used to treat rheumatoid arthritis—would topple the cholesterol drug Lipitor and the blood thinner Plavix to become the world's top-selling medicines that year.[11] In 2013 the Pharmaceutical Research and Manufacturers of America noted the current development of more than 900 biologic medicines and vaccines targeting more than 100 diseases.[12]

B. Making Biological Products versus Small-Molecule Drugs

Biologics and small molecules are now both routinely used in medical treatments, but the molecules have important differences.

Small-molecule drugs tend to have relatively few atoms and a correspondingly low molecular weight.[13] They are often synthesized from known starting materials pursuant to a series of more well-defined and more tightly controlled chemical reactions.[14] So long as a production facility follows good manufacturing practices, it should consistently produce a known, precisely defined, small molecule.[15]

By contrast, biologics include much larger, complex macromolecules and even larger cells or vaccines generated within a biological environment.[16] A typical biological product has layer upon layer of structure—including not only the basic chemical composition (such as the amino acid sequence of a protein) but also variable chemical modifications (such as phosphorylation and other post-translational modifications)[17] and complex three-dimensional organization. Biologics can be many orders of magnitude larger than a

---

9. CRS Report, *supra* note 4, at 3.

10. *Id.*

11. *Biologic Drugs Set to Top 2012 Sales*, 18 Nature Med. 636 (2012).

12. *More Than 900 Biologic Medicines and Vaccines in Development*, PhRMA, (Mar. 11, 2013), http://www.phrma.org/media/releases/biologics-research-promises-bolster-future-medicine.

13. *See* Steven A. Nash & Rebecca Workman, *A New Pathway for Follow-On Biologics*, 20 Fed. Cir. B.J. 193, 194–96 (2010–2011).

14. *Id.* at 194; *see also How Do Drugs and Biologics Differ?*, Biotechnology Indus. Org. (Nov. 10, 2010), http://www.bio.org/node/53 [hereinafter BIO].

15. *See* Nash & Workman, *supra* note 13, at 194.

16. *See* Brian F. McMahon, *The Biologics Price Competition and Innovation Act of 2009: Legislative Imprudence, Patent Devaluation, and the False Start of a Multi-billion Dollar Industry*, 100 Ky. L.J. 635, 657 fig.4 (2011/2012) (comparing the structure of aspirin (small molecule) and erythropoietin (biological product)).

17. The process that a cell uses to make a protein by building a chain of amino acids is called "translation." Chemical modifications can be made to proteins after they are built, and these are called "post-translational" modifications.

small-molecule drug.[18] That size alone adds to the complexity of developing and regulating these molecules, because there is greater potential that small changes in structure could have biological effects.

In addition to the size difference, the steps of biologic synthesis occur within a living organism, which is different from chemical synthesis of small molecules in a laboratory. It is difficult to regulate individual steps of the reaction/production in the same way that steps of a chemical synthesis can be regulated.[19] Living cells constantly carry out an enormous number of chemical and biological reactions that can, whether intended or not, lead to chemical modification of a product. Those changes are part of the metabolism of the biological system being used to manufacture the product and are, as a result, often sensitive to minor changes in the manufacturing environment. These minor changes can alter the nature of the biological product—often in ways that are difficult to monitor.[20] It is also possible that the living organism used to develop the biologic will develop mutations and internal changes over time that can change the product, again without anyone intending this result.[21]

In summary, biologic molecules have the potential to be substantially more amorphous and variable than small chemical molecules because: (1) their synthesis occurs within a living organism where every step can be affected by metabolism, small changes in manufacturing conditions, and even mutations that change the organism; and (2) biologics have many more atoms that can be affected by these various pathways. This raises a fundamental question that is central to the BPCIA: How do you, and who should, determine whether a follow-on biologic is really biosimilar to and/or interchangeable with the initial reference biologic? Congress was not oblivious to the fundamental differences between biologics and small chemical molecules, so this question proved to be challenging when Congress was trying to develop a legal framework for follow-on biologics.

III. Federal Regulation of Biologics

Most biologics are regulated by the FDA pursuant to its authority under Section 351 of the Public Health Service Act (PHSA).[22] In contrast, the

18. *See* McMahon, *supra* note 16, at 657 fig.4.
19. *See* BIO, *supra* note 14.
20. *Id.*
21. *Cf. id.*
22. Public Health Service Act, Pub. L. No. 78-410, 58 Stat. 682 (1944) (codified at 42 U.S.C. § 262). For a history of the Public Health Service Act and the evolution of biologics licensing, see Carver, *supra* note 6, at 681–84.

FDA regulates small-molecule drugs pursuant to its authority under the Federal Food, Drug, and Cosmetic Act (FDCA).[23] Although biologics are still considered drugs and subject to the safety and efficacy requirements of the FDCA, the statutory requirements for approval of a license to sell a biologic are set forth by the PHSA.[24]

Prior to, and unchanged by, the BPCIA is the requirement that the developer of a new biologic seek the FDA's approval to market it by filing a Biologics License Application (BLA).[25] A BLA is analogous to, and requires much of the same information as, a New Drug Application (NDA) filed under the Federal Food, Drug, and Cosmetic Act.[26] For example, a BLA must demonstrate that the biologic is safe, pure, and potent.[27] This generally requires the applicant to carry out and disclose the results of clinical trials.[28]

The costs of developing a new biologic have been estimated to be roughly similar to the cost of developing a new small-molecule drug—on the order of $1.2 billion (including time, clinical trials, and failed attempts).[29] The time it takes to develop and gain approval of a new biologic is also reportedly similar to that of a small-molecule drug (average of 97.7 months and 90.3 months, respectively).[30] Although biologics have a slightly higher probability of ultimate clinical success, those that fail often fail during Phase 3 clinical trials—after the majority of research-and-development money has already been invested.[31] Before the BPCIA provisions became effective, a company seeking to market a biological product had only one option—it had to file a full BLA. Unlike the legislation governing generic drugs introduced by the Hatch-Waxman Act in 1984, the PHSA had no

23. Drug Price and Patent Term Restoration Act of 1984, Pub. L. No. 98-417, 98 Stat. 1585 (the Hatch-Waxman Act).

24. *FDA 101*, *supra* note 8, at 2 ("[B]iologics are also subject to regulation under the [FDCA] because most biological products also meet the definition of 'drugs' cited within this Act."); *see also* 42 U.S.C. § 262(j). Historically, some biological products were approved under the FDCA. *See* Carver, *supra* note 6, at 684. In 1986, the agency decided to regulate recombinant versions of previously marketed products under the same statute that the originals were approved under. *Id.* at 685. By the 2000s, both Congress and the FDA had taken steps to harmonize the approval requirements under each statute, but to confirm that the PHSA did indeed govern biologics. *Id.* at 687–88.

25. 42 U.S.C. § 262(a).

26. 21 U.S.C. § 355(b); *see also* Richard A. Epstein, *The Constitutional Protection of Trade Secrets and Patents Under the Biologics Price Competition and Innovation Act of 2009*, 66 Food & Drug L.J. 285, 285 n.3 (2011).

27. 42 U.S.C. § 262(a)(2)(C)(i).

28. *See, e.g.*, 21 C.F.R. §§ 601.2, 601.20, 601.25, 601.27, 601.70 (2012).

29. CRS Report, *supra* note 4, at 3.

30. Henry Grabowski, *Follow-On Biologics: Data Exclusivity and the Balance Between Innovation and Competition*, 7 Nature Rev. Drug Discovery 479, 481 (2008).

31. *Id.* at 482.

express provision for expedited approval of a biological product based on an already approved reference product.

A. "Biosimilar" Drugs and "Interchangeable" Drugs

The BPCIA added Section 351(k) to the PHSA, allowing a company to seek approval of a follow-on biologic based on an already approved reference product.[32] The 351(k) applicant must demonstrate that its follow-on biologic is "biosimilar" to, or "interchangeable" with, the reference product.[33] The required showings are intended to be somewhat analogous to demonstrating bioequivalence under the Hatch-Waxman Act.

An applicant seeking expedited approval of a biologic under § 351(k) must provide information satisfying different criteria. First, the applicant must show that its product is *biosimilar* to the reference product.[34]

> The term "biosimilar" or "biosimilarity" . . . means—
>
> (A) that the biological product is highly similar to the reference product notwithstanding minor differences in clinically inactive components; and
>
> (B) there are no clinically meaningful differences between the biological product and the reference product in terms of the safety, purity, and potency of the product.[35]

This showing must be based on data from analytical, animal, and clinical studies,[36] although the Secretary may waive any studies that are, in his or her discretion, unnecessary.[37]

Second, the applicant must also show that its drug has the same mechanism of action, labeling, and route of administration/dosage as the reference product.[38]

Third, the applicant must show that the facility in which the biologic is manufactured meets standards to ensure the product remains safe, pure, and potent.[39] Because of the potential variability that arises from the production of biologics in living systems, as discussed earlier, it appears that

32. As codified, the subsections of the statute relevant to regulatory approval are 42 U.S.C. § 262(i) (definitions), 42 U.S.C. § 262(k) (regulatory pathway for follow-on biologics), and 42 U.S.C. § 262(m) (extension of reference product exclusivity periods by six months based on requested pediatric studies). 42 U.S.C. § 262(*l*) sets forth the procedures for resolving patent disputes, discussed *infra*, § IV.

33. 42 U.S.C. § 262(k)(2)–(6).

34. *Id.* § 262(k)(2)(A)(i)(I).

35. *Id.* § 262(i)(2).

36. *Id.* § 262(k)(2)(A)(i)(I)(aa)–(cc).

37. *Id.* § 262(k)(2)(A)(ii).

38. *Id.* § 262(k)(2)(A)(i)(II)–(IV).

39. *Id.* § 262(k)(2)(A)(i)(V).

the type (and amount) of data required to show similarity will depend on the characteristics of each individual product.[40]

Some commentators have predicted that, at least initially, the FDA will require clinical trials for the follow-on biologic, although they are not an explicit requirement.[41] This, of course, remains to be seen.

For approval of a follow-on biologic, establishing biosimilarity is a must. But an applicant may also, and optionally, submit additional information showing that its product is *interchangeable* with the reference product. There are, not surprisingly, advantages to doing so. According to the legislation, an interchangeable product "may be substituted for the reference product without the intervention of the health care provider who prescribed the reference product."[42] In other words, a pharmacist may dispense an interchangeable product even though a doctor has prescribed the reference product. A biosimilar product, by contrast, is a separate treatment option that a doctor might prescribe, but that cannot be substituted for a reference product by a pharmacist.

In order to show interchangeability, the 351(k) applicant must first show biosimilarity. The applicant must then also show that the biologic "can be expected to produce the same clinical result as the reference product in any given patient" and, for a product administered more than once to an individual, that there is no more risk in switching from the reference product to the follow-on product than there would be from continuing to administer the reference product.[43] The requirement that the follow-on product give the same clinical result in any given patient appears, by definition, to be something that will require clinical testing.

The statutory provisions of "biosimilar" and "interchangeable" in the BPCIA are at least superficially similar to the administrative designations of "pharmaceutical equivalent" and "therapeutic equivalent" in the Orange Book.[44] But because of the potential variability and complexity of a biologic, it remains to be seen whether such designations will be as useful in the biologics context as they are in the small-molecule context. A report by the Federal Trade Commission suggests that difficulties arising from the general lack of automatic substitution of biologics (unless they meet the strict requirements for interchangeability) and the setting where such drugs are usually dispensed (in a hospital or clinic rather than a pharmacy), as well as physicians' potential safety and efficacy concerns, might

40. CRS Report, *supra* note 4, at 13–14.

41. *Id.* at 13.

42. 42 U.S.C. § 262(i)(3).

43. *Id.* § 262(k)(4).

44. U.S. Dep't of Health & Human Servs., Approved Drug Products with Therapeutic Equivalence Evaluations, at vi–vii (33d ed. 2013) [hereinafter Orange Book].

hinder follow-on biologics from capturing market share to the same extent as generic small-molecule drugs.[45]

The complexity of biologics also raises the bar on the amount of data a follow-on applicant must show to demonstrate biosimilarity or interchangeability of the products and draws into question whether it would, as a practical and financial matter, be worthwhile for a follow-on manufacturer to seek approval of a 351(k) application. Although the cost of bringing a generic small-molecule drug to market averages around $5 million, the estimated average cost to bring a follow-on biologic to market is $100 million to $200 million.[46] Given that, a follow-on manufacturer might opt to apply for its own BLA rather than follow the BPCIA route. This is especially true if the follow-on manufacturer thinks it can promote its product as being "biobetter" rather than biosimilar by advertising purported improvements over the already approved product.[47]

B. Exclusivity Periods

Assuming that a follow-on manufacturer chooses to go the route of the BPCIA provisions, approval of a follow-on product is predicated on prior approval of the reference product. The BPCIA provides data and market exclusivity periods to the innovator of the reference product on which the follow-on biologic is based,[48] as well as to the first follow-on biologic that can demonstrate interchangeability with the reference product.[49] These exclusivity periods are substantially different from the exclusivity periods granted under the Hatch-Waxman Act for small-molecule drugs.[50]

The Hatch-Waxman Act provides some data exclusivity for the data in New Drug Applications, but the lion's share of market exclusivity for new drugs is often derived from the patents that cover them. Under the Hatch-Waxman Act, when the FDA approves a New Drug Application, the company that submitted the application receives five years of data exclusivity, during which the FDA will not consider a corresponding Abbreviated New Drug Application (ANDA).[51] That exclusivity period is extended by six months if the applicant provided the agency with the necessary pediatric study data. After expiration of the five-year or so exclusivity period, potential competitors may submit Abbreviated New Drug

45. Fed. Trade Comm'n, Emerging Health Care Issues: Follow-On Biologic Drug Competition 23 (June 2009) [hereinafter FTC Rep.].

46. *Id.* at 14.

47. CRS Report, *supra* note 4, at 18–19.

48. 42 U.S.C. § 262(k)(7).

49. *Id.* § 262(k)(6)(A).

50. 21 U.S.C. § 355(j)(5)(F)(ii).

51. *Id.*

Applications, and any further exclusivity for the pioneer comes from the patents that cover its product. Thus, under the Hatch-Waxman Act, the bulk of the incentive to innovate new drugs has traditionally appeared to come from the patent system.

By contrast, the BPCIA was intended to split the incentive to innovate between statutory exclusivity periods and patent protection. Under the BPCIA, the FDA is not permitted to consider a follow-on biologic license application that relies on an approved reference product's BLA within four years of approval of that reference product's BLA.[52] But, under the BPCIA, the agency may not *approve* a follow-on 351(k) application until 12 years after approval of the reference product.[53] Moreover, each exclusivity period is extended by six months if the original BLA applicant provided the agency with the necessary pediatric study data.[54] It has been estimated that these exclusivity periods will roughly approximate the period of exclusivity provided by most patents under the Hatch-Waxman Act.[55]

So why extend the statutory exclusivity as compared to the Hatch-Waxman provisions? According to some commentators, the statutory exclusivity period is designed to address concerns that the variable nature of biologic products renders them susceptible to slight modification that allows a follow-on to design around the innovator's patents.[56] Industry members argued that, if the patents protecting innovative biologics are potentially easy to design around, then using only the patent system to provide the exclusivity period to incentivize development is not sufficient. They argued that, as a practical matter for biologics, the exclusivity period during which an innovator could recoup its research costs might effectively disappear.

Congress may have implemented the BPCIA's statutory scheme of extended statutory exclusivity based on concerns that the patent system would not sufficiently protect the innovator company's investment, but this scheme potentially leads to other effects as well.

First, the BPCIA statute expressly prohibits an innovator from extending its exclusivity period based on minor changes and improvements to the product.[57] This is directed to what some critics have referred to as "evergreening" in the world of small-molecule patent practice.[58] Evergreening

52. 42 U.S.C. § 262(k)(7)(B).

53. *Id.* § 262(k)(7)(A).

54. *Id.* § 262(m)(2)(A).

55. Yaniv Heled, *Patents vs. Statutory Exclusivities in Biological Pharmaceuticals—Do We Really Need Both?*, 18 MICH. TELECOMM. & TECH. L. REV. 419, 443–47 (2012).

56. WENDY H. SCHADT & JOHN R. THOMAS, CONG. RESEARCH SERV., P.L. 111-148: INTELLECTUAL PROPERTY PROVISIONS FOR FOLLOW-ON BIOLOGICS 3–4 (May 25, 2010).

57. 42 U.S.C. § 262(k)(7)(C).

58. *See* Janet Freilich, *Patent Infringement in the Context of Follow-On Biologics*, 16 STAN. TECH. L. REV. 9, 23 (2012).

involves making minor changes to purportedly extend patent protection without the normal quid pro quo of innovation. Of course, the key (and contentious) question in that battle is this: what qualifies as a "minor" change? What some critics call evergreening may, in certain instances, actually encompass the development of new, improved, and innovative products. These fights are nothing new in the context of ANDA litigation, and the fights will be expected to continue here—but perhaps in the context of disputing how a modification should be viewed under the BPCIA statutory framework.

A second benefit that purportedly flows from the BPCIA's extended statutory exclusivity is that companies will not have to undertake expensive litigation to enforce their patent rights to recoup and protect their investment.[59] Because the FDA is statutorily prohibited from granting new applications that rely on existing applications for a set amount of time, the BLA holder does not have to do anything to protect its rights against a filer of an expedited application.[60] In a system of statutory exclusivity, there are no questions of whether the generic would infringe or whether it successfully designed around the patent—if a 351(k) applicant seeks to rely on another company's BLA, it has to wait 12 years regardless of any minor changes it made in relation to the reference product or the innovator's patents.

Congress also changed the triggering event for a follow-on manufacturer to potentially get its own period of exclusivity. Although the Hatch-Waxman Act focuses on patent exclusivity and rewards the first generic to successfully challenge the innovator's patents with a 180-day exclusivity period over other generics,[61] the BPCIA grants the first manufacturer of an interchangeable follow-on product at least 12 months of exclusivity during which no additional product can be found to be interchangeable with the reference product.[62] However, some commentators have questioned whether, due to the lengthy market exclusivity granted to the original reference product, exclusivity for the first interchangeable product is of any practical importance.[63]

---

59. *See* Heled, *supra* note 55, at 431–32.

60. *Id.*

61. 21 U.S.C. § 355(j)(5)(B)(iv).

62. 42 U.S.C. § 262(k)(6).

63. *See* McMahon, *supra* note 16, at 671–75. But there do seem to be substantial benefits to a product's classification as being interchangeable rather than merely biosimilar. *See* FTC REP., *supra* note 45, at 23 (noting that doctors are hesitant to prescribe a biosimilar treatment). The 351(k) procedure exists to provide a cheaper way for competitors to bring a follow-on product to market. It is not clear why a year of exclusivity of being the only interchangeable product, even when that year occurs after the innovator has been on the market for 12 years, is not valuable.

C. Challenges over Data Exclusivity

Not surprisingly, the BLA applicant will usually disclose confidential and proprietary information within its BLA in order to show that its product meets the relevant safety and efficacy requirements. The expedited approval process—which is based on showing that a follow-on biologic is "biosimilar" to an already-approved reference biologic—has been challenged as an improper use of the confidential data submitted in the BLA for the original reference product. How the FDA handles this challenge will determine when the expedited approval process will first be available and whether it will ever be available for many well-known and highly profitable biologic drugs.

The challenger, Abbott Laboratories, has made the following argument in a citizen petition to the FDA.[64] The BLA for Abbott's antibody product Humira includes data from preclinical and clinical testing and detailed manufacturing information for producing the biologic molecule. That information was submitted with the understanding that it would be kept confidential by the FDA. The confidentiality of those testing and manufacturing data, and their value in the field, make the data a trade secret belonging to Abbott—the BLA filer.[65] For any BLA filed before the enactment of the BPCIA in 2010, these trade secret data were submitted with no expectation that they could ever be used in a process for approval of a competitor's follow-on biologic.[66] If the data are used by the FDA in the expedited approval process under the BPCIA without providing just compensation to the filer of the original BLA (the owner of the trade secret), then that would be a taking in violation of the Fifth Amendment of the U.S. Constitution.[67] Considering the value of the confidential information here, providing just compensation does not look like a practical alternative. Abbott has urged the FDA to avoid the constitutional issue and refuse to consider any application for expedited approval of a biologic that refers to a reference compound approved based on a BLA filed before the BPCIA was enacted on March 23, 2010.[68]

The FDA has received comments regarding Abbott's petition, and some of those comments raise the following counterarguments. First, there is a question as to whether the process of expedited approval actually *uses* the

64. Abbott Laboratories Citizen Petition, Docket No. FDA-2012-P-0317 (Apr. 2, 2012) [hereinafter Abbott Petition].

65. *Id.* at 3–5.

66. *Id.* at 5, 21–26.

67. *Id.*

68. *Id.* at 1, 27–29. Abbott's drug Humira would be one of those compounds.

confidential information in the reference BLA.[69] Arguably, the FDA only needs to know (1) that the reference product is safe, pure, and potent, and (2) that the follow-on biologic has been shown to be "biosimilar" or "interchangeable" under the BPCIA. The information to demonstrate the second point comes from the follow-on application, not the original BLA. And the first point will be public knowledge, because the fact that the reference product is approved is not confidential. Thus, the counterargument goes, there is no taking because no trade secret is ever used. (Abbott and another commentator have argued that the data are used because they underlie the decision to approve the reference product.)[70]

Second, responders have argued there has never been an explicit guarantee that the FDA would not provide a pathway for expedited approval based on an already approved reference product.[71] In the Hatch-Waxman context, the FDA has stated that relying on the public fact that an approved product has been found to be safe and effective does not lead to Fifth Amendment issues.[72] The potential has always been there for Congress to enact legislation that puts in place a similar mechanism for biologics, and the biotechnology industry has been involved in the discussion leading up to the enactment of such a mechanism in the BPCIA. (Abbott's position is that, before the BPCIA was enacted, no BLA filer could have understood that its data would be used to benefit a competitor.)[73]

Third, at least one commentator has argued that even if there were a taking, the data exclusivity periods provided under the BPCIA and the ability to bring a patent infringement suit before a follow-on product is launched, which were both part of the negotiated balance of incentives, both provide just compensation.[74]

If the FDA agrees not to consider any application for expedited approval of a biologic that refers to a reference compound approved on a BLA filed on or before March 23, 2010, then: (1) because of the 12-year data exclusivity period under the statute, the earliest expedited approval

69. Comments of Generic Pharmaceutical Association (GPhA) to Abbott Laboratories Citizen Petition, Docket No. FDA-2012-P-0317, at 4–7 (Aug. 15, 2012) [hereinafter GPhA Comments]; Comments of Therapeutic Proteins International LLC (TPI) to Abbott Laboratories Citizen Petition, Docket No. FDA-2012-P-0317, at 2–4 (Nov. 16, 2012) [hereinafter TPI Comments]; Comments of Zuckerman Spaeder LLP on Citizen Petition, Docket No. FDA-2012-P-0317, at 4–7 (Dec. 24, 2012) [hereinafter Zuckerman Spaeder Comments].

70. Abbott Petition, *supra* note 64, at 16–20; Comments of the Washington Legal Foundation (WLF) to Citizen Petition, Docket No. FDA-2012-P-0317, at 8–14 (Feb. 13, 2013).

71. GPhA Comments, *supra* note 69, at 16–19; TPI Comments, *supra* note 69, at 4–5; Zuckerman Spaeder Comments, *supra* note 69, at 7–15.

72. GPhA Comments, *supra* note 69, at 17–19; TPI Comments, *supra* note 69, at 5; Zuckerman Spaeder Comments, *supra* note 69, at 9–15.

73. Abbott Petition, *supra* note 64, at 21–25.

74. Zuckerman Spaeder Comments, *supra* note 69, at 15–17.

will be sometime in 2022; and (2) there will never be an expedited approval under the statute for the biologics approved before 2010. That is not to say that there can be no follow-on drugs in those cases: eventually patent protection will either expire or be successfully challenged in litigation. But any such follow-on drug will need to be supported by a BLA.

D. The Continuing Role of Patent Protection under the BPCIA

Despite the incentives provided by the statutory exclusivity, patent protection will still play a major role for many biologics—especially the most profitable ones. First, patents can be asserted broadly—not just against those who seek expedited approval for a follow-on biologic. A competitor might decide that development of a particular product justifies the cost of filing its own independent BLA. In this situation, there is no exclusivity provided under the BPCIA, and patent protection becomes crucial to the developer of the reference product.

Second, patent protection can still potentially outlive the longer 12-year statutory exclusivity period for biologics.[75] This is likely to be particularly relevant at the end of the exclusivity period when a product is at its maximum profitability. During this important time window, expedited approval of follow-on biologics becomes available, along with the potential for resulting patent litigation under the statutory scheme of the BPCIA. That statutory scheme is where we turn next.

IV. Patent Litigation under the BPCIA

The BPCIA provides a scheme for resolving patent disputes between a BLA holder and a company seeking approval for a follow-on biologic.[76] Like the scheme of statutory exclusivity, the patent resolution process under the BPCIA differs from the process under the Hatch-Waxman Act. Instead of requiring a document like the Orange Book for small-molecule drugs, the BPCIA envisions a collaborative back-and-forth exchange between the reference product sponsor (RPS) and the follow-on manufacturer (FOM) to identify potentially relevant patents.[77] The statute provides for the timing of this exchange. Figure 1 depicts the statutory time periods between BLA filing and initial exchange of information. Figure 2 depicts the statutory time periods associated with beginning the statutory immediate infringement action.

75. *See* Heled, *supra* note 55, at 476 app. tbl.1.
76. 42 U.S.C. § 262(*l*).
77. *Id.* § 262(*l*)(3)–(5).

A. Exchange of Information

Within 20 days of filing its application, the FOM must provide the RPS with a copy of its application and information describing its manufacturing process.[78] Under specific provisions of the statute, disclosure of this information is limited to the RPS's outside counsel and one in-house counsel, so long as those attorneys do not engage in patent prosecution related to the reference product.[79] If the RPS has licensed the patent from a patentee who retained the right to participate in litigation, then the patentee's representative may also be provided with the information.[80] The information provided by the FOM is considered confidential and may only be used to determine whether the RPS could reasonably assert any of its patents.[81]

Figure 1 Initial Exchange of Information

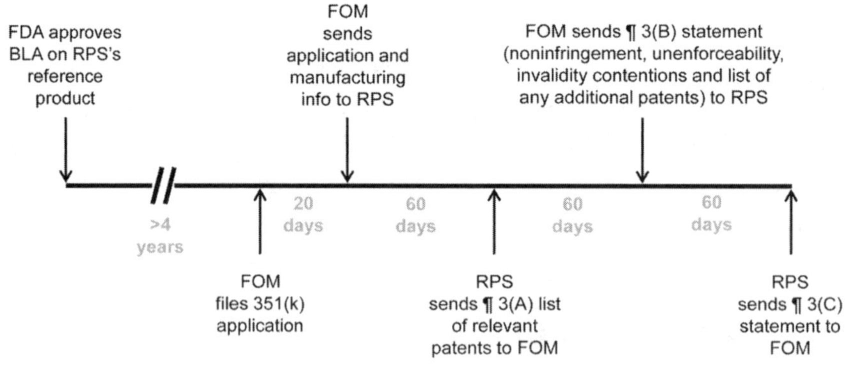

78. *Id.* § 262(*l*)(2).
79. *Id.* § 262(*l*)(1)(B)(ii).
80. *Id.* § 262(*l*)(1)(B)(iii).
81. The disclosure and confidentiality requirements of the BPCIA are enumerated more specifically than the disclosure requirement under the Hatch-Waxman Act. Under Hatch-Waxman, the ANDA applicant must provide an "offer of confidential access [that] shall contain such restrictions as to persons entitled to access, and on the use and disposition of any information accessed, as would apply had a protective order been entered for the purpose of protecting trade secrets and other confidential business information." 21 U.S.C. § 355(j)(5)(C)(i)(III).

Figure 2 Identifying Patents for Immediate Infringement Action

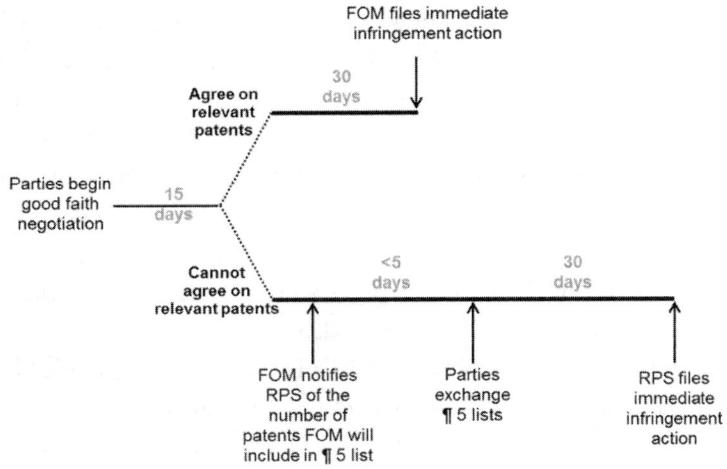

Unlike the Hatch-Waxman Act, the BPCIA does not provide for a database listing the patents that presumably cover the product of a BLA. In other words, there is no Orange Book in biologics regulation.[82] Accordingly, early in the process, the follow-on manufacturer need only provide a copy of its application and manufacturing details—it need not provide any noninfringement or invalidity contentions.

Within 60 days of receiving a copy of the FOM's information, the RPS must provide the manufacturer with a list of patents for which a claim of infringement "could reasonably be asserted" based on the biologic that is the subject of the follow-on application (this is the "paragraph 3(A) list," named after the provision of the statute).[83] If the RPS acquires a patent or a license to a relevant patent after providing the paragraph 3(A) list, it has 30 days to supplement its list and notify the FOM of the new patent.[84] The RPS must also identify any of these patents for which it would be willing to offer a license to the FOM. The paragraph 3(A) list is intended to cover essentially the same kinds of patents that would have been listed in the Orange Book, if such a listing existed. But the paragraph 3(A) lists may also include patents covering a method of manufacture.[85]

82. Orange Book, *supra* note 44.
83. 42 U.S.C. § 262(*l*)(3)(A)(i).
84. *Id.* § 262(*l*)(7).
85. *See id.* § 262(*l*)(3)(A)(i).

Then, within 60 days of receiving the paragraph 3(A) list, the FOM must respond. The response is analogous to the paragraph (iii) and (iv) certifications under the Hatch-Waxman Act: for each patent, the FOM must indicate that it does not intend to begin commercial marketing until after the patent expires, or provide its contentions as to why it believes that each claim is invalid, unenforceable, or not infringed.[86] The manufacturer must also respond to any licensing offers that the RPS made.[87] Finally, the FOM may, but is not required to, identify any additional patents on which it believes a claim of patent infringement could reasonably be asserted (this is the "paragraph 3(B) list").[88] And for any relevant patent identified by the RPS after provision of the paragraph 3(A) list, there is a 30-day period after receipt of that notice to supplement the paragraph 3(B) contentions.

The RPS responds, within 60 days of receipt of the paragraph 3(B) submission, with its opinion as to why each claim of each listed patent would be infringed ("paragraph 3(C) contentions").[89] The RPS must also provide its opinion as to why each claim identified as invalid or unenforceable by the applicant is actually valid and enforceable.[90]

B. Patent Resolution Negotiations

After the FOM receives the paragraph 3(C) contentions, the BPCIA mandates that the parties "shall engage in good faith negotiations to agree on which, if any, patents listed under paragraph (3)" shall be the subject of an immediate patent infringement action.[91] If the parties cannot agree on a list of patents for the immediate litigation, paragraph 5 of subsection (l) provides a mechanism for selecting the patents that are to be the subject of an immediate patent infringement action. Under paragraph 5, the FOM goes first and must notify the RPS of the number (but not the identity) of patents it thinks should be part of the immediate infringement action. The parties then agree on a day, within five days of the FOM's numerical notification, for a simultaneous exchange of lists containing the identity of patents that each party thinks should be the subject of the immediate patent infringement action. Each party's list may include up to the number of patents that the FOM determined.[92] If the FOM selected zero, however, the RPS may list

86. *Id.* § 262(*l*)(3)(B)(ii).
87. *Id.* § 262(*l*)(3)(B)(iii).
88. *Id.* § 262(*l*)(3)(B)(i).
89. *Id.* § 262(*l*)(3)(C).
90. *Id.*
91. *Id.* § 262(*l*)(4)(A). The immediate patent infringement action is a statutorily described action, codified at 42 U.S.C. § 262(*l*)(6).
92. *Id.* § 262(*l*)(5)(B)(ii)(I).

one patent.[93] This back-and-forth should be approached carefully, because there is the potential here for each side to broaden or limit the litigation.

Like Hatch-Waxman, the BPCIA also amended 35 U.S.C. § 271 to create an artificial act of infringement. Specifically, it is now an act of infringement, *with respect to any patent listed* by either party pursuant to paragraph (l)(3), to submit a 351(k) application seeking approval of a biologic.[94] And if the FOM failed to provide the RPS with a copy of its application and details of its manufacturing process, it is an act of infringement, with respect to any patent *that could have been listed* on a paragraph 3 list, to submit a 351(k) application.

C. Immediate Patent Infringement Action

Although the submission of the 351(k) application is now an artificial act of infringement with respect to certain patents, the statute details and limits the potential avenues for remedying that act of infringement. First, the statute provides for an "immediate patent infringement action." If the parties have agreed on patents to litigate, the scope of the immediate patent infringement action is simple—the litigation covers the agreed-upon patents and must be brought by the RPS within 30 days of the agreement.[95] If the parties fail to reach an agreement, the litigation must include all patents identified by either party in the lists exchanged under paragraph 5 of subsection (l), and the litigation must be commenced by the RPS within 30 days of the exchange of those lists.[96]

Unlike the analogous litigation in the Hatch-Waxman Act, the immediate patent infringement action under the PHSA does not automatically trigger a temporary stay of FDA approval.[97] As a practical matter, some commentators have suggested that this should rarely cause a problem because the RPS already enjoys 8 years of regulatory exclusivity beyond the time when a 351(k) applicant may file and these 12 years of exclusivity roughly overlap the exclusivity provided by patent protection.[98] But given

93. *Id.* § 262(*l*)(5)(B)(ii)(II).

94. 35 U.S.C. § 271(e)(2)(C)(i).

95. 42 U.S.C. § 262(*l*)(6)(A).

96. Because 35 U.S.C. § 271(e)(2)(C) makes it an act of infringement to submit the 351(k) application with respect to any patent listed under paragraph 3, some commentators have questioned whether the immediate patent infringement action is in fact limited to paragraph 5 listed patents. *See* Nash & Workman, *supra* note 13, at 208. But for the provisions of paragraph 5 to have any meaning, the immediate infringement action must be restricted to the patents listed in paragraph 5. A better interpretation is that the immediate patent infringement action is restricted to the paragraph 5 listed patents, and the broader scope of § 271(e) exists to give the federal courts jurisdiction over the declaratory judgment actions, *see* 42 U.S.C. § 262(*l*)(9), regarding other paragraph (3) listed patents.

97. 21 U.S.C. § 355(j)(5)(B)(iii).

98. *See* Heled, *supra* note 55, at 476 app. tbl.1.

the importance of market exclusivity, the RPS must consider whether it wants to seek a stay. To do so, it will need to seek a preliminary injunction through the declaratory judgment framework of the statute when the applicant's follow-on biologic is 180 days from commercial launch.[99]

Similar to the Hatch-Waxman Act, if the patentee prevails in the immediate patent infringement action, the court enjoins the FDA from approving the follow-on application until the relevant patents expire.[100] Similarly, monetary damages for infringement may only be awarded if there has been actual manufacture and use or sale/offer for sale.[101] The court also has discretion to award injunctive relief, just as it does in ANDA litigation.[102] But the BPCIA also added new subsection (D) to § 271(e)(4). Subsection (D) *requires* the court to enjoin the follow-on applicant from infringing the RPS's patents if the RPS prevailed in the specified immediate patent infringement action, the decision of that action is final (affirmed by the Federal Circuit or the time for appeal has expired), and the 12-year statutory exclusivity has not yet expired.[103]

The RPS's remedies are limited, however, if it fails to follow the notice and timing provisions of the statute. If the RPS fails to initiate the immediate patent infringement action within 30 days of the parties' agreement on which patents to litigate, or within 30 days of the exchange of paragraph 5 lists, it cannot seek injunctive relief and is limited to seeking a reasonable royalty on the applicant's ultimate biologic.[104] Perhaps, and some have suggested even more importantly, if the RPS fails to list a patent in its paragraph 3(A) list (or to supplement its list with a later acquired patent), it "may not bring an action under this section for infringement of the patent with respect to the biological product."[105] It is not clear if this language means that the patentee is barred from bringing an action for the artificial act of infringement defined under 271(e), or if it is barred from asserting any act of infringement under 271.[106] To err on the side of certainty, a patentee should list every potentially relevant patent in its paragraph 3 list or supplement.

D. Declaratory Judgment Suits and Preliminary Injunctions under the BPCIA

The BPCIA provides specific limitations on seeking declaratory judgment and preliminary injunctions. Not later than 180 days before the FOM will

99. 42 U.S.C. § 262(*l*)(8), (9). See Section IV.D.
100. 35 U.S.C. § 271(e)(4)(A).
101. *Id.* § 271(e)(4)(C).
102. *Id.* § 271(e)(4)(B).
103. *Id.* § 271(e)(4)(D).
104. *Id.* § 271(e)(6)(B).
105. *Id.* § 271(e)(6)(C).
106. *See* Nash & Workman, *supra* note 13, at 216.

first market its biological product, the applicant must give notice of that date to the RPS.[107] Upon receiving such notice (but before the actual date of first commercial marketing), the RPS may seek a preliminary injunction to prevent infringement of any patent that was included in any paragraph 3 list, but not included in the list subject to the immediate infringement action.[108]

As long as the follow-on applicant provided the statutorily required notice and information to the RPS within 20 days of first filing its application, neither party may file a declaratory judgment action prior to the 180-day notice of commercial marketing. This presumably reflects Congress's desire for the parties to follow the statutory framework set out for resolving patent disputes. If the FOM fails to take any action it is required to take after receiving the RPS's paragraph 3(A) list, however, then the RPS is free to bring a declaratory judgment action with respect to any patent it listed in its paragraph 3(A) list. And if the FOM failed to give the RPS notice of its application in the first place, the RPS is free to bring a suit for "declaration of infringement, validity, or enforceability of any patent that claims the biological product or a use of the biological product."[109]

V. Conclusion

The BPCIA regime is still in its infancy. The technical requirements for proving "biosimilarity" or "interchangeability" for large and complex biological molecules will likely be worked out case by case. There is also some uncertainty about whether and when the expedited approval process for biologics will become available, because of the potential that referring to the approval of a reference product would amount to an uncompensated taking.

But the statute does provide a detailed process that is intended to facilitate prompt resolution of patent disputes connected to the approval of a follow-on biologic, including a timed back-and-forth negotiation on the scope of potential patent litigation. Given the explosive growth of the biologics market, the BPCIA process for expedited approval will certainly be tested in litigation. And although there are similarities with Hatch-Waxman, the differences in the BPCIA process, outlined in this chapter, will be essential to keep in mind.

107. 42 U.S.C. § 262(*l*)(8)(A).

108. By providing a remedy for paragraph 3 patents excluded from paragraph 5 lists, this provision of the statute further supports the interpretation that the immediate patent infringement action is limited to the paragraph 5 listed patents. See note 96.

109. 42 U.S.C. § 262(*l*)(9)(C).

chapter 25

Pharmaceutical Regulation and Approval in Foreign Countries

I. Introduction

This chapter is a general overview of the regulatory steps and procedures related to the approval of branded and generic drugs in Australia, Brazil, Canada, the United Kingdom and the European Union, and Japan.

The reader should appreciate that the approval of a drug in each of these countries is a complex matter that involves many different levels of government and is subject to numerous laws and regulations that cannot all be included in this general review. Instead, this chapter focuses on the key regulations and legal considerations for drug approval requirements, pricing regulations, intellectual property issues (patents and trade-marks), and marketing considerations.

II. Australia—Pharmaceutical Regulatory Approval Strategy[1]

A. Introduction

In Australia, pharmaceuticals are governed by a comprehensive and complex regime of legislation, regulations, and industry codes spanning many different areas of law. The legislative and regulatory landscapes are also very dynamic, as patent laws are constantly under review and government authorities constantly update regulatory processes and policies. This

1. Grant Fisher of Ashurst, Melbourne, Australia, and Duncan Longstaff and Nadine Martino of Ashurst, Adelaide, Australia.

chapter provides an overview of each element in this complex Australian regime together with insights into the issues that might arise and how they might be navigated.

B. Australian Register of Therapeutic Goods

The Therapeutic Goods Administration (TGA), which is part of the Australian Government Department of Health (DoH), has responsibility for administering the *Therapeutic Goods Act 1989* (Cth) (TG Act), the *Therapeutic Goods Regulations 1990* (Cth) (TG Regulations), and the orders made under the TG Act. The TGA controls the supply of therapeutic goods in Australia through audit and assessment of the quality of manufacture, premarket assessment of goods, and postmarket monitoring of standards compliance. The TGA also maintains the Australian Register of Therapeutic Goods (ARTG) for the purpose of compiling information in relation to, and providing for evaluation of, therapeutic goods for use in humans.

A therapeutic good[2] (including a pharmaceutical medicine) can only be marketed, imported, manufactured, or supplied within Australia after it has been evaluated and approved by the TGA and included on the ARTG (unless specifically exempted by the TG Act). The sponsor, who is the person or company who wishes to export, import, or manufacture the therapeutic goods within Australia, or arranges for another party to do so, is responsible for applying to have its therapeutic good included on the ARTG. The sponsor must be a person or company who is resident of, or is carrying on business in, Australia.[3] This means that foreign companies must appoint a local person or company as the sponsor of their products,

---

2. Section 3 of the TG Act defines therapeutic goods as goods: "that are represented in any way to be, or that are, whether because of the way in which the goods are presented or for any other reason, likely to be taken to be for therapeutic use, or for use as an ingredient or component in the manufacture of therapeutic goods or for use as a container or part of a container for goods; or included in a class of goods the sole or principal use of which is, or ordinarily is, a therapeutic use or a use of a kind referred to in subparagraph (a)(ii) or (iii); and includes biologicals, medical devices and goods declared to be therapeutic goods under an order in force under section 7 of the TG Act (which allows for the Secretary to make a declaration that goods are not therapeutic), but does not include: goods declared not to be therapeutic goods under an order in force under section 7; or goods in respect of which such an order is in force, being an order that declares the goods not to be therapeutic goods when used, advertised, or presented for supply in the way specified in the order where the goods are used, advertised, or presented for supply in that way; or goods (other than goods declared to be therapeutic goods under an order in force under section 7) for which there is a standard (within the meaning of subsection 4(1) of the Food Standards Australia New Zealand Act 1991); or goods (other than goods declared to be therapeutic goods under an order in force under section 7) which, in Australia or New Zealand, have a tradition of use as foods for humans in the form in which they are presented."

3. See definition of "sponsor" in Section 3 of the TG Act.

which is typically done either as part of broader local licensing arrangements or under a specific agency agreement.

Patent protection is not required for a pharmaceutical medicine or other therapeutic good to be approved by the TGA. Similarly, even where one or more Australian patents or patent applications may relate to a pharmaceutical medicine or other therapeutic good, this does not affect the TGA's evaluation and approval processes. The Australian government's systems for the regulation of therapeutic goods and the enforcement of and challenges to patents (via the Patent Office or in court) are separate and formally unrelated, save for:

- the certificates generic pharmaceutical companies and patentees may be required to provide to the DoH under the TG Act (discussed in the following sections); and
- the extensions of patent term that patentees may obtain to account for the delays in exploitation while a product is being evaluated by regulators (also discussed in the following sections).

Therapeutic goods[4] (including medical devices)[5] intended for use only in clinical trials are exempt from registration or listing on the ARTG (subject to certain requirements). Such therapeutic goods and medical devices must, however, be registered or listed if they are to be used beyond clinical trials.

Individuals can also gain limited access to therapeutic goods not on the ARTG through three other avenues: the special access scheme,[6] authorised prescribers,[7] and importation for personal use.[8]

Therapeutic goods include medicines (prescription, nonprescription or "over-the-counter," and complementary medicines), medical devices, and biologicals. Products included in the ARTG are divided into four categories:

- **Registered goods:** Therapeutic goods that have to be "registered" on the ARTG are those for the treatment of diseases and illnesses and those containing higher-risk ingredients that are intended for supply in Australia. They include prescription and most nonprescription (over-the-counter) medicines. The TGA evaluates registrable goods for safety, quality, and efficacy. Registration is given in respect

4. See Section 18 or Section 19 of the TG Act, regulation 12 of the TG Regulations, and schedules 5 and 5A to the Section Regulations.

5. By either Section 41HA or Section 41HB of the TG Act and by part 7 of the Therapeutic Goods (Medical Devices) Regulations 2002 and schedule 4 to the Regulations.

6. *See Special Access Scheme*, Therapeutic Goods Admin., http://www.tga.gov.au/hp/access-sas.htm (last updated Apr. 16, 2014).

7. *See Authorised Prescribers*, Therapeutic Goods Admin., http://www.tga.gov.au/hp/access-authorised-prescriber.htm (last updated Mar. 26, 2014).

8. *See Personal Import Scheme*, Therapeutic Goods Admin., http://www.tga.gov.au/hp/access-personal-import-scheme.htm (last updated Oct. 1, 2004).

of the conditions and indications for which the clinical data support safe use. Generally speaking, nonprescription medicines are considered to have a lower risk than prescription medicines and therefore undergo a less onerous level of evaluation. Whether a medicine is deemed "high risk" and therefore required to be registered on the ARTG is determined by assessing: whether it contains a substance included in the Standard for the Uniform Scheduling of Drugs and Poisons (SUSDP); its dosage form(s); promotional/therapeutic claims made; the seriousness of the illness treated; and whether there are any significant side effects or adverse effects from prolonged use. If these factors indicate a high risk, sponsors are required to provide comprehensive safety, quality, and efficacy data from clinical trials (or, for well-known active ingredients, standard textbooks/guidelines) to obtain registration in one of two subcategories:

- o Prescription (high risk) registered—all prescription medicines and some specified products such as sterile injectables; or
- o Nonprescription (low risk) registered—considered to be lower risk than prescription medicines but still worthy of a high level of scrutiny to ensure adequate labeling and appropriate use, such as mild analgesics, cough/cold preparations, and antifungal creams.

All registered medicines must display an "AUST R" number on the label as proof of registration.

- **Listed goods:** Therapeutic goods that have to be "listed" on the ARTG are those that represent a low risk to patients. Listed goods contain only ingredients the TGA has evaluated as low risk, are manufactured by licensed manufacturers complying with good manufacturing practice, and carry indications only for health maintenance and/or enhancement, or certain indications for nonserious, self-limiting conditions. They include products such as medical devices, complementary medicines (such as vitamins, minerals, and sunscreens), and some nonprescription medicines. The TGA evaluates listable goods for safety and quality but not for efficacy. However, the sponsor of the listable good must certify to the TGA that it holds evidence to support all indications and claims for its products.[9] Therapeutic goods that are intended for export only are also required to be listed on the ARTG before export (not registered). Generally speaking, the requirements for "registered" goods are more onerous than those for "listed" goods.

9. See Sections 26 and 26A of the TG Act.

All listed medicines must display an "AUST L" number on the label as proof of listing and must *not* contain substances that are scheduled in the Poisons Standard.

- **Medical devices** that have a physical or mechanical effect on the body or are used to measure or monitor functions of the body[10] (discussed in further detail in the following sections).
- **Biologicals**[11] comprising, containing, or derived from human cells or human tissues and used to treat or prevent disease; diagnose the condition of a person; influence, inhibit, or modify a physiological process in persons; test the susceptibility of a person to a disease or ailment; or replace or modify parts in the anatomy of a person.

C. Registration of Prescription Medicines

An application to register a new prescription medicine must comply with the following regulatory documents:

- "Mandatory requirements for an effective application";[12]
- "Pre-submission planning form" (PPF);[13]
- Prescription medicines electronic facility lodgement or the form titled "Application for the registration, or to vary the conditions of registration, of prescription medicines";[14] and
- a number of common technical documents dealing with quality, efficacy, and safety.[15]

A sponsor must also follow the prescription medicines registration process that is set out in the Australian Regulatory Guidelines for Prescription Medicines (ARGPM).[16] The TGA also has detailed guidelines on its website

---

10. See Section 41BD of the TG Act.

11. Information about the regulation of biologicals is available at *Blood, Tissues & Biologicals*, Therapeutic Goods Admin., http://www.tga.gov.au/industry/btb.htm (last visited May 20, 2014).

12. Therapeutic Goods Admin., http://www.tga.gov.au/industry/pm-argpm-requirements.htm (last updated May 1, 2014).

13. Therapeutic Goods Admin., http://www.tga.gov.au/industry/pm-argpm-process-ppf.htm (last updated May 9, 2014) [hereinafter PPF].

14. Therapeutic Goods Admin., http://www.tga.gov.au/industry/pm-forms-registration.htm (last updated Oct. 3, 2007).

15. Being the CTD Module 1: Administrative information and prescribing information for Australia; ICH M4Q Common technical document for the registration of pharmaceuticals for human use—Quality (CPMP/ICH/2887/99 Rev 1 Quality); ICH M4S Common technical document for the registration of pharmaceuticals for human use—Safety (CPMP/ICH/2997/99 Rev 1 Safety); and ICH M4E Common technical document for the registration of pharmaceuticals for human use—Efficacy (CPMP/ICH/2887/99 Rev 1 Efficacy), which can be accessed at http://www.tga.gov.au/industry/pm-argpm-requirements.htm.

16. The ARGPM is available on the TGA's website at http://www.tga.gov.au/pdf/pm-argpm-process.pdf and applies to applications made from June 1, 2013.

about the registration or listing process of nonprescription medicines[17] and complementary medicines.[18] The fees and charges payable to the TGA for the registration or listing process are available at the TGA website.[19]

The process for a prescription medicine to obtain registration on the ARTG differs depending on whether the product is a new pharmaceutical product (i.e., a product not currently registered on the ARTG), or a generic version of a product already registered on the ARTG.

1. New Product

The prescription medicines registration process for new products is an eight-phase process with eight milestones, each of which must be completed before commencement of the next phase (see Table 1).

Table 1

Australian Regulatory Approval Milestones for Prescription Medicines

Phase	Milestone
1—Pre-submission	Outcome of pre-submission planning sent
2—Submission	Outcome of application consideration sent
3—First round assessment	Outcome of first round assessment and Section 31 request for information or documents sent
4—Consolidated Section 31 request response	End of Section 31 request response period
5—Second round assessment	Outcome of second round assessments sent
6—Expert advisory review	Outcome of expert advisory committee review sent
7—Decision	Decision made by delegate
8—Post-decision	Administrative and regulatory activities complete

The first step of the registration process is the pre-submission phase whereby a sponsor must provide the TGA with the PPF.[20] The PPF provides information on the scope and scale of a submission, including details

17. *See Australian Regulatory Guidelines for the OTC Medicines (ARGOM)*, Therapeutic Goods Admin., http://www.tga.gov.au/industry/otc-argom.htm (last updated Apr. 17, 2013).

18. *See Australian Regulatory Guidelines for Complementary Medicines (ARGCM)*, Therapeutic Goods Admin., http://www.tga.gov.au/industry/cm-argcm.htm (last updated Nov. 30, 2013).

19. *See* Office of Corporate Servs., Summary of Fees and Charges at 1 July 2013 (2013) [hereinafter Summary of Fees and Charges], *available at* http://www.tga.gov.au/about/fees-130701.htm.

20. PPF, *supra* note 13.

of the quality, nonclinical, and clinical evidence. The provision of this information allows the TGA to assign resources for the evaluation process and provide the sponsor with a timetable for the key dates of the process. The TGA's document "Information for applicants completing a pre-submission planning form"[21] provides guidance on the steps to be taken to complete a PPF. This phase of the process is completed when the applicant receives a planning letter from the TGA that sets out the key dates of the process.

The sponsor must then lodge a substantial regulatory submission, which the TGA terms the dossier, that to be effective must be lodged on the appropriate form, be lodged by the date set out in the planning letter received in the pre-submission phase, and meet all the relevant regulatory requirements. The dossier is often a considerable number of pages in length and will include overviews of the chemistry and manufacturing of the product, clinical and nonclinical data,[22] draft product information, and administrative information. This phase of the process is completed when the sponsor receives a notification letter that will advise whether or not the application is effective and whether it has been accepted for evaluation. If the application has been accepted for evaluation, the sponsor will be required to pay the TGA the evaluation fee specified in the notification letter.

During the first round assessment, the TGA evaluates the submission across the separate technical areas (i.e., chemistry and manufacturing, toxicology and clinical studies). During the registration process, the TGA will consider the safety, quality, and efficacy of the product but will not consider its cost effectiveness and total cost.[23] The TGA will undertake a request-for-information process with the sponsor during this phase. This process involves the TGA providing the sponsor with "consolidated section 31 [of the TG Act] questions," which is a single set of evaluation questions from all evaluation areas of the TGA that the sponsor will then be required to respond to with a consolidated Section 31 request response. The TGA will then undertake a second round of assessment in which it will consider consolidated Section 31 request responses and finalize its evaluation report.

On completion of the second round of assessment, the evaluation report will be considered by the TGA delegate who will approve the submission. The TGA delegate has the opportunity to seek independent advice on the submission from an independent expert advisory committee, such as the Advisory Committee on Prescription Medicine (ACPM) which includes external advisors, or more specific issues may also be referred to

21. Therapeutic Goods Admin., http://www.tga.gov.au/industry/pm-argpm-process-ppf-instructions-01.htm#.U3vxti_BLGU (last updated May 7, 2014).

22. There is no requirement that applications to the TGA to market medicines and medical devices must contain data from clinical trials conducted in Australia.

23. See Section 25(1) of the TG Act.

a subcommittee of the ACPM, the Pharmaceutical Subcommittee, or to the Advisory Committee on Safety of Medicines. Generally, new innovator products or new indications are referred to the ACPM. The sponsor will be provided with an opportunity to review the final report and advise the TGA of any perceived errors of fact or major omissions before the report is considered by the ACPM. Even if the final report is not referred to the ACPM, the sponsor will have 14 days to review the report and advise the TGA of any issues.

Finally, the TGA delegate will decide whether the application is to be approved (possibly modified or varied) or rejected. If approval is given, the product is entered on the ARTG. Information about the product and the decision made will be published on the TGA website—namely, the product information document, consumer medicines information, and an Australian public assessment report, which provides information about the evaluation process and why the TGA approved the application. The registration of a new prescription pharmaceutical product on the ARTG generally takes at least 12 months.

2. Generic Product

A generic product is also required to be registered pursuant to the prescription medicines registration process, and its application must also be in compliance with the key TGA regulatory documents. In fact, the registration of a new generic[24] product is very similar to the registration process of a new product in that it has the same phases and milestones (except for "Milestone 6: Expert advisory review," which is optional). The time frames for the process are also very similar, with the total assessment period by the TGA of either a new product or a generic being six months.

However, in some cases, the sponsor has the benefit of being able to submit an abridged application where, in lieu of providing detailed efficacy/clinical data, the sponsor can provide bioequivalence/bioavailability studies to demonstrate clinical equivalence between an existing Australian medicine and the generic.[25] Clinical equivalence can be established through appropriate bioequivalence studies, a Biopharmaceutics Classification System (BCS)–based biowaiver approach, or/and a robust scientific justification that has been prepared in accordance with the TGA's

24. A generic is defined in schedule 9 of the TG Regulations as a medicine when compared with a registered medicine: "has the same quantitative composition of therapeutically active substances—being substances of similar quality to those in the innovator product; has the same pharmaceutical form; is bioequivalent; and has the same safety and efficacy properties."

25. *See Mandatory Requirements for an Effective Application*, Therapeutic Goods Admin., http://www.tga.gov.au/industry/pm-argpm-requirements.htm (last updated May 1, 2014), and in particular Module 5—Efficacy (clinical study reports), for further detail.

"Biopharmaceutical studies—Justification for not submitting biopharmaceutic data."[26]

A generic may apply for registration by relying on its studies and data or those of an innovator. The generic will only be able to rely on the innovator's data after the five-year data exclusivity period has passed (discussed in further detail in following sections).

If the generic does rely on an innovator's data, it will need to comply with Section 26B of the TG Act, which requires a party that applies to register a therapeutic good and relies on evidence submitted by another party in the context of registration of another good to demonstrate safety or efficacy to provide to the TGA one of two certificates (26B Certificate):[27]

- First certificate—a certificate to the effect that the sponsor, acting in good faith, believes on reasonable grounds that it is not marketing and does not propose to market the product in a manner or in circumstances that would infringe a valid claim of a patent that has been granted in relation to the product; or
- Second certificate—a certificate to the effect that a patent has been granted in relation to the product, the applicant proposes to market the product before the end of the term of the patent, and the applicant has notified the patentee of the application for registration or listing of the product under Section 23 of the TG Act.

Under the TG Act it is an offense to give a certificate that is false or misleading in any material particular. The maximum penalty for that offense is "1,000 penalty units," which in monetary terms is approximately A$170,000. The TGA will not inquire into the correctness of the certificate.[28] Rather, the certificate is capable of enforcement by the Australian government following the outcome of any related court proceedings concerning patent infringement.

This certification scheme, which is comparable to the patent certification regime in the United States, was introduced as a result of reforms to the TG Act in 2005 as a result of the Australia-U.S. Free Trade Agreement (AUSFTA).[29]

---

26. Therapeutic Goods Admin., Australian Regulatory Guidelines for Prescription Medicines app. 15 (June 2004), *available at* http://www.tga.gov.au/pdf/pm-argpm-ap15.pdf. See *Mandatory Requirements for an Effective Application*, *supra* note 25, for further detail about the requirements for bioequivalence studies.

27. *Australia–United States Free Trade Agreement*, Therapeutic Goods Admin., http://www.tga.gov.au/about/international-usa-fta.htm (last updated Aug. 10, 2006).

28. See Section 26(1A)(c) of the TG Act.

29. Further details are available at *Australia–United States Free Trade Agreement*, *supra* note 27.

3. Medical Devices

The TGA's "Australian regulatory guidelines for medical devices" (ARGMD)[30] sets out the regulatory requirements for medical devices in Australia. Like other therapeutic goods, medical devices must be included on the ARTG before they can be supplied or marketed in Australia. The registration process for a medical device will depend on whether the product will be manufactured overseas or in Australia. The requirements in relation to the registration of medical devices will also depend on the classification of the device. There are five classes of medical devices: Class I, Class IIa, Class IIb, Class III, and Active Implantable Medical devices. The classification will be determined by considering: the manufacturer's intended use of the device; level of risk to patients, users, and other persons (the probability of occurrence of harm and the severity of that harm); degree of invasiveness in the human body; and duration of use.

Before a sponsor can make an application to the TGA for a medical device to be included in the ARTG, the manufacturer (if the product is being manufactured in Australia) must determine the classification of the device, ensure the procedures to be used meet the relevant Essential Principles, and apply to the TGA for a "conformity assessment certification." The conformity assessment certification relates to the procedures in place to demonstrate that the medical device and manufacture of the medical device conform to the Essential Principles of safety and performance defined in the TG Act, which are descriptively headed:

- use of medical devices not to compromise health and safety;
- design and construction of medical devices to conform to safety principles;
- medical devices to be suitable for intended purpose;
- long-term safety;
- medical devices not to be adversely affected by transport or storage; and
- benefits of medical devices to outweigh any side effects.

A further nine Essential Principles about design and construction apply to devices case by case.

If the application for a conformity assessment certification is successful, the manufacturer must prepare a Declaration of Conformity. If the product is being manufactured overseas, the manufacturer must either obtain a conformity assessment certification from the TGA or European notified body (accompanied by a declaration of its conformity with certain

30. Therapeutic Goods Admin., http://www.tga.gov.au/industry/devices-argmd.htm (last updated May 4, 2011).

Australian standards). Manufacturers of Class I devices must apply a conformity assessment procedure and prepare a Declaration of Conformity; however, the declaration does not need to be submitted to the TGA prior to submitting a device application. In order to lodge an application, the sponsor must complete the appropriate application form, pay the prescribed application fee,[31] ensure that if conformity assessment evidence is required for the device that appropriate evidence has been obtained, and ensure that the application does not contain information that is false or misleading. The sponsor is also required to certify that certain information is available and can be provided to the TGA on request, including, among other things, documents prepared by the manufacturer confirming that the devices are medical devices and are intended for a specified purpose, documents confirming that the conformity assessment procedure was selected and applied, and confirmation that the device is in compliance with the Essential Principles.[32]

D. Data Exclusivity

Section 25A of the TG Act provides a five-year data exclusivity regime that prevents the use of protected information about other therapeutic goods for the purpose of assessing an application for registration of a new therapeutic good.[33] Protected information will remain protected for five years from the date that a therapeutic good is registered. "Protected information" is:

- information supplied in relation to an application to register therapeutic goods;
- consisting of, or containing, an active component;
- where the information is not available to the public; and
- where no other therapeutic goods containing the same active component were already registered.

An "active component" is "a substance that is, or one of the substances that together are, primarily responsible for the biological or other effect identifying the goods as therapeutic goods."

If goods are registered and it becomes apparent that protected information was used when evaluating the goods for registration, the TGA must cancel the registration.

Section 25A of the TG Act (discussed earlier) includes the requirement that for data exclusivity to be granted, there can be no other therapeutic goods containing the same active component already on the register.

31. Current application fees are available SUMMARY OF FEES AND CHARGES, *supra* note 19.
32. Section 41FD of the TG Act.
33. *See* Section 25A(2) of the TG Act.

Consequently, data exclusivity will not be granted for new uses or new formulations of already approved drugs. The data exclusivity regime will apply despite a sponsor having no patent protection for its therapeutic good.

E. Orphan Drug Program

The Australian Orphan Drug Program came into effect on January 1, 1998, and is regulated by Part 3B of the TG Regulations. According to the TGA, the Orphan Drug Program "aims to bring orphan drugs to the market by reducing development costs."

TG regulation 16H(2) provides that to be an orphan drug, a medicine, vaccine, or in vivo diagnostic agent:

- must be intended to treat, prevent, or diagnose a rare disease; or
- must not be commercially viable to supply to treat, prevent, or diagnose another disease or condition.

The TG Regulations define a "rare disease" as a disease, or condition, likely to affect not more than 2,000 individuals in Australia at any time.[34]

To be an orphan drug the product must not have been rejected on safety grounds by the TGA; the Food and Drug Administration of the United States of America; the Medicines and Healthcare Products Regulatory Agency (formerly the Medicines Control Agency) of the United Kingdom; the Pharmacy Examining Board of Canada (formerly the Bureau of Pharmaceutical Assessment of Canada); the Medical Products Agency of Sweden; the Medicines Evaluation Board of the Netherlands; or the European Medicines Agency (formerly the European Agency for the Evaluation of Medicinal Products).[35]

The Orphan Drug Program has the following incentives:

- the waiver of application and evaluation fees;[36]
- no annual registration fees;[37] and
- priority evaluation.

F. Clinical Trials

Clinical trials in Australia are regulated by the TGA. The parties involved in the conduct of a clinical trial must comply with the TG Act and the TG Regulations.

As discussed, therapeutic goods intended for use in clinical trials are exempt from registration or listing under the TG Act. There are currently two routes that a sponsor can take for conducting a clinical trial of a new

34. *See* definition of "rare disease" in regulation 2 of the TG Regulations.
35. *See* regulation 16H(3) of the TG Regulations.
36. *See* regulation 45(12)(a) of the TG Regulations.
37. *See* regulation 45(12)(c) of the TG Regulations.

therapeutic good not entered on the ARTG or use of an existing therapeutic good beyond the conditions of its marketing approval—the Clinical Trial Notification Scheme (CTN Scheme) and the Clinical Trial Exemption Scheme (CTX Scheme).

Clinical trials in which therapeutic goods are used within the conditions of their marketing approval are not subject to CTN or CTX requirements, but they still need to be approved by a Human Research Ethics Committee (HREC) before the trial may commence.

The CTN Scheme and the CTX Scheme require an Australian "sponsor." The sponsor takes overall responsibility for the conduct of the trial and usually initiates, organizes, and supports a clinical study and carries the medicolegal responsibility associated with the conduct of the trial. Clinical trials can be sponsored by individuals, by bodies and organizations, or by companies.

It is important to note that an application or notification to conduct a clinical trial involving an unapproved therapeutic good is independent of an application for registration of that good on the TGA. A notification or application to conduct a clinical trial will be accepted while an application for registration of the same product is under review and vice versa.

The document "Access to Unapproved Therapeutic Goods—Clinical Trials in Australia"[38] sets out in detail the requirements of both the CTN Scheme and the CTX Scheme and the responsibilities of the parties involved in the clinical trial. A summary of each of these schemes follows.

1. Good Clinical Practice in Australia

Parties should be mindful that in conducting clinical trials in Australia they will be subject to standards of ethical conduct and good clinical practice. The TGA's *Australian Clinical Trial Handbook*[39] provides a practical guide to the conduct of clinical trials to international standards of good clinical practice in the Australian context. Two of the key standards that parties will need to adhere to (and that are discussed in further detail in the *Handbook*) include:

- the "National Statement on the Ethical Conduct of Research Involving Humans 2007" (National Statement);[40] and
- the "Note for Guidance on Good Clinical Practice (CPMP/ICH-135/95)" (GCP Note).[41]

---

38. Therapeutic Goods Admin., (Oct. 2004), *available at* http://www.tga.gov.au/pdf/clinical-trials-guidelines.pdf.

39. Therapeutic Goods Admin., (Mar. 2006), *available at* http://www.tga.gov.au/industry/clinical-trials-handbook.htm.

40. Austl. Res. Council (2007), *available at* http://www.nhmrc.gov.au/_files_nhmrc/publications/attachments/e72_national_statement_130207.pdf.

41. Therapeutic Goods Admin., http://www.tga.gov.au/industry/clinical-trials-note-ich13595.htm (last updated July 25, 2000).

The National Statement, which has been developed in accordance with the National Health and Medical Research Council Act 1992 (Cth), is the ethical standard against which all research in Australia involving humans, including clinical trials, is reviewed. It is required to be used by any researcher conducting research with human participants, any member of an ethical review body reviewing that research, those involved in research governance, and potential research participants. In particular, the National Statement also sets out the responsibilities, composition, function, and procedural and record-keeping requirements of HRECs in Australia. Several other TGA guidelines provide additional helpful advice with respect to clinical trial issues. One guideline of particular note is the "Note for guidance on clinical safety data management: Definitions and standards for expedited reporting," which describes the reporting processes for expedited reporting of adverse drug reactions in clinical trials.[42]

G. Marketing and Advertising of Therapeutic Goods

The marketing and advertising of therapeutic goods are subject to the requirements of the TG Act. Under the TG Act, an advertisement in relation to therapeutic goods includes any statement, pictorial representation, or design that is intended whether directly or indirectly to promote the use or supply of the goods.[43] Medicines that are not registered on the ARTG are not permitted to be marketed in Australia. Additionally, Section 22(5) of the TG Act specifies that a person will commit an offense if the person, by any means, advertises therapeutic goods that are included on the ARTG for an indication not accepted in relation to that good.

Generally, under the TG Act, advertising prescription medicines directly to consumers is prohibited, and advertising can only be directed to health professionals.[44] This advertising is regulated by a self-regulating scheme operated by Medicines Australia (the industry group representing innovator pharmaceutical companies in Australia).

Advertising for nonprescription medicines, including over-the-counter medicines and complementary medicines, can be directed to both consumers and health professionals (certain medicines in the Poisons Standard are an exception). The TGA, the Therapeutic Goods Advertising Code Council, the Australian Self-Medication Industry, and the Complementary

---

42. Therapeutic Goods Admin., http://www.tga.gov.au/industry/clinical-trials-note-ich37795.htm (last updated July 7, 2000).

43. *See* definition of "advertisement" in section 3 of the TG Act.

44. *See* Section 42DL of the TG Act and 5F and 5G of the TG Regulations for further information about how to apply for such approval. The relevant approval application forms are available at *Advertising Therapeutic Goods*, Therapeutic Goods Admin., http://www.tga.gov.au/industry/advertising.htm.

Healthcare Council regulate this advertising. Certain advertisements (for example, broadcast or print media) require approval from the relevant regulatory body prior to their publication.[45]

Advertisements for therapeutic goods directed to consumers must comply with the Therapeutic Goods Advertising Code 2007[46] (Advertising Code).[47] The Advertising Code aims to "ensure that the marketing and advertising of therapeutic goods to consumers is conducted in a manner that promotes the quality use of therapeutic goods, is socially responsible, and does not mislead or deceive the consumer." The code is generally consistent with the requirements of the World Health Organization's Ethical Criteria for Medicinal Drug Promotion 1988.

The advertising of therapeutic goods must also comply with the provisions of the Australian Consumer Law (ACL), as enacted in Schedule 2 to the Competition and Consumer Act 2010 (Cth) and equivalent Australian state and territory fair trading legislation (discussed in further detail later in this section). Section 18 of the ACL prohibits a person from, in trade or commerce, engaging in conduct that is misleading or deceptive, or that is likely to mislead or deceive. Conduct will generally be likely to mislead or deceive if there is a real or not remote chance or possibility that members of the relevant public would be misled or deceived. The absence of an intention to mislead or deceive is not a defense. Section 29(1) of the ACL also prohibits a person from making a false or misleading representation. There are equivalent provisions in the Fair Trading Act in each Australian state and territory. Sanctions that courts can impose against persons involved (including directors of companies in their personal capacity) for breaches of the ACL include pecuniary fines, damages, injunctions, adverse publicity (published statements), corrective advertising orders, and community services orders.

In Australia it is also unlawful for a person to pass off his or her products, services, or business as the products, services, or business of another person. "Passing off" is falsely representing that the products, services, or business are the products, services, or business of another person or that they are in some way connected with that other person or his or her products, services, or business.

Generic companies should be particularly careful to avoid falling foul of these laws and ensure that they and the doctors and pharmacists who are selling their products do not mislead, misrepresent their products as an innovator's product, or pass off their own products as an innovator's product.

45. *See* Section 42C of the TG Act.

46. Available at http://www.comlaw.gov.au/Details/F2007L00576.

47. *See* Section 3(1)(a) of the Advertising Code.

1. Counterfeit Products and Product Tampering

A person is guilty of an offense under the TG Act if the person intentionally manufactures, supplies, imports, or exports from or into Australia therapeutic goods that are counterfeit and the person knows that fact or is reckless as to whether that fact exists.[48] Goods will be considered counterfeit if the labels or presentations of the goods, documents/records relating to the goods or their manufacture, or any advertising contain a false representation in relation to the:

- identity or name of the goods;
- formulation, composition, or design specification of the goods or of any ingredient or component of them;
- presence or absence of any ingredient or component of the goods;
- strength or size of the goods (other than the size of any pack in which the goods are contained);
- strength or size of any ingredient or component of the goods; or
- sponsor, source, manufacturer, or place of manufacture of the goods.

The maximum penalty for the offense is seven years' imprisonment and 2,000 penalty units, which in monetary terms is approximately A$340,000. There are provisions under the TG Act that permit the counterfeit products to be forfeited by Australian Customs officials.[49]

It is also an offense for a person to supply or propose to supply therapeutic goods if the person knows that some or all of those therapeutic goods have been subject to actual or potential tampering and if the person fails to notify the TGA of having become aware of a substantial risk of, or the actual or potential, tampering of those goods (within 24 hours of becoming aware).[50] The maximum penalty for the offense is 12 months' imprisonment and 1,000 penalty units, which in monetary terms is approximately A$170,000.

2. Other Legislation

A sponsor should be aware that the TG Act and its regulations are not the only legislation which will need to be complied with to import into or export from Australia. These relate mainly to general prohibitions (subject to specific exceptions) under customs legislation on the importation or exportation of certain therapeutic goods or related substances without the license and permission of the Secretary of Health. Sponsors should ensure they are familiar with the requirements and obligations under the following legislation:[51]

---

48. *See* Section 42E of the TG Act.
49. *See* Sections 42F(1)–(2).
50. *See* Section 42T of the TG Act.
51. This may not be an exhaustive list.

- Customs Act 1901 (Cth);
- Customs (Prohibited Imports) Regulations 1956 (Cth);
- Customs (Prohibited Exports) Regulations 1958 (Cth);
- Quarantine Act 1908 (Cth)
- Poisons Standard 2013 (Cth), which is updated annually;
- Gene Technology Act 2000 (Cth), its regulations, and the corresponding state and territory legislation;
- Wildlife Protection (Regulation of Imports and Exports) Act 1982 (Cth);
- Commerce (Trade Descriptions) Act 1905 (Cth); and
- any state and territory legislative requirements relating to labeling and scheduling.

H. Pharmaceutical Benefits Scheme (PBS)

While TGA approval is all that is technically required for a pharmaceutical medicine to be marketed or supplied in Australia, most high-cost pharmaceutical products (such as prescription medicines) cannot be viably sold into the Australian market without the substantial government subsidies flowing from inclusion on the Pharmaceutical Benefits Scheme (PBS).

The PBS is a national scheme whereby the government subsidizes the cost of medicine for most medical conditions for all Australian residents. The PBS is part of the Australian government's broader National Medicines Policy, which aims to meet medication and related services needs, so that both optimal health outcomes and economic objectives are achieved.

The PBS is managed by the DoH and administered by the Department of Human Services (Medicare Australia) under the National Health Act 1953 (Cth) (NH Act) and the regulations and orders made under the NH Act, which include the National Health (Pharmaceutical Benefits) Regulations 1960 (Cth) (NH Regulations).

Under the PBS, a patient pays a fixed (subsidized) price of A$36.90[52] for most products listed on the Schedule of Pharmaceutical Benefits (PBS Schedule), regardless of the price at which the sponsor supplies that product to the market (but patients may pay extra if they are prescribed a brand that costs more than the least expensive brand of that medicine), with the Australian government paying the remaining cost (which it negotiates with each sponsor product by product).

Unlike the ARTG, the primary purpose of the PBS Schedule is to inform health care professionals of how medicines are indicated and priced

52. From January 1, 2014, noting that concession cardholders (students, senior citizens, persons with disabilities, etc.) pay A$6.00, and that these amounts are adjusted in line with inflation on January 1 each year.

in Australia. The PBS Schedule sets out pricing information, product strengths, and the therapeutic indication or restriction for each product. The PBS Schedule is published on the DoH website[53] and contains information on the conditions of use for the prescribing of PBS medicines.

Without listing on the PBS, a decision to supply prescription products to patients in the retail market will not be commercially practical as patients will be unlikely to pay the full, unsubsidized price. This differs in the hospital market where prescription products are supplied to hospitals before they are listed on the PBS. The purchase of medicines at the hospital level is funded by the states and territories under a tender process, as opposed to the Australian government under the PBS.

1. Formularies

Although all medicines included on the PBS are described as "listed," the PBS is divided into two parts:

- F1 "formulary" (for patented, single-brand medicines that are not interchangeable on an individual patient basis with drugs that have multiple brands and are not single-brand combination items); and
- F2 "formulary" (for multiple-brand, or generic, medicines or single-brand medicines that are interchangeable at the individual patient level with multiple-brand medicines).

A new patented pharmaceutical medicine will be recommended for inclusion in the F1 category if it is assessed by the Pharmaceutical Benefits Advisory Committee (PBAC) to have better efficacy than off-patent comparators for the same clinical indication.

A drug is classified into one formulary only. Once a drug is allocated to a formulary, all pharmaceutical items containing that drug (with the exception of combination items) attract the consequences of the drug's allocation to that formulary. Any drug listed on F1 will move to F2 upon listing of a new bioequivalent or biosimilar brand of the drug that is listed. Once a drug moves to F2, it cannot move back to F1, even if all other brands are delisted.

Following reforms effective since December 1, 2010,[54] the key consequences of a drug being allocated to the F2 formulary are mandatory price reductions (for example, a significant one-off price reduction at the time of PBS listing of 16 percent of the first generic brand of a drug)[55]

53. *See Pharmaceutical Benefits Scheme (PBS)*, DEP'T HEALTH, http://www.pbs.gov.au (last visited May 20, 2014).

54. Pursuant to a Memorandum of Understanding between the Commonwealth of Australia and Medicines Australia, September 28, 2010.

55. *See* Section 99ACB of the NH Act.

and price disclosure requirements.[56] The data that need to be submitted to the Australian government for medicines in the F2 formulary include the sales revenue for the brand, the volume sold, and the kind and value of sales incentives offered to wholesalers. These data must be collected monthly and disclosed to the department semiannually. The intention of price disclosure is to allow the Australian government to pay a price closer to the actual ex-manufacturer price, taking into account any discounting, rebates, and incentives offered by the manufacturer to wholesalers or directly to pharmacies.

2. Listing under the PBS

An applicant must complete a series of steps to obtain PBS listing for a new product. First, an innovator sponsor must make either a major or minor submission to the PBAC in accordance with the "Guidelines for preparing submissions to the Pharmaceutical Benefits Advisory Committee."[57]

PBAC is an independent statutory committee established under the NH Act. The primary role of the PBAC is to recommend to the minister for health which drugs should be subsidized under the PBS. The minister for health can only approve the subsidy of a medicine in line with PBAC' s independent recommendation.

Submissions to list new drugs on the PBS or to make substantial changes to current listings are classified as major submissions. Examples include the following additions or changes to the PBS:

- list a new drug (including a new fixed-combination product, a new nutritional product, a new vaccine, or a new orphan drug);
- substantially change the listing of a currently restricted drug (including a new indication or a de-restriction);
- enable a review of the comparative cost-effectiveness of a currently listed drug in order to change a PBAC recommendation to the PBPA on its therapeutic relativity or price advantage; and
- list a new form (or strength) of a currently listed drug for which a price advantage is requested.

---

56. Products in the F1 formulary are not subject to these requirements but may be if they are moved to the F2 category because they lose patent protection or other interchangeable medicines are introduced. *See Price Disclosure (EAPD)*, DEP'T HEALTH, http://www.pbs.gov.au/info/industry/pricing/eapd (last visited May 20, 2014), for further information about the PBS price disclosure requirements.

57. DEP'T OF HEALTH (Dec. 2008), *available at* http://www.pbs.gov.au/industry/listing/elements/pbac-guidelines/PBAC4.3.2.pdf. The relevant application forms are available at *PBS Forms*, DEP'T HEALTH, http://www.pbs.gov.au/info/industry/useful-resources/pbs-forms (last visited May 20, 2014).

A minor submission relates to new forms of previously listed products or changes to the conditions of use.

The PBAC meets three times a year[58] to assess applications for listing on the PBS based on the clinical benefit and cost-effectiveness of the proposed drug compared with other drugs already listed on the PBS for the same or similar indications. If there is no listed alternative, PBAC considers the clinical place, overall effectiveness, cost, and cost-effectiveness of the proposed drug compared with standard medical care. A major submission will be required to undergo an economic evaluation by the Pharmaceutical Evaluation Section and the Economic Sub-Committee of the PBAC before consideration by the PBAC. Minor submissions usually do not require an economic evaluation.

It is possible for ARTG and PBS applications to be filed in parallel (although a PBS listing will not occur until after the ARTG registration). These parallel ARTG and PBS processes are the result of a Memorandum of Understanding entered into between the Australian government and Medicines Australia dated September 28, 2010.[59] Sponsors should be aware that some unapproved indications and some other indications may not be subsidized under the PBS.

Once a major submission is received by the PBAC, it is forwarded to the Pharmaceutical Evaluation Branch (PEB). PEB uses external advisers and internal subcommittees to conduct a detailed evaluation of both the clinical benefit and the economic value of the drug compared with other treatments or products for the same condition or use.

On the basis of this evaluation, the PBAC may recommend that the Pharmaceutical Benefits Pricing Authority (PBPA) determine an appropriate pricing arrangement for new products containing the new drug based on a technical evaluation in which the cost and efficacy of the new drug are compared to those of a current drug or treatment. The decision to subsidize an item will be considered by the Cabinet if the net cost to the PBS is greater than A$10 million per year.

A major submission to the PBAC can be a time-consuming and expensive process. A major submission typically requires more than 12 months of preparation, and the process for obtaining listing of a new drug or indication can take at least nine months from the date of submission to the time of listing in the PBS Schedule plus an additional period for those

58. The PBAC cycle time frame for 2013 is available at *PBS Calendars*, DEP'T HEALTH, http://www.pbs.gov.au/info/industry/useful-resources/pbs-calendar (last visited May 20, 2014).

59. Available at http://www.pbs.gov.au/info/industry/useful-resources/memorandum.

drugs or indications that which require Cabinet approval. The submission fee for a major submission can be found on the PBS website.[60]

3. Listing a Generic Brand

The process for a PBS listing of a new generic brand of an already listed drug is much simpler than the process for listing a new drug or indication. Generally, no consideration by the PBAC or the PBPA is necessary as a full PBAC evaluation has already been carried out for the innovator's product. Instead, generic submissions are forwarded by the PBAC Secretary to the Listings Unit of the PEB for consideration. Unlike the sponsor of an innovative product, the sponsor of a generic product is not required to present a full submission and pays a small submission fee.[61] The process typically takes two and a half to four months depending on whether the listing causes a price change.

Significantly, the sponsor of the generic product is not required to submit a clinical evaluation or an economic evaluation. Instead, it relies on the information submitted by the sponsor of the innovative product to the PBAC. The Listings Unit does not undertake a further review of the cost-effectiveness of the generic product (although this will be a key consideration if the sponsor later applies for PBS subsidization, as discussed earlier).

The sponsor of any application to list a generic product on the PBS Schedule is required to guarantee to the department that, for a period of 24 months, it will be able to meet any pharmacy orders for the new product within a reasonable time.[62] This "guarantee of supply" requirement is significant in the context of urgent applications for interlocutory injunctions.

As discussed, the listing of a generic product may in certain circumstances trigger a price reduction of other listed brands of that pharmaceutical item and will cause the drug to be subject to the PBS price disclosure system.

I. Intellectual Property Considerations

Australia has separate registration systems for patents, trade-marks, designs, plant breeders' rights, and circuit layouts—although they are all managed by the same statutory body, IP Australia. Copyright and other rights akin to intellectual property (such as duties of confidence and trade secrets) are not capable of formal registration but are firmly established as

---

60. *See Cost Recovery, Fees and Charges—Frequently Asked Questions*, DEP'T HEALTH, http://www.pbs.gov.au/info/industry/listing/elements/fees-and-charges (last updated June 22, 2012).

61. *See id.*

62. *See* part VII, division 3C of NH Act. *See also Guarantee of Supply*, DEP'T HEALTH, http://www.pbs.gov.au/info/industry/listing/elements/guarantee-of-supply (last visited May 20, 2014).

part of Australia's law by statute and/or in common law and equity. Each of these areas of law can have potential relevance to therapeutic goods.

1. International Framework

Australia is a signatory to international agreements on intellectual property, including the Agreement on Trade-Related Aspects of Intellectual Property Rights (TRIPS), AUSFTA, the Patent Cooperation Treaty (PCT), Paris Convention for the Protection of Industrial Property, the Budapest Treaty on the International Recognition of the Deposit of Microorganisms for the Purposes of Patent Procedure, the Berne Convention, and the Madrid Agreement Concerning the International Registration of Marks, which lend to shaping the domestic Australian intellectual property law.

2. Patents

Patent law in Australia is governed by the Patents Act 1990 (Cth) (Patents Act) and the Patents Regulations 1991 (Cth) (Patents Regulations).

Australia has both standard patents and innovation patents:

- A *standard patent* typically provides 20 years of protection for the invention that is the subject of the registered patent, although the Commissioner may cap the term (for procedural reasons) or extend the term past 20 years (as happens in the cases of some pharmaceutical substances), which is discussed in further detail later in this section.
- An *innovation patent*[63] provides eight years of protection. It is an efficient way of providing fast protection without prolonged preregistration examination or the need for establishing an inventive step (instead, the lower standard of an innovative step applies). An innovation patent offers the same level of protection as a standard patent (i.e., a statutory monopoly for the subject invention), albeit for a shorter period.

When a patent is granted, the patentee may seek to prevent others from using or producing the invention without the patentee's permission by an infringement action in the federal or state supreme courts (although the Federal Court of Australia is used almost exclusively). It is also possible to apply to the courts to revoke a patent if the invention disclosed in the patent is not adequately described, lacks novelty, was obvious to a person skilled in the art (or, for an innovation patent, not innovative) as of

63. The Advisory Council on Intellectual Property (ACIP) is currently undertaking a review into the innovation patent system to consider the effectiveness of the system in stimulating innovation by Australian small- to medium-business enterprises.

the date of application, is not useful for its intended purpose or application, or is not patentable subject matter.[64]

Significant amendments have recently been made to the Patents Act with the passing of the Intellectual Property Laws Amendment (Raising the Bar) Act 2012 (Cth) (Raising the Bar Act). The Raising the Bar Act received Royal Assent on April 15, 2012, and most provisions took effect from April 15, 2013. Some provisions (e.g., with respect to correcting entitlement issues) apply retrospectively to patents filed, examined, and/or granted under previous incarnations of the Patents Act, while other provisions (e.g., with respect to the prior art base for assessing inventive step) only apply to patents filed, examined, or granted since April 15, 2013. The effects of the amendments are generally to raise the threshold requirements for patentable inventions in Australia and to raise the requirements for description and disclosure in Australian patent specifications. The current Patents Act (incorporating the Raising the Bar Act amendments) is discussed in the following sections.

3. Pharmaceutical Patents

As discussed, the term of a standard patent is generally for 20 years but may be extended in some instances for pharmaceutical patents. Pharmaceutical extensions of term provisions were first introduced in the Patents Act. According to IP Australia, "the pharmaceutical patent extension scheme was introduced in recognition that a patent owner is unable to commercially exploit a patent until regulatory approval from the TGA is given and to partially compensate pharmaceutical patentees for delays associated with seeking regulatory approval for new medicines in Australia."[65] The Explanatory Memorandum to the bill that introduced these amendments noted that the intention was to provide an effective patent term from the date of marketing approval that was "more in line with that available to inventions in other fields of technology." Australia is also obliged to retain a system of extensions for pharmaceutical patents under AUSFTA.

Section 70 of the Patents Act provides that the patentee of a standard patent may apply to the Commissioner of Patents for an extension of the term of the patent if the following conditions are met:

- One or more pharmaceutical substances per se must in substance be disclosed in the complete specification of the patent and in substance fall within the scope of the claim or claims of that

64. Section 138 of the Patents Act.

65. *See* IP Austl., Pharmaceutical Patents Review: Background and Suggested Issues Paper (2012), *available at* http://www.ipaustralia.gov.au/pdfs/Background_and_Suggested_Issues_Paper_PharmaReview.pdf.

specification, and/or one or more pharmaceutical substances when produced by a process that involves the use of recombinant DNA technology must in substance be disclosed in the complete specification of the patent and in substance fall within the scope of the claim or claims of that specification;

- goods containing, or consisting of, the pharmaceutical substance must be included in the ARTG;
- the period beginning on the date of the patent (effective date) and ending on the first regulatory approval date for the substance must be at least five years; and
- the term of the patent must not have been previously extended.

An application for an extension of term of a standard patent must be made within six months of the patent being granted or of the first inclusion of the pharmaceutical in the ARTG, whichever is later.[66] If a patent meets these requirements, it can be extended for a maximum period of five years, taking the term of the patent to 25 years. The term of the extension is equal to the period beginning on the date of the patent and ending on the earliest first regulatory approval date in relation to any of the pharmaceutical substances minus five years.[67]

In *H Lundbeck A/S v. Alphapharm Pty Ltd* (2009) 177 FCR 151, the Full Court (Emmett, J., Bennett, J., and Middleton, J.) held that a term extension should be removed for Lundbeck's Lexapro product (an isolated (+)-enantiomer of citalopram) on the basis that a product registered earlier on the ARTG [i.e., Cipramil, which consists of the racemate citalopram (a substance that consists of equal proportions of the (+)- and (−)-enantiomers)] also contained the relevant enantiomer.

Although an extension of a patent may be granted, the exclusive rights of the patentee will be limited if the extension is granted. The Patents Act provides that if the Commissioner grants an extension of the term of a standard patent, the exclusive rights of the patentee during the term of the extension are not infringed:

- by a person exploiting a pharmaceutical substance per se that is in substance disclosed in the complete specification of the patent and in substance falls within the scope of the claim or claims of that specification or a pharmaceutical substance when produced by a process that involves the use of recombinant DNA technology, that is in substance disclosed in the complete specification of the patent and in substance falls within the scope of the claim or claims of that specification for a purpose other than therapeutic use; or

66. Section 71 of the Patents Act.
67. Section 77 of the Patents Act.

- by a person exploiting any form of the invention other than a pharmaceutical substance per se that is in substance disclosed in the complete specification of the patent and in substance falls within the scope of the claim or claims of that specification or a pharmaceutical substance when produced by a process that involves the use of recombinant DNA technology, that is in substance disclosed in the complete specification of the patent and in substance falls within the scope of the claim or claims of that specification.

Chapter 11, Part 1 of the Patents Act also provides exemptions from infringement for obtaining regulatory approval for patented pharmaceuticals.[68] Under these exemptions, a patent will not be infringed by a person who exploits the invention solely for the purpose of obtaining inclusion in the ARTG or connected with obtaining similar regulatory approval under a law of a foreign country or of a part of a foreign country.

The Australian government is currently conducting a review of pharmaceutical patents in Australia. On October 15, 2012, the Parliamentary Secretary for Industry and Innovation, the Honourable Mark Dreyfus QC MP, announced a review of the pharmaceutical patents regime in Australia.[69] The key objective of this review was to evaluate whether the current system for pharmaceutical patents effectively balances the objectives of securing timely access to competitively priced pharmaceuticals, fostering innovation, and supporting employment in research and industry. In particular, the extension of term provisions of the Patents Act were reviewed.

Following a submissions period, a draft report of the pharmaceutical patents review was released in April 2013 and was the subject of substantial comment from across the Australian pharmaceutical and patent professional communities. The final report was due for release in May 2013, but has still not been published by IP Australia.[70]

4. Enforcement

Under Australian patent law, the patentee and/or an exclusive licensee (i.e., someone who is authorised to exploit the invention in Australia to the exclusion of the patentee and all other persons) may commence infringement proceedings. If the exclusive licensee commences infringement proceedings alone, the patentee must also be joined as a party. However, where the patentee and exclusive licensee are related companies, it is

68. Section 119A.

69. The terms of reference of the review are available at http://www.ipaustralia.gov.au/about-us/ip-legislation-changes/review-pharmaceutical-patents/.

70. As of July 8, 2013.

usual for them to commence proceedings as co-applicants, rather than joining the patentee after the proceedings have commenced.

Infringement and validity are dealt with simultaneously in the same proceeding. Liability is generally split from an assessment of any financial remedies, which is heard in a separate proceeding (as necessary). A case will typically come to trial 15 to 18 months after issuing proceedings, depending on the court's calendar and the attitude of the parties (i.e., the proceeding may be given a speedy timetable and come to trial within 12 months). In addition, a patentee must certify that the proceeding is commenced in good faith, has reasonable prospects of success, and will be conducted without unreasonable delay.

The relief that a court may grant for the infringement of a patent includes an injunction and, at the option of the applicant, either damages or an account of profits.[71] A patentee may now also claim exemplary and punitive damages in certain circumstances (although this is not the same as U.S.-style "treble damages").

- A permanent injunction prohibiting the generic from making, selling, importing, exporting, or stockpiling generic product until after the expiry of the patent is a remedy commonly awarded by the Federal Court of Australia upon making a finding of infringement of one or more valid claims of an Australian patent. An injunction may be subject to such terms, if any, as the court thinks fit.
- An applicant may also, in appropriate circumstances, obtain an interlocutory injunction at an early stage in infringement proceedings. Broadly, in Australia there is no presumption in favor of the grant of an injunction to a patentee. To obtain a preliminary injunction restraining infringement, a patentee must establish that:
  - o there is a serious question to be tried as to the patentee's entitlement to final relief;
  - o the patentee is likely to suffer loss for which damages will ultimately not be an adequate remedy, typically by that the alleged infringer's presence in the market will have effects that cannot be measured in money terms, such as non-recoverable price drop, or employees of the patentee losing their jobs (although convincing evidence, rather than mere assertion, would be required); and
  - o the balance of convenience favors the granting of a preliminary injunction, taking account of the hardship to either party or even a third party, the risk of irreparable damage or injury, the patentee's delay, how long the patents have stood undisturbed,

71. Section 122(1) of the Patents Act.

> the sufficiency of the patentee's undertaking to pay damages to compensate the generic for being kept out of the market if the patent is ultimately invalidated, and the need to protect the goodwill of the patentee.

More recent decisions in pharmaceutical patent cases have granted interlocutory injunctions in favor of patentees restraining potential infringers so as to preserve the status quo, but these have related mainly to the operation of the PBS price drop. This trend works in an innovator/patentee's favor as, where it applies for a preliminary injunction to restrain threatened or actual infringement of a pharmaceutical patent for a listed product, it will be seeking to prevent a downward-pricing trigger caused by the listing of a generic on the PBS (resulting in the innovator's product moving from the F1 to the F2 formulary). Therefore, if applied for and granted early enough, a preliminary injunction can potentially prevent a generic from even obtaining a PBS listing, and thereby prevent the PBS price reductions for the patentee's products, which are irreversible. The patentee must offer the court an undertaking as to damages (discussed next) as the "price" of the preliminary injunction. This undertaking extends not only to the generic but also, in theory, "to any person, whether or not a party, adversely affected by the operation of the interlocutory order," potentially including the Australian government.[72]

- If the TGA has been provided with a 26B Certificate as described earlier, the patentee must provide its own certificate before it can, for example, seek a preliminary injunction from a court that prohibits the generic from exploiting (for example, manufacturing or selling) the drug in Australia until the patent infringement proceedings have been decided.
- Section 26C of the TG Act requires the patentee to certify, prior to the commencement of any proceeding, that the proceedings:
  - are to be commenced in good faith; and
  - have reasonable prospects of success; and
  - will be conducted without unreasonable delay (Patentee Certificate).[73]

The TG Act provides that proceedings have reasonable prospects of success if the patentee had reasonable grounds in all the circumstances known to the patentee, or which ought reasonably to have been known to

72. Federal Court of Australia Practice Note CM14 – Usual Undertaking as to Damages, *available at* http://www.fedcourt.gov.au/law-and-practice/practice-documents/practice-notes/cm14/practice_notes_cm14.rtf.

73. *Therapeutic Goods Act 1989* (Cth) ss 26C(2) and (3).

the patentee (in addition to the fact of grant of the patent), for believing that he or she would be entitled to be granted final relief by the court for infringement of the patent, that the patent is valid, and that the proceedings are not otherwise vexatious or unreasonably pursued. Once issued, the patentee can then commence proceedings in the Federal Court or a state supreme court for a preliminary injunction against the generic.

Further, Section 26D of the TG Act affects a patentee who has received notice under Subsection 26B(1)(b) and who then obtains a preliminary injunction. If the principal proceedings are discontinued or dismissed, in the absence of reasonable grounds (in addition to the grant of the patent) to believe that the claims had a reasonable prospect of being held to be valid and that final relief would be granted for infringement, Subsection 26D(5) exposes the patentee to an award of damages to the Commonwealth (maximum of A$10 million), as well as to an award of compensation to the applicant on the basis of an account of gross profits, without requiring the applicant to quantify actual loss. To date, there have been no prosecutions for providing a false certificate.

5. Preliminary Discovery

Where an applicant does not have all the relevant facts it requires to commence proceedings, it may be possible to make an application for preliminary discovery. The Federal Court of Australia may make orders for preliminary discovery for the purpose of identifying a respondent[74] or determining whether there is a cause of action against a known party.[75] These type of orders are discretionary in nature and are often difficult to obtain.

Rule 7.23 of the *Federal Court Rules 2011* (Cth) sets out the requirements for making an application to obtain discovery from a prospective respondent. In summary, a prospective applicant may apply to the Court if it:

- reasonably believes that it may have the right to obtain relief in the Court from a prospective respondent whose description has been ascertained; and
- after making reasonable inquiries, does not have sufficient information to decide whether to start a proceeding in the Court to obtain that relief; and
- reasonably believes that the prospective respondent has or is likely to have or has had or is likely to have had in the prospective respondent's control documents directly relevant to the question of whether the prospective applicant has a right to obtain the relief

74. *See* rule 7.22 of the Federal Court Rules 2011 (Cth).
75. *See* rule 7.23 of the Federal Court Rules 2011 (Cth).

> and inspection of the documents by the prospective applicant would assist in making the decision.

In the case of *Wyeth v. Secretary, Department of Health and Ageing* [2009] FCA 313, Wyeth was unsuccessful in obtaining preliminary discovery as it did no more than establish a possibility of success at a final hearing. Wyeth's basic position in relation to the interests of justice was that unless it could obtain preliminary injunctive relief four to six weeks before a PBS listing of the generic, it may be exposed to irreparable harm for which damages would not be an adequate remedy, if the generic's product were listed. The nub of the issue was whether it was in the interests of justice to make an order for identity discovery (i.e., provide the names) or wait until the potential listing on the ARTG. The Federal Court held that a patentee, in those circumstances, must show that the making of the order is necessary to provide an effective remedy with respect to the "actionable wrong." This was not made out in this case as the delay was not considered long enough.

Comparable "preliminary discovery" actions are also available under the court rules of each Australian state and territory.[76]

6. Freedom of Information

An alternative to preliminary discovery that has been recently (unsuccessfully) tested is for patentees to seek to obtain information about generics applying for regulatory approval to the TGA via Australia's Freedom of Information Act 1982 (Cth) (FOI Act). However, the decision in *Secretary Department of Health and Ageing v. iNova Pharmaceuticals (Australia) Pty Ltd* [2010] FCA 1442 demonstrates that the FOI Act offers little assistance to innovators seeking to "open the window" between the innovator learning of an ARTG registration and the generic making an application for PBS listing, and obtain information about ARTG applications before they are granted and made public. The main issue in *iNova* was whether the DoH was also entitled to withhold the mere fact that exempt documents exist. Disclosure of this fact is useful information to innovators, because it informs them that an ARTG application has been made and enables them to start preparing for generic entry. Innovators may even be able to deduce the identity of the generic using their knowledge of the market. The Court found that the DoH is entitled to withhold the fact that exempt documents exist, if disclosing that fact in a notice issued to the innovator would make the notice an exempt document in one of the preceding categories. The

76. *See* Supreme Court Rules (NT) O 32; Uniform Civil Procedure Rules 1999 (QLD) pts 1, 2; Supreme Court Civil Rules 2006 (SA) r 146; Supreme Court Rules 2000 (TAS) r 403E, 403F; Supreme Court (General Civil Procedure) Rules 2005 (VIC) O 32, Rules of the Supreme Court (WA) O 26A.

Court held that disclosing the existence of documents could reasonably prejudice the enforcement or proper administration of law if, applying an objective test, a person receiving that information could identify the generic entrant and prevent PBS listing, thereby prejudicing the administration of the PBS.

The difficulty with obtaining information about pending ARTG applications makes it critical for innovator companies and their advisers to monitor regularly new registrations on the ARTG. Now that applications for PBS listing can occur in parallel with ARTG applications, it is also important for innovator companies to monitor the published agendas of PBAC. Generics continue to enjoy the confidentiality of their ARTG applications. However, if generics decide to take advantage of parallel ARTG and PBS processes, then they should be mindful that their impending market entry could be disclosed in a PBAC meeting agenda published on the PBS website.

7. Patent Revocation Proceedings

Under Section 138 of the Patents Act, any person may apply to a prescribed court for an order revoking a patent. Under Section 139 of the Patents Act, the patentee and any exclusive licensee are parties to any revocation proceedings.

As an alternative to commencing revocation proceedings, a cross-claim for revocation may be brought by an alleged infringer against the patentee for revocation of the patent, following the commencement of infringement proceedings. It is common for an alleged infringer to do so if the grounds for invalidity may exist. Procedurally, a cross-claim is a parallel proceeding to the principal proceeding and is heard and determined at the same time as the principal proceeding.

8. Other IP Considerations

In addition to patents, all pharmaceutical companies operating in Australia are entitled to the protection of names and other key intellectual property, such as copyright, designs, and trade-marks. Confidential information and trade secrets are also protected by common law action (the law evolved through decisions of the courts) for breach of confidence.

9. Trade-marks, Trading Style, and Trade Names

Rights in trading styles and trade names are governed by three areas of law in Australia:

- national company and business names legislation, including the Corporations Act 2001 (Cth), Business Names Registration Act 2011 (Cth), Business Names Registration (Transitional and

Consequential Provisions) Act 2011, and Business Names Registration (Fees) Act 2011;
- registration of trade-marks under the Trade Marks Act 1995 (Cth) and actions for infringement of trade-marks registered under that act; and
- action for misleading or deceptive conduct or false representations under the ACL, as well as the common law action for passing-off (as discussed earlier).

Registration of company and business names merely complies with legislative requirements to register a name under which business is conducted by a person or entity trading in Australia under a name that is not the name of that person or trading entity. Registration conveys no proprietary right to the name (although common law rights to the name arise from its use).

In addition to the preceding points noted in relation to the ACL and passing off, companies supplying therapeutic goods in Australia should be particularly careful in adopting brand names or trade-marks to ensure they are not infringing the registered or unregistered rights of a competitor. This is especially relevant in relation to generic pharmaceutical products, where generic brand names often mimic the trade-mark of the innovator product rather than the generic pharmaceutical compound name. Searches of the Australian Register of Trade Marks[77] and the Internet are straightforward and always advisable.

10. Copyright in Product Labels

The courts have held that copyright subsists in the product information that is supplied to the TGA as part of the registration process, but for acts after May 28, 2011, this does not amount to copyright infringement. For example, in the case of *Sanofi-Aventis Australia Pty Ltd v. Apotex Pty Ltd* [2011] FCA 846, Justice Jagot held that copyright subsisted in the product information for Arava (Sanofi-Aventis's own leflunomide product) and that Apotex had infringed Sanofi-Aventis's copyright (which Sanofi-Aventis submitted had been infringed by the generic Apotex reproducing the whole or substantial part of the Arava product information in its own leflunomide product information).

There is now an exception in the Copyright Act 1968 (Cth)[78] that allows the copyright in TGA-approved product information to be reproduced/copied without constituting an infringement of copyright. The exception provides

77. IP Austl., http://pericles.ipaustralia.gov.au/atmoss/falcon.application_start (last visited May 21, 2014).

78. Section 44BA(1) of the Copyright Act.

that the copyright subsisting in TGA-approved product information is not infringed by a direct or incidental act that is undertaken to register a medicine or restricted medicine under the TGA (done on or after May 27, 2011). In addition, the supply, reproduction, publication, communication, or adaptation of some or all of TGA-approved product information on or after May 27, 2011, does not constitute an infringement of the copyright subsisting in the product information, if the product information is used for a purpose related to the safe and effective use of a TGA-approved medicine.[79] Acts incidental to these acts also do not infringe copyright.[80]

As a result, generic pharmaceutical companies are able to use innovator product information that has already been approved by the TGA without infringing the copyright of the original product information. These provisions are recently new to the Copyright Act as they were only introduced with the enactment of the *Therapeutic Goods Legislation Amendment (Copyright) Bill 2011* (Cth).

III. Brazilian Drug Approval Considerations[81]

In Brazil, pharmaceuticals are governed by a comprehensive and complex regime of legislation and regulations spanning many different areas of law. The legislative and regulatory landscapes are also very dynamic, as patent laws are constantly under review and government authorities constantly update regulatory processes and policies.

On the Brazilian health regulatory system, introduced by Act No. 6.360/76, any drug may be marketed only if it has been previously registered with the pertinent authority of the Ministry of Health; such competence was attributed to Brazilian Agency of Sanitary Surveillance (ANVISA) by Law No. 9.782/99. The Brazilian patent system is ruled by the Brazilian Patent Office created by Federal Law No. 5.648/70 and its authority is defined by Brazilian Industrial Property Law (No. 9.279/96).

This chapter provides an overview of each element in this complex Brazilian regime together with insights into the issues that might arise and how they might be navigated.

79. Section 44BA(2) of the Copyright Act.

80. Section 44BA(3) of the Copyright Act.

81. Gustavo de Freitas Morais, Rodrigo Augusto Oliveira Rocce, and Vyctor Taddeucci de Araujo of Dannemann Siemsen Bigler & Ipanema Moreira, Sao Paulo, Brazil.

A. Patent Portfolio

1. Patentable Subject Matter in the Pharmaceutical Sector

Article 10, Sections VIII and IX, of the Brazilian Industrial Property Law (No. 9.279/96) establishes a prohibition on the patenting of: (1) operation or surgical techniques as well as therapeutic or diagnostic methods, for use on the human or animal body; and (2) natural living beings, in whole or in part, biological material including the genome or germ plasm of any natural living being, when found in nature or isolated therefrom, and natural biological processes.

There are three major types of pharmaceutical patents: Compound Patents, which protect the active ingredient; Formulation Patents, which cover the drug composition; and Process Patents, which protect the drug manufacturing process.

When a pharmaceutical patent application is directed toward "strategically important drugs and therapeutic destinations to the Brazilian Universal Healthcare System" as listed by the Ministry of Health, the patent is subject to the prior consent of the Brazilian health regulatory body called the Brazilian Health Surveillance Agency (ANVISA). If this occurs, acceptable patentable matter may be questioned and prior consent may be denied.

In these circumstances, the rejection of a patent application is grounded on an opinion that Swiss-type, second-use, polymorph, and selection patents present a health risk by creating barriers to the entry of generic drugs and thus compromising the Brazilian Universal Healthcare System (SUS).

2. Prior Consent and the Role of ANVISA in the Pharmaceutical Patent Application Process

The granting of patent applications for pharmaceutical processes and products requires prior consent from ANVISA.[82]

Resolution RDC 21/2013, which altered Resolution RDC 45/2008, regulates ANVISA's prior consent of pharmaceutical process and product patent applications. This Resolution is the result of the Public Consultation[83] by ANVISA.

As established in Resolution RDC 21/2013, ANVISA will not grant prior consent for pharmaceutical process or product patent applications when the applications contain a prohibited substance.[84] Furthermore, if the subject of the patent application is listed as a strategic drug for SUS

82. Article 229-C incorporated into the Brazilian Industrial Property Law by Brazilian Law No. 10,196/01.

83. Public Consultation 66/2012 by ANVISA and a meeting that was held by ANVISA on March 20, 2013.

84. Currently prohibited substances are those listed in List F of Ordinance 344/1998.

in Ordinance 3.089/2013,[85] then ANVISA will review patentability requirements according to Brazilian Intellectual Property Law.

For example, any antiviral or antiretroviral drug will be subjected to a substantive patentability analysis before ANVISA because antiretrovirals are a general category of drugs considered strategic under Ordinance 3.089/2013.

In summary, according to ANVISA's Resolution currently in force:

1. the prior consent for pharmaceutical patent applications relating to subject matter that represents a risk to the human health (i.e., drug substance)[86] will be denied;
2. pharmaceutical patent applications related to a substance or a therapeutic destination considered strategic for SUS[87] will undergo a substantive patent analysis by ANVISA based on the provisions of Brazilian Intellectual Property Law;[88] and
3. any pharmaceutical patent applications not related to a prohibited substance, to a substance, or to a strategic therapeutic destination for SUS will not undergo a substantive examination for the requirements of patentability by ANVISA.

B. Clinical Trials

1. Legal Framework

The legal framework concerning clinical trials in Brazil relates mainly, but not restrictively, to Resolution No. 466/2012 of the National Council of Health (CNS). This Resolution establishes the guidelines for conducting clinical trials. This legal framework emphasizes the main ethical aspects, the institutional ethics committee (CEP) attributes, and the National Commission for Ethics in Research (CONEP) attributes. It also lists the contents of Informed Consent Forms (ICF), protocols, and brochures.

Resolution RDC 36/2012 of ANVISA[89] describes the list of documents and procedures required for the approval of clinical research. All clinical Phase 1, 2, 3, and 4 trials must also show proof of registration with the Brazilian Registry of Clinical Trials (ReBEC) database. For cases filed prior to publication of the Resolution, it is acceptable to show proof of registration of the research that was previously recorded in other primary records of the International Clinical Trials Registry Platform (WHO ICTRP).

85. Ordinance 3.089/2013 may be updated from time to time.

86. These substances are listed in List F of Ordinance 344/1998 of the Brazilian Ministry of Health.

87. These are listed in Ordinance 3.089/2013 from the Brazilian Ministry of Health.

88. This analysis is a substantive examination of the requirements of patentability.

89. Resolution RDC 36/2012 amended Resolution RDC 39/2008.

2. Regulatory Pathway for Conducting Clinical Trials in Brazil

Brazilian regulatory approval follows a sequential process in which the first step is the translation of the study and/or its submission into Portuguese.[90]

The first ethical approval must be released by the CEP of the coordinating site. This release is required because it is one of the requirements for submission to CONEP. All trials supported by foreign sponsors require an additional ethical approval from CONEP, whose responsibilities include developing regulations for the protection of subjects in clinical trials.

In terms of coordinating the institutional CEP network, CONEP evaluates protocols relating to human genetics and reproduction, new drugs, procedures, devices, vaccines, and research that involves international cooperation. CONEP reviews the documentation from the coordinating site only. Once the approval is issued, it is extended to the other sites participating in the study.

Last, all clinical protocols carried out in Brazil must be approved by ANVISA. ANVISA is responsible for issuing the import license and the Special Communicate.[91] ANVISA also evaluates protocol methodological issues and the relevance of data for future submissions.

For studies sponsored by multinational companies, ANVISA analyzes the clinical trial information. ANVISA will only issue its approval following issuance of approvals from the CEPs and CONEP. Once satisfied that the approval should issue, ANVISA issues the Special Communicate and then the import license.

All therapeutic activities requested for the pharmaceutical product to be registered must be supported by clinical trial reports. Such clinical trials must be approved by the health authority of the country where the clinical trial was conducted. The clinical trials must also have been conducted with the finished pharmaceutical product presented for registration.

3. Regulatory Submission of Drug Application

In the Brazilian health regulatory system, Act No. 6.360/76 requires that a drug may only be marketed if it has been previously registered with the Ministry of Health:[92]

> Article 12—No product to which this Act refer[s], including the imported ones, may be manufactured, marketed or released before registration with the Ministry of Health[93]

90. The documentation refers to the translated dossier, including the protocol, investigator brochure, informed consent form, and sponsor and institutional declarations, which are sent to each site's institutional ethics committee (CEP) for review.

91. The Special Communicate is the official approval document.

92. This requirement is pursuant to ANVISA Bylaw No. 9.782/99.

93. Act No. 6.360/76.

The registration, or market approval, issued by ANVISA is the effective authorisation for manufacturing and marketing of a drug in Brazil. ANVISA issues market approval for the following kinds of drugs: (1) chemical drugs, which are divided into (a) reference drugs, (b) similar drugs, and (c) generic drugs; (2) biological products, which are divided into (a) new biological products and (b) biological products; and (3) herbal medicines.

4. Legal Framework

There are three market approval categories for chemical drugs: (1) reference drugs, (2) branded generic drugs, and (3) non-branded generic drugs.

A *reference drug*[94] is defined as the "innovative product registered with the Federal authority responsible for the health surveillance and marketed in the Country, whose efficacy, safety and quality were scientifically proven before the pertinent federal authority, by the time of the registration."

A *branded generic drug* is that which "contains the same active ingredient(s), has the same concentration, dosage form, administration route, dosage administration and therapeutic recommendation, it is equivalent to the drug registered with the federal authority responsible for the health surveillance, and it may differ only in characteristics relative to the product size and form, expiration term, packaging, labeling, excipients and vehicles, and it must always be identified by trade name or brand."

A *non-branded generic drug* is a "drug similar to a reference or innovative product, that is intended to be interchangeable, usually manufactured after the expiration or waiver of the patent protection or other exclusivity rights, with proven efficacy, safety and quality, and assigned by DCB or DCI, when the first one is absent." The similarity between the generic drug and reference drug is proven by pharmaceutical equivalence study results and relative bioavailability/bioequivalence study results.

Pharmaceutical equivalents[95] "are drugs that contain the same drug, that is, the same salt or ester of the same therapeutically active molecule, in the same quantity and dosage form, that may or may not contain identical excipients. They must comply with the same updated specifications of the Brazilian Pharmacopoeia and, in their absence, with those of other codes authorised by the legislation in force, or with other applicable quality standards related to the identity, dosage, purity, strength, content uniformity, time of decomposition and dissolution speed, if applicable."

94. Act No. 9.787/99.
95. ANVISA Resolution RDC 16/2007.

Bioequivalent drugs[96] are "pharmaceutical equivalents that, when given at the same molar dose, under the same experimental conditions, do not have statistically significant differences in bioavailability."[97]

The registration of biological products is regulated by ANVISA Resolution RDC 55/2010. Biological products are defined as drugs that have, as their active ingredient: (1) molecules extracted directly from microorganisms, organs, or tissues of animal origin, or cells or fluids of human or animal origin ("biological origin"); or (2) molecules produced by the process of genetic modification ("biotechnological origin").[98]

Drugs considered to be biological products are: (1) vaccines; (2) hyperimmune serum; (3) blood derivatives; (4) biodrugs, including (a) drugs obtained from biological fluids or animal tissues, and (b) drugs obtained from biotechnology procedures; (5) monoclonal antibodies; and (6) drugs containing live, attenuated, or dead microorganisms.[99]

RDC 55/2010 makes a distinction between "new biological products," which are biological products that have not previously been registered in Brazil, and "biological products," which contain a molecule with known biological activity that has previously been registered in Brazil.

For *new biological* drugs, the applicant must submit to ANVISA a dossier containing: details about the company; drug data (composition, technical information regarding the active ingredient, shelf life, quality control, etc.); and reports of preclinical and clinical trials (Phases 1, 2, and 3) to prove the quality, safety, and efficacy of the new biological drug.

There are two ways of obtaining market approval for *follow-on biological products*: (1) by the Individual Development route; and (2) by the Comparability route. By the Individual Development route, the applicant must submit reports of preclinical and clinical trials. The Phase 3 clinical trial must be comparative (non-inferiority, clinical equivalence, or superiority). The Phase 1 and 2 clinical trials need not be comparative. By the Comparability route, the applicant must present a report proving that its product is comparable to the comparator product. This report must contain information regarding the comparison of the two products at all development stages: manufacturing of the molecule, comparative analysis between the products, stability, purity, impurity profile, and so on. The applicant must also provide nonclinical trial reports designed to detect significant differences between the biological product and the comparator product. Furthermore, the applicant must file reports of: (1) pharmacokinetics studies;

96. *Id.*

97. According to Law No. 9.787—XXV—bioavailability indicates the velocity and extension of an active ingredient absorption in a dosage form, from its concentration/time curve in the systemic circulation or its excretion in the urine.

98. ANVISA Resolution RDC 55/2010.

99. *Id.*

(2) pharmacodynamic studies; and (3) pivotal studies of safety and efficacy. Such studies must also be comparative with the comparator product.

In order to obtain market approval in Brazil for both chemical and biological drugs, it is necessary to submit a dossier to ANVISA proving that the product meets the standards for quality, safety, and efficacy as defined by sanitary law. The basic requirements for obtaining market approvals in Brazil are discussed in a later section.

It is worth highlighting that only the reference and the non-branded generic drugs are interchangeable—that is, the patient is able to choose whether he or she will acquire the reference or the non-branded generic drug if the physician prescribes it by the reference brand or by its International Nonproprietary Name (INN). On the other hand, the branded generic product can be acquired only if it was prescribed by its brand. It cannot be changed at the point of sale for the non-branded generic.

Furthermore, although Resolution RDC 55/2010 does not address interchangeability, the Brazilian sanitary law states that only the chemical drugs (reference and non-branded generic) are interchangeable. Therefore, considering that a biological product cannot be considered a generic of a new biological product, both are not interchangeable.

5. Reference Drugs

The registration of new drugs is regulated by Resolution RDC 136/2003. An applicant for a new drug must submit to ANVISA a dossier containing: information related to the company; information related to the drug (composition of the drug, technical information regarding the active ingredient, shelf life, etc.); and reports of preclinical and clinical trials, in order to prove the quality, safety, and efficacy of the new drug.

Preclinical trials required by RDC 135/2006 include acute, subacute, and chronic toxicity studies; reproductive toxicity studies; mutagenic activity studies; and oncogenic potential studies. The applicant must then conduct Phase 1, 2, and 3 clinical trials. The applicant must also present a copy of the Good Manufacturing Process (GMP) Certificate issued by ANVISA to the manufacturing facility and, if applicable, the local labeling site.

6. Generic Drugs

The registration of non-branded and branded generics is regulated by Resolutions RDC 16/2007 and 17/2007, respectively.

It is not necessary to perform clinical trials in order to prove the safety and efficacy of branded or non-branded generics. The rationale is that clinical trials have already been performed on the reference drug.[100]

100. *See* Law No. 9.787/99 (generic drugs); RDC 16/2007, 17/2007.

Instead, the applicant must prove that its branded or non-branded generic drug is bioequivalent to the reference drug. If the applicant can prove bioequivalency, ANVISA will assume that the branded and/or non-branded generic drug is safe and effective by relying on the clinical data that was evaluated during the registration of the new drug.

An applicant seeking approval of a generic drug must therefore submit to ANVISA a dossier which includes information related to the company; details of the drug (composition of the drug, active ingredient, technical information, shelf life, etc.); and reports of relative bioavailability/bioequivalence studies. The applicant must also submit GMP Certificates for the manufacturing facility and local labeling site.

7. Biological Products

Biological drugs that have been manufactured in other countries will only be registered in Brazil if the products have market approval in those other countries.

To register a biological product, the applicant must prepare a dossier indicating: the name of the manufacturer; the country of manufacture of the active ingredient(s), the bulk biological product, the biological product in its primary package, and the finished biological product; the quality control tests conducted on the active ingredient, the bulk biological product, and the finished biological product batches; the site where the respective quality control tests will be conducted; and the product specifications.

The applicant must also indicate the active ingredient manufacturer name if the applicant is not the manufacturer. If the active ingredient manufacturer changes, the applicant must request a new registration and may not use the brand name of the previously registered biological product, except when:

- there is a technology transfer to the new company, which must be duly proven by providing the technology transfer agreement between the companies involved; or
- the applicant proves through comparative studies that the drug properties of the finished biological product are unchanged after the change of manufacturing facility.

In the same way, if the applicant is not the manufacturer of the biological product in its primary package, the manufacturer must be stated in the application. If the manufacturer of the biological product in its primary package changes, a new product registration must be requested, except when comparative studies can show that the properties of the finished biological product (safety and activity) remain unchanged after the change in manufacturer.

C. Strategies

1. Patent Enforcement

Once a generic is granted regulatory approval and is marketed or manufactured, if the drug infringes another person's patent, it is possible to file an ordinary action for infringement under Article 42 of the Industrial Property Law:

> Article 42. The patent grants to the holder the right to prevent a third party, without its consent, from producing, using, marketing, selling or importing for these purposes:
>
> I—product subject to patent;
>
> II—process or product obtained directly by patented process.

In addition to preventing the alleged infringer from marketing and/or manufacturing the drug, it is possible for the patent holder to obtain damages incurred by virtue of the infringement.[101]

It is worthwhile to emphasize, however, that a lawsuit for damages for industrial property (IP) rights expires after five years.[102]

In accordance with Article 209, Paragraph 1 of the Industrial Property Law (IPL), "the Judge may determine through a preliminary injunction, on the case records of the lawsuit, to stop the act or violation, before the defendant summons, upon deposit in cash or guaranty, as deemed necessary, to avoid irreparable damage or damage of difficult indemnification."

Accordingly, once the requirements necessary for granting a preliminary injunction are met,[103] the judge may order that the acts that violate the patent be halted.

2. Data Protection

There is no data protection for pharmaceuticals in Brazil. In practice, ANVISA will not respect any period of non-reliance after the issuance of a market approval for the innovator company. With regard to pharmaceutical products for human use, Brazilian law *does not* provide for specific RDE (Registration Data Exclusivity) periods or periods of non-reliance during which third parties cannot obtain a market approval by referring to the originator's data.

In any case, a registration dossier should only be presented to ANVISA with a notice that it is secret and should not be used or relied upon in order to facilitate the issuance of market approvals to unauthorised third parties.

---

101. Industrial Property Law Article 209.
102. Industrial Property Law Article 225.
103. This is provided for in Article 273 of the Brazilian Code of Civil Procedure.

Moreover, a lawsuit seeking to prevent the issuance of market approval to unauthorised third parties stands a good chance of success in view of: (1) Section 195.XIV of the IPL (which provides that "a crime of unfair competition is committed by he who . . . divulges, exploits or uses, without authorisation, the results of tests or other undisclosed data the elaboration of which involved considerable effort and which has been presented to government entities as a condition for approving the commercialization of products"); (2) the concept of unjust enrichment; and (3) the fact that data protection exists for veterinarian and agrochemical products under By-law No. 10,603/02.

3. Linkage Actions

In practical terms, there is no linkage system in Brazil. ANVISA, Brazil's regulatory agency, will issue market approvals to branded and non-branded generics even if there are patents in force in Brazil. If the generic product is launched (which, so far, rarely occurs), a patent infringement lawsuit may be initiated. There is no official register in which patents covering a given product are listed and will be checked by ANVISA.

However, there is at least one court case wherein the market approval of the generic manufacturer has been suspended through a preliminary injunction due to the existence of a patent in force in Brazil. Although it is an agrochemical case, its rationale would apply to pharmaceuticals. Article 43.VII of the IPL, which states that patent rights do not apply to "acts practiced by unauthorised third parties related to the invention protected by a patent, for the sole purpose of producing test results, information and data in order to obtain the commercialization registration in Brazil or abroad," was interpreted to relate only to the testing aimed at obtaining market approval. The issuance of the registration itself amounted to a violation of the patent. As a result, it may be able to prevent non-authorised generics from obtaining market approval.

4. Extrajudicial Measures

It may also be possible to solve a patent infringement dispute without the need of the judicial system through the use of an extrajudicial notice, which is provided for in Article 867 of the Brazilian Code of Civil Procedure:

> Article 867. Whoever wants to avoid responsibility, to provide conservation and reservation of his/her rights or formally express any intention, can make his/her complaint in writing, in a petition addressed to the judge, and request him/her to order the appropriate people.

Notices are used to warn the violator of the infraction and can be faster than and equally effective as a patent infringement action.

5. Possible Patent Term Extension for a Pharmaceutical Product

Brazilian IP Law[104] does not provide for patent term extension. The term of a regular patent is defined by Article 40 of the IPL as 20 years from the filing date or 10 years from grant, whichever is longer.

Only pipeline (or revalidation) patents, the corresponding applications of which were filed between May 1996 and May 1997 (under the provisions of Article 230 of the IPL), may have their term harmonised with that of the corresponding ground patent. In other words, if a Brazilian pipeline patent is grounded in a foreign patent, the patent is extended via a Supplementary Protection Certificate. The term of the pipeline patent may be harmonised with the new extended term in the foreign country. However, a lawsuit is required to implement such a harmonisation. Recent decisions issued by the Brazilian Superior Court have refused to harmonize a Brazilian pipeline patent with its corresponding foreign patent.

IV. Canadian Drug Approval Considerations[105]

Section IV is a general overview of the regulatory steps and procedures related to drug approval in Canada.

A. Introduction on Coordinating NDS/ANDS Filing Strategy in Canada

In Canada, a company seeking authorisation to sell a new drug product does so by filing a New Drug Submission (NDS) with the Health Products and Food Branch (HPFB) of Health Canada. An NDS contains scientific information regarding product safety, efficacy, and quality, including results of preclinical and clinical studies. A manufacturer wishing to sell a generic version of a previously approved drug product must file an Abbreviated New Drug Submission (ANDS). The ANDS provides data showing the generic drug to be of equivalent safety and efficacy as the reference product, usually by bioequivalence studies. Successful submissions will result in approval for a Notice of Compliance (NOC) issued by the minister of health. An NOC will allow the manufacturer to sell the generic drug in Canada.

The Patented Medicines (Notice of Compliance) Regulations (PM(NOC) Regulations) provides the link between patent protection and the NOC. An innovator must submit a patent list with respect to a new

104. Law No. 9.279/96—IPL.

105. Isabel Jaen Raasch, John E. Norman, and Scott E. Foster and Darren Blimkie of Gowling Lafleur Henderson LLP, Ottawa, Ontario, Canada with assistance from James Holtom (intern).

drug at the time of filing for an NDS or a Supplementary NDS (SNDS). Any issued patent covering the drug submitted after NDS filing will not need to be addressed by a generic company seeking NOC approval. A patent application filed prior to the NDS filing date that has yet to issue may still be added to the patent list with respect to that drug if it is filed within 30 days of patent issuance. A manufacturer wishing to market a generic version of an approved drug is able to "freeze" the Patent Register upon ANDS filing. Any innovator patents that have yet to issue and be added to the patent list with respect to that drug will not need to be addressed by the generic company seeking to gain marketing approval from the minister.

In addition, innovators may insulate against generic reliance on innovator-submitted data (data protection) in support of regulatory approval for new innovative compounds for a maximum period of eight years and six months from NOC issuance. This includes six years of data exclusivity, two years of market exclusivity, and a possibility of a six-month extension of market exclusivity for paediatric drugs. This period of innovator data protection operates independently from any patent protection.

B. Pre-Regulatory Submission Phase

1. R&D from Compound Identification to Preclinical Studies

a. Target Identification/Prediscovery Phase

Research and development toward a new drug begins with identification of an appropriate disease or condition "target." The timeline for this pre-drug discovery phase is highly variable and influenced by scientific advances, a motivation to cure the disease or condition, and the availability of funding.

b. Drug Discovery Phase

Scientists will search for a molecule or class of molecules that are active toward the target. Typically, thousands of compounds may be screened for activity as well as for attributes such as absorption into the bloodstream, distribution within the body, metabolism, excretion, and toxicity. The most promising "lead" compounds are then optimized prior to preclinical testing.

c. Preclinical Discovery Phase

Promising compounds are tested both to elucidate the drug's mechanism of action and its safety profile. Feasibility of scaling up for use in large clinical trials may also be evaluated. By the end of this process, the number of compounds of interest typically narrows. A few of these are then studied in clinical trials.

2. Clinical Trials

To seek approval for sale of a new drug in Canada, evidence of clinical effectiveness must be filed with the minister of health. The Food and Drugs Act[106] and Food and Drugs Regulations[107] authorise Health Canada to regulate the safety, efficacy, and quality of pharmaceuticals in Canada. This includes permitting drug sponsors to conduct clinical trials with drug candidates. Responsibility for this process rests with Health Canada's Health Products and Food Branch (HPFB). Directorates within HPFB include the Biologics and Genetic Therapies Directorate (BGTD), the Marketed Health Products Directorate (MHPD), the Natural Health Products Directorate (NHPD), and the Therapeutic Products Directorate (TPD). Clinical trials are governed by Part C, Division 5 of the Food and Drug Regulations.

a. Timing

Drug candidates that pass preclinical testing may then be eligible for human testing, as determined by Health Canada upon application by the drug sponsor to HPFB, specifically TPD's Office of Clinical Trials (OCT).

b. Preclinical Trial Application (PCTA) Consultation Meeting

Health Canada permits drug sponsors to request a preclinical trial application meeting where sponsors present relevant data and discuss any issues regarding drug development prior to submitting a clinical trial application (CTA). Health Canada will provide guidance on the acceptability of the proposed trial. This permits the sponsor to make any necessary modifications prior to submitting the full CTA. Guidance documents published by Health Canada indicate that the sponsor is expected to prepare and send to the appropriate directorate a written record of the discussions and conclusions of the consultation meeting within 14 days of the consultation date. This written record will be added to the Central Registry file for the drug.[108]

c. Clinical Trial Application (CTA)

A CTA must be filed with the TPD's OCT prior to initiation of the clinical trial.[109] CTAs are required for human clinical trials of drugs not authorised for sale in Canada, and for marketed drugs where the proposed use is outside the parameters of the Notice of Compliance (NOC) or the drug identification number (DIN). The application includes administrative and clinical information regarding the proposed trial, chemical and manufacturing summaries of the drug product to be used in the proposed trial, and

106. Food and Drugs Act, R.S.C. 1985, c. F-27.

107. Food and Drug Regulations, C.R.C., c. 870.

108. Health Can., Guidance Document for Clinical Trial Sponsors: Clinical Trial Applications 8–10 (Health Prods. & Food Branch 2013).

109. Food and Drug Regulations, C.R.C., c. 870, C.05.005.

any additional supporting quality information that is applicable to the trial.[110] The CTA also includes a copy of the protocol for the clinical trial and a copy of the statement of risks and anticipated benefits to the health of clinical trial subjects that will result from participation in the clinical trial. This is set out in the subject consent form.[111] Finally, the CTA must include a statement declaring that good clinical practices will be utilized.[112]

The OCT has set a target of seven days for review of CTAs for some Phase 1 trials and comparative bioavailability trials.[113] All other CTAs are reviewed by the OCT within 30 days of the date of receipt.[114] If the OCT does not issue a rejection letter within 30 days of the CTA submission, the trial may proceed by default.[115]

During the screening of CTAs, the OCT can request that the applicant clarify aspects of the application. The applicant has two calendar days to respond to the request.[116] If it is determined that information required under the Food and Drug Regulations Section C.05.005 or C.05.008 has not been included, or if no response to a request for clarification has been received within two days, a screening rejection letter may be issued. Further, a Not Satisfactory Notice (NSN) may be issued if significant deficiencies are identified during review of the application. Even if a screening rejection letter or an NSN is issued, the applicant has the option of resubmitting the information and materials at a future time for processing as a new CTA.

Successful sponsors are issued a No Objection Letter (NOL) during the 30-day review period. They are permitted to conduct the proposed clinical trial, which is completed in three to four phases and can last up to seven years.

No person is permitted to sell or import a drug for use in a clinical trial unless authorised to do so as outlined in Division 5 of the Regulations. Sponsors who wish to import a drug into Canada for the purpose of a clinical trial must produce a No Objection Letter to facilitate shipment and demonstrate compliance with the clinical trial regulatory process.

If a clinical trial sponsor wishes to change a trial protocol previously approved by the minister, it must submit a CTA Amendment application before implementing the desired changes.[117] This includes amendments to protocols that affect the evaluation of the clinical efficacy of the drug, and

110. *Id.*

111. *Id.*

112. *Id.*

113. Standing Senate Comm. on Soc. Affairs, Sci. & Tech., Canada's Clinical Trial Infrastructure: A Prescription for Improved Access to New Medicines 7 (Senate of Can. 2012).

114. Food and Drug Regulations, C.R.C., c. 870, C.05.006.

115. *Id.*

116. *Id.* at C.05.009.

117. *Id.* at C.05.008.

amendments to the chemistry and manufacturing information that may affect the safety or quality of the drug.[118] Experimentation in a clinical trial outside the approved CTA is not permitted without an approved amendment.

The clinical trial must be scientifically sound and clearly described in a protocol.[119] Furthermore, approval of a research ethics board must be obtained before the clinical trial begins.[120] The research ethics board will ensure that the trial is scientifically sound and has a clearly described protocol.

Studies using approved drugs whose dosage, method of administration, or indication for use differs from that of the corresponding NOC require clinical trial authorisation from the OCT.[121] Trials studying specific population groups for whom the drug is not currently approved for use also must obtain clinical trial authorisation prior to commencement of the study.[122]

In Phase 1 trials, the candidate drug is administered to a small number of subjects to test for the presence of adverse effects and to determine pharmacokinetic/pharmacological activity of the drug. Phase 1 trials are often conducted in healthy volunteers only. If the results are favorable, trials will proceed to Phase 2.

In Phase 2 trials, candidate drugs are given to a small number of subjects who suffer from the disease or condition to be treated, diagnosed, or prevented, in order to evaluate the efficacy of the drug. Side effects and risks associated with the drug are also studied. If all is satisfactory, drug candidates proceed to Phase 3 trials in a larger test population to yield statistically significant results. A drug candidate will be abandoned if the intended benefit is not demonstrated in human subjects in Phase 2.

In Phase 3 trials, drugs can be tested in several thousands of subjects, with or without control groups, to determine if the drug candidate is effective. This clinical trial stage can last several years to evaluate fully both toxicity and efficacy to the satisfaction of Health Canada.

Phase 4 trials occur after the drug has been approved for market. They are useful in optimizing the drug's use, including safety, mortality, morbidity, and epidemiological studies.

d. Experimental Use Patent Issues

Section 55.2(1) of the Patent Act states that it is not infringement of a patent for any person to make, construct, use, or sell the patented invention solely for uses reasonably related to the development and submission of information

118. *Id.* at C.05.008(2).
119. *Id.* at C.05.010(a).
120. *Id.* at C.05.010(d).
121. Standing Senate Comm. on Soc. Affairs, Sci. & Tech., *supra* note 113, at 6.
122. *Id.*

required under any law of Canada or a country other than Canada that regulates the manufacture, construction, use, or sale of any product.[123]

e. Funding

Clinical trials and the entire drug development process can be extremely costly, often requiring hundreds of millions of Canadian dollars to see through to completion.

As much as 80 percent of clinical research in Canada is publicly funded through the Canadian Institute of Health Research (CIHR).[124] The majority of the research is carried out in academic institutions and hospitals. However, clinical trials of investigational drugs are exclusively funded by pharmaceutical companies.[125] Approximately two-thirds of these clinical trials are conducted at privately owned and operated facilities, rather than in hospitals or academic settings.[126]

f. Confidentiality Issues

i. Clinical Trial Registration and Data Confidentiality

There is no requirement in the Food and Drugs Act or Division 5 of the Regulations to register clinical trials in Canada, despite the World Health Organization's support of a registration requirement through its International Clinical Trials Registry Platform.

Health Canada encourages clinical trial sponsors to make the data available via posting on publicly accessible registries, but acknowledges there is no current legislation that allows it to compel such transparency.

ii. Personal Health Information Confidentiality

Researchers conducting clinical trials are obliged to maintain the confidentiality of personal health information obtained. A physician may be liable for breach of confidence if she or he discloses confidential patient information. The same applies to organizations or institutions that perform clinical trials, including contract research organizations.[127] If personal health information is kept anonymous, disclosure to third parties can be made without breaching confidence.[128]

---

123. Patent Act, R.S.C. 1985, c. P-4, s. 55.2.
124. Standing Senate Comm. on Soc. Affairs, Sci. & Tech., *supra* note 113, at 8.
125. *Id.* at 8.
126. *Id.*
127. Mathieu Gagné, Drugs and Health Products Law in Canada 40 (2011).
128. *Id.*

3. Patent Considerations in the Pre-Regulatory Submission Phase

g. Timing

Canada uses a first-to-file patent system whereby the first inventor to file is entitled to the patent.

i. Public Disclosure

In Canada, a patent applicant has a one-year grace period after disclosing the invention to the public in which to file a Canadian patent application.[129] If disclosure occurred prior to the one-year grace period, in Canada or elsewhere, the patent is deemed anticipated, rendering the patent application invalid.[130]

ii. Convention Priority

Canada is a member of the Paris Convention for the Protection of Industrial Property.[131] This treaty allows an applicant to claim priority from an earlier filed application, provided the subsequent filing is done within one year of the original. Under this system, a patent can be filed in a Convention member country, and then filed in another member country within one year, and the earlier filing date will be the priority date for both applications.

Once a patentable invention has been disclosed publicly, a patent application should be filed within one year in Canada or another Convention country.

iii. Patent Register

Created under the PM(NOC) Regulations, the Patent Register is a list of patents with respect to drugs that have been issued a Notice of Compliance for marketing approval. The Patent Register is maintained by the minister of health. To list a patent on the Register, the applicant must submit the patent list at the time of filing an NDS or SNDS for the drug to which the patent relates.[132] If the Canadian filing date of the patent is after the NDS or SNDS filing date, the patent will not be eligible for listing. Generic manufacturers are required to address the patents on the Register upon submitting an ANDS. In this regard, a person who infringes a patent is liable to the patentee and to all persons claiming under the patentee for all damage sustained as a result of the infringement (i.e., making, constructing, using, or selling the patented invention, including by the act of importing material made through a patented process outside Canada).[133]

---

129. Patent Act, R.S.C. 1985, c. P-4, s. 28.2(1)(a).

130. *Id.* at s. 28.2.

131. Paris Convention for the Protection of Industrial Property, July 14, 1967, 828 U.N.T.S. 305.

132. Patented Medicines (Notice of Compliance) Regulations, SOR/93-133, s. 4(5).

133. Patent Act, R.S.C. 1985, c. P-4, ss. 42, 55(1), 55.1; Eli Lilly & Co v. Apotex Inc., [2009] F.C. 991.

h. Scope of Filings

In Canada, filing for a patent is governed by the Patent Act[134] and Patent Rules.[135] A patent is obtained by filing a patent application with the Canadian Intellectual Property Office (CIPO). CIPO will examine the application to ensure it meets with patentability requirements and complies with the Patent Act and Patent Rules. A patent may be obtained for any invention that is a new and useful art, process, machine, manufacture, or composition of matter, or a new and useful improvement thereof.[136]

To receive an official filing date in Canada, the applicant must submit a statement that a patent is sought, provide documentation describing the invention, provide the applicant's name and address along with the name and address of its patent agent (if applicable), and pay the filing fee.

i. Inventors

Section 27(1) of the Patent Act indicates that the inventor(s) of an invention may obtain a patent granting exclusive property in the invention, upon compliance with the application procedures set forth in the Patent Act and Regulations.[137]

There is no exact definition for "inventor" in the Patent Act. However, interpretation of "inventor" based on the definition of "invention" has been applied as the person who conceived of the inventive concept and reduced it to a definite and practical shape.

To satisfy the conception requirement of "inventor," one cannot simply identify a problem, one must also have a vision of the solution. The reduction to practice requirement is satisfied if the inventor takes the conception or discovery and adapts it to fit within the definition of "invention" (i.e., a new and useful art, process, machine, manufacture, or composition of matter, or improvement thereof).

Instances where a potential inventor did not conceive the solution, but did make a valuable and practical suggestion that assisted in making the invention operative or contributed to a piece of the invention that made the whole, may result in that person being considered a joint inventor. However, to be a " coinventor," one must participate in the inventive concept as opposed to merely the verification of the invention.[138]

When two or more persons are coinventors, they must jointly apply for a patent. If one of the inventors refuses to apply for a patent, the other inventors may still apply, and a patent may issue in their names alone,

134. Patent Act, R.S.C. 1985, c. P-4.
135. Patent Rules, SOR/96-423.
136. Patent Act, R.S.C. 1985, c. P-4, s. 2.
137. *Id.* at s. 27(1).
138. Apotex Inc. v. Wellcome Found. Ltd., 2002 S.C.C. 77.

if they are able to prove to the Commissioner that the joint inventor has refused to make application.[139]

j. Amount of Information Required for Specification/Disclosure

An application begins with an abstract that provides technical information and a concise summary of the matter contained in the application.[140] If a drug is being claimed, the chemical formula that characterizes the invention should be stated in the abstract.[141]

The applicant must also file a description or disclosure that describes the invention and its operation or use.[142] It must set out the steps in a process or method of constructing, making, compounding, or using the invention in terms such that a person skilled in the art may construct, make, compound, or use it.[143] For a process patent, the specification must explain the necessary sequence of the steps in the process, so as to distinguish the invention from other inventions.[144]

The application must contain claims that must define the subject matter of the invention in explicit terms to "fence off" the inventor's monopoly.[145]

An application may also contain drawings that are referred to in the description.[146]

k. Genus/Species Claims

A selection patent claims either a compound or small group of compounds (the "species") selected from a class of compounds (the "genus") previously claimed by an earlier Canadian patent. A selection patent must meet the following three requirements, originally outlined in *Re IG Farbenindustrie AG's Patents* (UK):

1. There must be a substantial advantage to be secured or disadvantage to be avoided by the use of the selected members.
2. The whole of the selected members must possess the advantage in question (a few exceptions may be permissible).
3. The special character must be peculiar to the selected group; the unselected compounds must not generally possess the same advantage; however, further research demonstrating a small number of unselected compounds possessing the same advantage will not invalidate the selection patent.[147]

139. Patent Act, R.S.C. 1985, c. P-4, s. 31(1).
140. Patent Rules, SOR/96-423, s. 79.
141. *Id.*
142. Patent Act, R.S.C. 1985, c. P-4, s. 27(3).
143. *Id.* at s. 27(3)(b).
144. *Id.* at s. 27(3)(d).
145. *Id.* at s. 27(4).
146. Patent Rules, SOR/96-423, s. 94(1).
147. Apotex Inc. v. Sanofi-Synthelabo Can. Inc., 2008 S.C.C. 61.

Additionally, the "species" claimed must not have been specifically disclosed or claimed in the "genus" patent.

The Supreme Court of Canada has unanimously ruled that "selection patents" are not inherently invalid.[148] Rather, selection patents are subject to validity attacks on the same grounds as any other patent.

In areas where advances are often obtained through experimentation, such as in the pharmaceutical industry, the court may employ an "obvious to try" or "ought to work" test to its obviousness analysis.[149] This test helps answer the question of whether the subject matter of the patent is truly inventive. The court will consider the following factors to determine if the claimed invention, such as selected compounds, were obtained by a process that would have been "obvious to try" to a person skilled in the art at the relevant time based on whether the purported invention "ought to work":

1. Is it more or less self-evident that what is being tried ought to work?
2. What is the extent, nature, and amount of effort required to achieve the invention (i.e., were the claimed compounds a result of routine trials or did they require prolonged experimentation)?
3. Is there a motive provided in the prior art to find the solution that the patent addresses?

l. Formulation Claims

Formulation claims in Canada typically involve an active pharmaceutical ingredient along with a pharmaceutically acceptable diluent or carrier.

m. Use Claims

A known compound may be deemed patentable subject matter if the inventor claims a use for the compound that is new, useful, and nonobvious, and hence satisfies the definition of "invention."[150]

A method of medical treatment is not patentable if it covers an area in which a physician's skill or judgment is normally exercised.[151] Thus, a claim for the delivery method of a drug is not a patentable invention as it interferes with an area normally reserved for a physician's professional judgment.[152] However, a claim for a specified dosage of a new drug may be patentable, as a physician can still decide to use that dosage or another

148. *Id.*
149. *Id.*
150. Shell Oil Co. v. Comm'r of Patents, [1982] S.C.R. 536.
151. Janssen Inc. v. Mylan Pharm. ULC, [2010] F.C. 1123.
152. *Id.*

unpatented dosage.[153] The issue is whether the patent would cover the doctor's behavior or a vendible product. Only the latter is patentable.

n. Process/Method of Manufacture Claims

In order to claim a process or method of manufacture, the disclosure must set out clearly the steps in a process or method of constructing, making, compounding, or using the invention in clear and exact terms such that any person skilled in the art may construct, make, compound, or use it.[154] A failure to comply may subject the patent to validity attacks for insufficient disclosure. The Patent Act further indicates that the specification must explain the necessary sequence of the steps in the process, so as to distinguish the invention from other inventions.[155]

Novel processes may be claimed even if they produce a compound that is already known. Furthermore, a known process or method of manufacture may be claimed provided that the resulting compound is both novel and nonobvious. Finally, novel products when produced by a novel chemical or biological process may sometimes be claimed in a product-by-process claim.

C. Naming the Compound—Trade-mark and Nonproprietary Name Issues

1. Typical Flow for Registration and Approval of Mark

a. NDS/ANDS Drug Submission

The Food and Drug Regulations require that a drug's brand name or identifying name be provided as part of the new drug submission to the minister.[156] Proposed proprietary drug names that are submitted are reviewed once the submission (NDS or ANDS) is filed. Health Canada then completes an initial review of the name within a 90-day target window.[157] Feedback is provided to the applicant by Health Canada. Specifically, names are reviewed for their potential to be confusing by looking like and/or sounding like the name of another drug, with the overall goal of preventing dispensing errors due to confusing similarities.

A naming issue may arise if another product with a conflicting name is approved after the initial review but prior to issuance of an NOC and/or a DIN of the applicant's drug. For this reason, Health Canada will conduct a second abbreviated name review within 90 days of the

153. Merck & Co. v. Pharmascience Inc., [2010] F.C. 510.

154. Patent Act, R.S.C. 1985, c. P-4, s. 27(3)(b).

155. *Id.* at s. 27(3)(d).

156. Food and Drug Regulations, C.R.C., c. 870, C.08.002(2)(b).

157. Health Can., Guidance for Industry: Drug Name Review: Look-Alike Sound-Alike (LA/SA) Health Product Names 7 (Health Prods. & Food Branch 2005).

anticipated day of NDS/ANDS approval. This will ensure no proprietary name has been approved by HPFB for another product between the initial review and drug submission approval.[158] If no conflicting name is found, Health Canada will proceed with approval. If it is found that two drug submissions have conflicting brand or identifying names, the sponsors of the second drug submission will have to change their name in order for drug submission approval to occur.

b. Trade-mark Registration

In Canada, trade-mark registration is separate from the drug approval process and is governed by the Trade-marks Act.[159] Hence, approval of a drug name in the trade-mark registration process or in the drug submission process does not guarantee acceptance of the name. Application for a trade-mark may occur at any time, but the process usually will not begin prior to the commencement of clinical trials.

Trade-mark applications are filed with the Canadian Intellectual Property Office. Once filed, the application is reviewed by an examiner for compliance with technical formalities and for potential objections based on the registrable criteria set out in the Trade-marks Act. The most common objections are based on descriptiveness and likelihood of confusion with prior registered marks. If the examiner discovers technical or substantive issues with the application, an Office Action will be issued. The applicant is provided with a definite period of time within which it must respond to the Office Action.

Once the examiner determines that the application is free from technical or substantive hurdles, CIPO will advertise the mark in the *Trade-marks Journal* to provide interested third parties with an opportunity to oppose the trade-mark registration. Opposition must be filed with CIPO within two months of publication.[160]

CIPO will issue a Notice of Allowance for applications that have passed both the examination and opposition processes. Once the registration fee is paid, the trade-mark will be registered for 15 years and may be renewed for an unlimited number of additional 15-year periods with payment of further registration fees.

158. *Id.*
159. Trade-marks Act, R.S.C. 1985, c. T-10.
160. *Id.* at c. T-10, s. 38(1).

2. Potential Pitfalls and How to Avoid Them

a. NDS/ANDS Drug Submission

The Food and Drug Regulations require that names of drugs cannot be confused with one another.[161] If HPFB believes confusion is likely, which may result in safety concerns, the agency may refuse to issue an NOC and/or a DIN.

To avoid confusion with another health product, Health Canada advises drug sponsors not to use abbreviations or suffixes in brand names, and encourages them to use a proprietary name that is only one word and that does not have qualifications by letters or numbers.[162] Further, Health Canada discourages the use of similarities between brand and generic names where the medicinal ingredients are different, as this may increase the likelihood of medication errors.[163]

In order to facilitate the name review process, Health Canada asks drug sponsors to provide detailed naming information as part of the drug submission, including:

1. a proposed proprietary name and an optional list of alternate name choices; and
2. a risk assessment of the product's proposed brand name supported by studies, data, and analysis, which aims to predict the likelihood of look-alike/sound-alike drug name errors.[164]

Risk assessments may include a drug/reference search record for similar proprietary and nonproprietary drug names, computer analysis of orthographic or phonetic similarities to other drugs, and assessment of the possibility of medication error arising from the name which may reference applicable medication error literature.[165]

b. Trade-mark Registration

As opposed to the drug approval process where the likelihood of medication error is assessed, the Trade-marks Office establishes whether or not use of the proposed trade-mark will cause confusion as to the source of the product.[166] If the proposed trade-mark is similar to the name of another drug and may lead consumers to believe that both drugs originate from the same source, the trade-mark registration may be refused.[167] Trade-marks

161. Food and Drug Regulations, C.R.C., c. 870, C.08.002.
162. Health Can., *supra* note 157, at 6.
163. *Id.*
164. *Id.* at 8.
165. *Id.*
166. Trade-marks Act, R.S.C. 1985, c. T-10, s. 6.
167. *Id.* at s. 37(1)(b).

for drug names also cannot be clearly descriptive or deceptively misdescriptive of the character or quality of the drug.[168] Trade-marks for drug names will be rejected for descriptiveness if the examiner deems them too similar to an international nonproprietary drug name. In order to avoid trade-mark rejection based on confusion or descriptiveness, the applicant should assess suitability for submission to the regulatory authorities prior to filing the application.

The required contents of a trade-mark application are dictated by the Trade-marks Act.[169] Applicants must enumerate the wares and services associated with the mark in ordinary commercial terms. The wares must be more specific than merely referring to "pharmaceutical preparations," which can be accomplished by either naming the disease, specifying the type of condition/disorder to be treated, or indicating the specific type of medication.[170]

D. Regulatory Approval Phase

Health Canada's Therapeutic Products Directorate (TPD) regulates pharmaceutical drugs for human use. The Biologics and Genetic Therapies Directorate (BGTD) regulates all biologic and radiopharmaceutical therapies.

1. NDS Types and Requirements (Including Typical Flowchart)

a. NDS

Any new drug must be authorised for sale by Health Canada before being sold in Canada. The manufacturer must submit a New Drug Submission (NDS) that details all scientific data that establishes the safety, efficacy, and quality of the drug. Health Canada will review the submission and issue a Notice of Compliance (NOC) and Drug Identification Number (DIN) if it feels the drug is suitable for marketing in Canada.

b. New Drug Submission Requirements[171]

New drugs typically require preclinical, clinical, chemistry, and manufacturing data, whereas generics typically only require bioequivalence and/or pharmaceutical equivalence data. Health Canada's TPD publishes guidance documents that break down the structure of a human NDS into five parts: Master Volume, Chemistry and Manufacturing, Comprehensive Summary, Sectional Reports, and Raw Data.

---

168. *Id.* at s. 12(1)(b).

169. *Id.* at s. 30.

170. Adrian Zahl, International Pharmaceutical Law and Practice 45 (Matthew Bender & Co. 2012).

171. Health Can., Therapeutic Products Program Guideline: Preparation of Human New Drug Submissions (Health Prot. Branch 1991).

i. Part 1: Master Volume

The Master Volume must contain:

1. Signed submission certification;
2. Completed NDS application form;
3. Table of contents;
4. Brief summary (300 to 500 words) detailing the name of the applicant, structural formula (if known), chemical name, other names, pharmacologic and chemical relationship to other drugs, therapeutic classification, route of administration, indications/contraindications, dosage and administration, and positive/negative features of the drug;
5. Product monograph;
6. Draft labeling;
7. List of prior related submissions; and
8. Non-Canadian package inserts.

ii. Part 2: Chemistry and Manufacturing

Required chemistry and manufacturing data include the following:

1. Drug substance: names, source, method of manufacture, structure elucidation and confirmation, physicochemical properties, impurities, reference standard, specifications and test methods, batch analyses, and stability;
2. Dosage forms: names, source, developmental pharmaceutics, formulation and method of manufacture, specifications for excipients and test methods, packaging, and stability; and
3. Addendum for biologics: description of manufacturing facilities, work flow, other drugs prepared in the same and adjacent areas, evidence establishing ability to produce a consistent product, proposed format of a protocol for lot release of the product, characteristics of the product (including safety, potency, pyrogenicity, identity, sterility, protein content, moisture content, particulate matter, and freedom from extraneous blood proteins), and active and passive immunization studies.

iii. Part 3: Comprehensive Summary

The comprehensive summary is generally between 20 and 200 pages and includes:

1. Investigational studies: pharmacology, toxicology, microbiology, published or unpublished investigational articles, and discussion and evaluation of investigational studies;

2. Clinical studies: clinical pharmacology, clinical trials, published or unpublished clinical articles, discussion and evaluation of clinical results, and overall conclusions from clinical studies; and
3. Status of research and development of the drug.

iv. Part 4: Sectional Reports

Sectional reports provide a detailed description of each available and completed preclinical or clinical study, organized in readable tables and graphs if possible. Sectional reports include:

1. Investigational studies: pharmacology, toxicology, microbiology, and published and unpublished investigational reports;
2. Clinical studies: clinical pharmacology, clinical trials, and published and unpublished clinical reports; and
3. Curriculum vitae of each investigator.

v. Part 5: Raw Data

Raw data are broken down into two sections:

1. Preclinical studies
2. Clinical studies

c. Steps in the NDS Process[172]

Step 1. Submission: NDS submitted to Submission and Information Policy Division (SIPD) of Health Canada.

Step 2. Processing: SIPD processes NDS and sends to TPD (Health Canada targets processing within ten calendar days).

Step 3. Screening: TPD screens application for completeness (Health Canada targets 45 calendar days).

- Priority review submissions and submissions accepted for advance consideration under the Notice of Compliance with Conditions (NOC/c) policy have a 25-calendar-day target for screening.

3a. Successful screening ↦ NDS is reviewed (step 4).

3b. Unsuccessful screening ↦ Screening Deficiency Notice (SDN) or Screening.

Rejection Letter (SRL) is issued.

Screening Deficiency Notice

172. Health Can., Guidance for Industry: Management of Drug Submissions (Health Prods. & Food Branch 2011).

- Sponsor corrects deficiency within 45 calendar days ↦ NDS is reviewed (step 4).
- Sponsor fails to provide all requested information within 45 calendar days ↦ SRL is issued.

Screening Rejection Letter

- The application process is terminated due to application rejection.

Step 4. Review: NDS is reviewed for safety, quality, and efficacy (Health Canada targets 300 calendar days for processing).

- Priority review submissions and submissions accepted for advance consideration under the NOC/c policy have a 180-calendar-day target for review.

4a. Successful review ↦ NOC and DIN issued (step 5).

4b. Unsuccessful review ↦ Notice of Deficiency (NOD), Notice of Non-compliance (NON), or Request for Clarification is issued.

Notice of Deficiency

- Issued if NDS has deficiencies or significant omissions.
- Review process cannot continue until deficiencies/omissions are corrected.
- Sponsor has 90 calendar days to submit all solicited information.
  - o Submitted information is screened by Health Canada within 45 days (25 days for priority or NOC/c submissions).
  - o Sufficient NOD response submission ↦ New review process commences (Health Canada target is 150 calendar days or 90 calendar days for priority NOC/c submissions).
  - o Insufficient NOD response submission ↦ Notice of Deficiency-withdrawal letter (NOD/w) is issued, NDS process ends.

Notice of Non-compliance

- Issued upon review completion if NDS does not comply with requirements of Food and Drugs Act or Regulations or is deficient.
- Sponsor has 90 calendar days to submit all solicited information.
  - o Submitted information is screened by Health Canada within 45 days (25 days for priority or NOC/c submissions).
  - o Sufficient NON response submission ↦ New review process commences (Health Canada target is 150 calendar days or 90 calendar days for priority or NOC/c submissions).
  - o Insufficient NON response submission ↦ Notice of Non-compliance-withdrawal letter (NON/w) is issued, NDS process ends.

Request for Clarification

- Can occur during screening or review.
- Sponsor must expand upon, add precision to, or reanalyze existing information in the submission within 15 calendar days.
- Upon response, review process resumes.

Step 5. NOC Issuance: NDS's that pass the review process are issued an NOC and a DIN and are thus permitted to be marketed in Canada.

NOC/c

- A Notice of Compliance with Conditions (NOC/c) may be issued by the Director and is restricted to products for serious, life-threatening, or severely debilitating diseases or conditions.
- The goal of the NOC/c is to provide physicians and patients early access to these types of medications.
- A sponsor under the NOC/c policy must commit to pursuing confirmatory studies and enhanced postmarketing surveillance.

d. NDS Flowchart

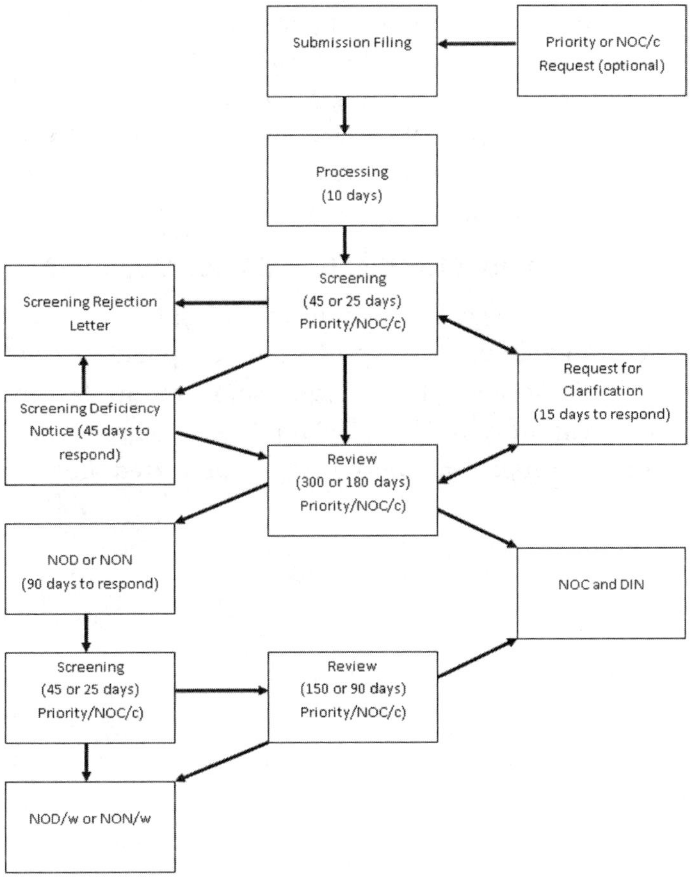

e. SNDS

A Supplemental NDS (SNDS) must be filed by the manufacturer in the event changes are made to drug products that have already been approved.[173] Changes include: a new description of the drug, brand name, specification of ingredients, method of manufacture, labeling, recommended route of administration, dosage, contraindications and side-effects, or claims made for the new drug.[174] Changes to a drug product are grouped into different "levels" based on the potential impact on safety and/or efficacy of the drug, as provided in Health Canada guidance documents.

i. Level 1—Major Quality Changes

Level 1 changes are those that have a substantial potential to have an adverse effect on the identity, strength, quality, purity, or potency of the drug product. The drug sponsor must file an SNDS, along with the recommended supporting data, and the change may not be implemented until an NOC has been issued.[175]

ii. Level 2—Notifiable Changes (Moderate Quality Changes)

Level 2 changes are those that have a moderate potential to have an adverse effect on the identity, strength, quality, purity, or potency of the drug product.[176] The same level of supporting data must be filed with the minister as in an SNDS filing, although it is filed as a notifiable change rather than as an SNDS. As with a Level 1 change, sponsors may not implement the change until a No Objection Letter (NOL) has been issued by the minister. Unlike a Level 1 change, if a response has not been received within 90 days, the change may be implemented.

iii. Level 3—Annual Notification (Minor Quality Changes)

Level 3 changes are those that have a minimal potential to have an adverse effect on the identity, strength, quality, purity, or potency of the drug product.[177] Changes in this category may be implemented prior to review of supporting data by Health Canada, although notification of the change is required. Supporting data are not submitted, but must be available upon request.

---

173. Food and Drug Regulations, C.R.C., c. 870, C.08.003(1).

174. *Id.* at C.08.003(2).

175. Health Can., Guidance Document: Post–Notice of Compliance (NOC) Changes: Quality Document 2–3 (Health Prods. & Food Branch 2011).

176. *Id.* at 3.

177. *Id.* at 3.

iv. Level 4—Record of Changes

All other changes to a drug product are designated as Level 4 and are those not expected to have an adverse effect on the product. Manufacturers are expected to maintain records of such changes in compliance with Good Manufacturing Practices required under the Food and Drug Regulations.

f. Steps in the SNDS Process

The SNDS application process (Level 1 changes) mirrors that of the NDS application process. It begins with submission, followed by processing, screening, review, and NOC issuance. Health Canada timeline targets are identical to those of the NDS process.[178] Sponsors may apply for priority review or NOC with conditions in the same manner as the NDS process.

2. ANDS Types and Requirements (Including Typical Flowchart)

Generic drug manufacturers can obtain marketing approval for a drug that has already been issued an NOC by filing an Abbreviated New Drug Submission (ANDS). An ANDS allows a generic company to base its request for approval on bioequivalence with the previously approved innovator drug product, thus avoiding the high cost of clinical trials that are required in the NDS process. The ANDS compares the generic drug with the innovative drug by reference to the scientific data submitted by the innovator in the original NDS.

a. ANDS Eligibility

Food and Drug Regulations Section C.08.002.1 defines the eligibility criteria for an ANDS as follows:

> (1) A manufacturer of a new drug may file an abbreviated new drug submission or an abbreviated extraordinary use new drug submission for the new drug where, in comparison with a Canadian resference product,
>
> (a) the new drug is the pharmaceutical equivalent of the Canadian reference product;
>
> (b) the new drug is bioequivalent with the Canadian reference product, based on the pharmaceutical and, where the minister considers it necessary, bioavailability characteristics;
>
> (c) the route of administration of the new drug is the same as that of the Canadian reference product; and

178. Health Can., *supra* note 172.

(d) the conditions of use for the new drug fall within the conditions of use for the Canadian reference product.[179]

b. Canadian Reference Product

For an ANDS to be filed, there must be an approved Canadian reference product that has been issued an NOC and is marketed by the innovator in Canada, or is no longer marketed in Canada but is still considered to be acceptable for the purposes of demonstrating bioequivalence or bioavailability by the minister.[180]

The applicant must also address any patents for the reference product that are listed on the Canadian Patent List, either declaring an acceptance that the NOC being applied for shall not be issued until patent expiry, or making an allegation against the patent (as will be discussed in Section IV.F.1, "Patented Medicine (Notice of Compliance) Regulations" later in this text).

c. Other Requirements

An ANDS application must include a product monograph, physicochemical characteristics, pharmacology data, pharmacokinetics data, a drug product classification, and an environmental assessment statement.[181]

d. Steps in the ANDS Process[182]

The ANDS application process mirrors that of the NDS application process. It begins with a submission, followed by processing, screening, and review, and culminates with the issuance of an NOC. Health Canada timeline targets are identical to those of the NDS process, with the exception of the review phase, which is reduced to 180 days from 300 days because bioequivalency data are less onerous to review than clinical trial data. The ANDS process does not permit a sponsor to obtain priority review or an NOC with conditions.

179. Food and Drug Regulations, C.R.C., c. 870, C.08.002.1(1).
180. *Id.* at C.08.002.1.
181. *Id.* at C.08.002.1.
182. HEALTH CAN., *supra* note 172.

e. ANDS Flowchart

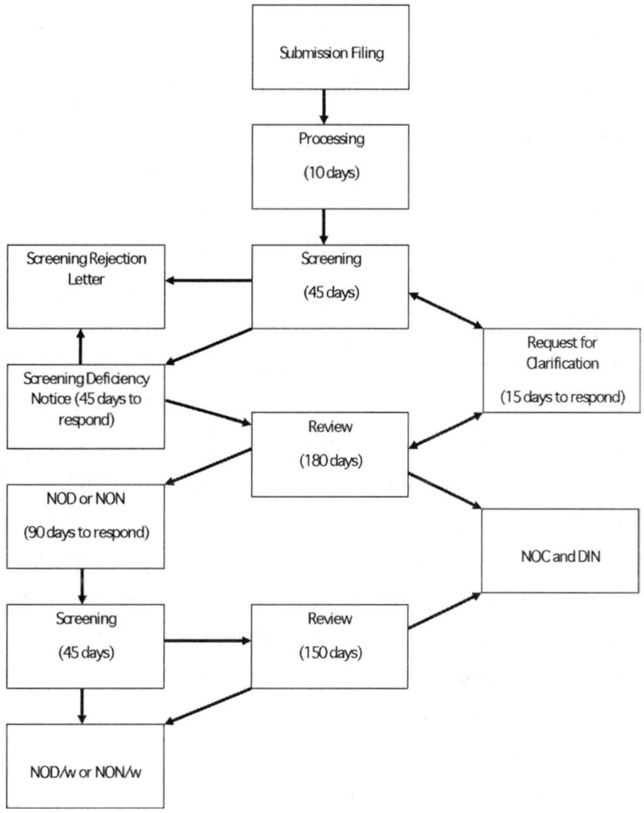

3. Biologic Drug Requirements (Including Typical Flowchart)

a. Innovator Biologic Drug Approval

A biologic is a type of drug produced from biologic sources. Biologics are inherently variable, and, as such, the structure can be sensitive to slight changes in manufacturing conditions.

Biologic drug approval is regulated by the Biologics and Genetic Therapies Directorate (BGTD) in accordance with Part C, Divisions 3, 4, and 8 of the Food and Drugs Regulations. Biologic drugs are listed on Schedule D of the Food and Drugs Act.

Similar to non-biologic drug approval, the sponsor of a biologic drug must receive an NOC and a DIN in order to market the drug in Canada. The application process is similar; however, the scope of information required for approval of a biologic drug is broader in that the sponsor must include more detailed chemistry and manufacturing information. A biologic drug sponsor must still meet NOC application requirements for non-biological drugs to establish safety and efficacy, which includes clinical trial data. The BGTD also imposes more stringent inspection requirements for production facilities and manufacturing processes, and may test biologic drugs periodically to ensure consistency of manufacturing.

b. Subsequent Entry Biologic Drug Approval

A subsequent entry biologic (SEB) is a biologic drug that enters the market subsequent to a version previously authorised in Canada and that has demonstrated similarity to a reference biologic drug. An SEB is not considered a generic version of a biologic because the active component of an SEB can differ from that of the reference product; thus, approval for marketing cannot be acquired by filing an ANDS. In an ANDS, the sponsor must prove the generic product is bioequivalent to the reference product, whereas an SEB requires more rigorous testing to establish that the drug is sufficiently equivalent to its reference product in terms of its effectiveness and safety. Health Canada indicates that an SEB sponsor may be required to provide both clinical and nonclinical data to illustrate the similarity between the SEB and the reference product.[183] If sufficient similarity can be established with the reference product, the SEB submission may proceed with a reduction of required clinical and preclinical data. As opposed to a generic approval, SEB approval is not a declaration of pharmaceutical equivalence with the reference product.

SEBs must be approved through the same NDS process that applies to non-biologic drugs and innovator biologic drugs. The target time for review of an SEB is the same as that for an NDS.[184]

c. Reference Product[185]

Health Canada publishes a comprehensive guidance document outlining its requirements for appropriate SEB reference products. An SEB sponsor must name the reference biologic drug authorised for sale in Canada to which the SEB is subsequent. The sponsor must be able to demonstrate that the chosen reference biologic drug is suitable to support the submission. The same reference biologic drug should be used throughout the studies supporting safety, quality, and efficacy of the product, and the dosage form, strength, and route of administration should be the same for both drugs. Furthermore, the active medicinal ingredient of the reference drug and the SEB must be shown to be similar. Finally, the reference biological drug should have adequate safety, quality, and efficacy data accumulated

---

183. HEALTH CAN., GUIDANCE FOR SPONSORS: INFORMATION AND SUBMISSION REQUIREMENTS FOR SUBSEQUENT ENTRY BIOLOGICS 13 (Health Prods. & Food Branch 2010).

184. *Id.* at 6.

185. *Id.* at 5.

such that the demonstration of similarity will bring into relevance a substantial body of reliable data.

Foreign reference products may be used if they are marketed by the same company in the foreign country as in Canada and if the countries have comparable standards. The foreign reference product should also be authorised for sale in Canada.

E. Regulatory Exclusivity

1. Innovative Drug Data Protection

The Food and Drug Regulations provide legal protection to innovative drugs from generic competition for a period of eight years through the protection of innovator data.[186] Generic manufacturers must wait six years from the date of NOC issuance before filing an NDS, SNDS, or ANDS with respect to the innovative drug.[187] The minister may not issue an NOC for the generic version of the drug for a full eight years from the date of NOC issuance for the innovative drug.[188]

a. Definition of Innovative Drug

An "innovative drug" is defined as a drug that contains a new medicinal ingredient not previously approved in a drug by the minister and that is not a variation of a previously approved medicinal ingredient, such as a salt, ester, enantiomer, solvate, or polymorph.[189] The definition of "innovative drug" prohibits innovators from obtaining subsequent data protection terms for variations of medicinal ingredients, even if it is considered a "new" drug for which an NOC was issued. Data protection guidance documents published by Health Canada indicate that a combination drug in which at least one of the ingredients is an innovative drug for which a data protection term is still in effect will still receive data protection for the eight-year period.[190] However, a combination of previously approved medicinal ingredients, all of which have exhausted their eight-year data protection period, will not be eligible for an additional data protection period.[191]

186. Food and Drug Regulations, C.R.C., c. 870, C.08.004.1.

187. *Id.* at C.08.004.1(3)(a).

188. *Id.* at C.08.004.1(3)(b).

189. *Id.* at C.08.004.1(1).

190. Health Can., Guidance Document: Data Protection under C.08.004.01 of the Food and Drug Regulations 3 (Health Prods. & Food Branch 2011).

191. *Id.* at 3–4.

b. Registration of Innovative Drugs

In order to be eligible for data protection, a drug must be listed on the Register of Innovative Drugs maintained by the minister of health.[192] Innovative drugs can be added to the Register after NOC issuance.

2. Paediatric Drug Protection

In addition to the eight-year term of data protection, an innovator can gain a six-month extension if its NDS, or any supplement to the NDS filed within the first five years of the eight-year protection period, contains results of clinical trials that were designed and conducted with the purpose of increasing knowledge about the use of the drug in paediatric populations and that will lead to a health benefit for children.[193] Health Canada maintains it must be clear that the goal of such studies was to increase knowledge about the use of the drug in paediatric populations that will assist health professionals, parents, caregivers, and patients in making informed choices about drug therapy.[194]

3. Orphan Drug Protection

On October 3, 2012, the Canadian government announced initiatives to ameliorate access to drugs for rare diseases in Canada.[195] A new framework has been suggested to encourage research and innovation of drugs for small patient populations. Although no legislation has been passed, the Standing Senate Committee on Social Affairs, Science and Technology is encouraged by the government's commitment to this issue and recommends that an extension to market exclusivity is provided to orphan drugs in Canada.[196]

F. Patented Medicine Considerations in Canada

1. Patented Medicine (Notice of Compliance) Regulations

a. Patent Listability Requirements

The PM(NOC) Regulations require the minister of health to maintain a Patent Register that lists patents for drugs that have been issued an NOC.[197]

Section 4 of the PM(NOC) Regulations provides the requirements that must be met before a patent can be listed on the Patent Register. These include timing of filing of patent lists, required information to be provided on a patent list, the type of drug submissions for which a patent list may be

192. Food and Drug Regulations, C.R.C., c. 870, C.08.004.1(9).

193. *Id.* at C.08.004.1(4).

194. Health Can., *supra* note 190, at 9.

195. *Harper Government Takes Action to Help Canadians with Rare Diseases*, Health Can. (Oct. 3, 2012), http://www.hc-sc.gc.ca/ahc-asc/media/nr-cp/_2012/2012-147-eng.php.

196. Standing Senate Comm. on Soc. Affairs, Sci. & Tech., *supra* note 113, at 28.

197. Patented Medicines (Notice of Compliance) Regulations, SOR/93-133, s. 3(2).

filed, and more substantive eligibility requirements relating to the claims of the patent.[198]

i. Timing Requirements

A first person who wishes to file a patent list for a drug must do so at the time of filing the NDS or SNDS to which the patent relates.[199] Patent lists that are submitted separately from the drug submission will be refused as not meeting the timing requirements according to Health Canada guidance documents.[200]

However, if the Canadian filing date of the patent precedes the drug submission filing date *and* the patent list is submitted to the Office of Patented Medicines and Liaison (OPML) within 30 days of the grant of the patent, a first person may submit a patent list with respect to a previously filed drug submission.[201]

ii. Patent List Information Requirements[202]

A first person must identify the following information on each patent list:

1. The submission to which the patent list relates (NDS or SNDS);
2. The medicinal ingredient, brand name, dosage form, strength, route of administration, and use set out in the NDS or SNDS to which the list relates;
3. The patent number, filing date of the patent application in Canada, date of grant of the patent, and expiry date of the patent;
4. A statement that the first person who filed the NDS or SNDS to which the list relates is the owner of the patent or has an exclusive license to the patent, or has obtained the consent of the owner of the patent to its inclusion on the list;
5. The address in Canada for service, of the first person, of a notice of allegation, or the name and address in Canada of another person on whom service may be made; and
6. A certification that the information submitted is accurate and that each patent meets the eligibility requirements of the PM(NOC) Regulations.

198. *Id.* at s. 4.

199. *Id.* at s. 4(5).

200. Health Can., Guidance Document: Patented Medicines (Notice of Compliance) Regulations 5 (Health Prods. & Food Branch 2012).

201. Patented Medicines (Notice of Compliance) Regulations, SOR/93-133, s. 4(6).

202. *Id.* at s. 4(4).

2. Drug Submissions Eligible for Filing a Patent List

A patent on a patent list in relation to an NDS is eligible to be added to the register if the patent contains a claim for the medicinal ingredient, formulation, dosage form, or use of the medicinal ingredient, provided they have been approved through issuance of an NOC with respect to the submission.[203]

A patent on a patent list in relation to an SNDS is eligible to be added to the register if the supplement is for a change in formulation, dosage form, or use of the medicinal ingredient, and the change has been approved through issuance of an NOC with respect to the supplement.[204]

a. Claim for the Medicinal Ingredient

A patent on a patent list that claims a medicinal ingredient, which has been issued an NOC with respect to the NDS submission, is eligible to be added to the register.[205] Product-by-process patents, patents claiming biological drugs, and patents claiming different polymorphs of the medicinal ingredient are included in this group.[206] Patents claiming a combination of medicinal ingredients are not eligible for listing with respect to a drug that contains only one of the claimed medicinal compounds.[207] However, a patent that claims, as a compound, a single medicinal ingredient is eligible for listing with respect to a drug that contains that medicinal ingredient in combination with other medicinal ingredients, despite the medicinal ingredient being in a combination of medicinal ingredients on the NOC.[208] A patent that claims an enantiomer is not eligible to be listed with respect to a medicinal ingredient that is its racemate.[209]

(a) Claim for Formulation[210]

To be added to the register, a claimed formulation in a patent must include the medicinal ingredient of the drug for which an NOC has been issued. The formulation claimed in the patent must correspond to the formulation approved in the relevant drug submission.

(b) Claim for the Dosage Form[211]

A claimed dosage form must correspond to the dosage form identified on the NOC and approved in the relevant drug submission. An eligible dosage

---

203. *Id.* at s. 4(2).
204. *Id.* at s. 4(3).
205. *Id.* at s. 4(2)(a).
206. *Id.* at s. 2.
207. HEALTH CAN., *supra* note 200, at 8.
208. *Id.* at 8–9.
209. *Id.* at 9.
210. Patented Medicines (Notice of Compliance) Regulations, SOR/93-133, s. 4(2)(b).
211. *Id.* at s. 4(2)(c).

form patent must include a claim that includes within its scope the medicinal ingredient or formulation for which an NOC was issued.

(c) Claim for the Use of the Medicinal Ingredient[212]

Use patents must contain a claim for use of the medicinal ingredient that has been approved through issuance of an NOC. The patent claim cannot have a substantially different scope from the use described in the NOC. Patents that claim the use of a combination of medicinal ingredients usually are not eligible for listing against a drug that contains only one component of the combination.

b. Challenges to Listing

The minister of health may refuse to add a patent to the Register and may delete any patent or other information that does not meet the requirements of the PM(NOC) Regulations.[213] Removal may also be made on request by an interested third party. The minister cannot be compelled to remove a patent from the Register by means of judicial review.[214]

c. Freezing Patent Register

A second person wishing to obtain an NOC with respect to a drug and who directly or indirectly compares or makes reference to a drug already marketed in Canada under an NOC issued to a first person and which has been submitted to a patent list must declare acceptance that the NOC will not be issued until the patent expires.[215] Otherwise, the second person must allege that the patent has expired, that the patent is not valid, or that no claim in the patent for the medicinal ingredient, formulation, dosage form, and use of the medicinal ingredient would be infringed by granting an NOC.[216]

However, the PM(NOC) Regulations indicate that a second person is not required to address any patent listed on the Patent Register with respect to the first person's drug on or after the date of filing of the second person's submission.[217] The result is the Patent Register being "frozen" with respect to the first person's patent list as of the date of filing of the second person's ANDS. The Patent Register is frozen only in respect to the second person's submission or supplement. Subsequent second persons who wish to obtain an NOC for the same drug may not benefit from the

212. *Id.* at s. 4(2)(d).
213. *Id.* at s. 3(2).
214. Apotex v. Canada, [1998] F.C.J. 1096.
215. Patented Medicines (Notice of Compliance) Regulations, SOR/93-133, s. 5(1).
216. *Id.*
217. *Id.* at s. 5(4).

original second person's date of filing; they may freeze the Register only as of their own submission filing date.[218]

Incomplete or invalid submissions by a second person will not freeze the Patent Register until the HPFB receives either the requisite missing information or a new version of the submission. A second person must address all patents that were added to the Register prior to the date of filing of his or her submission.

d. Preparing Notice of Allegation

A second person filing a submission for an NOC, or supplement thereto, may address a listed patent by alleging that:

- The statement made by the first person under Section 4(4)(d) of the PM(NOC) Regulations that he or she is the owner or exclusive licensee of the patent is false,
- The patent has expired,
- The patent is not valid, or
- No claim for the medicinal ingredient (or use thereof), formulation, or dosage form will be infringed by making, constructing, using, or selling the drug for which the second person's submission pertains.[219]

These allegations are part of the submission to the HPFB for NOC issuance, and notice of the allegations must be provided to the first person.[220] The notice of allegation (NOA) must include a description of the medicinal ingredient, dosage form, strength, route of administration, and use of the drug with respect to which the submission or supplement has been filed, and a detailed statement of the legal and factual basis for the allegation.[221] For example, if the NOA alleges noninfringement of a listed patent, the alleged differences between the claims and the product/process of the second person must be clearly stated. If invalidity is alleged, prior art that is relied upon to invalidate the listed patent should be disclosed. Once the NOA has been filed, the second person may not add additional prior art or new grounds for noninfringement or invalidity.[222]

The NOA must also include a certification by the minister of the date of filing of the submission or supplement, and proof of service of the documents to the first party must be provided to the minister.[223]

218. Health Can., *supra* note 200, at 21.
219. Patented Medicines (Notice of Compliance) Regulations, SOR/93-133, s. 5(1).
220. *Id.* at s. 5(3)(a).
221. *Id.* at s. 5(3)(b).
222. Zahl, *supra* note 170, at 73.
223. Patented Medicines (Notice of Compliance) Regulations, SOR/93-133, s. 5(3)(c)(d).

In the event the first person's drug is considered an "innovative drug" under Subsection C.08.004.1(1) of the Food and Drug Regulations, the second person may not file an ANDS for a generic version of the drug until six years after the first person's NOC issuance, as previously described. Thus, the second person may not serve an NOA on the first person until the six-year data exclusivity period has lapsed.[224]

If the second person cancels her or his submission or supplement to which the allegation relates, or the minister deems the submission or supplement to not be in compliance with the requirements of the Food and Drugs Act, the second person must retract the NOA and serve notice of the retraction to the first person within 90 days of cancellation or notification of noncompliance.[225]

e. Preparing Application and Responding to Notice of Allegation

Once served with an NOA, a first person has 45 days to apply to the Federal Court to issue an order prohibiting the minister from issuing an NOC for the drug until after the expiration of the listed patents.[226] If the first person does not own the patent(s), the patent owner(s) must be made a party to the application. If no application is submitted within 45 days, the minister is free to issue an NOC for the generic product.

Once an application of prohibition is filed, the minister is unable to issue an NOC to the generic manufacturer for 24 months, unless the proceedings terminate before the end of the 24-month period.[227] If a prohibition proceeding is not complete within this 24-month period after receipt of proof of the application, the minister may issue the NOC.[228]

In order to support the application and meet the requisite burden of proof, the innovator responds to the NOA with evidence in the form of expert affidavits or reports, and arguments. The generic company will respond with its own evidence and arguments, after which both sides have an opportunity to cross-examine the other side's affiants. If the innovator's application is withdrawn, discontinued, or dismissed, the generic company may bring an action under Section 8 of the Regulations to recover its losses (damages) incurred as a result of being kept from participating in the market due to the inability of the minister to issue an NOC as a result of the filing of the application.

224. *Id.* at s. 5(5).
225. *Id.* at s. 5(6).
226. *Id.* at s. 6(1).
227. *Id.* at s. 7(1).
228. *Id.*

i. Infringement

The applicant bears the burden of establishing that the allegations contained in the NOA are not justified and that a prohibition order should be granted such that an NOC is not issued until after patent expiry. The innovative drug manufacturer bears the legal burden of demonstrating that the generic's evidence supporting its allegation of noninfringement is not justified.

However, if the generic manufacturer alleges that the process used to make its product does not infringe a patented process, the burden of proof is on the generic company.[229]

In the application process, the patent holder must produce evidence that its patent will be infringed by the generic product. The ANDS filed by the generic manufacturer can be used to provide evidence of infringement. A patent holder may make a motion for the generic company to produce any portion of the submission or supplement filed with the minister that is relevant to the disposition of the issues in the proceeding.[230]

ii. Validity

Allegations of invalidity in an NOA must have sufficient evidence to rebut the presumption of validity found in Subsection 43(2) of the Patent Act, and as such the burden of proof of invalidity lies with the generic company. However, despite the presumption of validity, the court has taken the stance that evidence rebutting patent validity presented by the generic company will be sufficient if it is not clearly incapable of establishing the allegations of invalidity.[231] Thus, once the second person submits any evidence capable of rebutting the presumption of validity, the burden essentially shifts back to the innovator to establish validity of its patent(s) being challenged. The relevant date for determining validity of the allegations is the date of the hearing, not the date the NOA was filed.

iii. Other Motions

The generic company may bring a motion to dismiss the application for being redundant, scandalous, frivolous, or vexatious or otherwise being an abuse of process with respect to one or more patents.[232] Motion may also be made to dismiss the application with respect to patents that are not eligible for inclusion on the register.[233] The innovator may bring a motion seeking disclosure of the generic's ANDS.

---

229. Pfizer Can. Inc. v. Novopharm Ltd., [2005] 42 C.P.R. 4th 502, para. 13.
230. Patented Medicines (Notice of Compliance) Regulations, SOR/93-133, s. 6(7).
231. Pfizer Can. Inc. v. Canada, [2007] F.C.A. 209, para. 109.
232. Patented Medicines (Notice of Compliance) Regulations, SOR/93-133, s. 6(5).
233. *Id.*

iv. Appeal Process

A Federal Court decision under the PM(NOC) Regulations may be appealed to the Federal Court of Appeal. However, if an appeal is not made and an NOC is issued by the minister, the patentee may not seek judicial review against the minister's decision to issue an NOC.

2. Patented Medicines Price Review Board Requirements

The Patented Medicines Price Review Board (PMPRB) is an independent tribunal authorised under Sections 79 to 103 of the Patent Act to monitor prices of patented medicines and order price reductions where it is deemed necessary. The board reviews the price charged by the patentee to hospitals, pharmacies, and wholesalers, but cannot control prices charged to consumers by these resalers.

a. Listing Requirements

Section 82 of the Patent Act requires a patentee who intends to sell a medicine in Canada to notify the PMPRB of its intention to sell and the date the medicine will be first offered for sale.[234]

Patentees are required to provide specific information to the PMPRB, including:

- the identity of the medicine;
- the price at which the medicine is sold in Canada or elsewhere;
- the cost of making and marketing the medicine;
- the factors used to determine what constitutes an excessive price (see Section IV.F.2.a.ii, "Price," later in this text); and
- any other related matter.[235]

i. Identity

The Patented Medicines Regulations require the patentee to submit the following information regarding the identity of the medicine:

- name and address of the patentee or former patentee and the address for correspondence in Canada;
- information as to whether the reporting patentee is the patent holder, a license holder, or any other person referred to in the definition of patentee in Subsection 79(1) of the Patent Act; the generic name and brand name of the medicine;
- whether the medicine is for human or veterinary use;
- the therapeutic use of the medicine approved by the minister of health;

234. Patent Act, R.S.C. 1985, c. P-4, s. 82(1).
235. *Id.* at s. 80(1).

- the date on which the first notice of compliance was issued to the patentee or former patentee with respect to the medicine;
- the DIN assigned to each strength and dosage form of the medicine under the Food and Drug Regulations; and
- the patent number of each invention of the patentee or former patentee pertaining to the medicine, the date on which each patent was granted, and the date on which each patent will expire.[236]

All information pertaining to the identification of the medicine must be provided within seven days of NOC issuance or the date the patented medicine is first offered for sale within Canada, whichever is earlier.[237]

ii. Price

The Patented Medicines Regulations require the patentee to submit the following information concerning the price of the medicine:

- the quantity of medicine sold in final dosage form and either
  - o the average price per package, or
  - o the net revenue from sales with respect to each dosage form, strength, and package size in which the medicine was sold by the patentee or former patentee to each class of customer in each province or territory;
- the publicly available ex-factory price for each dosage form, strength, and package size in which the medicine was sold by the patentee or former patentee to each class of customer in each province and territory; and
- if the medicine is being sold in one or more of the countries set out in the schedule, the publicly available ex-factory price for each dosage form, strength, and package size in which the medicine was sold to each class of customer in each of those countries.[238]

The preceding pricing information must be provided for the first day of sales in Canada and for each six-month period beginning January 1 and July 1.[239] The patentee has 30 days after each required date to provide the information to the PMPRB.

In calculating average price per package of medicine, the actual price of the patented medicine after any reduction provided as a rebate, discount, refund, free goods, free services, gifts, or any other like benefit and after deducting sales tax shall be used.[240] In a similar fashion, calculation

236. Patented Medicines Regulations, SOR/94-688, s. 3(1).
237. *Id.* at s. 3(2).
238. *Id.* at s. 4(1)(f).
239. *Id.* at s. 4(2).
240. *Id.* at s. 4(4)(a).

of net revenue from sales of each dosage form, strength, and package size shall include the preceding deductions in determining the final value.[241]

Pricing information pertaining to veterinary medicines or over-the-counter patented drugs need not be provided. However, the information should be maintained and must be offered to the board upon request.[242]

b. Pricing Restrictions

A patentee is responsible for reporting and submission of certain sales, pricing, and research and development (R&D) data to the PMPRB. The PMPRB reviews the information obtained from the patentee and will determine if the price of the patented medicine is within the PMPRB Guidelines. If the investigation determines the price to be excessive, the PMPRB may direct the patentee to reduce the price to a level deemed not excessive as well as levy fees and fines.[243]

To determine what constitutes an excessive price, the board must consider:

- the price the medicine has been sold for in the relevant market;
- the price other medicines in the same therapeutic class have been sold for in the relevant market;
- the price the medicine and other medicines in the same therapeutic class have been sold for in other countries;
- changes in the consumer price index; and
- any other factors specified in the Patented Medicines Regulations.[244]

The board may not consider research costs other than the Canadian portion of the world costs related to the research that led to that medicine or to the development and commercialization of that invention, calculated in proportion to the ratio of sales by the patentee in Canada of that medicine to total world sales.[245]

c. Compendium of Policies, Guidelines and Procedures

The PMPRB outlines its price review process in the Compendium of Policies, Guidelines and Procedures.[246] The board will apply the guidelines to determine if the price of a medicine is excessive. The allowable introductory price of a new patented drug is dependent on the level of therapeutic improvement,

241. *Id.* at s. 4(4)(b).
242. *Id.* at s. 4(3).
243. Patent Act, R.S.C. 1985, c. P-4, s. 83.
244. *Id.* at s. 85(1).
245. *Id.* at s. 85(3).
246. *Compendium of Guidelines, Policies and Procedures*, PATENTED MED. PRICES REV. BOARD, http://www.pmprb-cepmb.gc.ca/english/View.asp?x=1733 (last updated June 2012).

which is categorized as breakthrough, substantial improvement, moderate improvement, or slight/no improvement over existing therapies.

i. Breakthrough Drugs

A breakthrough drug may be priced at the median of prices charged for the same drug in other industrialized countries listed in Schedule 5 of the Regulations.[247]

ii. Substantial Improvement Drugs

A drug providing a substantial improvement over existing therapies may be priced at the higher of the median international price and the highest non-excessive priced product in the same therapeutic class.[248]

iii. Moderate Improvement Drugs

A drug that provides a moderate improvement over existing therapies may be priced at the midpoint between the median international price and the highest non-excessive priced product in the same therapeutic class.[249]

iv. Slight or No Improvement Drugs

A drug that provides slight or no improvement over existing therapies will be reviewed using a "therapeutic class comparison test" to ensure the price is not in excess of comparable drugs.[250]

G. Other Patent Actions

1. Actions and How They Differ from Applications under the PM(NOC) Regulations

a. Parties

An action for infringement may be commenced in either Federal Court or a provincial court by a patentee or a person claiming under a patentee. Both exclusive and nonexclusive licensees are considered to be a person claiming under a patentee.[251] The patentee must be made a party to any infringement proceeding, unless otherwise specified by court order.[252]

An action for infringement begins with a statement of claim alleging the defendant(s) directly or indirectly (inducement) infringed the claims in a patent. This differs from an application under the PM(NOC) Regulations, which is an application to the court for an order prohibiting the

247. *Id.*
248. *Id.*
249. *Compendium of Guidelines, Policies and Procedures*, *supra* note 246.
250. *Id.*
251. Armstrong Cork Can. Ltd. v. Domco Indus. Ltd., [1982] 1 S.C.R. 907.
252. Patent Act, R.S.C. 1985, c. P-4, s. 55(3).

minister of health from issuing an NOC to the manufacturer wishing to gain approval for a generic version of a drug.

b. Timing

i. Limitation Period

An action for infringement in Federal Court may be commenced within six years of the infringing act.[253] Limitation periods to commence an action for infringement in provincial courts vary depending on the applicable provincial legislation. The vast majority of patent actions are brought in Federal Court, as it has jurisdiction to invalidate a patent in rem whereas a provincial court does not. For this reason, only patent actions arising in Federal Court will be discussed further.

ii. Filing of Documents

Once a patentee files a statement of claim with the court alleging infringement, it must be served on the defendant within 60 days.[254] The defendant may then file a statement of defense and counterclaim, indicating that it is not infringing and/or alleging that the patent is invalid. The statement of defense and counterclaim must be served on the plaintiff and filed with the court within 30 days of service of the statement of claim, provided the defendant was served in Canada.[255]

The patentee may then file a reply and defense to the counterclaim within 30 days of service of the statement of defense and counterclaim, indicating that the defense is not warranted and the patent is, in fact, valid.[256] The defendant may then file a reply to the defense to the counterclaim filed by the patentee. If the defendant files a statement of defense without a counterclaim, the plaintiff may reply to the statement of defense within ten days after service.[257]

iii. Motions

The action then proceeds with motions, often requesting particulars,[258] to strike pleadings,[259] or for summary judgment or a summary trial.[260]

iv. Discovery

Often the parties commence document discovery, whereby both parties must serve the opposing party and the court an affidavit of documents

253. *Id.* at s. 55.01.
254. Federal Court Rules, SOR/98-106, s. 203.
255. *Id.* at s. 204.
256. *Id.* at s. 192.
257. *Id.* at s. 205.
258. *Id.* at s. 181(2).
259. *Id.* at s. 221(1).
260. *Id.* at s. 213.

containing schedules of all documents relevant to the proceedings. Every party to the action must serve an affidavit of documents on every other party within 30 days after the close of pleadings.[261] There is an obligation to correct any inaccuracy or deficiency in an affidavit of documents throughout the duration of the proceedings once the party becomes aware of such deficiencies.[262] Examination for discovery follows document discovery, where one party will conduct an oral examination of the other party regarding the contested matters of the action.[263]

v. Expert Reports

Expert reports (if necessary as an aid to the court) may be exchanged. Expert reports are only admissible at trial if the expert witness is available for cross-examination.[264] There is no expert "discovery" in Canada or cross-examination on the expert report other than during the trial or hearing before the court.

vi. Settlement Discussions and Pretrial Conference

Within 60 days after the close of pleadings, the solicitors for both parties must engage in discussions to settle any or all issues in the action.[265] If a trial is necessary, either party may request a pretrial conference, which the court will schedule within 60 days of requisition.[266] Offers made in writing can affect the ultimate award as to costs if the final award is less than the offered amount.[267]

vii. Trial

Once pleadings, discovery, settlement discussion, and pretrial conference are complete and all evidence has been submitted, the trial may begin. The Federal Court strives to provide trial dates within two years from the commencement of litigation.

viii. Appeal

A final judgment of the Federal Court of Canada may be appealed to the Federal Court of Appeal.[268] Within 30 days of judgment, the applicant for appeal will file a notice of appeal that clearly and concisely sets out the grounds for appeal and the relief sought.[269] If an appeal of a Federal Court

261. *Id.* at s. 223(1).
262. *Id.* at s. 226(1).
263. *Id.* at s. 234.
264. *Id.* at s. 279.
265. *Id.* at s. 257.
266. *Id.* at s. 258–59.
267. *Id.* at s. 420.
268. Federal Courts Act, R.S.C. 1985, c. F-7, s. 27(1).
269. *Id.* at s. 27(2).

of Appeal decision is desired, an application must be made to the Supreme Court of Canada for leave to appeal. Unlike the Federal Court of Appeal, there is no right of appeal to the Supreme Court. The Supreme Court will consider, among other factors, whether the issues in dispute are of "national importance" in deciding whether to grant leave.

c. Available Remedies

i. Injunction

(a) Interlocutory Injunction

In an action for infringement, a patentee may bring a motion for interlocutory injunction either before or during trial that restrains the defendant from any further use, manufacture, or sale of the patented item.[270] An injunction is an equitable remedy that is very rare. The Federal Court of Canada generally finds that the required element of "irreparable harm" in the absence of injunction has not been met; any harm is reparable with an award of damages.

(b) Permanent Injunction

If a patentee is successful in the action for infringement, the court can issue a permanent injunction forbidding the infringing party from making, constructing, using, or selling the patented invention until patent expiry.[271]

ii. Delivery Up or Destruction

The court may order the party who was found to have infringed a patent to deliver up all infringing goods that remain in that party's possession or to destroy them under oath.

iii. Damages

Any person who infringes a patent is liable to the patentee and all persons claiming under the patentee for damage sustained by reason of infringement after the patent is issued.[272] The successful patentee is generally entitled to select between damages or an accounting of the infringer's profits. Punitive damages are also available under certain circumstances.

iv. Costs

The successful party in a civil case, such as a patent infringement action, may be able to recover a portion of its legal costs. The amount of costs that are recoverable will depend on a number of factors to be reviewed by the court.[273]

---

270. Patent Act, R.S.C. 1985, c. P-4, s. 57(1).
271. *Id.*
272. *Id.* at s. 55(1).
273. Federal Courts Rules, SOR/98-106, s. 400(3).

2. Impeachment Actions

The attorney general of Canada or any interested party may bring an application to the Federal Court to have a patent or any claim in a patent declared invalid.[274] As opposed to an action for infringement, in an impeachment action the interested party is the plaintiff and the patentee is the defendant.

In addition to seeking a declaration of invalidity or voiding of a patent, any person who reasonably believes he or she may be sued under a patent may bring an action in Federal Court to seek a declaration of noninfringement.[275]

A person who seeks to impeach a patent or seeks a declaration of noninfringement must post security for costs of the patentee in an amount determined by the court.[276]

A successful plaintiff does not receive a monetary award for a successful impeachment action. Rather, such plaintiff will receive the desired invalidation of claims or voiding of the patent, or the desired declaration of noninfringement.

A licensee may be estopped from challenging the validity of the patent being licensed,[277] absent an appropriate clause in the license agreement.

H. Marketing Considerations

No person may sell or advertise a new drug unless the requisite conditions from Section C.08.002 of the Food and Drug Regulations have been satisfied. These conditions include:

- Filing the appropriate new drug submission with the minister;
- Issuance of an NOC to the manufacturer of the new drug with respect to the submission;
- Ensuring the NOC has not been suspended; and
- Submission of label specimens to the minister, including package inserts, product brochures, and file cards, and a statement setting out the proposed date on which the labels will first be used.[278]

1. Advertising

The Food and Drugs Act and Regulations govern the packaging, labeling, and advertising of drug products in Canada. Health Canada's Therapeutic Products Directorate (TPD) regulates drug advertising to ensure information

274. Patent Act, R.S.C. 1985, c. P-4, s. 60(1).

275. *Id.* at s. 60(2).

276. *Id.* at s. 60(3).

277. Dableh v. Ontario Hydro, [1993] F.C.J. 924, 50 C.P.R. 3d 290, 355–56 (FCTD) (reversed on other grounds).

278. Food and Drug Regulations, C.R.C., c. 870, C.08.002(1).

contained in advertisements are not false, misleading, or deceptive. The TPD sets standards for advertising and delegates some authority to independent agencies, such as Advertising Standards Canada (ASC) and the Pharmaceutical Advertising Advisory Board (PAAB), the latter of which is responsible for advertising of health products directed to health professionals.[279]

The Food and Drugs Act defines "advertisement" as any representation by any means whatever for the purpose of promoting directly or indirectly the sale or disposal of any food, drug, cosmetic, or device.[280] If the purpose of a message is to promote the sale or disposal of a drug, then it will be subject to the legislative regulation, whereas a message that is deemed not to promote sale or disposal of the drug will not. The Food and Drug Regulations restrict advertisements for prescription drugs to mentioning the brand name, proper name, common name, price, and quantity of the drug.[281] Thus, an advertisement in which the name of a prescription drug is mentioned without the name of the disease is permitted. However, depictions of easily recognizable product packages, including blister packs or inhalers, that could lead to the identification of the therapeutic indication of the prescription drug being advertised are not permitted. Schedule D drugs, including vaccines and insulin, are not subject to the "name, price, and quantity" restriction.

No drug may be promoted or advertised until market authorisation for the drug has been approved from the minister in the form of an NOC.[282] Further, no person may label, package, treat, process, sell, or advertise any drug in a manner that is false, misleading, or deceptive or that is likely to create an erroneous impression regarding its character, value, quantity, composition, merit, or safety.[283]

Although it is not mandatory to have drug advertisements reviewed prior to release, Health Canada strongly encourages preclearance by independent advertising preclearance agencies.[284]

---

279. Health Can., Overview of Health Product Advertising (Health Prods. & Food Branch 2006).

280. Food and Drugs Act, R.S.C. 1985, c. F-27, s. 2.

281. Food and Drug Regulations, C.R.C., c. 870, C.01.044.

282. *Id.* at C.08.002(1).

283. Food and Drugs Act, R.S.C. 1985, c. F-27, s. 9(1).

284. Health Can., Overview of Health Product Advertising (Health Prods. & Food Branch 2006).

2. Labeling

a. Principal Display Panel

The Food and Drug Regulations indicate specific requirements that must be present on the principal display panel of the inner and outer labels of a drug. These include:

- the proper name of a drug immediately preceding the brand name of the drug (if any);
- the common name of the drug if there is no proper name;
- if prescribed standards in Division 6 of Part C apply, a statement that the drug is a Canadian Standard Drug (CSD may be used as an abbreviation);
- where no prescribed standard in Division 6 of Part C exists, but a standard is contained in a publication mentioned in Schedule B to the Food and Drugs Act, the name of the publications containing the standard used or its abbreviation;
- if a manufacturer's standard is used, a statement setting forth the fact that such a standard is used; and
- the notation "sterile" in both official languages if the drug is required to be sterile under the regulations.[285]

b. Any Panel

In addition to the requirements of the principal display panel, the label of a drug must show on any panel:

- the name and address of the manufacturer of the drug;
- the lot number of the drug;
- adequate directions for use of the drug;
- a quantitative list of the medicinal ingredients by their proper names or common names (if no proper name exists);
- the expiration date; and
- if the new drug was approved for extraordinary use, a prescribed statement indicating this authorisation.[286]

c. Outer Label

In addition to the preceding requirements, the outer label must show:

- the net amount of drug in terms of weight, measure, or number;
- for drugs intended for parenteral use, a quantitative list of preservatives by their proper names or common names; and

285. Food and Drug Regulations, C.R.C., c. 870, C.01.004(1)(a).
286. *Id.* at C.01.004(1)(c).

- a quantitative list of all mercurial preservatives by proper names or common names.[287]

I. Trade-mark Considerations

Innovative pharmaceutical companies sometimes seek trade-mark protection for the appearance of the drug products sold. This appearance is often referred to as a product's "trade dress." Generic pharmaceutical companies will often oppose these applications to ensure they may sell their generic version of the same drug using the same trade dress as the innovator. The Supreme Court of Canada has held that color alone cannot constitute the basis for a registrable trade-mark.[288] However, the Federal Court has since acknowledged that a trade-mark for a particular color applied to a particular size and shape of tablets is valid.[289] The Trade-marks Office generally indicates that while color alone is not registrable, it may be acceptable if it is claimed as a feature of a trade-mark.

A trade-mark consisting of a combination of product, shape, and color is inherently registrable, but there is a high evidential burden on the applicant to establish that the mark serves to distinguish its goods and services from those of others. While combinations of color, shape, and size of a product may together constitute an acceptable trade-mark, the resulting mark is generally seen as weak.[290]

A three-dimensional (shape) mark of a drug tablet, capsule, or device can be protected under trade-mark law as a distinguishing guise. A distinguishing guise is registrable only if it has been used in Canada such that the drug product is rendered distinctive as of the filing date, and exclusive use of the distinguishing guise will not unreasonably limit the development of the drug industry.[291]

J. Conclusion

In summary, a company seeking authorisation to sell a new drug product in Canada must file an NDS with HPFB. The NDS must contain data sufficient to establish that the product is safe, effective, and of acceptable quality. A manufacturer seeking authorisation to sell a generic version of a previously approved drug must file an ANDS with HPFB. The ANDS must contain data sufficient to establish that the generic drug has equivalent safety, efficiency, and quality.

287. *Id.* at C.01.004(2).

288. Henry K Wampole & Co. v. Hervay Chem. Co., [1930] S.C.R. 336.

289. Smith Kline & French Can. Ltd. v. Canada (Registrar of Trade-marks), [1987] 2 F.C. 633 (FCTD).

290. Novopharm Ltd. v. Bayer Inc., [2000] 9 C.P.R. 4th 304 (FCA).

291. Trade-marks Act, R.S.C. 1985, c. T-10, s. 13(1).

Before obtaining marketing approval, a generic manufacturer may have to address patents listed on the patent list submitted by the innovator pursuant to the PM(NOC) Regulations or await expiry of the data protection period.

V. The United Kingdom and European Union Approval Considerations[292]

After a brief introduction to the legal structure of the European Union, this section discusses the patentability of medicinal products in the European Union and the regulation of clinical trials as well as the framework regulating the placing of medicinal products for human use on the market (licensing procedure). It also includes a discussion of trade-marks for pharmaceuticals, supplementary protection certificates for patented products, regulatory data protection, and various regulatory/trade issues that arise once the product has been authorised.

A. Introduction to the European Union Legal Structure

With the accession of Croatia on July 1, 2013, the European Union (EU) is comprised of 28 member states,[293] being a single market with free movement of goods, services, persons, and capital. The laws of the EU are governed by regulations and directives.[294] In the case of directives, each EU member state must take its own steps to implement the directive into national law. In doing so, it must ensure that implementation delivers the outcomes required by the directive, although there is an element of discretion in deciding how to go about achieving the result, to the extent that the directive does not prescribe a particular approach. This section of the chapter explains many of the topics covered at the level of the provisions of European legislation, without reference to implementation at the member state level. Where more specific aspects of procedures and practices are examined, reference is made to the United Kingdom (UK). It is important to note that other EU member states will have adopted their own legal instruments to implement the relevant directives and may not

292. Jennifer Jones and Sarah Faircliffe of Bird & Bird LLP, London, United Kingdom.

293. Austria (1995), Belgium (1952), Bulgaria (2007), Croatia (2013), Cyprus (2004), Czech Republic (2004), Denmark (1973), Estonia (2004), Finland (1995), France (1952), Germany (1952), Greece (1981), Hungary (2004), Ireland (1973), Italy (1952), Latvia (2004), Lithuania (2004), Luxembourg (1952), Malta (2004), The Netherlands (1952), Poland (2004), Portugal (1986), Romania (2007), Slovakia (2004), Slovenia (2004), Spain (1986), Sweden (1995), and United Kingdom (1973).

294. EU legislation extends to three non-EU countries by way of the European Economic Agreement—being Iceland, Liechtenstein, and Norway.

have an identical approach. In addition, certain areas, such as the pricing and reimbursement of medicines, are effectively dealt with at the member state level, and each will have its own system. Regulations are described as having "direct effect" as legal instruments, which means that their provisions are binding at EU member state level without the need for implementing legislation.

Although national courts apply the laws of the EU, the judicial authority and highest court in the EU is the Court of Justice of the European Union (CJEU). Seated in Luxembourg, the CJEU is comprised of the Court of Justice (CJ) and the General Court (GC). The General Court hears appeals over the registration of community trade-marks, for instance, with an appeal available to the CJ. The CJ is composed of 28 judges, one from each member state, and eight advocates general, who provide an independent opinion in cases assigned to them. In order to provide, as much as possible, harmonisation across the EU, if a particular area of EU law is unclear, a member state court has the option of referring a question or questions to the CJ. The CJ will then provide a ruling, which is binding across the EU. Referrals to the CJ are available in all areas of EU law governed by regulations or directives.[295] Reported decisions of the CJ and GC are found in the European Court Report (ECR). However, many decisions are not reported and can be found at the European Court website, Curia (http://curia.europa.eu/juris/recherche.jsf?cid=11466).

B. Patents in the European Union

1. European Patent Convention (EPC)

In the EU, a patent can be prosecuted centrally under the European Patent Convention (EPC), undergoing a single examination at the European Patent Office (EPO), located in Munich and The Hague. Once granted, however, the "European patent" becomes a bundle of national patents. The EPC dates to 1973 and was revised in 2000. It is an international convention with 38 signatories, rather than just the 28 member states of the EU.[296] Patent applications can also be prosecuted nationally in national intellectual property (IP)

295. As will be discussed briefly as patent law is largely harmonised under the European Patent Convention (EPC), which is an international agreement, rather than a regulation or directive, no reference to the CJ is available. Also the EU is set to welcome a Unitary Patent and Unified Patents Court.

296. Besides the 28 member states, Iceland, Monaco, Norway, Switzerland, (including Liechtenstein) and Turkey are also included. In addition, Albania, Bosnia, and Herzegovina, the former Yugoslav Republic of Macedonia and Serbia recognise patents granted by the EPO.

offices of each of the EPC member states, and many member states provide the option of protecting inventions by utility model.[297]

2. Enforcement of Patents in National Courts and the EPO

Once patents grant via the EPO route, they may be enforced and challenged in the national courts of the member states. Their validity can also be challenged by third parties centrally at the EPO nine months after grant in opposition proceedings, whereby the patentee can seek to amend the patent centrally, with the effect that all national patents become so amended. Similarly, at any time during the life of the patent, a patentee can seek to amend a national patent in the national IP office or centrally at the EPO.

The first stage of proceedings at the EPO is before the Opposition Division, which consists of three examiners (and often will include the examiner who prosecuted the patent). After the nine-month opposition period in which oppositions have been filed, the patentee is given an opportunity to respond in writing. During an oral hearing, each of the invalidity arguments is considered, with the patentee provided an opportunity to advance amended claims in order to overcome an adverse ruling. Each ground of invalidity is then considered again for the new claims, until the amended claims are found valid or until the patentee has no further claims to advance and the patent is found invalid. An appeal to the Technical Board of Appeal (TBA) can be filed by either party. The TBA is comprised of three members, typically two of whom are technically, rather than legally, qualified. Typically, the TBA reverses half of the decisions of the Opposition Division, and any decision is final. A point of law may be referred to the Enlarged Board of Appeal, which is comprised of five legally qualified and two technically qualified members. If an EPO decision is not reported in the EPO Official Journal, it can be found at the "Case law and appeals" section of the EPO website (http://www.epo.org/law-practice/case-law-appeals.html).

Given that the EPC is an international agreement, the Court of Justice does not interpret patent law. Harmonisation of patent law throughout Europe has occurred by way of the EPC as well as extensive case law from the EPO Boards of Appeal. Although national courts are not bound by decisions of the Boards of Appeal, courts particularly situated in the most often litigated countries of the UK, Germany, France, and The Netherlands are keenly aware of and consider the corresponding EPO decisions as well as decisions from other national courts. Courts specializing in intellectual property disputes are found in the UK, Germany, The Netherlands, France, and Italy.

297. Utility models differ from patents in that they are not generally examined before grant and they provide more limited terms of protection than patents.

Patent litigation in all but Germany and Austria proceeds with infringement and validity heard together. Germany and Austria have what is referred to as a split system whereby infringement and validity are heard separately, with the validity trial generally taking twice as long as infringement.

3. Unitary Patent and Unified Patents Court

On December 17, 2012, two regulations were adopted that provide for a Unitary Patent, being a single patent granted centrally that will be effective in all EU member states except Italy and Spain. Two months later, in February 2013, all EU member states except Spain and Poland signed the Unified Patent Court Agreement, which provides for the new Unitary Patent and existing and future European patents (bundle patents) to be litigated centrally, in a new court system.

The Unitary Patent essentially adopts the framework of the European Patent Convention. Although it appears a sensible approach for Europe, it has taken over 40 years to achieve, amid struggles over language and translations. The final agreement with respect to translation arrangements envisages only one translation filed, into English if the language of the EPO prosecution is French or German and into French or German if the language of the proceedings is English. Therefore, there will always be an English text. Italy and Spain were implacably opposed to these linguistic arrangements as the scheme did not accord their languages the same status, and, as a result, they chose not to become parties to the agreement.

Although the Unitary Patent regulations entered into force on January 20, 2013, they will apply from January 1, 2014, or the date of entry into force of the Unified Patent Court Agreement, whichever is later. The agreement will only enter into force upon ratification from at least 13 contracting states, with ratification required from France, Germany, and the United Kingdom. As of November 2013, Austria had ratified, with other member states expected to follow.

The Unified Patent Court (UPC) will comprise a Court of First Instance and a Court of Appeal. The Court of First Instance will be composed of a central division as well as local and regional divisions. The seat of the central division is Paris, with two "thematic clusters"—one in Munich for mechanical engineering, and the other in London for pharmaceuticals and "human necessities." Regional divisions will be comprised of two or more contracting states and may hear cases in multiple locations. A contracting state may also set up a local division; however, a second local division will require that state to hear over 100 patent cases in a calendar year.

Draft Rules of Procedure were published for the UPC in July 2013 with the purpose of soliciting comments, due by October 2013. As of summer

2013, the Preparatory Committee[298] had estimated that the UPC will be established by mid-2015. Importantly, once established, there will be a transitional period of seven years, which is extendible by a further seven years, whereby actions for infringement or revocation may still be brought before national courts, and proprietors of, or applicants of, a European patent granted or applied for will have the option to opt out of the UPC, provided an action has not already been brought.

4. Patentability Requirements

That an invention be new, involve an inventive step, and be capable of industrial application is provided for in Article 52(1) EPC:

> European patents shall be granted for any inventions, in all fields of technology, provided that they are new, involve an inventive step and are susceptible of industrial application.

Exclusions from patentability that are relevant to pharmaceuticals and biotechnology are found under Article 52(2)(a) EPC (as qualified by Article 52(3) EPC) for "discoveries"[299] and under Article 53 EPC, which provides for exclusions for:

> (a) inventions the commercial exploitation of which would be contrary to "ordre public" or morality; such exploitation shall not be deemed to be so contrary merely because it is prohibited by law or regulation in some or all of the Contracting States;
>
> (b) plant and animal varieties or essentially biological processes for the production of plants or animals; this provision shall not apply to microbiological processes or the products thereof;
>
> (c) methods for treatment of the human or animal body by surgery or therapy and diagnostic methods practised on the human or animal body; this provision shall not apply to products, in particular substances or compositions, for use in any of these methods.

That a patent must disclose the invention in a manner sufficiently clear and complete for it to be carried out by a person skilled in the art, and

298. Composed of each of the member states party to the UPC designed to oversee five work streams: legal framework, financial aspects, information technology, facilities, and human resources and training. *See* UNIFIED PAT. CT., http://www.unified-patent-court.org/ (last visited May 22, 2014).

299. The "mere discovery" objection under Article 52(2)(a) EPC is in practice rare, as it does not exclude from patentability the practical application of a discovery.

the requirement that any amendment to claims must not add matter, are provided for by Article 100(b) EPC (and Article 83 EPC during patent prosecution) and Article 138(1)(d) EPC, respectively.[300] Unity of invention and sufficiency are each provided for by Articles 82 and 84 EPC, respectively.

5. Pharmaceutical Patents

Under the EPC, a pharmaceutical patent may relate to a new compound, or a new class of compounds,[301] an enantiomer for which the racemate is known,[302] and to known compounds where the invention lies in the process, formulation, combination,[303] or medical use.[304] Discussed in following sections are patents to formulations and new medical uses of known compounds.

a. Exclusions from Patentability for Pharmaceuticals

The "method of treatment" exclusion, as provided for in Article 53(c) just discussed, was based on the desire to not impede medical practitioners in providing treatment. However, "medical use" claims, discussed in a later section, achieve the same result as a claim to a method of medical treatment.

The last sentence of Article 53(c) makes clear that although patent protection cannot be obtained for products that are used in the surgical and

300. A patent cannot be challenged during EPO opposition proceedings on the grounds of added matter, as there is no scope for the claims of a European patent that has only just been granted to have been extended by amendment. Claims can be amended during the opposition proceeding as long as the amendments do not add matter, as per Article 123(3) EPC. Article 138 sets out the only grounds under which a national court may revoke a European patent under the law of a contracting state.

301. Selection patents may be allowed if when selecting a group of compounds from the prior-disclosed wider class of compounds, the selected group has a certain advantageous property or unknown technical effect, which is caused by the structural features that distinguish the selected group from the prior-disclosed group. *See* T-939/92 Triazoles/AGREVO [1996] OJEPO 309; Dr. Reddy's Labs. v. Eli Lilly, [2008] EWHC (Pat) 2345; [2009] C.A. 1362.

302. Although claims to an enantiomer are novel over its racemate, inventive step is more difficult given the routine testing of the individual enantiomers in a racemate. The claim may be inventive, however, if the compounds have more than one asymmetrical carbon atom, presenting a number of possible isomers, or if the resolution of the enantiomers proves particularly difficult. *See* T-296/86 enantiomers/HOECHST; Generics (UK) Ltd. v. H Lundbeck A/S, [2007] EWHC (Pat) 1040, [2008] C.A. 311, [2009] RPC 13 (H.L.).

303. To be patentable, a combination of known compounds must provide a "synergistic effect" and not just an effect that is merely additive or that is merely a bonus effect attributable to testing what is already an obvious combination. *See* T-1429/04 UNILEVER/Biocidal Activity, 2006 EPOR 50. Patentability of a combination of known compounds may also occur if the combination is the result of overcoming a technical prejudice to use the compounds together, although a technical prejudice can be difficult to demonstrate. *See* T-0794/05 ASTRA ZENECA/Asthma-Combination; T-1028/05 ASTRA ZENECA/New Therapeutic Use.

304. Not discussed here are "product-by-process" claims, which are rarely granted and provide little or no benefit over a process claim, and "reach through claims," which are also rarely granted and after a decision by the EPO TBA in T-1063/06 Reach-Through Claim/ BAYER SCHERING PHARMA AG, 2009 OJEPO 516, are likely to be found invalid for insufficiency.

therapeutic treatment of humans and animals, the new chemical entity or pharmaceutical itself is patentable.

What is a "therapy" for the purposes of Article 53(c) has been considered by the Boards of Appeal of the EPO. A treatment for relieving discomfort associated with menstruation was considered a "therapy,"[305] as was treatment of external parasites on pigs used on a large scale by farmers.[306] Both preventative and curative treatments are considered within "therapy" as both have the same goal of maintenance or restoration of health.[307] Although a cosmetic treatment is not caught by the exception, whether a condition is considered therapeutic or cosmetic can be difficult to predict. For instance, treatment for suppression of appetite was considered cosmetic,[308] whereas treatment for dental plaque was held to be a "therapy."[309]

If one step in a diagnostic process is considered surgical, the exclusion applies, as was the case for a claim to an imaging method for a diagnostic purpose whereby the surgeon injected polarized ^{129}Xe into the heart of the patient to evaluate blood flow. The exclusion from patentability may be avoided by disclaiming the offending embodiment, if possible without offending the other patentability criteria.

b. Formulation Patents

As many of the approaches to formulating pharmaceuticals are well established, patents to formulations are often found invalid for lack of inventive step. Two exceptions were for a patent held by Novartis[310] for a formulated microemulsion of cyclosporine and a patent held by Gedeon Richter for contraceptive formulations.

In the first instance, claim 1 was to:

> A pharmaceutical composition comprising a cyclosporine as active ingredient, (1) a hydrophilic phase, (2) a lipophilic phase, and (3) a surfactant which composition is an "oil-in-water microemulsion pre-concentrate."

Although microemulsions were known, they had not previously been used in the pharmaceutical field and, therefore, did not form part of the common general knowledge of the skilled person.

305. T-0081/84 RORER/Dysmenorrhea, 1988 OJEPO 297.
306. T-0116/85 DUPHAR/Pigs I, 1989 OJEPO 13, 1988 EPOR 1.
307. T-0019/86 DUPHAR/Pigs II, 1989 OJEPO 24, 1988 EPOR 10.
308. T-144/83 DUPONT/Appetite Suppressant, 1987 EPOR 6.
309. T-290/86 ICI/Cleaning Plaque, 1991 EPOR 157.
310. Novartis AG v. IVAX Pharm. UK Ltd., 2006 EWHC (Pat) 2506.

The UK High Court, upheld on appeal, found a formulation patent valid in *Gedeon v. Richter Plc v. Bayer Schering Pharma AG*.[311] The claims at issue concerned two steroidal hormones that regulate the female menstrual cycle and are used as contraceptives. The claims were directed to a tablet formulated so as to rapidly dissolve in the stomach, as defined by an in vitro dissolution test. Although it was agreed between the parties that the in vitro steps involved were routine and would provide the skilled person with the rate of dissolution of the formulation, there was disagreement between the experts over what decision would be taken after such testing. In the opinion of the expert for the patentee, the skilled researcher would test only an enterically coated tablet (which would not meet the claimed dissolution test), whereas in the opinion of the claimant's expert, testing would occur for both an uncoated and coated formulation. The court held that it would not be routine to do animal tests on an uncoated formulation; such a decision would be for the skilled formulator and would be taken without the necessary "fair expectation of success."

c. Medical Use Patents

A first, second, and subsequent medical use of a known compound may be patentable, as per Articles 54(4) and 54(5) EPC.

> 54(4) Paragraphs 2 and 3 [which provide for what constitutes the state of the art] shall not exclude the patentability of any substance or composition, comprised in the state of the art, for use in a method referred to in Article 53(c), provided that its use for any such method is not comprised in the state of the art.
>
> 54(5) Paragraphs 2 and 3 shall also not exclude the patentability of any substance or composition referred to in paragraph 4 for any specific use in a method referred to in Article 53(c), provided that such use is not comprised in the state of the art.

Thus, Article 54(4) provides for a claim to a first medical use of a known compound, and Article 54(5), to second and subsequent uses. Article 54(5) was introduced with the EPC 2000 amendments, which came into

311. 2011 EWHC (Pat) 583. At the Opposition Division at the EPO the patent was revoked (EP 1 380 301) as the amendments were held to add matter, and, therefore, the question of whether the formulation was obvious was not considered. The decision was appealed to the TBA, and an oral hearing took place in November 2013. However, as the TBA will only consider the issues raised in the OD decision, only if the OD decision is reversed will the patent be sent back to the OD to consider novelty and inventive step. In any event, the OD's decision was upheld by the TBA.

force in December 2007. Previously, such second and subsequent medical uses were claimed in what was referred to as "Swiss form" claims, which were worded thus:

> The use of a [known substance or composition] for the manufacture of a medicament for a specified new and inventive therapeutic application.

Following a precedent set in Switzerland, the claim was allowed as it referred to the process of manufacture, regardless of whether the process was known.

Following a decision of the EPO in 2010 discussing the introduction of Article 54(5),[312] the UK IP Office provided guidance as to how second medical use claims will be examined, such that only claims of the form "substance X for use in the treatment of disease Y" will be allowed, with any "Swiss-type" claims being rejected for lack of clarity. Amendment of a patent application to replace a Swiss-type claim with the new form of second medical use claim will be allowed; however, post-grant amendment is unlikely to succeed as the new form may have a broader scope.

A new medical use claim may be allowed for a compound known for treating different patient populations. For instance, in T-0019/86 DUPHAR/Pigs II,[313] treatment of a known condition in seropositive pigs was considered a new therapeutic application despite the same compound being used for the same treatment in seronegative pigs (lacking maternal antibodies). Similarly, a claim for treatment of an expanded patient population may be patentable—for instance, if the pharmaceutical can be administered for a known therapy in combination with substances previously contraindicated for that group of patients.

New modes of administration may also be allowable, as in T-0051/93 SERONO/HCG.[314] Human chorionic gonadotropin, hCG, used for treating male sexual disorders, was previously administered only intramuscularly as it was understood that administration subcutaneously would require an increase in dosage to achieve the same therapeutic effect, which was known to lead to undesirable side effects. The invention was the administration of hCG subcutaneously. Although this provided lower hCG blood levels, it produced the same therapeutic levels as intramuscular delivery.

A claim that differs from the prior art only as to its dosing regime (provided it is not a mere selection within the teaching of a broader prior

---

312. *See Nicotinic Acid Compositions for Treating Hyperlipidemia*/Abbott Respiratory LLC, G 0002/08, 19 February 2010, Enlarged Board of Appeal, 2010 OJEPO 456.

313. T-019/86 DUPHAR/Pigs II, 1989 OJEPO 24.

314. (unpublished) (8 June 1994), *available at* http://www.epo.org/law-practice/case-law-appeals/recent/t930051eu1.html.

disclosure) may also be patentable as a new medical use, as held by the Enlarged Board of Appeal in February 2010.[315] The claimed dosage regime was "once per day prior to sleep" with the prior art dosing being a sustained release medicament for administration at any time of the day. Administration before sleep had the result of reducing or eliminating side effects (being an elevation in liver enzymes, causing liver damage). The suggestion in the prior art to reduce side effects was to reduce the dosage or stop taking the drug altogether. The Enlarged Board of Appeal was asked to answer the legal question of whether under Article 54(5) dosing regime claims were patentable, which was answered in the affirmative. The UK Court of Appeal has also held valid a patent claiming a new dosing regime (from 5 mg per day to 0.05–1.0 mg per day) for finasteride, used for male pattern baldness.[316]

6. Biotechnology/Biopharmaceutical Patents

Biopharmaceuticals are biological molecules used as medicines, such as synthesized insulin or the breast cancer antibody drug Herceptin. The preceding discussion for pharmaceuticals applies to biopharmaceuticals, with specific issues relating to biotechnology discussed next.

a. Directive on Legal Protection of Biotechnological Inventions

The directive was meant to provide a harmonised position across the EU with respect to patenting biotechnology, although by the time it was implemented into national law in July 2000 much of what the directive provides had already been achieved by virtue of a succession of decisions of the EPO. The scope of the directive is limited to "biological material," which is defined to mean "any material containing genetic information and capable of reproducing itself or being reproduced in a biological system." Therefore, the directive has no direct application to proteins as, unlike DNA, proteins contain amino acids and are not capable of reproducing themselves or being reproduced. The directive is discussed in following sections where applicable.

b. "Product of Nature"

EPO guidelines express the extent to which a "product of nature" can be claimed:

> To find a previously unrecognised substance occurring in nature is also mere discovery and therefore unpatentable. However, if a substance found in nature can be shown to

---

315. *See* G 0002/08, n312.

316. Actavis UK Ltd. v. Merck & Co., [2008] EWCA (Civ) 444. In Germany, the Bundespatentgericht decision also held that a dosing regime can be patentable, but in this particular case found the claim lacked novelty, 26 June 2008, Ni 58/06 (EU).

> produce a technical effect it may be patentable. An example of such a case is that of a substance occurring in nature which is found to have an antibiotic effect. In addition, if a micro-organism is discovered to exist in nature and to produce an antibiotic, the micro-organism may itself be patentable as one aspect of the invention. Similarly, a gene which is discovered to exist in nature may be patentable if a technical effect is revealed, *e.g.,* its use in making a certain polypeptide or in gene therapy.[317]

Hence, biological materials, such as genes and proteins, may be patentable regardless of the fact they already occur in nature. The simple discovery of a gene or protein (or any element of the human body) will not be patentable, but the isolated gene or protein may be. As the identification, purification, and classification of the isolated gene or protein is the result of technical processes and human intervention, if the gene or protein is capable of industrial application, it may be patentable.

The directive on the legal protection of biotechnological inventions addresses the "product of nature" issue in Articles 3 and 5:

> 3.1 For the purposes of this Directive, inventions which are new, which involve an inventive step, and which are susceptible of industrial application shall be patentable even if they concern a product consisting of or containing biological material or a process by means of which biological material is produced, processed or used.
>
> 3.2 Biological material which is isolated from its natural environment or produced by means of a technical process may be the subject of an invention even if it previously occurred in nature.

For biological material found in the human body, the directive provides at Articles 5.1 through 5.3 provisions addressing novelty, the "discovery" exception, and industrial applicability, but does not deal with inventive step:

> 5.1 The human body, at the various stages of its formation and development, and the simple discovery of one of its elements, including the sequence or partial sequence of a gene, cannot constitute patentable inventions.[318]

317. EPO Guidelines for Examination in the European Patent Office, at pt. G, ch. II, 3.1 (June 2012).

318. Articles 5.1 and 6 (which provides that uses of human embryos are unpatentable) were recently considered by the CJ in Oliver Brüstle v. Greenpeace eV, Case C-34/10, see

> 5.2 An element isolated from the human body or otherwise produced by means of a technical process, including the sequence or partial sequence of a gene, may constitute a patentable invention, even if the structure of that element is identical to that of a natural element.
>
> 5.3 The industrial application of a sequence or a partial sequence of a gene must be disclosed in the patent application.

With respect to an isolated DNA sequence encoding the V28 protein, the Opposition Division held:

> 19(i) Although nucleic acid encoding the V28 protein exists as a segment of the human genome and thus it is a part of nature, the purified and isolated nucleic acid having that sequence does not exist in nature and thus, cannot be discovered. The purified and isolated polynucleotide encoding V28 protein is *de facto*, not a discovery.[319]

With respect to claims to cDNA (complementary DNA that codes for a protein but does not contain the introns of genomic DNA), the Opposition Division stated:

> 4.3.1 It is common ground amongst the parties that until a cDNA encoding human H2-relaxin and its precursors was isolated by the proprietor, the existence of this form of relaxin was unknown. It is established patent practice to recognise the novelty for a natural substance which has been isolated for the first time and which had no previously recognised existence.[320]

Curia website: http://curia.europa.eu/juris/document/document.jsf?text=&docid=111402&pageIndex=0&doclang=EN&mode=lst&dir=&occ=first&part=1&cid=418152), in which the court held that a human ovum as soon as fertilized is regarded as a "human embryo" for the purposes of the directive. The question of whether pluripotent cells were unpatentable was left to national courts to decide; if they are capable of commencing the process of the development of a human being, they are unpatentable. Induced pluripotent cells therefore remain patentable. In April 2013 the UK High Court made a further reference to the CJ requesting that the term "human embryo" in Article 6(2)(c) of the directive by clarified, in particular whether it includes unfertilized human ova whose division and further development have been stimulated by parthenogenesis, and that, in contrast to fertilized ova, contain only pluripotent cells that are incapable of developing into human beings. *International Stem Cell Corporation*, Case C-364/13.

319. ICOS Corporation's Patent, 2002 OJEPO 293 (EP0630405).

320. HOWARD FLOREY INSTITUTE/Relaxin [1995] OJEPO 388 (upheld on appeal to the TBA, T-0272/95, 23 October 2002, 1999 OJEPO 590).

c. Susceptible of Industrial Application

To be patentable, a biotechnological product or process must have industrial application—for instance, as a biopharmaceutical. For DNA sequences, the sequence must possess a known utility or code for proteins with known utility. The sequences must possess a "technical effect."[321]

In a recent and significant case on this issue, both the EPO and UK courts considered whether a claim to DNA and amino acid sequences for neutrokine-α was capable of industrial applicability. Although no specific practical use for the protein had been established by the patentee at the time of filing, bioinformatics indicated that neutrokine-α belonged to a family of proteins known as the "TNF ligand superfamily," which are proteins known to act as intercellular mediators of inflammation and immune responses. Based on this, the patent identified a long list of biopharmaceutical uses for the protein. The Opposition Division, the UK High Court, and the Court of Appeal each found the patent invalid. The Technical Board of Appeal reversed the Opposition Division and upheld the patent, and the UK Supreme Court followed, laying down the principles as extracted from EPO case law with respect to sufficiency of patent claims to biotechnological products: a "speculative" use will not suffice; the patent must disclose a "practical application" and "some profitable use" for the claimed substance with "a real as opposed to a purely theoretical possibility of exploitation." Although experimental evidence is not necessary, merely identifying the structure of a protein without attributing to it a "plausible" or "reasonably credible" use is fatal. Where in this case the protein belonged to a superfamily of proteins, and all members of the superfamily had the same understood role, assigning a similar role to the claimed protein sufficed.[322]

C. Regulation of Clinical Trials in the European Union

1. Legislative Framework

Since being adopted in 1964, the Declaration of Helsinki[323] has provided a level of consensus among the international community on how medical research involving human subjects should be conducted. The current version (dated October 2008) provides a set of principles that apply to all medical research. Although primarily addressed to physicians, it encourages other participants in medical research involving human subjects to adopt the principles, which include those relating to scientific integrity,

321. T 939/92 Agrevo/Triazoles, 1996 OJEPO 309.

322. Human Genome Scis. Inc. v. Eli Lilly & Co., [2012] Report of Patent, Design and Trade-mark Cases 6, 2 November 2011.

323. *World Medical Association Declaration of Helsinki*, World Med. Ass'n, http://www.wma.net/en/30publications/10policies/b3/ (last visited May 23, 2014).

research ethics committee approval, and informed consent. This sentiment is echoed in the second Recital of European Directive 2001/20 (Clinical Trials Directive),[324] which explains that the accepted basis for the conduct of clinical trials in humans is founded on the protection of human rights and the dignity of the human being as reflected in the Declaration of Helsinki. All references to articles in this section refer to the Clinical Trials Directive unless otherwise stated.

The Clinical Trials Directive provides the legal framework for the regulation of clinical trials for medicinal products in the 28 member states of the European Union. As discussed in the introduction to Section V of this text, European Union directives are not directly effective in the member states, but are implemented through national law. The national execution measures for implementing the Clinical Trials Directive in the member states can be found on the European Union's legal database.

The Clinical Trials Directive is further enhanced by Directive 2005/28, which lays down principles and detailed guidelines for good clinical practice as regards investigational medicinal products for use, as well as the requirements for authorisation of the manufacturing or importation of such products (the GCP Directive).[325] The GCP Directive is not discussed in this section; however, all clinical trials, including bioavailability and bioequivalence studies, must also be designed, conducted, and reported in accordance with the principles of good clinical practice.[326]

2. Scope of the Clinical Trials Directive

The Clinical Trials Directive applies to the conduct of clinical trials for medicinal products carried out on humans. However, if clinical trials are conducted outside the European Union, but the application for marketing authorisation is submitted in the European Union, the principles of the directive must be complied with.[327]

"Clinical trials" are specifically defined as being "any investigation in human subjects intended to discover or verify the clinical, pharmacological and/or other pharmacodynamic effects of one or more investigational

324. Directive 2001/20/EC, of the European Parliament and of the Council of April 4, 2001 on the Approximation of the Laws, Regulations and Administrative Provisions of the Member States Relating to the Implementation of Clinical Practice in the Conduct of Clinical Trials on Medicinal Products for Human Use, 2001 O.J. (L 121/34).

325. Commission Directive 2005/28/EC of April 8, 2005, laying down principles and detailed guidelines for good clinical practice as regards investigational medicinal products for human use, as well as the requirements for authorisation of the manufacturing or importation of such products (Text with EEA relevance), 2005 O.J. (L 91/13).

326. Article 1(2)–(4), Clinical Trials Directive.

327. For further information, see *Medicinal Products for Human Use*, European Comm'n, http://ec.europa.eu/health/human-use/clinical-trials/#geninf (last visited May 23, 2014).

medicinal products, and/or to identify any adverse reactions to one or more investigational medicinal products and/or to study absorption, distribution, metabolism and excretion of one or more investigational medicinal products with the object of ascertaining its (their) safety and/or efficacy."[328] This includes all phases of clinical trials required for market authorisation and limited bioavailability and bioequivalence studies for generic authorisation.

Thus, the first question to consider in determining whether the Clinical Trials Directive is relevant to clinical study is whether the test substance is a "medicinal product." A medicinal product is any substance or combination of substances that is (a) "presented as having properties for treating or preventing disease in human beings" or (b) "which may be used in or administered to human beings either with a view to restoring, correcting or modifying physiological functions by exerting a pharmacological, immunological or metabolic action, or to making a medical diagnosis."[329]

The scope of the Clinical Trials Directive does not extend to clinical studies involving medical devices, food supplements, other non-medicinal therapies, or non-interventional trials. Non-interventional trials are studies "where the medicinal product is prescribed in the usual manner in accordance with the terms of the marketing authorisation" and where no additional diagnostic or monitoring procedures are applied to patients.[330]

3. General Considerations Prior to Commencing a Clinical Trial

The procedure for commencing a clinical trial is set out in Article 9 of the Clinical Trials Directive. Clinical trials require: (1) a favorable ethics opinion from an ethics committee;[331] and (2) authorisation from the competent authority of the relevant member state.[332] These applications can be simultaneously submitted and are discussed in further detail in a later section of this text.

As discussed, a fundamental principle of the Clinical Trials Directive is to safeguard the health and dignity of the human subjects. In that regard, Article 3 imposes certain criteria that must be satisfied before a clinical trial can be undertaken. Additional conditions need to be satisfied in relation to clinical trials involving children and incapacitated adults as set

328. Article 2, Clinical Trials Directive.

329. Article 1(2) of Directive 2001/83/EC of the European Parliament and of the Council.

330. Article 2, Clinical Trials Directive.

331. Ethics Committee is defined in Article 2 as "an independent body in a Member State, consisting of health care professionals and non-medical members, whose responsibility is to protect the rights, safety, and well-being of human subjects involved in a trial and to provide public assurance of that protection, by, among other things, expressing an opinion on the trial protocol, the suitability of the investigators and the adequacy of facilities, and on the method and documents to be used to inform trial subjects and obtain their informed consent."

332. Articles 3(2) and 9(1), Clinical Trials Directive.

out in Article 4 and Article 5, respectively. Given the vulnerability of such patients, the Clinical Trials Directive makes it clear that their interests always prevail over science and society.[333]

For all clinical trials, considerations must be given to (among other factors):

- the balance between the foreseeable risks and inconveniences and the anticipated benefit for the individual trial subject and other present and future patients. The trial is only permitted if the ethics committee and/or the competent authority reaches the conclusion that the anticipated benefits justify the risks;
- informed consent (discussed in further detail in a later section);
- the subject's rights to physical and mental integrity, privacy, and data protection; and
- insurance or indemnity to cover the liability of the investigator[334] and sponsor.[335]

4. Clinical Trial Authorisation

Applications for authorisation are made by the sponsor of the clinical trial.[336]

The Clinical Trials Directive provides a time limit of 60 days in which the competent authority must process the application. However, member states can legislate for shorter periods in accordance with local practice. The period is generally non-extendable, but can be extended with respect to medicinal products for gene therapy, somatic cell therapy, and all medicinal products containing genetically modified organisms. No time limits apply to clinical trials for xenogeneic cell therapy.[337]

Clinical trial authorisation in the United Kingdom is obtained from the Medicines and Healthcare Products Regulatory Agency (MHRA). The other member states of the EU will have their own national bodies that act as the competent authority.[338] Therefore, in instances where clinical

333. Articles 4(i) and 5(h), Clinical Trials Directive.

334. Investigator: "the doctor or person following a profession agreed in the Member State for investigations because of the scientific background and the experience in patient care it requires. The investigator is responsible for the conduct of a clinical trial at a trial site. If a trial is conducted by a team of individuals at a trial site, the investigator is the leader responsible for the team and may be called the principal investigator." Article 2.

335. The "sponsor" is defined in Article 2 as being a natural or legal person (e.g., a company or institution) that takes responsibility for the initiation, management, and/or financing of the clinical trial.

336. Article 9(1), Clinical Trials Directive.

337. Article 9(4), Clinical Trials Directive.

338. For a list of the national competent authorities as of March 13, 2013, see European Meds. Agency (Mar. 13, 2013), https://eudract.ema.europa.eu/docs/msca/MSCA_Contact_List.pdf.

trials are conducted in more than one member state, an application for authorisation will need to be submitted in each country in which the trial is conducted. This can potentially be time-consuming and expensive.

5. Ethics Committee Approval

The sponsor must apply for approval from the appropriate ethics committee in the relevant member state before conducting a clinical trial. Each member state may have a number of ethics committees, and some consideration will be needed to ensure that approval is sought from the correct body.

In preparing its opinion, the ethics committee has to attend to certain issues, which are set out in Article 6(3) of the Clinical Trials Directive:[339]

a. The relevance of the trial and its design;
b. Evaluation of the anticipated benefits and risks and whether the conclusions are justified;
c. The "protocol"—namely, the document that describes the objectives, design, methodology, statistical considerations, and organization of the trial;
d. Suitability of the investigator and supporting staff;
e. The investigator's brochure (i.e., a compilation of the clinical and nonclinical data on the investigational medicinal product[340] or products that are relevant to the study of the product or products in human subjects;
f. The quality of the facilities;
g. The adequacy and completeness of the written information to be given and the procedure to be followed for the purpose of obtaining informed consent and the justification for the research on persons incapable of informed consent with regard to the specific restriction under Article 3;
h. Provision for indemnity or compensation in the event of injury or death attributable to the clinical trials;
i. Any insurance or indemnity to cover the liability of the investigator and sponsor;
j. The amounts and, where appropriate, the arrangements for rewarding or compensating investigators and trial subjects and the

339. Article 6(4) specifies that the member state can decide that the competent authority is responsible for the matters described in Article 3(h), (i), and (j), Clinical Trials Directive.

340. Investigational medicinal product: "a pharmaceutical form of an active substance or placebo being tested or used as a reference in a clinical trial, including products already with a marketing authorisation but used or assembled (formulated or packaged) in a way different from the authorised form, or when used for an unauthorised indication, or when used to gain further information about the authorised form." Article 2, Clinical Trials Directive.

relevant aspects of any agreement between the sponsor and the site; and

k. Arrangements for the recruitment of subjects.

If the clinical trial is to be undertaken in children, an ethics committee must not endorse a protocol unless it has paediatric expertise or has taken advice in clinical, ethical, and psychological problems in the field of paediatrics.[341] Similarly, the ethics committee can only endorse a protocol for a clinical trial in incapacitated adults if it has expertise in the relevant disease and patient population or after taking advice in clinical, ethical, and psychological problems in the relevant disease and patient population concerned.[342]

Once the sponsor has submitted a request for approval, the ethics committee will have up to 60 days to give its opinion to the applicant and the competent authority. This period can be extended for potentially up to 120 days (subject to certain conditions) for clinical trials of gene therapy, somatic cell therapy, or medicinal products containing genetically modified organisms. No time limits apply to xenogeneic cell therapy.

Where multicenter clinical trials are to be conducted in a single member state, it is sufficient for the ethics committee to issue a single opinion for that member state. However, if the multicenter clinical trials are to be carried out in more than one member state, a single opinion for each member state is required.[343]

6. Informed Consent

An interaction between a health care professional and a patient without informed consent is unlawful. Given the potential risks involved in clinical trials, it is very important that informed consent of the subject is obtained and recorded.

Informed consent under the Clinical Trials Directive is defined as a "decision, which must be written, dated and signed, to take part in a clinical trial, taken freely after being duly informed of its nature, significance, implications and risks appropriately documented, by any person capable of giving consent or, where the person is not capable of giving consent, by his or her legal representative; if the person concerned is unable to write, oral consent in the presence of at least one witness may be given in exceptional cases, as provided for in national legislation."[344] Article 3(2) allows informed consent to be withdrawn at any time.

341. Article 4(g), Clinical Trials Directive.
342. Article 5(g), Clinical Trials Directive.
343. Article 7, Clinical Trials Directive.
344. Article 2, Clinical Trials Directive.

Consent by proxy with respect to adults who are not incapacitated is not permitted. However, specific requirements for obtaining informed consent in children and incapacitated adults are provided in Articles 4 and 5, respectively. In the case of children, informed consent is given by the parents or the legal representative and must represent the child's presumed will, which can be revoked at any time. The child should also be given information pitched at his or her level of understanding (from staff with experience with children) setting out the risks and benefits of the clinical trial. The explicit wish of the child not to take part in the trial or to withdraw from it must be considered by the investigator and, if appropriate, the principal investigator. Additional provisions exist with respect to an incapacitated adult; however, informed consent is given through his or her legal representative. The form of the legal representative will depend on national law and can include a court.

7. Conduct of a Clinical Trial

Article 10 deals with the conduct of clinical trials. The sponsor is able to make amendments to the protocol after the trial has started. If the amendments are substantial and are likely to have an impact on the safety of the trial subjects or to change the interpretation of the scientific documents in support of the conduct of the clinical trial, the sponsor has to inform the relevant competent authorities of the reasons for the change and notify the relevant ethics committees.

The Clinical Trials Directive imposes a monitoring and reporting duty. The investigator has a duty to report all serious adverse events to the sponsor.[345] In turn, the sponsor has a duty to inform the competent authorities and the ethics committees of all serious, unexpected adverse events that are fatal or life-threatening as soon as possible and no later than seven days after the sponsor becomes aware of the case.[346] All other suspected serious, unexpected adverse reactions are reported within 15 days. There is also a further annual reporting obligation.

Member states can suspend or prohibit a clinical trial if it no longer meets the conditions in the request for authorisation submitted to the competent authority or if it has doubts about the safety or scientific validity of the clinical trial.[347]

The sponsor must also notify the competent authorities of the relevant member states and the ethics committee when the clinical trial ends or if

345. Article 16, Clinical Trials Directive.
346. Article 17, Clinical Trials Directive.
347. Article 12, Clinical Trials Directive.

it is terminated early. The purpose of this is to inform these bodies that no further regulatory surveillance of the clinical trial is required.

8. Access to Information about Clinical Trials

Exchange of information between member states is essential for the safe conduct of clinical trials. Under Article 11, the Clinical Trials Directive provides that member states have to enter certain information relating to a clinical trial in a European database, which is only accessible by the competent authorities of member states, the European Medicines Agency, and the European Commission.

EudraCT[348] is a restricted-access database that contains information about clinical trials from May 1, 2004, onward. Some of the information on this database can be viewed by any member of the public through the EU Clinical Trials Register,[349] which allows users to view the description of:

- Phase 2 to 4 adult clinical trials with trial sites in the European Union and European Economic Area; and
- any paediatric clinical trial with trial sites in the European Union and any trials that form part of a paediatric investigation plan irrespective of the location of the trial site.

The clinical trial description will include the design of the trial, the identity of the sponsor, the investigational medicine, the therapeutic areas, and the status of the trial.

9. Future Regulation of Clinical Trials

The lack of harmonisation, along with the complexity and expense of the current system, led to criticism of the Clinical Trials Directive, and on July 17, 2012, the European Commission adopted a proposal for a regulation on clinical trials on medicinal products for human use that would repeal Directive 2001/20/EC. This move appeared to be a direct response to the "unfavourable regulatory framework for clinical research" that contributed to a decrease of 25 percent in clinical trials conducted between 2007 and 2011.[350]

The proposed regulation will introduce simplified application and reporting procedures. There is also the possibility of the European Commission conducting controls in member states and other countries to ensure that the rules are being properly supervised and enforced.

---

348. *See EudraCT*, European Meds. Agency, https://eudract.ema.europa.eu/ (last visited May 23, 2014).

349. *See Clinical Trials*, European Meds. Agency, https://www.clinicaltrialsregister.eu (last visited May 23, 2014).

350. *See* Press Release, European Comm'n, Fostering EU's Attractiveness in Clinical Research: Commission Proposes to Revamp Rules on Trials with Medicines (July 17, 2012), http://ec.europa.eu/health/files/clinicaltrials/2012_07/press-releases/ip-12-795_en.pdf.

As regulations are directly effective, the new regulation (which is expected to come into force in 2016), will mean that the law governing clinical trials will be fully harmonised across the European Union, simplifying the process of conducting multicountry clinical trials.

10. Further Information

General information relating to the requirements and conduct of clinical trials can be found on the MHRA's and the European Commission's websites.[351]

Specific guidance on medicinal products and clinical trials can be found in the European Commission's Volume 10 clinical trials guidelines.[352]

D. Authorisation of Medicinal Products for Human Use

1. EU Regulatory Framework

The legal framework governing the placing on the market of pharmaceuticals (referred to in the legislation as "medicinal products") in all EU member states is contained in two key pieces of European legislation (Directive 2001/83/EC[353] and Regulation (EC) No. 726/2004[354]), together with a number of associated directives and regulations[355] that govern specific aspects, such as orphan drugs and requirements for paediatric studies. Directive 2001/83/EC and Regulation (EC) No. 726/2004 have both been amended several times by subsequent updating legislation. In addition, there are numerous guidelines that, together with implementing legislation, supplement and explain the key provisions. The European Commission's "Notice to Applicants"[356] is one such key guideline, but national regulatory

351. MHRA, http://www.mhra.gov.uk (last visited May 23, 2014); *Clinical Trials*, European Comm'n, http://ec.europa.eu/health/human-use/clinical-trials/ (last visited May 23, 2014).

352. *Volume 10 Clinical Trials Guidelines*, European Comm'n, http://ec.europa.eu/health/documents/eudralex/vol-10/ (last visited May 23, 2014).

353. Directive 2001/83/EC, of the European Parliament and of the Council of November 6, 2001 on the Community Code Relating to Medicinal Products for Human Use, 2001 O.J. (L 311) 67, as amended. For a consolidated version of the text incorporating amendments, see http://ec.europa.eu/health/files/eudralex/vol-1/dir_2001_83_consol_2012/dir_2001_83_cons_2012_en.pdf.

354. Regulation (EC) No. 726/2004 of the European Parliament and of the Council of March 31, 2004 Laying Down Community Procedures for the Authorisation and Supervision of Medicinal Products for Human and Veterinary Use and Establishing a European Medicines Agency (Text with EEA Relevance), 2004 O.J. (L 136) 1, as amended. For a consolidated version of the text incorporating amendments, see http://ec.europa.eu/health/files/eudralex/vol-1/reg_2004_726_cons/reg_2004_726_cons_2013_en.pdf.

355. For a list of regulations and directives, see *EudraLex—Volume 1—Pharmaceutical Legislation Medicinal Products for Human Use*, European Comm'n, http://ec.europa.eu/health/documents/eudralex/vol-1/index_en.htm (last visited May 23, 2014).

356. *EudraLex—Volume 2—Pharmaceutical Legislation Notice to Applicants and Regulatory Guideliens Medicinal Products for Human Use*, European Comm'n, http://ec.europa.eu/health/documents/eudralex/vol-2/index_en.htm (last visited May 23, 2014).

authorities and the European Medicines Agency (EMA) also produce many guidelines of their own. These guidance documents can be found on the websites of the regulatory bodies listed in Annex I of the directive.

This regulatory system dates to 1965 and essentially provides that, subject to certain limited exceptions, no medicinal product may be placed on the market in the EU unless it has received a marketing authorisation in accordance with one of the prescribed procedures (for details of which see a later section). The key aim of the legislation is that of "protection of public health," but as explained in Recitals 2 and 3 of the key Directive 2001/83/EC, this is balanced against the need to support the European pharmaceutical industry:

> (2) The essential aim of any rules governing the production, distribution and use of medicinal products must be to safeguard public health;
>
> (3) However, this objective must be attained by means which will not hinder the development of the pharmaceutical industry or trade in medicinal products within the Community.

Directive 2001/83/EC (as amended) contains the key definitions (including the definition of what constitutes a medicinal product) and provides that (subject to limited exceptions) no product may be placed on the market in a member state of the EU unless a marketing authorisation has been granted according to one of the possible procedures (which are explained in a later section).[357] A marketing authorisation will stipulate the condition(s) and patient population(s) for which the medicine is authorised, together with any conditions imposed on its use. Directive 2001/83/EC also provides general governance of national, "mutual recognition," and "decentralised" marketing authorisation procedures (and even though the "centralised" procedure is governed by Regulation (EC) No. 726/2004, for many elements, such as the application dossier content, it too follows the directive's provisions).

The directive sets out the following key aspects of the marketing authorisation framework:

357. Article 6(1) Directive 2001/83/EC: "No medicinal product may be placed on the market of a Member State unless a marketing authorisation has been issued by the competent authorities of that Member State in accordance with this Directive or an authorisation has been granted in accordance with Regulation (EC) No. 726/2004, read in conjunction with Regulation (EC) No. 1901/2006 of the European Parliament and of the Council of 12 December 2006 on medicinal products for paediatric use (OJ L 378, 27.12.2006, p.1) and Regulation (EC) No 1394/2007."

(l) The "dossier requirements" (i.e., the administrative information and clinical, preclinical, and quality data, together with other information required to support the marketing authorisation application

(m) The "Mutual Recognition" procedure (mutual recognition procedure) and the "Decentralised" procedure (DCP) to be used when seeking marketing authorisations in more than one member state (unless the "Centralised" procedure is used)

(n) The data and market exclusivity framework, together with the concept of "global marketing authorisation" (see Section V.F, "Regulatory Exclusivity for Medicinal Products in the EU," later in this text)

(o) The "referral" procedures for harmonising product authorisations throughout the EU and resolving authorisation-related divergences between member states

(p) Requirements for manufacturing authorisations and importation of medicines

(q) Provisions concerning homeopathic and traditional herbal medicines (not covered in this text)

(r) Pharmacovigilance provisions

(s) Provisions on wholesale distribution of medicines

(t) Advertising of medicines

(u) Product information—details of the Summary of Product Characteristics (SmPC), labeling, and patient information leaflet

(v) Supervision and sanctions

(w) Classification of products

(x) A technical annex provides further detail on the content of marketing authorisation application dossiers.

Each EU member state has taken its own steps to implement the directive in national law; the provisions governing marketing authorisations (and other related topics) in the UK are contained in the Human Medicines Regulations 2012 (SI 2012 No. 1916).

Regulation (EC) No. 726/2004 contains the provisions that govern the establishment and operation of the EMA, including its various scientific committees, as well as the "centralised procedure" route to authorisation, which results in a single marketing authorisation valid throughout the EU. As a regulation, it has "direct effect" in all EU member states and does not need to be implemented into national law.

Given the complexity of the EU system and the volume of both legislation and (in particular) guidelines, this text can only provide an overview of the regulatory framework. For each product authorised, there will be issues specific to the nature of the product, the disease area, the applicant, and so forth, some of which may be of critical importance in that instance but which cannot be covered here.

a. The Regulatory Agencies

Each EU member state has one or more national regulatory agencies, often referred to as "competent authorities." In the UK, the Medicines and Healthcare Products Regulatory Agency (the MHRA) is responsible for the authorisation of medicinal products for human use and for medical devices, and has other related tasks, such as authorising clinical trials, granting manufacturing, wholesale dealing, and parallel import licenses, and carrying out inspections to ensure compliance with "good clinical practice," "good manufacturing practice," "good pharmacovigilance practice," and the like.

The European Medicines Agency (EMA) was established in 1995 under European legislation and is based in London. Its role has expanded over time, and it is now home to seven specialist scientific committees (most of the members of which are nominated by and represent the EU member states), as well as numerous additional working groups, specialist advisory groups, a management board, and a permanent administrative staff. The seven committees are as follows:

1. Committee for Medicinal Products for Human Use (CHMP)
2. Pharmacovigilance Risk Assessment Committee (PRAC)
3. Committee for Medicinal Products for Veterinary Use (CVMP)
4. Committee for Orphan Medicinal Products (COMP)
5. Committee on Herbal Medicinal Products (HMPC)
6. Paediatric Committee (PDCO)
7. Committee for Advanced Therapies (CAT)

The EMA works closely with the EU national regulatory agencies (as well as international regulators) and the European Commission's DG-SANCO (Directorate General for Health and Consumers).

The national regulatory agencies themselves work together through the framework of the "Heads of Agencies" and the CMDh (the Co-ordination Group for Mutual Recognition and Decentralised Procedures—Human), which has an important role in some of the procedures described in Section V.C, "Regulation of Clinical Trials in the European Union."

b. Definition of "Medicinal Product"

The regulatory regime for medicines is one of several, generally mutually exclusive European regulatory frameworks governing products that in some way either have or are alleged to have an effect on health. The question of what constitutes a product falling within the scope of the medicines framework is not always straightforward, but it is important as the regulatory requirements for medicines are more rigorous than those set out under the European legislation governing food, medical devices, or cosmetics. The starting point is the definition of "medicinal product" in Article 1(2) of Directive 2001/83/EC:

> Any substance or combination of substances presented as having properties for treating or preventing disease in human beings; or
>
> Any substance or combination of substances which may be used in or administered to human beings either with a view to restoring, correcting or modifying physiological functions by exerting a pharmacological, immunological or metabolic action, or to making a medicinal diagnosis.

This definition is broadly construed, and the two limbs are alternatives. No two regulatory frameworks apply to the same product, so overlap cannot occur. The supremacy of the medicines regulatory regime over the other regimes is made clear by Article 2(2) of Directive 2001/83/EC:

> In cases of doubt, where, taking into account all its characteristics, a product may fall within the definition of a "medicinal product" and within the definition of a product covered by other Community legislation, the provisions of this Directive shall apply.

The scope of the "medicinal product" definition has been the subject of a number of European Court judgments, providing some clarification and a degree of harmonisation between the EU member states, although this is an area where it is possible to find different member states reaching different conclusions on product classification in the case of a borderline product. In the UK, the MHRA's Borderline Section offers advice on the status of a product in cases of doubt. Each product is considered on its merits and on any information that may have a bearing on the product's status—for example, the claims made for the product, the pharmacological properties of the ingredients, whether there are any similar licensed products on the market, and how it is presented to the public through labeling, packaging, promotional literature, and advertisements. The MHRA's Guidance Note 8, "A Guide to What Is a Medicinal Product," provides more detailed information.[358]

2. Routes to Authorisation

a. The Four Routes to Authorisation

There are four possible routes to obtaining a marketing authorisation in the EU:

1. The National procedure—where a marketing authorisation is required in only one EU member state
2. The Mutual Recognition procedure (MRP)—where a marketing authorisation has already been granted in one member state and is then sought in one or more other member states
3. The Decentralised procedure (DCP)—where a marketing authorisation has yet to be granted and the applicant is seeking marketing authorisations in more than one member state
4. The Centralised procedure—where a single application is made to the EMA for an EU-wide marketing authorisation, valid in all 28 EU member states. Norway, Iceland, and Liechtenstein have, through the European Economic Area agreement, adopted the complete Community acquis on medicinal products and are consequently parties to the centralised procedure. However, legally binding acts from the Community, including commission decisions, do not directly confer rights and obligations in Norway and Iceland, but first have to be transposed into legally binding acts in these states. In Liechtenstein, however, they are directly applicable, according to national legislation. According to Decision No. 74/1999 of the EEA Joint Committee, when decisions on approval of medicinal products are taken by the Community, Norway, Iceland, and Liechtenstein will take corresponding decisions on the basis of the relevant acts.

The applicant sometimes has a choice as to which procedure to use, and sometimes the nature of the authorisation sought dictates that the centralised procedure must, or alternatively may not, be used. The mandatory and optional scope of the centralised procedure is explained in a following section. Broadly speaking, applications for generic/hybrid products use the decentralised procedure or the mutual recognition procedure, and the centralised procedure is used primarily for innovative products.

358. (Nov. 2012), *available at* http://www.mhra.gov.uk/home/groups/is-lic/documents/publication/con007544.pdf.

Whichever route is chosen, assessment standards are the same, and an authorisation will be granted or refused based on a review of the quality, safety, and efficacy of the product; only if the risk-benefit profile[359] is found to be positive (following assessment of the data submitted) and the product is of sufficiently high quality in terms of its production may an authorisation be granted.

b. Which Procedure to Use

A number of factors influence the choice of procedure, such as the plans for marketing the product, the applicant, and the nature of the product itself, but in some instances the application must go via the centralised procedure. The centralised procedure is mandatory for medicinal products that are specified in the Annex to Regulation (EC) No. 726/2004 (see Article 3(1) of the regulation). For certain other medicinal products, the centralised procedure is optional.

i. Centralised Procedure—Optional Scope

According to Article 3(2) of Regulation (EC) No. 726/2004, for the following types of medicines for human use, the centralised procedure may be used, but is not compulsory:

(cc) medicinal products that contain a new active substance that was not authorised in the Community before November 20, 2005, and that do not fall under the mandatory scope of the regulation;[360]

(dd) medicinal products that constitute a significant therapeutic, scientific, or technical innovation, or where the granting of a marketing authorisation is in the interest of patients at the Community level; and

(ee) generic medicinal products of centrally authorised products ("automatic" access under Article 3(3) of Regulation EC No. 726/2004).

---

359. The term "risk-benefit balance" is defined in point 28a, Article 1 of Directive 2001/83/EC as "an evaluation of the positive therapeutic effects of the medicinal product in relation to the risks as defined in point 28, first indent."

360. A new active substance is defined in Annex I of Chapter 1, Volume 2A of the European Commission's Notice to Applicants (rev. 4, June 2013) guidance as follows: "A new chemical, biological, or radiopharmaceutical active substance includes: a chemical, biological, or radiopharmaceutical substance not previously authorised as a medicinal product in the European Union; an isomer, mixture of isomers, a complex or derivative or salt of a chemical substance previously authorised as a medicinal product in the European Union but differing significantly in properties with regard to safety and efficacy from that chemical substance previously authorised; a biological substance previously authorised as a medicinal product in the European Union but differing in molecular structure, nature of the source material, or manufacturing process; or a radiopharmaceutical substance which is a radionuclide, or a ligand not previously authorised as a medicinal product in the European Union, or the coupling mechanism to link the molecule and radionuclide has not been authorised previously in the European Union."

(ff) Multiple/duplicate or "informed consent" (see a later section) applications for a specific medicinal product with an active substance already authorised via the centralised procedure have automatic access to the centralised procedure.

For the purpose of determining whether "a significant therapeutic, scientific, or technical innovation" is established, the agency will consider whether:

(gg) the medicinal product provides a new alternative to patients in treating, preventing, or diagnosing a disease, or
(hh) the medicinal product development is based on significant new scientific knowledge or on the application of new scientific knowledge, or
(ii) a new technology or a new application of technology is used for the development or the manufacture of the medicinal product.

A product that does not fulfill any of the preceding criteria can be found to be one for which authorisation under the centralised procedure is "in the interest of patients at the Community level" when it addresses a special health issue, allows access to medicines, or provides another type of contribution to patient care in the Community.

Applications for certain medicinal products for paediatric use may also be eligible for evaluation through the centralised procedure in accordance with the Paediatric Regulation (EC) No. 1901/2006.

ii. Centralised Procedure—Mandatory Scope

As set out in the Annex to Regulation (EC) No. 726/2004, the centralised procedure must be used for:

(a) medicinal products developed by means of one of the following biotechnological processes:

- recombinant DNA technology
- controlled expression of genes coding for biologically active proteins in prokaryotes and eukaryotes, including transformed mammalian cells
- hybridoma and monoclonal antibody methods

"Biosimilar" medicinal products (see elsewhere in this chapter) developed by one of the preceding biotechnological processes also fall within the mandatory scope of the centralised procedure. Examples of such products include the following:

(i) products intended for gene therapy
(ii) vaccines from strains developed by means of recombinant DNA technology, including gene deletion

(iii) any medicinal product for which a monoclonal antibody is used at any stage in the manufacturing process
(iv) cell therapy products

(b) "advanced therapy medicinal products" (as defined in Regulation (EC) No. 1394/2007—see elsewhere in this section for more detail);
(c) medicinal products for human use containing a new active substance[361] for which the therapeutic indication is the treatment of the following:

(i) acquired immune deficiency syndrome (HIV/AIDs)
(ii) cancer
(iii) neurodegenerative disorders
(iv) diabetes
(v) auto-immune diseases and other immune dysfunctions
(vi) viral diseases

Clarifications on the working definitions of these diseases are to be found in the guideline "Scientific aspects and working definitions for the mandatory scope of the centralised procedure."[362]

(d) medicinal products that are designated as orphan medicinal products.

3. How the Various Application Procedures Work

Marketing authorisation applicants must be "established in the Community" in order to submit an application (Article 8(2), Directive 2001/83/EC; Article 2, Regulation (EC) No. 726/2004). Section 2.8, Chapter 1, Volume 2A of the European Commission's Notice to Applicants (Revision 4, June 2013) guidance provides a discussion of the concepts of "applicant" and "marketing authorisation holder." As described, the "national" procedure will apply if the product is only authorised in one member state; the main features of the various multistate application procedures are outlined later in this section.

a. Mutual Recognition Procedure

The applicant first obtains a marketing authorisation in one member state (Directive 2001/83/EC will govern this) and then selects further member states in which he or she wishes to apply for a marketing authorisation, asking them to recognise the initial authorisation. All such applications are based on identical dossiers, to ensure harmonisation of product authorisations throughout the EU. The "reference member state" (RMS) prepares

361. *Id.*

362. Scientific Aspects and Working Definitions for the Mandatory Scope of the Centralised Procedure, EMA/CHMP/121944/2007.

a report for the additional "concerned member state(s)" (CMS). The CMS authorisations are based on "mutual recognition" of the RMS assessment.

b. Decentralised Procedure

The decentralised procedure is available when a marketing authorisation has not yet been granted in any member state, but in other respects it is procedurally similar to the mutual recognition procedure, with an RMS preparing an initial draft assessment report and product information on behalf of itself and the other CMSs.

The provisions governing the mutual recognition procedure and the decentralised procedure are set out in Chapter 4 of the directive. In both these procedures, once the CMSs have reviewed the assessment report and product information, national authorisations based on these will be granted in the CMS territories. Only if a CMS considers that there is a "potential serious risk to public health" may the competent authority of a CMS refuse to approve the authorisation or assessment report prepared by the competent authority of the RMS. In such cases, the directive provides for a procedure to resolve the differences, ultimately referring the matter for a decision by the main scientific committee of the EMA, if this proves necessary. A "potential serious risk to public health" is defined as a situation where there is a significant probability that a serious hazard resulting from a human medicinal product in the context of its proposed use will affect public health.[363]

c. Centralised Procedure

A successful application through the centralised procedure leads to a European marketing authorisation, which automatically allows the marketing authorisation holder to market its medicinal product throughout the EU. Applications for the centralised procedure are submitted to the EMA. The EMA's Committee for Medicinal Products for Human Use (CHMP) conducts the scientific evaluation, with appropriate input from other committees (Committee on Advanced Therapy Products, Pharmacovigilance Risk Assessment Committee, etc.) and other expert groups as necessary. The assessment procedure is led by a rapporteur and a co-rapporteur, who are members of, and appointed by, the CHMP. Upon receiving the EMA's scientific opinion, the European Commission grants a marketing authorisation for the entire EU, if this is appropriate. When the EMA issues a positive opinion in favor of granting a marketing authorisation, then the European Commission will usually accept this (although it is not bound to do so) and grant an authorisation.

363. Guideline on the Definition of a Potential Serious Risk to Public Health in the Context of Article 29(1) and (2) of Directive 2001/83/EC, 2006 O.J. (C 133/5).

The marketing authorisation holder must, subject to certain limited exceptions, use a single name for a medicinal product authorised through the centralised procedure[364] Applicants wishing to use an "invented name" (trade name) should identify at an early stage a name that is suitable for use in all member states, and guidelines have been published to explain the factors that will be taken into account by the EMA's Invented Name Review Group when deciding whether to approve a proposed invented name.[365] In exceptional cases, the Commission may authorise the use of a different invented name in a member state or states where the invented name has been cancelled, opposed, or objected to on the grounds of trade-mark law.

Potential applicants to the centralised procedure are encouraged to contact the EMA well in advance of their intended application submission date, and pre-submission discussions with the agency are common. The EMA also encourages applicants to apply for scientific advice[366] to guide them with respect to any potential issues regarding, for example, clinical trials or quality concerns, in order to discuss and resolve possible problems at an early stage. Scientific advice will often be appropriate long before the authorisation is applied for—for example, at the stage of designing clinical trials.

A potential applicant should submit a request for confirmation of eligibility to use the centralised procedure regardless of whether the product falls within the mandatory or optional scope. The standard length for the assessment procedure is 210 days, although if the applicant is required to provide additional information (as is often the case), then there is a "clock-stop" period (typically three months) to allow the applicant to gather this information. There is also a provision for an accelerated procedure of 150 days where the applicant can establish that there is a "major interest from the point of view of public health and in particular from the viewpoint of therapeutic innovation" (Article 14(9), Regulation (EC) No. 726/2004).

If the applicant does not receive the outcome hoped for, there is a provision for a "reexamination" of the CHMP's opinion. This is still performed by the CHMP, but a different rapporteur is appointed. Details of the reexamination procedure are explained in a guideline.[367]

---

364. Article 6(1) Regulation (EC) No. 726/2004: "otherwise than in exceptional cases relating to the application of the law on trade marks, [the application particulars] shall include the use of a single name for the medicinal product."

365. See the EMA's "Guideline on the Acceptability of Names for Human Medicinal Products Processed Through the Centralised Procedure" (CHMP/328/98, Dec. 11, 2007).

366. European Meds. Agency, EMEA-H-4260-01-REV.6, Guidance for Companies Requesting Scientific Advice and Protocol Assistance, 21 May 2010.

367. Procedural Advice on the Re-examination of CHMP Opinions, EMEA/CHMP /50745/2005 (rev. 1).

Chapter 4, Volume 2A of the European Commission's Notice to Applicants provides details on various aspects of the operation of the centralised procedure.

In the framework of the centralised procedure, only one marketing authorisation may be granted to an applicant for a specific medicinal product, although Article 82(1), second subparagraph of Regulation (EC) No. 726/2004 provides that the same applicant can submit more than one application for the same medicinal product in certain circumstances:

> Only one authorisation may be granted to an applicant for a specific medicinal product.
>
> However, the Commission shall authorise the same applicant to submit more than one application to the Agency for that medicinal product when there are objective verifiable reasons relating to public health regarding the availability of medicinal products to health-care professionals and/or patients, or for co-marketing reasons.
>
> As regards medicinal products for human use, Article 98(3) of Directive 2001/83/EC shall apply to medicinal products authorised under this Regulation.

The European Commission decides whether multiple applications may be justified in any particular case and has produced a Note on the Handling of Duplicate Marketing Authorisations;[368] there seems to be a move in recent years toward limiting the circumstances in which permission for multiple applications will be granted.

4. Contents of the Application Dossier

The contents of the application dossier will to some extent depend on the legal basis for the application, particularly in relation to the safety and efficacy data required and whether the applicant is able to cross-refer to such data in another product's dossier. There may also be particular product-related aspects to take into account when compiling the dossier (for example, for products containing genetically modified organisms, or other specific types of products such as vaccines)—these detailed requirements areas are outside the scope of this chapter, but the regulatory authorities and the European Commission have produced numerous guidelines for applicants, which are available on their websites.

368. Handling of Duplicate Marketing Authorisation Applications, SANCO/D3/RSR /iv(20011)ddg1.d3.1137738.

a. Full Dossier/New Active Substance

A "full application" dossier contains manufacturing information, results of pharmaceutical and preclinical tests as well as the results of clinical trials and other supporting data, such as a summary of the pharmacovigilance system, the risk management plan and an evaluation of the potential environmental risks posed by the medicine, a Summary of Product Characteristics (SmPC), and other information relevant to the application, including administrative data. For some types of products, such as radionuclide generators, further information will be required. For new medicines authorised after July 28, 2008, results of all agreed paediatric studies must also be included, unless a deferral or waiver has been granted (see Section V.D.7, "Paediatric Studies," later in this text). Economic issues are not considered during the assessment of the dossier—those are a matter for the local health authorities, which may or may not choose to reimburse the cost of a particular medicine.

Article 8(3) of Directive 2001/83/EC outlines the required contents of a full application for a marketing authorisation (regardless of the route to authorisation followed)—see Annex II for the full text of Article 8. The application and accompanying documents must be presented in accordance with Annex I of Directive 2001/83/EC, and must follow the guidance published by the European Commission in the Notice to Applicants, Volume 2B, Common Technical Document (CTD).

For active substances, certificates of suitability (CEPs) may be issued. CEPs are certificates issued by the European Directorate for the Quality of Medicines (EDQM) to confirm that a certain active substance is produced according to the requirements of the relevant monograph of the European Pharmacopoeia or of the monograph on transmissible spongiform encephalopathy (TSE). CEPs can be used by companies to replace much of the documentation required for the active substance in the dossier. The qualitative and quantitative particulars of all the constituents of the medicinal product (i.e., the active substance(s) and the excipients) must always be provided. Inactive ingredients may have a CEP certificate to show compliance with the monograph on TSE.

b. Abridged Applications/Old Active Substances

Directive 2001/83/EC also provides a simplified "abridged" procedure for certain types of products, whereby the applicant is able to provide a reduced data package in terms of safety and efficacy data, relying on alternative sources of data.

i. Generic/Hybrid Applications

The most commonly used abridged route is that available to obtain authorisation for a generic (or for a "hybrid") version of an innovative product;

according to Article 10(1) of Directive 2001/83/EC, the applicant is not required to provide the results of preclinical and clinical tests and trials for applications for generic products, where the data exclusivity period of the reference product has expired (see Section V.F, "Regulatory Exclusivity for Medicinal Products in the EU," for a full explanation of the data/market exclusivity framework). The generic applicant must identify an appropriate reference product, the data exclusivity period for which has expired, and it may then submit a generic application cross-referring to the preclinical and clinical data contained in the dossier of the reference product, instead of having to provide its own data package. Administrative and quality data are still required as for any other application.

The important definitions are set out in Article 10(2) as follows:

> For the purposes of this Article:
>
> "reference medicinal product" shall mean a medicinal product authorised under Article 6, in accordance with the provisions of Article 8;
>
> "generic medicinal product" shall mean a medicinal product which has the same qualitative and quantitative composition in active substances and the same pharmaceutical form as the reference medicinal product, and whose bioequivalence with the reference medicinal product has been demonstrated by appropriate bioavailability studies. The different salts, esters, ethers, isomers, mixtures of isomers, complexes or derivatives of an active substance shall be considered to be the same active substance, unless they differ significantly in properties with regard to safety and/or efficacy. In such cases, additional information providing proof of the safety and/or efficacy of the various salts, esters or derivatives of an authorised active substance must be supplied by the applicant. The various immediate-release oral pharmaceutical forms shall be considered to be one and the same pharmaceutical form. Bioavailability studies need not be required of the applicant if he can demonstrate that the generic medicinal product meets the relevant criteria as defined in the appropriate detailed guidelines.

Bioequivalence studies[369] provide the regulator with the assurance it needs that it is appropriate to allow cross-reference to data contained in the originator's dossier when assessing a generic application, as the two

369. See the EMA's "Guideline on the Investigation of Bioequivalence" (CPMP/EWP/QWP/1401/98 rev. 1).

products have essentially the same safety and efficacy profile. The studies are thus a necessary component of a generic application dossier (except in certain limited situations where bioequivalence may be presumed).

A "hybrid" application can be made based on Article 10(3) where the product for which authorisation is sought is similar, but not identical, to the reference product; "bridging data" are required to provide the link between the product to be authorised and the product cross-referred to:

> In cases where the medicinal product does not fall within the definition of a generic medicinal product as provided in paragraph 2(b) or where the bioequivalence cannot be demonstrated through bioavailability studies or in case of changes in the active substance(s), therapeutic indications, strength, pharmaceutical form or route of administration, vis-à-vis the reference medicinal product, the results of the appropriate pre-clinical tests or clinical trials shall be provided.

Section 5.3.1.1, Chapter 1, Volume 2A of the European Commission's Notice to Applicants (Revision 4, June 2013) guidance explains which type of application can provide the basis for a reference medicinal product:

> Reference can be made to the dossier of a reference medicinal product for which a marketing authorisation has been granted in the Union in accordance with Articles 8(3), 10a, 10b or 10c of Directive 2001/83. The application form in Module 1 of the dossier for the generic product should clearly identify the reference product in order for the reference Member State, in case of mutual recognition procedure/decentralised procedure, to prepare the assessment report.
>
> On the contrary, reference cannot be made to the dossier of a medicinal product for which a marketing authorisation has been granted in the Union in accordance with Article 10(1). However, in those cases where a medicinal product authorised under Article 10(1) has been developed through an application submitted in accordance with Article 10(3) of Directive 2001/83/EC leading to a new indication, strength, pharmaceutical form, a marketing authorisation application of a subsequent generic of this medicinal product can include the new indication, strength, pharmaceutical form, etc., To this effect, it will also be possible to refer to the data submitted to support the development.
>
> Reference must be made to a product which is or has been authorised in the Union, (i.e., a marketing authorisation

> has been granted for the reference medicinal product, but it may have ceased to exist) and in accordance with the Union law.[370]

A product can be used as a reference medicinal product if three cumulative criteria are met:

- The reference medicinal product is (or has been) authorised in the EU (through a centralised or national procedure);
- The reference medicinal product is or has been authorised in accordance with the pharmaceutical acquis;
- The relevant data exclusivity period has expired.

A generic application may be made even if the reference medicinal product was never authorised in the member state in which such application is submitted, provided that a reference product authorised in another member state can be identified.[371] The data exclusivity period for the reference product must have expired in the member state where it is authorised.

ii. "Biosimilar" Applications

Where the proposed reference product is a biological product, the traditional route used for a generic application has no real application since, due to the nature of the product and the manufacturing process, it will not be possible to guarantee sufficient similarity (identicality) between the reference product and the "copy," as the product is so process dependent. Most biological medicinal products are complex molecules or mixtures of molecules (such as blood products, monoclonal antibodies, or vaccines) produced by manufacturing processes (such as expression by a cell line maintained by fermentation) that inevitably introduce a degree of heterogeneity in the product.

Thus, in the field of biologicals, a special "biosimilars" framework exists, which may be relied on by applicants wishing to cross-refer to data contained in the dossier of a biological reference product. Article 10(4) of Directive 2001/83/EC provides that where a biological medicinal product is similar to a reference biological product, results of appropriate tests and trials (as required by the Annex to Directive 2001/83/EC and related detailed guidelines) must be included in the application dossier:

> Where a biological medicinal product which is similar to a reference biological product does not meet the conditions in the definition of generic medicinal products, owing to, in particular, differences relating to raw materials or differences in

370. Case C-527/07, The Queen, Generics (UK) Ltd. v. Licensing Authority, Shire Pharm. Ltd. & Janssen-Cilag AB, 2009 E.C.R. I-5259.

371. Directive 2001/83/EC, Article 10(1), third paragraph.

> manufacturing processes of the biological medicinal product and the reference biological medicinal product, the results of appropriate pre-clinical tests or clinical trials relating to these conditions must be provided. The type and quantity of supplementary data to be provided must comply with the relevant criteria stated in Annex 1 and the related detailed guidelines. The results of other tests and trials from the reference medicinal product's dossier shall not be provided.

The assessment of biosimilar applications requires much more of a case-by-case approach than that for traditional generics, and the extent of "comparability data" required will depend on the nature of the product. There is a clear legal basis for the establishment of guidelines, according to which biosimilar authorisations may be granted, and these address the data necessary to support such applications (which can be quite extensive). The CHMP has issued both general and product-specific guidance covering the requirements for biosimilar applications, and these can be found on the EMA website (see Annex I).

The first biosimilars approved (from 2006 onward) through this process were for relatively simple biological products (somatropin, filgrastim, epoetin alpha/zeta), but the system is evolving to cope with applications concerned with more complex molecules, and in June 2013 the EMA issued positive opinions for approval of the first monoclonal antibody biosimilars.

iii. "Well-Established Use" (or "Bibliographic") Applications

Although this type of application is rather infrequently used in practice, Article 10a of Directive 2001/83/EC allows the applicant to provide a data package with references to scientific literature where the active substance(s) can be shown to have "well-established use" as a medicine within the EU/EEA for at least ten years:

> By way of derogation from Article 8(3)(i), and without prejudice to the law relating to the protection of industrial and commercial property, the applicant shall not be required to provide the results of pre-clinical tests or clinical trials if he can demonstrate that the active substances of the medicinal product have been in well-established medicinal use within the Community for at least ten years, with recognised efficacy and an acceptable level of safety in terms of the conditions set out in Annex I. In that event, the test and trial results shall be replaced by appropriate scientific literature.

The "well-established medicinal use" concept is defined in Part II, Annex I to Directive 2001/83/EC, which provides that "the period of time required for establishing a well established medicinal use . . . must not be less than one decade from the first systematic and documented use of that substance as a medicinal product in the [EU]."

iv. Authorisations for Fixed Combination Products

A fixed combination medicinal product is one containing a combination of two or more active substances within a single pharmaceutical form of administration.

Article 10(b) of Directive 2001/83/EC sets out the data required in the marketing authorisation dossier in such cases:

> In the case of medicinal products containing active substances used in the composition of authorised medicinal products but not hitherto used in combination for therapeutic purposes, the results of new pre-clinical tests or new clinical trials relating to that combination shall be provided in accordance with Article 8(3)(i), but it shall not be necessary to provide scientific references relating to each individual active substance.

The individual substances must have been authorised within the EEA. The use of the fixed combination must be justified, based on valid therapeutic principles such as improvement of the benefit—risk profile or simplification of therapy (improved patient compliance, for example).

v. "Informed Consent" Applications

It is possible for a marketing authorisation holder to give consent to another applicant to cross-refer to the data in the dossier of the authorised product, thereby allowing the second applicant to make an abridged application (Article 10c, Directive 2001/83/EC):

> Following the granting of a marketing authorisation, the authorisation holder may allow use to be made of the pharmaceuticals, pre-clinical and clinical documentation contained in the file on the medicinal product, with a view to examining subsequent applications relating to other medicinal products possessing the same qualitative and quantitative composition in terms of active substances and the same pharmaceutical form.

Such applications would typically be made in situations where duplicate applications are sought for marketing purposes. Section 5.6, Chapter 1, Volume 2A of the European Commission's Notice to Applicants (Revision 4, June 2013) guidance explains the scope of the provision as follows:

> As Article 11 of Directive 2001/83/EC refers only to application on the basis of Article 10, an informed consent application does not have to cover all presentations/indications of the medicinal product with regard to which consent is given. Consent may be given to use the documentation contained in the file of the relevant medicinal product for a given presentation/indication provided that the application relies on that consent as regards all three modules of the dossier. It is a prerequisite for the use of article 10c that consent has been obtained for all three modules containing the pharmaceutical, preclinical and clinical date. It is not possible to use Article 10c a legal basis for an application consisting of the applicant's own module 3 and for which consent has been given for modules 4 and 5. In such cases the legal basis for the application is Article 8(3).
>
> An informed consent application cannot cover more presentations or indications than the medicinal product with regard to which consent is given.

An Article 10c application can only be submitted in a member state where the medicinal product with regard to which consent is given is authorised, and the second applicant must have permanent access to the documentation in order to fully perform his or her responsibilities as marketing authorisation holder.

5. Advanced Therapy Medicinal Products

In order to address the specific regulatory challenges of authorising "advanced therapy" products, namely

(a) gene therapy products,
(b) somatic cell therapy products, and
(c) tissue engineered products,

The Regulation on Advanced Therapy Medicinal Products (EC) No. 1394/2007[372] introduced a number of amendments into Directive 2001/83/EC and Regulation (EC) No. 726/2004. It established a specialised Committee for Advanced Therapies within the EMA and introduced special regulatory requirements, such as traceability, for such products. Part IV of Annex I to Directive 2001/83/EC contains detail on the specific requirements for authorisation of advanced therapy products, as well as definitions.

372. Regulation (EC) No. 1394/2007 of the European Parliament and of the Council of 13 November 2007 on Advanced Therapy Medicinal Products and Amending Directive 2001/83 /EC and Regulation (EC) No. 726/2004 (Text with EEA Relevance), 2007 O.J. (324/121).

One important provision to note is the "hospital exemption" as provided for in Article 3(7) of Directive 2001/83/EC, as amended:

> This Directive [2001/83/EC] shall not apply to:
>
> . . .
>
> (7) Any advanced therapy medicinal product, as defined in Regulation (EC) No. 1394/2007, which is prepared on a non-routine basis according to specific quality standard, and used within the same Member State in a hospital under the exclusive professional responsibility of a medical practitioner, in order to comply with an individual medical prescription for a custom-made product for an individual patient.

6. Orphan Drugs

a. Incentives for Orphan Drug Development

The EU institutions have provided a number of incentives to encourage the development of orphan drugs to treat rare diseases; without special provisions, the pharmaceutical industry would have little interest in developing and marketing such products under normal market conditions. The most important of these incentives is the period of market exclusivity provided by the Orphan Drug Regulation (EC) No. 141/2000[373] (Orphan Drug Regulation). As stated in the Explanatory Memorandum accompanying the proposal for this regulation:

> Society cannot accept that certain individuals be denied the benefits of medical progress simply because the affliction from which they suffer affects only a small number of people.

b. Orphan Designation

In order to benefit from orphan drug status, orphan designation must be applied for at any stage prior to an application for marketing authorisation being made. The designation criteria are set out in Article 3 of the Orphan Drug Regulation, as follows:

> A medicinal product shall be designated as an orphan medicinal product if its sponsor can establish:
>
> that it is intended for the diagnosis, prevention or treatment of a life-threatening or chronically debilitating condition

373. Regulation (EC) No. 141/2000 of the European Parliament and of the Council of 16 December 1999 on Orphan Medicinal Products, 2000 O.J. (L 18/1).

> affecting not more than five in 10 thousand persons in the Community when the application is made, or
>
> that it is intended for the diagnosis, prevention or treatment of a life-threatening, seriously debilitating or serious and chronic condition in the Community and that without incentives it is unlikely that the marketing of the medicinal product in the Community would generate sufficient return to justify the necessary investment; and
>
> that there exists no satisfactory method of diagnosis, prevention or treatment of the condition in question that has been authorised in the Community or, if such method exists, that the medicinal product will be of significant benefit to those affected by that condition.

The prevalence criterion (i.e., a condition affecting not more than 5 in 10,000 people in the Community) is by far the most commonly used as a route to designation (the insufficient return on investment route having been used on only a few occasions). The EMA's Committee for Orphan Medicinal Products (COMP) has produced guidance on how designation applicants should address the prevalence issue and the recommended sources of prevalence data.[374]

Where "significant benefit" needs to be established, then Article 3 of the Orphan Drug Regulation's accompanying implementing Commission Regulation 847/2000[375] provides that "a clinically relevant advantage or major contribution to patient care" must be shown; the demonstration of "significant benefit" may be based on reasonable assumptions at the time of designation supported by "proof of principle" evidence in experimental studies or early clinical data.

The COMP will assess the designation application, and, if the designation criteria are met, then an opinion in favor of designation is granted, and the European Commission will adopt a decision designating the product. The product will then be entered into the Community Register of orphan products and be eligible for a number of incentives, such as free "protocol assistance" (scientific advice) from the EMA and reductions/waivers in marketing authorisation application and renewal fees. The EU member states

374. Points to Consider on the Calculation and Reporting of the Prevalence of a Condition for Orphan Designation, COMP/436/01; Relevant Sources for Orphan Disease Prevalence Data, EMA/452415/2012.

375. Commission Regulation (EC) No. 847/2000 of April 27, 2000 Laying Down the Provisions for Implementation of the Criteria for Designation of a Medicinal Product as an Orphan Medicinal Product and Definitions of the Concepts "Similar Medicinal Product" and "Clinical Superiority," 2000 O.J. (L 103/5).

also have their own programs of incentives to encourage orphan drug development, such as R&D and tax incentives.

Around 20 percent of products now being authorised through the centralised procedure are designated orphans; many of them are products to treat rare cancers, with metabolic diseases also well represented. The "exceptional circumstances" and "conditional approval" routes to authorisation (see later in this chapter) are frequently relied on to help address the challenge of providing sufficient clinical data to support the application when patient numbers are very low.

7. Paediatric Studies

Regulation (EC) No. 1901/2006[376] (the Paediatric Regulation) entered into force on January 26, 2006, establishes a system of requirements, rewards, and incentives to ensure that medicines are researched and, where possible, authorised in paediatric populations. This legislation aims to address the acknowledged problem of a scarcity of clinical data on use of many medicines in children and the resulting need for physicians to have to resort to off-label prescribing in many cases when treating paediatric patients.

The regulation established a specialist Paediatric Committee, the PDCO, within the EMA, responsible for various tasks, including providing opinions on the development of medicines for use in children; assessing and approving proposed "paediatric investigation plans," waivers, and deferrals (see later in this chapter); assessing compliance with approved PIPs (Paediatric Investigation Plans); and providing opinions on the collection and assessment of paediatric data and establishing an inventory of paediatric medicinal product needs.

The Paediatric Regulation also requires companies holding data on the safety and efficacy of authorised products in the paediatric population to submit these to the regulatory authorities.[377] These are made available in an EMA database, while general information on paediatric clinical trials can be found in the European Clinical Trials Register.

a. Requirements for Paediatric Studies

A Paediatric Investigation Plan (PIP) defines a research and development program aimed at ensuring that the necessary data are generated determining the conditions in which a medicinal product may be authorised to treat the paediatric population. It provides details of the timing

376. Regulation (EC) No. 1901/2006 of the European Parliament and of the Council of December 12, 2006 on Medicinal Products for Paediatric Use and Amending Regulation (EEC) No. 1768/92, Directive 2001/20/EC, Directive 2001/83/EC and Regulation (EC) No. 726/2004 (Text with EEA Relevance), 2006 O.J. (C 378/1).

377. Article 45, Regulation (EC) No. 1901/2006. Article 46 requires companies to submit newly generated paediatric data.

and studies proposed by a marketing authorisation holder or applicant in relation to the various paediatric subsets affected by the condition (and, in some instances, there may be studies proposed in one or more subsets while a waiver or deferral of studies is agreed for other subsets). The paediatric population is defined as "that part of the population aged between birth and 18 years" (Article 2(1)), but for the purposes of the PIP, different patient population subsets can be considered individually.

Article 7 of Regulation (EC) No. 1901/2006 requires all marketing authorisation applications for new medicines to include the results of a completed PIP, unless there is a PIP waiver or deferral in place:

> (1) An application for marketing authorisation under Article 6 of Directive 2001/83/EC in respect of a medicinal product for human use which is not authorised in the Community at the time of entry into force of this Regulation shall be regarded as valid only if it includes, in addition to the particulars and documents referred to in Article 8(3) of Directive 2001/83/EC, one of the following:
>
> (a) the results of all studies performed and details of all information collected in compliance with an agreed paediatric investigation plan;
>
> (b) a decision of the Agency granting a product-specific waiver;
>
> (c) a decision of the Agency granting a class waiver pursuant to Article 11;
>
> (d) a decision of the Agency granting a deferral.
>
> For the purposes of point (a), the decision of the Agency agreeing the paediatric investigation plan concerned shall also be included in the application.
>
> (2) The documents submitted pursuant to paragraph 1 shall, cumulatively, cover all subsets of the paediatric population.

In establishing whether or not an application concerns a product that "is not authorised in the Community," the "global marketing authorisation" concept (discussed elsewhere in this chapter) applies.

Article 8 provides that the same requirement applies when a marketing authorisation is extended or varied to include a new indication, pharmaceutical form, or route of administration and the medicine is protected either by a Supplementary Protection Certificate (SPC) or a patent that qualifies for the grant of an SPC. For the purposes of Article 8, the documents referred to previously in Article 7(1) must cover both the existing and the new indications, pharmaceutical forms, and routes of administration.

The Paediatric Regulation also provides that the requirements of Articles 7 and 8 not apply where the legal basis for the application is the generic/hybrid, biosimilar, or "well-established use" route, nor in the case of homeopathic or traditional herbal medicinal products.[378]

The procedure for requesting agreement of a PIP is set out in Section 1, Chapter 3 of Regulation (EC) No. 1901/2006:[379]

> Article 15
>
> 1. Where the intention is to apply for a marketing authorisation in accordance with Article 7(1)(a) or (d), Article 8 or Article 30, a paediatric investigation plan shall be drawn up and submitted to the Agency with a request for agreement.
>
> 2. The paediatric investigation plan shall specify the timing and the measures proposed to assess the quality, safety and efficacy of the medicinal product in all subsets of the paediatric population that may be concerned. In addition, it shall describe any measures to adapt the formulation of the medicinal product so as to make its use more acceptable, easier, safer or more effective for different subsets of the paediatric population.

It is possible to apply for modifications to the PIP as studies progress, and this is commonly done in practice.[380]

b. PIP Waivers and Deferrals

The need to agree to a PIP and perform the agreed studies can be waived in certain circumstances, as set out in Article 11 of the regulation:

> (1) Production of the information referred to in point (a) of Article 7(1) shall be waived for specific medicinal products or for classes of medicinal products, if there is evidence showing any of the following:
>
> (a) that the specific medicinal product or class of medicinal products is likely to be ineffective or unsafe in part or all of the paediatric population;

378. Article 9, Regulation (EC) No. 1901/2006: "Articles 7 and 8 shall not apply to products authorised under Articles 10, 10a, 13 to 16 or 16a to 16i of Directive 2001/83/EC."

379. *See also* Communication from the Commission—Guideline on the Format and Content of Applications for Agreement or Modification of a Paediatric Investigation Plan and Requests for Waivers or Deferrals and Concerning the Operation of the Compliance Check and on Criteria for Assessing Significant Studies (2008/C 243/01), 2008 O.J. (C 234/1).

380. European Medicines Agency Policy on Changes in Scope of Paediatric Investigation Plan (PIP) Decisions, EMA/472551/2012.

> (b) that the disease or condition for which the specific medicinal product or class is intended occurs only in adult populations;
> (c) that the specific medicinal product does not represent a significant therapeutic benefit over existing treatments for paediatric patients.
>
> (2) The waiver provided for in paragraph 1 may be issued with reference either to one or more specified subsets of the paediatric population, or to one or more specified therapeutic indications, or to a combination of both.

Thus the waiver may cover either the whole paediatric population or just part of it; where a total waiver is granted, then none of the rewards for performing paediatric studies are available. A waiver can be granted on a product-specific basis (by applying to the PDCO), or under the PDCO's agreed list of products for which a waiver is appropriate (a "class waiver").

In many cases, the agreed studies can be deferred (according to a time frame agreed with the PDCO and specified in the PIP), as provided for by Article 20(1) of the regulation. This ensures that research in children is conducted only once it is safe and ethical to do so and prevents delays in authorising the product in question for adult use:

> At the same time as the paediatric investigation plan is submitted under Article 16(1), a request may be made for deferral of the initiation or completion of some or all of the measures set out in that plan. Such deferral shall be justified on scientific and technical grounds or on grounds related to public health.
>
> In any event, a deferral shall be granted when it is appropriate to conduct studies in adults prior to initiating studies in the paediatric population or when studies in the paediatric population will take longer to conduct than studies in adults.

c. Rewards for Completion of Paediatric Studies

The reward for successfully completing the studies contained in a PIP is, where that product is covered by an SPC or a patent eligible for an SPC, a six-month extension to the term of the SPC. This reward is available even when the results of the studies do not support the grant of a paediatric indication; the important things are that the agreed PIP has been complied with (there are procedures in place for the regulatory authorities to perform "compliance checks") and that the results are made available in the product. In order to benefit from this reward, the marketing authorisation must contain a statement that the PIP has been complied with, the product must be authorised in all EU member states, and the product must not have obtained the "+1 year" extra market exclusivity for a significant new indication under the "8+2+1" rule (described in the section on data exclusivity later in this chapter). Article 36 of Regulation (EC) No. 1901/2006 is the relevant provision:

> Where an application under Article 7 or 8 includes the results of all studies conducted in compliance with an agreed paediatric investigation plan, the holder of the patent or supplementary protection certificate shall be entitled to a six-month extension of the period referred to in Articles 13(1) and 13(2) of Regulation (EEC) No. 1768/92.
>
> The first subparagraph shall also apply where completion of the agreed paediatric investigation plan fails to lead to the authorisation of a paediatric indication but the results of the studies conducted are reflected in the summary of product characteristics and if appropriate in the package leaflet of the medicinal product concerned.
>
> The inclusion in a marketing authorisation of the statement referred to in Article 28(3) shall be used for the purposes of applying paragraph 1 of this Article.
>
> Where the procedures laid down in Directive 2001/83/EC have been used, the six-month extension of the period referred to in paragraph 1 shall be granted only if the product is authorised in all Member States.
>
> Paragraphs 1, 2 and 3 shall apply to products that are protected by a supplementary protection certificate under Regulation (EEC) No. 1768/92 or under a patent which qualifies for the granting of the supplementary protection certificate. They shall not apply to medicinal products designated as orphan medicinal products pursuant to Regulation (EC) No. 141/2000.
>
> In the case of an application under Article 8 which leads to the authorisation of a new paediatric indication, paragraphs 1, 2 and 3 shall not apply if the applicant applies for, and obtains a one-year extension of the period of marketing protection for the medicinal product concerned on the ground that this new paediatric indication brings a significant clinical benefit in comparison with existing therapies in accordance with Article

> 14(11) of Regulation (EC) No. 726/2004 or the fourth subparagraph of Article 10(1) of Directive 2001/83/EC.

For authorised orphan drugs, rather than obtaining an SPC extension, the 10-year period of orphan market exclusivity provided for in Regulation 141/2000 is extended to 12 years (Article 37 of the regulation):

> Where an application for a marketing authorisation is submitted in respect of a medicinal product designated as an orphan medicinal product pursuant to Regulation (EC) No. 141/2000 and that application includes the results of all studies conducted in compliance with an agreed paediatric investigation plan, and the statement referred to in Article 28(3) of this Regulation is subsequently included in the marketing authorisation granted, the ten-year period referred to in Article 8(1) of Regulation (EC) No. 141/2000 shall be extended to twelve years.
>
> The first paragraph shall also apply where completion of the agreed paediatric investigation plan fails to lead to the authorisation of a paediatric indication, but the results of the studies conducted are reflected in the summary of product characteristics and, if appropriate, in the package leaflet of the medicinal product concerned.

For products authorised under a Paediatric Use Marketing Authorisation (PUMA—see the following section), Article 38 of the regulation provides that data and market exclusivity will apply to the paediatric test and trial data contained in the PUMA dossier (in accordance with the "8+2+1" framework described elsewhere in this chapter).

Introduction of the Paediatric Regulation raised the question of whether it is possible to have negative term SPCs, since a negative term SPC[381] of less than six months duration can become a (short but potentially valuable) positive term SPC when extended by six months. Court decisions have confirmed that it is possible to grant a negative term SPC (see the section on SPCs later in this chapter for a full discussion).

d. Paediatric Use Marketing Authorisation (PUMA)

Article 30 of Regulation 1901/2006 also introduced a new type of marketing authorisation, the Paediatric Use Marketing Authorisation (PUMA), although this has been little used in practice:

381. A negative term SPC is possible in accordance with the SPC legislation but was not foreseen at the time of its adoption, as there was at that stage no possibility of extending a negative term SPC to give a positive, meaningful outcome.

(1) Submission of an application for a paediatric use marketing authorisation shall in no way preclude the right to apply for a marketing authorisation for other indications.

(2) An application for a paediatric use marketing authorisation shall be accompanied by the particulars and documents necessary to establish quality, safety and efficacy in the paediatric population, including any specific data needed to support an appropriate strength, pharmaceutical form or route of administration for the product, in accordance with an agreed paediatric investigation plan.

The application shall also include the decision of the Agency agreeing the paediatric investigation plan concerned.

(3) Where a medicinal product is or has been authorised in a Member State or in the Community, data contained in the dossier on that product may where appropriate be referred to in accordance with Article 14(11) of Regulation (EC) No. 726/2004 or Article 10 of Directive 2001/83/EC, in an application for a paediatric use marketing authorisation.

(4) The medicinal product in respect of which a paediatric use marketing authorisation is granted may retain the name of any medicinal product which contains the same active substance and in respect of which the same holder has been granted authorisation for use in adults.

8. Authorisations in Special Circumstances

a. Expedited Procedure

Article 14(9) of Regulation 726/2004 allows for an applicant to request assessment of his or her application under the expedited, or fast-track, procedure if the product is "of major interest from the point of view of public health and in particular from the viewpoint of therapeutic innovation" (which the applicant must substantiate). If accepted, this results in a review time of 150 days, rather than the usual 210 days.

Any request for accelerated assessment should be made as early as possible before the actual submission of the marketing authorisation application. Applicants must substantiate the request and, in particular, justify their expectation that the medicinal product is of major public health interest, particularly from the point of view of therapeutic innovation. There is no single definition of what constitutes major public health interest; this should be justified by the applicant case by case.

The key items to be described in the justification, and the appropriate level of detail, should be evaluated case by case, but the following key items would normally be addressed:

- The unmet needs and the available methods of prevention, diagnosis, or treatment;
- The extent to which the medicinal product is expected to have major impact on medicinal practice, its major added value, and/or how it addresses the greater unmet needs; and
- A brief outline of the main available evidence on which the applicant bases its claim of major public health interest.

b. Conditional Authorisation

Where there is scientific data to demonstrate a positive benefit-risk profile (pending confirmation), but the clinical part of the dossier is incomplete, a conditional authorisation may be granted (Article 14(7), Regulation (EC) No. 726/2004):

> Following consultation with the applicant, an authorisation may be granted subject to certain specific obligations, to be reviewed annually by the Agency. The list of these obligations shall be made publicly accessible.

A Commission regulation[382] provides an explanation of how this system works. Article 4 sets out the requirements for granting a conditional marketing authorisation:

> (1) A conditional marketing authorisation may be granted where the Committee [the CHMP] finds that, although comprehensive clinical data referring to the safety and efficacy of the medicinal product have not been supplied, all the following requirements are met:
>
> (a) The risk-benefit balance of the medicinal product, as defined in Article 1(28a) of Directive 2001/83/EC, is positive;
>
> (b) It is likely that the applicant will be in a position to provide the comprehensive clinical data;
>
> (c) Unmet medical needs will be fulfilled;

382. Commission Regulation (EC) No. 507/2006 of March 29, 2006 on the Conditional Marketing Authorisation for Medicinal Products for Human Use Falling Within the Scope of Regulation (EC) No. 726/2004 of the European Parliament and of the Council (Text with the EEA Relevance), 2006 O.J. (L 92/6).

> (d) The benefit to public health of the immediate availability on the market of the medicinal product concerned outweighs the risk inherent in the fact that additional data are still required.
>
> In emergency situations as referred to in Article 2(2), a conditional marketing authorisation may be granted, subject to the requirements set out in points (a) to (d) of this paragraph, also where comprehensive pre-clinical or pharmaceutical data have not been supplied.
>
> (2) For the purposes of paragraph 1(c), "unmet medical needs" means a condition for which there exists no satisfactory method of diagnosis, prevention or treatment authorised in the Community or, even if such a method exists, in relation to which the medicinal product concerned will be of major therapeutic advantage to those affected."

The conditional authorisation is valid for one year and may be renewed following an annual review of the specific obligations attached to such authorisation. The specific obligations include a time frame for completion of the additional studies required. It is normally possible to lift the conditional status once the studies have been completed and the conditions satisfied, so most conditional authorisations revert to "normal" status at this time.

c. Authorisation under "Exceptional Circumstances"

Article 14(8) of Regulation (EC) No. 726/2004 makes provision for authorisations to be granted under "exceptional circumstances," where comprehensive clinical data cannot be provided but it is nevertheless in the interests of public health for a marketing authorisation to be granted:

> In exceptional circumstances and following consultation with the applicant, the marketing authorisation may be granted subject to certain conditions, in particular relating to the safety of the medicinal product, notification to the competent authorities of any incident relating to its use, and action to be taken. The marketing authorisation may be granted only when the applicant can show that he is unable to provide comprehensive date on the efficacy and safety of the medicinal product under normal conditions of use, for objective, verifiable reasons and must be based on one of the grounds set out in Annex I to Directive 2001/83/EC. Continuation of the marketing authorisation shall be linked to the annual reassessment of these conditions.

Article 22 of Directive 2001/83/EC provides in the same terms for the possibility of "exceptional circumstances" authorisations for mutual recognition procedure/decentralised procedure/nationally authorised products.

The specific grounds justifying the grant of such an authorisation are set out in the legislation (Annex I of Directive 2001/83/EC):

> (d) Where the indication is encountered so rarely that the applicant cannot reasonably be expected to provide comprehensive evidence;
> (e) Where, in the present state of scientific knowledge, comprehensive information cannot be provided;
> (f) Where it is contrary to generally accepted principles of medical ethics to collect such information.

As for conditional authorisations, specific obligations are attached to the marketing authorisation, which is subject to an annual reassessment to examine the product's risk-benefit profile. However, in such cases, the expectation is that the authorisation will remain under "exceptional circumstances" rather than revert to a normal marketing authorisation, as it will often not be possible to gather a full set of data (given the reasons previously listed). Orphan drugs quite often need to rely on the "exceptional circumstances" route to obtain a marketing authorisation.

E. Pharmaceutical Trade-marks

Many pharmaceutical trade-marks are word marks, but they can also include figurative logos, colors, or the "get-up" of the product—for example the well-known, diamond-shaped, blue tablet of Pfizer's Viagra.

A trade-mark provides a company with the means to protect the monopoly created by a patent after the expiry of that patent. However, the scope of protection provided by a pharmaceutical trade-mark is limited in the European Union by regulatory restrictions on advertising and by obligations always to associate use of the trade-mark with its nonproprietary name.[383] There is also pressure on physicians from public health authorities (that pay for pharmaceuticals) to prescribe using the generic or nonproprietary name rather than the trade-mark. In some jurisdictions within the EU, when a trade-mark is used on a prescription, the pharmacist fulfilling the prescription is at liberty to substitute a generic equivalent, if available.[384]

---

383. *See* Article 54(a) of Directive 2001/83/EC of November 6, 2001 on the Community Code Relating to Medicinal Products for Human Use, 2001 O.J. (L 311), 67, as amended.

384. *See* Case 266/97 and Case 267/87, R v. Royal Pharm. Soc'y of Great Britain *ex parte* the Ass'n of Pharm. Imps., [1989] 2 C.M.L.R. 751, [1989] 2 All E.R. 758.

1. Double Hurdle for Pharmaceutical Names/Trade-marks

In the EU, before commercialization of a pharmaceutical product, the name of the product must overcome two hurdles: (1) registration of the trade-mark at a national IP office of a member state or centrally at the Office for Harmonisation in the Internal Market (OHIM),[385] and (2) examination of the name as part of the process of obtaining a marketing authorisation, either nationally or via the centralised procedure.

A granted marketing authorisation has no influence on the examination of the trade-mark at national IP offices or OHIM, as explicitly stated by the General Court:[386]

> [A]s regards the reference by the applicant to the fact that the European Agency for the Evaluation of Medicinal Products has granted it authorisation to market its product using the trade mark TRAVATAN . . . it is irrelevant in the present case, since any such authorisation has no bearing on the assessment of likelihood of confusion in connection with the application of Regulation No. 40/94.[387]

The reverse is also true at the European Medicines Agency (the EMA, formerly the EMEA).

> It is evident that even where an invented name has been registered in Member States/Community for a medicinal product, safety considerations must determine whether that invented name may be used for the medicinal product . . . It should be highlighted that the issue of whether a particular invented name will or may constitute an infringement of another entity's intellectual property rights cannot be one of the EMEA's concerns and is therefore not taken into account by the EMEA in its consideration of the acceptability of a proposed invented name.[388]

2. Approval of a Pharmaceutical Name by the European Medicines Agency

The (Invented) Name Review Group (NRG) within the EMA is responsible for reviewing proposed names from a safety/public health point of

385. There is currently a proposal to rename OHIM to European Union Trade Marks & Designs Agency.

386. For a description of the General Court see the Introduction. The General Court was known as the Court of First Instance since its inception until November 2009.

387. Case T-130/03, Alcon Inc. v. OHIM & Biofarma, 2005 E.C.R. II-3861, 3883.

388. Guideline on the Acceptability of Names for Human Medicinal Products Processed Through the Centralised Procedure, CPMP/328/98, at 2–3 (rev. 5).

view. The NRG is chaired by an EMA representative and is composed of representatives from member states of Southern, Central, Northern, and Eastern Europe to ensure representation of different language groups. The members are representatives from national competent authorities[389] having scientific and regulatory responsibilities and experience.

The proposed invented name and all the background information provided by the applicant is sent by the NRG to the national competent authorities of the member states, the European Commission, and the World Health Organization, and each must then, within 30 days, inform the NRG of any objections to the proposed name on grounds of safety or other concerns.

Up to four invented names per marketing authorisation can be proposed to the NRG for consideration, essential given that between 40 and 50 percent of invented names are rejected. Importantly, regardless of the route to registration of the marketing authorisation, use of a single trade-mark throughout the European Union for the same medicinal product is desirable, as per Article 6 of Regulation (EC) No. 726/2004:

> [E]ach application for the authorisation of a medicinal product . . . otherwise than in exceptional cases relating to the application of the law on trade marks, shall include the use of a single name for the medicinal product.

Only in exceptional cases where the proposed trade-mark has been canceled, opposed, or objected to by a national IP office will the existence of more than one name for a single product be accepted. There has been only one reported case in which a different trade-mark was authorised for use in Germany because of an objection by a third party to use of the name in Germany.[390] In reality, it is still easier and faster to propose a new mark for all of the EU than to challenge an adverse decision in the courts.

The name of the medicinal product "may be either an invented name not liable to confusion with the common name, or a common name or scientific name *accompanied* by a trade mark or the name of the marketing authorisation holder."[391]

A "common name" is the "international non-proprietary name [INN] recommended by the World Health Organisation, or, if one does not exist, the usual *common* name."[392] The WHO administers INNs internationally (although some countries, such as the United States with its U.S. Adopted

389. For a definition of "competent authority," see section 4.1.2.

390. Case T-123/00, Dr. Karl Thomae GmbH v. Comm'n, 2002 E.C.R. II-5193, under Regulation (EC) No. 2309/93.

391. Article 1(20) of Directive 2001/83/EC.

392. Article 1(21) of Directive 2001/83/EC.

Names, provide their own nonproprietary name system).[393] INNs should not in principle be used as trade-marks as doing so endangers the principle that INNs are public property, can frustrate the rational selection of further INNs for related substances, and can ultimately compromise the safety of patients by causing confusion in drug nomenclature.

The invented name should not be liable to cause confusion in print, handwriting, or speech with the invented name of another medicinal product. When assessing potential confusion, the NRG will take into account: the indication(s); patient population(s); pharmaceutical form(s); route of administration(s); strength(s); setting for dispensing and use; whether a prescription is required; orphan status; and, perhaps most importantly, the potential for harm to the patient in case of a mix-up. The invented name should not convey misleading therapeutic and/or pharmaceutical connotations and should not be misleading with respect to the composition of the product.

After authorisation, if prescription errors occur due to confusion with the trade-mark of another medicinal product and there is an adverse drug reaction, this should be reported within the pharmacovigilance systems established by the marketing authorisation holder. If the confusion does not result in an adverse drug reaction, the prescription error should be reported to the EMA. Such reports will be reviewed by the NRG with possible outcomes being a change to the trade-mark or notification of the error to the relevant patients and health care professionals.

3. Registration of Pharmaceutical Trade-marks

Applicants have the choice of pursuing registration of a Community trade-mark (CTM),[394] having unitary effect throughout the European Union, via OHIM located in Alicante, or national trade-mark registrations via national IP offices. The national offices, and national courts, apply a harmonised trade-mark law under Trade Mark Directive 2008/95/EC.[395] This directive closely parallels Regulation (EC) 207/2009,[396] the Community Trade Mark Regulation, which is applied to CTMs by OHIM and, on appeal, by the General Court and ultimately by the Court of Justice (as well as by national courts when CTMs are enforced nationally).[397]

393. *See Guidance on INN*, World Health Org., http://www.who.int/medicines/services/inn/innquidance/en/index.html (last visited May 23, 2014); Guidelines on the Use of International Non-proprietary Names (INNs) for Pharmaceutical Substances (World Health Org. 1997).

394. There is a current proposal to rename the CTM to "European trade mark."

395. Directive 2008/95/EC of 22 October 2008 to Approximate the Laws of Member States Relating to Trade Marks, O.J. (L 299/25), 8.11.2008.

396. Council Regulation (EC) No. 207/2009 of February 26, 2009, O.J. (L 78/1), 24.3.09.

397. For a description of regulations and directives, national and European courts, see the Introduction.

a. Absolute Grounds of Objection and Invalidity

The absolute grounds throughout the EU for refusing to register a trade-mark or for establishing its invalidity are set out in Article 3 of the Trade Mark Directive and Article 7 of the Community Trade Mark Regulation. The relevant provisions of the regulation are those set out in the first eight provisions of Article 7(1), which corresponds to Article 3(1) of the directive.

> 7(1) The following shall not be registered:
>
> (a) signs which do not conform to the requirements of Article 4 [signs of which a Community trade-mark may consist];
>
> (b) trade marks which are devoid of any distinctive character;
>
> (c) trade marks which consist exclusively of signs or indications which may serve, in trade, to designate the kind, quality, quantity, intended purpose, value, geographical origin or the time of production of the goods or of rendering of the service, or other characteristics of the goods or service;
>
> (d) trade marks which consist exclusively of signs or indications which have become customary in the current language or in the bona fide and established practices of the trade;
>
> (e) signs which consist exclusively of:
>
> (i) the shape which results from the nature of the goods themselves;
>
> (ii) the shape of goods which is necessary to obtain a technical result;
>
> (iii) the shape which gives substantial value to the goods;
>
> (f) trade marks which are contrary to public policy or to accepted principles of morality;
>
> (g) trade marks which are of such a nature as to deceive the public, for instance as to the nature, quality or geographical origin of the goods or service;
>
> (h) trade marks which have not been authorised by the competent authorities and are to be refused pursuant to Article 6 . . . of the Paris Convention.

A trade-mark will be refused registration on the grounds set out in paragraph (1) even if the grounds pertain only to a part of the European Union, as per Article 7(2).

If the sign designates in any of its possible meanings, whether now or in the future, the characteristics of the goods or service in which registration is sought, then as per Articles 7(b) and 7(c), the signs are incapable of fulfilling the indication-of-origin function, unless the trade-mark has acquired distinctive character through use, as per Article 7(3).

For example, the mark "5 HTP" for pharmaceutical products was canceled for being descriptive, because it is an abbreviation of the chemical 5-hydroxytryptophan, with the abbreviation differing only in that it did not contain the hyphen.[398]

Similarly, the General Court found that English consumers would immediately understand the mark "Passion for Better Foods" to mean that the products offered were of superior quality, which included pharmaceutical products, as they are consumed orally and can be ingested with food. The mark was considered an advertising slogan, was devoid of distinctive character, and did not therefore identify the commercial origin.[399]

Customary use of a mark as per Article 7(d) will bar registration, as was the case for the mark "BSS" with respect to "ophthalmic pharmaceutical preparations: sterile solutions for ophthalmic surgery." The General Court and the Court of Justice held that "BSS" was at the time of registration a generic reference to "balanced salt solution" among the relevant public, being ophthalmologists and ophthalmic surgeons, and no distinctive character had been acquired through use.[400]

Words, although descriptive themselves, may be combined to produce a mark that is not itself descriptive if it produces an impression that is sufficiently removed from that produced by the sum of the elements that compose it. Therefore, if the new word is more than a sum of its parts, it may be registrable.

This was not the case for the fused word "SupplementPack," for which registration was sought for chemical and biochemical products, despite arguments from the applicant that the juxtaposition of the words does not exist in the English language, is not grammatically correct, and is not a technical term used in the relevant scientific circles. The General Court held that the word was descriptive, as there was not a perceptible difference between the combined word and the mere sum of its parts. In fact,

398. Case T-190/09, Longevity Health Prods. Inc. v. OHIM, General Court, 2011 E.C.R. II-0044; Case C-222/11, Longevity Health Prods. Inc. v. OHIM, Order of the Court (not published in English, available in German and French at Curia website), http://curia.europa.eu/juris/document/document.jsf?docid=117141&doclang=FR}}.

399. Case T-251/08, Vion NV v. OHIM (Sept. 23, 2011) (not published in English, available in German and French at Curia website).

400. Case T-273/01, Alcon v. OHIM, 2003 E.C.R. II-411; Case C-192/03, P Alcon v. OHIM, 2004 E.C.R. I-8993.

use of capital letters for each word diminished the juxtaposition of the words, and the capital letter within the term emphasized that the mark is composed of two elements.[401]

Nor was the word "Vektor-Lycopin" registrable, as it was descriptive of goods that may transmit lycopene or that may contain lycopene. The meaning of the German words *vektor* and *lycopin* are found in dictionaries and on websites and would be understood by the relevant public (in English, "lycopene," meaning a carotenoid with antioxidant properties).[402]

"Celltech" was held registrable by the General Court as, although one meaning of the word is "cell technology," the relevant public would not associate it with pharmaceutical products.[403] In contrast, the mark "BioGeneriX" was refused registration for "pharmaceutical biotechnology" goods as being descriptive of those goods, particularly as aurally the mark is identical to "biogenerics." The argument by the applicant that, in fact, biological products are not truly "generic" but, rather, "biosimilar," not surprisingly failed. Replacing *c* and *s* with *X* and capitalizing *G* and *X* did not sufficiently distinguish.[404]

Trade-marks that incorporate descriptive elements may be able to overcome the absolute grounds of examination prior to publication, but may be of limited value when trying to oppose a later trade-mark application or in subsequent trade-mark infringement litigation.

This was the case in the opposition proceedings between the registered mark Echinacin and the word for which registration was sought, Echinaid.[405] Because the common prefix "echina" would be understood by the relevant public as referring to the properties of the Echinacea plant and not the commercial origin of the goods, the General Court found that it was descriptive and that it would be disregarded by the consumer. The comparison was, therefore, between the endings of the words, with "cin" and "id" being distinctive and dominant elements that attract the attention of consumers. Consequently, there was no likelihood of confusion between the two marks.

In contrast, the mark Botox, although initially derived from the name of the protein botulinum toxin, had acquired a highly distinctive character and substantial reputation such that registration of BotuMax, Botolist, and Botocyl were each canceled or refused on the grounds that their use would

401. Case T-113/09, PromoCell Bioscience Alive Biomedizinische Produkte GmbH v. OHIM, 2010 E.C.R. II-0017.

402. Case T-85/08, Exalation Ltd. v. OHIM, 2010 E.C.R. II-3837.

403. Case T-260/03, Celltech R&D Ltd. v. OHIM, 2005 E.C.R. II-1215, upheld on appeal to the CJ; Case C-273/05 OHIM v. Celltech R&D Ltd., 2007 E.C.R. I-2883.

404. Case T-47/07 Ratiopharm GmbH v. OHIM, 2008 E.C.R. II-0179; Case T-48/07 Ratiopharm GmbH v. OHIM, 2008 E.C.R. II-0181.

405. Case T-202/04, Madaus AG v. OHIM & Optima Healthcare Ltd., 2006 E.C.R. II-1115.

take unfair advantage and dilute the distinctive character of Botox. This was even the case for Botolist, where the goods in question were dissimilar.[406]

b. Relative Grounds of Objection and Invalidity

The relative grounds for refusing to register a trade-mark or for establishing its invalidity are set out at Article 8(1) of the Community Trade Mark Regulation, which corresponds to Article 4(1) of the Trade Mark Directive.

> 8(1) Upon opposition by the proprietor of an earlier trade mark, the trade mark applied for shall not be registered or, if registered, shall be liable to be declared invalid:
>
> (d) if it is identical with the earlier trade mark and the goods or services for which registration is applied for are identical with the goods or services for which the earlier trade mark is protected;
>
> (e) if because of its identity with, or similarity to, the earlier trade mark and the identity or similarity of the goods or services covered by the trade marks there exists a likelihood of confusion on the part of the public in the territory in which the earlier trade mark is protected; the likelihood of confusion includes the likelihood of association with the earlier trade mark.

Examination of relative grounds for a CTM takes place for each of the member states of the EU, and registration will be refused based on relative grounds in a single member state.

The risk that the public might believe that the goods or services in question come from the same undertaking or from economically linked undertakings constitutes a likelihood of confusion, which must be assessed globally, according to the relevant public's perception of the signs and goods or services in question. A likelihood of confusion will be found where the marks in issue are identical or similar for goods or services that are also identical or similar. A lesser degree of similarity between the goods designated (for example, due to their different therapeutic indications) may be offset by a greater degree of similarity between the marks and vice versa.

c. The Relevant Public

Where the goods at issue are pharmaceutical products requiring a doctor's prescription prior to their sale to end users in pharmacies, the relevant

406. Case T-131/09, Farmeco AE Dermokallyntika v. OHIM & Allergan Inc. (BOTUMAX), 2010 E.C.R. II-0243; Joined Cases T-345/08 (BOTOLIST) and T-357/08 (BOTOCYL), Helena Rubinstein SNC & L'Oreal SA v. OHIM & Allergan, 2010 E.C.R. II-0279. Decided under Article 8(5) of Regulation No. 40/94, now Article 8(5) of Regulation 209/2009 in relation to "famous marks."

public comprises both end users and health professionals, being doctors who prescribe and pharmacists who sell the products. Despite the fact that the choice of those products is influenced or determined by the health professionals, consumers as end users are still included in the relevant public.

The average consumer is deemed to impart a high degree of attentiveness when purchasing pharmaceutical products. This higher level of attention applies both when the consumer purchases the pharmaceutical product without a prescription and when a prescription is necessary.

Given the different levels of medical knowledge, there can be a likelihood of confusion for end consumers where there is none for medical professionals. For example, as between the words Respicur and Respicort, although above-average consumers are able to understand the descriptive "respi" to refer to respiratory illnesses, their limited medical knowledge would prevent them from discerning the conceptual references of the components "cur" and "cort." German medical professionals, on the other hand, would understand "cur" to mean "heal" or "cure" and "cort" to mean "corticoids"; thus, these elements had a weak distinctive character. However, as the relevant public was composed of both professionals and end users, the mark was canceled.[407]

d. Comparison of Goods

In assessing the similarity of the goods or services at issue, the following factors are taken into account: their nature, their intended purpose, their method of use, and whether they are in competition with each other or are complementary, as well as the distribution channels of the goods concerned.

Although the category of "pharmaceutical preparations" (class 5 of the Nice Classification of Goods) is sufficiently broad to allow subcategories to be considered independently,[408] the courts take a broad view of what goods and services are similar. This is because products within class 5 are of the same nature (pharmaceutical preparations), have the same function or intended purpose (treatment of health problems), are directed at the same consumers (health care professionals and patients), and use the same distribution channels (typically pharmacies).

An exception is sometimes made for goods that have different therapeutic indications. The marks Famoxin and Lanoxin were found not to be confusingly similar in part because of their different therapeutic indications, with the former being used for the treatment of metabolic disorders

407. Case T-256/04, Mundipharma AG v. OHIM, 2007 E.C.R. II-0449.

408. Joined Cases T-493/07, T-26/08, T-27/08, GlaxoSmithKline SpA, Laboratorios Wellcome de Portugal, Lda & Wellcome Found. Ltd. v. OHIM & Serono Genetics Inst. SA, 2009 E.C.R. II-0175.

and the latter for cardiovascular illnesses.[409] Similarly, Tolposan and Tonopan were found not to be confusingly similar as the signs had an average degree of similarity and the uses for the goods—pain alleviation and muscle relaxant medicines, respectively—were only slightly similar.[410]

However, differing therapeutic indications are generally not enough to outweigh other similarities, especially where the signs present a high degree of similarity. Citracal and Cicatral were found confusingly similar despite having different therapeutic indications ("dietary supplements, including dietary supplements inclusive of calcium salt" and "healing salves") and methods of application (ingestion as supplements to daily nutrition versus application of the healing salve to wounds and sores).[411] Similarly, there was a likelihood of confusion between Galzin and Calsyn despite having different therapeutic indications: "treatment of Wilson's disease" (a genetic disorder in which copper levels accumulate in tissues) and "pharmaceutical and medical preparations, more specifically calcium-based preparations." In *Alcon Inc. v. OHIM and Biofarma SA*,[412] the marks were considered confusingly similar despite a difference in the method of administration, with one being a tablet taken orally and the other, eye drops.

Some subcategories within class 5 pharmaceutical preparations have been considered similar:

- "veterinary preparations" was identical to "pharmaceutical preparations" (the marks being ZydU.S. and ZimbU.S., respectively);[413]
- "vitamin preparations" was highly similar to "pharmaceutical preparations" (the marks being Eviol and Evion, respectively);[414] and
- "insecticides for killing dust mites" provided a low degree of similarity to "pharmaceuticals" (the marks being a figurative sign "astex Technology" and Astex, respectively).[415]

Goods from other classes may also be considered similar to "pharmaceutical preparations" in class 5:

409. *Id.*

410. Case T-331/09, Novartis AG v. OHIM & Sanochemia Pharmazeutika AG, 2010 E.C.R. II-5967.

411. Case T-277/08, Bayer Healthcare LLC v. OHIM & Uriach-Aquilea OTC, SL, 2009 E.C.R. II-0214.

412. Case T-130/03, Alcon Inc. v. OHIM & Biofarma SA, 2005 E.C.R. II-3861.

413. Case T-288/08, Cadila Healthcare Ltd. v. OHIM & Novartis AG (2012), not published, available at Curia website; http://curia.europa.eu/juris/document/document.jsf?docid=120423&doclang=EN.

414. Decision of the Cancellation Division of 07.07.2009, 2890 C, *available at* http://oami.europa.eu/LegalDocs/Cancellation/en/C002997856_1630.pdf.

415. Case T-48/06, Astex Therapeutics Ltd. v. OHIM & Protec Health Int'l Ltd., 2008 E.C.R. II-0161.

- "surgical, medical and veterinary apparatus and instruments" in class 10 were considered similar (the marks being a figurative sign "AMS Advanced Medical Services" and AMS, respectively);[416]
- hair care products with anti-hair loss properties within the category "hair lotions" in class 3 had a low degree of similarity to pharmaceutical products despite the fact that hair care products may contain active ingredients to treat and prevent hair loss (the marks being Neodys and Neo Dyl, respectively);[417] and
- "clothing for professional use and protective clothing for staff and patients in hospitals" showed a "certain degree of similarity" to "medical and surgical plasters; compresses; materials for dressings" given that the two products could be used simultaneously and that the manufacturers and distribution channels could overlap (the marks being Mesilette and Medinette, respectively).[418]

e. Comparison of Signs

Two marks are considered similar when, from the point of view of the relevant public, they are at least partially identical as regards one or more of the visual, phonetic, and conceptual aspects.

For identical or very similar goods, a single letter change was not sufficient to prevent a likelihood of confusion, as was the case with the signs Seroslim and Serostim.[419] With respect to visual comparison, courts consider that consumers focus on the prefix of words, disregarding identical suffixes, particularly if they are descriptive. This was the case for the sign applied for Biovisc, for which there was a clear visual difference when compared to the trade-marks Provisc and Duovisc.[420]

However, although it is understood that the consumer focuses on the prefixes of words, even if the first letter of the trade-mark is different from the sign to be registered, this will not always prevent a likelihood of confusion. This was the case for Resverol and Lesterol, on the basis that the consumer would consider the first syllable and, although the first letter was different, the following two letters within the first syllable were

416. Case T-425/03, AMS Advanced Med. Servs. GmbH v. OHIM & AMS, 2007 E.C.R. II-4270.

417. Decision of the Opposition Division, 30.09.2009, *available at* http://oami.europa.eu/LegalDocs/Opposition/2009/en/000965949.pdf.

418. Case T-342/10, Paul Hartmann AG v. OHIM & Mölnlycke Health Care AB, 2012, not published, available at Curia website; http://curia.europa.eu/juris/document/document.jsf?text=&docid=123791&pageIndex=0&doclang=EN&mode=lst&dir=&occ=first&part=1&cid=153237.

419. Case T-201/08, Market Watch Franchise & Consulting Inc. v. OHIM & Ares Trading SA, 2010 E.C.R. II-0203.

420. Case T-106/07, Alcon v. OHIM & Acri.Tec AG, 2008 E.C.R. II-0177.

identical.[421] Similarly, although Sorvir and Norvir were conceptually dissimilar, the visual and phonetic similarity between the marks was held to create a likelihood of confusion.[422]

Again, with Prazol and Prezal, the different vowel did not outweigh the visual similarity.[423] The presence of the character "-" in E-plex, which indicated that the mark was composed of two elements whereas the earlier mark Epilex consisted of one word, was not enough to differentiate the marks, particularly as the marks were phonetically similar.[424]

Any figurative elements of a mark must be considered in the comparison if those elements form a significant component in the overall impression produced by the mark. As between Hitrechol, with a figurative element of two Chinese characters immediately above the word, and the word Hypochol, although the goods in question were identical, the Chinese characters introduced an element of conceptual differentiation and the General Court found no likelihood of confusion.[425]

The stylized character of the graphical representation of the sign oli provided a distinctive component and, for that reason, was not similar to Olay.[426] Similarly, the stylized apomea had only a very low degree of visual and aural similarity to APO, and there was no likelihood of confusion.[427]

F. Regulatory Exclusivity for Medicinal Products in the EU

Regulatory exclusivity, or regulatory data protection, was first established in the European Union in 1987, by amendments to Directive 65/65/EC, well before it was mandated by Article 39.3 of TRIPS.[428] The resulting Directive 87/21/EEC was replaced in 2001 by a codifying measure,

---

421. Case T-363/09, Longevity Health Prods. Inc. v. OHIM & Gruppo Lepetit SpA, 2010 E.C.R. II-0286.

422. Case T-149/08, Abbott Labs. v. OHIM & aRigen Inc., 2010 E.C.R. II-0198.

423. Case T-95/07, Aventis Pharma SA v. OHIM & Nycomed GmbH, 2008 E.C.R. II-0229.

424. Case T-161/10, Longevity Health Prods., Inc. v. OHIM & Tecnifar-Industria Tecnica Farmaceutica, SA, 2011 E.C.R. II-0156.

425. Case T-517/10, Pharmazeutische Fabrik Evers GmbH & Co. KG v. OHIM & Ozone Labs. Pharma SA—available at Curia website: http://curia.europa.eu/juris/document/document.jsf?text=&docid=125004&pageIndex=0&doclang=EN&mode=lst&dir=&occ=first&part=1&cid=153364

426. Case T-240/08, Procter & Gamble Co. v. OHIM & Labs. Alcala Farma, SL, 2009 E.C.R. II-0114.

427. Decision of the Opposition Division, OHIM (Jan. 15, 2009), *available at* http://oami.europa.eu/LegalDocs/Opposition/2009/en/001211176.pdf.

428. Directive 65/65/EEC of January 26, 1965 on the Approximation of Provisions Laid Down by Law, Regulation or Administrative Action Relating to Proprietary Medicinal Products, 1965 O.J. (22) 0368-0373, as amended by Directive 87/21/EEC of December 22, 1986, 1987 O.J. (L 15) 36–37.

Directive 2001/83/EC,[429] and this has been amended effective from November 2005[430] to establish the current system of regulatory exclusivity in the EU. Regulatory exclusivity for centrally authorised products is governed by Regulation 726/2004.

Regulatory exclusivity provides innovator companies with a term of protection for their medicinal products[431] that is commercially more reliable than patent protection, as patents are always liable to be found invalid. Also, although regulatory exclusivity is shorter than the patent term, the period of protection begins only when a marketing authorisation has granted.

As discussed in Section V.D.2, "Routes to Authorisation," marketing authorisations in the EU can be obtained via the centralised, national, decentralised, or mutual recognition procedures. The voluminous data required to be provided by the applicant under these procedures, to establish that the drug is safe and efficacious, are protected via regulatory exclusivity by limiting the degree to which subsequent applicants can rely upon them. Although the protection used to be variable throughout the EU,[432] all applications for marketing authorisations submitted from November 2005 benefit from the same regulatory protection.

1. Regulatory Exclusivity

As discussed in Section V.D.4, "Contents of the Application Dossier," an applicant for a marketing authorisation that has not itself carried out the necessary preclinical and clinical testing can obtain a marketing authorisation either by consent of the marketing authorisation holder via the "generic" application route or by the "well-established use" route, as per Article 10 of Directive 2001/83/EC and Article 14(11) of Regulation 726/2004. Authorisations for fixed combination products, as per Article 10(b), are also available, as discussed in Section V.D.4.

429. Directive 2001/83/EC of November 6, 2001 on the Community Code Relating to Medicinal Products for Human Use, 2001 O.J. (L 311) 67.

430. Directive 2004/27/EC of March 31, 2004 Amending Directive 2001/83/EC on the Community Code Relating to Medicinal Products for Human Use, 2004 O.J. (L 136) 34. Although there have been subsequent amendments, they do not affect the regulatory exclusivity provisions.

431. For definition of "medicinal product," *see* Section D.1.b.

432. For applications for a marketing authorisation filed before November 2005, the term of protection varied depending on whether the marketing authorisation was applied for centrally or nationally. If obtained via the centralised procedure, the term of protection was ten years, and if nationally, it depended on the member state. If obtained in Belgium, Germany, France, Italy, The Netherlands, Sweden, or the United Kingdom, the term of protection is ten years. For all other member states, it is six years.

a. Duration of Protection

The "generic" application, whether made centrally or nationally, allows an applicant to obtain a marketing authorisation on the basis of the originator's data (for which a marketing authorisation has granted), provided the medicinal product is a generic of the originator's product and eight years have elapsed from the originator's first marketing authorisation in the EU. However, the generic entrant cannot market the generic medicinal product for a further two years, although this period can be extended by a further year if within the first eight years a new indication is introduced by the originator. This gives rise to the often quoted "8+2+1" formula and is provided for at paragraph 1 of Article 10(1) of Directive 2001/83/EC:

> 1. By way of derogation from Article 8(3)(i), and without prejudice to the law relating to the protection of industrial and commercial property, the applicant shall not be required to provide the results of pre-clinical tests and of clinical trials if he can demonstrate that the medicinal product is a generic of a reference medicinal product which is or has been authorised under Article 6 for not less than eight years in a Member State or in the Community.
>
> A generic medicinal product authorised pursuant to this provision shall not be placed on the market until ten years have elapsed from the initial authorisation of the reference product."

The additional one year for a new indication was introduced in the amendments that took effect in 2005 and requires that:

> during the first eight years . . . the marketing authorisation holder obtains an authorisation for one or more new therapeutic indications which, during the scientific evaluation prior to their authorisation, are held to bring a significant clinical benefit in comparison with existing therapies.[433]

The same periods of protection apply in the case of centrally authorised products pursuant to Article 14(11) of Regulation (EC) No. 726/2004.

433. Paragraph 4 of Article 10 of Directive 2001/83/EC as amended. See http://www.ema.europa.eu/ema/index.jsp?curl=pages/regulation/document_listing/document_listing_000211.jsp&mid=WC0b01ac0580031b0a, Guidance on Elements Required to Support the Significant Clinical Benefit in Comparison with Existing Therapies of a New Therapeutic Indication in Order to Benefit from an Extended (11 Year) Marketing Protection Period (European Commission—November 2007) and the European Commission's Notice to Applicants (rev. June 4, 2013), Volume 2A, Chapter 1, Section 6.2.

For applications made for a new indication for a well-established substance, a period of one year of exclusivity shall be granted, as per Article 10(5):[434]

> In addition to the provisions laid down in paragraph 1, where an application is made for a new indication for a well-established substance, a non-cumulative period of one year of data exclusivity shall be granted, provided that significant pre-clinical or clinical studies were carried out in relation to the new indication.

The exclusivity period is noncumulative to other periods of protection in that it refers only to the data concerning the new indications. An application for a new indication is assessed by the competent authority,[435] which will determine whether the new indication for a well-established substance is based on significant preclinical or clinical studies.

There is also a one-year period of protection for significant preclinical or clinical trials carried out for the purpose of substantiating an application for a change of classification, as per Article 74a of Directive 2001/83/EC:[436]

> Where a change of classification of a medicinal product has been authorised on the basis of significant pre-clinical tests or clinical trials, the competent authority shall not refer to the results of those tests or trials when examining an application by another applicant for or holder of marketing authorisation for a change of classification of the same substance for one year after the initial change was authorised.

For fixed combination products, the Notice to Applicants, Volume 2A, Chapter 1, Section 5.5 addresses the question of the data and market exclusivity from which the data on the combination may benefit:

> Strictly speaking, any "combination" of active substances within the same pharmaceutical form is a new and unique medicinal product requiring a separate marketing authorisation and SPC. Therefore a new "combination" medicinal product will have an independent period of data exclusivity and market protection from its first authorisation within the Union. Consequently applications in accordance with Article 10 and 10c referring to "combination" dossiers are acceptable.

434. See Notice to Applicants (rev. June 4, 2013), Volume 2A, Chapter 1, Section 6.3.
435. For an explanation of the term "competent authority," see Section D.1.a.
436. See Notice to Applicants (rev. 4, June 2013), Volume 2A, Chapter 1, Section 6.4.

> An authorisation for a "combination" medicinal product is not considered to fall within the scope of the global marketing authorisations of the individual active substances as described in Article 6(1) of Directive 2001/83/EC.

As of mid-2013, the scope of this position was being challenged in an action in the European Courts, in relation to a refusal by the EMA to validate an application dossier for a generic version of a fixed-dose combination of two known actives (abacavir/lamivudine).[437] The EMA's refusal to validate was based on the fact that the regulatory exclusivity period for the reference combination product had not expired, but the applicant claims that providing regulatory exclusivity in such cases is contrary to the legislation in situations where the reference product is a fixed-dose combination of two known actives, and that the correct interpretation is that the combination product falls within the same "global marketing authorisation" (see the following section) as the earlier individual actives and so does not enjoy a further independent period of protection.

b. Global Marketing Authorisation

Also introduced by the 2005 amendments was the concept of a "global marketing authorisation," as per the second paragraph of Article 6, Directive 2001/83/EC, as amended:

> When a medicinal product has been granted an initial marketing authorisation in accordance with the first subparagraph, any additional strengths, pharmaceutical forms, administration routes, presentations, as well as any variations and extensions shall also be granted an authorisation in accordance with the first subparagraph or be included in the initial marketing authorisation. All these marketing authorisations shall be considered as belonging to the same global marketing authorisation, in particular for the purpose of the application of Article 10(1).

It follows that the global marketing authorisation comprises all marketing authorisations held by the same company for the same active substance, including any additional strengths, pharmaceutical forms, variations, or extensions authorised through separate procedures and under a different name. Consequently, the holder of the initial marketing authorisation cannot obtain any new period of regulatory exclusivity for

437. Teva Pharma and Teva Pharmaceuticals Europe v. EMA (Case T-547/12) (2013/C 46/38), 2013 O.J. (C 46/21), action brought on December 18, 2012.

that medicinal product unless, as noted, a new therapeutic indication "of significant clinical benefit" is established, providing a one-year extension.

The initial marketing authorisation and any other marketing authorisations that fall within the global marketing authorisation are held by the same company if they belong to the same mother company or group of companies, the product is under license, or the companies exercise a concerted practice concerning the placing on the market of the relevant medicinal product. All such companies are considered one entity, and the ten-year period of protection commences with the first marketing authorisation granted.

i. New Active Substance

Importantly, the marketing authorisations within the global marketing authorisation must all refer to the "same active substance," which has the "same qualitative and quantitative composition in active substances," and which includes "different salts, esters, ethers, isomers, mixtures or isomers, complexes or derivatives . . . unless they differ significantly in properties with regard to safety and/or efficacy."[438] Not surprisingly, controversy has arisen as to when a compound can be considered a "new active substance"[439] such that it is worthy of its own global marketing authorisation and ten-year period of regulatory exclusivity.

Two such instances have involved a previously authorised racemate and the new, more active enantiomer. If the enantiomer is considered a new active substance distinct from the racemate, then it is entitled to a marketing authorisation separate from the global marketing authorisation for the racemate. If, instead, the enantiomer is considered the same active substance, then the ten-year period of regulatory exclusivity for the enantiomer is the same as for the racemate.

In *Escitalopram*, the generic entrant applied for a marketing authorisation for the enantiomer escitalopram with reference to the data provided under the marketing authorisation of the racemate Citalopram (which contains the S(+) enantiomer, escitalopram, as well as the R(−) enantiomer). The Dutch competent authority granted the marketing authorisation. Lundbeck, the original marketing authorisation holder for the racemate, appealed to the Dutch court, arguing that escitalopram was a "new active substance" and, therefore, was deserving of its own marketing

438. Article 10(2)(b) of Directive 2001/83/EC, as amended.

439. A definition of "new active substance" is provided in the Notice to Applicants (Volume 2A, Chapter 1, Annex III) that provides some instruction with respect to biological substances: "a new biological substance, . . . a biological substance previously authorised as a medicinal product in the EU, but differing in molecular structure, nature of the source material or manufacturing process."

authorisation and its own period of regulatory exclusivity. As a result, a generic application for escitalopram should only reference data for escitalopram and not the racemate (and as the regulatory exclusivity period was still in place for the enantiomer, any generic application must be refused). The Dutch court found no significant clinical differences with regard to safety and efficacy between the enantiomer and racemate as there were no clinically relevant differences in efficacy, the profiles of adverse events were similar, and the toxicity profiles comparable. Escitalopram was, therefore, not a "new active substance" and thus formed part of the same global marketing authorisation as Citalopram, and a generic application for the enantiomer could therefore refer to the regulatory data for the racemate. However, a subsequent opinion of the EMA Committee for Medicinal Products for Human Use (CHMP) concluded that there was insufficient evidence to establish that the generic escitalopram had comparable efficacy and safety to the racemate, and the generic marketing authorisation was suspended (the CHMP opinion reviewed only the publicly available data relating to Citalopram).[440]

In contrast, the CHMP rejected the enantiomer eszopiclone as a "new active substance" when comparing it to its racemate, having concluded that the enantiomer was not significantly different in safety and efficacy. Sepracor Pharmaceuticals marketed the racemate zopiclone for insomnia and had applied for a marketing authorisation for the enantiomer eszopilcone, also for insomnia. Although limited pharmacokinctic data suggested that the absorption of the enantiomer was faster by 30 minutes when compared to the racemate (therefore inducing sleep faster) and the plasma concentrations were below the therapeutically effective level after eight hours compared to over nine hours for the racemate (providing an advantage to the patient of next-day, unwanted side-effects), the CHMP concluded that the lack of comprehensive data between eszopiclone and the racemate made the evaluation of relative clinical benefit impossible. The CHMP therefore recommended that the marketing authorisation be granted, but that the medicine should not be granted "new active substance" status. Following a reexamination process at the request of Sepracor, the CHMP confirmed its previous opinion, and the Commission (responsible for actually granting the marketing authorisation) indicated in a letter several months later that it was inclined to follow that opinion. As a result, Sepracor withdrew its application for a marketing authorisation, shortly before grant, stating that the CHMP opinion compromised the

440. Decision of the Dutch Medicines Evaluation Board (Apr. 29, 2009), *available at* http://www.cbg-meb.nl/NR/rdonlyres/925652E3-845D-4A28-899E-DF6D0AD050B9/0/2522010escitalopram.pdf.

commercial viability of launching the product in Europe.[441] Sepracor then sought to annul the Commission's decision by appeal to the General Court. However, the court, upheld on appeal, found the letter no more than an opinion without legal effect, which could not therefore be the subject of an annulment.[442] Although the enantiomer is marketed in the United States by Sepracor (under the name Lunesta), it is not marketed in the EU.

Following these decisions, in October 2012, the EMA published a Reflection Paper[443] that outlines the scientific approach that will be taken when evaluating whether enantiomers, complexes, derivatives, or different salts or esters of an existing active will be considered a "new active substance." Although the EMA will take each decision case by case, the default position is that such actives are not different from the reference unless demonstrated otherwise. In order to demonstrate a significant difference, the potential new active must be directly compared to the reference, with indirect, non-comparative evidence being unlikely to persuade. Such direct comparisons will be most compelling if presented in head-to-head clinical studies in order to demonstrate clinically relevant differences in safety and/or efficacy. However, preclinical data or pharmacological, microbiological, and toxicological studies may also be convincing on their own.

The Reflection Paper sets out the evidence that is more likely to support a finding of a "new active substance":

- Significant changes to the dosing frequency (e.g., twice daily to once daily) or another route of administration mandated by significant differences in safety and/or efficacy properties;
- Changes to the overall efficacy at clinically relevant doses (e.g., clinically and statistically significant difference in the primary endpoint);
- Clinically relevant changes that result in differences to contraindications, warnings, or clinically significant adverse reactions;
- Clinically relevant changes that affect significantly drug:drug interactions such that the population able to take the drug is significantly different;

441. Press Release, EMA, Sepracor Pharmaceuticals Ltd. Withdraws Its Marketing Authorisation Application for Lunivia (Eszopiclone), EMEA/303056/2009 (May 15, 2009).

442. Case T-275/09, Sepracor Pharm. (Ir.) Ltd. v. European Comm'n, 2011 E.C.R. I-1753; Case C-477/11, P Sepracor Pharm. (Ir.) Ltd. v. European Comm'n (May 14, 2012), not published, see Curia website; http://curia.europa.eu/juris/document/document.jsf?text=&docid=107561&pageIndex=0&doclang=EN&mode=lst&dir=&occ=first&part=1&cid=156626 .

443. Reflection Paper on Considerations Given to Designation of a Single Stereo Isomeric Form (Enantiomer), a Complex, a Derivative, or a Different Salt or Ester as New Active Substance in Relation to the Relevant Reference Active Substance, EMA/651649/2010 (Oct. 18, 2012).

- Clinically relevant changes that allow the product to be used in a wider patient population within the current indication or previously excluded sub-groups (e.g., the product is to be used in the patient population which previously was excluded due to significant safety concerns or lack of efficacy); or
- Compelling preclinical data where it is not feasible to conduct head-to-head clinical studies (e.g., differences in reproductive toxicity or carcinogenicity, or the reference active substance is not authorised for the proposed indication).

The Reflection Paper also provides the evidence that is unlikely to be sufficient to support a finding of a "new active substance":

- Changes to pharmacokinetics alone (that do not alter elements as previously set out);
- Preclinical differences that are inconclusive or unlikely to result in significant changes in clinical efficacy or safety;
- Extrapolation between studies (i.e., the claim for significantly different efficacy is based on a study or studies with the reference active substance and a different study or studies with the proposed form); or
- The widening of the patient population (either within the current indication or to a different indication) to groups not previously studied for the reference active substance.

ii. Initial Marketing Authorisation

A marketing authorisation granted other than under European Directive 2001/83/EC is not considered a marketing authorisation for the purposes of Articles 6 and 10 and will not therefore form part of the global marketing authorisation. Therefore, a product granted a marketing authorisation under the national laws of a country before it became a member state of the European Union cannot be the initial marketing authorisation as per Articles 6 and 10.

This was clarified by the General Court in June 2009.[444] In 1963, the Austrian authorities granted Waldheim a marketing authorisation for galantamine, under the brand name Nivalin, for the treatment of poliomyelitis. When Austria joined the EU in 1995, the marketing authorisation was not updated in accordance with the requirements of the EU directive. Then in 2000, a full marketing authorisation for galantamine under the EU directive was granted to Waldheim (and Janssen-Cilag as a result of

444. Case C-527/07, The Queen, Generics (UK) Ltd. v. Licensing Authority, Shire Pharm. Ltd. & Janssen-Cilag AB, 2009 E.C.R. I-5259.

cooperation agreements) under the brand name Reminyl. In 2005, Generics applied for a generic marketing authorisation referencing the earlier product Nivalin as the reference medicinal product. Although regulatory exclusivity continued to apply to Reminyl, no such protection applied to Nivalin. However, the court held that only those medicinal products benefiting from a marketing authorisation issued in accordance with the EU directive (thus providing the certainty that the authorisation was the result of rigorous testing and review) can be considered to be a reference medicinal product, and the generic application was rejected.[445]

c. Well-Established Medicinal Use

As discussed in Section V.D.4.b.iii., "'Well-Established Use' (or 'Bibliographic') Applications," Article 10a of Directive 2001/83/EC provides for a simplified procedure for grant of a marketing authorisation by allowing an applicant to forgo preclinical and clinical trials and replace them with references to published scientific literature, if it can be demonstrated that the active substances of a medicinal product have been in well-established medicinal use within the EU for at least ten years, with recognised efficacy and an acceptable level of safety.[446]

Generic applicants have in the past abused this provision, applying for a generic marketing authorisation before the exclusivity period for the reference medicinal product has expired, by referring to the results of the clinical trials for the reference product as published by the marketing authorisation holder in publicly available scientific journals. One such example was for Clopidogrel (Plavix) where, although the German authorities had granted the marketing authorisation, the European Commission indicated that the well-established use provision had been incorrectly applied and threatened proceedings before the Court of Justice if the German authorities did not change their practice.[447]

It follows that the ten-year exclusivity period cannot be circumvented via Article 10a by reference to scientific literature concerning the preclinical and clinical trials of the reference medicinal product published before expiry of exclusivity.

---

445. This is reflected in Volume 2A, Chapter 1, Section 5.3.1.1 of Notice to Applicants, (June 2013).

446. *See* Notice to Applicants (June 2013), Volume 2A, Chapter 1, Section 5.4.

447. "Health: Germany receives final warning to comply with EU rules on well established medicinal use," Press Release, IP/10/536 (May 5, 2013), stating: "Information available to the Commission suggests that when calculating the period of ten years, German authorities take less extensive use of the active substance into account, such as clinical trials. This inaccurate calculation of the time period infringes the terms of Directive 2001/83/EC and obscures the clear separation between different authorisation procedures."

2. Orphan Medicinal Product Exclusivity

As discussed in Section V.D.6, "Orphan Drugs," pharmaceuticals intended to treat small patient populations receive special treatment under the Orphan Medicinal Products Regulation.[448] The most valuable incentive from which orphan products can benefit is the ten years of market exclusivity provided for by Article 8(1) of the Orphan Drug Regulation, once a marketing authorisation has been obtained. The orphan designation does not of itself constitute any right to place the product on the market, and a marketing authorisation application must be made separately (via the centralised procedure); the usual standards of quality, safety, and efficacy must be met before an authorisation will be granted. The ten years of market exclusivity apply from the date of grant of the marketing authorisation, as follows:

> Where a marketing authorisation in respect of an orphan medicinal product is granted pursuant to Regulation (EEC) No. 2309/93 or where all the Member States have granted marketing authorisations in accordance with the procedures for mutual recognition laid down in Articles 7 and 7a of Directive 65/65/EEC or Article 9(4) of Council Directive 75/319/EEC of 20 May 1975 on the approximation of provisions laid down by law, regulation or administrative action relating to medicinal products, and without prejudice to intellectual property law or any other provision of Community law, the Community and the Member States shall not, for a period of 10 years, accept another application for a marketing authorisation, or grant a marketing authorisation or accept an application to extend an existing marketing authorisation, for the same therapeutic indication, in respect of a similar medicinal product.

If the applicant has successfully completed paediatric studies in accordance with an agreed paediatric investigation plan, then this ten-year exclusivity period is extended to 12 years.

The Orphan Drug Regulation's implementing regulation (Commission Regulation 847/2000) provides the definition of what constitutes a "similar medicinal product" in Article 3(3)(a)–(c):

> For the purposes of the implementation of Article 8 of Regulation (EC) No. 141/2000 on orphan medicinal products the following definitions shall apply:

448. Regulation (EC) 141/2000 of December 16, 1999 on Orphan Medicinal Products.

> (a) "active substance" means a substance with physiological or pharmacological activity;
> (b) "similar medicinal product" means a medicinal product containing a similar active substance or substances as contained in a currently authorised orphan medicinal product and which is intended for the same therapeutic indication;
> (c) "similar active substance" means an identical active substance, with the same principal molecular structural features (but not necessarily all of the same molecular structural features) and which acts via the same mechanism;
> (d) a "similar active substance" is defined as "an identical active substance or an active substance with the same principal molecular structural features (but not necessarily all of the same molecular structural features) and which acts via the same mechanism.

The definition goes on to give examples. In light of this definition, it can be appreciated that the Orphan Drug Regulation provides broad protection of the authorised orphan against market entry of other, similar products for the same indication. There are, however, circumstances in which the market exclusivity period may be "broken." These are set out in Article 8(3) of the Orphan Drug Regulation; by way of derogation from the market exclusivity provided in Article 8(1), a marketing authorisation may be granted, for the same therapeutic indication, to a similar medicinal product if:

> (a) the holder of the marketing authorisation for the original orphan medicinal product has given consent to the second applicant, or
> (b) the holder of the marketing authorisation for the original orphan medicinal product is unable to supply sufficient quantities of the medicinal product, or
> (c) the second applicant can establish in the application that the second medicinal product, although similar to the orphan medicinal product already authorised, is safer, more effective, or otherwise clinically superior.

The concept of "clinical superiority" is explained in the implementing regulation (Commission Regulation 847/2000), Article 3(3)(d):

> (d) "clinically superior" means that a medicinal product is shown to provide a significant therapeutic or diagnostic advantage over and above that provided by

an authorised orphan medicinal product in one or more of the following ways:

1. greater efficacy than an authorised orphan medicinal product (as assessed by effect on a clinically meaningful endpoint in adequate and well controlled clinical trials). Generally, this would represent the same kind of evidence needed to support a comparative efficacy claim for two different medicinal products. Direct comparative clinical trials are generally necessary however comparisons based on other endpoints, including surrogate endpoints may be used. In any case, the methodological approach should be justified; or
2. greater safety in a substantial portion of the target population(s). In some cases direct comparative clinical trials will be necessary; or
3. in exceptional cases, where neither greater safety nor greater efficacy has been shown, a demonstration that the medicinal product otherwise makes a major contribution to diagnosis or to patient care.

The European Commission has also produced a guideline on aspects of the application of Articles 8(1) and 8(3): "Assessing similarity of medicinal products versus authorised orphan medicinal products benefiting from market exclusivity and applying derogations from that market exclusivity,"[449] which gives further detailed explanation of this area. An earlier Commission Communication on Regulation (EC) No. 141/2000[450] provides further elaboration on a number of points relating to application of the Orphan Drug Regulation's provisions.

There is also a provision that the ten-year exclusivity term may be reduced to six years if, at the end of the fifth year, it is established that the criteria laid down in Article 3 are no longer met; inter alia, where it is shown on the basis of available evidence that the product is sufficiently profitable not to justify maintenance of market exclusivity (Article 8(2), Orphan Drug Regulation). There is a guideline to explain the application of this provision but, as of July 2013, it had not yet been used.

449. 2008 O.J. (C 242/12).
450. 2003 O.J. (C 178/2).

G. Supplementary Protection Certificates

1. Background

Although legislation in the United States and Japan concerning patent term extension was introduced in 1984 and 1988, respectively, it was not until April 1990 that the European Commission submitted its proposal for a Council Regulation concerning the creation of supplementary protection certificates (SPCs) for medicinal products.[451] The proposal stated that the aim of the regulation was to narrow the expanding research gap perceived to exist between Europe and the United States and Japan, as only 40 percent of all new chemical entities reaching the research and development stage worldwide were of European origin as opposed to 65 percent ten years earlier.

The result was the Medicinal Products SPC Regulation (SPC Regulation),[452] which came into force throughout most of the European Union on January 2, 1993, and has applied throughout most of the European Economic Area[453] since July 1, 1994. Also introduced was the Plant Protection Products SPC Regulation,[454] which came into force on February 8, 1997, and is in similar terms to the SPC Regulation. Despite one regulation's relating to medicinal products and the other to plant products, the former is to be read in association with the latter, as although the latter does not amend the earlier, it clarifies certain problems of interpretation that have arisen with respect to the SPC Regulation.[455]

2. The Medicinal Products SPC Regulation—Grant

As discussed in the introduction to Section V of this text, a regulation has direct effect and is therefore harmonised law among all member states of the EU. Just as a "European patent" once granted becomes a bundle of national patents, each national IP office interprets the SPC Regulation and grants a national SPC with effect in that member state only.

451. COM (90) 101 final (Apr. 11, 1990).

452. Regulation (EEC) No. 1768/92 of June 18, 1992 Concerning the Creation of a Supplementary Protection Certificate for Medicinal Products, 1992 O.J. (L 182) 1, codified and repealed by Regulation (EC) 469/2009, 2009 O.J. (L 152/1).

453. The Agreement of the European Economic Area entered into force on January 1, 1994 and brings together the 28 EU member states and the 3 EEA EFTA states—Iceland, Liechtenstein, and Norway—in a single market, referred to as the "Internal Market."

454. Regulation (EC) No. 1610/96 of July 23, 1996 Concerning the Creation of a Supplementary Protection Certificate for Plant Protection Products, 1996 O.J. (L 198) 30.

455. Recital 17 of the Plant Products SPC Regulation reads: "Whereas the detailed rules in recitals 12, 13 and 14 and in Articles 3(2), 4, 9(1)(c) and 17(2) of this Regulation are also valid, mutatis mutandis, for the interpretation in particular of recital 9 and Articles 3, 4, 8(1)(c) and 17 of Council Regulation (EEC) No. 1768/92."

The SPC Regulation distinguishes between "medicinal product"[456] (or pharmaceutical) and "product," the product being the "active ingredient" or "combination of active ingredients" within the medicinal product.[457] The grant of the SPC centers around the "product," in that before an SPC is granted, the "product" must be the subject of a marketing authorisation and must be protected by a "basic patent" in force in the member state in which the SPC is sought.[458] Although the medicinal product that is the subject of the marketing authorisation may encompass active ingredients other than the "product,"[459] the marketing authorisation must be the first for the "product" in the EU, and the SPC application must be the first for that "product" within the member state.[460]

Although this appears to be a circuitous route to grant, because SPCs are national rights, this ensures that only one SPC per "product" is granted in each national jurisdiction and that national SPCs granted for the same "product" expire on the same day throughout the EU.

Article 3 of the SPC Regulation sets out the conditions for grant:

> 3. A certificate shall be granted if, in the Member State in which the application referred to in Article 7 is submitted and at the date of that application:
>
> (a) the product is protected by a basic patent in force;
> (b) a valid authorisation to place the product on the market as a medicinal product has been granted in accordance with Directive 2001/83/EC or Directive 2001/82/EC, as appropriate;
> (c) the product has not already been the subject of a certificate; and
> (d) the authorisation referred to in point (b) is the first authorisation to place the product on the market as a medicinal product.

456. For the definition of "medicinal product," see Section D.1.b.

457. Article 3 of the SPC Regulation.

458. What constitutes "protected by a basic patent" in relation to combination products is discussed in Section G.3 below in light of the CJ decisions in Case C-322/10, Medeva BV v. Comptroller Gen. of Patents, Designs & Trade Marks, November 24, 2011, not published, available on Curia website; http://curia.europa.eu/juris/document/document.jsf?text=&docid=115209&pageIndex=0&doclang=EN&mode=lst&dir=&occ=first&part=1&cid=177182, and Case C-422/10, Georgetown Univ., Univ. of Rochester, Loyola Univ. of Chi. v. Comptroller Gen. of Patents, Designs & Trade Marks, November 24, 2011, not published, available on Curia website; http://curia.europa.eu/juris/document/document.jsf?text=&docid=115204&pageIndex=0&doclang=EN&mode=lst&dir=&occ=first&part=1&cid=177283 .

459. *See* Case C-322/10, note 458.

460. For grant in one member state, it is not relevant if the product has been the subject of a certificate in another member state.

3. Scope

Just as the grant of the SPC is to the "product," the protection conferred by an SPC is limited to the "product." In this way, the SPC system differs considerably from the U.S. and Japanese patent term extensions. By limiting the grant of the SPC to the "product," the patent is not itself extended but, rather, the term for the specific "product" protected by the "basic patent" (and not, for example, to other compounds within the scope of the basic patent).

The scope of the SPC includes any other use of the "product" as a "medicinal product" authorised before expiry of the basic patent, such as an authorisation for a new indication, a new formulation, or a combination of that active with other active ingredients.[461] The scope of the SPC is, of course, limited by the scope of the patent; if the basic patent is a process patent, then the SPC is also similarly limited.[462]

For example, an SPC to A will provide the same protection as a patent claim to A, encompassing all medicinal products that contain A, such as A+B, A+B+C, and so on. An SPC to A+B will likewise protect all medicinal products containing A+B, such as A+B+C, and so on.

The SPC can be invalidated on the same grounds as would be effective against the basic patent and additionally on the ground that the SPC should not have been granted.

4. Term

The SPC comes into force upon expiry of the basic patent. After the first five years after filing the basic patent, the applicant is compensated by way of the SPC for the time it takes to obtain a marketing authorisation, up to a maximum of five years.

The term of the SPC is calculated as per Article 13:

> 13(1) The certificate shall take effect at the end of the lawful term of the basic patent for a period equal to the period which elapsed between the date on which the application for a basic patent was lodged and the date of the first authorisation to place the product on the market in the Community reduced by a period of five years.

461. Article 4 of the SPC Regulation and Case C-442/11, Novartis AG v. Actavis UK Ltd. February 9, 2012, not published, available on Curia website; http://curia.europa.eu/juris/document/document.jsf?text=&docid=120021&pageIndex=0&doclang=EN&mode=lst&dir=&occ=first&part=1&cid=177413.

462. Article 5 of the SPC Regulation.

> 13(2) Notwithstanding paragraph 1, the duration of the certificate may not exceed five years from the date on which it takes effect.

It follows from Article 13 that an SPC will only provide a term of protection if the date of the marketing authorisation is more than five years from the filing date of the basic patent. Although it seems sensible that no SPC would be applied for if the marketing authorisation is granted within less than five years from the filing of the basic patent, because of the paediatric extension of six months, discussed next, the effective protection of an SPC can be up to five and a half years. Therefore, an applicant will apply for an SPC from four and a half years from the filing date of the basic patent, such that the SPC has a zero or negative term, which can then be extended by the six-month paediatric extension.[463]

Therefore, subject to the paediatric extension, the SPC cannot last for more than five years from the date it takes effect (patent expiry) or for more than 15 years after the first marketing authorisation in the EEA, if earlier.

5. Paediatric Extensions for Medicinal Product SPCs

The Medicinal Products for Paediatric Use Regulation[464] was introduced in 2007 to extend the duration of a medicinal product SPC by a single six-month period if the medicinal product has been tested in accordance with an agreed paediatric investigation plan. The aim was to increase the number of new medicines available for children rather than rely on off-label use.

As discussed in Section V.D.7, "Paediatric Studies," the paediatric investigation plan (PIP) is agreed between the applicant and the Paediatric Committee of the European Medicines Agency, specifying the timing and measures proposed to assess the quality, safety, and efficacy of the medicinal product in all subsets of the paediatric population that may be concerned. In addition, it will describe any measure to adapt the formulation of the medicinal product so as to make its use more acceptable, easier, safer, or more effective for different subsets of the paediatric population.[465]

463. Case C-125/10, Merck Sharp & Dohme Corp. v. Deutsches Patent und Markenamt, not published, available on the Curia website; http://curia.europa.eu/juris/document/document.jsf?text=&docid=116141&pageIndex=0&doclang=EN&mode=lst&dir=&occ=first&part=1&cid=177602.

464. Regulation (EC) No. 1901/2006 of December 12, 2006 on Medicinal Products for Paediatric Use and Amending Regulation (EEC) No. 1768/92, Directive 2001/20/EC, Directive 2001/83/EC and Regulation (EC) No. 726/2004, 2006 O.J. (L 378) 1.

465. Article 15(2) Regulation (EC) No. 1901/2006 of December 12, 2006 on Medicinal Products for Paediatric Use.

A six-month extension in the context of a 20-year patent term may not appear significant, but given that the profitability of a pharmaceutical is almost always at its highest at the end of the patent term, a six-month extension to an SPC can be very valuable.

6. Application Procedure

The SPC application is made to the national IP offices. It must be applied for by the holder of the patent or his or her successor in title[466] within six months of the grant of the first marketing authorisation for the medicinal product effective in that member state[467] or, where the authorisation has granted before grant of the basic patent, six months from grant of the patent.

Therefore, in order to apply for an SPC, a patentee must choose, in each member state in which a marketing authorisation for a medicinal product is granted (a centralised marketing authorisation having automatic effect in all member states), a suitable national patent in force that protects the product of the authorised medicinal product. This patent becomes the "basic patent."

The applicant must also provide the relevant IP office with documentary evidence of the first marketing authorisation,[468] although this was modified by way of the Plant SPC Regulation to require only details of the authorisation without the actual documents. Should the applicant and owner of the patent differ from the holder of the marketing authorisation, and the latter refuse to cooperate in providing the marketing authorisation, the national IP office can intervene by requesting a copy of the marketing authorisation from the national issuing authority.

With each round of enlargement of the European Union, deadlines for applying for an SPC have been relaxed, by Article 20, to provide for the fact that no SPC application would have been possible in the acceding states within the deadline.

7. Past Issues Relating to SPCs[469]

a. Article 3(b): Can a Valid Marketing Authorisation Include Active Ingredients Other Than the "Product"?

An SPC is intended to protect the "product," as specified in the claims of the basic patent, and not the medicinal product. Therefore, as affirmed by the

466. Article 6 of the SPC Regulation.

467. Article 7(1) of the SPC Regulation.

468. See Section G.7.b for a discussion of what constitutes a "first marketing authorisation."

469. Not discussed in detail is Case-422/11, Novartis v. Actavis, which concerned the scope of an SPC, see Section G.3, and Case C-125/10, Merck v. Deutches, which concerned negative or zero term SPCs, see Section G.4.

Court of Justice[470] in *Medeva*,[471] it follows that an SPC can be granted for a product, specified in the claims of the patent, where the marketing authorisation contains not only that product but also other active ingredients.

For instance, an SPC application to A+B will be granted if the basic patent specifies in the claims the combination of A and B, and the marketing authorisation relied upon for the application contains A+B+C. As long as the marketing authorisation contains the "product" A+B, the SPC will grant.

b. Article 3(d): What Is the First Marketing Authorisation to Place the Product on the Market?

As the first marketing authorisation is linked to both the grant of the SPC as per Articles 3(b) and (d) and the term of protection as per Article 13, controversy has arisen as to what is considered the "first" marketing authorisation.

In order to obtain SPC protection, the marketing authorisation relied upon must be granted under the relevant EU Directive. However, as in *Synthon v. Merz*,[472] if such a product has already been on the market under a marketing authorisation granted other than under the EU Directive (obtained in a country before it joined the EU), the SPC will be refused. It is contrary to the policy of the SPC Regulation to extend the term of protection for a product that has already been on the market before being subject to the EU authorisation procedure.

A second controversy arose with respect to whether a marketing authorisation granted in Switzerland could be considered a "first marketing authorisation." Switzerland is not part of the EEA,[473] but it is part of the EFTA,[474] and Swiss authorisations are automatically recognised in Liechtenstein, which is part of the EEA. Therefore, marketing authorisations granted in Liechtenstein are pursuant to the EU Directive and recognised under the SPC Regulation. The issue was whether the Swiss marketing authorisation should be recognised such that the SPC term was calculated from the earlier Swiss authorisation pursuant to Article 13. Because the European Economic Area Agreement (the EEA Agreement) recognises that Liechtenstein can have two coexisting types of marketing authorisation, one from the Swiss authorities and one under the EU Directive, the Court of Justice held that a Swiss authorisation that has been automatically

470. For an explanation of European Courts see the Introduction.

471. See note 459.

472. Case C-195/09, Synthon BV v. Merz Pharma GmbH & Co. KGaA, 2011 E.C.R. I-7011; Case C-427/09, Generics (UK) Ltd. v. Synaptech Inc., 2011 E.C.R. I-7099.

473. See note 293.

474. The European Free Trade Association (EFTA) is an intergovernmental organization set up for the promotion of free trade and economic integration for the benefit of its four member states: Iceland, Liechtenstein, Norway, and Switzerland.

recognised in Liechtenstein will be regarded as a first marketing authorisation under Article 13 of the SPC Regulation.[475] The decision was no doubt influenced by a desire to limit the term of SPC protection to no more than 15 years, which would be exceeded if the Swiss authorisation was not recognised as the "first" for the purposes of Article 13.

The court followed this decision in November 2013 in finding a marketing authorisation granted in Switzerland and recognised in Liechtenstein as the first marketing authorisation under Article 13, despite the fact that the EMA refused the marketing authorisation based on the same data and the Swiss marketing authorisation was subsequently suspended. Therefore, the SPC application based on a subsequently granted marketing authorisation by the EMA for the same product, which would have provided a term of five years, was granted a term of just less than three years based on the Swiss marketing authorisation.[476]

The third issue arose in France. Although a marketing authorisation had granted, before the medicinal product could be introduced to the French market, agreement had to be reached with the national authorities with respect to pricing and reimbursement. Despite what was therefore in effect a useless marketing authorisation, the Court of Justice held in *Hässle*[477] that the term of the SPC is calculated from the grant of this first marketing authorisation, which effectively reduces its term by the time required to agree on pricing with the national authorities.

c. Article 3(d): What Is the Meaning of "Product" When Determining What Constitutes the First Marketing Authorisation?

As "product" is defined in Article 1(b) as "the active ingredient or combination of active ingredients of a medicinal product" and as Article 3(d) permits grant of an SPC only to a "product" that has not already been the subject of a marketing authorisation, there have been attempts to differentiate two arguably similar medicinal products to allow grant of an SPC to both.

---

475. Joined Cases C-207/03, Novartis AG v. Comptroller, available on Curia; http://curia.europa.eu/juris/document/document.jsf?text=&docid=58140&pageIndex=0&doclang=EN&mode=lst&dir=&occ=first&part=1&cid=178498, and C-252/03, Ministre de l'Economie v. Millennium Pharm. Inc., 2005 R.P.C. 33.

476. Case C-617/12, Astrazeneca AB v. Comptroller Gen. of Patents, Designs & Trade Marks, (Nov. 14, 2013).See Curia website: http://curia.europa.eu/juris/document/document.jsf?text=&docid=144849&pageIndex=0&doclang=EN&mode=lst&dir=&occ=first&part=1&cid=5927

477. *Hässle AB v. Ratiopharm GmbH*, decision of the CJ, December 11, 2011, Case C-127/00, 2011 E.C.R. I-14781, considering the phrase "first authorisation to place the product on the market" as found in Article 19 of Council Regulation (EEC) No 1768/92 of 18 June 1992 concerning the creation of a supplementary protection certificate for medicinal products (Regulation 1768/92/EEC). Article 19 has been amended in SPC Regulation 469/2009, but the decision is relevant to Article 13(1) of 469/2009.

For instance, a medicinal product that contained a highly cytotoxic active ingredient for the treatment of brain tumors, in combination with an inactive matrix that enabled the active ingredient to be released slowly and gradually, was not distinguishable over the earlier authorised medicinal product containing the active ingredient alone. The matrix, although contributing to the efficacy of the medicinal product, did not have any therapeutic effect of its own and was not therefore considered an "active ingredient." Consequently, the cytotoxic active ingredient plus matrix was not a separate "product" per Article 1(b), and therefore the marketing authorisation was not the first to the "product," and no SPC was granted.[478]

Similarly, an earlier product containing 80 percent of an active ingredient with what was effectively 20 percent impurities was considered the same product under Article 1(b) as a compound containing 90 percent of the active ingredient and only 10 percent impurities, despite the fact that the latter compound was the result of a patented process to remove impurities. Therefore, two products that differ only in the proportion of the active chemical compound to the impurity, one having a greater percentage of the impurity than the other, are regarded as the same "product" within the meaning of Article 1(b), and only one SPC can be granted to the earlier marketed product per Article 3(d).[479]

Conversely, a single enantiomer was considered a new "product" in comparison to its racemic mixture. The English Court of Appeal held that the enantiomer was a compound with its own distinct activity, bioavailability, and toxicity and was not a minor variant of the racemate. Importantly, the earlier marketing authorisation to the racemate did not allow the marketing of the new enantiomer, and research and testing of the enantiomer was of the scale seen for new medicines, which was wholly within the policy reasons for the grant of an SPC.[480]

An adjuvant that has no therapeutic effect on its own, but which enhances the therapeutic effect of an antigen when combined with that antigen, is a new "active ingredient," and therefore the combination is not a new "product," and no SPC will grant to the combination.[481]

478. Massachusetts Institute of Technology, decision of the CJ, Case C-431/04, 2006 E.C.R. I-04089

479. Case C-258/99, BASF AG v. Bureau voor de Industriële Eigendom (BIE), 2001 E.C.R. I-3643.

480. Generics (UK) Ltd. v. Daiichi Pharm. Co. & Daiichi Sankyo Co., [2009] EWCA (Civ) 646.

481. Case C-210/13, Glaxosmithkline v. Comptroller of Patents, Designs and Trade Marks, November 14, 2013 by Reasoned Order. See Curia website: http://curia.europa.eu/juris/document/document.jsf?text=&docid=144850&pageIndex=0&doclang=en&mode=lst&dir=&occ=first&part=1&cid=324383

A new use for an already authorised product is seen as a variant of the earlier product and will not be considered a new "product" under Article 1(b) where the basic patent protects the second medical use of the compound. As the newly authorised product falls within the scope of any earlier SPC, there is no policy reason to extend protection to the second medical use of the product.[482]

However, in the case of a known active ingredient for which the basic patent is limited to a therapeutic use, a second patented therapeutic use of that known active ingredient can be the subject of a valid SPC, as was held by the Court of Justice in *Neurim*.[483] The use of the known active ingredient melatonin (which was never itself patented) was patented in 1987 for regulating the seasonal breeding activity in sheep, and the product was put on the market thereafter. Then in 1992, melatonin was patented for use in humans for insomnia, and the applicant subsequently applied for an SPC. Because the first product and relevant marketing authorisation were limited in scope by the basic patent to use as a medicine for sheep, the new use in humans of the same compound was treated as a distinct "product" per Article 1(b). Therefore, no marketing authorisation had been granted to this new product, and the SPC was valid.

8. Current Issues Relating to SPCs

a. Article 3(a): When Is a Product Protected by a Basic Patent in Force?

The issue of when the "product is protected by a basic patent in force" arose in the context of combination products.[484] Presented with the same two combination SPC applications, national IP offices and courts in different member states applied the Article 3(a) criteria differently, resulting in the grant of a combination SPC in some member states while others were refused. The IP offices and courts of France, Belgium, and Germany applied the well-known infringement test from patent law to assess whether the combination product was "protected by the basic patent" and granted the combination SPCs. For instance, in this case, because X, X+Y, and X+Y+Z are each separate "products" under Article 1(b), they are each "protected" by a claim to X in the "basic patent" and may each be entitled to separate SPC protection.

482. Case C-202/05, Yissum Research & Dev. Co. of the Hebrew Univ. of Jerusalem v. Comptroller-Gen. of Patents, 2007 E.C.R. I-2839.

483. Case C-130/11, Neurim Pharm. (1991) Ltd. v. Comptroller-Gen. of Patents, decision of the CJ, July 19, 2012, not published, see Curia website: http://curia.europa.eu/juris/document/document.jsf?docid=125216&doclang=EN.

484. There is no controversy that an SPC may be granted to both a single "product" and a "combination of products," as per the definition of "product" in Article 1(b).

The corresponding combination SPC applications were, however, refused by the UK IP Office and the High Court of England and Wales, which favored an alternative test, referred to as the "scope of protection" test. In this case, unless the basic patent disclosed the combination, the combination was not "protected," and the SPC was denied. Recognizing that the law was not "acte clair," the UK court referred the issue (which arose in two separate cases) to the Court of Justice, and in November 2011, this resulted in the decisions of *Medeva*[485] and *Georgetown*.[486]

Medeva and *Georgetown* both concerned multi-disease vaccines, which by law must be marketed as combination products to provide infants and young children vaccination against multiple diseases. Essentially, the combination "product," the subject of the SPC applications, provided for more active ingredients than were identified in the wording of the claims of the "basic patent."

The Court of Justice agreed with the UK court in finding that the active ingredients within the combination product must be "specified" in the wording of the claims of the basic patent.[487] Consequently, for an SPC application to succeed for a combination product with two active ingredients, both actives must be specified in combination in the claims of the basic patent. Similarly, if a patent claims a combination product only, such that there is no claim to either of the individual active ingredients, an SPC will be granted for the combination only (provided the marketing authorisation also contains that combination).[488]

Advocate General Trstenjak's Opinion in *Medeva* provides further clarification to the conclusions adopted by the Court of Justice. The advocate general implied that it is the inventive advance that will be protected

485. Case C-322/10, Medeva BV v. Comptroller Gen. of Patents, decision of the CJ, 24 November 2011, not published, see Curia website: http://curia.europa.eu/juris/document/document.jsf?text=&docid=115209&pageIndex=0&doclang=EN&mode=lst&dir=&occ=first&part=1&cid=359844.

486. Georgetown Univ., Univ. of Rochester, Loyola Univ. of Chi. v. Comptroller-Gen. of Patents, Designs & Trade Marks, Case C-422/10, decision of the CJ, November 24, 2011, not published, see Curia website.

487. In decisions delivered one day later, November 25, 2011, *University of Queensland and CSL Ltd. v. Comptroller General of Patents, Designs and Trade Marks*, C-630/10, not published, see Curia website, and *Daiichi Sankyo Company v. Comptroller General of Patents, Designs and Trade Marks*, C-6/11, not published, see Curia website, the Court stated that the product must be *identified* in the wording of the claims. Some commentators suggest that "identified" is more limiting than "specified." See: http://curia.europa.eu/juris/document/document.jsf?text=&docid=116322&pageIndex=0&doclang=EN&mode=lst&dir=&occ=first&part=1&cid=324569

488. Case C-518/10, Yeda Research & Dev. Co. & Aventis Holdings Inc. v. Comptroller Gen. of Patents, Designs and Trade Marks, decision of the CJ, November 25, 2011, not published, see Curia website: http://curia.europa.eu/juris/document/document.jsf?text=&docid=116161&pageIndex=0&doclang=en&mode=lst&dir=&occ=first&part=1&cid=324254.

by the basic patent when considering Article 3(a). Therefore, a patent that claims one active ingredient only does not protect (when considering Article 3(a)) that active ingredient plus another active ingredient as the patent has contributed no more than the former. The combination of the two actives does not embody the inventive advance of the patent, and so no SPC to the combination will be granted based on that patent. However, if the inventive feature of a patent is in fact a combination (and it is claimed), then the combination is protected by that patent for the purposes of Article 3(a), and an SPC to the combination can be obtained.[489]

The ambit of "specified" in the claim of the patent has been tested in the national courts, with the English courts rejecting the contention that inclusion of the term "comprising" in the claim is sufficient to include other active ingredients.[490] The Court of Appeal of The Hague held that a patent claim to a method for preparing escitalopram and the nontoxic acid addition salts thereof "specified" escitalopram oxalate, even though the literal wording of the claim did not include that particular salt.[491] However, the claim did specify a narrow class of compounds structurally defined, and the specification identified the oxalate and the combination was a preferred embodiment.

A claim to a technique did not specify the antibody ranibizumab, according to the UK court. Although ranibizumab was identified using the technique claimed (being the selection of an antibody by screening a library of antibodies), the lengthy claim provided for the technique only, without identifying the antibody itself. Ranibizumab was not therefore protected by the patent, and the SPC was invalid.[492]

National courts in Paris,[493] Dusseldorf,[494] and The Hague[495] came to differing decisions when asked whether the combination product "irbesartan + hydrochlorothiazide" was specified in the wording of a claim to irbesartan

489. What was not clear was whether the AG and the CJ in *Medeva* held that only one SPC can be obtained per patent, a question that has been answered by the CJ in *Actavis v. Sanofi*, Case C-443/12. See Section G.8.b.

490. Medeva BV v. Comptroller Gen. of Patents, [2012] EWCA (Civ) 523. The claim was to "A method for the preparation of an acellular vaccine, which method *comprises* preparing . . . [antigen A and] . . . [antigen B] . . . to provide a synergistic effect."

491. Joined Cases 200/044/332/01, 200/081/980/01, 200/081/996/01, H. Lundbeck A/S v. Alfred E. Tiefenbacher GmbH (Jan. 24, 2012).

492. Novartis Pharm. UK Ltd. v. Medimmune Ltd. & Med. Research Council, [2012] EWHC (Pat) 181.

493. Case RG 12/55806, Sanofi v. Sandoz, Tribunal de grande instance [TGI] [Court of First Instance] Paris, Aug. 10, 2012.

494. Case 4a O 109/12, Sanofi SA v. Actavis Deutschland GmbH & Co KG, Dusseldorf Landesgericht [District Court] Aug. 15, 2012.

495. Case 425814/KG ZA 12-905, Sanofi SA v. Pharmachemie BV, Rechtbank's-Gravenhage [District Court of The Hague] Sept. 14, 2012.

"in association with a diuretic." Nowhere in the specification of the patent was hydrochlorothiazide mentioned, but it was recognised by court experts that hydrochlorothiazide was commonly used as a diuretic in such combinations. The courts of The Hague and Dusseldorf held that hydrochlorothiazide was sufficiently specified by reference to "a diuretic," and the SPC was upheld; however, the Paris court disagreed. The UK court was then asked the same question in the corresponding case and expressed its frustration with the lack of clarity despite five Court of Justice references on this point so far,[496] and referred a further question asking that the Court of Justice clarify the criteria for deciding when a "product is protected by a basic patent."[497] Although the court answered this reference on December 12, 2013, it found it unnecessary to clarify this position as the SPC at issue was found invalid for other reasons.

However, on the same day, December 12, 2013, the court softened its position in *Eli Lilly and Company* in holding that it is not necessary for the active ingredient to be identified in the claims. In *Eli Lilly*, the claim was to antibodies defined in terms of their binding characteristics to a target protein, which was found sufficient provided the claim relates "implicitly but necessarily and specifically, to the active ingredient in question."[498] Although the court noted that the interpretation of "necessarily and specifically" was a matter to be determined by national courts, there is no doubt that future references will ask for clarification.

b. One SPC per Patent?

Prior to *Medeva* it was generally understood that it was possible to obtain one SPC per product, per basic patent, where one patent covered multiple different medicinal products. The decision in *Medeva* plainly stated that no more than one SPC could be granted per patent, but commentators and

496. "In my judgment it is necessary to seek further guidance from the CJ as to the correct interpretation of Article 3(a) of the Regulation in order to resolve this dispute. To reiterate, the reason why this is necessary is because the Court [the CJ] did not answer question 1 in the previous five references and the rulings it did make does not provide a clear test which can be applied to cases such as the present." Actavis Grp. PTC EHF & Actavis UK Ltd. v. Sanofi & Sanofi Pharma Bristol-Myers Squibb SNC, [2012] EWHC (Pat) 2245 at para. 72 (Arnold, J.).

497. Case C-443/12, Actavis Grp. PTC EHF & Actavis UK Ltd. v. Sanofi & Sanofi Pharma Bristol-Myers Squibb SNC, [2012] EWHC (Pat) 2545. The question referred was: "What are the criteria for deciding whether 'the product is protected by a basic patent in force' in Article 3(a) of Regulation 469/2009/EC ('the Regulation')?" *Id.* at para. 76 (Arnold, J.) (expressing opinion, which is in line with the Advocate General in Medeva).

498. C-493/12, Eli Lilly & Co. v. Human Genome Scis. Inc., [2012] EWHC (Pat) 2857. Relevant claim 13 of the Patent reads: "An isolated antibody or portion thereof that binds specifically to: (a) the full length Neutrokine-α polypeptide (amino acid sequence of residues 1 to 285 of SEQ ID NO:2); or (b) the extracellular domain of the Neutrokine-α polypeptide (amino acid sequence of residues 73 to 285 of SEQ ID NO:2)."

judges have been wary of accepting this at face value as it is unclear if the Court of Justice properly appreciated that more than one invention (or more than one "product") can be contained within the same basic patent.

Following the decision in *Medeva*, national IP offices have interpreted this issue differently, with the UK IP Office granting more than one SPC for the same basic patent (arguably justified on the basis that the SPCs are based on different inventions within the same basic patent), and the Dutch Patent Office conversely prohibiting grant of more than one SPC per patent regardless of the number of products (or inventions) claimed in the basic patent. The court answered the references on December 12, 2013, ruling that more than one SPC can be obtained per patent.[499] If there is more than one invention within the same basic patent, then an SPC may be granted for each.

However, with respect to combination products, it now appears difficult to obtain a combination SPC that is not in itself innovative and claimed in its own patent. No SPC will be granted to a combination product when based on the same basic patent as the single product and a later marketing authorisation unless the combination is itself innovative (i.e., not relying on the inventive single active ingredient to confer innovation). Importantly, the court did make clear that if a combination consisting of an innovative active ingredient in respect of which an SPC has already been granted and another active ingredient is the subject of a new basic patent "in so far as it covered a totally separate innovation,"[500] this would confer entitlement to an SPC for the new combination that is subsequently placed on the market.

Whether more than one SPC can be granted per basic patent was the question referred by both the Dutch and English courts to the Court of Justice in October 2012.[501]

---

499. However, the Rechtbank's-Gravenhage [District Court of The Hague] favored the approach of the UK IP office in Case 425814/KG ZA 12-905, Sanofi SA v. Pharmachemie BV Sept. 14, 2012.

500. Case C-443/12, Case C-484/12 and Actavis Group. PTC EHF and Actavis UK Ltd, Sanofi, Sanofi Pharma Bristol-Myers Squibb SNC,. para. 42.

501. GeorgeTown (Dutch case and CJ reference) and Actavis (English High Court case and CJ reference; Case AWB 10/4769, Georgetown Univ. v. Octrooicentrum Nederland (July 11, 2012); Case C-484/12 and Actavis Grp. PTC EHF and Actavis UK Ltd. v. Sanofi & Sanofi Pharma Bristol-Myers Squibb SNC, [2012] EWHC 2545 (Pat); Case C-443/12. The question referred in the latter: "In a situation in which multiple products are protected by a basic patent in force, does the Regulation, and in particular Article 3(c), preclude the proprietor of the patent being issued a certificate for each of the products protected?"

H. Post-authorisation Issues

Once a marketing authorisation has been obtained, it is important to ensure that it is actively managed. In particular, use of the product must be monitored to ensure that the positive risk-benefit profile concluded during the assessment procedure, based on the data available at that point, is found to be true on an ongoing basis. The conditions under which clinical trials are conducted do not necessarily reflect the way that a medicine is used once marketed. Placing the product on the market so that it is available to the general patient population (which will lead to wider-scale use than was the case with the clinical trials undertaken prior to authorisation, over a longer period, and with exposure to a wider selection of patients, such as those taking other medicines at the same time) may reveal new safety- or efficacy-related issues that require regulatory follow-up. The regulatory authorities have various tools at their disposal to gather post-authorisation data and to take action in relation to the granted marketing authorisation where this proves necessary, and these are outlined next.

Post-authorisation measures can be attached to a marketing authorisation requiring, in addition to the obligations imposed on all marketing authorisation holders, additional specific obligations (in the case of conditional authorisations and those under exceptional circumstances—see earlier in this text), additional pharmacovigilance activities, or other obligations.

This section addresses various post-authorisation administrative points concerning the ongoing management of the authorisation and a number of practical issues relating to placing the product on the market (pricing and reimbursement, SmPC harmonisation, and parallel imports). The legislation also covers other post-marketing aspects, such as advertising and promotion of medicinal products, the classification for supply of the product (e.g,. prescription only) and changes to this, and wholesale distribution. These topics are outside the scope of this chapter, but information can be found on the websites listed in Annex I.

1. Post-authorisation Management of the Marketing Authorisation

a. Validity of the Marketing Authorisation

Unless subject to the special arrangements for authorisations granted under the conditional/exceptional circumstances provisions described in Section V.D.8, "Authorisations in Special Circumstances," once granted, a marketing authorisation is valid for five years, at which point it may be renewed and will then normally be valid indefinitely. The renewal procedure will involve a reassessment of the product's risk-benefit profile based on use of the product in practice; if necessary "on justified grounds relating to pharmacovigilance, including exposure of an insufficient number of patients to the medicinal product concerned," the regulatory authority may decide to require one additional five-year renewal, after which the authorisation may be renewed for an unlimited period.[502]

There are also requirements in place (referred to as the "sunset clause") to make sure that the authorised product is actually placed (and maintained) on the market within a certain time frame, if the validity of the authorisation is to be maintained. Article 24(4)–24(6) of Directive 2001/83/EC provides as follows:

> Any authorisation which within three years of its granting is not followed by the actual placing on the market of the authorised product in the authorising Member State shall cease to be valid.
>
> When an authorised product previously placed on the market in the authorising Member State is no longer actually present on the market for a period of three consecutive years, the authorisation for that product shall cease to be valid.
>
> The competent authority may, in exceptional circumstances and on public health grounds grant exemptions from paragraphs 4 and 5. Such exemptions must be duly justified.

Article 14(4)–14(6) of Regulation EC No. 726/2004 makes the same provision with respect to centrally authorised products.[503] The start of the three-year period from grant of the marketing authorisation is determined taking into account the date on which it actually becomes possible to place the product on the market without infringing market exclusivity or other protection periods that need to be respected.

b. Changes to the Marketing Authorisation

The marketing authorisation will evolve over time, and there will be circumstances in which the authorisation holder will want (or be required) to make changes to it. The product may be further developed, and, in any event, the marketing authorisation holder is required to take into account current scientific knowledge and update the authorisation accordingly.[504]

---

502. Article 24(1)–(3), Directive 2001/83/EC; Article 14(1)–(3), Regulation EC No. 726/2004.

503. See Questions and Answers on notification to the EMA of actual marketing and cessation of placing on the market for centrally authorised products, EMEA/180078/2005, and Questions and Answers" on the application of the so-called "sunset clause" to centrally authorised medicinal products, EMEA/180079/2005.

504. Article 16(3) of Regulation (EC) No. 726/2004 and Article 23(3) of Directive 2001/83 /EC.

Whether administrative changes are made, or changes to the product information are required, or new quality, safety, or efficacy data need to be submitted for assessment, if new strengths, indications, pharmaceutical forms, and so on of the product are applied for, then these need to be reported to, or approved by, the regulatory authority.

The procedures in place for varying a marketing authorisation follow a risk-based approach and are set out in Commission Regulation (EC) No. 1234/2008. This covers centrally authorised products and authorisations obtained via the decentralised and mutual recognition procedures and has more recently been extended (with transitional provisions) also to purely national authorisations. Variations may be "minor" (e.g., tightening of specification limits) or "major" (e.g., addition of a new therapeutic indication). The most significant changes, such as changes to the active substance, strength, pharmaceutical form, and route of administration, are referred to as "extensions." The regulation requires the European Commission to publish and update guidelines on the details of the various categories of variations, as well as guidelines on the operation of the procedures and the documentation to be submitted as part of the variation application.

Variations classed as types IA and IB involve minor changes, and type II variations are all other variations not classified as an "extension." Type I variations require notification (the procedure differing slightly between types IA and IB), while type II variations require approval before they can be implemented. A special procedure is provided for in cases where an "urgent safety restriction" is needed. "Extensions" of marketing authorisations, which fulfill the conditions set out in Annex II to the regulation, do not fall within the scope of the regulation and are, in effect, applications for a new marketing authorisation (although see the discussion on the concept of "global marketing authorisation" elsewhere in this chapter, as for data exclusivity purposes the extension is treated as part of the original marketing authorisation).

Where marketing authorisations are granted in accordance with Article 10 or 10c of Directive 2001/83/EC, the marketing authorisation holder should swiftly introduce variations whenever the marketing authorisation of the reference or "original" medicinal product is changed to address a safety or efficacy concern.

Changes to the labeling or package leaflet that do not affect the SmPC may be implemented by means of a simpler "Article 61(3) [of Directive 2001/83/EC] notification procedure."

There are also separate specific procedures in place to transfer an authorisation from one legal entity to another. For centrally authorised products, Commission Regulation (EC) No. 2141/96 governs such transfers.

2. Post-authorisation Pharmacovigilance, Safety, and Efficacy

a. Pharmacovigilance

In July 2012, important changes in Directive 2001/83/EC and Regulation (EC) No. 726/2004 came into effect, introducing new pharmacovigilance requirements into the legislation with the aim of strengthening the system for post-authorisation monitoring of medicinal products. Requirements for putting in place pharmacovigilance systems to include risk-management systems (and, where appropriate, risk-minimization plans), for submitting post-authorisation data on products placed on the market, and for establishing procedures to ensure appropriate regulatory follow-up based on such data were all addressed as part of the changes to the legislation. Directive 2010/84/EU[505] and Regulation (EU) No. 1253/2010[506] were the legal instruments that introduced the changes into Directive 2001/83/EC and Regulation (EC) No. 726/2004, respectively.[507] Title IX (Articles 101–108) of Directive 2001/83/EC and Articles 21–29 of Regulation (EC) No. 726/2004 impose general obligations on member states regarding pharmacovigilance,[508] as well as obligations on marketing authorisation holders.[509] The legislation is accompanied by an implementing regulation, published by the European Commission in June 2012, which provides details on the operational aspects of the new pharmacovigilance requirements (Commission Regulation No. 520/2012).

A new EMA scientific committee, the Pharmacovigilance Risk Assessment Committee (PRAC), was established and given a mandate including risk-management issues and communication relating to the risk of adverse reactions, as well as the design and evaluation of post-authorisation safety studies and pharmacovigilance audits. The PRAC focuses on assessing and monitoring safety issues relating to human medicines, and its recommendations are used by the existing Committee for Medicinal Products for Human Use (CHMP) when adopting opinions for centrally authorised medicines and by the Coordination Group (CMDh) for nationally authorised medicines.

---

505. 2010 O.J. (L 348) 74

506. 2010 O.J. (L 348) 1. *See also*, for centrally authorised products, Commission Regulation (EC) No. 540/95 of March 10, 1995 Laying Down Arrangements for Reporting Suspected Unexpected Adverse Reactions Which Are Not Serious, 1995 O.J. (L 55/5).

507. The pharmacovigilance legislation was again amended in October 2012 by Regulation EU No. 1027/2012 (applies from June 5, 2013) and Directive 2012/26/EU (applied from October 28, 2013).

508. Article 101(1), Directive 2001/83/EC: "Member States shall operate a pharmacovigilance system for the fulfilment of their pharmacovigilance tasks and their participation in Union pharmacovigilance activities."

509. When necessary, stricter requirements can be imposed as a condition of grant of the marketing authorisation. *See* Article 21a, Directive 2001/83/EC.

In order to support the implementation of the new legislation for pharmacovigilance, a new set of guidelines—the Good Pharmacovigilance Practices (GVP) modules—for the conduct of pharmacovigilance in the EU is being developed, which will replace the current set in Volume 9A of the Rules Governing Medicinal Products in the EU (Volume 9A remains the reference as applicable until the transition period ends or until a specific GVP module is published as final). Final GVP modules can be found on the EMA website. Full detail on the pharmacovigilance framework is beyond the scope of this chapter, but some key points are noted next.

i. PSURs and ADRs

The legislation requires the submission of post-authorisation data both to report adverse drug reactions (ADRs) and in the form of Periodic Safety Update Reports (PSURs). The EMA has published module VI on ADRs and module VII on PSURs as part of the GVP Guideline. ADR reporting is done electronically through EudraVigilance, the EMA-managed database and data processing network that collates pharmacovigilance information. EudraVigilance contains information on suspected ADRs arising from use of the product both within and outside the terms of the marketing authorisation, and ADR reporting also covers, for example, medication errors and overdoses. The EMA also collates PSURs. All these reports are made available to the regulatory authorities, EMA committees, and the European Commission and are used for evaluating medicines during their development and monitoring their safety following authorisation. Health care professionals and the public also have "appropriate" access to the EudraVigilance database.

The obligations on marketing authorisation holders regarding the recording and reporting of suspected adverse reactions are set out in Article 107, and those on member states are contained in Article 107a of Directive 2001/83/EC. Patients have a right to report suspected side effects of medicines directly to national medicines regulatory authorities. They apply equally to centrally authorised products by virtue of Article 28(1) of Regulation (EC) No. 726/2004.

Concerning PSURs, Article 107b of Directive 2001/83/EC sets out the requirements on marketing authorisation holders:

> 1. Marketing authorisation holders shall submit to the Agency periodic safety update reports containing:
>
> (a) summaries of data relevant to the benefits and risks of the medicinal product, including results of all studies with a consideration of their potential impact on the marketing authorisation;

> (b) a scientific evaluation of the risk-benefit balance of the medicinal product;
> (c) all data relating to the volume of sales of the medicinal product and any data in possession of the marketing authorisation holder relating to the volume of prescriptions, including an estimate of the population exposed to the medicinal product.
>
> The evaluation referred to in point (b) shall be based on all available data, including data from clinical trials in unauthorised indications and populations.
>
> The periodic safety update reports shall be submitted electronically.

PSURs are submitted by marketing authorisation holders at defined periods during the post-authorisation phase. Article 107c of Directive 2001/83/EC provides the framework for this. The default position (the marketing authorisation can require a different submission schedule) is that PSURs must be submitted:

- Immediately upon request;
- Where the product has not yet been placed on the market, at least every six months following authorisation until the placing on the market; or
- Where the product has been placed on the market, at least every six months during the first two years following initial placing on the market, once a year for the following two years, and at three-yearly intervals thereafter.

Equivalent provisions apply to centrally authorised products by virtue of Article 28(2) of Regulation (EC) No. 726/2004.

Products authorised under the generic and "well-established use" routes (Articles 10(1) and 10a of Directive 2001/83/EC, respectively) are exempt from routine submission of PSURs unless an obligation to provide them has been laid down as a condition in the marketing authorisation or a competent authority requests them on the basis of concerns relating to pharmacovigilance data or due to the lack of PSURs relating to an active substance after the marketing authorisation has been granted.

ii. Post-authorisation Safety and Efficacy Studies (PASS/PAES)

The new legislation introduced a strengthened legal basis for requiring post-authorisation safety and efficacy studies (PASSs/PAESs) from marketing authorisation holders as a means of improving the ongoing risk-benefit monitoring of medicines. A PASS is a study of an authorised medicine that identifies, characterizes, or quantifies a safety hazard, confirms the safety profile of the medicine, or gauges the effectiveness of risk-management measures during its lifetime. A PAES is a study aimed at verifying the efficacy of a medicine on the market, including efficacy in everyday medical practice. The purpose of the information in a PASS/PAES is to support decision making on the safety and benefit-risk profile of a medicine to enhance its safe and effective use. The EMA published a GVP module (VIII) on PASS in June 2012 and was preparing a separate guideline on methodological aspects of PAES in 2013.

iii. Provision of Key Information on Medicinal Products

Marketing authorisation holders must provide (and maintain) key information on all authorised medicines in the EU to the EMA, using, in accordance with Article 57 of Regulation (EC) No. 726/2004, a dedicated format for the electronic submission of such information. This information is then available to be used to support all activities related to the ongoing safety monitoring of medicines.

b. Regulatory Actions When Issues Arise

The legal framework for post-authorisation monitoring of medicinal products is backed up by a number of provisions that provide a legal basis for the regulatory authorities to take appropriate action with respect to the authorisation if required.

EU member states are, according to Article 101(2) of Directive 2001/83/EC, required to take regulatory action in response to pharmacovigilance "signals":

> (2). Member States shall, by means of the pharmacovigilance system referred to in paragraph 1, evaluate all information scientifically, consider options for risk minimisation and prevention and take regulatory action concerning the marketing authorisation as necessary.

Regarding assessment of PSURs, Article 107d provides that:

> [t]he national competent authorities shall assess periodic safety update reports to determine whether there are new risks or whether risks have changed or whether there are changes to the risk-benefit balance of medicinal products.

Article 107f goes on:

> Following the assessment of periodic safety update reports, the national competent authorities shall consider whether any action concerning the marketing authorisation for the medicinal product concerned is necessary.

> They shall maintain, vary, suspend or revoke the marketing authorisation as appropriate.

A "pharmacovigilance referral" procedure is provided for, to ensure harmonised implementation of safety recommendations across the EU.[510]

Where, as a result of evaluating pharmacovigilance data, urgent action is considered necessary, a member state or the European Commission can initiate the "Urgent Union procedure,"[511] which is designed to rapidly assess significant emerging safety issues linked with a medicine or a class of medicines available in the EU, regardless of its initial authorisation route (i.e., centralised or national authorisation). A scientific assessment of the particular medicine or class of medicines is conducted through the PRAC and/or the CHMP, in conjunction with the CMDh in the case of nationally authorised products. On the basis of the scientific assessment, regulators can adopt a decision to vary, suspend, revoke, or refuse renewal of the marketing authorisation, in the interest of protecting public health. This procedure should be initiated where the member state or Commission considers urgent action to be required in any of the following cases:

(f) it considers suspending or revoking a marketing authorisation;
(g) it considers prohibiting the supply of a medicinal product;
(h) it considers refusing the renewal of a marketing authorisation;
(i) it is informed by the marketing authorisation holder that, on the basis of safety concerns, he or she has interrupted the placing on the market of a medicinal product or has taken action to have a marketing authorisation withdrawn, or that he or she intends to do so;
(j) it considers that a new contraindication, a reduction in the recommended dose, or a restriction to the indications is necessary.

Article 116 of Directive 2001/83/EC provides that the competent authority shall suspend, revoke, or vary a marketing authorisation when, inter alia:

> the view is taken that the medicinal product is harmful or that it lacks therapeutic efficacy, or that the risk-benefit balance is not favourable, or that its qualitative or quantitative composition is not as declared. Therapeutic efficacy shall be considered to be lacking when it is concluded that therapeutic results cannot be obtained from the medicinal product.

Article 117 goes on to define the situations in which, without prejudice to the provisions of Article 116, member states must take appropriate

---

510. Article 107g, Directive 2001/83/EC.
511. Articles 107i–107k, Directive 2001/83/EC.

steps to ensure that the supply of medicinal product is prohibited and the medicinal product withdrawn from the market (which action may be limited to the batches of product in dispute)—namely, where "the view is taken that":

- The medicinal product is harmful, or
- It lacks therapeutic efficacy, or
- The risk-benefit balance is not favorable, or
- Its qualitative and quantitative composition is not as declared, or
- Proper controls have not been carried out or some other obligation relating to the manufacturing authorisation has not been fulfilled.

For products authorised through the centralised procedure, Article 20 of Regulation (EC) No. 726/2004 allows a member state to suspend, on its own initiative or at the European Commission's request, the use in its territory of the product "where urgent action is essential to protect human health or the environment." In such cases, a procedure to reach a final decision applicable throughout the EU is initiated.

3. Practical Issues Concerning Placing the Product on the Market

a. Pricing and Reimbursement

Save for the provisions of the Transparency Directive,[512] the provisions governing the pricing and reimbursement of medicinal products for human use are not harmonised at EU level, and each member state has its own system in place. Although issues of pricing are not considered as part of the marketing authorisation assessment procedures, this point is in practice a "fourth hurdle" (after quality, safety, and efficacy) to commercialization of the product due to the price controls and reimbursement systems in place, imposed by national governments. The mechanisms in any one-member state typically differ between patented (normally branded) products and generic (normally unbranded) products; special systems may operate for special categories of medicines, such as orphan drugs, depending on the member state in question.

i. The Transparency Directive

The differences in the systems can lead to significant price differences between member states (hence the existence of the parallel import trade, as discussed later in this chapter), but such systems remain largely unharmonised due to the member states' well-established freedom to organize their own social security systems and retain national competences in the

512. Council Directive 89/105/EEC of December 21, 1988 Relating to the Transparency of Measures Regulating the Prices of Medicinal Products for Human Use and Their Inclusion in the Scope of National Health Insurance Systems, 1989 O.J. (L 40) 8–11.

area of public health. This being the case, the Transparency Directive lays down harmonised provisions to ensure the transparency of national provisions regulating the pricing and reimbursement of medicinal products for human use. It defines a series of procedural requirements designed to enable interested parties to verify that national pricing and reimbursement decisions do not create obstacles to pharmaceutical trade within the European Union, in accordance with Article 28 of the EC Treaty. The provisions of the directive do not affect national policies on the setting of prices, except as far as is necessary to achieve the transparency required by the directive.

Prior to the adoption of the Transparency Directive, the European Commission issued a communication addressing the legality of various aspects of price controls and reimbursement mechanisms,[513] accepting that national measures adopted with a view to offering the best possible treatment to all citizens without excessive cost to the public purse were in practice compatible with Article 34 TEFU, provided that the restrictive effect was not disproportionate to this legitimate aim. However, national measures could not lay down different systems for imported and domestic products and thus hinder the sale of imported ones. On the question of reimbursement, the Commission stated that a national scheme excluding certain products or categories of products from reimbursement under a country's national health care scheme ("a negative list") or one that only listed certain products ("a positive list") was not necessarily incompatible with EU law as long as in deciding what products were to be excluded only objective criteria were involved and there was no discrimination as to the origin of the product.

The Transparency Directive lays down three major requirements with respect to individual pricing and reimbursement decisions:

- decisions must be made within a specific time frame (90/180 days);
- decisions must be communicated to the applicant and contain a statement of reasons based on objective and verifiable criteria; and
- decisions must be open to judicial appeal at the national level.

Several proceedings have come before the CJEU under the directive,[514] mostly in relation to the alleged breach of Article 4 in the context of attempts

513. Commission Communication 86/C310/08 on the Compatibility with Article [34TFEU] of Measures Taken by Member States Relating to Price Controls and Reimbursement of Medicinal Products, 1986 O.J. (C 310) 7.

514. *See also* Case C-62/09, The Queen, on the Application of the ABPI v. MHRA, Apr. 22, 2010, See Curia website: http://curia.europa.eu/juris/document/document.jsf?docid=79658&doclang=EN

to challenge compulsory price freezes or price reductions[515] or of Article 6 by those countries that cover a medicinal product by their national health insurance system only after a decision to include it in a positive list.[516]

The Transparency Directive was, as of mid-2013, under review by the European legislators, with a view to repealing it and replacing it with a new directive "relating to the transparency of measures regulating the prices of medicinal products for human use and their inclusion in the scope of public health insurance systems."[517] The aim is to speed up the placing on the market of authorised medicines, so a reduction in the duration of national pricing and reimbursement decisions is proposed. In the future, such decisions should be taken within 120 days for innovative medicines (180 days for more complex procedures) and for generic medicinal products within only 30 days, instead of the current 180 days. Strong enforcement measures in case the decisions do not comply with the time limits are also proposed.

ii. The PPRS in the UK

In the UK, the Pharmaceutical Price Regulation Scheme (PPRS) operates as a voluntary agreement between the government and the pharmaceutical industry. It covers all branded National Health Service (NHS) medicines.

---

515. Namely, Case E-2/98, Fed'n of Icelandic Trade v. Gov't of Iceland & Pharm. Pricing Comm. (EFTA Advisory Court Nov. 24, 1998); Joined Cases C-352/07 to C-356/07, C-365/07 to C-365/07 to C-367/07 and C-400/07, Menarini Industrie Farmaceutiche Riunite Srl & Others v. Ministero della Salute & Agenzia Italiana del Farmaco (Apr. 2, 2009); Joined Cases C-471/07 and C-472/07, Ass'n Generale de l'Industrie du medicament (AGIM) ASBL & Others v. Belgium (Jan. 14, 2010). For CJ decisions see Curia website: Joined Cases C352/07 to C356/07, C365/07 to C367/07 and C400/07 (http://curia.europa.eu/juris/document/document.jsf?text=&docid=73797&pageIndex=0&doclang=en&mode=lst&dir=&occ=first&part=1&cid=329512) and Joined Cases C471/07 and C472/07 (http://curia.europa.eu/juris/document/document.jsf?text=&docid=75517&pageIndex=0&doclang=en&mode=lst&dir=&occ=first&part=1&cid=329704).

516. Namely, Case C-424/99, Commission v. Austria (Nov. 27, 2011) (See Curia webite: http://curia.europa.eu/juris/document/document.jsf?text=&docid=47334&pageIndex=0&doclang=EN&mode=lst&dir=&occ=first&part=1&cid=360951), Case C229/00, Commission v. Finland (June 12, 2003) (See Curia website: http://curia.europa.eu/juris/document/document.jsf?text=&docid=47923&pageIndex=0&doclang=EN&mode=lst&dir=&occ=first&part=1&cid=361225), Case C-245/03, Merck Sharp & Dohme BV v. Belgium (Jan. 20, 2005) (See Curia Website: http://curia.europa.eu/juris/document/document.jsf?text=&docid=49849&pageIndex=0&doclang=EN&mode=lst&dir=&occ=first&part=1&cid=361324); Case C296/03, GlaxoSmithKline SA v. Belgium (Jan. 20, 2005) (see Curia Website: http://curia.europa.eu/juris/document/document.jsf?text=&docid=49851&pageIndex=0&doclang=EN&mode=lst&dir=&occ=first&part=1&cid=361531), Case C-317/05, G Pohl Boskamp GmbH & Co KG v. Gemeinsamer Bundesauschuss (Oct. 26, 2006) (See Curia website: http://curia.europa.eu/juris/document/document.jsf?docid=63936&doclang=en), and Case C-311/07, Commission v. Austria (July 17, 2008).

517. COM (2012) 84 final 1 March 2012.

The PPRS sets a ceiling on companies' profits on NHS sales. The system is complex, and full details can be found on the UK government's website.[518]

Any supplier of NHS medicines may join the PPRS. A company may choose not to become a member, or it may be excluded by the Secretary of State for Health if, for example, it has failed to comply with the requirements of the scheme. In such circumstances, a statutory scheme prevails.[519] The Health Service Branded Medicines (Control of Prices and Supply of Information) (No. 2) Regulations 2008 limit the maximum price of prescription-only, branded medicines supplied to the NHS and require manufacturers and suppliers of branded pharmaceutical medicines to provide the Department of Health with information on sales income and discounts. Members of the PPRS are exempt from such statutory requirements.

The PPRS is due to be replaced by a "value-based" pricing mechanism in 2014. The statutory scheme is also due to be revised.

b. EU-wide SmPC Harmonisation

One of the aims of the EU's regulatory framework for medicinal products is to ensure that the key terms of the marketing authorisation for any particular product are harmonised in all member states in which the product is placed on the market. This is clearly not an issue for products authorised via the centralised procedure because a single marketing authorisation covers all EU member states, but where products (and particularly older ones) have been authorised by the national regulatory bodies, there may be national differences in the authorisations for a particular product. The resulting disparities between the Summaries of Product Characteristics (SmPCs) in different countries provide a particular challenge for generic companies wishing to obtain authorisation for a generic product in more than one member state by means of the mutual recognition or decentralised procedures, as these require a harmonised SmPC.

The European Commission thus initiated a program of SmPC harmonisations, using the "referral" procedures set out in Directive 2001/83/EC. Details on the Community Referral Procedures are set out in Chapter 3, Volume 2A of the Notice to Applicants (Revised September 2007).

Article 30(2), Directive 2001/83/EC provides the legal basis for the SmPC harmonisation exercise:

---

518. The Pharmaceutical Price Regulation Scheme 2014, found at https://www.gov.uk/government/publications/pharmaceutical-price-regulation-scheme-2014.

519. Details of the statutory scheme can be found at https://www.gov.uk/government/publications/statutory-scheme.

> In order to promote harmonisation of authorisations for medicinal products authorised in the Community, Member States shall, each year, forward to the coordination group a list of medicinal products for which a harmonised summary of product characteristics should be drawn up. The coordination group shall lay down a list taking into account the proposals from all Member States and shall forward this list to the Commission.
>
> The Commission or a Member State, in agreement with the Agency and taking into account the views of interested parties, may refer these products to the Committee in accordance with paragraph 1 [which sets out the "divergent decisions" referrals procedure].

The CMDh website provides a tracking table showing the progress being made with the SmPC harmonisation exercise.

c. Parallel Imports

The pharmaceutical market within the EU is fragmented by different national pricing for the same product. Price differences between member states occur because of disparities in national prosperity and differing government price controls intended to limit public health expenditure, resulting in a medicinal product being sold for less in some member states and more in others.

Price differences for the same medicinal product across member states allow parallel traders to profit by buying low and selling high. The trade is "parallel" to the extent that it takes place outside, and in parallel with, the distribution network that the manufacturer or original supplier has established for its products. Although the term "import" is used for historical reasons, arguably it is no longer relevant when referring to trade within the EU, with its policy of free movement of goods and people.

This principle gives rise to the principle of "exhaustion"—that is, once a product has been placed on the European market with the owner's consent, the IP rights in those goods are said to be "exhausted." Although the medicinal products are the subject of national patents (a "European patent" once granted becomes a bundle of national patents), neither patent rights nor trade-mark rights can be asserted to prevent parallel trade within the EU. Therefore, once a marketing authorisation has been granted for a medicinal product and that product has been placed on the European market, parallel trade in that product cannot be prevented, unless justified to protect the health and life of humans.

Nationally authorised and centrally authorised medicinal products are treated differently. Centrally authorised products are by definition authorised for the whole of the European market, and, therefore, any products distributed by the manufacturer and parallel trader are identical as they are covered by the same marketing authorisation. Thus, once the marketing authorisation holder has placed the medicinal product on the market, the parallel trader is free to enter the market of any member state with that product regardless of the national market into which the marketing authorisation holder introduced the product. Although a trader is free to distribute the product in any member state without applying for a license, notification must be made to the EMA before doing so. Further, as with nationally authorised products, the labeling and package leaflet must be amended so as to comply with national legislation in the member state of importation.[520]

i. Nationally Authorised Medicinal Products

Once a medicinal product has been placed on the market under a national marketing authorisation, a parallel importer can apply for a license to market that product rather than apply for a separate marketing authorisation. The competent authorities[521] grant such a license (a modified form of a marketing authorisation) to parallel importers under what is referred to as a "simplified" procedure that requires the applicant to provide far less information than is required for a marketing authorisation.[522] This reflects the underlying policy that repetition of preclinical and clinical testing is unnecessary and disproportionate.

In order to be awarded a license under the simplified procedure, the following conditions must be met:

---

520. Not discussed here are the changes that parallel importers may introduce to the packaging of the imported medicinal products, being those which are necessary to market the product in the member state of destination (e.g., relating to the language of the labeling and the instructions or the size of the packaging). Repackaging the product and reaffixing the trade-mark or replacing it with a different trade mark used for the same product in the member state of destination may be necessary to allow the parallel product to be marketed in the destination member state. The changes introduced must be strictly necessary to allow the parallel product to be marketed in the destination member state. If the condition of the parallel goods is changed or impaired, then the principle of exhaustion of trade-mark rights will not apply and the proprietor of the trade-mark can oppose further commercialization of the goods. *See* Case 1/81, Pfizer v. Eurim-Pharm, [1981] E.C.R. 2913; Joined Cases C-427/93, C-429/93, C-436/93, Bristol-Myers Squibb & Others, [1996] E.C.R. I-3457; Case C-379/97, Upjohn, [1999] E.C.R. I-6927; Case C-443/99, Merck, Sharp & Dohme, [2002] E.C.R. I-3703; Case C-143/00, Boehringer, [2002] E.C.R. I-3759.

521. For a definition of "competent authority," *see* Section D.1.a.

522. Case 104/75, Officier van Justitio v. Adriann de Peijper, [1976] E.C.R. 00613; *see also Commission Communication on Parallel Imports of Proprietary Medicinal Products for Which Marketing Authorisations Have Already Been Granted*, COM (2003) 839 final (Dec. 30, 2003).

- the imported product has been granted a marketing authorisation in the member state of exportation; and
- the imported product is sufficiently similar to a product that has already received marketing authorisation in the member state of destination.

(a) "Sufficiently Similar"

The parallel importer must therefore establish that the medicinal product is the subject of a marketing authorisation in both the member state of export and of import and that the product being exported is "sufficiently similar" to that authorised in the member state of destination. Essentially, there should be no difference to the patient between the parallel imported product and the reference product already on the market; the two products should be interchangeable. Small differences between the two products are acceptable provided that the active ingredient is identical and that any differences do not have a therapeutic effect.

A product is considered "sufficiently similar" if it has been manufactured according to the same formulation, using the same active ingredient, and has the same therapeutic effects.[523] If the medicinal product is marketed in both member states by the same entity, it is likely that the product is identical, so no issue arises. Complications can arise, however, when the medicinal product is marketed in the two member states by different companies.

Previously, to be considered "sufficiently similar," the manufacture of the two medicinal products also required a "common origin." For instance, although two products were manufactured by two unrelated companies, they were considered "sufficiently similar" if the products were each manufactured under separate licenses from the same licensor.[524] However, this criterion no longer applies. If the assessment of safety and efficacy can be performed with reference to the authorised product on the market of the member state of import without any risk to public health, then the fact that the two products do not have a common origin is irrelevant, although it may help in establishing the similarity of the two products.[525]

---

523. Case C-201/94, R v. Meds. Control Agency ex parte Smith & Nephew Pharm. Ltd. & Primecrown v. Meds. Control Agency, [1996] E.C.R. I-5819.

524. *Id.*

525. Case C-112/02, Kohlpharma GmbH v. Germany, [2004] E.C.R. I-3369. In the United Kingdom, whether the products are manufactured by the same or different companies influences whether the licensing application falls into the "simple" or "standard" category. If the UK product and the product to be imported are manufactured by companies in the same group of companies or are made under license from the same licensor, the application for a license falls into the "simple" category. If the UK and parallel product do not have a common

A product manufactured using a different manufacturing process and with different excipients will also be deemed "sufficiently similar" to the reference product provided that the difference does not influence the therapeutic effect, as stated by the General Court in *Rhône-Poulenc Rorer*:

> [I]t is permissible . . . to place the imported product upon the market . . . if . . . [the imported product] has the same active ingredients and therapeutic effect as [the domestic reference product], but does not use the same excipients and is manufactured by a different manufacturing process.[526]

However, as a small difference in excipients may cause a difference in bioavailability, a change of excipients is particularly relevant to controlled-release products. For this reason, at least the Dutch competent authority requires that for controlled-release tablets with different excipients to the reference product, the applicant must demonstrate that the two products are interchangeable.[527]

(b) Withdrawal of Marketing Authorisation in Member State of Import

Parallel importation of a medicinal product is still possible when the marketing authorisation for the reference product has been withdrawn in the member state of import, provided it was not withdrawn based on health and safety reasons. For instance, a new version of a medicinal product may be introduced into one member state and the marketing authorisation withdrawn for the old version, while the old version continues to be marketed in the member state of export. In such a case, the old version parallel product will enter the member state of import with reference to the marketing authorisation for the new version of the product. The national

origin but are in all other respects the same, then the "standard" application applies. Further, if there is no common origin and the imported products contain a new excipient, the product is made by a different route, is a controlled release preparation, is a sterile product which is sterilized in a different way from the UK product, is a sterile product whereby the container is made from a different material to the container of the UK product, is an influenza vaccine, is a metered dose inhaler, is a powder for inhalation, the bioequivalence of the UK and imported product cannot be demonstrated through bioavailability studies, or the sole or primary evidence for the safety and efficacy of the imported product consists of published scientific literature, the application for a license falls into the "complex" category. For applications within the "standard" and "complex" category, pharmacoviligence plans must be approved. *See* Parallel Import Licences, MHRA, http://www.mhra.gov.uk/Howweregulate/Medicines/Licensingofmedicines/Parallelimportlicences/index.htm#l5 (last updated May 19, 2014).

526. Case C-94/98, R v. Licensing Auth. ex parte Rhöne Poulenc Rorer, [1999] E.C.R. I-8815, 8835–36.

527. Meds. Evaluation Bd., MEB-14-3.2, Parallel Import: Authorisation and Maintenance (May 29, 2012).

competent authorities in the member state of import must request documentation from the competent authority in the member state where the old version of the medicinal product continues to be marketed in order to verify that the old version has the same therapeutic effect as the new version.[528] If in fact there is a risk to the health of humans if the old version of the product is allowed to remain on the market at the same time as the new version, no license will be granted.[529]

In *Ferring*, the new version medicinal product differed from the old version only with respect to modified excipients, leading to improved temperature stability and thus making refrigerated storage unnecessary. Ferring argued that given the difference in storage requirements between the new and old versions, this may place public health at risk. The Court of Justice held in this respect that:

> while it is not open to the Court to rule on the question relating to the existence and reality of a risk to public health linked to the coexistence of the two versions for the medicinal product in question on the German market, it is conceivable that the risk mentioned by Ferring may be of such a nature that it cannot be averted satisfactorily by appropriate labeling.[530]

VI. Japanese Drug Approval Considerations[531]

This section introduces an overview of the regulatory schemes and procedures regarding the regulatory approval process of branded and generic drugs in Japan.

A. Building a Drug and Patent Portfolio: Considerations for Innovators

1. Application for a New Prescription Drug

To obtain marketing approval for a prescription drug in Japan, an entity must submit an application to the Ministry of Health, Labor, and Welfare under the Pharmaceutical Affairs Law of Japan, Article 14. This application must comply with one of the categories stipulated in the

528. Case C-172/00, Ferring Arzzneimittel GbbH v. Eurim-Pharm Arzneimittel GmbH, [2002] E.C.R. I-6891.

529. Case C-15/01, Paranova Lakemedel AB & Others v. Lakemedelsverket, [2003] E.C.R. I-4175.

530. *Ferring*, [2002] E.C.R. I-6891, para. 45.

531. Junko Iyoda of Taiyo, Nakajima & Kato, Tokyo, Japan.

Pharmaceutical Affairs Law. Each category requires specific types of documents to be submitted with the application. The categories of new drugs as well as the types of documents to be submitted are as follows:

Prescription drugs with new active ingredients refers to a drug with an active ingredient that has never been used in drugs already approved for manufacture/marketing or not specified in the Japanese Pharmacopoeia ("approved drugs, etc.");

New combination prescription drugs refers to a drug with active ingredients or ratios of active ingredients different from those of combination drugs specified in the Japanese Pharmacopoeia or combination drugs already approved for manufacture/marketing as prescription drugs;

Prescription drugs with new administration routes refers to a drug that has the same active ingredient as an approved drug, etc., but has a different route of administration (oral, subcutaneous, intramuscular, intravenous, percutaneous, per-rectal, transvaginal, eye drops, nasal drops, inhalation, etc.);

Prescription drugs with new indications refers to a drug that has the same active ingredient and route of administration as an approved drug, etc., but for a different indication;

Prescription drugs with new dosage forms refers to a drug that has the same active ingredient, route of administration, and indication as an approved drug, etc., but has a new dosage form, such as sustained release instead of immediate release; and

Prescription drugs with new doses refers to a drug that has the same active ingredient and route of administration as an approved drug, etc., but has different dosage strength.

Tables 2 and 3 illustrate the documents for requiring marketing approval in each category. Table 2 is a list of documents specified by the Minister of Health, Labor, and Welfare. Table 3 illustrates documents required for marketing approval in each category described earlier.

An applicant seeking marketing approval for the "prescription drugs with new active ingredients" is required to file "origin or background of discovery, conditions of use in foreign countries," "manufacturing methods, standards and test methods" "Stability,","Pharmacological action," "Absorption, distribution, metabolism, and excretion (except for bioequivalence)," "Acute, subacute, and chronic toxicity, teratogenicity, and other types of toxicity," and "Clinical study."

Table 2
Documents Specified by MHLW

	A 1 2 3	B 1 2 3	C 1 2 3	D 1 2 3	E 1 2 3 4 5 6	F 1 2 3 4 5 6 7	G
(1) Prescription drugs with new active ingredients	○○○	○○○	○○○	○○Δ	○○○○×Δ	○○○Δ○ΔΔ	○
(2) New combination prescription drugs	○○○	×○○	○○○	○ΔΔ	○○○○×Δ	○○×××Δ×	○
(3) Prescription drugs with new administration routes	○○○	×○○	○○○	○ΔΔ	○○○○×Δ	○○×Δ○ΔΔ	○
(4) Prescription drugs with new indications	○○○	×××	×××	○××	ΔΔΔΔ×Δ	×××××××	○
(5) Prescription drugs with new dosage forms	○○○	×○○	○○○	×××	○○○○×Δ	×××××××	○
(6) Prescription drugs with new doses	○○○	×××	×××	○××	○○○○×Δ	×××××××	○

○: Date required × : Data not required Δ: Data required depending on individual cases

Table 3
Documents Required for Marketing Approval

a. Origin or background of discovery, conditions of use in foreign countries	1. Origin or background of discovery 2. Conditions of use in foreign countries 3. Special characteristics, comparisons with other drugs, etc.
b. Manufacturing methods, standards and test methods	1. Chemical structure and physicochemical properties, etc. 2. Manufacturing methods 3. Standards and test methods
c. Stability	1. Long-term storage tests 2. Tests under severe conditions (stress tests) 3. Accelerated tests
d. Pharmacological action	1. Tests to support efficacy 2. Secondary pharmacology, Safety pharmacology 3. Other pharmacology
e. Absorption, distribution, metabolism and excretion	1. Absorption 2. Distribution 3. Metabolism 4. Excretion 5. Bioequivalence 6. Other pharmacokinetics
f. Acute, subacute, and chronic toxicity, teratogenicity, and other types of toxicity	1. Single dose toxicity 2. Repeated dose toxicity 3. Genotoxicity 4. Carcinogenicity 5. Reproductive toxicity 6. Local irritation 7. Other toxicity
g. Clinical studies	Clinical trial results

An applicant seeking marketing approval for the "new combination prescription drugs," is required to file "origin or background of discovery, conditions of use in foreign countries," "manufacturing methods, standards, and test methods (except for chemical structure and physicochemical properties)," "Stability," "Pharmacological action," "Absorption, distribution, metabolism, and excretion (except for bioequivalence)," "Acute, subacute, and chronic toxicity, teratogenicity, and other types of toxicity (except for genotoxicity, carcinogenicity, reproductive toxicity, and other toxicity)" and "Clinical study."

An applicant seeking marketing approval for the "prescription drugs with new administration routes" is required to file "origin or background of discovery, conditions of use in foreign countries", "manufacturing methods, standards and test methods (except for Chemical structure and physicochemical properties)", "Stability", "Pharmacological action", "Absorption, distribution, metabolism, and excretion (except for bioequivalence)", "Acute, subacute, and chronic toxicity, teratogenicity, and other types of toxicity (except for genotoxicity)" and "Clinical study."

An applicant seeking marketing approval for the "prescription drugs with new indications" is required to file "origin or background of discovery, conditions of use in foreign countries," "Pharmacological action (except for secondary pharmacology, safety pharmacology and other pharmacology," "Absorption, distribution, metabolism, and excretion (except for bioequivalence)" and "Clinical study".

An applicant seeking marketing approval for the "prescription drugs with new dosage forms" is required to file "origin or background of discovery, conditions of use in foreign countries", "manufacturing methods, standards and test methods (except for Chemical structure and physicochemical properties)", "Stability", "Absorption, distribution, metabolism, and excretion (except for bioequivalence)" and "Clinical study".

An applicant seeking marketing approval for the "prescription drugs with new doses", is required to file "origin or background of discovery, conditions of use in foreign countries", "Pharmacological action (except for secondary pharmacology, safety pharmacology and other pharmacology", "Absorption, distribution, metabolism, and excretion (except for bioequivalence)" and "Clinical study".

The documents described in Tables 2 and 3 must be prepared in the Common Technical Document (CTD) format established by the International Conference on Harmonisation of Technical Requirements for Registration of Pharmaceuticals for Human Use (ICH).

2. Regulatory Exclusivity of Prescription Drugs

In April 1980, the Ministry of Health, Labor, and Welfare introduced a reexamination system for the purpose of preventing drug-induced damages, such as those caused by thalidomide and clioquinal (SMON, or subacute myelo-optico-neuropathy).[532] In the reexamination system, the indications and safety of an approved new drug are evaluated again after a certain period has passed.

532. Article 14-quarter of the Pharmaceutical Affairs Law.

The main reasons for the introduction of the reexamination system include: (1) unknown side effects that may occur when an approved drug is used by a large number of patients; (2) a limited number of cases studied in the clinical trials; and (3) differences in usage of the drug between actual clinical practice and the clinical trials (in the clinical trials, dosage, co-administered drugs, symptoms, age, and preexisting diseases of the patients are controlled).

The Ministry of Health, Labor, and Welfare does not accept an application for marketing a generic drug until the reexamination period of a reference drug expires. This is because efficacy and safety of a reference drug are not assured. Accordingly, a new drug is protected from generic competition during the reexamination period.

Thus, although both Japan and the United States have a regulatory exclusivity period after the approval of a new drug, the underlying basis for regulatory exclusivity in Japan is different from that of the United States.

Prescription drugs have the following durations of regulatory exclusivity:

- Prescription drugs with new active ingredients: eight years
- New combination prescription drugs: six years (four years in some cases)
- Prescription drugs with new administration routes: six years
- Prescription drugs with new indications: four years
- Prescription drugs with new dosage form: four years
- Prescription drugs with new doses: four years
- Prescription drugs with orphan drug: ten years

3. Application for Drugs Manufactured Overseas

An overseas manufacturer (a manufacturer who wishes to produce a drug overseas for the purpose of importation into Japan) may directly apply for marketing approval of a drug under the manufacturer's own name, provided that the manufacturer conducts the necessary tests (clinical trials) concerning the quality, efficacy, and safety of the drug and follows procedures required by the Minister of Health, Labor, and Welfare.[533]

a. Accreditation of Overseas Manufacturers

Similar to the case of a domestic manufacturer, an overseas manufacturer must receive accreditation from the Minister of Health, Labor, and

533. Overseas Special Approval: Article 14-ter(ii) of the Pharmaceutical Affairs Law.

Welfare.[534] The accreditation is also required for an overseas manufacturer who produces only a drug ingredient for a drug to be exported to Japan.

An overseas manufacturer can apply for the accreditation through a representative of the manufacturer or an agent thereof. The application must be filed before the submission of an application for overseas special approval. If an overseas manufacturer has not obtained the accreditation, the manufacturer must indicate in the application form for overseas special approval that the application for accreditation is pending. The Minister of Health, Labor, and Welfare will not grant the overseas special approval unless the accreditation is obtained.

For an application for accreditation, an overseas manufacturer must submit a document providing a detailed summary of the structure and facility of its production plant. Also, the overseas manufacturer must have a site inspection carried out by the Minister of Health, Labor, and Welfare. Other details related to the application are described on the website of the Pharmaceuticals and Medical Devices Agency (PMDA).[535]

b. Appointment of Marketer

When an overseas manufacturer applies for overseas special approval, the overseas manufacturer must appoint a marketer who has received a drug marketing business license in Japan. After the overseas manufacturer obtains overseas special approval, the overseas manufacturer can market the imported drug only via the appointed marketer. This system has been adopted mainly to prevent the occurrence of health and hygiene-related hazards that may be caused by a drug receiving overseas special approval.

c. Manufacturing License

In addition to obtaining the overseas special approval, an overseas manufacturer must conduct certain production-related procedures, such as labeling, examination, and storage of the drug products in Japan. Therefore, the overseas manufacturer must obtain not only a marketing license but also a manufacturing license. The marketing license refers to a license required for shipping a drug product to the Japanese market.[536] A holder of the license has the final responsibility for a sold product and ships a drug product to the market under the holder's name. Production is not permitted with this license alone. The manufacturing license refers to a license required for manufacturing drug products.[537] A drug manufacturer must

534. Accredited Foreign Manufacturer: Article 13 of the Pharmaceutical Affairs Law.

535. *Accreditation of Foreign Manufacturers*, http://www.pmda.go.jp/english/service/acc_foreign.html.

536. Article 12 of the Pharmaceutical Affairs Law.

537. Article 13 of the Pharmaceutical Affairs Law.

obtain a class of license for each manufacturing plant that corresponds to its manufacturing category. Shipment to the market is not permitted with this license alone.

d. Handling of Data Obtained by Stability Tests and Clinical Trials

It is stipulated that the stability tests on drugs carried out in foreign countries are accepted, provided that the stability tests conform to the regulations and guidelines of Japan.[538]

Clinical trials conducted in foreign countries can also be accepted for approval review, provided that: (1) the trial data are gathered and prepared according to the guidelines stipulated in the Ministerial Ordinance Concerning the Guidelines for Practicing Clinical Trials of Drugs and the provisions of Article 43 of the implementing rules of the Pharmaceutical Affairs Law; and (2) the documents are confirmed to satisfy the requirements of (1) based on the written explanation according to Article 14(5) of the Pharmaceutical Affairs Law or the site inspection. However, data from clinical trials conducted in Japan are also requested for the purposes of judging whether the foreign clinical trial data can be used to evaluate the efficacy and safety of the drug in the Japanese population.

Of the documents to be attached to the application for marketing approval, those originally written in English need not be fully translated if a Japanese abstract is attached to the original texts.[539] Further, if modules 3, 4, and 5 of an application for approval in the CTD format are originally written in English, the Japanese abstract is unnecessary.

4. Patent Term Extensions

Provisions similar to the extension of patent term defined in 35 U.S.C. § 156 also exist in the Japanese Patent Law. Specifically, the Japanese Patent Law stipulates that a patentee can recover up to five years of the patent term during which a patented invention could not have been practiced due to a regulatory period under the Pharmaceutical Affairs Law (hereinafter referred to as “marketing approval”).[540]

The requirements for an application for registration of an extension are described next.

a. Application

An applicant for registration of an extension of the term of a patent right (hereinafter also referred to as “application for registration of an

538. Communication from the chief of the Pharmaceutical Affairs Bureau (Feb. 15, 1991), http://www8.cao.go.jp/kisei-kaikaku/oto/otodb/english/faq/qa/q3-1.html}}.

539. Notification No. 256 of the PMSB (Mar. 18, 1998), for Japanese see http://www.rsihata.com/guidance/pdf/031898IYS265A.pdf}}.

540. Article 67(2) of the Patent Law.

extension") is limited to the patentee.[541] When a patent is owned jointly, each of the joint owners may not, except jointly with the other owners, apply for registration of an extension of the patent term.[542] Further, marketing approval may only be obtained by the patentee or a person who has an exclusive license or a nonexclusive license on the patent right.[543]

b. Timing

An application for registration of an extension must be filed within three months of the date on which the market approval is obtained. Moreover, the application may not be made after the expiration of the term of the original patent right.[544] Nevertheless, in a case in which, due to reasons outside his or her control, the applicant is unable to file the application within three months of the market approval date, the applicant may file the application within two months (provided the applicant is a resident abroad) from the date when the reasons ceased to be applicable (but not later than nine months from the date on which the marketing approval was obtained).[545]

When an applicant anticipates that the marketing approval cannot be obtained by the day before six months prior to the expiration date of the term of the original patent right, the applicant desiring to apply for registration of a patent term extension must file a specified document.[546] Without submission of the specified document by that date, an application for the registration of a patent term extension may not be made after six months prior to the expiration date of the term of the original patent.[547] Here, the date on which the marketing approval was obtained means the date on which a written approval reached the applicant for the approval or, in a case in which the applicant for the approval knew of the approval before the written approval was delivered, the date on which the applicant actually knew of the approval.

c. Term

A term of extension applied for means a period during which a patented invention cannot be practiced, but is limited to five years. The date that the period during which the patented invention cannot be practiced starts is the later of: (1) the date on which the clinical trial was started (for example, the date on which the clinical trial regimen was submitted to the

541. Article 67ter(1)(iv) of the Patent Law.
542. Article 67bis(4) of the Patent Law.
543. Article 67ter(1) of the Patent Law.
544. Article 67bis(3) of the Patent Law.
545. Article 4 of Order for Enforcement of the Patent Law.
546. Article 67bis-bis(1) of the Patent Law.
547. Article 67bis-bis(2) of the Patent Law.

Ministry of Health, Labor, and Welfare); or (2) the registration date of the patent right.

If a clinical trial was carried out in a foreign country and if marketing approval was subsequently obtained by carrying out a clinical trial or a bridging test in Japan and if the registration date of the patent right in Japan is before the commencement date of the clinical trial in the foreign country, the term in which the clinical trial was carried out in the foreign country may be added, provided that the requirements, such as proving the commencement date of the clinical trial in the foreign country, are satisfied.[548]

The termination date of the term during which the patented invention could not be practiced is the date on which marketing approval was obtained.

d. Conditions

An application can be filed for registration of a patent term extension when a patented invention cannot be practiced due to the necessity of obtaining the marketing approval. The relevant categories include patents directed to materials, uses, and production methods. In Japan, unlike in the United States, when there are multiple patent rights (patents directed to a material, a use, and a production method) corresponding to one marketing approval, the registration of a patent term extension can be individually approved for all of the patent rights related to the single marketing approval so long as the marketing approval is considered necessary to practice the patented invention.

Other than those just described, a more detailed explanation of the application for registration of an extension is described in the Examination Guidelines concerning registration of an extension of a patent term, revised on December 28, 2011.

e. Important Points to Consider

The following two points should be remembered when filing an application for registration of an extension.

First, in the past, in the case of a patent term extension of a patent for producing a pharmaceutical product, the Japan Patent Office (JPO) compared only the final product obtained by the production method and the product that had already received approval for production in determining eligibility for the patent term extension, not the patented production method and the production method specified in the Ministry of Heath, Labor, and Welfare (MHLW) approval. However, according to the revised guidelines issued by JPO, JPO compares the patented production method and the production method specified in the MHLW approval in considering an application for registration of a patent term extension, and if the

548. Examination Guidelines for Patent and Utility Model, ch. VI, s. 3.1.1.(2).

production method specified in the MHLW approval for production is not within the scope of the patented production method, the application for registration of a patent term extension will be rejected.

Second, JPO grants registration of a patent term extension with respect to a patent as a whole, rather than for individual claims in the patent. Therefore, the following case may occur. A patent includes claim A, which recites an indication of a pharmaceutical product (such as an analgesic containing substance X as an active ingredient), and claim B, which depends on claim A and recites a dosage form (such as an injectable solution) of the pharmaceutical product, and approval for production and marketing has already been granted for the active ingredient in a tablet dosage form. In such a case, a patent term extension based on a subsequently granted approval for production and marketing of the active ingredient in an injective solution dosage form will not be granted. While this might seem somewhat inconsistent, the reason given by the JPO is that the pharmaceutical product claimed in claim A could already have been practiced after the prior approval of the tablet dosage form, and since claim B includes all of the limitations of claim A, the scope of claim B is within the scope of claim A. Therefore, when both an indication of a pharmaceutical product and a dosage form of the pharmaceutical product are to be claimed, it is advisable to avoid inclusion of both claims in a single patent application if a patent term extension is likely to be sought. This may be achieved, for example, by filing a divisional application including a claim reciting a dosage form, or by filing a patent application including a claim reciting an indication and a separate patent application including a claim reciting a dosage form on the same day.

5. Trade-marks and Nonproprietary Names

a. Nonproprietary Names

i. INN And JAN Application and Approval

In Japan, an application for marketing approval must use the Japanese Accepted Name (JAN) as a generic name. In the past, there were cases in which a name according to JAN differed from a corresponding name according to International Nonproprietary Names (INN). However, around 2006, the Ministry of Health, Labor, and Welfare made attempts to make JAN consistent with INN, and JAN and INN have since been modified to a unified nomenclature.

The Ministry of Health, Labor, and Welfare permits an applicant to apply for an INN before a name according to JAN is decided. Further, the procedures for incorporating the items included in rINN (Recommended International Nonproprietary Names) into JAN are specified in

Notification No. 0331001 of the Pharmaceutical and Food Safety Bureau (PFSB), MHLW, dated March 31, 2006.

The application for a JAN may be made before the filing of an application for marketing approval of a drug and registration of the drug in the drug master file (MF), provided that sufficient basic trials of the drug have been carried out and clinical usability is sufficiently proved.

b. Pharmaceutical Trade-marks

i. Registration with the Japan Patent Office

The period from the filing of an application to the issuance of the decision to grant from the Japan Patent Office is about 18 months. Once a trade-mark is registered, the trade-mark right can be retained semipermanently by repeating renewals, unless opposition is raised.

ii. MHLW Approval of Trade-marks

In an application for marketing approval of a new drug in Japan, a brand name is required to satisfy that: (1) the brand name drug shall not cause health or hygienic hazards; (2) the brand name shall have a character of a prescription drug; and (3) the information about the dosage form and the content (or concentration) of an active ingredient shall be appended to the brand name. The Ministry of Health, Labor, and Welfare has not provided particular guidelines with respect to these requirements. This is quite different from the U.S. Food and Drug Administration (FDA) practice, which has a system for approving trade-marks given to drugs.

With respect to a generic drug for which an application for marketing approval is made after September 2005, the Ministry of Health, Labor, and Welfare prohibits the use of its own commercial name and stipulates that the commercial name of the generic drug shall be a combination of an established name, a dosage form, a content, and a company name (shop name). This provision is intended to prevent medical accidents, facilitate use of generic drugs, and promote prescription using established names.

iii. Limited DTC Advertising

To ensure proper use of drugs, the Ministry of Health, Labor, and Welfare prohibits advertisements for prescription drugs to general consumers (direct consumer advertising) in Japan.[549] This regulation is quite different from the practice in the United States.

549. Notification No. 1339 of the PAB, MHW (Oct. 9, 1980).

B. Considerations for Generic Drug Companies

1. Identifying and Evaluating Patents on Brand Products

In Japan, for the purpose of ensuring a stable supply of generic drugs, the Ministry of Health, Labor, and Welfare will not approve an application for a generic drug before patents for the active ingredient and the first approved indication of the reference drug have expired. However, a generic manufacturer may obtain approval for other indications of the reference drug if there are no patents directed to the other indications. This is somewhat similar to the carve-out provision in the United States.

The Ministry of Health, Labor, and Welfare does not have a system similar to the Orange Book provided by the Food and Drug Administration in the United States. Further, the Ministry of Health, Labor, and Welfare is not involved in checking whether patents exist that protect the active ingredients, indications, or dosage/administration of reference drugs.

Therefore, a generic drug applicant must carry out its own search to check whether patents protecting the reference product exist, and their effective terms, before marketing the generic product in Japan. Further, the generic applicant is required to submit a certification relating to a patent directed to the substance or the use of the reference product.

Further, if an approval holder of a brand product has a patent protecting a brand product (more specifically, a patent directed to the material or the use of the brand product), it is recommended that the approval holder submit patent information conforming to a specified form to the general drug examination division of the Pharmaceuticals and Medical Devices Agency (PMDA).[550] Such patent information should be submitted before the expiration of the patent directed to the material or the use, and the approval holder should be aware of the time at which generic drugs are expected to come into the market. Without the submission of patent information, PMDA may approve a generic drug before the expiration of the patent.

2. Evaluating Patent Validity and Enforceability

a. Product-by-Process Claims

Under Japanese patent practice, there are three major categories of subject matter that may be eligible for a patent: (1) an invention of a product; (2) an invention of a method; and (3) an invention of a method for producing a product. It is recommended that an invention of a product be specified and defined by the structure or properties of the product. However, it is permissible to define a product by a method of production where it

550. Notification No. 0115001 of the Health Policy Bureau (Jan. 15, 2009).

is impossible or difficult to describe or define the product by its structure or properties (case A). A claim written in this manner is called a product-by-process claim in Japan, similarly to the patent practice in the United States. When a claim is written in the product-by-process format, the specific production method recited in the claim is viewed as being described only for the purpose of specifying the product, and the technical scope of the invention recited in the claim is not limited to products that are actually produced by using the specific production method. Thus, the technical scope is generally considered to cover products.

Many patented inventions that have claims written in the product-by-process format do not correspond to the cases just described. That is, the product-by-process format is often used even in cases where it would have been possible to define a product by its structure or properties (case B). A judgment concerning interpretation of the technical scope of such claims was delivered by the full bench of the Intellectual Property High Court in 2012.[551]

In the judgment, the full bench of the Intellectual Property High Court classified product-by-process claims as *genuine product-by-process claims* that correspond to case A, and *pseudo product-by-process claims* that correspond to case B. In the judgment, it was acknowledged that the technical scope of a genuine product-by-process claim that recites a specific production method is not limited to products that are actually produced by using the specific method, and that the technical scope generally covers products. On the other hand, the judgment clarified that the technical scope of a pseudo product-by-process claim is limited to products that are actually produced by using the production method recited in the claim. This case is currently under appeal to the Supreme Court in Japan.

Accordingly, the interpretation of product-by-process claims in Japan is different from the judgment handed down in *Abbott Laboratories v. Sandoz, Inc.*[552] Therefore, when interpreting a product-by-process claim in a Japanese patent, it is important that the claim is evaluated to determine whether it is a genuine product-by-process claim or a pseudo product-by-process claim, and the technical scope of the claims should be determined accordingly.

3. Application to Manufacture/Market a Generic Drug

The following data must be submitted with an application to manufacture or market a generic drug.

551. Intellectual Property High Court See http://www.ip.courts.go.jp/hanrei/pdf/20120313110328.pdf Jan. 27, 2012, Patentee: TEVA Gyogyszergyar Zartkoruen Mukodo Reszvenytarsasag.

552. Fed. Cir. (2007-1400), May 18, 2009, see http://cafc.uscourts.gov/images/stories/opinions-orders/07-1400.pdf.

a. Standards and Test Method

Standards are defined as a list of test methods, analytical methods used in the tests, and criteria for acceptance. The standards provide measures for quality that a drug manufacturer should follow in order to ensure the quality and consistency of drug products.

The standards and test method document shall describe the actual measured values of the purity test and moisture content specified in the ICHQ6A guideline, and the document serves as the basis for setting the test conditions, such as those related to analytical validation demonstrating the properness of the actual measured values.

Actual measured values obtained by measurement of three samples per lot for three lots (nine samples in total) must be submitted. Further, a generic applicant must submit a document that demonstrates that the residual solvents, impurities, and analogues in the production process are properly controlled. These documents must be prepared according to the ICHQ3A, ICHQ3B, and ICHQ3C guidelines.

b. Accelerated Study

Accelerated studies are tests in which drug quality is evaluated after being subjected to various environmental factors, such as temperature, humidity, and light. These tests are used to set the retest period of the drug substance, the effective life of the drug, and the storage conditions for the drug.

Accelerated studies must be carried out according to ICHQ1A and ICHQ1B guidelines.

c. Bioequivalence

The purpose of conducting a bioequivalence study is to guarantee the therapeutic equivalence of a generic product as compared to a reference product. Therefore, in the bioequivalence study, the bioavailability of a reference product and the bioavailability of a generic product are compared.

When comparison of bioavailability is difficult, or when the measurement of bioavailability does not serve as an indicator of therapeutic effects, a reference product and a generic product are compared with respect to pharmacological effects supporting the efficacy or therapeutic effects. In the case of an orally administered drug, since the dissolution behavior exerts an influence on bioequivalence, a dissolution test needs to be conducted. Bioequivalence studies must be conducted according to the Guidelines for Bioequivalence Studies on Generic Drugs.[553]

553. Notification No. 229(10) of the Evaluation and Licensing Division, PFSB (Feb. 29, 2012).

4. Required Items to Be Matched with Brand Products

a. Salts, Esters, Crystal Forms, Hydrate/Non-hydrate Forms

In Japan, a generic product is a drug product that contains the same active ingredient as that of a reference product in the same amount and has the same dosage form and dosage/administration as that of the reference product.

A difference between crystal form or hydrated form and anhydrous form ("crystal forms") is not deemed a difference in an active ingredient, unlike a difference between salts or esters. Therefore, it is permissible to use a drug substance that differs in crystal form from that of a reference product, provided that the equivalence of the bioavailability of the reference product and generic product is ensured and that information about the crystal form based on literature is submitted.[554]

In the case of salts or esters, the drug substance of a generic product must be identical to that of the reference product.

b. Indications

Similar to the Section viii statement under Section 505(j)(2)(a)(viii) in the United States, the labeling carve-out is also permitted in Japan. Specifically, although a patent exists with respect to an indication and dosage/administration of a reference product, if another indication outside the scope of the patent is possible, a generic product directed to such an indication may be approved.[555]

In Japan, in an attempt to promote the usage of generic products, the Ministry of Health, Labor, and Welfare requires a generic company to provide the generic product in the same active ingredient strengths that are on the market for the reference product. This provision was introduced on May 15, 2006, in order to account for a possibility that a reference product, rather than a generic product, is selected due to the lack of a generic product of a particular strength when a dosage may be changed in accordance with the patient's symptoms.

An example following the classification of dosage forms stipulated in Notification No. 0215007 of Health Insurance Bureau dated February 15, 2006, is provided next. When a reference product has tablet forms (50 mg tablet, 100 mg tablet, 200 mg tablet) classified to oral medicine-1 and powdered forms (grain 10 percent, grain 20 percent) classified to oral medicine-2 in its product line, a generic product does not have to cover both tablet and powdered forms that differ in dosage form classification;

554. Notification No. 0616(1) of the Evaluation and Licensing Division, PFSB (June 16, 2011).

555. Notification No. 0605001 of the Health and Policy Bureau, Notification No. 0605014 of the Evaluation and Licensing Division, PFSB (June 5, 2009).

however, if a generic company comes to market with a tablet, the company must have a 50 mg tablet, a 100 mg tablet, and a 200 mg tablet in its tablet product line.

C. Drug Pricing in Japan

1. Determination of Reimbursement Prices for New Drugs

Unlike U.S. citizens, all Japanese citizens are legally required to be insured by their employer's health insurance or the national health insurance. Although the proportion of the medical costs directly borne by Japanese citizens varies depending on the type of the applicable health insurance and age, most of the medical cost is borne by the insurers. Therefore, the compensation for medical procedures (reimbursement of medical fees) and the prices of prescription drugs (drug prices) are decided by the Minister of Health, Labor, and Welfare.

In Japan, a list of prescription drugs that can be used in medical treatment designated by medical insurances along with the price of the prescription drugs designated by the medical insurances (drug prices) are published in official gazettes and are generally referred to as the "drug price list." The basic calculation of prices of new drugs depends on the prices of drugs having similar indications. Notification No. 0212(1) of the Health Insurance Bureau dated February 12, 2010, stipulates, for example, that a drug price may be set to be from 70 percent to 120 percent higher than that for drugs having similar indications, provided that a new drug has a clinically useful new mechanism of action, has been shown to have greater efficacy and safety than existing drugs in the same class, and has been shown to improve treatment of the indicated disease or trauma. This is referred to as a premium for innovativeness.

Even after the drug price is included in the drug price list, re-pricing is undertaken for prescription drugs that showed at least a two-fold increase in the market scale as compared to that originally expected at the time of listing and for which annual sales exceed fifteen billion yen. Further, drug prices are decreased for drugs that have passed a certain period after entry onto a drug price list, or drugs for which generic products have entered the drug price list. The review of drug prices is usually carried out every two years.

2. Entry of Generic Drugs in the Drug Price List

Prices for generic products are entered in the drug price list twice a year, in June and December. Similarly to reference products, the method for calculating the drug price is also clearly defined for generic products.[556]

556. Notification No. 0210(4) of Health Policy Bureau (Feb. 10, 2012).

To have a generic product placed on the drug list in June, the generic product is required to have received marketing approval by February 15. To have a generic product placed on the drug list in December, the generic product is required to have received marketing approval by August 15. However, the marketing approval for a generic product will not be granted if an active ingredient patent or a first approved indication patent protecting the reference product exists. For generic companies, timing for placing generic drugs on the drug price list and the remaining effective terms of patents directed to active ingredients or first approved indications for reference drugs are to be considered in Japan.

Appendix A

Confidential Invention Disclosure Form

[Company name]
[Address]

I. Title of Invention:

____

II. Invention

A. Background

(Please provide a description of background of the invention: what was the state of the art before the invention was made? What problem does the invention solve?)

B. Description of the invention

(Please provide a detailed description of the invention and attach additional information and supporting documents if necessary. In the description, please address the following questions: what are the technical features of the invention? is there an alternative to each of the technical features? is (are) there a formula(e), diagram(s), or figure(s) that describes the invention? what does the invention do? how can the invention be used? how was the invention made? what is the significance of the invention? what problem does this invention solve? how does the invention solve the problem? what were the earlier solutions, if any? what advantages does this invention provide over the earlier solutions? how is the invention an improvement over the state of the art?)

C. Description of the drawings

(Please provide a detailed description of the drawings and explain what each component in the drawings is and how it functions.)

D. How was the invention made?

- When did you first think of the invention?
- Laboratory Notebook Pages on which this disclosure is based:
- Date of first drawing, if any, illustrating the invention:
- Date of first written description of the invention:
- Date of disclosure:

- To whom was the invention first disclosed?
- When did you first do experimental work toward carrying out the invention?
- When were you satisfied that you had solved the problem toward which your invention was directed?
- Who has observed the progress of your experimental work?
- List all prior patents (including patents of Company) and published articles of which you are aware that are related to the invention:
- Has the invention been described in a printed publication or sold or offered for sale?

III. Inventors

A. Please list the name, residence address, and citizenship of each inventor.

B. Please briefly describe each inventor's contribution to the invention.

IV. Signatures

Inventor 1

Name **Signature** **Date**

The above was read and understood by me on this ______ day of _________, 20_____.

Name: ______________________ Signature: ______________

Inventor 2

Name **Signature** **Date**

The above was read and understood by me on this ______ day of _________, 20_____.

Name: ______________________ Signature: ______________

Inventor 3

Name **Signature** **Date**

The above was read and understood by me on this ______ day of _________, 20_____.

Name: ______________________ Signature: ______________

Index

About the ABA Section of Intellectual Property Law

From its strength within the American Bar Association, the ABA Section of Intellectual Property Law (ABA-IPL) advances the development and improvement of intellectual property laws and their fair and just administration. The Section furthers the goals of its members by sharing knowledge and balanced insight on the full spectrum of intellectual property law and practice, including patents, trademarks, copyright, industrial design, literary and artistic works, scientific works, and innovation. Providing a forum for rich perspectives and reasoned commentary, ABA-IPL serves as the ABA voice of intellectual property law within the profession, before policy makers, and with the public.

ABA Section of Intellectual Property Law
Order today! Call 1-800-285-2221
Monday-Friday, 7:30 a.m. – 5:30 p.m., Central Time
or Visit the ABA Web Store: www.ShopABA.org

Qty	Title	Regular Price	ABA-IPL Member Price	Total
_____	ADR Advocacy, Strategies, and Practice in Intellectual Property Cases (5370195)	$139.95	$114.95	$_____
_____	ANDA Litigation (5370199)	$299.00	$249.00	$_____
_____	Careers in IP Law (5370204)	$24.95	$16.95	$_____
_____	Computer Games and Virtual Worlds (5370172)	$69.95	$55.95	$_____
_____	Copyright Remedies (5370208)	$89.95	$74.95	$_____
_____	Distance Learning and Copyright (5370163)	$89.95	$79.95	$_____
_____	Fundamentals of Intellectual Property Valuation (5370143)	$69.95	$49.95	$_____
_____	The Intellectual Property Handbook (5620116)	$110.00	$90.00	$_____
_____	IP Attorney's Handbook for Insurance Coverage in Intellectual Property Disputes, Second Edition (5370210)	$129.95	$103.95	$_____
_____	A Lawyer's Guide to Section 337 Investigations before the U.S. International Trade Commission, Second Edition (5370203)	$119.95	$89.95	$_____
_____	A Legal Strategist's Guide to Trademark Trial and Appeal Board Practice, Second Edition (5370200)	$159.95	$129.95	$_____
_____	Music & Copyright in America (5370201)	$97.95	$67.95	$_____
_____	New Practitioner's Guide to Intellectual Property (5370198)	$89.95	$69.95	$_____
_____	The Patent Infringement Litigation Handbook (1620416)	$149.95	$129.95	$_____
_____	Patently Persuasive (5370206)	$129.95	$99.95	$_____
_____	Patent Obviousness in the Wake of *KSR International Co. v. Teleflex Inc.* (5370189)	$129.95	$103.95	$_____
_____	Trademark and Deceptive Advertising Surveys (5370197)	$179.95	$134.95	$_____
_____	Practitioner's Guide to the PCT (5370194)	$139.95	$109.95	$_____
_____	Practitioner's Guide to Trial Before the Patent Trial and Appeal Board (5370209)	$139.95	$114.95	$_____
_____	Settlement of Patent Litigation and Disputes (5370192)	$179.95	$144.95	$_____
_____	Starting an IP Law Practice (5370202)	$54.95	$34.95	$_____
_____	The Tech Contracts Handbook (5370188)	$34.95	$29.95	$_____
_____	Trademark and Deceptive Advertising Surveys (5370197)	$179.95	$134.95	$_____
_____	Trademark Surveys (5370207)	$269.95	$239.95	$_____

*** Tax**
DC residents add 6%
IL residents add 9.50%

Payment
❑ Check enclosed payable to the ABA
❑ VISA ❑ Mastercard ❑ American Express

* Tax $_____
** Shipping/Handling $_____
TOTAL $_____

****Shipping/Handling**

Up to $49.99	$5.95
$50 to $99.99	$7.95
$100 to $199.99	$9.95
$200 to $499.99	$12.95
$500 to $999.99	$15.95
$1,000 and above	$18.95

Name____

Firm/Organization_______________________________________

Address____

City_________________________ State_____ Zipcode__________

Phone____________________ E-mail______________________
(in case of questions about your order)

Please allow 5 to 7 business days for UPS delivery. Need it sooner? Ask about overnight delivery. Call the ABA Service Center at 1-800-285-2221 for more information.

Guarantee: If – for any reason – you are not satisfied with your purchase, you may return it within 30 days of receipt for a complete refund of the price of the book(s). No questions asked!

Please mail your order to:
ABA Publication Orders, P.O. Box 10892, Chicago, Illinois 60610-0892
Phone: 1-800-285-2221 or 312-988-5522 • Fax: 312-988-5568
E-mail: orders@abanet.org

Thank you for your order!

Section of Intellectual Property Law
AMERICAN BAR ASSOCIATION